Table of Contents

Part II

Completion of the project (January 1958-November 1965)

Foreword

The Eurochemic company was in many respects an original approach to international co-operation. It is the only company to have carried out research and industrial operations, on an international basis, in the particularly complex and sensitive area of nuclear fuel recycling. For most of the thirteen western European countries that took part in Eurochemic from 1959 onwards, it was their first experience in the field, and the research laboratory and prototype chemical reprocessing plant built at Mol in Belgium provided them with the means of training their specialists-to-be.

The operation of the plant led to significant progress in nuclear chemical engineering, but this scientific and technical success contrasts sharply with Eurochemic's failure to make its mark on the industrial scene, due largely to the fact that the limited market for reprocessing was very soon shared out between a small number of national plants. In this way the opportunity to lay the foundations for a European nuclear industry was missed.

The Eurochemic company subsequently found a new and equally advanced vocation in 1977 when it changed over to managing the radioactive wastes the operation of the plant had produced: it was in fact the first reprocessing plant to embark upon a comprehensive programme to decontaminate and decommission its installations, as well as to condition all the different types of radioactive wastes accumulated on the site.

Finally, the life of the Eurochemic company was marked by the original nature of its statute. It was set up under a Treaty signed by the participating governments as an "international shareholding company", the intention being to involve private industry in public sector activities and to allow its management structure a considerable degree of independence. As things turned out, on completion of the research and development phase, the lack of commercial prospects ultimately placed the burden of funding and decision-making squarely on the governments. However its unusual status did nevertheless encourage a very open form of co-operation on a European scale, both amongst the company's team of researchers and engineers and between the design offices and firms that became involved in its work.

The Eurochemic company itself, prior to its liquidation in 1990, made the necessary arrangements for its history to be written. A publishing committee was established, all the members of which had played a leading role in the company; it was chaired by Pierre Huet, former Director-General of the OEEC European Nuclear Energy Agency. The book was written by Jean-Marc Wolff, an historian of contemporary science and technology, and was also submitted as a doctoral thesis, being accepted with the distinction "*magna cum laude*". It is intended mainly for managers and experts concerned with the nuclear fuel cycle and for all who take an interest in international technical co-operation and the history of nuclear energy, for whom it is a very rich source of references.

This book is published under the responsibility of the Secretary-General of the Organisation but does not necessarily reflect the views of the OECD countries that took part in the Eurochemic company.

Part III

The company's first life: reprocessing (December 1965-January 1975)

Part V

Final considerations

SOURCES, BIBLIOGRAPHY AND REFERENCES

INDEX OF TABLES AND FIGURES

N° of table

N° of figure

"In the wintertime you must have seen those unfortunate people scratching for bits of coal amongst the ashes in the dustbins.

This is surely the most basic way of reprocessing a spent fuel. But if it were possible for these people to continue their task amongst not coal ashes but in atomic waste, they would soon be rich, because alongside the unburned material they would find the equivalent of gold nuggets."

Robert Sartorius, The Processing of Irradiated Fuels, *Industry and Nuclear Energy, papers given to the second Information Conference on Nuclear Energy for Managers, Amsterdam, 24 to 28 June 1957.*

"Those who criticise the 'technical system', who condemn the inherent power they see in technology as the embodiment of an ancient destiny or as a curse, should always remember the principle of the 'concealing hand' conceived by Albert Hirschman – a principle which, to my mind, has as much impact in the social sphere as that of the 'invisible hand' in the economic: it is possible for an undertaking to succeed not because all the risks have been carefully worked out but because in fact they have been underestimated. There again, if we could know and evaluate every potential difficulty in advance, what would become of human adventure and progress?"

Jean-Jacques Salomon, *Le destin technologique.*

Introduction

This history of Eurochemic is the first book devoted to an undertaking at the back end of the nuclear fuel cycle. A product of European co-operation in the field of the chemistry of radioactive materials, Eurochemic had a short but double life.

The period that elapsed between the conception and ultimate demise of Eurochemic covered thirty-five years of the second half of the twentieth century. It began eleven years after the explosion of the first atomic bomb, at a time when civilian nuclear power was emerging in the euphoria of the early days of high growth. It ended as the nineties were about to begin. By then the development of civilian nuclear power was profoundly affected by fifteen years of economic crisis, by the transformation of the link between growth and energy, by changes in political attitudes and decision-making about questions of high technology and, finally, by the first signs of changes in the military nuclear sector arising from the radical international developments accompanying the collapse of the Soviet Union.

The European Company for the Chemical Processing of Irradiated Fuels (Eurochemic) was set up on 27 July 1959. From 1956 onwards, governments, nuclear agencies and private firms from fourteen different countries had collaborated to create the company under the auspices of the European Organisation for Economic Co-operation, the OEEC.

The establishment was officially inaugurated at Mol in Belgium on 7 July 1966. Reprocessing activities went on for nearly ten years until January 1975. This date marked the end of the first life of the undertaking, which was then converted to the management of the radioactive waste it had produced.

Eurochemic went into liquidation in 1982. The site and its installations were transferred in 1985 to the Belgian state, which entrusted its management to Belgoprocess, a subsidiary of the National Radioactive Waste and Fissile Material Organisation, ONDRAF. International co-operation at Eurochemic terminated when the company was dissolved on 28 November 1990.

The principal feature of this study of Eurochemic is that it places the emphasis, in a history of nuclear energy that has developed substantially in the contemporary period, on civilian nuclear power and more particularly on the activities of the back end of the nuclear fuel cycle. From the characteristics of this nuclear undertaking it is possible to evaluate the potential of and constraints affecting three aspects of international technical, political and industrial co-operation, always tightly interwoven.

The sources and the historical approach they dictate are such that the story to be told has to be primarily chronological in format, with the topical taking second place, and covering – in an attempt to combine these together – the political, economic and technical aspects that are essential for an understanding of the development of that small part of the "atomic complex" known as the reprocessing and management of radioactive waste.

The history of the activities at the back end of the fuel cycle necessitates a new approach in the nuclear story

The history of Eurochemic occupies a special place in any history of the rapidly developing nuclear energy

A rapidly developing story

The process of writing the history of nuclear energy began during the sixties when work started, at the request of the public authorities, on books devoted to the national nuclear agencies of the United States[1] and the United Kingdom[2]. In 1980 the first comprehensive work was published, focusing essentially on the international

---

1 Initially ordered by the USAEC and carried out for thirty years by R. G. Hewlett, assisted in turn by O. E. Anderson, F. Duncan and J. M. Holl, this primarily political history currently consists of three volumes and covers the period from 1939 to 1961: *The New World, 1939-1946*, published in 1962; *Atomic Shield, 1947-1952*, published in 1969; and finally *Atoms for Peace and War, 1953-1961*, published in 1989.

2 GOWING M. (1964), *Britain and Atomic Energy 1939-1945*, followed by GOWING M. (1974), *Independence and Deterrence, Britain and Atomic Energy 1945-1952.*

aspects of military and civilian atomic developments from their beginnings up to 1979, *Le complexe atomique, histoire politique de l'énergie nucléaire* [The Atomic Complex, a Political History of Nuclear Energy], by Bertrand Goldschmidt[3]. In recent years the history of nuclear energy has been considerably enriched first by the opening of the archives at the end of the 30-year embargo period and, secondly, by the launching of national and international research programmes.

Considerable progress was made in the military field, through the international programme known as "Nuclear Forces in the Evolution of Postwar American-European-Soviet Relations: A Four Nations Program", also known as the Nuclear History Program (NHP), covering the period 1987-1992[4].

Less progress was made with regard to civilian nuclear power which, it is true, developed later than the military side. The opening of the archives – essential if the historian is to do his job – was very recent and very limited[5]. As regards the international undertakings, it is important to stress the pioneering role of the history of CERN[6], produced by an international team of science historians, the first two volumes of which were published in 1987 and 1990. However in most cases published works have been produced by researchers in political science or law, whose approach is different from that of historians[7]. However the bibliography is now in the process of evolving as the archives are opened[8].

The history of the Eurochemic company covers a particular branch of nuclear technology, the so-called "activities at the back end of the fuel cycle", which has hitherto not been the subject of historical research[9]

"The nuclear fuel cycle" (Figures 1 and 2) is the term usually applied to the industrial activities that extend from the mining of uranium through to the final burial of nuclear waste – a stage which, although planned, has not yet been carried out on an industrial scale. The mid-point of the cycle is the nuclear generation of electricity. Seen from this mid-point, a distinction is drawn between the activities at the so-called "front end of the cycle" which extend from mining through refining and isotopic separation to the fabrication of the fuel, and those at the so-called "back end of the cycle" which cover reprocessing and waste management. The reprocessing activity is essential in justifying this terminology. During reprocessing, fissile material is in fact recovered before being recycled in the isotopic separation or fuel fabrication plants. Thus the reprocessing plants are said to "close the cycle".

As a general rule, bibliographical references are given in abbreviated form as footnotes. A bibliography by subject is included after the general conclusion. Comprehensive references are set out in alphabetical order at the end of the book.

[3] GOLDSCHMIDT B. (1980) is still today an essential reference work, despite the highly regrettable lack of any list of sources or bibliography. Its author is not an historian, but one of the leading French personalities in the nuclear field. The book is in two parts, "The Explosion", devoted to the military aspects, and "Combustion" which covers the civilian aspects. Born in 1912, a graduate engineer of the Ecole de physique et de chimie de Paris, Bertrand Goldschmidt worked in the Curie laboratory from 1934 to 1940. During the second World War he contributed to the British and American atomic research and together with Jules Guéron, also one of the "CEA Canadians", was at the origin of the development of the chemistry of nuclear materials in the Commissariat à l'énergie atomique (CEA) [French Atomic Energy Commission], before going on to be its Director of External Relations. In this role he played a part in the history of Eurochemic.

[4] This project, financed by German and American foundations and involving university staff and researchers, under the direction of Uwe Nerlich (Stiftung Wissenschaft und Politik) and George Quester (Maryland University), resulted in the publication of "Occasional Papers" as from 1989. The French branch was directed by Maurice Vaïsse, together with the Groupe d'études français pour l'histoire de l'armement nucléaire (GREFHAN). See *Vierteljahreshefte für Zeitgeschichte* (1988), pp. 373 et seq., VAISSE M. (1991), L'Histoire de l'armement nucléaire, *Vingtième siecle*, n° 32, October- December 1991, pp. 93-94.

[5] Although the historians of CERN have had free access to that organisation's archives, the EURODIF historian had to work on fragmented and partial documentation because of the lack of historical records at Pierrelatte: he basically had to construct his own history – even though it was ordered by the organisation itself – from the results of interviews. See DAVIET J. P. (1993), p. 367. However he did have access to "a certain number of files" in the historical archives now being established at the CEA.

[6] HERMANN A., KRIGE, J., MERSITS U., PESTRE D. (1987 and 1990).

[7] For example the only general works on EURATOM available at the moment are books written in the sixties by researchers in political science: POLACH J. G. (1964), SCHEINMAN L. (1967), which is in fact a long article, and that by PIROTTE O., GIRERD P., MARSAL P., MORSON S. (1988), who are lawyers.

[8] This move seems more advanced in the United States, as a result of the DOE policy for the massive declassification of the USAEC archives.

[9] However a rapid introduction to the back end of the fuel cycle and its characteristics was recently provided, WALKER W. (1992).

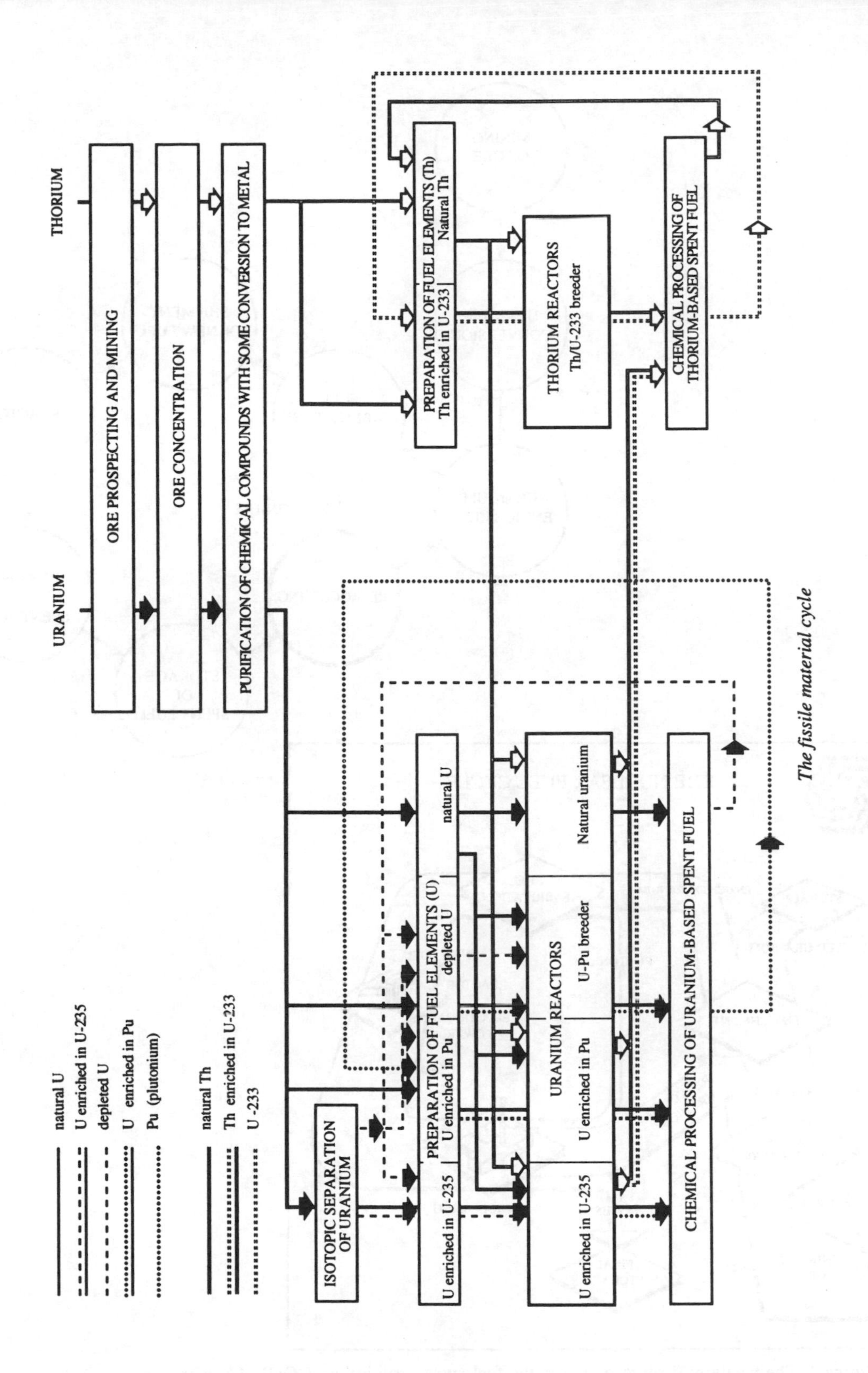

Figure 1. The “fissile material cycle” seen by the OEEC in 1957. At that time two cycles, based upon uranium and thorium respectively, were being considered. The waste aspect is ignored (Source: OEEC/EPA (1957), p. 309).

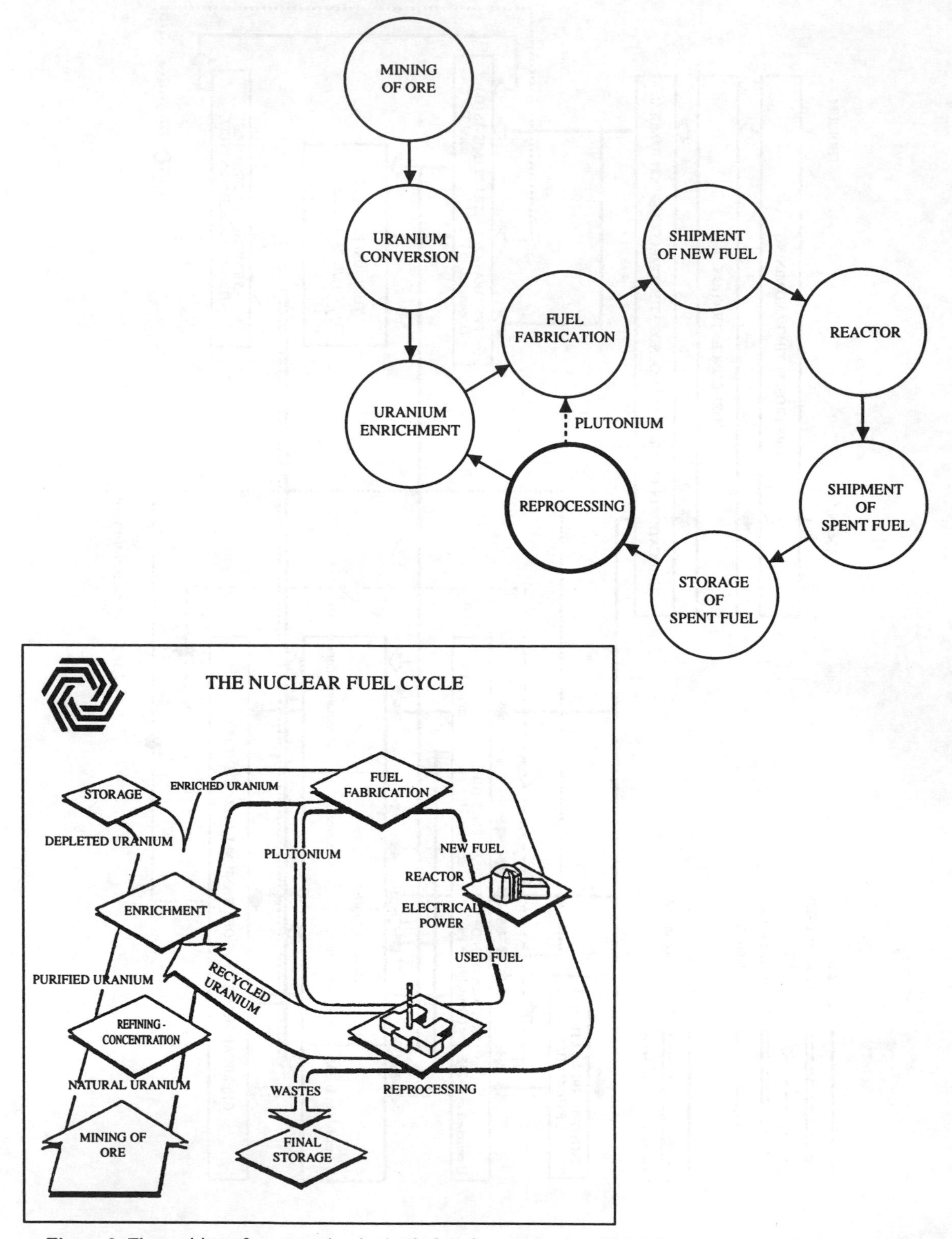

Figure 2. The position of reprocessing in the fuel cycle, seen by the COGEMA management in 1979 and by the Communications Directorate of the COGEMA La Hague establishment in 1992. Although the head-end of the cycle, from the mine to the conversion process, is shown in 1979, the diagram does not mention the management of the nuclear wastes resulting from reprocessing, which means that the cycle is in fact not completely closed. Although reprocessing does allow the uranium and plutonium to be recycled, this too produces waste. Hence there is a tail-end of the cycle, which is taken into account in exploded fashion on the 1992 diagram, where it is shared between the storage of the spent uranium and the storage – currently temporary for the more active substances – of the reprocessing wastes (Source for upper diagram: RGN (1979), p. 160. Source of lower diagram: booklet of diagrams "The Reprocessing of Spent Fuel at La Hague", Figure n° 1, Tourlaville, handed out to visitors to the establishment in 1992).

Nuclear reprocessing and waste management are particularly sensitive activities that are being hotly debated today for reasons concerned with problems of nuclear proliferation, safety and security, and the environment. Thus any study utilising high grade archives can contribute to elucidating the current debate. Such a study may cover the situation of reprocessing in the nuclear industry (even though the United States abandoned reprocessing in 1977 and Germany in 1989-1990, France, Japan, the United Kingdom and India are continuing to invest in it), the risks of proliferation (the function of reprocessing is to isolate uranium, plutonium and fission products), or even the production and management of nuclear wastes (particularly as regards the quantities to be processed, the costs and methods of conditioning). However the subject of final storage on the surface or in deep geological formations is outside the field of the history of Eurochemic.

From the standpoint of the history of technology, Eurochemic was a place of learning, and of the creation, transfer and diffusion of knowledge.

Eurochemic handled, separated and conditioned uranium, plutonium and fission products. It developed special reprocessing techniques, applied a comprehensive programme of conditioning to its radioactive wastes and began the decommissioning of its site.

The history of Eurochemic is also that of two kinds of technology transfer: geographical and sectoral. The initial steps of the undertaking were guided largely by the United States, the pioneers in the field of reprocessing. Eurochemic also relied on French experience. The European scientists and engineers took over American and French know-how, recasting and adapting it to the needs of European co-operation. The plant was built jointly by European contractors and involved, besides an outstanding degree of co-operation, the adaptation to a nuclear establishment of systems from other industries such as chemicals, metallurgy and electronics.

One of the aims of the undertaking was to diffuse knowledge to the shareholders and to train scientists and engineers. In this regard, the company produced as much information and know-how as it did radioactive materials. Through its activities it trained several hundred specialists who can now be found throughout Europe in reprocessing, waste management and other sectors of the nuclear industry. Accordingly this aspect of Eurochemic has contributed to a better understanding of how scientific and technical networks function.

As a locus of European co-operation, Eurochemic was both an actor on the international scene as well as an entity that was worked and remodelled by the international forces and the way in which they developed from the middle of the nineteen-fifties

There is no doubt that the history of international relations is at present experiencing a new beginning. This development is characterised by an approach paying more attention to the interplay of underlying forces, to the links between the different decision-making levels and to the convergent and divergent movements between the different cultures, and is beginning to bear fruit in the history of European and international nuclear relations[10]. The co-operation in the Eurochemic framework mainly involved the countries of western Europe, but the United States did play an important role. This co-operation began contemporaneously with the emergence of European organisations in the 1950s, in the nuclear field as well as in others. Its development parallels that of international relations in general.

A study of how the undertaking was created can make a contribution to the history of international decision-making in matters of high technology[11]. The project that gave birth to the undertaking was successful because the national interests of the different creating countries coincided at a particular moment in time, and because this favourable situation was turned to account by the determined action of a group of men from a variety of countries and professional backgrounds, led by Pierre Huet who was to become the first Director-General of the OEEC European Nuclear Energy Agency.

Using a method of "successive approximations", which in the space of a few months made it possible to move from an abstract concept of a joint undertaking to a precise project for a reprocessing plant, a technical co-operation system which met the technical and political aspirations of most OEEC countries was devised and internationally accepted. The Eurochemic project, through close co-operation in a new type of joint venture, was to involve an "international semi-public company", public bodies and leading European industries. In this way the Eurochemic experiment was an early example of international co-operation not only in R & D but also in the industrial field.

---

10 See for example n° 68 and 69 of *Relations internationales*, winter 1991 and spring 1992, devoted to "nuclear questions in international relations". As regards nuclear co-operation in Europe, see the historiographical review by Maurice Vaïsse on the occasion of the conference held on 18 and 19 November 1991 at Louvain-la-Neuve, VAISSE M. (1992).

11 See PESTRE D. (1986) and KRIGE J. ed. (1992), and the lecture given by John Krige on a comparative study of CERN and ELDO at the Summer School held at the Cité des sciences et de l'industrie at La Villette in July 1994.

The interests of the participants were highly diverse and, as a result, the consensus became unstable as time progressed. Continuing this experiment in co-operation involved seeking to maintain a dynamic and fluctuating equilibrium in a changing environment. The political balance needed for continuing the experiment was maintained, or rather continuously recreated, for complex and sometimes unexpected reasons which can be identified from an analysis of the records. The decision to put an end to the reprocessing work marked the end of the equilibrium, although not immediately that of co-operation, particularly because of the powerful constraints resulting from the presence of nuclear waste on the site.

What happened to Eurochemic in the period 1974 to 1985 is also a unique case of the "nationalisation" of an international undertaking. It reflects the links that formed between a sovereign state – Belgium – and an international structure located on its territory. This situation had special legal, economic and environmental implications, particularly because the company was governed by an international Convention with national law playing only a secondary role, applying notably as regards relations between the company and its staff, the procedure for reaching agreement with the national safety and security authorities, as well as industrial and commercial activities.

Eurochemic, a creature of the OEEC, was also in touch with other international organisations in the nuclear field, such as the OEEC European Nuclear Energy Agency (ENEA and Eurochemic were both offshoots of the OEEC and ENEA made a powerful contribution to setting up the company), EURATOM (of the six signatory countries of the Treaty of Rome, only Luxembourg took no part in the experiment) and the International Atomic Energy Agency, a United Nations body (in particular for security control). Eurochemic was the site of special links between these different international organisations, especially concerning the system of guarantees related to the supply of American fuel and then to the Non-Proliferation Treaty.

Finally, the Eurochemic experiment, a unique international reprocessing venture, provides food for thought about how fuel cycle activities can be internationalised, an approach which from time to time is suggested as one way of reducing the risks of proliferation[12].

Eurochemic was a nuclear undertaking which, during its short life, underwent profound changes both in its structures and its objectives

Eurochemic was also an enterprise, an international shareholding company whose objectives were applied research and the supply of reprocessing as an industrial service. International co-operation in advanced and costly technologies has taken a variety of forms ranging from a specialist international organisation, like CERN, to the Groupement d'intérêt économique (GIE) [Economic Interest Grouping] such as Airbus Industrie[13]. The form chosen for Eurochemic was highly original and played a major role in the development of the company. As such its history forms part of what is now the fertile field of the history of enterprises, the core of a history of emerging organisations[14]. Since Alfred D. Chandler, the history of enterprises has paid special attention to the links that form between strategies and structures; moreover the history of organisations, following the example of the management researcher Henry Mintzberg, relates the rational character usually attributed to the formulation of strategies in enterprises to a changing and sometimes surprising environment. Changes in the nature of the enterprise are inseparable from those in its structures and reflect, from time to time, the constraints suffered or the strategies drawn up. The instability of Eurochemic's structures during its first life, contrasting with the immobility of the second period, is very striking.

The dual vocation of the undertaking, expressed in the definition of it as an "industrial pilot", resulted at the time of its creation in a compromise which somewhat encumbered its future. Its R & D vocation – as far as internal operations were concerned – led to the establishment of highly specific links between the Research Division and the Technical Division during the phases of planning and construction. Its industrial vocation linked it to the reprocessing market which has unique features because of the type of industrial service provided – the extraction of uranium and plutonium, regarded as materials with "high added political value".

Because of the nuclear nature of the activities, work was organised in a highly specific way, with job definitions and strict rules of procedure and surveillance which it is possible to precisely describe and analyse. In this field it appears that the safety, risk prevention and control policies were introduced at a very early stage.

---

12 Starting with the 1946 Lilienthal-Acheson plan and including the past and present proposals made by the IAEA.

13 For the forms taken by co-operation in the aviation and space field, the "mutual organisations", see KOENIG C., THIETARD R. A. (1988).

14 For the history of organisations and their links with management sciences, see the seminal article by Patrick Fridenson in *Les Annales,* FRIDENSON P., (1989). Eurochemic was not "a true company" but a public-private hybrid, at least in its early life, but is certainly a subject in the history of organisations as understood by Patrick Fridenson.

Potential incidents and accidents were analysed as early as 1965 and actual incidents covered in detailed reports[15] describing both their progress and their radiological effects.

However the history of Eurochemic is also one of rapid change, which took the undertaking from reprocessing to radioactive waste management. These changes were naturally accompanied by changes in its structures and retraining of its staff. The modest size of the undertaking allows this metamorphosis to be followed very closely.

However the possibility of placing this new object in an historical perspective depended strongly on the available sources and how these were approached.

A history emerging from a systematic comparison of the written sources and the memories of those who were involved

The written sources

The unrestricted opening of the archives of the Eurochemic company provides a rare opportunity to apply conventional historical methods to a contemporary subject.

As early as 1985, the members of the company's Board of Liquidators[16] had envisaged having the history of the Eurochemic enterprise written, regarding it as the final phase of the co-operative venture which had lasted thirty-five years. The idea was to "learn the lessons of experience", but also to save the company from being forgotten[17]. At the final meeting of the Board in November 1990, the practical ways and means of producing a "history" were decided upon. The Board set up a Publishing Committee involving the principal players and other senior figures[18] and an historian was selected in February 1991.

As that historian my first task was to construct an historical archive from the abundant administrative papers which had built up over the years and had been distributed between the offices of the organisation which set up the company, the OECD Nuclear Energy Agency in Paris[19], the company's own offices and those of the Belgoprocess company which had taken over management of the site, both of these at Mol in Belgium[20]. These archives are of outstanding quality and richness, owing to the need to keep the many people involved in any international co-operative venture fully informed, and further enhanced by one of the particular aims of this co-operative venture which was to provide information to thc participants. Also a substantial effort of popularisation was made in the documents, in that experts had to explain technical subjects to non-specialists so that they might take political or financial decisions. From the outset this state of affairs facilitatcd understanding of the technical issues, even if the technical arguments are always somewhat one-sided[21].

However the mass of documentation was such that it was essential to devise a strategy for reading and utilising it.

It was necessary to decide which archives would be used as principal written sources for the present work. Indeed it was impossible in the time available – and in any cvent inappropriate for a general history of the undertaking – to read all the historical archives, covering several tens of mctrcs of shelving.

---

15 In this way the history of Eurochemic can help improve understanding of the actual management of industrial hazards in a sector usually regarded, particularly after the Chernobyl disaster, as particularly delicate and "incident-prone", see KERVERN G. Y, RUBISE P. (1991).

16 The Secretary of the Board of Liquidators, Otto von Busekist, had then started preparatory work and begun to put the archives in order. He was kind enough to give me access to his files.

17 The termination of another joint undertaking of the OECD European Nuclear Energy Agency, the Dragon project, resulted in a history of the project being written by a British scientific journalist, SHAW E. N. (1983).

18 See below for its membership.

19 The archives of the ENEA are in the process of being transferred to the European archives in Florence and will be accessible after the 30-year embargo period.

20 It would have been impossible to construct the historical archive without the assistance of Pierre Strohl and of the OECD/NEA documentalist, Sally Godwin; or without the contribution in Mol, as regards the Eurochemic papers, of Otto von Busekist, former legal adviser to the OECD/NEA and former Secretary of the Board of Liquidators of the Eurochemic company; Willem Drent, former Head of Eurochemic's Documentation Section, and Oscar Martinelle. Resources belonging to the Belgoprocess company were made available to me by Els Delande, Head of Documentation and Public Relations, with the agreement of J. Claes, Managing Director of the company and with the agreement of ONDRAF. It is intended that all Eurochemic's records will be incorporated in those of the OEEC/OECD at the European Community Archives in Florence, as from 1995.

21 This fact makes it virtually impossible to develop "analyses of controversies" from the Eurochemic archives alone.

Although there were very few internal sources for the beginnings of the project, making my documentation very sparse, they exploded in volume as the project developed. Having reviewed the company's information and decision-making mechanisms, I focused, for the lifetime of the company, on the records of the Board of Directors and its heir, the Board of Liquidators, and on those of the Special Group, which expressed the views of governments within the OECD on the company. In fact all the information necessary for supervising the project and taking technical decisions was referred to the Board of Directors. The Technical Committee and the Managing Director reported regularly to the board. The Special Group representing the governments, for its part, debated the major strategic orientations and settled the financial problems, on the basis of summary documents that were usually very well produced. From the minutes of the meetings it is possible, despite their often smooth style, to follow the reviews and discussions and even the disputes between the partners in this co-operative venture[22]. However, in addition to these sources, many other documents were used according to needs or discoveries.

The use of this type of source does of course have an impact on the type of history produced and does mark out its limitations. The internal story and the one seen "from on high", i.e., from the managers' point of view, are rather particular, to the detriment of social history and the determination of the local or regional impact of the undertaking[23]. In order to avoid a blinkered approach and to explain certain allusions, an attempt has been made to put the subject in perspective and take a broad view by invoking the few existing review works and above all by systematically examining the relevant technical journals for the whole of the period of interest[24]. Here most use was made of two nuclear sources in two principal member countries of Eurochemic. The German journal *Atomwirtschaft*[25], has been published in an unbroken series since 1956. For France, use was made of the series *L'âge nucléaire* (1956-1958), *Energie nucléaire* (1957-1974)[26] and the *Revue générale nucléaire* since 1975[27]. These references provided a chronological framework which may be useful for understanding changes internal to the company[28].

Systematic comparison with the memories of those involved

Of course the contemporary historian fortunate enough to have archives at his disposal also has the privilege of being able to discuss the results of his perusal of written sources and his hypotheses not only with his colleagues but also with those directly involved.

In an article published in *Annales* in 1935[29], Lucien Fèvre expressed his long-felt desire to see the historical task approached using the techniques of a "working co-operative" consisting of real technicians and one historian, the latter being charged with "finally wielding the pen to identify, attune, arrange and, if necessary, explain the results obtained by all".

The way in which the co-operative work with the Publishing Committee was organised bears some analogy with this model.

The Committee was chaired by Pierre Huet, former Director-General of the ENEA, honorary member of the French Council of State, and included Emile Detilleux, former Managing Director of Eurochemic, and former Managing Director of ONDRAF, Marcel Frérotte, former Director for Energy in the Belgian Ministry for

---

[22] However it is difficult or even impossible to make any fine analysis of the decision-making process. The preserved text in fact consists of summaries, approved subsequent to the discussion and decision processes.

[23] Quite apart from the time constraint, producing such a history calls for special linguistic capabilities. It is true that the languages of decision were French and English, but most of the social life of the undertaking, i.e., that involving the staff and the local environment, was conducted in Dutch.

[24] The greatest use was made of the collections in the Centre de documentation of the CEN located at the Saclay CEN (with its wealth of books, reports and journals this is an unrivalled source of documentation for any nuclear historian in France), together with the resources of various other libraries and documentation centres, the Bibliothèque de documentation internationale contemporaine [Library of International Contemporary Documentation] (BDIC) at Nanterre, the German Historical Institute and the Médiathèque d'histoire des sciences et des techniques [Multimedia Library of the History of Science and Engineering] at the Cité des sciences et de l'industrie at La Villette, and the library of the OECD.

[25] Abbreviated as *AtW*.

[26] Abbreviated as *EN*.

[27] Abbreviated as *RGN*.

[28] A brief list of events in chronological order showing the most important aspects of this matrix is included at the end of each part of the book.

[29] Thoughts on the history of technology, *Annales,* 1935, pp. 531-535. Republished in the collection *Pour une histoire à part entière,* Paris, SEVPEN, Bibliothèque générale de l'EPHE, VIth section, 1962, 1852 p., pp. 559-664. The quotation was drawn from the latter, p. 664.

Economic Affairs and former Chairman of ONDRAF, Oscar Martinelle, former Financial Director of Eurochemic, Jean-Michel Pictet, former Chairman of the Eurochemic Special Group, Rudolf Rometsch, former Managing Director of Eurochemic, formerly a Deputy Director-General of the IAEA and former Chairman of CEDRA/NAGRA, Yves Sousselier, former Chairman of Eurochemic's Technical Committee, former scientific adviser to the CEA, and Pierre Strohl, former Secretary-General of Eurochemic and former Deputy Director-General of the OECD Nuclear Energy Agency (NEA)[30]. The first drafts of the different chapters were forwarded to the members of the committee as I worked through the archives. The committee members provided me with their views and comments either in writing or orally, at three general meetings and during regular meetings with two technical groups, one focused on legal and political problems, the other on more specifically chemical issues.

The contemporary nature of the subject and the method used also allowed a number of iterations between the draft and the memories of those concerned.

A schedule of individual interviews was prepared[31]. This took place in two stages. The first, pre-drafting phase, concerned those people involved with the company over a long period and dealt with general topics. The aim was to obtain their overall view of the history of the company and of its different phases. The second stage took place in parallel with the drafting process and referred to more particular issues which were unclear or controversial.

This process of comparing past events – as evoked in the work of an historian[32] – with the memories of those concerned was extremely valuable. It not only stimulated the thought processes of the historian but also those of the people consulted with regard to the nature of their own memories, concerned about their own place in the history of the company and hence the "viewpoint" they had about its evolution. Any written history, including this one, is a construction rather than a reconstruction of facts, in the commonly used sense of an "historical reconstruction". The particular feature of the historical approach used here – which has proved its worth but which also has its limitations – is the attempt to give meaning to what happened, a meaning that was not perceived by those directly involved who, to use their own words, felt that they were "moving forward in a fog". This meaning, constructed *post hoc*, should not obscure the fact that decisions were taken in a context of great uncertainty. This is particularly clear as regards the management of nuclear wastes. However, although the process of picking out the "relevant information" – what was "useful" for the story – may attract teleological reproaches, not to do so incurs the risk of incompleteness, of continuous questioning, quite apart from the danger of drowning in the documentation. The process had to be finite and a book produced, so a particular approach had to be chosen. Of course others are possible and probably as legitimate as the one adopted. Indeed it is strongly to be hoped that the records lodged in the European Union Archives in Florence will stimulate further research about an undertaking whose history certainly deserves preserving and whose life could thus be extended as a "subject of history".

The pattern of the story: essentially chronological and only secondarily topical

There are a number of reasons why a chronological breakdown was essential: the great mass of documentation, which had to be broken down into subsets; the procedure of comparing two major types of source – written and oral; the requirements inherent in the basic task which had to be both monographic and

---

30 L. A. Nøjd, former chairman of the Special Group and Teun J. Barendregt, former Technical Director, who were originally members of the publishing committee, died before it held its first meeting.

31 The interviews involved Maurits Demonie, a technician and subsequently an engineer at Eurochemic and at Belgoprocess since 1962, Emile Detilleux, Managing Director of Eurochemic from 1973 to 1984, Louis Geens, technician and subsequently engineer at Eurochemic and then at Belgoprocess, Pierre Huet, Director-General of the ENEA from 1959 to 1963, legal adviser to Eurochemic from 1963 to 1990, Yves Leclerq-Aubreton, Eurochemic's Director of Administration from 1960 to 1969, Managing Director of Eurochemic from 1969 to 1972, Michel Lung, engineer at SGN, Oscar Martinelle, head of the Finance Division of Eurochemic, Alain Mongon, engineer responsible for instrumentation from 1962 to 1964, André Redon, engineer at the CEA, Rudolf Rometsch, Managing Director of Eurochemic from 1964 to 1969, Earl Shank, chemical engineer at ORNL, American adviser to Eurochemic from 1962 to 1969, Yves Sousselier, chairman of Eurochemic's Technical Committee from 1957 to 1990, Pierre Strohl, Deputy Director-General of the OECD Nuclear Energy Agency (NEA) from 1982 to 1992, and Walter Schüller, a former Section Head at Eurochemic and former Managing Director of the German WAK pilot plant. The dates and venues of these interviews are given in the references to sources.

32 An initial version of this study was prepared under the supervision of Patrick Fridenson and submitted as a doctoral thesis to the Ecole des hautes études en sciences sociales, Paris, in 1994, before Messrs Jean-Jacques Salomon (Chairman), Dominique Pestre, Maurice Vaïsse (rapporteurs) and Pierre Strohl. Jean-Pierre Daviet, John Krige, Michel Margairaz and Alan S. Milward also gave me the benefit of their encouragement and advice.

general; and finally the features that were specific to the evolution of an organisation which changed its principal activity during its lifetime.

The scheme adopted for the study therefore involves four main parts covering half a century of history, the first and last each covering a period of 15 years (1940-1955 and 1975-1990), the second and third each covering ten years. A fifth part seek to place the entire period into a topical perspective.

Part I covers the origins of a project which emerged from the conjunction – at first sight unexpected – between the Organisation for European Economic Co-operation (OEEC) and the reprocessing of spent nuclear fuel for civilian purposes. It shows how and why a number of different European countries were led to consider co-operation in this field within the OEEC framework, and how this desire to co-operate led to the Convention of 20 December 1957, giving birth to the company.

Part II covers the accomplishment of the project. This was difficult and the project almost foundered for financial and political reasons, but the arguments for technical co-operation, which governed the transfers of nuclear technology as well as co-operation between the European firms responsible for construction, led to the commencement of reprocessing activities in 1966.

Part III retraces the first life of the company and lays emphasis on the conditions in which reprocessing was conducted at Eurochemic for ten years, and upon the reasons why it was decided that these activities should come to an end. It terminates with an assessment of the industrial and technical inheritance, part of which largely determined the second life of the undertaking.

This second life lasted 15 years from the end of reprocessing to the liquidation of the company in 1990. This period was marked by the recovery of the site and by the execution of a virtually complete programme of nuclear waste management.

A concluding part seeks to place the undertaking in a broader perspective, first concerning the technical history of reprocessing, and secondly that of European co-operation.

Postscript on units

From 1950 to 1958 the accounting unit used by the OECD and Eurochemic was the European Payments Union Unit of Account (EPU/UA) and then the European Monetary Agreement Unit of Account (EMA/UA) up to 1973. The values of the EPU/UA and the EMA/UA were tied to that of the American dollar until the end of the period of fixed parities and was equivalent to 50 Belgian francs (BF). As from 1971 the unit of account used by Eurochemic was made equivalent to the value of 50 BF.

The main types of radioactivity and their measurement

Type of radioactivity	*Nature*	*Range in increasing order*
Alpha	emission of a helium nucleus	a few cm in air
Beta +	explusion of a positron	
Beta - (natural beta)	expulsion of an electron	
Gamma[33]	emission of a photon	several cm of lead

Units for measuring radioactivity[34]

Two systems of measurement units were used during the period covered by the history of Eurochemic:

Type of unit	*Emission: activity of source*	*Impingement energy received (absorbed dose)*	*Impingement: biological effect (EBD: equivalent biological dose);(EDE: effective dose equivalent)*[35]
Old unit and definition	CURIE (Ci) 1 Ci is the activity of 1 gramme of the radium isotope 226	RAD[36] 1 Ci generates 1 rad at a distance of 1 metre	REM 1 rem corresponds to 1 rad
Unit used in the international system (SI) and definition	BECQUEREL (Bq) 1 Bq is the activity that corresponds to 1 disintegration per second	GRAY (Gy) Energy given up by radiation to 1 unit of mass of the exposed material. 1 Gy = 1 joule per kg or 10^7 ergs per kg	SIEVERT (Sv) 1 Sv corresponds to 1 Gy
Conversion factors	1 Ci = 3,7 10^{10} Bq 1 Bq = 2,7 10^{-11} Ci	1 RAD= 10^{-2} Gy 1 Gy = 100 RAD	1 REM = 10^{-2} Sv (1 REM = 10 mSv) 1 Sv = 100 REM

---

[33] Gamma radiation involves the emission of a photon, which is an X-ray of very short wavelength (10^{-10} to 10^{-14} metres, frequency of 1018 to 1022 and energy 104 to 108 eV) which can penetrate deep into matter.

[34] Source: TEILLAC J. (1988), TUBIANA M., BERTIN M. (1989).

[35] The conversion of received energy into biological dose equivalents involves weighting coefficients that take into account the type of radiation and the nature of the biological tissue involved, involving a "quality factor" or "Q factor", ranging from 1 to 20. When calculating dose per unit time, a "dose rate" or an "equivalent dose rate" is obtained, expressed as Sv/year or as rem per year, day, hour, etc.

[36] The Roentgen (R) is also a unit outside the international system; it measures exposure to X-ray or gamma radiation as well as the quantity of electrostatic electricity produced by ionisation in 1 cm^3 of air: 1 R = 0.88 rad; 1 rad = 1.14 R. It was virtually never used at Eurochemic.

Units used for the study

In the sources, the international units came into use only at a very late stage.

The conversion of REMs into mSv allows easy translations. *The millisievert (mSv) is therefore used as the unit of dose in the study.*

However for *measuring activity* we preferred to retain the *Curie (Ci)* rather than the Becquerel (Bq). Since the Ci is a much larger unit than the Bq, the quantities of radioactivity present in nuclear plants or in wastes are easy to appreciate intuitively when expressed in Ci, i.e. in thousands or millions, while if the unit Bq is used the values are in high powers of 10 (for example a million Ci corresponds to $3.7 \cdot 10^{16}$ Bq).

PART I

The origins and birth of the project

February 1940 – December 1957

Introduction

On 24 October 1956 the "Working Party for the Creation of a Joint Undertaking for the Chemical Processing of Irradiated Fuels" held its first meeting at the headquarters of the Organisation for European Economic Co-operation (OEEC), in the Château de la Muette, Paris, beginning the process which was to lead to the establishment of the Eurochemic company.

At first sight, this encounter between an international economic organisation and a highly specialised branch of the nuclear chemical industry seems odd, and deserves some explanation.

To provide one, it is important to trace two historical paths that are initially very distinct but which gradually converge to meld in the Eurochemic project.

The first is the history of the chemistry of nuclear materials, which began in the military sphere, in connection with the American programme for producing the atomic bomb, and was "demilitarised" some ten years later, as international relations improved following the death of Stalin. This history is the subject of the first chapter.

The second path is the development of European co-operation from the end of the 1940s, which is covered in the next chapter. The treatment shows first how the OEEC came to wrestle with the problems of technical co-operation and to suggest appropriate structures to the countries of Europe and, secondly, the reasons why fourteen countries of the OEEC decided to work together in the field of nuclear energy.

Finally the third chapter explains how this conjunction of the chemistry of nuclear materials and the OEEC led to the Eurochemic project and why this particular project, focused on reprocessing, was successful while its fellow, concerned with the joint construction of an isotopic separation plant, came to nought.

Chapter 1

The chemical processing of irradiated fuel: a brief technical and diplomatic history, 1940 to 1956

In February 1941 at Berkeley, California, scientists found that when uranium had been bombarded by neutrons it contained traces of a new element; only a month later they showed it to be fissile, and realised that it was easier to produce than the fissile isotope of uranium, U-235. The new element, subsequently named "plutonium," was to become the subject of intense research as part of the atomic bomb development project.

Hardly three years after plutonium had been discovered, its chemical properties had been sufficiently investigated in the laboratory for its production to move on to the industrial stage. However it was only in 1954 that the process of separating out the plutonium reached technical maturity with the commissioning of the military plant at Savannah River (South Carolina) which employed the PUREX extraction process. Subsequent to December 1953, the PUREX method was applied for civilian purposes, like most of the techniques the nuclear industry originally developed for military applications. The process for extracting plutonium was adapted for the chemical reprocessing of spent nuclear fuel, whereupon it came within the ambit of the international organisations, particularly the OEEC in Europe, which made it a subject for technical co-operation.

Until 1954 therefore, the history of the chemical processing of spent fuels was an integral part of military atomic developments, specifically that of the plutonium bomb, one of the two types of weapon covered by the Manhattan Project. In the early days the main objective of reprocessing was to extract the plutonium from the uranium fuel within which it had been formed: the fission products were separated out at the same time since it was important to remove any troublesome radioactive elements. The recovery of the uranium itself was considered only after the war.

The emergence of civilian reprocessing was one result of the *Atoms for Peace* speech given by United States President Dwight D. Eisenhower in December 1953. The change to a civilian context in 1954 and 1955 refocused reprocessing endeavours on *all* the products: plutonium, uranium and fission products. However the presence of the plutonium in the spent fuel required special guarantees and control procedures that were inseparable from the specifically chemical processes. It was in this way that reprocessing found itself embroiled in "plutonium diplomacy".

Technically speaking, these developments took place in two stages. During the war, time pressure forced the choice of a discontinuous co-precipitation process in a simple configuration, that of the "canyon plant". The post-war period saw the development of continuous processes, in which the nuclear fuels were dissolved to form an aqueous solution, with counterflow extraction using organic solvents not miscible with the water which together brought about the separation. This development arose partly from the desire to recover not only the plutonium from the spent fuel but also the uranium, owing to fears of an impending shortage. In the early 1950s came the discovery of the properties of tributyl phosphate (TBP), the basis of the process selected for Eurochemic, known as PUREX.

Thus diplomacy and engineering were closely allied in the development of reprocessing. Its invention stemmed from the war; the immediate post-war period and the Cold War saw parallel developments in a partitioned world; the early improvements in world relations brought about a gradual lifting of the curtain of secrecy with the PUREX process winning the day.

From Berkeley to Nagasaki: the crucial role of reprocessing in the race to make the plutonium bomb (December 1940 - August 1945)[1]

The industrial facilities for producing plutonium were designed and built in a very short time as part of the "Manhattan District" project. This was revealed immediately after the war in the Smyth report[2] which until 1955 was the only public source of information about activities till then kept under the seal of military secrecy.

Less than four years passed between the discovery of plutonium in the United States and its military use; the first reprocessing plant at Hanford (Washington) was put into service only 13 months after its construction had started. This great speed was of course made necessary by military urgency, but rendered possible by the very special way in which the project was organised. The different phases of research, pilot unit development and full scale construction, which would normally follow one another in sequence, in fact took place simultaneously. Moreover the change of scale was enormous, since there was no time to try out the intermediate stages.

From the laboratory to the pilot plant: Berkeley, Chicago, Oak Ridge (1940-1943)

From the discovery of neptunium to the first production of plutonium in the laboratory: 1940 - 1942

The research phase[3] began in February 1940 when E. M. McMillan bombarded a thin film of uranium with neutrons in an experiment aimed at determining the penetrative power of certain fission products. When McMillan analysed his results, he also found a type of radioactivity different from that of the fission products used. P.H. Abelson joined him to identify a possible unknown element. Together they succeeded in separating the first transuranium element, of atomic number 93[4], which they named neptunium[5]. Its radioactivity was due to its disintegration with a period of 2.3 days, into another transuranium element of atomic number 94[6].

This research was continued[7] during the winter of 1940–1941 by Glenn T. Seaborg, A. C. Wahl, J. W. Kennedy and E. G. Segré, who produced the new element by bombarding uranium with deuterons and then by neutrons, using the Berkeley 60-inch cyclotron. Half a microgramme of the 239 isotope of element 94 was isolated in this first experiment, but even this tiny quantity allowed information to be gathered about its chemical properties[8]. In mid-May 1941 the researchers were able to show that the new element could be made to fission by slow neutrons. This discovery caused them to look for ways of producing more because its fissile nature could make it possible, as with U-235, to produce an atomic bomb. Some 500 microgrammes were produced by the end of 1942 through the prolonged bombardment of uranyl nitrate and provided a better understanding of the chemistry of the element which had been named plutonium[9]. Laboratory production was improved and repeated in Berkeley and then at Washington University in Saint Louis. By December 1943, two milligrammes had been produced in the laboratory. The irradiated uranium was separated by a co-precipitation process with lanthanum fluoride. However the use of this chemical, which is highly corrosive, made it very difficult to move to an industrial scale.

---

1 This section is based upon GOLDSCHMIDT B. (1980), GROVES L. (1962), HEILBRON J.L., SEIDEL R.W. (1989), HOUNSHELL D.A., SMITH J.K. Jr. (1988), UN (1956), RADVANYI P. and BORDRY M. (1992), SEABORG G.T. (1958), SMYTH H.D.W. (1948), SOUSSELIER Y. (1960), USAEC (1955), and on the trilogy HEWLETT R.G., ANDERSON O.E. (1962), HEWLETT R.G., DUNCAN F. (1969) and HEWLETT R.G., HOLL J.M. (1989).

2 This report, the full title of which is *General Account of the Development of Methods of Using Atomic Energy for Military Purposes,* was made public six days after Hiroshima, see HEWLETT R.G., ANDERSON O.E. (1962), p. xi, and published the same year in the *Review of Modern Physics* XVII, October 1945, pp. 351-471, for reasons internal to the United States, on the instructions of the project director, Leslie H. Groves. If the project had failed, Congress would have required explanations for expenditure exceeding three billion dollars, undertaken outside its control. The report would have provided the justification. It was turned into a book in 1948.

3 SEABORG G.T. (1958), notes provided by Rudolf Rometsch. HEILBRON J.L., SEIDEL R.W. (1989) describes the European search for transuranium elements since 1934, pp. 430-441, and gives a detailed account of the parallel research leading to the discovery of neptunium and plutonium, pp. 456-464.

4 ${}_{92}{}^{238}U + {}_{0}{}^{1}n \rightarrow {}_{92}{}^{239}U \rightarrow {}_{-1}{}^{0}e + {}_{93}{}^{239}Np$

5 Named after Neptune, the planet whose orbit around the sun lies immediately outside that of Uranus. It will be recalled that the discoverer of uranium, Klaproth, named it in memory of the discovery of the planet Uranus by Herschel in 1781.

6 ${}_{93}{}^{239}Np \rightarrow {}_{-1}{}^{0}e + {}_{94}{}^{239}Pu$

7 HEWLETT R.G., ANDERSON O.E., (1962), pp. 34 *et seq.*

8 Analysis of such small quantities was based on methods used in biology and pharmacy.

9 Using the same approach as for naming element 93. Element 94 was named only in 1942, HEWLETT R.G., ANDERSON O.E. (1962), p.89.

From the reactor to the bomb: the essential stage represented by the chemical processing of irradiated fuel

Laboratories at the X10 pile and at the Clinton (Oak Ridge) pilot plant

To produce sufficient plutonium for manufacturing the bomb, a change of scale was needed in the separation process. Since Enrico Fermi's achievement of the first chain reaction in a natural uranium-graphite pile on 2 December 1942, it had become clear that it would be relatively simple to produce large amounts of plutonium in a "pile" of this type.

In the pile, the neutrons emitted during fission of the uranium 235 isotope[10] are in fact mostly absorbed by the predominant isotope, uranium-238, a process that converts it to uranium-239. This is then transformed in two successive beta emissions first into neptunium-239 and then into plutonium-239.

The fissile plutonium produced in this way can be separated from the uranium by chemical means, being a different element, in proportions measurable in kilograms per tonne of uranium. By now the critical mass – in other words the minimum mass necessary to sustain the fission of plutonium 239 – had been calculated[11].

Accordingly it was decided in January 1943 to build a 1000 kW pile, known as X10, for producing plutonium, at the Oak Ridge Clinton Engineer Works, near Knoxville (Tennessee). The plutonium was to be extracted and refined in a pilot reprocessing plant which had to be near the pile[12] and which was therefore the world's first reprocessing unit.

Plutonium, the "Metallurgical Project" and Du Pont de Nemours

Plutonium research had been incorporated into the Manhattan project and had become part of the "Metallurgical Project" developed at the University of Chicago. The plutonium chemistry group was led by Glenn T. Seaborg, who had left Berkeley for Chicago in April 1942. From the summer of 1942 his team sought to identify suitable industrial separation processes[13]. The deadline subsequently laid down by the industrial operator was 1 June 1943. Four methods of chemical separation were explored[14]. The precipitation process was chosen, with preference given to solvent extraction. This was a well-known approach that was frequently used in radioactive chemistry[15]. It had technical advantages which were carried over into the costs of the process. The operation could be broken down into separate cycles, allowing systems to be used more than once, thus limiting the overall scale[16]. However the best precipitation agent had to be found. A systematic exploration of the properties of heavy metal phosphates revealed the qualities of bismuth phosphate, which was then adopted as the basis for the industrial process.

In early November 1942, Leslie H. Groves, who had been directing the project since September 1942, made the chemical firm Du Pont de Nemours responsible for the industrial production of plutonium, comprising the piles and treatment plants[17].

The Du Pont management was not overly enthusiastic about the project, which risked damaging its image and diverting its best researchers away from its commercial and private work. However the company agreed to contribute to the war effort, on condition that the fact that it would make no profit would be made public[18]. For this reason the contract negotiated between the government and the company was of the "cost plus fixed fee" type, but the profit was a merely symbolic one dollar. In return the government indemnified the company against loss and assumed full liability in the event of an accident. This situation was tenable only in wartime and as

---

10 Natural uranium contains 0.71% of uranium-235, which is fissile.

11 The theoretical value of this mass for plutonium metal containing over 90% of the isotope 239 is 5 kilogrammes, compared with 21 kilogrammes of highly enriched uranium (containing over 80% of U-235). However practical problems – isotopic composition, purity of the metal, inadequate focusing of the initial neutron flux, and so on – make the critical mass higher in bombs.

12 See HEWLETT R.G., ANDERSON O.E. (1962), Insert between pp. 120-121. For X10, pp. 193-198. See also the ORNL REVIEW (1992), Chapter 1, *Wartime Laboratory*, especially pp. 8-28, on the Metallurgical Laboratory, X10 and the pilot unit for plutonium extraction.

13 See HEWLETT R.G., ANDERSON O.E. (1962), pp. 182-185.

14 "Volatility, absorption, solvent extraction and precipitation". See also HEWLETT R.G., ANDERSON O.E. (1962), p 204.

15 SMYTH H. DW (1948) p. 137.

16 The discontinuous nature of the process, regarded at the time as an advantage, was one of the reasons why it was later abandoned in favour of organic solvent processes, usually continuous.

17 See HOUNSHELL D.A., SMITH J.K. Jr. (1988), pp. 331-346, GROVES L.H. (1962) p. 46 *et seq* and HEWLETT R.G. ANDERSON O.E. (1962), pp. 105 *et seq.* and 186-187.

18 "They [i.e. the Du Pont management] wanted the company to meet its patriotic responsibilities but not at the cost of being viewed again as a merchant of death". HOUNSHELL D.A., SMITH J.K. Jr. (1988), p. 333.

soon as hostilities ended Du Pont made it clear that it wished to disengage from the nuclear field both at Oak Ridge and Hanford in order to return to its primary vocation[19].

Du Pont therefore set up a special section in its Explosives Division known as TNX, with the task of building – together with the Chicago "Metallurgical Laboratory" and the firm Stone and Webster[20] – the Clinton pile and the pilot reprocessing unit located nearby, and also working on the plans for the industrial unit to be located at Hanford. At the end of 1942, when the chemical process to be used had not yet been selected, the main construction options for the first generation of American reprocessing plants, known as "canyon plants" were established. To carry through this massive change of scale and provide liaison between the academic researchers and the industrial engineers, Du Pont appointed Crawford H. Greenewalt to be head of the TNX R & D section; Greenewalt had been responsible for the successful industrial development of nylon which Wallace H. Carothers had invented in the company in 1936[21].

On 4 January 1943 Leslie Groves awarded the contract for the construction of the Clinton pilot plant to Du Pont de Nemours.

The "canyon" type pilot plant (Figures 3 to 6), the foundations of which were poured in March 1943, was about 35 metres long[22] and was two-thirds underground. Adjoining were the control rooms, the analytical laboratories and a plutonium purification laboratory. The pile began operation on 4 November 1943 and the first experimental separation of plutonium began on 20 December. By 1 February 1944, 190 mg of plutonium had been produced. The separation efficiency improved because by June of the same year the recovery rate was around 80-90%. By the summer of 1944, plutonium produced at Oak Ridge had been shipped to Los Alamos where the basic weapon research was carried out.

The experience accumulated at Clinton was transferred as quickly as possible to Hanford where work was proceeding at full speed.

Hanford and the first three reprocessing plants (1943-1945)

Development of the site

The government had bought land at Hanford at the beginning of 1943[23]. The construction of Hanford Engineer Works began on 7 June 1943 and the first pile was in service by September 1944.

Three separation plants of the "canyon" type were built, on two sites at least ten miles from the piles, to the south of Gable Mountain (Figure 7): the T and U plants were built on the "200 West" site, and plant B on the "200 East" site. Work began during the summer of 1943, but was then suspended until 1944 because priority had to be given to building the piles. Construction of the separation building for the first plant – T – commenced in January 1944 and the building was completed in September. Plants U and B followed three months later.

The first reprocessing process

The highly radioactive uranium rods which had been irradiated in the pile were left in the ponds for a period of time to allow their activity to decay somewhat, then taken in lead flasks by rail to the chemical plant where they were dissolved in nitric acid.

The resulting solution was then passed through three stages[24]: the plutonium was first separated out by co-precipitation with bismuth phosphate, then concentrated by lanthanum fluoride, using the laboratory process, and finally purified using hydrogen peroxide.

The final product, an aqueous solution of plutonium nitrate, was shipped to Los Alamos for reduction into metal. The uranium and fission products remained in the high activity liquid wastes that were stored in tanks nearby. The other effluent was simply discharged into the enormous areas of dessert surrounding the site.

---

19 The nuclear field was therefore, to use the expression of HOUNSHELL D.A., SMITH J.K. Jr (1988), p. 342, a "road not taken". In fact until the end of the war it was uncertain whether nuclear activities would be included in the company's future business. Finally it was decided that nuclear engineering was not chemistry, but rather came under the energy sector. Significantly it was General Electric that inherited the Hanford site (see below).

20 HOUNSHELL D.A., SMITH J.K. Jr. (1988) p. 338.

21 HOUNSHELL D.A., SMITH J.K. Jr. (1988) op. 339-340.

22 100 feet.

23 HEWLETT R.G., ANDERSON O.E. (1962) gives extensive treatment to Hanford p. 212-226.

24 HEWLETT R.G., ANDERSON O.E. (1962) p. 222.

Figure 3. General view of the Clinton Laboratories after the end of the Second World War. At centre right, the building containing the "Clinton pile" and immediately behind it on the left, the pilot extraction plant with the ventilation ducts on the roof and, to the side, the stairways giving access to the semi-underground cells (Source: ORNL REVIEW (1992) p. 28).

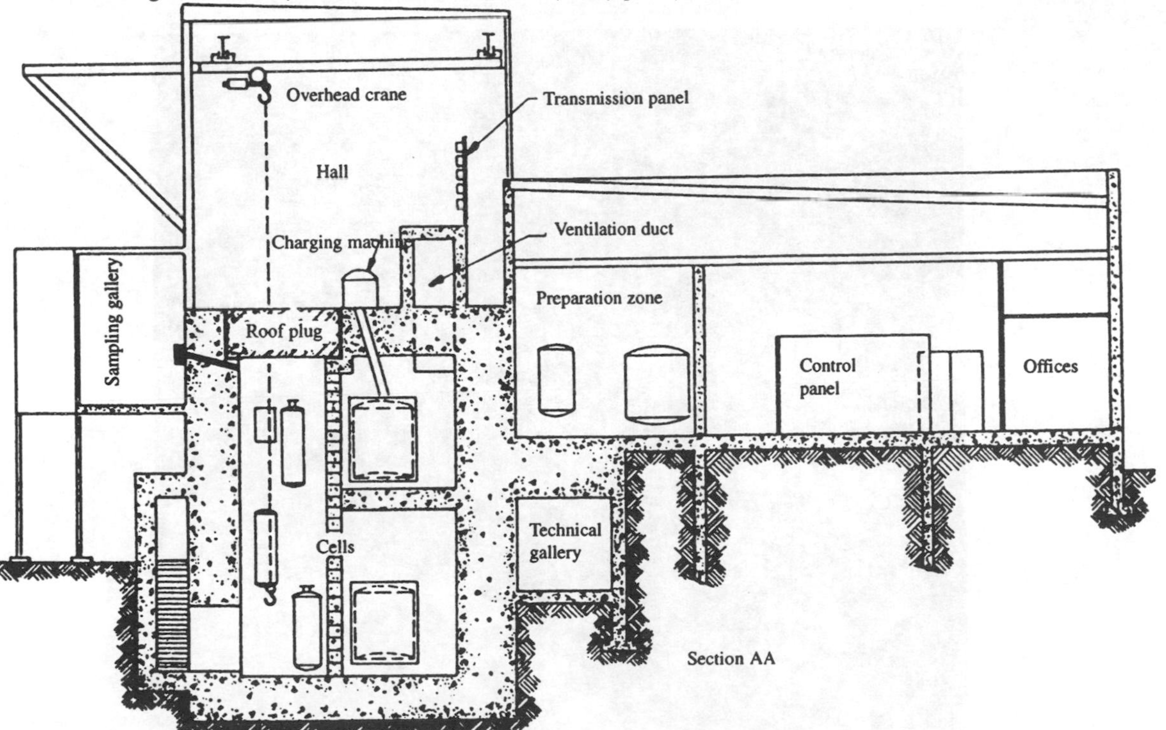

Figure 4. Sectional drawing of the Thorex facility at Oak Ridge, installed in the building which had housed the first pilot processing plant at Oak Ridge. The overhead crane facility was built after 1945 (Source: BRUCE F.R., SHANK E.M. et al (1955), p.61).

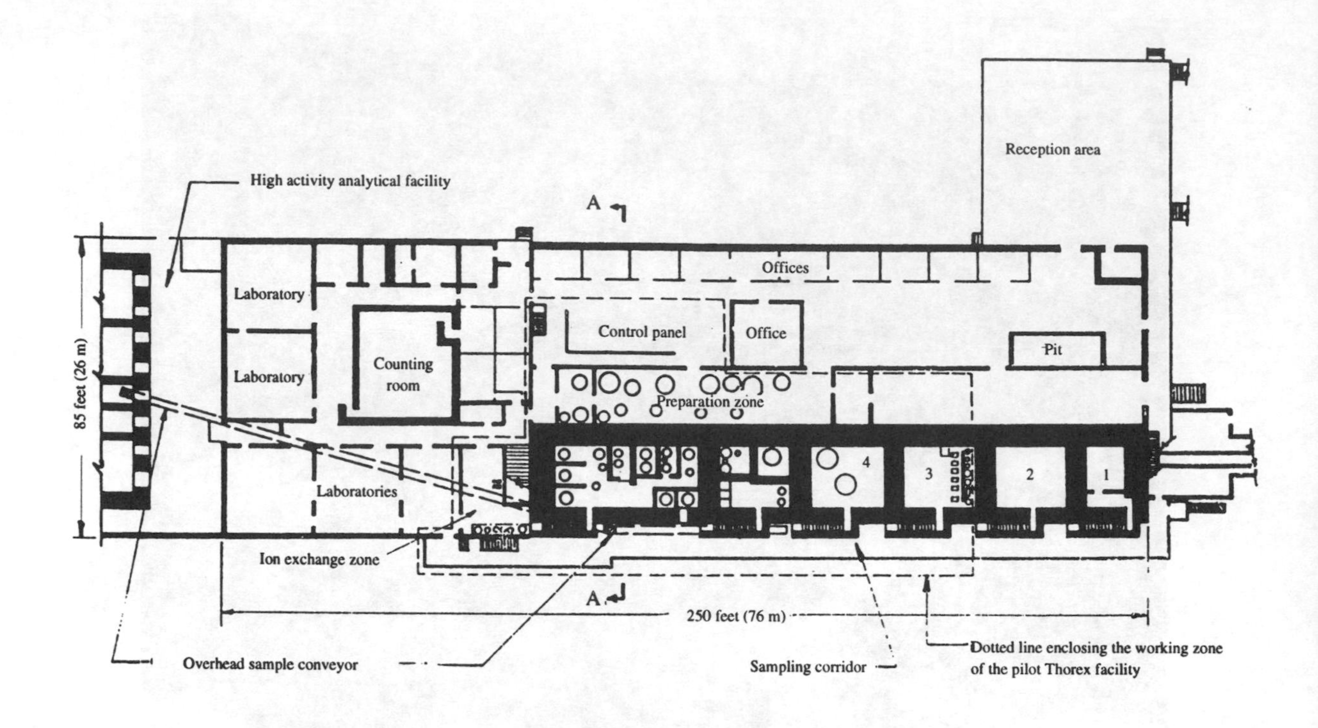

Figure 5. Plan view of the Thorex facility at Oak Ridge, installed in the building that had housed the first pilot reprocessing unit at Oak Ridge. The original installation included six semi-underground cells, connected to the X-10 pile by a channel the dimensions of which can still be seen on the right-hand side of the drawing (Source: BRUCE F.R., SHANK E.M. et al (1955), p.60.).

Figure 6. View of the control panels on the outer walls of the cells at the pilot Oak Ridge plant in 1944 (Source: ORNL REVIEW (1992), p. 23).

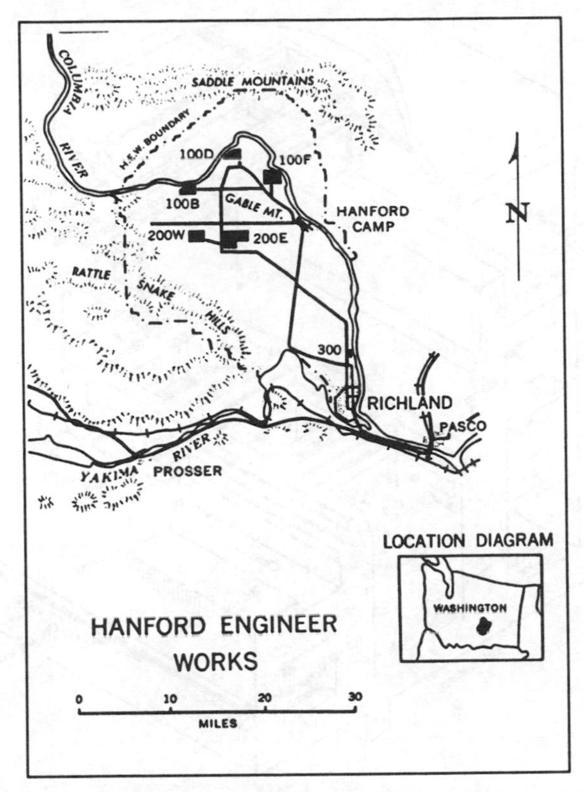

Figure 7. Simplified diagram of the Hanford site at the end of the Second World War. Of the three reprocessing plants, units T and U were at 200 W and unit B at 200 E (Source: HEWLETT R.G., ANDERSON O.E. (1962), Insert between pages 224 and 225).

A remotely maintained canyon plant (Figure 8)

The various stages of the process were carried out behind thick radiation shielding and were entirely remotely controlled. The fact that the cells were inaccessible during operation meant that the design had to be simple. The systems used required only slight maintenance, involved the smallest possible number of moving parts and used components that could easily be replaced remotely.

The building in which separation took place consisted of a row of 40 concrete cells, each separated from its neighbour by nearly 2 metres of concrete and forming a "canyon" 240 metres long, enclosed in a building 20 metres wide[25]. Each cell was covered by a heavy removable concrete block. An overhead crane ran above the cells, under the concrete roof which was at a height of 24 metres, and was used to lift off the cell covers and carry out maintenance work or replacement of tanks and equipment. The crane driver's cab was separated from the

---

[25] Description of the 221 T plant taken from HEWLETT R.G., ANDERSON O.E. (1962), pp. 219 - 221: "Each separation plant was designed to include a separation building, [...] a ventilation building, [...] a waste storage area. The building itself, more than 800 feet long, 65 feet wide, and 80 high". "Inside [...] a row of forty concrete cells, most of them about 15 feet square and 20 feet deep, ran the length of the building. Each cell was separate from its neighbour by 6 feet of concrete and would be covered by concrete blocks 6 feet thick. [...] Along one side of the cell row and separated from it by 7 feet of concrete were the operating galleries on three levels, the lowest for electric controls, the intermediate for piping and remote lubrication equipment, and the upper for operating control boards. The entire area above the cells was enclosed by a single gallery 60 feet high and running the length of the building. Its 5 foot concrete walls and 3 foot roof slabs were designed to prevent the escape of radiation when the cell covers were removed. [...] Radiation meant remote control, which in turn placed a premium on simplicity of design, mechanical perfection, maintenance free operation, and interchangeability of parts. To avoid servicing pumps and valves, steam jets were developed to transfer process materials from one tank to another. [...] Once the plant was operating, the only access to the cells would be by means of the huge bridge crane which travelled the length of the building. From the heavily shielded cab behind a concrete parapet above the gallery, operators could look over the canyon with specially designed periscopes and television sets. They could use the seventy-ton hook to lift off the cell covers and lighter equipment to work within the cells."

See also GROVES L. (1962), p. 85.

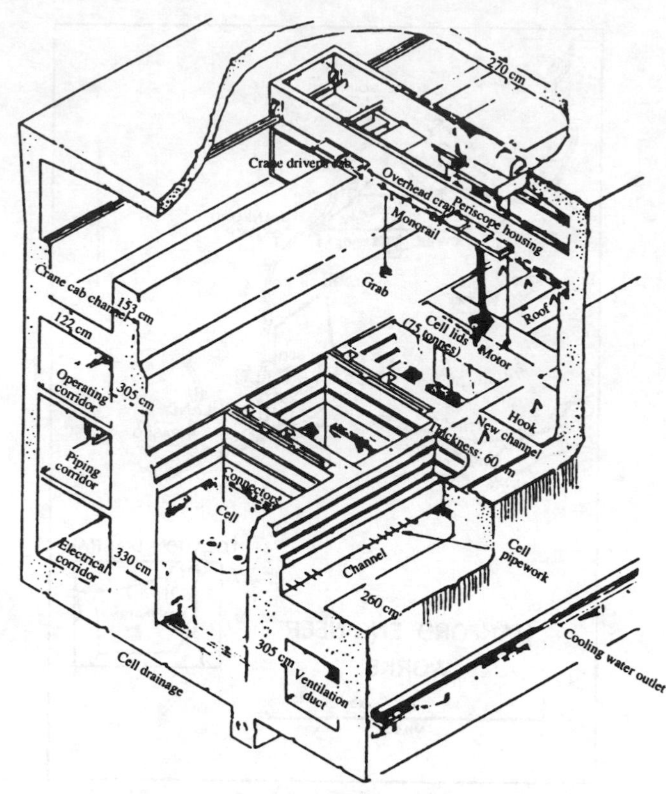

Figure 8. Perspective sectional view of one of the Hanford bismuth phosphate treatment plants showing the structure of a remotely controlled and maintained "canyon plant" (Source: SHWENNESEN J.L. (1958), p. 321).

crane gallery by a concrete wall. The operators worked remotely using periscopes and television cameras. Under the cab gallery there were three operating galleries for instruments, valves, pumps and the steam ejectors used for transferring liquids from one tank to another. There were separate buildings for the ventilation system, the filters and effluent storage.

Plutonium production for the bombs

The first batch of fuel elements from pile B entered the dissolving unit of plant T on 26 December 1944, and the first ever industrial plutonium reprocessing cycle was completed at the end of January. On 2 February 1945 the first load of plutonium nitrate arrived at Los Alamos by truck. Others rapidly followed and provided the fissile material needed for producing the two bombs which were exploded at Alamogordo on 16 July and at Nagasaki on 8 August[26]. The other two plants at Hanford entered production in the summer of 1945 and all three remained in service until 1952.

The period of secrecy. Parallel discoveries of organic solvents. 1945 - December 1953

From the industrial standpoint, the bismuth phosphate process was not the most efficient. It was used at Hanford as a laboratory process transposed to the industrial scale in record time with no regard to cost, for reasons of urgency and because of the need to minimise the technical risks in view of the little experience available with plutonium chemistry.

After the war, research into plutonium extraction focussed on continuous processes using organic solvents which held out hope of greater efficiency. New plants were commissioned, using new processes, as shown in Table 1:

26 The bomb dropped on Hiroshima on 6 August, known as "Little Boy", was a uranium weapon. That dropped on Nagasaki was known as "Fat Man".

Table 1. **Industrial reprocessing processes: 1945-1954**

Name of process	Principal extraction product	Chemical formula	Plant	Date commissioned
PHOSPHATE CO-PRECIPITATION	Bismuth phosphate	$Bi P O_4$	Hanford (United States)	1945
ALL ACETATE CO-PRECIPITATION	Sodium acetate?	?	Pu extraction plant at the Mayak chemicals Combinat (Cheliabinsk 40)	1948
TRIGLY[27]	Triglycol dichloride	$Cl(C_2H_4O)_2.C_2H_4Cl$	Chalk River (Canada)	1949
REDOX	Hexone or methyl isobutyl ketone	$CH_3CO.C_4H_9$	Hanford (United States)	1952
BUTEX	Dibutylcarbitol (DBC)	$C_4H_9.O(C_2H_4.O)_2.C_4H_9$	Windscale (United Kingdom)	1952
PUREX	Tributyl-phosphate (TBP)	$O{=}P(OC_4H_7)_3$	Savannah River (United States)	1954

This research was carried out separately in the United States, Canada, the United Kingdom, France and the Soviet Union. The curtain of military secrecy which had enveloped nuclear research from 1939 onwards owing to the war, lifted only slightly for domestic reasons by the Smyth report, was very soon lowered again owing to the worsening relations between the Allies. This distrust naturally involved the two great powers but also had an influence on relations between the Western Allies. In 1943 the British were excluded from working on the bomb following the Quebec Agreements of 19 August[28]. The MacMahon Act of August 1946 reinforced the policy of secrecy in the United States[29]. As far as France was concerned, the fact that the French Atomic Energy Commission was headed by a communist scientist, Frédéric Joliot-Curie, who had worked on uranium fission before the war, made very strict precautions necessary in the eyes of thc Americans[30]. However information gleaned during the war and data obtained from the Smyth report provided the basis for research and are a partial explanation of the extent to which work went on in parallel. However it was in the United States that the most effective organic solvent was discovered.

Competition between two organic solvents – hexone and TBP – in the United States (1946-1954)

As soon as the war ended, Du Pont de Nemours informed Leslie H. Groves that it wished to withdraw from the management of Hanford. General Electric was approached and took over the site in May 1946[31]. There were many management difficulties and a delicate transitional period began which lasted until May 1947[32].

There were two problems with regard to reprocessing itself. First of all it was necessary to find a more effective process than the one used during the war; the particular need was for one that would make it possible to recover not only plutonium but also uranium from the spent fuel (in the immediate post-war period in the United States there were considerable fears of an impending shortage of uranium). The second problem was to recover the uranium contained in the high activity liquid wastes arising from the bismuth phosphate reprocessing at Hanford. Du Pont had drawn the USAEC's attention to the importance of this recovery. Plutonium research which, post-war, had been conducted in a fairly disorganised way[33], was refocused on these two problems, and produced results that demonstrated the advantages of organic solvents.

---

27 This was in fact a semi-industrial process, included here for completeness.

28 See below for the Canadian research.

29 However as far as the United Kingdom and its dominion, Canada, were concerned, this isolationism underwent a change following the appointment of Marshall as US Secretary of State in January 1947, when the two powers found a certain *modus vivendi*. See GOWING M. (1974), Vol. 1, Chapter 8, pp. 241 *et seq*. Exchanges of information took place in 1948 on nine specific aspects of nuclear technology, the seventh aspect being extractive chemistry. However the impact of these exchanges was limited: see below.

30 WEART S. (1979).

31 HEWLETT R.G., ANDERSON O.E. (1962), pp. 628-630.

32 HEWLETT R.G., DUNCAN F. (1969), pp. 142-145.

33 HEWLETT R.G., DUNCAN F. (1969), pp. 142-143.

Given the task of developing a new reprocessing method, General Electric concentrated on a process using the solvent hexone, which had been explored by the team of Glenn T. Seaborg in 1945. Preparations for building a new plant using the hexone REDOX process began at the Knolls Atomic Power Laboratory, managed by General Electric, and in a pilot unit at Hanford[34]. Construction began at Hanford early in 1950, the plant commenced operations in August 1951 and entered into production in 1952. At the same time at Idaho Falls the hexone process was adapted to the reprocessing of highly enriched uranium fuel from a nearby reactor. In February 1953 a pilot plant – the Idaho Chemical Processing Plant (ICPP) – came into service and was to play an important role in the history of Eurochemic.

However it was the search for answers to the second problem which led to the discovery of the most effective organic solvent. The Kellex Corporation had been charged with recovering the uranium present in the tanks containing the liquid wastes arising from bismuth phosphate reprocessing[35]. Laboratory research with this objective was carried out at the Oak Ridge National Laboratory (ORNL) where the properties of other organic solvents were also being investigated. At the end of 1949, ORNL reported the advantageous properties of tributyl phosphate or TBP, which were such that plans were made to connect a TBP unit to the bismuth phosphate plant still in service – thus threatening the future of REDOX – and the Kellex corporation was given the job. However General Electric's research was already well advanced when Kellex was still tackling the problems of raising a laboratory method to the industrial scale as well as being highly preoccupied with the industrial requirements of the Korean war. Work on the Kellex plant began in the autumn of 1950, but was only half completed when the REDOX plant began production in 1952[36] and in fact came into service only at the beginning of 1953[37].

In the meantime Du Pont de Nemours had begun working on a comprehensive spent fuel reprocessing unit using TBP near the Savannah River piles, having been recalled to its patriotic duty at the beginning of the Korean war[38]. The first industrial military unit for reprocessing with TBP was commissioned in 1954. This plant reproduced the general layout of the canyon plant and was fully remotely controlled.

In the United States all these activities were tightly protected by the McMahon Bill of 20 December 1945, which became the 1946 Atomic Energy Act which was voted at the end of July 1946[39]. This Act established the United States Atomic Energy Commission (USAEC) and enshrined America's nuclear isolationism. The objective was to continue the technical and military monopoly for as long as possible, but the American "closed door" policy did not prevent nuclear proliferation either in the USSR or in the United Kingdom.

Reprocessing in support of the Soviet nuclear bomb[40]. The Mayak (Cheliabinsk 40) extraction plant

After the Nagasaki explosion and the publication of the Smyth report, Soviet nuclear research was concentrated on plutonium. The "plutonium programme" directed by Igor Kourtchatov was planned to be completed by 21 December 1949 for Stalin's seventieth birthday. The research on the reprocessing method was led first by G. N. Yakovlev then by B. A. Nikitin and A. P. Ratner at the Radium Institute.

The decision was taken to build a major nuclear research centre. A site was chosen in the Urals at Kasli, near Kychtym, a city with a long tradition of ammunition production, located about 120 kilometres south of Sverdlovsk. Work on the "Urals Centre"[41] began in November 1945, involving 70 000 workers from 12 forced labour camps. In 1947 the construction of the first plutonium-generating reactor began within the secret military complex, now known by its postal address: Cheliabinsk 40[42]. This first graphite-moderated reactor, known as reactor A, was commissioned in June 1948. It appears that the first plutonium extraction plant, RT1, began to operate a few months later[43]. The first Soviet plutonium bomb was exploded on 29 August 1949 at the Semipalatinsk range in Kazakhstan. This put an end to the American atomic monopoly and also changed the whole pattern of international diplomacy.

---

34 GROVES L. (1962), p. 88 and HEWLETT R.G., DUNCAN F. (1969), pp. 549-551.

35 HEWLETT R.G., DUNCAN F. (1969), p. 174.

36 HEWLETT R.G., DUNCAN F. (1969), p. 550.

37 HEWLETT R.G., HOLL J.M. (1989), p. 162.

38 HOUNSHELL D.A., SMITH J.K. Jr. (1988), p. 345.

39 The text of the Act is given in CANTELON P. L., HEWLETT R. G., WILLIAMS R. C. (1991), pp. 77-91.

40 Confirmation of the information given here should be found in HOLLOWAY D. (1994).

41 When the Centre was opened to foreign visitors in 1989, it covered 200 km and had a direct workforce of 10 000. COCHRAN T. B. et al. (1993), p. 508.

42 Known as Cheliabinsk 65 since 1990. The first Soviet reactor went critical there at the end of December 1946.

43 It appears that two plants were in service at Cheliabinsk 40, but in 1989 there was only one.

The plutonium production process consisted of an initial phase of dissolution in nitric acid, followed by co-precipitation using a so-called "all acetate" process[44]. The effluent contained up to 100 g/l of sodium nitrate and up to 80 g/l of sodium acetate[45] as well as uranium, fission products and residual plutonium. Until September 1951 all the effluent, without exception, was diluted and discharged into the nearby river Techa, a tributary of the Irtych. Thereafter, and until 1953, it was discharged into Lake Karachay, with catastrophic effects on the environment and the nearby population. It was only in 1953 that the first storage facilities were commissioned, and then only for the fission product precipitates mixed with nitrates and acetates[46].

The original process was modified several times but then abandoned at some time unknown in favour of PUREX. The RT1 plant which had an estimated capacity of 400 t/year in 1989, was used for producing military plutonium until 1 November 1990. By then, 25 tonnes of plutonium were stored on the site.

Reprocessing in Canada: a short history (1944-1954)[47]

The Anglo-American Quebec Agreement of 19 August 1943 banned any transfer of technology concerned with plutonium chemistry and any communication on this subject to other countries[48], but the Americans had agreed on 8 June 1944 to supply the Anglo-Canadian laboratory in Montreal with the so-called "hot dogs"[49] – fuel rods which had been irradiated in the Oak Ridge pile and which contained plutonium.

In August 1944, an industrial chemistry section had been set up under A. Cambron in order to build a pilot separation unit. The choice of extraction method was entrusted to Bertrand Goldschmidt, who had worked with Glenn T. Seaborg in Chicago for a few months in 1942-1943. Goldschmidt chose to explore the extraction process using organic solvents which was one of those investigated in Chicago. The approach was empirical and systematic[50]: "to pick the best of the 300 organic solvents commercially available in the United States: half a dozen of these were chosen and using one of them we succeeded in developing a new plutonium extraction process in Montreal in 1945." The TRIGLY process was based upon triglycol dichloride. This process was discontinuous, worked well in separating plutonium but much less well for uranium.

Early in 1949, a small extraction unit was started up to extract plutonium from the Chalk River NRX reactor which had gone critical in July 1947. This unit was not very successful and in 1948 and 1949 modifications were made to the original process by combining the TRIGLY and hexone methods[51]. Modified in this way, the process worked satisfactorily in the spring of 1950. However on 13 December 1950 there was a fatal accident in the reprocessing waste management sector which followed the extraction process. An explosion in an evaporator containing radioactive ammonium nitrate killed one worker outright; four others required hospital treatment. The plant was then shut down until a replacement for ammonium nitrate could be found. Production resumed only in autumn 1952 and the plant continued to operate until September 1954. By then the Canadians were considering developing a plutonium separation plant using ion exchange. The construction of such a unit was still being considered in September 1955 but it was never built, for two reasons: first because in September 1954 Canada had concluded a reprocessing agreement with the United States for fuel irradiated in NRU; secondly and most important, because the fear of supply problems for nuclear fuel had been eliminated when large deposits of uranium were found in Ontario. The fear of a shortage had been the principal prime mover of Canadian reprocessing[52]. The fact is that Canadian plans for plutonium reprocessing came to an end one month after the first Geneva Conference.

---

44 "All-acetate precipitation scheme", COCHRAN T. B. et al., (1993) p. 520.

45 Composition confirmed in 1953.

46 It was one of the specially built storage tanks which exploded in September 1957, discharging 20 million Ci, or about 40% of the activity discharged during the Chernobyl accident.

47 EGGLESTON W. (1965), pp. 192 *et seq,* devotes a short chapter to this topic. BOTHWELL R. (1988) gives only a few lines and makes no allusion to the accident.

48 The text is given in FRUS (1970), pp. 1117-1119 and repeated in CANTELON P. L., HEWLETT R. G., WILLIAMS R. C. (1991), pp. 31-33. The third of five sections in the Agreement states: "[...] we will not either of us communicate any information about Tube Alloys to third parties without each other's consent".

49 This was the name given by the researchers to the fuel irradiated in the first piles.

50 GOLDSCHMIDT B. (1980) p. 69 and (1987) p. 328.

51 The latter had probably been suggested by the Americans.

52 It was for this reason that Canada was also developing a method for reprocessing thorium. In fact the situation is a little more complex than this, because although plutonium production work was quickly successful, a parallel facility for reprocessing thorium known as "laboratory 23" languished until 1951 and operated until 1958. This reprocessing "Cinderella" as it was called by Bothwell, was the poor relation of the plutonium extraction unit. The intensification of

The output of Canadian reprocessing was modest: 17 kilogrammes of plutonium had been extracted from 517 fuel elements. However out of the brief Canadian experience was born British reprocessing.

The United Kingdom, Windscale and BUTEX

British reprocessing began in 1946 when the Anglo-Canadian team at Chalk River discovered the qualities of an organic solvent other than triglycol dichloride. This was dibutyl carbitol which formed the basis of the BUTEX process.

Plans for a British plutonium extraction plant began with the decision taken in December 1945 (but not announced publicly until January 1947) to build a reactor for producing the plutonium needed for manufacturing a bomb[53]. The British "plutonium plan" was in four stages: construction of the experimental BEPO pile at the Harwell Research Centre, fabrication of the fuel at Springfields, irradiation of the fuel in the Windscale piles and, on the same site, extraction of the plutonium. Priority was naturally given to building the natural uranium air-cooled piles. The first Windscale pile came into service in October 1950 and the second eight months later.

Design and construction of the Windscale extraction plant was a co-operative venture involving three groups. The first two were public bodies which came under the Controller of Atomic Energy in the Ministry of Supply, Lord Portal; they involved scientists from the chemistry division of the Harwell Research Centre, directed by Robert Spence[54], and engineers and planners from the Risley Industrial Group under the responsibility of Christopher Hinton[55]. The third group was private: the Imperial Chemical Industries (ICI) company and more precisely its General Chemical and Metals Division, which had laboratories at Widnes and Springfields. ICI was the industrial operator of the project, analogous to Du Pont in the United States.

At the end of 1946, the team of Robert Spence, who was still working in Canada on the plutonium extracted from the "hot dogs" from the American Oak Ridge pile[56], was given the task of developing a process that could be lifted to the industrial scale. They opted for an organic solvent, dibutyl carbitol (DBC). This was more stable than TRIGLY when exposed to radiation and heat. Moreover it required no "salting out" agent, a fact which substantially reduced the amount of liquid wastes. Finally, unlike TRIGLY, this was a continuous process. Robert Spence also recommended the use of counter-flow extraction columns which had been tried out on a small scale in a pilot unit that operated until 1950 at Chalk River.

In the autumn of 1947 Christopher Hinton took over the Risley Industrial Group. He laid down the schedule and negotiated the contracts with the various firms responsible for building the different parts of the plant. On the ICI side, the project was directed by J. P. Baxter. The firm tested the extraction columns at full scale at Springfields and devised the process details in its industrial laboratories at Widnes. ICI was responsible for all the purely chemical part of the plant.

Construction work began on 1 January 1949. The Windscale treatment system involved six separate main buildings, each with a specialist task: the main plant[57], where the spent fuel was treated by DBC in five extraction columns and which was remote-controlled; the plutonium purification plant, the plutonium metal fabrication plant, the uranium purification plant, the solvent recovery plant and the radioactive waste treatment plant. The most active liquid waste was stored in tanks, the remainder being diluted and discharged into the Irish Sea from the adjoining shore.

The first non-active tests began in June 1951 but these resulted in a few unpleasant surprises, notably in August with the "virtual explosion" of Butex in an evaporator, which necessitated last minute modifications. In December a further problem appeared[58]. The plant went active on 22 February 1952 and the first plutonium metal was produced on 31 March 1952. On 3 October of the same year the first British bomb was exploded, this too being a plutonium type. The Windscale plant using the BUTEX process remained in service for 12 years.

By 1955 therefore, the British had a new plant which covered all their needs. The French, for their part, were just beginning to build their own.

---

the Cold War in 1949 resulted in Canada considering building a plutonium generating reactor, the output of which would have been sold to the United States. BOTHWELL R. (1988), p. 105-107.

53 There was also a uranium plan.

54 The Centre was directed by Sir John Cockroft.

55 Hinton had started his career with ICI before moving to the Ministry of Supply during the war, see GOWING M. (1974), Vol. 2, p. 7-10.

56 Spence returned to Harwell in the autumn of 1947.

57 Primary separation plant.

58 See GOWING M (1974), Vol. 2, p. 417.

Reproccessing in France: from Fort de Châtillon to the beginnings of Marcoule (1949-1955)[59]

By 1955 France had already made considerable progress in developing its "plutonium project".

The project had been made possible by the extraction of the first milligramme of plutonium on 20 November 1949 from fuel irradiated in ZOE, the first French pile. Extraction was carried out in the old military arsenal at Le Bouchet, madc available to the CEA[60] by the Ministry of Defence. The new chemistry department of the CEA, set up in 1949 and directed by Bertrand Goldschmidt, had in fact taken delivery of the spent fuel from ZOE the previous month and had put it through the Canadian TRIGLY extraction process. However research continued on extraction processes that might be capable of extension to the industrial scale, with systematic investigation of the properties of organic solvents, which the CEA ordered from Union Carbide[61].

Bertrand Goldschmidt was put on the track of TBP early in 1951, following a discussion with the Deputy Director of the Harwell Centre, who told him that the Americans had discovered a new solvent. Goldschmidt then carried out a thorough literature search "of all recent American publications covering solvent extraction in inorganic chemistry"[62], which led to the discovery of TBP by the French[63]. TBP was investigated at Le Bouchet and its qualities demonstrated in September or October: "For the last three months, a new solvent has been investigated in detail: its extraction and decontamination capability are such as to make it much superior to all those considered so far, and a number of versions of a process which seems full of promise are being investigated"[64]. During December, 40 mg of plutonium had been produced but the path now seemed open to industrial reprocessing. The TBP process was selected in 1952 for the construction of the plutonium plant.

The laboratory discovery very quickly found its way into the first nuclear Five-year Plan, also known as the "Gaillard" Plan (1952-1956). This was outlined in September 1951 and adopted by parliament in July 1952. The Plan gave priority to the development of plutonium production: 40 billion francs were set aside for the construction of two graphite-moderated piles and the corresponding plutonium extraction plant. In the meantime, the "Plutonium Section", directed by Pierre Regnaut, had moved from Le Bouchet to Fort de Châtillon, premises vacated as the Saclay Centre developed. Contacts were made with industry as from December 1951 and in January 1952 Bertrand Goldschmidt selected Saint-Gobain to construct a reprocessing plant in conjunction with the CEA. This was the beginning of a co-operative venture in reprocessing which continued until the company left the nuclear sector in 1978[65]. Saint-Gobain began by setting up a Nuclear Chemistry Laboratory, followed by an Organic Plutonium Products Division (known as "POP") which operated from 1955 to 1956, when it was joined by a young engineer from the Ecole de Chimie in Bordeaux, Michel Lung, who was also to play a major part in the history of Eurochemic[66].

---

59 General sources: GOLDSCHMIDT B. (1987), the annual Reports of the CEA, and notes taken by the author from B. Goldschmidt's presentation, *"Chemistry in the CEA, from the Laboratory to the Factory"*, at the Seminar arranged by Dominique Pestre at the Koyré Centre on 19 April 1991.

60 The French Atomic Energy Commission had been created on 18 October 1945 by an order of General de Gaulle, who was then head of the provisional government of the French Republic (GPRF).

61 As soon as he returned from Canada, Bertrand Goldschmidt continued his research on 50 solvents ordered from Union Carbide: he selected six. He was directed towards dibutyl carbatol in 1947 by the firm's salesman (Goldschmidt; private communication 1991).

62 GOLDSCHMIDT B. (1987), p. 421.

63 It was used as a separation agent for rare earths.

64 CEA (1951), p. 29.

65 Bertrand Goldschmidt relates how he came to select the Manufacture des glaces et produits chimiques of Saint-Gobain, Chauny and Cirey, in preference to Rhône-Poulenc, for reprocessing in GOLDSCHMIDT B. (1987), pp. 454-456.

The book by Jean-Pierre Daviet on Saint-Gobain has only a few lines on the firm's nuclear activities, DAVIET J. P. (1989) p. 246. In fact the chemical group's nuclear activities were never more than marginal.

66 The POP Division formed the nucleus from which the specialist subsidiary, Saint-Gobain nucléaire (SGN), was created; in 1965 it became Saint-Gobain techniques nouvelles. SGN played a very important role in the construction of the Eurochemic plant. In the early 1970s SGN came under the management of Pont-à-Mousson in the group and then in 1978 was sold to the CEA subsidiary specialising in the head-end and tail-end activities of the fuel cycle, the Compagnie générale des matières nucléaires (COGEMA), which became the majority shareholder, and to the engineering consultants (mainly in the petroleum field), TECHNIP. It is still active today under the original logo SGN, now meaning Société générale des techniques nouvelles, and constitutes the main focus of the EURISYS engineering consultancy network set up in 1989 by some 50 firms involved in the construction of the UP3 and UP2-800 plants at La Hague.

Building T at the Fort de Châtillon was chosen to accommodate a pilot plant. Experiments on a ZOE fuel rod were done in 1952 in the original Le Bouchet installation. As from 1952, one of those seconded to work for the CEA at Le Bouchet by the army technical section was André Redon, a young graduate engineer from the Conservatoire national des arts et métiers (CNAM), who was to play an important part in the history of Eurochemic and who soon joined the plutonium team of the CEA[67]. Designs for the pilot plant were ready by February 1953. Until early 1955, the pilot plant project was directed by Yves Sousselier, a petroleum engineer recruited by Bertrand Goldschmidt for his experience in industrial extraction processes[68]. Construction took about 8 months. The pilot plutonium extraction unit at Châtillon (Figures 9 and 10) was commissioned in 1954 and processed spent fuel from ZOE (renamed EL1) and, after undergoing modifications in 1955, fuel from the EL2 reactor at Saclay. It provided sufficient material for plutonium fuel production tests and for work on alloys with a view to manufacturing the bomb[69]. However the unit was shut down in 1957, mostly because of the problems raised by the management of its wastes. It was dismantled in 1961 and the premises taken over by hot laboratories the design of which was later to inspire those of Eurochemic.

The development of the pilot unit led to the construction of a full-scale plant. On 4 December 1952 the Marcoule site was chosen for the construction of the French plutonium producing piles and the extraction plant[70]. An Industrial Division under Pierre Taranger was set up to develop the site. The preliminary plans were drawn up by Saint-Gobain in 1954 and civil engineering work commenced on the site in June 1955.

The Marcoule plant (Figure 11), later named UP1, was commissioned in 1958 and the first reprocessing campaign took place at the end of 1959. The first French bomb was exploded on 13 February 1960.

A month after work began at Marcoule, the first International Conference on the Peaceful Uses of Atomic Energy was opened in Geneva. This put an end to the systematic policy of secrecy by unveiling certain nuclear techniques, particularly those related to reprocessing[71], and opened the way to international co-operation in the nuclear industry by demilitarising certain aspects. This Conference marked the culmination of a process that had begun two years earlier – inseparable from the overall changes in international relations following the death of Stalin, during the period of transition between the worst of the Cold War and the beginnings of *détente* – which might be denoted "the end of the "hot" Cold War"[72].

The end of secrecy and the beginning of civilian reprocessing (December 1953 - August 1955)

The metamorphosis of American nuclear policy in 1953[73], announced by President Eisenhower's "Atoms for Peace" speech to the UN General Assembly on 8 December 1953, and made manifest by the amendment to the MacMahon Act in July 1954, allowed international nuclear co-operation to begin in the field of reprocessing. Initially this was bilateral in form, but multilateral projects were subsequently reactivated.

---

67 Born in 1924, André Redon spent a year studying chemistry at university, then worked in a dye factory while continuing his studies in his special subject but also in industrial electricity and nuclear physics at the Conservatoire national des arts et métiers (CNAM).

68 Born in 1919, Yves Sousselier was preparing for the competitive entry examination to the Ecole polytechnique when war broke out. Having received his degree in mathematics in 1945 he entered the Ecole nationale du pétrole which was then in Strasbourg. On completion of his education, he worked as a petroleum engineer from 1948 to 1953, particularly on extraction processes. In 1953 he was introduced to Bertrand Goldschmidt who was looking for an engineer familiar with industry and solvent extraction. Following his period with the Fontenay pilot unit he worked for the CEA on the industrial aspects of ore processing. His experience at Fontenay made him one of the experts on the Eurochemic project and he was soon appointed chairman of the Technical Committee.

69 In fact in the meantime the French military nuclear programme had taken shape, as the recent work by Dominique Mongin shows. On 4 November 1954 the government of Pierre Mendès-France took a secret decision to set up a Nuclear Explosives Committee, chaired by a former classmate (École Polytechnique – X) of Pierre Guillaumat, General Jean Crépin. On 28 December 1954 the "General Research Bureau" the direct ancestor of the Military Applications Division (DAM) was created within the CEA under the direction of Colonel Albert Buchalet. On 20 May 1955 the Gaillard plan was extended and given a much more military connotation under a secret protocol between the delegate minister to the presidency of the cabinet Gaston Palewski, the Minister of Defence and the Minister of Finance. See MONGIN D. (1993), p. 18.

70 CEA (1953) p. 39.

71 However the veil of secrecy was lifted only partly: techniques considered to be strategically or economically vital, such as those of enrichment, remained confidential.

72 To use the expression of GIRAULT R., FRANK R., THOBIE J. (1993), p. 190.

73 HEWLETT R. G., HOLL J. M. (1989), especially Chapter VIII, Atoms for Peace, building American policy.

Figure 9. View of the mimic diagram indicator panel (in the foreground) and the control panel (in the background) of the Châtillon pilot plant (Source: *L'âge nucléaire* (1958), p. 333).

Figure 10. View of part of the plutonium solution concentration unit at the Châtillon pilot plant (Source: *L'âge nucléaire* (1958), p. 333).

Une vue de l'usine d'extraction de plutonium en voie d'achèvement. La conception technique et la réalisation de cette usine entièrement télécommandée ont été confiées par le C. E. A. à la Cie de St -Gobain.
St-GOBAIN - MARCOULE

Figure 11. Advertising by Saint-Gobain, designer and builder of the Marcoule plant (Source: *Energie nucléaire* (1957), n° 3, July-September, advertising brochure, p. X).

The failed attempts at post-war co-operation. The work of the UN Atomic Energy Commission

An initial approach to co-operation had been sketched out by the Atomic Energy Commission set up in January 1946 by the first General Assembly of the United Nations[74], meeting in London. The Commission continued its work until the spring of 1948. The committee established by the American Under-Secretary of State, Dean Acheson, on the initiative of James F. Byrnes, to set out the American position on the international control of atomic energy they were considering, formulated a proposal in the so-called "Lilienthal-Acheson" report[75] which was published in March 1946. The report largely took up the position of Robert Oppenheimer, who was then one of the five members of the group of consultants appointed by the committee. It proposed that the "hazardous" activities, i.e., any that could lead to the manufacture of atomic weapons – in fact all stages of the fuel cycle from the uranium mine through to reprocessing – should be under international management – not the control of an international authority overseeing national installations – and be entrusted to an "International Authority". The committee believed that the only way completely to eliminate atomic weapons was to internationalise nuclear activities.

The Lilienthal-Acheson report was revised by the American representative on the Commission, the financier Bernard Baruch, and the "Baruch Plan" was presented to the UN on 14 June 1946[76]. This proposed the establishment of an "Atomic Development Authority" (ADA). The third of the twelve concrete measures set out in the plan provided for the ADA to manage and control the plants producing fissile materials[77]. The Baruch plan served as a basis for discussions which continued until the spring of 1948. In the meantime however, the

---

[74] See GOLDSCHMIDT B. (1980), p. 83 *et seq.*

[75] David Lilienthal was the Director of the Tennessee Valley Authority. The development of the Oak Ridge nuclear centre had been dependent on hydro power from the TVA.

[76] FRUS (1972), 1946, Vol. 1, pp. 846-851, reproduced in CANTELON P. L., HEWLETT R. G., WILLIAMS R. C. (1991), pp. 92-96.

[77] CANTELON P. L., HEWLETT R. G., WILLIAMS R. C. (1991), pp. 93: "3. Primary Production Plants – The Atomic Development Authority should exercise complete managerial control of the production of fissionable materials. This means that it should *control and operate all plants* producing fissionable materials in dangerous quantities and *own and control the production* of these plants".

international scene had changed considerably and this first proposal for the international management of nuclear materials and reprocessing for peaceful purposes remained a dead letter. The concept of international management was then replaced by the idea of international control, put up by the Soviet Union, to sweeten their rejection of the Baruch plan.

However at that time such a system of control would have been applied only to the Americans, because the Soviet Union had not officially acknowledged that it had a nuclear programme. The rapidly deteriorating relations between the Soviet Union and the USA as from 1947 soon put an end to all discussion of these questions, and the issue of setting up an international organisation with responsibility for nuclear matters did not come up again until the end of 1953, when both the Soviet Union and the United Kingdom had achieved a nuclear capability and a new phase was opening in post-war international relations.

The year of "Atoms for Peace"

Fears that US technology was lagging behind

The shock caused in the United States by the explosion of a Soviet thermonuclear device on 12 August 1953 was only to be expected: this explosion took place only a few months after that of an American device and the inauguration of the Republican Dwight D. Eisenhower as President of the United States. The Americans feared that the Soviet Union had stolen a march on them[78]. Analysis of the explosion cloud by the American "radioactive monitoring" systems concluded that the bomb was transportable and had contained lithium deuteride. The Americans were intending to use solid lithium deuteride for their superbomb which had not yet been tested. Of the two devices exploded at the Eniwetok atoll in May 1951 and November 1952, the first was a bomb doped with a mixture of deuterium and tritium and the second "a very large device using liquefied deuterium kept at very low temperature as the nuclear fuel"[79], in other words, a non-transportable device. It was therefore against this background of fear that the Soviet Union had a technological lead that the idea of international nuclear co-operation was raised once again. However the death of Stalin on 5 March 1953 and the Korean armistice on 27 July 1953 gave new hope for the success of the American initiative which the Soviet Union did not oppose.

However the new American position also reflected the desire of private American firms to gain access to new markets which had been closed to them by the government monopoly and by legal restrictions.

Promoting private initiatives in the nuclear field

Republican policy was to develop nuclear activities of a new type, compatible with the market economy, and to reduce the powers – considered to be excessive – of the USAEC (an organisation regarded at the time as an "island of socialism") and to link it to the "mainland of free enterprise" and to put an end to the "gilded days of privilege and isolation"[80]. One step in this direction was the appointment of Lewis L. Strauss, a financial autodidact, as chairman of the USAEC on 2 July 1953[81]. The nuclear field held out opportunities for – or at least some hope of – profit both at home and abroad. To exploit this potential it was necessary to burst the chains applied by the MacMahon Act[82]. It was also essential to promote a new, civilian and peaceful image of nuclear energy. Private industry wanted to be ready to grasp the opportunity and in April 1953 set up the first "Atomic Industrial Forum".

---

[78] See GOLDSCHMIDT B. (1980), p. 124-126. It was not until 1 March 1954 that the United States was able to explode an air transportable bomb for the first time in Bikini, and the Soviet Union did not explode its own until November 1955.

[79] GOLDSCHMIDT B. (1980), p. 125.

[80] HEWLETT R. G., HOLL J. M (1989), p. 114, with regard to the Amendment to the MacMahon Act: "The golden days of privilege and isolation, however, were beginning to fade in 1953. The rising interest in nuclear energy within American industry, the determination of the Eisenhower administration to reverse the trend toward greater governmental control of the economic system, [..] these efforts would in part establish at least some bridges between the 'island of socialism' and the mainland of the nation's 'free enterprise system'". The Republicans were afraid that the USAEC, with the likely expansion of nuclear electricity, would become a sort of nuclear Tennessee Valley Authority.

[81] He had already been a member of the USAEC from 1946 to 1950. He retained his post until June 1958. For a review of his career, see HEWLETT R. G., HOLL, J. M. (1989), p. 429.

[82] By the autumn of 1952, the Joint Committee for Atomic Energy (JCAE) had emphasized the need for the Act to be amended in an information booklet *Atomic Power and Private Enterprise*, HEWLETT R. G., HOLL J. M. (1989), p. 22.

The birth of the idea of the "peaceful atom" and the speech of 8 December 1953

However it was the President who imposed his concept of the "peaceful atom". This idea, sketched out during the summer of 1953[83], led to the proposal that the military stocks of fissile material should be diverted and offered for peaceful purposes at international level. Before presenting his plan to the United Nations, Eisenhower had briefed his British and French allies at the Bermuda Conference.

In his "Atoms for Peace" speech[84], Eisenhower suggested that international discussions should be started in order to speed up the transition from the "atom of fear" to the "atom of peace":

The United States is "instantly prepared ... to seek 'an acceptable solution' to the atomic armaments race which overshadows not only the peace, but the very life of the world [...]. The United States would seek more than the mere reduction or elimination of atomic materials for military purposes.

It is not enough to take this weapon out of the hands of the soldiers. It must be put into the hands of those who will know how to strip its military casing and adapt it to the arts of peace.

[...] The United States knows that peaceful power from atomic energy is no dream of the future.

[...] Who can doubt, if the entire body of the world's scientists and engineers had adequate amounts of fissionable material with which to test and to develop their ideas, that this capability would rapidly be transformed into universal, efficient, and economic usage"[85].

In concrete terms, he proposed that the countries possessing fissile materials should progressively hand these over to an international body of the United Nations, an "Atomic Energy Agency", responsible for using them for peaceful purposes. The Agency would safeguard, store and protect the material and would primarily be responsible for developing civilian applications, particularly in agriculture, medicine and electricity generation.

As far as the terms of reference of the international organisation were concerned, this proposal fell short of the Baruch Plan since the production of fissile material remained a national issue. However it did give the necessary impetus to the emergence of "civilian nuclear power".

The birth of civilian reprocessing, from the amendment of the MacMahon Act to the first Geneva Conference (1954-1955)

The first concrete actions

This announcement of a change in the American position was followed in 1954 by concrete measures such as the declassification and publication of a number of scientific and technical data on nuclear energy[86]. On 30 August 1954, Congress voted a new Atomic Energy Act[87] which amended the MacMahon Act, relaxed the policy of secrecy and laid down rules governing the export of technology and equipment. It permitted bilateral co-operative agreements on highly specific topics, including the transfer of small amounts of American fissile material exclusively for research and peaceful purposes. The United States also reserved the right to monitor its use. This then raised the problem of monitoring, which had two possible solutions: national control by the United States, or control by an international organisation.

On 23 September 1954 John Foster Dulles proposed to the ninth UN General Assembly that an international agency be set up and an international conference convened on the peaceful uses of atomic energy. This was the origin of the UN International Atomic Energy Agency and preparations began for the 1955 Geneva Conference. Bilateral co-operative treaties were prepared and some signed as early as June 1955[88] (Table 2).

This exercise in *détente* resulted in the UN Secretariat, then directed by Dag Hammarskjoeld, and under the responsibility of Walter G. Whitman, an MIT chemical engineer, organising the International Conference on the Peaceful Uses of Atomic Energy. It was chaired by the Indian nuclear scientist Homi J. Bhabba and was held

83 HEWLETT R. G., HOLL J. M. (1989), pp. 65 *et seq.*

84 Reproduced in CANTELON P. L., HEWLETT R. G., WILLIAMS R. C. (1991), pp. 96-104.

85 Ibid. pp. 101-102.

86 See for example the publication dates of the American sources in the bibliography of the paper by B. Goldschmidt: A Solvent Process for Extracting Plutonium and Uranium Irradiated in Piles, in UN (1956), Vol. 9, p. 567 *et seq.*

87 An extract is given in CANTELON P. L., HEWLETT R. G., WILLIAMS R. C. (1991), pp. 331-335, which refers to the complete version in Laws of Eighty-third Congress – Second session, pp. 1098 *et seq.*

88 Bilateral agreements between the United States and the Eurochemic member countries, arising from the 1954 Amendment to the MacMahon Act. Six countries signed before the first Geneva Conference was held, and 11 before the Convention establishing Eurochemic was signed on 20 December 1957. The agreements are shown in chronological order.

in Geneva from 8 to 20 August 1955[89] (Figure 12). Some 1400 delegates from 75 countries heard more than a thousand papers at the Conference. The Conference proceedings were published the following year in English and in French[90]. To a great extent these proceedings broke down the wall of secrecy although a few very substantial blocks did remain standing, particularly as regards the methods of isotopic separation and data on uranium resources and production.

Disclosures at the 1955 Geneva Conference concerning reprocessing

Sessions 21 and 22.B.I of the Conference[91], held on 19 and 20 August, were devoted to "The Chemical Processing of Irradiated Active Fuel Cartridges". Twenty-two papers were given during these sessions by scientists from the British Harwell Centre, the American centres of Oak Ridge, Hanford, Argonne and Idaho Falls, by Norwegians and Swedes from JENER at Kjeller[92] and French scientists from the CEA.

The TBP process was brought out into the open both by Bertrand Goldschmidt[93] and by Floyd L. Culler and J. R. Flanary of the Oak Ridge laboratory as regards the American method[94]. In the same session, Erik Haeffner, a Swede working at JENER on the project for a small pilot unit which it was planned to add to the JEEP pile and who later was to join Eurochemic, presented a highly significant paper. R. B. Lemon and D. G. Reid described the operation of the Idaho Chemical Processing Plant (ICPP), operated at that time by the Phillips Petroleum Company and designed to process aluminium-uranium fuel rods from materials testing reactors. During the Conference, the United States arranged for several volumes of "Selected Reference Material" to be printed in Geneva and supplied to the participants.

Table 2. **Bilateral co-operation agreements**

Country	Date of signature	Period of validity	Type A : *research* B : *research and power reactors*
Turkey	10 June 1955	10 years	A
Switzerland	18 July 1955	10 years	B
Belgium	21 July 1955	10 years	B
Portugal	21 July 1955	7 years	A
United Kingdom	21 July 1955	10 years	B
Denmark	25 July 1955	10 years + 3	A
Sweden	18 January 1956	10 years + 2	A
France	20 November 1956	10 years	B
Norway	10 June 1957	10 years	B
FR of Germany	7 August 1957	10 years	B
Netherlands	8 August 1957	10 years	B
Spain	12 February 1958	10 years	A
Italy	15 April 1958	10 years + 10 years	A
Austria	25 January 1960	10 years	A

89 See FERMI L. (1957).

90 UN (1956).

91 UN (1956), Vol. IX.

92 The Joint Establishment for Nuclear Energy Research was the first ever international co-operative organisation in the civilian nuclear field. It emerged in 1950 from the coming together of Norwegian heavy water and a stock of uranium bought pre-war by a Dutch university and which the British agreed to refine, see GOLDSCHMIDT B. (1980), pp. 263-265. Thus Europe's first joint reactor, JEEP, diverged in July 1951. In the early 1950s, other Scandinavian countries joined this co-operative venture, followed by Germany. In 1953 a project for a small pilot reprocessing unit was initiated, which came to fruition in 1961 and continued until 1966. This pilot unit enabled the "small countries of northern Europe" to gain practical experience in reprocessing. See Part I, Chapter 2.

93 According to Bertrand Goldschmidt, the French took the initiative of publishing their research in the hope that other countries would do the same.

94 «Séparation du plutonium des produits de fission: méthode d'extraction par solvant au moyen du TBP», ibid, Vol. IV, p. 606 *et seq.*

Figure 12. Symbol of the "Atoms for Peace" operation used during the 1955 Geneva Conference. It was mounted above the door to the building housing the American swimming pool reactor and decorated the documentation distributed during the Conference and the commercial fair. Around a planetary representation of the atom showing its nucleus and orbiting electrons are symbolised the four areas of activity in which civil applications of atomic energy were to be beneficial: scientific research, medicine, industry and agriculture. Two olive branches highlight the peaceful dimension of the operation (Source: USAEC (1955), cover page).

The sixth volume, entitled "Chemical Processing and Equipment"[95], contains for example an extended presentation of the ICCP in forty-four pages illustrated by photographs and diagrams to supplement and explain the paper by Lemon and Reid.

The volume of the same collection devoted to instrumentation is nothing less than a catalogue for purchasing American equipment. Indeed commercial concerns were far from absent during the Conference which was in fact a shop window for American nuclear technology. Many other countries also took the opportunity at the Conference exhibition to present the results of their work and their successes in this new field of technology and industry.

In 1955 therefore the objectives of reprocessing were refocused on civilian applications. It was no longer a question of extracting plutonium and uranium to produce weapons, but of recovering uranium from fuel elements irradiated in power reactors for re-use in the same reactors, and plutonium to supply the future fast breeders. To do this it was necessary to separate the fissile materials – uranium and plutonium – from the fission products, which are reactor "poisons". They also had to be stored in an appropriate manner96.

A consensus had now emerged in favour of the industrial use of organic solvents for these purposes, one of the leaders being TBP[97], itself at the centre of the PUREX process.

95 USAEC (1955).

96 See session 7.3 in UN (1956), Vol. IX.

97 In the mid-1950s reprocessing methods fell into two major types: first the aqueous processes, subdivided into co-precipitation and separation by organic solvents (DBC, hexone, TBP); secondly the non-aqueous processes encompassing three different approaches being investigated in the laboratory: the pyrometallurgical processes, oxide scorification and the volatilisation of halides.

The PUREX process for reprocessing spent nuclear fuels (Figure 13)

A few definitions: The fissile material, cladding and structural components of nuclear fuel elements.

Before a nuclear fuel element is inserted into a reactor it consists essentially of two parts: first the fissile material core and, secondly, the cladding and structural components.

It is in the fissile core, consisting of uranium which can be in a variety of forms – metal, alloy or oxide – that the fission reaction and the accompanying secondary nuclear reactions take place during the time the element spends in the reactor.

The cladding and structural components, made of metals or alloys based upon aluminium, magnesium, iron (stainless steel), zirconium (zircaloy) or beryllium, have the respective functions of isolating the fissile material from the environment and facilitating the dispersion of the heat generated by the nuclear reactions occurring in the fuel.

The chemistry of spent fuel reprocessing: principles, objectives and general characteristics

Irradiated fuel: a complex cocktail of chemical elements

The chemical composition of the fissile material in a fuel element which has produced fission energy in a reactor is complicated, and depends in particular on how long the element spent in the reactor and on the time which has elapsed since it was unloaded.

The fission process breaks up the nuclei of the elements originally present and may create heavier elements by the absorption of protons. It causes a large number of transmutations because heavy nuclei can be broken up in a large number of ways. Thus the fissile core of irradiated fuel will contain about a third of all the chemical elements of the periodic table. These can be grouped into two main categories:

- heavy elements such as uranium and transuranium elements such as neptunium, plutonium, americium and curium;
- elements from the middle of the period table, mostly resulting from the breaking down of the nuclei of heavy elements by fission, hence their generic name of fission products. They are about 30 in number: rare gases such as krypton and xenon, metalloids including the halogens – bromine and iodine – alkaline and alkaline-earth metals such as rubidium, caesium, strontium, barium, etc., miscellaneous metals such as silver, cadmium, zirconium, niobium, molybdenum, ruthenium, rhodium and palladium, and finally a series of rare earths, from lanthanum to gadolinium.

The capture or emission of neutrons also increases the number of isotopes of these various elements present. Consequently the chemical and isotopic composition of irradiated fuel is extremely complex.

The objectives of civilian reprocessing

In a "military" plant, the objective is to recover and purify the plutonium for use in weapons[98], but in a "civilian" plant the aim is to recover and purify both the uranium and plutonium so that they can be recycled in new nuclear fuel. For this purpose they have to be separated from the cocktail of elements present in spent fuel, the so-called "fission products", which are useless and even harmful for the production of recycled fuel.

Very thorough separation under special conditions

The fission products are equivalent to 4 to 5% by weight of the uranium and plutonium contained in the irradiated fuel, but they act as a poison to the fission reaction and are a substantial source of radiation which complicates the business of recycling. Accordingly if recycling is to be possible, the greatest possible proportion of the fission products must be removed. The desirable degree of decontamination depends on the concentration of the fission products which is itself related to the burn-up of the fuel. In the early 1960s this concentration was of the order of one part of fission products to one thousand parts of uranium[99].

---

[98] "Military grade" plutonium must consist almost exclusively of the isotope 239 and has to be extracted from fuels which have spent only a short time in the reactor.

[99] The sharp rise in burn-up values during the period that Eurochemic was in operation increased this concentration to 3% for the fission products. The decontamination requirements thus also rose substantially and could reach a factor of 6 to 7 million. This means that after decontamination only one part of fission products should remain out of the 6 or 7 million originally present.

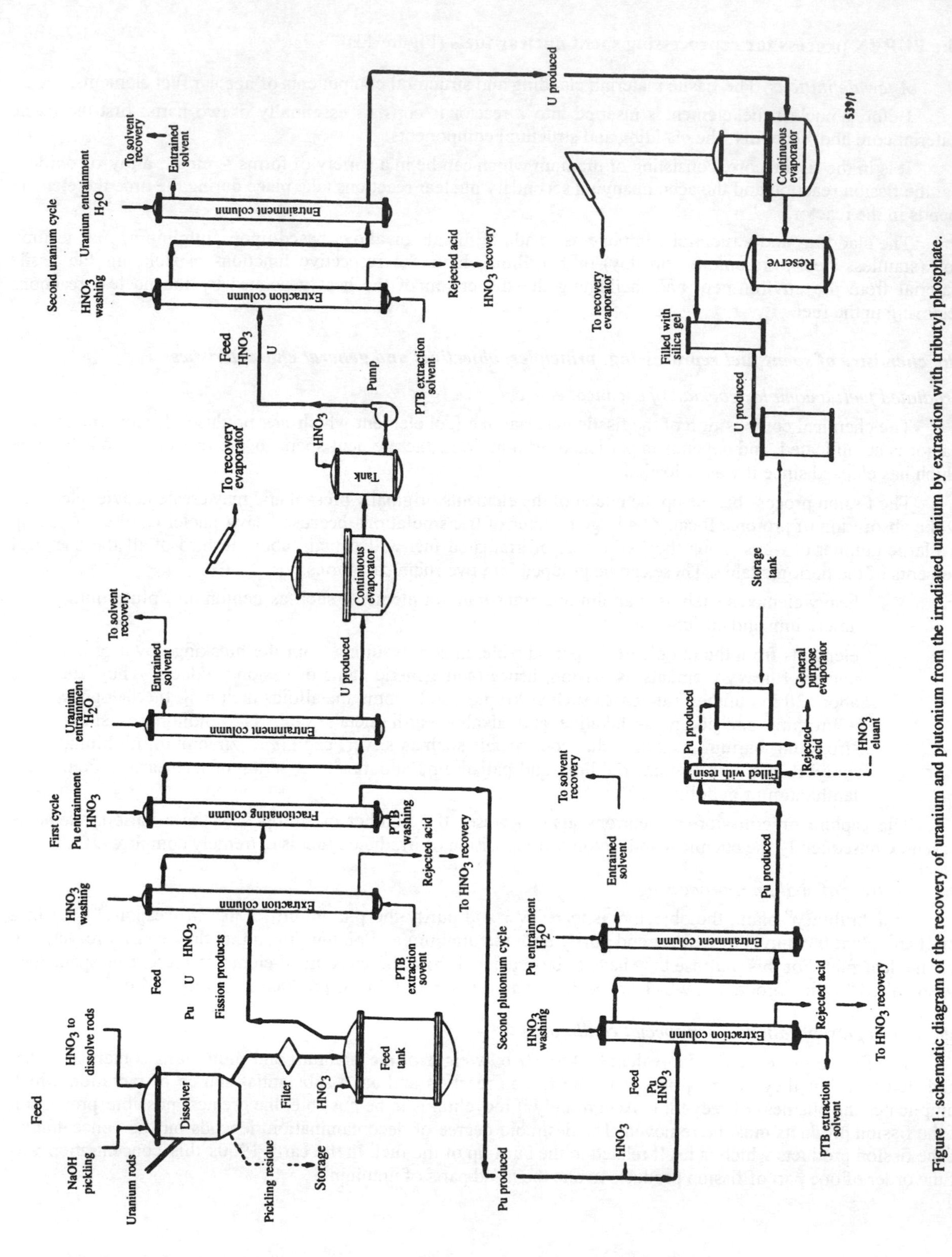

Figure 13. The first schematic diagram of the two-cycle PUREX process presented by ORNL to the 1955 Geneva Conference. Both cycles employ extraction columns. Final uranium purification is done using silica gel, and that of plutonium by ion exchange resins (Source: UN (1956), p. 607).

The uranium and plutonium, whose proportions in the fuel also depend on the burn-up, possibly varying from one per thousand to 1%, also have to be separated so that their concentration in the recycled fuel can be adjusted.

It is necessary to recover about 99% of the uranium and plutonium present in the fuel sent for reprocessing. This high value is motivated essentially by the need to minimise the quantity of fissile material – primarily plutonium – present in the various types of waste leaving the plant.

The processes of chemical separation described below, although relatively simple in principle, are complicated by the presence of a very high level of radiation and by the absolute necessity to avoid any uncontrolled build-up of fissile material, especially plutonium.

In fact there is no difference between the fissile materials dealt with in reprocessing plants and those which permit a chain reaction inside a reactor. Consequently it is absolutely essential, during the chemical separation processes, to prevent any fissile substances from accumulating in terms of mass or shape in such a way as to trigger a critical reaction through the emission of neutrons. Such a build-up could result at best in a criticality excursion causing the emission of a substantial amount of radiation and, at the worst, in a major accident, in other words an explosion that would destroy the installation and eject radioactive substances outside the plant.

There are various ways of guarding against criticality accidents, which represent an important aspect of the safety[100] of a reprocessing facility. These include using special materials – "neutron absorbers", also known as "neutron poisons" – to prevent the reaction from starting; by monitoring the concentration of fissile materials in quantitative and qualitative terms and keeping it at a sufficiently low level; and ensuring "safe geometry", by designing containers in such a way that neutrons escape from their surface before they are able to interact with fissile nuclei (in this case systems are designed in such a way as to maximise the ratio of surface area to volume). In practice, safety is assured by a combination of these methods and requires strict control at the design phase as well as during operations.

Protection against ionising radiation, which is the second important aspect of safety, is achieved in particular by locating equipment and systems inside thick-walled cells that are inaccessible during normal plant operations and accessible only with difficulty during a shutdowm because of the residual radioactivity due in particular to the contamination of the inner walls of the equipment. Consequently the operation of the plant must be monitored remotely.

The prevention of bodily contamination by inhalation and the control of gaseous effluents are achieved by two separate ventilation systems that share a single principle. This involves establishing a cascade of higher and higher vacuum as one moves from one zone to another that is more active. Since the air comes from outside, it cannot increase the activity of the zone through which it passes. Once it has reached the most active zones it is extracted through a system of filters. There is also an independent ventilation and filtration system for the tanks.

The method of solvent extraction: simple in principle but complicated to apply

A simple principle

The principle of the solvent extraction method is based upon two properties of organic solvents. First of all, they are not miscible with aqueous solutions. Secondly they complex nitrates of the heaviest elements in the periodic table – thorium and beyond – and retain them in solution: thus they can extract nitrates from an aqueous solution. This does not occur with lighter elements such as fission products and, in the case of plutonium, takes place only when it is oxidised to valencies of 4^+ or 6^+. Plutonium of valency 3^+ remains in the aqueous solution or passes through it.

The method we are concerned with therefore involves placing the irradiated fuel material itself into an aqueous nitric solution and then bringing this solution into contact with a counterflow of TBP, a particularly effective complexing agent, diluted in an organic solvent. This causes the uranium and the plutonium which is oxidised to valencies 6 and 4 to pass into the organic solution. The fission products remain in the aqueous nitric phase and are removed with it. This is the co-decontamination or co-extraction stage.

When the organic solution is brought into contact with an aqueous solution containing a reducing agent, the plutonium is reduced to valency 3 and then passes into the aqueous solution. In this way the plutonium is separated from the uranium, which is why this operation is called the separation stage.

---

100 In today's nuclear installations, a distinction is made between safety and security. The term "safety" is applied to internal technical aspects – for example technically safe plant and equipment; procedures designed to prevent the system from being jeopardized by operating errors – while the term "security" essentially concerns protection against criminal acts.

The reprocessing method mostly involves continuous operations that are fairly easy to monitor and operate remotely. However it is relatively complicated in practice, particularly for the reasons already mentioned. Some further explanation is appropriate.

A complex procedure

The "head-end" of the process

Mechanical or chemical decladding

For the fuel material itself to be placed in solution it must be brought into contact with the dissolving reagent, here nitric acid. The cladding surrounding the fuel material must first be removed. This can be done either by dissolving the cladding using a suitable chemical reagent – chemical decladding – or by peeling off the cladding by mechanical means – mechanical decladding.

In chemical decladding, the cladding is selectively dissolved by a reagent which attacks the fuel material only slightly or not at all. While some of the cladding metals used, such as aluminium and magnesium, are easy to dissolve, others like zirconium and stainless steel are very resistant to chemical attack. Also if the cladding were to be dissolved along with the fuel material, the cladding metal would be present with the fission products in the aqueous phase at the end of co-decontamination, increasing the amount of high activity liquid wastes and complicating their storage and subsequent treatment[101].

Hence there are good reasons for dissolving the material in two stages – first decladding and then dissolving the fuel material – by varying the acid concentration since the fuel is more resistant than the cladding to attack. However it is inevitable that some of the fuel is dissolved during decladding and then enters the decladding solution, generating large quantities of intermediate activity liquid waste.

Mechanical decladding is used either to remove the cladding entirely or to penetrate it so that selective attack of the fuel material is possible. This can be done remotely prior to dissolving, if the fuel consists of a rigid metal rod. For fuels consisting of an assembly of pins, consisting of stacks of fuel pellets[102], the pins are sheared mechanically, producing segments open at both ends. The segments are placed in a tank and the fuel pellets dissolved by nitric acid without the cladding being attacked. When the fuel has been completely dissolved, the remains of the cladding, known as "hulls", are removed from the acid bath and processed as waste. The shearing operation also produces a certain amount of finally divided solid waste, known as "fines", which must also be separated out. Mechanical decladding systems are usually adapted to the composition and particular shape of the fuel being reprocessed.

Dissolving the fuel material

The fuel itself is dissolved using boiling nitric acid. The resulting solution contains uranyl ions UO_2^{++} in the form of uranium nitrates, a mixture of Pu^{4+} and Pu^{6+} together with the non-gaseous fission products.

The dissolving reaction generates nitrogen oxides, the so-called "nitrous vapours": these are recombined into nitric acid which is then returned to the tank. The purpose here is to clean up the dissolution gases rather than to recover the nitric acid. The dissolving process also releases gaseous or volatile fission products, mostly rare gases such as xenon or krypton, together with iodine, and these are discharged through the plant stack unless subjected to appropriate treatment.

Adjusting and purifying the solution

The solution contains a certain amount of suspended solids, fines, insoluble elements such as molybdenum and silicon which are alloyed with uranium in certain types of fuel, fission products of limited solubility, and so on. Before proceeding to the extraction stage, these solids are separated out by filtration or centrifuging in order to give a clear solution. Also the degree of acidity, the uranium concentration in the solution, and the valency of the plutonium are adjusted to the conditions required for extraction.

Solvent extraction

The solvent: TBP

Solvent extraction is the decisive phase of the reprocessing method. As already indicated, three organic solvents have been used on an industrial scale: hexone, dibutyl carbitol (DBC) and tributyl phosphate (TBP). TBP is now used everywhere in the PUREX process, in the form of a solution containing about 30% by volume in a saturated hydrocarbon such as certain kerosenes or dodecane.

---

101 And ultimately its solidification, which was not considered in the 1950s.

102 Made for example of sintered uranium oxide which today is the commonest form of fuel for light water reactors.

The extraction system: mixer-settlers or pulse columns

The system used is the conventional solvent extraction apparatus used in the chemical industry: mixer-settlers or pulse columns (Figure 14)[103] which are both highly effective.

The former are more stable in operation and easier to control than pulse columns. However their operation involves a relatively long contact time between the highly radioactive aqueous solution and the solvent, with a consequently higher risk of degrading the solvent due to the effect of radiation. In addition the relatively large dead space in the mixer-settler makes the prevention of criticality incidents more complicated.

Compared with the mixer-settler, pulse columns are smaller in volume and require a shorter contact time between the active solution and the solvent. They also have no mechanical parts in the active zone that require regular or frequent maintenance.

The extraction operations

The first extraction cycle

Co-decontamination or fission product separation. In the first extraction operation the uranium and plutonium enter the solvent, leaving the great majority of the fission products in the aqueous dissolving solution. The solvent charged with uranium and plutonium is then washed by nitric acid solutions of different concentrations in order to eliminate some of the small fraction of fission products that has entered the organic solution, mainly zirconium-niobium, ruthenium and caesium.

The recovery rate of the uranium and plutonium is of the order of 99.9% and their decontamination as regards fission products can achieve a factor ranging between 1000 and 10 000. These high separation rates are obtained by using a large number of stages in counterflow columns.

Separation of uranium and plutonium. The organic solution is then brought into contact with an aqueous solution containing a reducing agent capable of giving the plutonium a valency of 3^+, causing it to pass into the aqueous solution while the uranium remains in the organic solution. The reducing agent first used was ferrous sulfamate, followed by uranyl sulphate, which had the disadvantage of introducing unwanted ions into the process: iron or the sulphate anion[104].

Re-extraction of uranium. By washing the organic solution (from which the fission products and plutonium have been removed) using an aqueous solution of very low acidity, the uranium is returned to the aqueous solution. This is known as re-extraction. The efficiencies of plutonium separation and uranium re-extraction are of the same order of magnitude as when they are separated from the fission products. However the overall decontamination factor achieved on completion of the above operations, which constitute the so-called "first extraction cycle" is too low to meet the optimal conditions of uranium and plutonium recycling. The purification operations must therefore be continued after this cycle.

The "second extraction cycles"

By applying extraction processes, similar to those described above, to the uranium and plutonium separated out during the first cycle, the overall decontamination factor of the uranium (with respect to plutonium and fission products) and of the plutonium with respect to fission products and uranium are increased. These further extraction operations are known as the "second extraction cycles", for uranium and plutonium respectively. The additional decontamination factor obtained for each of these second cycles is of the order of 100.

The "tail-end" of the process. Final purification of uranium and plutonium

Following the two extraction cycles, it is usually necessary to remove further traces of fission products, especially zirconium and niobium, from the uranium. This final purification stage involves either a third extraction cycle or, more simply, allowing the solution to percolate through absorption columns containing silica gel.

With regard to the plutonium, it must usually be decontaminated by a further factor of 10 with respect to the fission products. It is also often necessary to further eliminate traces of uranium. For this purpose, either ion

---

[103] In the 1950s, filled columns were also used.

[104] Eurochemic played a significant role in adopting the present-day process using uranyl nitrate. In current plants the plutonium is reduced by uranyl nitrate produced by electrolysis in the extraction column itself.

Types of contactor used for solvent extraction

The main types of continuous contactor used for the liquid-liquid extraction of fissile and fertile substances are gravity columns, including filled columns and those containing perforated plates, and contactors with forced circulation, including pulse columns and mixer-settlers.

The filled column (Fig. 13) contains stainless steel Raschig rings for example, forcing the dispersed phase droplets to follow a sinuous route through the continuous phase, improving contact between the two phases. For a given application, the type of filling used depends a great deal on the chemical nature of the system and the required treatment conditions, its height depending on the desired degree of separation.

It is possible to use mixer-settler systems that give high efficiency. Horizontal mixer-settlers usually require less vertical clearance but take up more horizontal space than the vertical solvent extraction columns of the same capacity. Types of mixer-settlers with a pump have been developed (Fig. 14) [14]. In this compact system, the contact chambers are separated from the pumping chambers by partition walls.

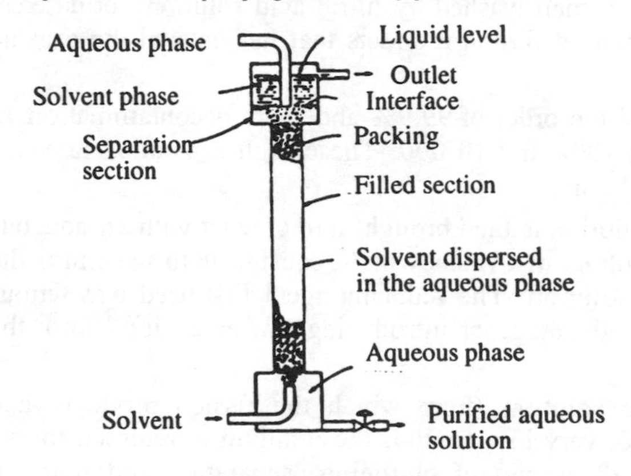

Figure 13. Filled solvent extraction column

Inside each contact chamber, a centrifugal stirrer pump draws the two incoming phases into a vertical pipe: they are intimately mixed and then discharged radially into the adjoining settling zone, thus producing the head necessary for flow between the plates and maintaining the liquid-liquid contact at the desired position on each plate. In each settling zone the two phases flow in one direction but flow in opposite directions between the plates. There are neither seals nor submerged bearings; the motor-pulser system can be removed or replaced by remote control.

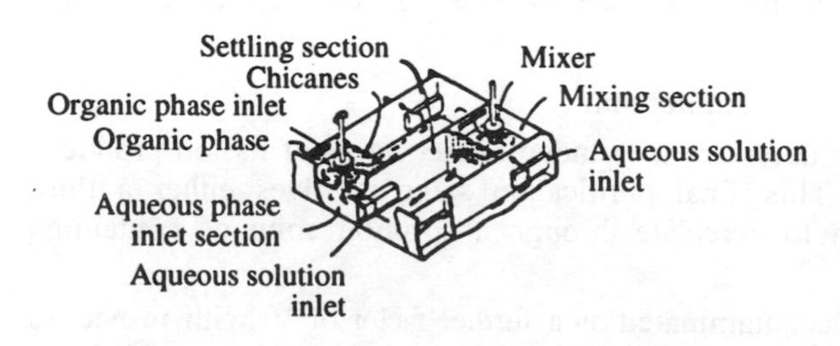

Figure 14. Schematic cross-section of a mixer-settler (pump type) for solvent extraction

The pulse column with perforated plates (Fig. 15) was first described by Van Dijk [15]; other data on pulse column developments have been published since then [16-19]. With this system the column height can be appreciably reduced - often by more than 50% - compared with the conventional filled column. In the pulse column, a vertical reciprocating pulsation is superimposed on the opposed flows of the liquid phases through a series of fixed perforated horizontal plates, all equidistant from one another. The perforated plates can be replaced by Raschig rings or another form of packing.

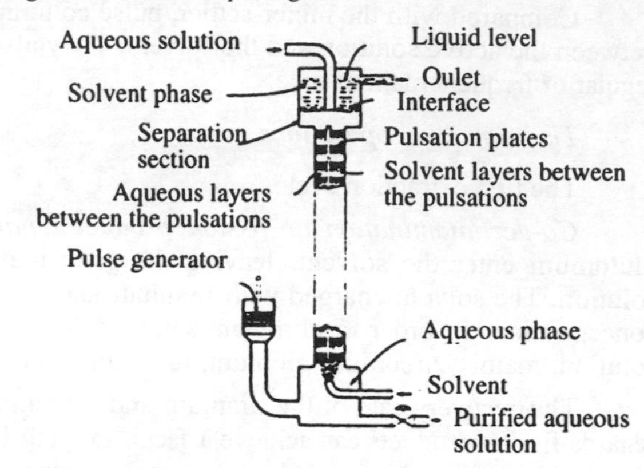

Figure 15. Pulse column for solvent extraction

In columns containing perforated plates, the pulsation plays another important role in facilitating the opposing flows of the aqueous and organic phases. The difference in density between the two phases is usually not enough to produce any significant reverse flow through the small holes in the perforated plates. Consequently the upward movement of the light phase and the downward movement of the heavy phase towards the bottom of the column are almost entirely due to the action of the pulse generator and the circulating pumps.

The greatest advantage of pulse columns over filled columns is their much smaller height for a given extraction result. Other advantages are that the extraction efficiency depends less on flow rate and, for columns containing perforated plates, it is easier to halt the installation temporarily and restart it, particularly as the flow through the plates is very slight in the absence of pulsations. On the other hand the need for a pulse generator is a drawback owing to its initial cost and the maintenance required.

14. Coplan B.V., Davidson J.K. and Zebroski E.I., *The Pump-Mix Mixer Settler, A New Liquid-Liquid Extractor*, Chem. Eng. Progress, 1934, 403-8.
15. Van Dijck W.J.D., U.S. Patent 2.011.187, 13 August 1935.
16. Nege G. and Woodfield F.W., *Pulse Column Variables*, Chem. Eng. Progress, 1934, 40-396-1.
17. Nege G. and Woodfield F.W., *A Lower-Plate Redistributor for Lurye-Diameter Pulse Columns*, Chem. Eng. Progress Symposium Series, 1954, n° 13.
18. Thornton J.D., *Recent Developments in Pulse Column Techniques*, Chem. Eng. Progress Symposium Series, 1954, n° 13.

Figure 14. Description of the three main types of contactor used for solvent extractıon ın the United States in 1955: mixer-settler, filled column and the pulse column with perforated plates (Source: CULLER F. L., Processing of Nuclear Fuel and of the Recovery Layer by Solvent Extraction, UN (1956), pp. 552-553).

exchange resins are used or the plutonium is precipitated in the form of plutonium oxalate[105]. The oxalate is filtered, washed and then calcined at a high enough temperature to break it down into plutonium dioxide. Another advantage of precipitation is that it can be done continuously, whereas purification on resins is a discontinuous process.

Some ancillary operations

Fission product concentration

As indicated above, the great majority of the fission products remain in the initial solution after the extraction of uranium and plutonium by the solvent in the first cycle. This solution is the most highly radioactive liquid waste produced in reprocessing, containing over 99.8% of the beta-gamma activity present in the spent fuel. At the end of the first extraction cycle this effluent is concentrated by evaporation in order to reduce the volume to be stored[106].

Processing the solvent

The TBP solution is recycled after each operation. However it does degrade to some extent under the effect of radiation and the chemical reagents, producing small amounts of dibutyl and monobutyl phosphates together with butyl alcohol. The presence of these degradation products is harmful to the extraction process so the solvent must be purified before being recycled. Also, traces of fission products as well as of zirconium and niobium which are difficult to remove make the solvent slightly active, necessitating certain precautions.

Practical problems

All these processes would be similar to those used in the conventional chemical industry if they were not complicated by the very high radioactivity of the substances being treated and by the presence of large amounts of fissile materials.

In particular the mixture of fission products is a powerful source of beta and gamma radiation which requires thick shielding to be used: concrete walls, lead glass observation windows, lead or thick steel cells, and so on.

The alpha radiation from plutonium makes it extremely harmful if it is inhaled or ingested into the body so its handling calls for special precautions and the use of sealed enclosures such as gloveboxes.

The problems of radioactivity impose a strict hierarchy in the way the plant is arranged: the active areas are inaccessible during operations and so must be remotely controlled. They are separated from the non-active areas, where restrictions on personnel are minimal, by semi-active areas where workers may spend only a strictly limited time.

Accordingly the construction of a reprocessing plant was not and is still not comparable with that of a conventional chemical works.

To build such a plant in Europe capable of reprocessing fuels from future reactors was a technical challenge. This challenge was accepted as a result of changes in international relations which, from the mid-1950s, became favourable to international co-operative initiatives in the nuclear field and particularly as regards civilian reprocessing. It was in the OEEC framework that the first of these took shape in the form of the Eurochemic company.

---

105 This latter technique is the one most frequently used today because it is an unavoidable step in the preparation of plutonium oxide, which is the form in which the Pu leaves the civilian reprocessing plants. However the military plants require metallic Pu, so the plutonium oxide is converted by hydrofluoric acid into PuF_4 which is then converted into the metal by calcination in an inert atmosphere.

106 The liquid wastes from today's most modern plants are vitrified.

Chapter 2

The OEEC, the countries of Europe and international co-operation in the nuclear field

1948-1956

It is reasonable to ask how it came about that an organisation of European countries which, in 1948, was given the task of assisting the reconstruction of Europe and distributing Marshall Aid, found itself eight years later heading a co-operative project on the chemistry of radioactive materials[1]. Again, how was this organisation's administrative experience converted to the management of a project for building a joint plant for the chemical reprocessing of irradiated fuel? These are the questions underlying the story of the somewhat surprising conjunction between the OEEC and nuclear power, a conjunction which led in June 1955 to the establishment of a working group with the task of preparing a Nuclear Energy Committee, the embryo of the future European Nuclear Energy Agency (OEEC/ENEA), within the Organisation.

Amongst the firm co-operative projects that the OEEC proposed in the nuclear field between June 1955 and October 1956, the one for a reprocessing plant was regarded as the least difficult to carry through. Accordingly a Study Group was set up to examine the proposal further.

Of course an essential prerequisite was that the countries of Europe should find the co-operation proposed by the Organisation attractive. A review of the contemporary status of nuclear activities in the member countries of the Study Group shows that their expectations, apart from the apparent consensus that belonging to the Group reflected, were extremely varied and sometimes contradictory.

The OEEC and nuclear power: an unexpected conjunction (April 1948-May 1955)

The nuclear commitment of OEEC stemmed from three concerns of the Organisation in the early 1950s.

The first preoccupation was to find an answer to the problem of energy supplies during those early years of "high growth". The danger of Europe's becoming increasingly dependent on imported energy, demonstrated by the investigations carried out by the Organisation's "vertical committees", caused the OEEC to take an interest in the prospects for nuclear power development and proposing, with a view to speeding up this development, the establishment of a European Nuclear Energy Agency together with projects for multilateral co-operation. In the beginning these were very diverse, covering not only most of the fuel cycle, from isotopic separation to the reprocessing of spent fuel and the development of power and research reactors, but also ship and aircraft propulsion and the production of heavy water. However the range of proposals was gradually narrowed down and in the end only three projects were successfully accomplished: two experimental reactors – Halden and Dragon – and the Eurochemic reprocessing plant. The way in which projects were selected and organised were closely related to the structures and "enterprise culture" of the OEEC. In this way there emerged an original project for international co-operation that was doubly hybrid, being halfway between the laboratory and the factory and halfway between a private company and a public organisation.

The second concern of the OEEC was to contribute to modernising European industry, particularly by arranging transfers of American knowledge and technology to western Europe. By 1949 appropriate structures had been put in place, and these activities were being steadily intensified at the beginning of the 1950s. One of the primary sectors ripe for modernisation appeared to be the chemical industry: the United States had a substantial technological lead in chemistry, with the chemistry of radioactive substances playing a leading role.

Finally the OEEC had to find new areas of activity following the internal crisis that occurred when the period of reconstruction – and the Marshall Plan – came to an end. One aspect of OEEC redeployment post-1951 was the development of a branch of the Organisation devoted to nuclear questions, resulting in 1957 in the

---

[1] For the history of the OEEC, see BOSSUAT G. (1992a and b), GIRAULT R., LEVY-LEBOYER M. dir (1993), MELANDRI P. (1980), MILWARD A. S. (1984). Use has also been made of the Acts of the OEEC Council, referred to in this chapter as *Acts,* volume n°, decision n°, date.

creation within the OEEC of the European Nuclear Energy Agency (ENEA) and of a joint undertaking involving 12 and subsequently 13 member countries of the Organisation, the Eurochemic company.

Elements of an OEEC "organisation culture" from 1948 to 1951

The establishment of the Organisation and the vertical technical committees, 1948-1949

The Organisation for European Economic Co-operation (OEEC)[2] was created by the Convention of 16 April 1948 signed by ministers from 16 European countries and the commanders-in-chief of the British, American and French occupied zones of Germany[3]. Its primary objective was "the planning and execution of a joint recovery programme" in the period from 1 July 1948 to 30 June 1949, whereby the European countries in the short term "could maintain a satisfactory level of economic activity without undue external aid"[4]. It also had the task of distributing Marshall Aid amongst its members.

Marshall Aid was first offered on 5 June 1947, and was worked out and approved by the United States Congress on 2 April 1948. The Economic Co-operation Act released an initial tranche of $5.3 billion for the first year of the five-year European Recovery Program (ERP)[5]. The OEEC's opposite number in America was the European Co-operation Administration (ECA), directed by Paul J. Hoffman. The new Organisation quickly established its structures: a Council of Ministers, chaired by Paul-Henri Spaak (Belgium), an Executive Committee with seven members, headed by Sir Edmund Hall-Patch, and a general secretariat, directed by Robert Marjolin (France)[6] were the prime movers[7]. Although decisions of the OEEC Council had to be unanimous, the Secretary-General had some power to make proposals. The Council could set up technical committees or other bodies as necessary.

For example in August 1948[8], eight vertical technical committees[9] were charged with preparing studies, each in their particular area, with a view to formulating the Long-term Programme which was to supplement the first recovery programme. These committees were frequently extensions of special technical committees formed in July 1947 as sub-groups of the European Economic Co-operation Committee (EECC) whose task was to prepare the European response to the Marshall offer.

Three of these committees took care of energy problems which were of course vital for reconstruction, covering coal, oil and electricity respectively.

Thus the Electricity Committee dealt with problems of electricity generation and exchanges of power between member countries. Accordingly it was given the task in August 1949 of looking into the possibilities of co-ordinating investment in the electrical industry[10], in order to overcome Europe's shortage of generating capacity as quickly as possible. On this occasion the committee's tasks were as follows:

---

2 See STROHL P. (1959).

3 The signatory countries were Austria, Belgium, Denmark, France, Greece, Iceland, Ireland, Italy, Luxembourg, the Netherlands, Norway, Portugal, Sweden, Switzerland, Turkey and the United Kingdom.

4 Convention of the OEEC, Article 1.

5 Or First Interim programme. According to R. Marjolin, the ECA provided Europe with $12.4 billion – or 1.2% of United States GNP – in 45 months from 1 April 1948 to 30 December 1951. This figure bears comparison with the target of 0.7% of GNP of the member countries of the OECD Development Aid Committee (DAC) in the 1970s for aid to the Third World...GIRAULT R. and LEVY-LEBOYER M. ed. (1993) suggests different figures: $13.5 billion under the Plan including $8.6 billion in the Counterpart Fund provided to countries.

6 See R. MARJOLIN (1986). Trained as a philosopher and economist, Robert Marjolin was Jean Monnet's right-hand man for the French Plan from 1945 to 1948. He had married an American woman whom he had met during his stay in the United States during the war. Robert Marjolin was Secretary-General until 31 March 1955 when he was replaced by another Frenchman, René Sergent. *Acts* (1955) 1327, 25 February.

7 The presence of representatives from Belgium, Britain and France reflects a political balance. Spaak was supposed to defend the "small countries" against the two major European powers, linked by their "cordial distrust".

8 *Acts* n° 80, 12/08/1948.

9 Council decision n° 173 of 12 April 1949 lists the technical committees with their terms of reference and modes of operation. The committees could be of three kinds: vertical, horizontal and special. The terms of reference of the vertical committees cover "specific products" and they are "charged with investigating certain problems using an analytical method". The horizontal committees "cover certain aspects of economic activity", and "are charged with employing the studies carried out by the former in order to bring these together and identify the general principles".

10 *Acts* n° 229, 19 August 1949. This task led to the creation in May 1951 of the UCPTE, the "Union for the Co-ordination of Generation and Transmission of Electricity", which still exists. See BANAL M., «Le Plan Marshall et l'électricité en France», GIRAULT R. and LEVY-LEBOYER M. ed. (1993), pp. 257-258.

- to assess the needs for electric power through surveys in member countries and by updating the data held by the Organisation;
- to submit within four months a report on the priority investment needed having regard to the constraints laid down by the Council: first, projects should be economically justified and capable of being carried through quickly, secondly the location of new power plants should be considered in such a way as to allow "a suitable balance between national considerations" and the potential of existing or planned interconnections;
- to estimate the cost;
- to form links with a horizontal committee, "the Supply Committee, charged with verifying the extent to which these needs may be met by the electrical and mechanical engineering industries in member countries".

Between 1948 and 1951, the OEEC also co-ordinated European reconstruction by means of annual development programmes and by relaunching intra-European trade on a multilateral basis through "annual multilateral compensation agreements". These prefigured the establishment of the European Payments Union (EPU) under the agreement of 19 September 1950 and the signature of 18 August 1950 of the Code of Free Trade. Specialist bodies had multiplied within the Organisation, and new links had been formed with the United States and Canada which were more closely associated with the work of the Organisation after June 1950[11].

The "intangible Marshall Plan"[12]*, technical assistance, OEEC projects on research, productivity and scientific & technical co-operation: 1950-1951*

Increasing production was inseparable from improving productivity. The "productivity gap" which had formed between the United States and Europe had to be bridged: in this connection the OEEC technical committees provided one channel for disseminating the technical information necessary for improving the productivity of the European economic system. The links of economic assistance between Europe and the United States were soon joined by those of technical assistance, which was the subject of a special amendment to the 1948 Act of Congress which had set up the Marshall Plan[13]. The aim of European leaders at the time was to attain the American level of excellence. As Robert Marjolin wrote, describing the state of mind in the OEEC throughout the 1950s, "we were mesmerized by America, and American material success was our ideal; practically our only aim was to bridge the gap between European industry and that of the United States"[14].

In 1950 the number of visits to the USA in connection with research and productivity questions increased under the auspices of a Technical Assistance Group. For example there was a series of visits concerned with the chemicals and oil sectors, dealing with chemical plants, the problems of water and air pollution caused by industry, the links between research laboratories and industry, and petrochemicals. Two American specialists were also invited to visit Europe on the occasion of the Twenty-third Annual Industrial Chemicals Conference held from 17 to 23 September 1950 in Milan. The visit was prepared jointly by the Chemicals Committee and the ECA[15].

The first clause in the document arranging the first visit[16] is extremely revealing as to their function: the Council recognises "the essential importance for the European chemical industry to get a precise idea of the progress made in the United States as regards the plant and equipment used in this industry so that European

---

11 *Acts,* n° 384, 23 June 1950.

12 The term used by Christian Stoffaes, STOFFAES C., «La Révolution invisible: une mise en perspective de l'expérience des missions de productivité», GIRAULT R. and LEVY-LEBOYER M. ed. (1993), pp. 755-762. The Bercy conference included interesting papers on productivity within the Marshall Plan. In his introductory report to the part on the Plan's industrial spin-off in France, François Caron affirms that "although it did not in fact create the move to modernisation", the Marshall Plan "added impetus to this growth in productivity which continued throughout the 1950s". As regards transfers of technology, it was "the instrument of a major technical breakthrough in most sectors". The paper by Richard F. Kuisel, "The Marshall Plan in Action: Politics, Labour, Industry and the Program of Technical Assistance," pp. 335-358, reviews how the ECA productivity projects were executed in France. In the Netherlands, the upturn in productivity took place in 1950 and is related to Marshall aid, see the paper by Erik Bloemen, "Technical Assistance and Productivity in the Netherlands", ibid. pp. 503-514.

13 Under Article 111a.3 concerning "the acquisition and supply of information and technical assistance".

14 MARJOLIN R. (1986), p. 228. As regards France, see also KUISEL R. F. (1988), «L'American way of life et les missions françaises de productivité», *Vingtième siècle*, n° 17 (January-March 1988), pp. 21-38.

15 The arrangements for the OEEC technical assistance projects are set out in the Council Resolution of 4 May 1950, and a selection of projects is included in the recommendation of the same day, *Acts,* n° 352 and 353. The visits mentioned were covered by resolutions 355, 356, 358 and 434.

16 *Acts* (1950) 355.

plants may be modernised in the best possible way, this being the only means of putting the European chemical industry in a position to cope with world-wide competition [...]". The experts were proposed by member countries and appointed by the technical committees. They were endorsed by the ECA and produced a preliminary report on their return. These reports were discussed by the technical committee which then made proposals, in further reports, as to how the data collected in the United States could be put to practical use. These reports were provided to member countries who passed them on to the appropriate organisations. In July 1951, the ECA decided to reinforce its productivity policy and announced, just as the one-thousandth French expert was leaving on mission, a special programme known as the "Production Assistance and Productivity Program" (PAP)[17].

The OEEC also began to take an interest in European scientific and technical co-operation, and in June 1949 the Council established an appropriate working group. As from January 1950 its work was extended into three new groups: the Working Group on Co-operation in Applied Sciences, that of Productivity and the Advisory Group for Increasing Productivity Through Standardisation (AGIPS). In February 1951 a new scientific and technical committee was set up with three sub-committees which took over from the working groups.

The looseness of the organisation had had undesirable effects and the increasing numbers of specialist bodies had resulted in duplication of effort and the usual problems of responsibility in an expanding administrative organisation. The completion of the Marshall Plan in October 1951 was to lead to a crisis from which the Organisation was to emerge transformed.

The crisis and reorientation of the OEEC. The European Productivity Agency (EPA): 1951-1953

The worsening in international relations, the deterioration of the Cold War and the Korean war which began in June 1950 brought military concerns to the forefront and, as Robert Marjolin wrote, "led to a profound crisis in European co-operation, even if attempts were made to conceal this as much as possible"[18]. In this situation, with Marshall aid expected to come to an end in October 1951, the future of the Organisation appeared threatened[19].

The rising tide of American military aid, which amounted to 10 billion dollars between June 1950 and June 1952, compared with 3.3 billion in economic and technical aid, was not without an impact on the OEEC. When NATO was established, it appeared superfluous or even harmful to keep in being an independent economic organisation, the civilian side of "containment", at a time when the worsening of international relations appeared to call for a new joint organisation of a war economy. In July 1950 the Americans began to consider merging the civilian and military forms of aid. This was done on 1 January 1952 when the ECA was replaced by the Mutual Security Agency (MSA). In 1951 the British did not hide the fact that they favoured the disappearance of the OEEC, with only the European Payments Union being retained. Robert Marjolin explains that in British eyes the Organisation had been valuable only as "a channel for American aid. Co-operation (for them) had no value in itself"[20]. Also the multilateral nature of the Organisation, with militarily neutral countries amongst its members, ran contrary to their desire for it to be dominated by the British and the Americans. It seemed that this Anglo-American domination would be easier to exercise within NATO.

France and Germany both wanted the OEEC to survive, although for different reasons. Those aspects of the OEEC seen by the British as defects were regarded by France as advantages. As to Germany[21], she would have been excluded from any European co-operation apart from the coal and steel sector, precisely as Konrad Adenauer's strategy for the country's re-entry to international life involved enhancing her international links. Also German participation was a matter of some reassurance to the French. Finally France and Germany obtained the support of the United States and Canada for their views and the OEEC survived[22]. However this was at the price of a severe slimming cure.

---

[17] For the ups and downs of the PAP, see KUISEL R. F., GIRAULT R. and LEVY-LEBOYER M. ed. (1993), pp. 345-347.

[18] R. MARJOLIN (1986), p. 238.

[19] With regard to the OEEC at the end of Marshall aid, see BOSSUAT G. (1992a) Chapter 9.

[20] R. MARJOLIN (1986), p. 238. These problems and disputes contributed to the decision that Marjolin should leave in December 1953.

[21] For a few months after the creation of the Federal Republic of Germany in September 1949, the new state took part in the work of the OEEC before being officially admitted following the signature of the Bilateral Treaty of Adhesion to the Marshall Plan on 15 December 1949.

[22] Pierre Huet remembers that Spaak wanted the Organisation to have a political role going beyond its function of co-operation, which he thought could be left to the diplomats. This resulted in the chairman of the Council being given

The Council set up a working group to "review the Organisation's current and future activities". Its conclusions were embodied in the Council Resolution of 28 March 1952[23]: "It is appropriate for member countries to focus their attention and their efforts on specific problems, particularly on the freedom of trade and payments, internal financial stability and *the expansion of production*. [...]. To achieve these objectives the Organisation must be made as efficient as possible, eschewing any expenditure not essential to the efficient conduct of its work". The "productivity" attitude invaded the organisation itself: the number of committees was reduced and their structure revised, and staff numbers fell by a third under the effect of what the international civil servants soon called the "axe committee".

As early as August 1951, Robert Marjolin had suggested that the aims of the OEEC should be immediately reorientated to the preparation of a European Plan for expanding production by 25% over the coming five years. Although the Plan remained a dead letter, with the Council doing no more than stating an objective, the refocusing of the Organisation was reflected in a number of measures adopted in 1952 and 1953 covering not only trade but also payments and monetary stability, expansion and the organisation of production.

Thus in March 1952 the Council decided to conduct a joint annual review of the economy and its prospects in each member country and associate countries (United States and Canada). Accordingly the OEEC began to exchange information, act as a forum for discussing economic policy, and move towards a convergence in statistical analysis[24]. The Council discussed the first general report on the economic situation in Europe in October 1952.

A year later, on 24 March 1953, the European Productivity Agency (OEEC/EPA) was created[25] as the first Agency of the OEEC, preceding the European Nuclear Energy Agency. The OEEC/EPA inherited the complex structures of technical assistance and scientific and technical co-operation that predated the reorganisation of 1951-1952. A degree of simplification had been introduced in May 1952 with the creation of a Committee for Productivity and Applied Research, to direct co-operative activities in research, productivity and standardisation, working through three sub-committees and centralising the activities of technical assistance.

The main result of the change from a committee to an agency was that the latter had its own budget, with two special sources of funds: the OEEC (in the form of grants), and the United States government (as "contributions to the Organisation for the Agency"). In this way, during the first three years of the operation of the Agency, the United States provided a quarter of its resources, $2.5 million out of 10 million.

The aim of the Agency was "to identify, develop and encourage the most appropriate and rational methods for increasing productivity in the different firms belonging to the various sectors of economic activity in member countries, and in all parts of their economy"[26].

A good idea of the functions of the OEEC/EPA can be obtained from its budget and, more precisely, from its running costs: thus for the financial year 1954-1955 the latter amounted to 800 million FF under twelve headings. The section on "industrial technology and applied research" (with 90 million FF), came in fourth position after "employment problems in industrial and commercial firms" (120 million FF), the "operation of industrial and commercial firms" (115 million FF), and the "organisation and management of agricultural undertakings" (95 million FF). However marketing and distribution, covered in two sections of which one was agricultural, the other not, amounted to 120 million FF. Thus the OEEC/EPA focused its efforts on highly concrete and immediately applicable problems such as organisation and management.

The Agency played a significant role in spreading the American management model. It also extended the productivity visits to the United States, assisted the development of management institutes, organised summer schools, provided research scholarships and organised productivity and information conferences for managers of firms. The first such conference was held from 14 to 19 September 1953 at the Administrative Staff College at Henley-on-Thames in the United Kingdom, and covered "the training of company managers". The first productivity conference was held in London in October 1953.

The Agency's role in the dissemination of technical advances was more modest. It had to make the European industrial elite aware of new American technology and in this context was one of the two channels by which nuclear activities penetrated the OEEC. Thus the Agency organised a conference for managers of nuclear

---

a mandate for this purpose. Spaak saw himself as the person to play this role. However although the planned "Spaakistan" did not come to fruition, an embryo of a permanent political structure did appear in the OEEC.

23 *Acts*, n° 782 of 28 March 1952. Author's underlining.

24 These annual reviews and general reports are still being published today.

25 *Acts*, n° 976, 24 March 1953. With regard to the EPA, see OEEC (1958), CAREW A. (1987), and *Notes et etudes documentaires* (1959), n° 2604. The EPA did not survive the transformation of the OEEC into the OECD in 1960.

26 Statutes of the Agency.

energy firms, which was held in April 1957 in Paris[27]. However by that time specialist bodies were being set up within the OEEC and soon took over[28].

The other channel through which nuclear power entered the Organisation was the problem of energy. The economic growth which followed European reconstruction appeared to have a great appetite for energy, while the region's resources were not increasing[29].

The advent of nuclear energy in the OEEC. The role of the Armand Report: December 1953-May 1955

The entry of nuclear energy took place in three stages:

– the energy problems facing Europe as it began the period of high growth caused the Secretary-General of the OEEC, Robert Marjolin, to concentrate on restarting co-operation in this field at the end of 1953;

– in the late summer of 1954 civil nuclear power, the development of which had been rendered politically possible by the changed American attitude at the end of 1953, appeared as one way of relaunching European construction after the failure of the European Defence Community. Two approaches were tried:

  The first, within the OEEC and therefore including the United Kingdom (at that time the most advanced country in Europe for nuclear power), was to lead to the joint creation of the ENEA and the Eurochemic company.

  The second approach began with the member countries of the European Coal and Steel Community (ECSC). Jean Monnet, Paul-Henri Spaak and Louis Armand played a crucial role in utilising nuclear power as an instrument of the "relaunch", which finally had its diplomatic result in the Messina Conference held in June 1955. This was to lead to the joint creation of two new "European Communities", one focused on the development of the nuclear industry, the other on establishing a common market in Europe[30].

  At the outset these two approaches were in competition with one another, but as they progressed they became in fact more complementary and parallel;

– at the end of 1954 Robert Marjolin asked Louis Armand[31] to prepare a report on the European energy problem. The Armand Report laid particular stress on the new possibilities for co-operation in this field, which in fact went beyond the OEEC alone; one of its products was the Eurochemic company, others being the European Nuclear Energy Agency (OEEC/ENEA) and the European Atomic Energy Community (EAEC or EURATOM).

The energy problem in Europe

The growth of energy consumption in Europe had slowed down in 1951 and stagnated in 1952; however in 1953 it resumed its growth at a rate exceeding that of production, resulting in a widening import gap (Figure 15). This worsening situation, which the OEEC forecasters thought would persist, caused the OEEC Secretary-General, Robert Marjolin, to submit to the Council on 14 December 1953[32], "a memorandum drawing the attention of member countries to the problem of rising energy costs in Europe, the dangers this trend could represent to the development of member countries and to the methods he thought could be used for resolving this problem, particularly through co-operation between member countries"[33].

---

27 OEEC/EPA (1957).

28 The second Conference held in Amsterdam in June 1957 was published by the OECD/ENEA and no longer by the EPA.

29 For energy developments and their links to initiatives in nuclear co-operation within the Europe of the Six, see POLACH (1964), pp. 30 *et seq.*

30 For the Messina Conference and the development of the negotiations leading to the Treaty of Rome, see SERRA E. ed. (1989). This reference is the proceedings of the Rome conference held from 25 to 28 March 1987. For EURATOM, see pp. 513-545.

31 With regard to Louis Armand and his role in the development of nuclear power in Europe, see the two biographies by TEISSIER DU CROS H. (1987), pp. 225 *et seq.*, and ASSOCIATION DES AMIS DE LOUIS ARMAND (1986), pp. 97 *et seq.*

32 i.e., six days after the Atoms for Peace speech.

33 OEEC (1955), p. 7.

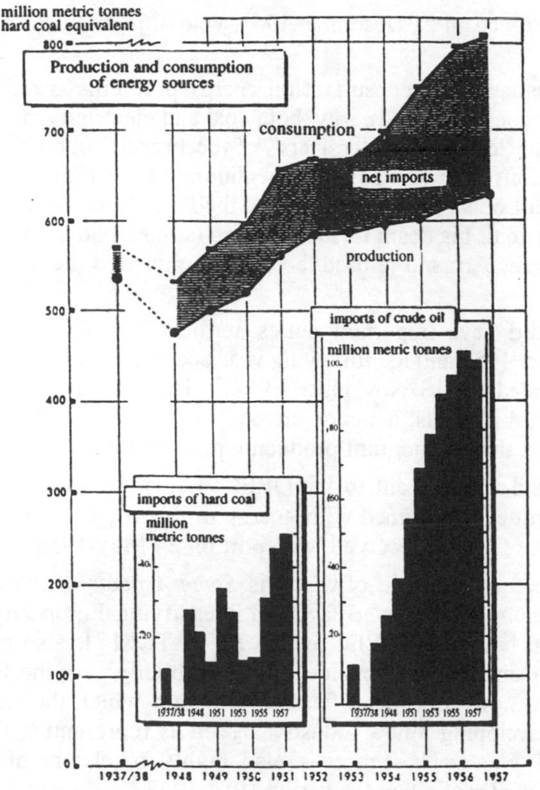

Figure 15. The energy problem in western Europe according to the OEEC in 1958: the stagnation in output and the rise in consumption led to growing reliance on imports of coal and, especially, oil (Source: OEEC (1958), p. 113).

The Council examined the memorandum on 11 January 1954, then "acknowledged the existence of an energy problem in Europe", and requested the Secretary-General to collect more information and make concrete proposals. This was done officially in December 1954. In the meantime, nuclear power had moved to the centre of the concerns of those who favoured intensified European co-operation.

The choice of nuclear power as an instrument for the European "relaunch"

During 1954, nuclear power had become a preferred topic for the "Europeans" close to Jean Monnet. The failure of the European Defence Community in August, owing to the French refusal to ratify, resulted in a search for new topics for the necessary "relaunch". This found official expression at Messina in June 1955 but the ground was prepared in "monetarist" circles in the summer of 1954[34].

Nuclear power enters the Château de la Muette: the Armand Report

Robert Marjolin chose a consultant from the "leading figures with access to the relevant government and professional circles"[35]. It was Louis Armand, at that time chairman of French railways (SNCF)[36], recruited by the

34 Sources say little and are fairly contradictory about the choice of nuclear power as a topic for the relaunch, in parallel with the Common Market, and on the role played by Louis Armand in making this choice. According to BROMBERGER S. and M. (1968), Chap. IX, it was Armand who had the idea soon after August 1954, but the book does make several errors of chronology. MONNET J. (1976), p. 590, brings Armand into the affair only in 1955. The available OEEC archives say nothing on the subject, as do the two biographies of Armand. Apparently Jean Monnet consulted Armand about the suitability of relaunching Europe by pooling transport and electricity. The latter then apparently preferred nuclear power, starting from the idea of a nuclear EDC. Monnet mentioned nuclear power only during a conversation he had with Paul-Henri Spaak "in the Ardennes whitened by winter" in 1954-1955. However it very soon became clear that if the project was to be successful, the civilian and military sides would have to be completely separated, and the relaunch was based only on the economic and civilian dimension of atomic power. It must be pointed out however that the military aspects of co-operation were also investigated elsewhere (see Part I, Chapter 3).

35 It will be recalled that Robert Marjolin was a close colleague of Jean Monnet at the Plan.

36 He had been appointed as managing director of SNCF in May 1949.

Organisation on the proposal of France on 21 December 1954, who brought nuclear power into the Château de la Muette.

The choice of a railways manager as a consultant on energy problems is only apparently paradoxical. To begin with, the railways are large consumers of energy, both coal and electricity, and of machines[37], and secondly because Louis Armand, one of the leading "modernisers" of the French economy in the post-war period, had been involved with nuclear questions in France since the beginning of the 1950s. Raoul Dautry[38], who had been to the Paris-Orléans Railway what Louis Armand had been to the Paris-Lyon-Méditerranée Railway, was the first general administrator of the CEA until his death in 1951. Félix Gaillard had then quite logically suggested that Louis Armand should succeed him. Armand refused, because he preferred the railways, and Raoul Dautry was succeeded by Pierre Guillaumat.

However Louis Armand did have important duties at the CEA. He was appointed a member of the Scientific Council on 21 December 1951 and the following year became chairman of the CEA Industrial Supply Committee which had been created on 18 November 1952[39]. He was therefore closely involved with the development of the CEA industrial projects, a major element of which in 1954 was the construction of the Marcoule nuclear complex with its three plutonium-producing piles and its reprocessing plant.

Once he had been appointed a consultant to the OEEC, Louis Armand worked in conjunction with the Secretariat and the vertical committees concerned with energy problems, and he took over certain projects and studies that were in progress[40]. The Council received his report on 24 May 1955.

The "Armand Report", the complete title of which is *Some Aspects of the European Energy Problem*[41], reflects the prevailing attitude amongst those who favoured intensifying European co-operation. It served as a reference text for European actions both in the OEEC and in EURATOM. It also enhanced the reputation of its author as a specialist, and he became the first President – albeit briefly – of the EURATOM Commission in 1958[42]. A study of the report shows the conceptual framework within which the Eurochemic project developed, as part of an overall scheme for developing a new industry[43], seen as representing the future of European energy supplies as well as its political future. The report is also highly revelatory of the technical optimism and enthusiasm that characterised the nuclear field in the period 1954-1958.

The Armand Report is divided in three topical sections preceded by a brief review of its origins and objectives, and followed by a concluding chapter setting out concrete proposals for European co-operation on energy questions. The very structure of the text places particular emphasis on the importance of nuclear power. The first chapter reviews the "*Economic Aspects of the Energy Situation*", the second analyses "*The New Situation: Atomic Power*", and the last chapter is concerned with "*Present and Future Aspects of Certain Problems Concerning Traditional Energy Sources*".

As far as Louis Armand was concerned, it was a question of "rethinking energy problems in terms of the overall economy, so that technological progress could make the best possible contribution to improving the conditions for everyone in a Europe that was ever more united" (p. 52). It was also a matter of breaking out from the traditional approach, reflected in the very existence of the OEEC vertical committees, in other words an approach based upon each type of energy[44]. The report adopts a highly "Schumpeterian" approach in its analysis of technological progress. It is also a plea for a controlled and international energy policy.

---

[37] Modernisation of the SNCF covered not only the rolling stock, with the call to import American locomotives, but also the track which was to be electrified. On 24 June 1952 the newly electrified Paris-Lyon line was opened with much ceremony. See RIBEILL G., «La SNCF au temps du Plan Marshall», GIRAULT R. and LEVY-LEBOYER M. ed. (1993), pp. 319-330.

[38] For Raoul Dautry, see the biography extracted from the thesis by Rémi Baudouï. BAUDOUI R. (1992).

[39] He retained this post until his appointment to EURATOM, and returned to it after his resignation in 1958.

[40] For example the transport of gas over long distances.

[41] OEEC (1955).

[42] Louis Armand hardly took up his post. He was experiencing personal difficulties which led him to resign. He was replaced in 1959 as president of EURATOM by Etienne Hirsch.

[43] In Armand's eyes this overall scheme went beyond the OEEC framework alone. However at the outset it appeared to him that the Organisation was the inevitable host owing to the presence of the British, who at the time held a substantial lead in the European nuclear field.

[44] Committees for oil, coal and electricity.

The question of Europe's lagging behind the United States permeates the report. The report was issued simultaneously with the publication of the British "White Book"[45] on the development of nuclear power, which provided the document with quotations and quantitative estimates.

The first chapter opens on an historical note: the energy field was about to change from the "quantitative and technical" to the "qualitative and economic", resulting in the appearance of competition between the different forms of energy. It was thus entering a phase through which the transport industry had already passed and which led to chaos in the absence of national and international regulation. The "structural analogies" and the "parallel evolution of problems"[46] between transport and energy are used to "identify the leading concepts that should steer the development of Europe's energy structure".

These observations are followed by a review of energy balances and prices, which emphasize the increase in production costs and prices, particularly when the latter are compared with rates of pay, where there was a considerable difference with the United States. The investment strategy during the reconstruction phase was governed by the requirement for "ever more energy": during the growth phase, a further requirement, that of "cheap energy first of all" was needed[47]. Similarly preference should be given to "productive energy" compared with "convenient energy" as regards both investment and pricing.

From all these standpoints Armand saw nuclear power as the touchstone for a Europe which had missed out on oil[48], which he saw as one of the reasons for its economic sluggishness.

The second chapter begins with remarks on the "turning points in the history of energy", with successive periods dominated in turn by steam, electricity and oil. A new era was beginning – that of nuclear power – the onset of which was comparable "quite reasonably...with the discovery of fire"[49]. "It is when such developments appear, as turning points in economic history, that it is necessary to review the situation and determine a policy for the future"[50]. After briefly sketching out the principles of fission, Armand stresses that the development of this technology will "call for an enormous effort of technological research, the success of which will determine whether nuclear power expands quickly or otherwise".

Europe appeared to have many advantages in this energy revolution: local uranium resources were abundant, the price of atomic power, estimated on the basis of British and American data at an average of 0.7 US cents per kWh within two years[51], allowing it "already to be regarded as competitive"[52]. Capital costs should be "very approximately somewhere between thermal and hydro plants"[53]. With atomic power, there would be no further need to site power plants near the source of fuel supply since "the costs of transporting fissile materials are negligible"[54]. It would free Europe from "the fear of energy shortages in the future" and "the idea that energy prices in Europe are irretrievably bound up with the price of coal"[55]. Even though the figures were estimates, concluded Armand, the risk had to be taken. He was strengthened in his conviction by the fact that the United States were embarking on nuclear development and that the British were also doing so "with a confidence in the future that deserves admiration".

It appeared to Armand that the time was ripe to establish special structures for international co-operation, on the grounds that this was "easier in a new system than in the old one"[56]. At the time, this idea was widespread in the circles around Jean Monnet, where the problems of the ECSC and the failure of the EDC were attributed to resistance from the existing structures.

---

[45] Published on 15 February 1954, the White Book on Electricity provided for the construction of 12 power stations within ten years, whereby the contribution of nuclear power to British energy needs would be raised to 5%.

[46] OEEC (1955), p. 16

[47] Ibid. p. 35.

[48] Which accounted for 16% of total primary energy consumption in 1953 compared with 38% in the United States, p. 27.

[49] This view was common amongst nuclear scientists: S. WEART (1980), p. 71 quotes Jean Perrin in 1920: "Compared with this discovery (that of radioactivity), that of fire will be insignificant in the history of mankind".

[50] OEEC (1955), p. 35.

[51] Within a range of 0.5 US cents in 1945 to 0.9 US cents in 1955. DEBEIR J. C., DELEAGE J. P., HEMERY D. (1986), p. 267 gives an initial 1947 estimate of 3.5 to 8 cents per kWh, by Sam SCHURR.

[52] OEEC (1955), p. 29.

[53] $250 per kWe compared with $180 for thermal power plants and $270 for hydro plants, p. 32.

[54] OEEC (1955), p. 32.

[55] Ibid., p. 34.

[56] Ibid., p. 35.

While it appeared to Louis Armand that exchanges of scientific knowledge were more a matter for the European Nuclear Research Organisation[57] or the European Atomic Energy Society[58], and the supply of fissile material for the IAEA which was gestating at the time, the task of the OEEC "being to keep abreast of developments in these areas"[59], he believed that the "exchanges of technical data, the construction and joint funding of industrial installations" were a preferred field of action for the OEEC. Only the "combination of the industrial potential of all member countries", and the presence in the OEEC of "countries that were very advanced in nuclear achievements, like the United Kingdom" – and Armand points out that the UK was prepared for broad co-operation with other countries, as confirmed in Article 39 of the White Book – would allow Europe "to possess an industry comparable with that of the United States"[60].

In more concrete terms the author, in his conclusions, recommends that "the different European countries should work together in the supply of raw materials (fissile materials and special metals) and pool their technological potential in the many fields necessary for the new technology (production and shaping of new materials, reactor technology and the corresponding chemical processes)"[61]. "The Organisation could play a fundamental role in [technical and industrial] co-operation", in "a great variety of forms: government agreements, constitution of industrial consortia from different countries, and so on". A "study group for industrial nuclear power" should be "set up within the Organisation without delay".

Chapter 3 examines traditional forms of energy which, within the OEEC, were matters for the vertical committees, while stressing possible areas of co-operation, and the "transverse" forms of energy: Armand proposes joint oil prospecting in Africa, develops a plea for "gas-productivity" by pointing out the American achievements as well as those in Germany and Belgium, recalls that the transport of gas over long distances as mastered in the United States would make it possible to link Europe to Iraq or Saudi Arabia. With regard to electricity supplies, he proposed that "energy-productivity" should be developed by creating new downstream sites for "European industrial areas" later denoted "international industrial areas" following joint investigations of sites for low cost production facilities (in Norway, Yugoslavia, Austria and also Central Congo). As regards coal which, for six members of the Organisation, already came under the ECSC, the problem of prices – in Europe twice as expensive at the mine as in the United States – meant that modernisation and a "creaming-off" policy involving abandoning certain pits to improve production costs were inevitable. In conclusion, the report envisages maximum utilisation of coal and gas "until atomic power was able to play a leading role". In Armand's view, this transitional phase would be of short duration.

With regard to the role of the OEEC concerning these traditional forms of energy, Armand proposed that the Organisation should set up a horizontal Committee for Energy Economics, encompassing gas. This committee would be requested to investigate the different proposals, prepare studies and "promote the spirit of co-operation".

The Armand Report was submitted to the OEEC Council on 24 May 1955, a few days before the opening of the Messina Conference which was to "relaunch" the construction of the Europe of the Six. Its approach and its conclusions, together with Armand's role in European construction led to the development – both parallel and in competition – of initiatives in nuclear co-operation in the two frameworks which have come to be called little and big Europe[62]. The creation of Eurochemic is linked to the development of the different nuclear projects which were agitating Europe from 1955 to 1957 and in particular, to use the expressions of Louis Armand, to the "two sauces", "Luxembourg" and "Muette"[63], prepared in Brussels and Paris, from which would emerge not only the European Atomic Energy Community (EAEC or EURATOM) but also the European Nuclear Energy Agency (ENEA).

Reprocessing with the "Muette sauce": June 1955 - October 1956

On 10 June 1955[64] the OEEC Council of Ministers adopted the conclusions of the Armand Report and set up a working group (n° 10) of three people "charged with examining the extent, form and methods that

57 i.e., CERN.

58 Set up in 1954.

59 OEEC (1955), p. 33.

60 Ibid., p. 34.

61 Ibid., p. 51.

62 For the emergence of the EAEC, see WEILEMANN P. (1983).

63 The OEEC premises were in the Château de la Muette in Paris. Jean Monnet who was behind the Messina relaunch, had been President of the High Authority of the ECSC, based in Luxembourg.

64 *Acts*, XV, 1371 of 10 June 1955.

co-operation [on nuclear matters within the Organisation] could take and to report to the Council on this subject as quickly as possible". On the same day the Council set up an Energy Committee to investigate the other proposals for co-operation in non-nuclear areas contained in the Armand Report. The two bodies submitted their reports, the first in December 1955, the other in April 1956. The Nicolaïdis Report and the Hartley Report identified the co-operative frameworks within the OEEC in the energy field and marked a new step forward in the formulation of plans for the reprocessing plant.

Working Group n° 10 and the "Nicolaïdis Report". June - December 1955. The first proposal for co-operation on reprocessing

The Working Group

Working Group n° 10 was chaired by Leander Nicolaïdis, Head of the Greek delegation to the Council, assisted by Roger Ockrent, the permanent delegate of Belgium and W. Harpham, Deputy Head of the British delegation. Between 26 July and 10 November the group visited 12 member countries, made contacts with the members of the United States and Canadian delegations at the first Geneva Conference, and submitted its report entitled "*Possibilities of Action in the Field of Nuclear Energy*" to the Council on 15 December 1955. The Nicolaïdis Report was published in January 1956.

The recipe for "Muette sauce" and the "Swiss article"

The Nicolaïdis Report recommended that the OEEC should explore a series of co-operative actions which differed fairly substantially from other contemporary forms of international co-operation. The "style" of OEEC co-operation may be called "Muette sauce", an allusion to the name of the château that houses the Paris offices of the Organisation. The "Muette sauce" was characterised by its flexibility and by the possibility of creating projects, programmes or specialist bodies involving only interested countries, under the "restricted agreements" made possible by Article 14 of the OEEC Convention, known as the "Swiss article"[65]:

"Decisions. Unless the Organisation decides otherwise in special circumstances, decisions shall be taken by mutual agreement between all members. If one member states that it has no interest in a particular question, its abstention shall not hinder decisions which are binding on the other members".

Thanks to the "Swiss article", the OEEC was able to develop "variable geometry" projects involving only certain member countries, under "restricted agreements"[66].

Paragraph 56 of the Nicolaïdis Report is highly significant of this approach as regards the nuclear field, which is fundamentally different from that which was to prevail in EURATOM[67]. The organisation rejected a supranational approach, preferring flexible arrangements on specific projects: "We do not believe that activities such as the management of plants or laboratories, the definition of trade arrangements, the harmonisation of public health legislation, and the control of the use of fissile materials can be handled by a single institution – no more than the manufacture of motor vehicles, the construction of motorways and the highway police in a particular country come under the same organisation".

This pragmatic approach is typical of the OEEC and is profoundly different from that of Monnet. George Ball, Under-Secretary in the Department of State in Washington, sketched out two portraits of Jean Monnet and Robert Marjolin, which could be applied to the two institutions[68]:

"Monnet and Marjolin tackled Europe's problems in different ways: Monnet continually set himself objectives which he could approach but never reach; Marjolin, whose job was to translate general ideas into institutions that would work, was necessarily aware of the boundaries of the possible and of the need to make compromises".

The Muette spirit was of course entirely supported by those responsible for nuclear policy in the Organisation. Pierre Huet, in his reviews of the Organisation's activities at the first two Information Conferences on Nuclear Energy for Industrial Managers, held at the OEEC in Paris in April 1957 and in Amsterdam in

---

[65] This article had its origins in the requirements of Switzerland which wished to adhere to the Marshall Plan but was concerned about the effects on its neutrality. For this reason it is known as the "Swiss article". On this subject, see FLEURY A., «Le Plan Marshall et l'économie suisse», GIRAULT R. and LEVY-LEBOYER M. ed. (1993), p. 555, which refers to BAUER G., «L'adhésion de la Suisse à l'OECE: ses conditions principales», *Relations internationales*, n° 30, summer 1982, p. 215.

[66] In a way, these prefigured the "à la carte agreements" of the 1960s, see Part V, Chapter 2.

[67] For EURATOM's approach to nuclear integration from a legal standpoint see PIROTTE O., GIRERD P., MARSAL P., MORSON S. (1988), pp. 44-87.

[68] Quoted by MARJOLIN R. (1986).

June 1957, put it thus[69]: "The Organisation has not sought to implement a comprehensive scheme of joint action, but it has tried to set up projects to meet the immediate needs of the countries of Europe and thereby likely to be interesting to them to the extent that they may decide to contribute[70]...This empirical method stands out clearly when one examines how the Organisation's programme was gradually developed, using a method that could be called one of 'successive approximations'".

The OEEC therefore considered itself as a service provider, a forum in which projects could be proposed, and not a place for spending all allocated funds, unlike the situation at EURATOM. It was in this manner that the project for a reprocessing plant gradually emerged, by increasingly precise brush strokes, starting from a large number of possible projects.

The report's proposals

The major types of institution

The three authors of the Nicolaïdis Report considered three types of institution[71] that might be suitable for nuclear co-operation in the OEEC framework:

- a Steering Committee for Nuclear Energy reporting to the OEEC Council, with the task of "comparing national programmes, promoting joint undertakings, harmonising legislation, promoting education and standardisation, and devising proposals for international trade". These various functions were to be handled by bodies subordinate to the Committee:
- a Control Bureau responsible for the security control of fissile material;
- "companies", or "joint undertakings" set up as and when required, which would be independent of the Organisation, with the task of "joint projects concerning production and applied research", having "their own management and not being under the authority of the Steering Committee".

The first series of joint undertakings to be envisaged included an isotopic separation plant, *one or more plants for the chemical separation of irradiated fuels*[72], joint undertakings for the production of heavy water, a joint centre for fabricating fuel elements, joint nuclear power stations, semi-industrial laboratories, such as testing centres and others carrying out fundamental research into metallurgy, materials testing reactors and prototype reactors.

These joint undertakings might benefit from appropriate extra-territorial status, like the provisions of the Franco-Swiss Convention of 4 July 1949 covering the construction of the Basle-Mulhouse airport.

The initial outline of a joint reprocessing undertaking

The report discussed a number of different projects including one for a reprocessing plant. Section 57 was the first document of the Organisation to consider the construction of a plant for reprocessing irradiated fuel as a new "joint undertaking". Annex V of the Report was devoted to the "brief description of an installation for the chemical treatment of irradiated uranium". It evaluated the cost of an installation of capacity 500 tonnes/year at $40 million, "so long as the plant is located not too far from a reasonably industrialised area". The type of plant envisaged in December 1955 was an automatic plant with no direct access to the cells[73], similar to the large American military plants.

The Energy Committee and the Hartley Report: September 1955 - May 1956

The Energy Committee was set up on 10 June 1955; its eight members "chosen on a personal basis for their knowledge of general energy problems and their standing within the Organisation or in their respective countries" were appointed on 23 August 1955[74]. The Committee met from September 1955 to April 1956. On 6 December 1955, a few days before the Nicolaïdis Report was submitted, it decided that it should encompass

[69] OEEC/EPA (1957a), p. 288.

[70] Pierre Huet was to add, shortly after the Amsterdam Conference, OEEC/EPA (1957b), p. 23, that this method arose "from circumstances rather than as the result of a deliberate choice".

[71] § 57.

[72] Ibid. Author's italics.

[73] § 29.

[74] *Acts* XV, 1370 of 10 June 1955, 1383 of 29 June 1955. Chaired by Sir Harold Hartley, its other members were Jacques Desrousseaux, Henning Daniel Fransén, Francesco Giordani, Henri Niez, Gustaaf Adolf Tuyl Schuitemaker, Pierre Uri and Friedrich Wilhelm Ziervogel.

the issues of nuclear energy requirements and supplies. The "Hartley Report", entitled *"Europe's Growing Needs of Energy"* was published in May 1956.

The Armand Report had restricted the preferred field of OEEC action to technical matters. The Nicolaïdis Report widened the horizon. The Hartley Report had a refocusing effect. It put the role of nuclear power in perspective for the near future and stressed that energy co-operation in the OEEC should be concentrated on the traditional forms of energy. Consequently nuclear co-operation policy should be reorientated towards applied research and technical investigations. This refocusing approach added strength to the project for a reprocessing plant.

The report argued as follows. It was true that "the advent of nuclear fuel had considerably enhanced energy reserves" (§ 36); it was true that nuclear power lifted the location constraints specific to conventional forms of energy owing to its low transport costs. However the Committee's experts believed that "it was unlikely that nuclear power would supply more than 8% of western Europe's total energy needs by 1975" (§ 39).

Further, the report criticised the adverse effects of "the public enthusiasm raised by this new form of energy and the excessive publicity given to the relatively insignificant progress made in this new field" (§ 37). Hence the role of conventional resources had been underestimated. Moreover "the undue optimism of certain statements about nuclear power had led many people to believe that coal was an outdated fuel of no importance" (§ 40), which had "devastating consequences on worker recruitment, investment and scientific progress in the coal industry".

Concluding the short section devoted to nuclear energy, the report ranked nuclear co-operation with advanced scientific and technical developments, suggesting that it had no importance in "the energy economy of the near future". The report's conclusions (§ 141/viii) stated that "there was no doubt that after 1975 nuclear energy would increasingly replace coal in thermal power stations", but that this coal would be needed for other purposes. The nuclear field should be made the subject of "an immediate and substantial technological effort".

The thinly veiled criticism of the nuclear forecasts given in the Armand Report was stiffened by the estimates given for the capital cost of new power plants. For thermal plants these costs were reduced to $145/kWe, those of hydro plants went up to $315. The cost of an installed nuclear kWe ranged from 250 to 400 dollars[75].

While it is difficult to assess the impact of the Hartley Report on the work of the special committee for nuclear energy questions which had been set up in the meantime as recommended by the Nicolaïdis Report, its preferences were certainly similar[76]. By proposing co-operation on reprocessing and the investigation of new reactor types, the Committee did demonstrate its intention to work in the longer term.

The Special Committee of the Council on Nuclear Energy, the Secretariat of the committee and Working Group n° 1 on joint undertakings: February - June 1956

On 29 February 1956[77] the OEEC Council created the Special Committee of the Council for Nuclear Energy, and requested it to submit proposals within three months. The primary objectives of co-operation between member countries were given as "the establishment of security control, the creation of joint undertakings and the conservative measures to be taken in intra-European trade with regard to nuclear materials and systems". The Committee was also asked to make preparations for the creation of a Steering Committee for Nuclear Energy and to propose ways and means of working together with the existing organisations.

The Committee was chaired by Leander Nicolaïdis with Pierre Huet as secretary[78]. The latter gathered round himself a small group of colleagues who in 1956 and 1957, under his direction, gave considerable impetus

---

75 See Appendix 7 of the report.

76 On Pierre Huet for example, this impact was zero, but it did alter the perception of the OEEC Council.

77 *Acts,* XVI, 1956, 119.

78 Pierre Huet, born in 1920, member of the French Council of State, had been involved in the establishment of the OEEC in 1947 alongside Robert Marjolin as Deputy Secretary-General of the CEEC whose task was to draw up the report presenting the European plan to the American government. When the Organisation came into being, he became a close colleague of the Secretary-General, Robert Marjolin, then as a civil servant seconded to the Organisation was secretary of the Council, then legal adviser. From January 1956 he was asked to examine the legal problems involved in the proposals for nuclear co-operation set out in the Nicolaïdis Report. Pierre Huet became the first Director-General of the European Nuclear Energy Agency and played a very important part in the process of setting up Eurochemic, of which he is regarded as one of the "founding fathers", see ETR 318 (1984). Following his departure from ENEA, he remained a consultant to the Organisation and closely followed the development of the undertaking, while continuing his career in the Council of State whose Secretary-General he was from 1966 to 1970. Pierre Huet was also president of the Association technique de l'énergie nucléaire (ATEN) from 1965 to 1975.

to the nuclear initiatives of the OEEC and formed the nucleus of the future ENEA. This small team was made up of engineers and scientists. Roland Perret was a highly imaginative Swiss engineer with leanings towards physics. He had come from the OEEC Directorate for Industry. Lew Kowarski, a former colleague of Frédéric Joliot, Hans Halban and Bertrand Goldschmidt, regarded as the father of ZOE, the first French nuclear pile, was the group's scientific adviser and was particularly interested in the prospects for developing reactors, especially those of the heavy water type. There was also a Norwegian chemical engineer, Einar Saeland, who had worked at the Norwegian-Dutch centre at Kjeller (JENER)[79]. The team also involved lawyers such as the British advocate Jerry Weinstein, the German Karlfritz Wolff and the young French lawyer Pierre Strohl[80].

The Committee was empowered to create "subordinate bodies" to help it in its work. Therefore on the proposal of the secretariat it set up four working groups covering joint undertakings (n° 1)[81], security control (n° 2)[82], harmonisation of legislation (n° 3)[83] and co-operation in education (n° 4)[84]. At the same time a working group of the Electricity Committee was looking into co-operation related to nuclear power stations (Working Group n° 2 of the Electricity Committee).[85]

Working Group n° 1 had thirty-two members from nineteen countries, some of whom already sat on the Special Committee. It was requested to take a preliminary look at the establishment of joint undertakings. The group's members represented national delegations to the OEEC, research institutions (nuclear and other), and private companies. The United Kingdom had delegated an under-secretary from the Atomic Energy Office, Michael I. Michaels. On the Italian side Professor Felice Ippolito, Secretary-General of the CNEN, accompanied the Italian representative to the OEEC, Achille Albonetti. The Swiss delegation consisted of experts from three companies, Elektrowatt S.A. and Reaktor S.A. in Zurich and Charles Gränacher, head of department in the Basle chemical firm CIBA. France was represented by Bertrand Goldschmidt, Director of External Relations and Chemistry at the CEA, Canada by J. D. Babbitt from the Canadian National Research Council and the United States by Howard A. Robinson[86]. Four possible areas of co-operation were considered: a plant for separating uranium isotopes, a plant for the chemical separation of irradiated fuels, a heavy water production plant and, finally, prototype and test reactors. The group's task was to "prepare a brief note on the technical problems facing each undertaking and the financial resources required".

Information under the following headings was collected for each of the three types of plant:

- a list of feasible processes and the relevant achievements of each country;
- technical problems yet to be resolved;
- expected design and construction times;
- potential problems concerning the necessary raw materials, plant and equipment, power supplies and staff;
- expected scale from the economic standpoint;
- cost of building the installations;
- production costs.

---

[79] He became Deputy Director-General of ENEA and succeeded Pierre Huet as Director-General in 1964.

[80] Pierre Strohl was then aged thirty. Entering the OEEC in 1948 as a doorman to pay for his legal studies, he was an assistant in the documents section of the OEEC conferences service in 1956 and chairman of the Staff Committee. Pierre Huet, who had known Strohl in the latter function, brought him into his team in 1956. He was very closely involved in the development of the company, as secretary of the Board of Directors, and then as a member of the Board of Liquidators. Seconded to Eurochemic from 1959 to 1964 as General Secretary of the company, he then returned to ENEA where he spent his entire career, to be Deputy Director-General during the last ten years of his working life.

[81] Its chairman was R. Sontheim, a director of Reaktor AG, Zurich, and the vice-chairman Bo Aler, a research officer in the Swedish Research Institute for National Defence.

[82] Chaired by Bertrand Goldschmidt from the French CEA.

[83] Chaired by Miss I. M. Collaço, member of the permanent delegation of Portugal.

[84] Chaired by R. Major from the Royal Norwegian Council for Science and Industrial Research.

[85] The group's report was submitted to the Electricity Committee in May 1956. In its conclusions the Committee stressed the widely varying positions of the different countries and suggested that the status of "joint undertaking" could be valuable for power plants located near national frontiers, which could serve two or more nearby countries. See EL (56) 5 and 6, 26 and 30 May 1956. It is worth recalling here that the only two truly international joint undertakings created in the EURATOM framework until that of JET in 1978 were the Franco-Belgian nuclear power station at Chooz in 1961 (SENA: Société d'énergie nucléaire des Ardennes) and the Franco-Belgian power plant at Tihange in 1974 (SEMO: Société d'énergie nucléaire Mosane).

[86] NE/WP.1(56)2.

Working Group n° 1 met for the first time on 10 and 11 April 1956 and submitted its report to the Special Committee on 25 May[87]. The British expert[88] who, with the help of the CEA, prepared a preliminary note on the reprocessing plant on the basis of which the Group made its initial choices.

The report was adopted as a document of the Special Committee on 8 June 1956[89]. It began with an estimate of expected needs in the year 1970. As regards the plans for a chemical separation plant, the working group concluded, following a survey of member countries, "that it was not possible to make meaningful predictions in this field", since the quantities to be processed depended on a number of factors which were still not well known: the nature, power output and operating regime of the reactors, and the number and size of national chemical treatment plants. France, for its part, "planned to cover its own needs as regards the processing of low-enriched spent fuel until at least 1970", but "might consider having all her highly enriched uranium processed in this joint plant". Despite the uncertainties, the working group came up with an estimate of 3500 tonnes a year, mostly of natural or slightly enriched uranium.

A comparative analysis of the different plant projects developed by the working group shows that the processing plant was the joint undertaking whose short term completion in the OEEC framework appeared most straightforward (Table 3).

Table 3. **Matrix of the main arguments put forward by Working Group n° 1**

Project	Process	Capital cost (million $)	Special problems
Isotopic separation plant	2 competing processes	65 to 70	Techniques, equipment and viability
Heavy water plant	Various but secret	50 to 60	Lack of information
Reprocessing plant	A single type of solvent	35.5 to 45	No problems
Research reactor	Various: high temperature, fast breeders, ship reactor, boiling light water reactor	7 to 14	No problems

It was only in respect of the reprocessing plant project that the group was able to make an immediate choice of process. For the isotopic separation plant, the choice between the industrial enrichment process using gaseous diffusion (supported by France), and the very recent German process which had been tested in the laboratory[90], was said to depend on the American supply situation, but in fact revealed the extent to which the two countries disagreed about the project. For the heavy water plant, the report listed the many processes used or envisaged in the different industries, without making a choice; the Norwegian rapporteur in his enquiry had encountered problems of industrial confidentiality in a field that was regarded as highly promising at the time.

The reprocessing plant also seemed to be the least expensive: 35.5 million dollars – plus a maximum of 25%, making $44.37 million – for a plant with a capacity of 500 tonnes a year, by comparison with the heavy water plant. The isotopic separation plant appeared even more costly at 65-70 million dollars – 50 million for a plant processing 500 tonnes of natural uranium a year and producing an additional 1 to 1.5 tonnes of U-235, 10 to 15 million for the power plant and 5 million dollars for the unit producing hexafluoride.

Reprocessing was put forward as having the fewest known material or human problems, which was not the case for heavy water production. At the time it was considered that the shortage of manpower – seen as the only difficulty facing reprocessing – could be resolved by an appropriate training policy. The isotopic separation plant, besides the manpower problems affecting any nuclear undertaking, raised two substantial questions. The first of these concerned the diffusion membranes which were difficult to develop and install because of fluorine corrosion and, being made of nickel, were likely to lead to European shortages of that metal. The second question concerned the possibility of shortages in the supply of compressors, which were needed in very large numbers.

It appeared above all that the isotopic plant could never be viable in competition with American prices. The report quoted a price range of $34 to 38 per additional gramme of uranium-235 produced in the European

87 OEEC (1956c).

88 The United Kingdom had made it clear at the outset that she would participate in the working group but not in the plant.

89 NE(56)12 of 8 June 1956 (annexes dated 12 June).

90 The "Trenndüse" jet separation process developed at the Institute of Physics of the University of Marburg under the direction of Erwin Becker, see MÜLLER W. D. (1990), pp. 505-506.

plant. When the report was published in June 1956, the United States was selling each additional gramme of uranium-235 in a 20%-enriched fuel at $25, and since 26 February the United States had virtually removed any limitations on quantity[91].

As regards heavy water, the report by the French expert suggested that no action should be taken immediately since "in the next 18 months a considerable amount of results and experimental data should become available in Europe about the main processes for producing heavy water".

The study of research reactors suggested much lower capital costs – 7 to 14 million dollars – together with running costs of 2 to 5 million dollars. In this field the report suggested either setting up a joint research centre or implementing a programme for co-ordinating national research. The report by the Dutch expert examined four possible areas of co-operation, the last one being appreciably more costly: a high temperature reactor, a ship propulsion reactor, a boiling water reactor and a fast breeder.

A note prepared by the Norwegian expert[92] "on the possible follow-up to our work" drew some conclusions from the report on 9 June 1956. With regard to the different plants, the note stressed that only the reprocessing plant was immediately feasible.

"Any decision about the isotopic separation plant depends upon the conditions governing the export of U-235 by the United States".

"With regard to heavy water [...] the demand is at present met by American exports, and European private industry is working hard to develop methods of production capable of competing with American prices". Therefore it was unnecessary to make any joint effort in this field since a co-operative venture would not have any apparent advantages. However the expert declared himself in favour of developing the project for a chemical separation plant "as soon as possible". The expected advantages were both economic and technical: the design studies and "perhaps even the construction of a plant" would be a learning process and generate savings by avoiding duplication of effort. "Only when we know the cost of processing shall we be able to determine what will be the cost of nuclear power in Europe".

Thus by June 1956 the essential features of the plant first proposed in December 1955 were becoming clear.

The choice fell on an extraction process using organic solvents "used effectively" which should "not raise any particular difficulties".

The plant was intended to process natural or slightly enriched uranium but did not exclude the possibility of processing highly enriched fuel, "which was not examined further owing to lack of time".

Its capacity would be 500 t/year of natural or slightly enriched uranium so long as the range of viability was estimated at 250 to 500 t/year. The variety of fuels to be processed would make 500 tonnes the upper limit on capacity and would have an adverse effect on viability.

Working Group n° 1 stressed that the hardware problems were not very significant, solvent production being within the capability of a modern chemical industry, that the systems needed were either available or could be developed and that the power requirements were not very high. The problem would be that of recruiting chemists, chemical engineers and physicists.

It was estimated that design and construction of the plant would take a total of four years (two years plus two years) but that the design time could be "considerably reduced if assistance from countries with practical experience in this field could be obtained". Thus overtures were made both to France and the United Kingdom.

The cost of designing and building a plant for reprocessing natural or slightly enriched fuel was estimated to be 35.5 million dollars, but could be reduced with the technical assistance of countries having experience in the field. An additional cost of under 25% would be needed for a polyvalent installation. The processing price would be $14 per kilogramme of uranium.

Creation of the Steering Committee for Nuclear Energy (SCNE) and study groups: July-October 1956

The Council embodied the Committee's conclusions in its decision of 19 July 1956[93] and laid down the guidelines for the Organisation's activities in the nuclear field. These activities were in two groups, involving two different types of body: the first was the Steering Committee for Nuclear Energy (SCNE) and the working groups, which were OEEC bodies. The second consisted of the study groups already working within the

---

91 For the American policy on enriched fuel prices, see Part I, Chapter 3.

92 NE/WP(56)11.

93 *Acts,* XVI, 1956, n° 208.

Organisation but more independently. In fact membership was not compulsory for all OEEC members, in line with Article 14 of the Convention.

The Special Committee became the Steering Committee for Nuclear Energy. Leander Nicolaïdis remained Chairman, the Vice-Chairman being Pierre Guillaumat, then Administrator-General of the CEA, and Friston C. How, Secretary of the UKAEA. Pierre Huet and his team provided the secretariat for the Steering Committee.

The Committee was given the task of preparing a security control system and the terms of reference for the future European Nuclear Energy Agency. In co-operation with the Trade Committee, it was to attempt to "make intra-European trade in products of particular importance for the production and peaceful uses of nuclear energy as free as possible". Another task was to prepare various measures for harmonising regulations applying in Europe to the new nuclear industry.

The feasibility of joint undertakings was to be investigated within the OEEC by two study groups: the first considered prototype, testing and research reactors, the other "nuclear power stations". The Steering Committee and working groups were requested to submit their report within six months.

Although the Annex to the Council's decision of 18 July referred to three study groups taking up the three plant projects examined by Working Group n° 1, the heavy water project, despite the extreme enthusiasm of Lew Kowarski, the Steering Committee's scientific adviser, was in fact no more than an empty vessel, and the other two had different centres of gravity: EURATOM for the isotopic separation plant project, and OEEC for the reprocessing plant project.

Denmark, Sweden and Switzerland indicated that they would like to join the working party on the isotopic separation project which had just been created by the Six following the plans produced by the expert group set up in February 1956 and chaired by Bertrand Goldschmidt. This group continued its work until May1957, in other words after the signature of the Treaty of Rome establishing the EAEC, but without reaching a conclusion. The reasons for this failure throw some light on the success of Eurochemic, and will be described in the next chapter.

As far as the reprocessing plant was concerned, the approach was a symmetrical mirror image: the Six, who had just set up a working party chaired by Erich Pohland, stated that they would like to take part in a study group to be set up within the OEEC, along with Austria, Denmark, Norway, Portugal, Sweden, Switzerland, Turkey and the United Kingdom.

This was in fact the beginning of a sharing out of tasks between the two nuclear co-operation organisations that was already taking shape in mid-1956.

Before going on to consider the work of the "Study Group on the Plant for Processing Irradiated Fuels", which is dealt with in Part I, Chapter 3, it is important to get a better idea of what motivated each member country of the study group in 1956. These factors were extremely varied, and reflected the great disparity between nuclear programmes and the structures set up for their implementation. These differences make it easier to understand the compromise which made possible the signature of the Convention in 1957.

Of the fourteen European countries that took part in the work of the study group on the processing plant and in the company from 1956 to 1958, two – France and the United Kingdom – were already involved in the reprocessing industry. Their achievements were described in the previous chapter, and we shall return to them here only in order to try to explain the extent of their involvement in international co-operation. Four others – Denmark, the Netherlands, Norway and Sweden – had some knowledge of the field through their participation in a small reprocessing pilot project which was the logical sequel to their co-operation in the reactor field at JENER[94]. The other countries had very limited knowledge of reprocessing, but all were keen to take part. We must try to understand why.

The nuclear position of the European countries involved in the Eurochemic experiment in 1956-1957[95]

After the Geneva Conference, most of the European countries tried to speed up their nuclear energy development programmes. The situation however remained very uneven (Figure 16).

[94] The aim of the project was to reprocess fuel from the JEEP pile in a pilot unit with a capacity of 3 tonnes a year, using 20% TBP. The estimated cost in 1958 was £30,000. See BARENDREGT T. J., KOREN LUND L. (1958).

[95] See the bibliography for each country in an Annex. Research into national policies is not very uniform, a state of affairs that is reflected in the present attempt at a synthesis.

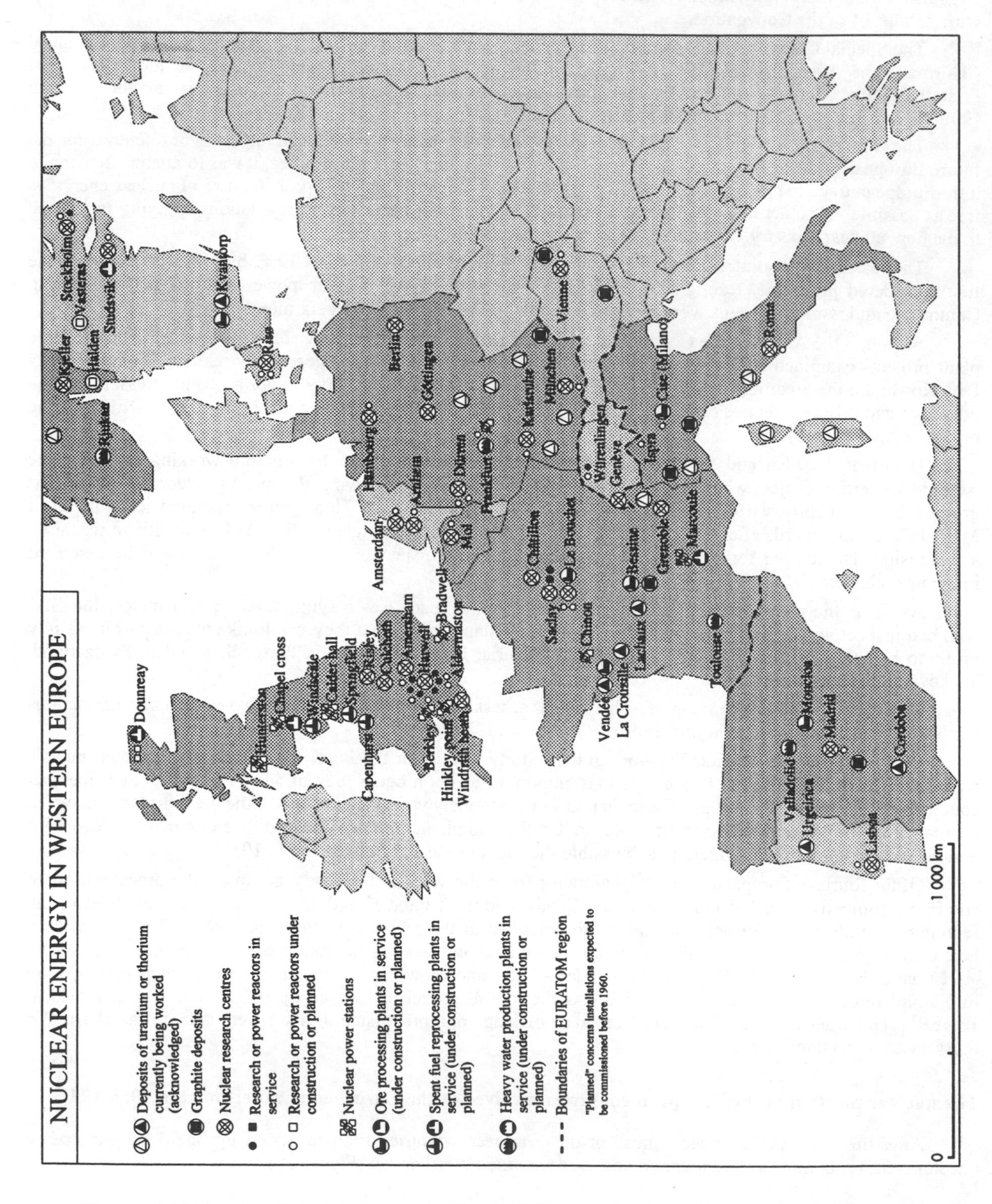

Figure 16. Nuclear installations in Europe in May 1957 (Source: *Die Atomwirtschaft,* May 1957, p. 150).

Wide variations in funding

A first approach is to determine how much was spent on nuclear energy. Tables 4 and 5 show the enormous differences that existed, first between the United States and the countries of Europe, and then between France and Britain compared with the other countries participating in Eurochemic.

Two factors play an important and complementary role: first the length of time a country has had a nuclear policy and, secondly, its economic strength. By applying these criteria to the fourteen European countries that took part in the Study Group or in the company it is possible to divide them into four groups in differing situations: two countries, France and the United Kingdom, dominated nuclear Europe with structures reflecting more than ten years of development; the total estimated cost of Eurochemic was equivalent to about a third of their annual expenditure. Nine countries had recently established a nuclear programme or possessed research structures. They were aware that their nuclear development would depend upon co-operation. The total estimated cost of Eurochemic was equivalent to ten times their annual expenditure. Two countries, Turkey and Portugal, had a very low nuclear budget and only embryonic structures. Finally the Federal Republic of Germany is in a class of its own.

A Franco-British nuclear condominium in Europe with differing interests. The French commitment to co-operation

During the war, scientists from France and the United Kingdom had contributed to the construction of the atomic bomb. Government research and development structures were set up at a very early stage and large sums were devoted to nuclear science and technology.

The United Kingdom, having kept its teams in place since the war, had a lead of several years over its OEEC partners in the mid-1950s. This fact throws some light on the country's low degree of interest in co-operation from which it would have received only limited benefit, particularly as it had special bilateral links with the United States.

However the United Kingdom was involved in European co-operation in the OEEC and took an active part in the work of the Study Group until December 1957. However the UK did not sign the Convention creating Eurochemic. The reasons for this half-hearted support are closely bound up with the status of co-operative projects between the Six, and are reviewed in the conclusion to the next chapter. This progressive detachment is in strong contrast to the French commitment.

The participation of France, already with ten years of experience through the CEA, therefore appears paradoxical, and indeed the idea of co-operation in reprocessing had not received universal support in the CEA. It seems clear that there was a gulf within the Commissariat between those who favoured purely national nuclear development and the partisans of European co-operation on nuclear technology. The special identity of reprocessing – "upstream" of the bomb – and the strong military influence within the CEA certainly contributed significantly to this hesitation, which was expressed at the highest level through the hostility of Pierre Guillaumat, the initial scepticism of Bertrand Goldschmidt[96] and later by the displeasure of Pierre Taranger when he found out that engineers who had worked at Marcoule would be seconded to Eurochemic[97]. However the CEA did have some real enthusiasts for international co-operation, such as Jules Guéron, one of the CEA "Canadians", who was shortly to leave the CEA General Programmes Directorate to become head of the EURATOM Directorate-General for Research and Education, and Jacques Mabile, who was to play a major role in French participation in Eurochemic, and also Yves Sousselier himself.

Finally however it was more advantageous for France to co-operate than not to co-operate, and it was in this realistic but unenthusiastic spirit that Pierre Guillaumat supported the project.

The expected advantages were primarily economic. For example participating in the project would make it possible for Saint-Gobain rapidly to turn its knowledge and experience to account and to sell equipment[98]. International joint funding would mean that France would not have to bear alone the cost of experiments that would have to be conducted in any event.

---

[96] Yves Sousselier opened the first session of the 1983 conference on Eurochemic experience by telling the following significant story: "One afternoon in 1956, my boss called me into his office and said: Goldschmidt is sending me to an OEEC meeting tomorrow about building a reprocessing plant. Neither of us believes it will ever come to anything. It's really of no interest but in any case we have to be represented. I haven't got the time, so can you go?" ETR 318, introduction to the first session, no page numbers.

[97] Communication from Yves Sousselier.

[98] One of the conditions for French participation laid down at the outset was that Saint-Gobain should be selected as the principal industrial architect.

Table 4. **Expenditure on nuclear energy (orders of magnitude) in 1956-1958 (millions of $)**[99]

Eurochemic project, 1956 estimate (*pro mem*)	40
USAEC civilian expenditure 1956-1957	400
UKAEA civilian credits 1956-1957	136
French CEA budget	110
Federal Republic of Germany	20
Belgium	12
Italy	7
Denmark	3
Switzerland	3
Norway	2.3
Sweden	20
Netherlands	2
Turkey	0.2
Austria	0.04
Portugal	very low
Spain	very low

Table 5. **Estimated expenditure on R & D**[100]

Country	Expenditure on nuclear R & D	Number of employees
United States (civilian)	441	119 368
United Kingdom (including military aspects)	304.5	30 000
France	80	5070
Germany (Federal Republic)	17.7	1100 (1958)
Belgium	11.6	916
Italy	6.4	1390
Netherlands	6	418

There were also technical advantages. Through Eurochemic, France could count on benefiting from new information from the United States. Thus the construction and operation of a new plant would be a further source of experience and know-how.

They were also political. Co-operation with Germany should allow a degree of "control by integration" and was thus entirely in line with the policy France had followed with regard to Germany since the creation of the ECSC.

Finally, French involvement in reprocessing might prepare the way for co-operation in building an enrichment plant, something strongly desired at the time by the CEA owing to the costs involved[101], which in the beginning was one of the main motivations for French participation in the negotiations that led to the EAEC. We shall develop this point further in the final part of this chapter.

Germany's search for nuclear power: the role of the chemical industry

The central role of Germany in Eurochemic co-operation, together with France and Belgium, merits careful consideration.

In 1956 Germany was in a special situation. The development of nuclear science and technology in this country, which not only had national resources but also industrial and scientific potential favourable to the introduction of a nuclear industry, had been held up for nearly ten years owing to the results of the 1945 defeat.

---

[99] The data are indicative only; and taken from contemporary sources, some of which are only approximate. Source for the three figures concerning the USAEC, UKAEK and CEA: GUERON J. , OEEC/EPA (1957), Vol. 1, p. 235. Basis for conversion 1$ = 500FF. Source for the remainder of the table: NICOLAIDIS L., OEEC/EPA (1957), Vol. 1, pp. 237-247. These are estimates made in 1957 and rounded by the author.

[100] NAU H. R. (1975) p. 626, using a EURATOM document of 1958, gives fairly similar estimates for six European countries, except for the Netherlands.

[101] And strongly contested by the British.

The country had been officially banned from engaging in any nuclear activity until 1952, and was unable fully to move into civil nuclear power until 1955, after it had renounced the production of atomic weapons. In 1956 therefore, German industry began to construct a nuclear industry[102].

Engaging in international co-operation allowed Germany to save time and gain experience, and at the same time set up its administrative structures, a feature of which is the extensive involvement of private interests. Although the country had a Ministry for Atomic Affairs, its decisions were to a large extent prepared by the German Atomic Commission (Deutsche Atomkommission, DAtK), in which industrial interests predominated. Thus German decisions as regards reprocessing were guided by one of the country's main chemical firms, Farbwerke Hoechst, arising from the ashes of the Konzern IG Farben.

A delayed return to the nuclear scene

As the world's second industrial power in the 1930s, Germany had played a pioneering role in atomic theory since the 1920s through the work done by the university of Göttingen and the discovery of the properties of radioactive substances by Otto Hahn and Lise Meitner. During the war the possibility of developing an explosive based upon fission had been investigated: Werner Heisenberg led a uranium project and was appointed Director of the Berlin Kaiser Wilhelm Institut. At a meeting held at the Kaiser Wilhelm Gesellschaft in June 1942, attended by Albert Speer, the Minister for Armaments, Werner Heisenberg explained that it was in principle possible to produce bombs based upon uranium fission, but that this would be extremely expensive in terms of both money and time. Although research did continue in Germany, it was fairly scattered and had fairly modest financial support, entirely incommensurate with the funding committed to the "Manhattan Engineering District" project, which in fact involved German scientists who had fled racist persecution and Nazi policies.

The leading nuclear scientists who had not fled the country were spirited away by the Allies when the war ended and some of them interned at Farm Hall in the United Kingdom[103]. On their return to Germany they pressed for a resumption of theoretical work and reactor experiments. The "Göttingen Group", focused on the Max Planck Institut für Physik, and involving Werner Heisenberg, Carl Friedrich von Weizsäcker and Karl Wirtz, was one of the leading pressure groups. It was a centre of scientific know-how and urged the resumption of experiments. It was supported by industry, particularly the chemical sector, which during the war had developed a certain experience in this field. At Hoechst, the group was able to rely on the support of the Director, Karl Winnacker[104].

However the resumption of research was vetoed by the Allied armed forces which continued to run the country until 1949, and the establishment of the new West German state was quickly followed by a reaffirmation of strict control on nuclear activities. However it appeared that the first slackening of the bonds would be possible with the CED project.

In May 1952 the Federal Republic of Germany secured the right to produce plutonium, but not more than 500 grammes a year, and to build an experimental reactor not exceeding 1500 kWe in power[105]. Work on the EBR1 reactor expanded in 1953 and the German atomic scientists were able the same year to take part in the Conference on the Kjeller reactors. By the spring of 1953 the chemical industry – led by Hoechst, since Bayer and BASF had less interest at the time – approached the authorities through their Employers' federation, the Verband der Chemischen Industrie, for authority to carry out research in the nuclear field. Besides the chemical firms, those involved in the steel industry and the non-ferrous metals trade, such as DEGUSSA, also played a role.

The acceleration of 1954-1956

The situation evolved quickly from 1954 onwards. On 2 October, Konrad Adenauer removed one obstacle by renouncing, on behalf of his country, the production of nuclear weapons, in the WEU negotiations which came to an end on 23 October. However the country was not given complete freedom of civilian development until after the Treaty of Paris of 5 May 1955, under which the country recovered full sovereignty.

---

102 The creation of the journal *Die Atomwirtschaft* in 1956 was highly significant.

103 For the debate about the "German bomb" and the controversy about the attitude of German scientists, see the writings of Mark Walker, particularly WALKER M. (1990) and WALKER M. (1993), notably the references on p. 519 concerning the publication of transcripts of conversations in "Operation Epsilon" at Farm Hall by FRANK Sir C., intr. (1993).

104 See WINNACKER K., WIRTZ K. (1975).

105 On these points, see ECKERT M. (1989).

On 8 November 1954, the Physikalische Studiengesellschaft Düsseldorf mbH was created, in which sixteen German firms supported scientific work on the peaceful uses of atomic energy. The establishment of this company typifies the close links in Germany between industry and research.

The German delegation to the August 1955 Geneva Conference included a large number of scientists and industrialists. This pattern of forces, so different from the situation in France, proved to be the lynchpin in 1955 and 1956 when the administrative structures for nuclear energy were set up in the federal state. On 16 October 1955 Franz-Joseph Strauss (CSU) was appointed to the new ministry for atomic energy, the Bundesministerium für Atomenergie. Adenauer had sought the creation of this small independent ministry so as not to add to the powers of the Ministry for Economics and thus give the Chancellor more elbow room in the negotiations leading to the Treaty of Rome. In February 1957 Franz-Joseph Strauss was replaced as head of this small ministry by Siegfried Balke, who came from the chemical industry – a subsidiary of Hoechst – and who had defended the interests of the chemical industry in Bavaria as president of the Verein der Bayrischen Chemischen Industrie before embarking on a political career. The ministry was assisted by a Consultative Committee, the Deutsche Atomkommission (DAtK), which held its first meeting on 26 January 1956, in which industrial interests were strongly represented in addition to those of the scientific institutions and the government. In the beginning this Committee played a predominant role in determining the technical and economic choices in the nuclear field[106].

The nuclear research centres were established in 1956. Although Karlsruhe was chosen as the location for the German FR1 reactor following the personal intervention of Konrad Adenauer in July 1955, the Kernreaktor Bau- und Betriebsgesellschaft mbH was created on 19 July 1956, shortly after the opening of the Geesthacht centre (GKSS) in early 1956, focused on a ship reactor, and a little before the GFKF in Jülich (Juliers), which started its life with British reactors. The Karlsruhe Research Centre (KfK) involved the Land of Baden-Wurtemberg, the federal government and industry.

The first German nuclear programme known as the 500 MW programme or the "Eltviller Programm", named after the small Rhineland town where its founders met under the auspices of Hoechst, was formulated during 1957 and submitted to the parliament in early 1958.

The role of the German chemical industry: Erich Pohland, Hoechst and reprocessing

Farbwerke Hoechst played a central role in the chemistry of nuclear materials and especially in reprocessing, and Germany's participation in the Eurochemic project was closely bound up with past and present members of the firm's staff, some of whom, such as Karl Winnacker or the minister Siegfried Balke, occupied a number of key posts. Erich Pohland, who was the first Managing Director of Eurochemic, is typical of the organic links in Germany between industry and the government[107].

Born in Elberfeld in 1898 and trained as a chemist, Erich Pohland was an assistant in the Kaiser Wilhelm Gesellschaft from 1922 to 1926, and then worked in parallel for the Karlsruhe Technische Hochschule and for IGFarben. He joined the civil service in 1937 and held a post in the economic administration of the Third Reich[108]. The collapse of the Nazi state led him briefly in 1945 to accept a post in the agriculture department at Hanover, but in 1946 he became Director of the Chemistry Section of the Zentralamt für Wirtschaft in the Minden dual zone. In this post he came into contact with the allies and efficiently defended the interests of the chemical industry in matters of production and dismantling licences. When the Federal Republic was set up in 1949 he was charged with problems of chemistry within the Ministry of Economics, as Referatsleiter Chemie am Bundeswirtschaftsministerium. His job also involved him in co-operation with the OEEC. After being a ministerial adviser (Ministerialrat) in the ministry of finance, he left along with Joachim Pretsch to join the atomic ministry when it was set up in 1956, as Leiter des Referats für Strahlennutzung und Isotopentechnik. He became chairman both of the EURATOM Working Party and the OEEC Study Group responsible for work on a project for a joint reprocessing undertaking and ended his career as the first Managing Director of Eurochemic from 1959 to 1963.

In 1956, Farbwerke Hoechst had established a laboratory to work on the chemistry of radioactive substances[109], at Griesheim, near Frankfurt. Its Director was a former assistant of Otto Hahn, Hans Götte, who had entered private industry after the second world war and was to become a director of the Eurochemic company.

---

[106] RADKAU J. (1983) gives a good description of its function.

[107] And this is well documented in the printed sources, MÜLLER W. D. (1990), AtW (1959), p. 151, and AtW (1961), Wer ist wer.

[108] Amt für Roh-und Werkstoffe, and then in the Reichsamt für Wirtschaftsausbau.

[109] Radiochemisches Laboratorium.

In 1956 however, German knowledge of reprocessing was no more than had been revealed at Geneva the previous year. For Germany, joining in an international reprocessing project was wholly advantageous. It would allow the country to catch up in a field where there was no urgency[110], owing to the slow development of German nuclear power. It provided reassurance in a sensitive area to the other European countries[111]. Eurochemic experience was systematically recycled in the design and construction of the German pilot unit which started up in Karlsruhe[112] in 1971, for which the first milestones were laid down in 1962.

In 1956 therefore, Germany was very attracted by international co-operation in reprocessing. The more liberal approach offered by co-operation within the OEEC was preferred by most of those involved, not only the government but the industrial representatives in DAtK[113], since a national experiment was unthinkable at the time owing to the young age of the German nuclear industry.

However the lack of any urgency in the development of reprocessing in Germany made it difficult for firms to participate immediately and spontaneously, despite the Ministry's declared determination and the support of Hoechst.

It cannot be ruled out that this ministerial desire to co-operate was also inspired by the government's wish to keep open the doors that might possibly allow the country access to technology which might have military applications. As Colette Barbier has pointed out[114], it appears that Franco-German co-operation in the military field, which later embraced Italy, might have been instituted as from 1955 and was not really removed from the agenda until General de Gaulle came to power, marking "the end of Bonn's [military] nuclear ambitions"[115].

The peripheral countries of the south

By contrast with these nuclear "prime movers" in Europe, there were countries that were interested but lacked resources owing to their overall level of development. They can be divided into two groups. Since the early 1950s Italy and Spain had been trying to develop structures, dominated by major firms in Italy and by the authoritarian government in Spain. Portugal and Turkey had no infrastructure except for what had been specially created.

Nuclear Italy: caught between the government and industry and between European co-operation and transatlantic bilateralism

The development of government structures and of applied nuclear sciences in Italy[116] was late in coming[117]. The government had to share the initiative with major firms, both private and public. In fact in 1946 a private research centre, the Centro Italiano Studi e Esperienze (CISE) was set up in Milan on the initiative of northern Italian firms, particularly the chemical firm Montecatini and the public petroleum conglomerate ENI. Research was carried on there, particularly in the field of uranium refining, but with few resources. The 1947 peace treaty restricted research to the civilian field. However in August 1951 an inter-university research institute was set up to pool the efforts of the physics departments of the universities of Milan, Padua, Rome and Turin, in the Istituto Nazionale di Fisica Superiore. The Italian government really came on the scene only on

---

[110] A DAtK memorandum of 9 December 1957, quoted by MÜLLER, p. 516, emphasizes Germany's lack of need for a national reprocessing plant prior to 1965: "Aus den Leistungsreaktoren des 500 MW-Programms werden etwa ab 1965 verbrauchte Brennelemente zur Aufarbeitung anfallen. Diese Mengen sind zumindest in den ersten Jahren zu gering, um die Rentabilität einer eigenen Anlage zu sichern". Since the firm had little immediate interest, the private companies were to some extent led by the nose by the ministry to buy shares.

[111] And it may also have reassured the Germans. On 12 April 1957, eighteen atomic scientists reacted with the "Göttingen Declaration" to the possibility raised in the NATO framework of providing the Bundeswehr with "tactical" atomic weapons.

[112] The Managing Director of WAK, Walter Schüller, was a former member of Eurochemic staff. For the transfer of technology between Eurochemic and Germany, see Part V, Chapter 1.

[113] "Keine Atomkommandowirtschaft, sondern...Ermutigung der privaten Unternehmerinitiative in der Atomwirtschaft", in a "Tätigkeitsbericht des BMf Atomfragen für das Jahr 1956", quoted by MÜLLER W. D. (1990), p. 154 and repeated by STAMM-KUHLMANN (1992), p. 46.

[114] BARBIER C. (1990).

[115] Ibid., p. 126.

[116] AtW (1956), p. 189-190. The first part of the article by NUTI L. (1990), pp. 134-139 devotes a few pages to "the situation of nuclear research in Italy after the second world war".

[117] According to Edoardo Amaldi the leading Italian scientists, or at least those who had not left fascist Italy, would have voluntarily directed their research to fields that could not be used for building atomic weapons. One cannot avoid being struck by the similarity of the arguments put forward by the German and Italian scientists.

26May1952 with the establishment of a nuclear agency, the Comitato Nazionale per le Ricerche Nucleari (CNRN)[118], which had many problems in making its way, especially owing to its lack of administrative and financial independence[119]. Accordingly the first three-year co-operative programme it proposed remained a dead letter.

The first concrete project[120] did not come into being until after the "Atoms for Peace" speech, and the 1955 Geneva Conference awakened enthusiasm. It had been preceded in July 1955 by the conclusion of a standard bilateral agreement with the United States. The Italian government supported the efforts of the major firms who then declared their intention of acquiring American research reactors; in fact this was done by FIAT, Edison and Montecatini. At the end of 1955 the electricity producers grouped in ANIDEL announced the creation of Societa Elettronucleare Italiana, shortly to become SENN, with the objective of building a power reactor in the Mezzogiorno, which turned out to be the Latina power plant. In 1956 FIAT and Montecatini combined in the Societa Richerche Impiandi Nucleari (SORIN) in order jointly to establish a specialist centre at Saluggia (Piedmont) in the vicinity of which a small reprocessing plant inspired by Eurochemic was to be built many years later. In July the CNRN decided to build a reactor at Ispra on the banks of Lake Maggiore. However in November 1956 Italy had no reactor in service, even though three had been ordered including the future Ispra 1 reactor[121] and a draft "five-year nuclear plan" had been announced in April. In fact according to Leopoldo Nuti, the Italian situation was characterised by discord between the government and business, particularly those involved in electricity production, who did not wish to see the government get involved in nuclear generation just when it was preparing to nationalise the electricity supply industry. There was also a degree of consensus on the need for Italy to make up for lost time in nuclear energy "by making use of American support to avoid being in a weak situation when EURATOM came into being"[122].

In these circumstances the Eurochemic project was exactly in tune with the wishes and desires of both government and business in Italy, particularly as it had the full support of the United States.

Spain: nuclear power, the touchstone for modernisation in a government framework

In Spain[123] the Franco government and the army controlled all initiatives in the nuclear field. The first signs of interest appeared in 1948 but these were not translated into institutional structures until 1951. The establishment of a national agency was a very early sign of the new technocratic and modernist tendencies that appeared in Franco's entourage under the impetus of Opus Dei[124]. In fact on 22 October 1951 the Junta de Energia Nuclear (JEN) was set up, under the direct authority of admiral Luis Carrero Blanco, a Franco minister since July 1951 and a member of Opus Dei. In 1956 the JEN was chaired by General Hernandez Vidal, a member of the army High Command; its vice-president was a scientific specialist in military optics, José Maria Otero Navascués. The JEN, which held the valuable title to the uranium resources that had been discovered in the south of the country, had established a research centre at la Moncloa, near Madrid, which employed about 150 people in 1957, including 60 chemists. The latter worked essentially on ore extraction processes. However Spain was also preparing to acquire its first reactor. In May 1956 General Vidal went to the United States to negotiate the purchase from General Electric of a 30 MW swimming pool reactor[125] and the JEN announced a reactor development plan.

Spain was not yet a member of the OEEC[126], but had been associated with the Organisation since May1954. Its participation in the co-operative undertaking on reprocessing satisfied not only its modernist aspirations but also its desire to be integrated into the international community and, perhaps, raised hopes of access to a technology likely to please the military.

---

[118] In 1956 its president was Professor Francesco Giordani, who in 1957 became the co-author, along with Louis Armand and Frantz Etzel, of EURATOM's "Report of the Three Wise Men". Its Secretary-General was Felice Ippolito.

[119] Thus Italy became a member of CERN before it had set up its own research structure.

[120] An accelerator at Frascati, near Rome.

[121] Purchased from American Machine and Foundry Co.

[122] NUTI L. (1990), p. 139.

[123] AtW (1956), p. 313.

[124] The student riots in February 1956 gave support to the modernist trend; Luis Carrero Blanco became Secretary to the Presidency.

[125] The JEN 1 reactor was installed in 1958 under the Standard Agreement signed with the United States on 12 February.

[126] In 1950 the reintegration of Spain began (the country had been banned from the international organisations five years earlier) with its admission to the FAO in November 1950 and then to UNESCO in November 1952. Spain became a member of the UN on 14 December 1955. See HERMET G. (1992), *Spain in the Twentieth Century,* PUF, Paris, p. 203-216.

In Portugal, the announcement of the Atoms for Peace policy and the discovery of ore deposits in the north of the country had led on 31 March 1954 to the creation of an atomic commission, the Junta de Energia Nuclear (the Portuguese JEN). In Turkey there were only the academic activities at the university of Ankara.

Those responsible for the indispensable modernisation of these countries, both fairly short of energy resources, believed that the development of nuclear activities was also required. Participation in the joint undertaking for reprocessing was a long-term investment, and a gamble on modernisation.

The "small countries": nuclear development through co-operation

Mid-way between the major European powers and the two countries just discussed are those countries whose GNP is limited by their small size or low population. They constitute the biggest group of countries involved in the history of Eurochemic: two of the Benelux countries, the Netherlands and Belgium, the Scandinavian countries, Denmark, Sweden and Norway and the Alpine countries, Austria[127] and Switzerland. These intermediate countries were very aware not only of the importance of nuclear power for their future, but also of the impossibility of their developing it alone. For them international co-operation was a necessity. In fact some of these countries had already embarked upon a small co-operative project in the field of reprocessing.

Nuclear co-operation in reprocessing amongst the countries of northern Europe: the role of JENER at Kjeller-Lillestrom

Some of these "small states" in northern Europe already had limited experience of reprocessing problems through their involvement in the Joint Establishment for Nuclear Energy Research (JENER) at Kjeller-Lillestrom[128], where a small pilot project involved an international team of scientists and technicians, some of whom were to play an important role in the history of Eurochemic, such as Sven G. Terjesen (Norway), Teun J. Barendregt (Netherlands) and Erik Haeffner (Sweden). The Scandinavian countries in general were kept informed of research going on in Norway through a "nuclear contact committee" set up by the Nordic Council[129]. As far as these countries were concerned, the Eurochemic project was an extension to their own joint project, and the small pilot unit commissioned in 1961 did in fact provide some technical services to the OEEC joint undertaking.

Because of their subsequent activities, Sweden and Denmark merit further consideration.

Sweden

The central nuclear policy body in Sweden, Aktiebolaget Atomenergi (AB Atomenergi) was set up in 1947[130]. It was not a government agency but a semi-public company, 4/7 of its capital of 14 million crowns being held by the government, the rest being shared by twenty-four private companies. These were in the electrical industry, iron and steel (Bofors), equipment suppliers (ASEA) and shipping lines (Johnson Line) and so on. AB Atomenergi, which in 1956 had about 300 employees, co-ordinated its shareholders' nuclear work. Through AB Atomenergi Sweden had attempted to exploit ore deposits on its own territory and had developed its own 300 kW reactor, known as R1, a natural uranium heavy water reactor, located in a cave excavated near Stockholm. In 1956, having announced a moderate nuclear development programme at the end of the previous year, Sweden built a research centre at Studsvik to house a second reactor, R2. Under the standard bilateral agreement signed in January 1956 with the United States, plans for the national reactor were abandoned in favour of a light water swimming pool reactor imported from the USA using 20% enriched uranium.

Denmark

Sweden had a nuclear policy at an early stage, but in the country of Niels Bohr the administrative structures for nuclear energy were not set up until after the definition of the "Atoms for Peace" policy, which opened the way to imports of fissile material. A nuclear engineering committee was established in February 1954 at the Academy of Technical Sciences (Akademiet for de Tekniske Videnskaber). A 24-member Atomic Energy Commission (Atomenergikommissionen) came into being on 21 December 1955 following preparations by the Ministry of Finance commencing on 8 March. Two bilateral treaties with the United States and the United Kingdom were then signed during the summer. Naturally Niels Bohr was the president of the commission. Continuity was maintained by an executive committee of seven members chaired by H. H. Koch. The

---

[127] For the general background to Austrian policy concerning European integration, see the recent article by WEISS F. (1994). No reference is made to Eurochemic.

[128] See Part I, Chapter 1 for the origins of this early European co-operation.

[129] The other members being Finland and Iceland. AtW (1957), p. 376-378.

[130] AtW (1956), p. 380-383.

Commission's immediate objective was to build the Research Centre at Risø, some 30 kilometres to the west of Copenhagen. Construction work began in 1956, contracts for the supply of research reactors being concluded the same year with the American firms Atomics International (for DR1), Foster Wheeler Co. (DR2) and with the British (DR3).

Switzerland

Amongst the small countries the case of Switzerland is both typical and particular[131]. Its status of "armed neutrality"[132] is the reason for its early interest in the possibilities of the defensive use of atomic weapons[133]. On 5 November 1945 the federal government had established an Atomic Energy Research Commission, a body which had a consultative role as well as managing federal nuclear energy funds.

In 1956 the country was in the middle of its nuclear "industrial phase". The major companies, which until then had shown no interest in an activity dominated by the scientists[134] under the auspices of the Federal Military Department, in 1953 grasped the opportunities provided by changes in the international situation and sought to establish a national reactor system using natural or slightly enriched uranium and heavy water.

In this country devoid of coal resources, the hydro sites had mostly been harnessed during the first half of the twentieth century and the electricity supply companies were wondering how to meet the expected growth in demand. The engineering firms, including Brown Boveri (BBC), then led by Walter Boveri, soon joined by Sulzer Frères SA and Escher-Wyss, banks and electricity suppliers, totalling 150 companies, began a process in 1953 which on 1 March 1955 led to the formation of Reaktor AG. This group's task was to build and operate a privately-owned National Research Centre at Würenlingen, with the primary aim of building a Swiss natural uranium, heavy water reactor, DIORIT[135]. This private company nevertheless maintained special links with the Confederation, which approved its statutes, research programme and budget and had a consultative vote in the board of directors and in appointing the chairman[136]. The construction of the Würenlingen Centre began in 1955. Meanwhile the Geneva Conference had taken place, which enabled Switzerland to obtain the first American reactor exported to Europe, in this case a demonstration light water, enriched uranium "swimming pool pile" which had been one of the main attractions at the exhibition forming part of the conference. It was bought for $180Ê000 (770 000 Swiss francs), transported to Würenlingen and renamed SAPHIR. It was recommissioned by Swiss engineers on 30 April 1957. The sale was accompanied by a co-operation agreement signed with the United States on 18 September 1955, which was negotiated jointly by the two leading figures in Swiss nuclear energy at the time, the physicist Paul Scherrer and the industrialist Walter Boveri. The agreement covered the supply of fuel and provided for an increase in the thermal power of the reactor.

Switzerland thus appeared to have a bright nuclear future. The strategy of building a national reactor system was accompanied by many attempts to weave a network of international links for the supply of uranium and heavy water. In 1952 overtures were made to Norway, which in that year liberalised its heavy water supply policy. Switzerland declared an interest in the JENER co-operative experiment but encountered an Anglo-American veto, the two Allies fearing leaks to the eastern countries. France offered its natural uranium in 1951-1952, subject to recovering the plutonium, which led to the negotiations being suspended for several years. Complicated three-way negotiations resulted in September 1954 in the delivery of 10 tonnes of Belgian uranium, conditioned in Britain.

Participation in Eurochemic was in line with the country's overall nuclear strategy, and would allow the chemical firms, whose leaders included CIBA in Basle, to take part[137] in the domestic effort in an international framework.

---

[131] See HUG P. (1991), Schweizerische Gesellschaft der Kernfachleute (1992), both referring to FAVEZ J. C., MYSYROWICZ L. (1987), which could not be located. Peter Hug suggests that nuclear energy in Switzerland was developed in four stages. The industrial phase covers the period 1954-1959, when the prevailing nuclear optimism suggested that Switzerland could build a power reactor on its own.

[132] Adopted in 1815 following the Napoleonic war.

[133] Switzerland formally renounced nuclear weapons when it acceded to the NPT in 1969.

[134] Led by Paul Scherrer, president both of the Zurich Federal Polytechnic and of the Swiss AERC.

[135] It went critical on 15 August 1960.

[136] These links prepared the way for the nationalisation of the centre, which took place in 1960 when it became an "institute attached to" the Zurich Federal Polytechnic. It was then known as the Federal Reactor Research Institute (IFR in French, EIR in German). It was merged in 1988 with another "attached institute" which was its immediate neighbour, the Villigen research centre, and is now known as the Institut Paul Scherrer (IPS).

[137] There were plans for developing a heavy water industry in Switzerland but these were torpedoed by the United States in 1955. See Part I Chapter 3.

The announcement of the project resulted in the formation of a group involving CIBA, BBC, Sulzer and Escher-Wyss. Rudolf Rometsch, a young researcher in CIBA, was his firm's representative in the group.

Belgium

Finally we must deal with the nuclear situation in Belgium[138], where Eurochemic was established.

In terms of the proportion of total electricity generated in nuclear plants, Belgium is today in third place in the world, after France and Lithuania. Between the two wars the country was the world's largest producer of radium[139], which came from the uranium deposit at Shinkolobwe in the Congo. This deposit was discovered in 1915 and was exploited until April 1960 by the Union minière du Haut-Katanga (UMHK). The ore was shipped to Belgium and from 1922 onwards the radium was extracted at Olen, a small town to the east of Antwerp.

In the beginning therefore, Belgium's nuclear history was closely linked to the supply of radium by UMHK or rather its American subsidiary African Metals, and then of uranium to the Americans[140]. From 1942 on, a series of commercial contracts was concluded between the firm and the managers of the Manhattan project. In this way, ore from the Congo apparently provided 72% of the uranium that went into the Hiroshima bomb.

On 26 September 1944 an agreement was signed between Belgium, represented by its government in exile in London, the United Kingdom and the United States, guaranteeing these two countries exclusive deliveries of Belgium uranium. The agreement was extended several times. Revenue from these sales went partly to fund the development of Belgian nuclear research in the 1950s[141].

Belgium had hoped that supplying fissile material would give it privileged access to American scientific and technical information[142]. This hope was disappointed, a fact which probably played a role in the country's support for European co-operation in the nuclear field and in the relatively late establishment of its own structures. It is true that an organisation for co-ordinating fundamental research had been set up in 1947[143], but real development did not begin until 1950.

On 31 December 1950 in fact, the function of Commissioner for Atomic Energy (overseeing the Belgian CEA), attached to the Ministry for Foreign Affairs, was created by Royal Order. In January 1951 the post was given to Pierre Ryckmans, former Governor-General of the Congo, who held it until his death in February1959. The CEA was responsible for setting up on 9 April 1952 the Research Centre for the Applications of Nuclear Energy (CEAN), the objective of which was to build a nuclear reactor. The CEAN was established at Mol, not far from Olen, on land belonging to ex-King Léopold III and Marie-José. An area of 184 hectares was acquired in December 1953 and in June 1954 work began on the BR1 reactor[144]. This reactor started up in May1956. In May 1957 the CEAN became the Centre for Nuclear Research (CEN). It then covered some 570 hectares of land.

In the meantime, industry and the electricity suppliers had begun to take a serious interest in nuclear power, and in the background was emerging an overall development scheme driven largely by the Société générale de Belgique (SGB) which held a large shareholding in UMHK and controlled most of the companies

138 This part is based upon preparatory notes made for VANDERLINDEN J. (to be published).

139 However it must be remembered that radium never played more than a marginal role in the turnover of UMHK compared with copper and other non-ferrous metals, and that at the time uranium was a mining waste product.

140 For the uranium supply policy, see HELMREICH J. E. (1986) and HEWLETT R. G., ANDERSON O. E. (1962), p.287 for the negotiation of the 1944 agreement.

141 The arrangements were complicated and involved the governments which signed tripartite agreements in 1951 and 1956, the American procurement agency, the Combined Development Agency (CDA), the Belgian Atomic Energy Commissioner and UMHK. The CDA paid a levy of 60 cents on every pound of uranium oxide to UMHK, which paid it into a public account to which only the Commissioner had access, under the control of the President of the Court of Auditors. The amount of this levy was to remain confidential since it could have been used to estimate the size of the American stocks of fissile material. It provided $9.5 million between 1951 and 1956 and 700 million Belgian francs from 1951 to 1960, of which 234 million went to the inter-university institutes, the remainder to the CEN. The costs of building the Mol Centre were for example entirely funded in this way until 1955.

142 Belgium was relying on its own interpretation of Article 9a of the agreement of September 1944, which in fact is not very clear: if uranium from the Congo was to be used as a source of energy for commercial purposes, the Americans or the British would allow Belgium "to participate in such utilisation on equitable terms". The only advantage gained by the Belgians was their participation in training courses such as those on radioisotopes held at Oak Ridge as from 1949.

143 The Inter-University Institute for Nuclear Physics (IIPN) was a simple body with no legal personality, emanating from the Board of Directors of the National Scientific Research Fund (FNRS) which in September 1951 became the Inter-University Institute for Nuclear Science (IISN), an establishment of public utility devoted exclusively to research. In 1956 its budget was 42 million BF.

144 Reactor of the British BEPO type, resulting from Anglo-Belgian co-operation which had continued since 1951.

involved in nuclear development, such as the Société générale métallurgique de Hoboken (SGMH), les Ateliers de construction électrique de Charleroi (ACEC)[145] and the Fabrique nationale d'armes de guerre (FN).

At the initiative of Union minière and a group of electricity suppliers who formed for the purpose the Syndicat d'études des centrales atomiques (SYCA), a Syndicat d'études de l'énergie nucléaire (SEEN) [Nuclear Energy Working Party] was set up in 1954. The Chairman was Herman Robiliart of UMHK, assisted as Vice-Chairman by Marcel De Merre, of the Société générale métallurgique de Hoboken.

SGMH had originally managed the iron and steel installations on the banks of the Escaut in the southern suburbs of Antwerp; its entry into the nuclear sector was linked to the takeover of the Société le radium belge, located in Olen and which hitherto had been a subsidiary of UMHK. The SGMH for its part had set up a specialist subsidiary, Métallurgie et mécanique nucléaire (MMN), on land purchased to the north of the Mol CEN in order to build a fuel fabrication plant.

The Chief Executive of SEEN was Jean Van der Spek. The company's immediate objective was to build a PWR, the intention at the time being to provide power for the 1958 Brussels Universal Exhibition[146]. However its broader goal was to foster the development of a comprehensive nuclear industry.

The SEEN began its negotiations with Westinghouse in the corridors of the 1955 Geneva Conference. In January 1957 the activities of SEEN were to be taken over by Société belgonucléaire (BN) and extended, notably to embrace reprocessing. In fact that year saw the launch of a large number of projects based upon reactors: Cockerill introduced a ship reactor project, Vulcain, and licensing agreements were signed between Westinghouse and ACEC, Cockerill, SGMH and FN.

In this context, participation in a reprocessing undertaking appeared desirable and logical, since it would give Belgium a presence throughout the fuel cycle, extending from ore in the Congo to reactors manufactured under licence, and to reprocessing.

Conclusion

The wide variety of motives for taking part in the Eurochemic project are summarised in Table 6. The preparations for the project by the Working Party were aimed at drawing up a Convention under which all the countries, whatever their own interests, could come together to build a joint plant.

Table 6. **Converging reasons for participation in Eurochemic**

Country and characteristics	Motives and objectives
France, most advanced of all the participants	Range of economic, political and technical motives.
Germany, then constructing its nuclear industry	To acquire knowledge and experience in the field of reprocessing as quickly as possible, possibly to exploit them itself later. Influence of the chemical lobby, central role of Hoechst.
Belgium, major uranium producer with its resources in the Congo.	To master all the fuel cycle operations. Influence of the SGB through the network of its subsidiaries and holdings.
Italy, just beginning its nuclear development.	Comparable to Germany. The role of Montecatini.
Netherlands, Norway, Sweden, Denmark, already involved in nuclear co-operation at Kjeller.	Continuation and extension of existing co-operation.
Spain, Switzerland, Austria, Portugal, Turkey.	To learn, and to build a joint plant in the OEEC framework.

145 See the contribution by C. Gérard in VANDERLINDEN J. ed. (1994).

146 The project was abandoned, but the 1958 Exhibition went ahead under the sign of the atom, symbolised by the "Atomium".

Chapter 3

Preparation of the project

The Study Group and the Convention of 20 December 1957

October 1956 - December 1957

The "Study Group for the Setting Up of a Joint Undertaking for the Chemical Separation of Spent Fuels" was established on 28 September 1956 and met for the first time on 24 October. Over an 11-month period, the Study Group refined the project while the Convention and Statutes of the future joint undertaking were under preparation by the OEEC.

The Study Group worked closely with the Steering Committee for Nuclear Energy, while negotiations went on between six of the twelve members of the Study Group which were ultimately to lead to the Treaty of Rome that created the European Economic Community (EEC) and EURATOM. The result was a degree of complexity in the organisation which must be considered insofar as it affected the results obtained.

Some crucial decisions were taken during this period, concerning a number of technical options, the choice of site, and the adoption of an original legal formula governing the project, which is expressed in the Convention signed by twelve OEEC countries on 20 December 1957.

Finally it is important to try to explain why the Eurochemic reprocessing project succeeded and the plans for a European isotopic separation plant failed. The two attempts at co-operation were not only conducted in parallel, but also influenced one another.

Organisation of the Study Group

The Study Group and the OEEC nuclear co-operation projects

The Study Group on experimental reactors and the Halden and Dragon joint undertakings

Of the three Study Groups covered by the decision of 19 July 1956, two came into being and worked in parallel: the Study Group for the joint reprocessing undertaking and the Study Group on experimental reactors[1], chaired by S. Eklund (Sweden), with Lew Kowarski as rapporteur, and Roland Perret as secretary.

The second Study Group formulated a programme of research and development on experimental reactors of the types it believed most appropriate to European conditions and having regard to the expected technological developments over the next five years.

It proposed five specific projects, ranked at three levels of priority. The two projects of highest priority concerned a test reactor and a boiling water reactor. Next came plans for an aqueous homogeneous reactor. The projects for a liquid metal fuel reactor and the projects for a fast breeder[2] were relegated to a third stage.

The activities of the Study Group on experimental reactors led in 1958 and 1959 to the creation of two of the three joint undertakings of the OEEC, through the signature of two agreements, one covering the Halden boiling water reactor, the other a high temperature reactor known as Dragon. In both cases it was a question of internationalising existing national projects, Norwegian and British respectively, unlike the Eurochemic project.

Negotiations opened in 1957 between twelve participants – Austria, Denmark, Norway, Sweden, Switzerland and the United Kingdom, with EURATOM acting on behalf of the Six, and led to the Agreement of 11 June 1958 on the joint operation of the boiling heavy water reactor built at Halden, 120 kilometres to the south of Oslo, by the Norwegian Institute for Atomic Energy (IFA). This Agreement has been extended and remodelled many times. Co-operation at the Halden reactor still continues today.

In March 1958, following the failure of a proposal for co-operation on a light water reactor, the United Kingdom suggested the internationalisation of the construction and operation of a high temperature, gas-cooled

1 Steering Committee for Nuclear Energy (1958), ENEA report (1958).

2 Literally translated into French this becomes "couveuses rapides".

reactor, a project on which the Harwell centre had been working since December 1955. The Agreement on the Dragon reactor, which was to be built at the new British research centre at Winfrith Heath[3] was signed on 23 March 1959. There were twelve participants: the UKAEA which owned and managed the reactor, and the same members as Halden. The reactor diverged in 1964 and the project continued until 1976[4].

Table 7. **Participation in the three OEEC joint nuclear undertakings**

Country	Eurochemic	Halden and Dragon
Germany	X	X
Austria	X	X
Belgium	X	X
Denmark	X	X
Spain[5]	X	
France	X	X
Italy	X	X
Luxembourg[6]		X
Norway	X	X
Netherlands	X	X
Portugal	X	
United Kingdom		X
Sweden	X	X
Switzerland	X	X
Turkey	X	

Comparing the participants in the OEEC joint undertakings shows how the "small countries" supported co-operative ventures and highlights the originality of Turkish and Portuguese participation in the reprocessing project. The fact that the United Kingdom and France did not co-operate directly together in any of the projects reflects a degree of rivalry, but also differences in perception of international co-operation. The United Kingdom joined in experiments which had national origins[7], France took part in a co-operative venture that started from scratch. These positions correlate very well with the attitudes of the two countries as regards European integration, for which the way was being prepared elsewhere.

The OEEC-EAEC Study Group on chemical processing

The Steering Committee for Nuclear Energy and some of its subsidiary bodies co-operated with the Study Groups, for example the Working Group on the administrative and financial arrangements for the joint undertakings, and the Legal Working Party, which had a joint sub-group with the Study Group on chemical processing which was charged with looking into specific aspects of the Convention and the future Statutes.

Because of the parallel nature of the nuclear projects of the OEEC and those of the EAEC, then under negotiation, the Organisation set up a liaison group to work with the Brussels Conference, chaired by Jean Renou, at that time Deputy Director of the CEA External Affairs Directorate and a close colleague of Bertrand Goldschmidt. The Eurochemic project and the EURATOM negotiations were linked by the fact that Erich Pohland was chairman of both the Reprocessing Working Party of the Intergovernmental Conference for the Common Market and EURATOM, and of the OEEC Eurochemic Study Group. The EURATOM negotiations came to an end on 25 March 1957 with the signature of the Treaty of Rome. Thus negotiations concerning Eurochemic ran parallel to those of the isotopic separation plant. However, while the latter did not come into being[8], the efforts of the Study Group led to the signature of the Convention on 20 December 1957 and the creation of the undertaking in July 1959.

3 Work on this site had begun in November 1957.

4 For the history of the DRAGON project, see SHAW E. N. (1983).

5 In 1959.

6 Through EURATOM.

7 And was itself the originator of Dragon.

8 For reasons given in the third part of this chapter.

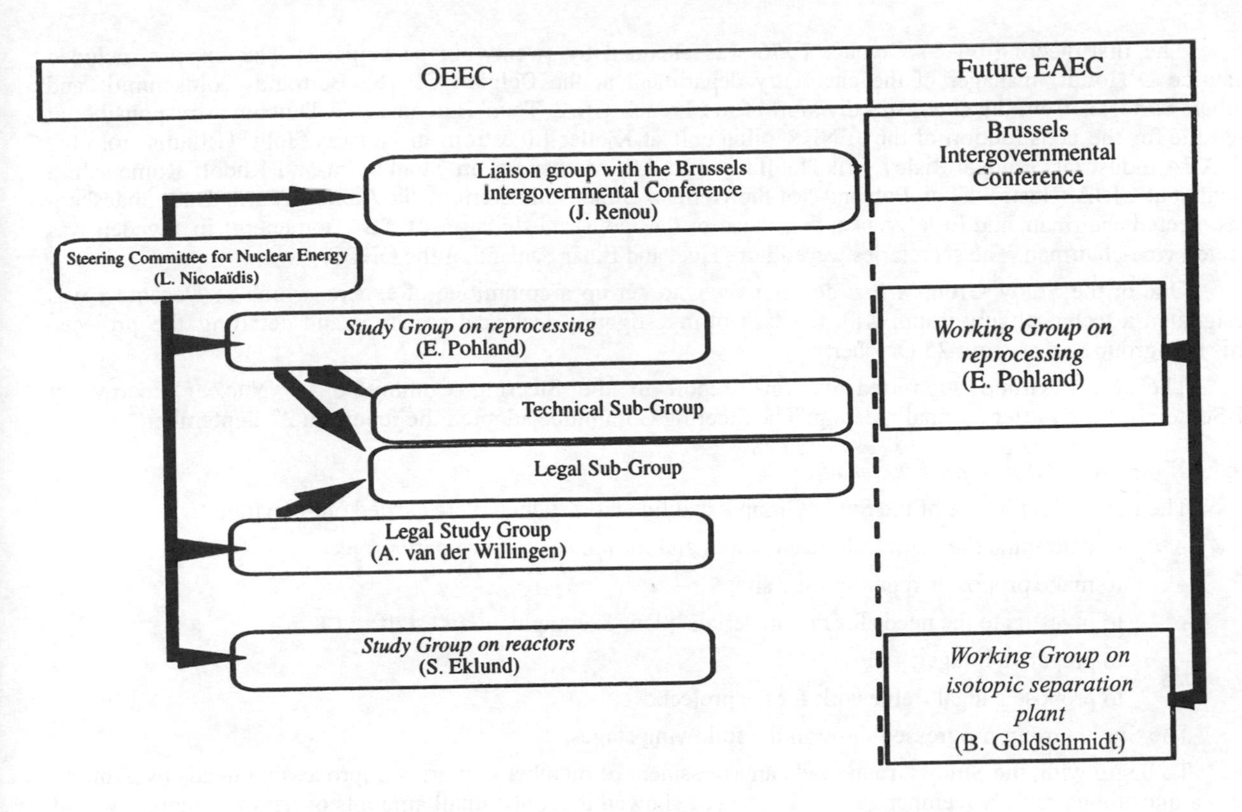

Figure 17. Study groups and groups involved in preparing projects.

Terms of reference and tasks of the Eurochemic Study Group

Terms of reference

The terms of reference of the Study Group set up by the OEEC with a view to the creation of joint undertakings in the field of nuclear energy were defined by the Council at its three-hundred fortieth meeting on 28 September 1956. Those pertaining to the "Study Group for the Setting Up of a Joint Undertaking for the Chemical Separation of Irradiated Fuels" were approved by the Council on 7 December 1956, by which time it had already held its first meeting.

"The task of the Study Group [was] to proceed with the technical studies, to draw up the financial estimates, and to prepare the legal instruments needed for the setting up of a joint undertaking for the chemical processing of irradiated fuels"[9]

Make-up of the Study Group and the technical sub-group

The Study Group was made up of "qualified experts appointed by the participating countries", and came in fact from thirteen countries. The United States, although not a member of the Study Group, did, in its capacity as an associate country, often send one or two representatives from its delegation to the OEEC, the United States Representation Office (USRO) or from the USAEC liaison office. Decisions were taken unanimously, but there was the possibility, if disagreement about technical options should arise of concern to at least two countries, to set up a second Study Group – this did not happen, but demonstrates the very flexible nature of co-operation in the OEEC.

The Study Group held nine meetings between 24 October 1956 to 17 September 1957, the number of members present varying from 28 to 32. A total of 94 people took part in the meetings[10]: they were a mixture of diplomats, senior civil servants from a variety of national ministries, and experts from national nuclear agencies or nuclear firms such as Belgonucléaire, Agip Nucleare or Sorin. However the Federal Republic of Germany and Switzerland did send a few experts from private chemical companies such as Farbwerke Hoechst, Bayer-Leverkusen and CIBA.

9 Statutes, C(56)233 of 7.12.1956.

10 A list of the members of the Study Group is given in the annexes.

The first meeting on 24 October 1956 was attended by twenty-six participants. The experts included Maurice D'Hondt, manager of the chemistry department at the Belgian CEAN, Bertrand Goldschmidt and Robert Sartorius from the chemistry division of the French CEA, Teun Barendregt, a Dutchman responsible at the time for the construction of the JENER pilot unit at Kjeller-Lillestrom in Norway, John Gillams from the UKAEA Industrial Group at Risley, Erik Haeffner from AB Atomenergi in Stockholm and Rudolf Rometsch, a chemist at CIBA (Basle). Erich Pohland[11] of the German Bundesministerium für Atomfragen in Bad Godesberg was elected chairman, and Erik Svenke, then head of the industrial division of AB Atomenergi in Sweden was elected vice-chairman. The secretaries were Pierre Huet and Einar Saeland of the OEEC.

One of the Study Group's first decisions was to set up a committee of experts which in December was designated a technical sub-group, with the task of investigating technical scenarios and detailing the projects. This sub-group met as from 25 October.

The Study Group submitted its final report to the Steering Committee for Nuclear Energy on 17 September 1957 after its final meeting. The Steering Committee adopted the report on 27 September[12].

Tasks of the Study Group: progress made

The terms of reference of the Study Group listed five tasks, which were carried out in parallel:

- to determine the main technical choices and formulate a preliminary project;
- to make proposals regarding the site;
- to investigate the needs for raw materials, plant & equipment and staff;
- to prepare a budget;
- to propose a legal framework for the project.

The Study Group progressed through the following stages.

To begin with, the Study Group made an assessment of member countries' reprocessing needs by sending them a questionnaire in November 1956. This survey showed that only small amounts of natural uranium would need to be reprocessed up to 1960[13], but that there would be an upward trend as from 1964-1965. The total quantity estimated for the period 1964-1970 was of the order of 3500 tonnes. However the Study Group was careful to stress how imprecise these estimates were[14].

It was decided to get in touch with the US authorities: a request for information on specific technical problems was sent on 26 December 1956 to the American delegation to NATO and to the European regional organisations. In response, ORNL sent a volume of data in March 1957 but, and most important, the Brussels Symposium on Chemical Processing was arranged on 20 May 1957[15], which provided the European specialists with much information on which their subsequent work was based.

The technical investigations were initially based on two schemes, according to the type of fuel to be processed – natural or low-enriched uranium (LEU) or highly enriched uranium (HEU). In February 1957 an initial option was adopted for a plant with a capacity of 50 tonnes of natural uranium or LEU, also capable of processing HEU, at a cost of about 18.5 million dollars with the potential for expansion to a major plant with a capacity of about 1000 tonnes of LEU. The final report came out in favour of a non-expandable installation with a capacity of 100 tonnes and an estimated cost of 12 million dollars, to come into service in 1961.

Consideration of the possible location, based on a British report, resulted in ten countries reaching an agreement in principle. Four countries – Belgium, Denmark, Italy and Norway – submitted specific proposals. These proposals led to the adoption of the Mol site in Belgium.

The legal form the project should take was examined by the joint sub-group of the Study Group and the Working Group on the Administrative and Financial Arrangements for Joint Undertakings, which began by considering three options: a commercial company, a government establishment or an intergovernmental organisation that might subsequently entrust its activities to an industrial group. On 31 January 1957 the option of an international company was adopted[16], as being "preferable (in order to) disseminate the technical

[11] France supported the chairmanship of Pohland to encourage the Germans to support the choice of Saint-Gobain as industrial architect.

[12] This report was published as Annex A of the first report of the Steering Committee for Nuclear Energy in March 1958, OEEC (1958).

[13] No more than 100 tonnes a year from 1961 to 1964, OEEC (1958), Steering Committee for Nuclear Energy, p. 35.

[14] Ibid. p. 33.

[15] USAEC (1957).

[16] SEN/CHEM(57)3.

knowledge arising from the operation of the plant". A first draft of the company's Statutes was issued on 12 April 1957, which made the first mention of the "European Company for the Chemical Processing of Irradiated Fuels". This was adopted in December. The only significant modification stemmed from a Swiss request[17]: a third paragraph added to Article 3 of the Statutes, left open the possibility of the subsequent joint construction of a larger plant.

During its life the Study Group took three principal types of decision that were essential to the future of the project.

Major decisions in the preparation of the project

The decision-making processes as regards technical, geographical and legal aspects merit discussion.

The main technical options

The technical sub-group and project variants

The membership of the technical sub-group comprised six different nationalities: Erich Pohland (Germany), Teun Barendregt (Netherlands), Yves Sousselier (France) who succeeded Robert Sartorius after the first few meetings, John Gillams (United Kingdom), Erik Haeffner (Sweden) and Rudolf Rometsch (Switzerland). Some were more knowledgeable about reprocessing, others were familiar with industrial practices. This expert group drew up proposals for the Study Group which took up most of its conclusions and formed the nucleus of the future Technical Committee of the company.

Estimating requirements in terms of quantity and quality, the sub-group determined the maximum capacity of the plant as 100 tonnes a year. However various approaches were explored as the Study Group's work proceeded. In an initial approach, two variants were explored for essentially technical reasons according to the type of fuel – natural uranium/LEU or HEU; this was followed by design considerations of a plant that would be easily extendable later to 1000 tonnes (denoted "variant A") or a non-extendable plant (denoted "variant B"), which satisfied criteria that were political – linked to the parallel process of selecting a site (see below) – as well as economic. Finally the sub-group concentrated on a multipurpose natural uranium/LEU or HEU type B project.

Assembling information: co-operation inside and outside the Study Group. The roles of France, Britain and the United States

France and the United Kingdom, which took part in the Study Group, supplied information from their research centres and reprocessing plants: Marcoule, through the CEA industrial division, and Windscale through the Chemical Plant Design Office of the UKAEA Industrial Group Headquarters.

The USRO also arranged access to USAEC data through an exchange of letters[18] but mainly by organising a Symposium on reprocessing which was held in Brussels from 20 to 25 May 1957 and which was comparable, for the European reprocessing chemists, with the Geneva Conference.

This Symposium heard papers from Oak Ridge National Laboratory (ORNL, Tennessee), Hanford Atomic Products Operation (HAPO), also the ICPP, Argonne National Laboratory (Lemont, Illinois), Knolls Atomic Power Laboratory (Schenectady, New York), Atomics International, North American Aviation Inc. (Canoga Park, California) and Brookhaven National Laboratory (Long Island, New York). Papers came from all the government and private research groups involved in reprocessing in the United States, but there was no direct contribution from the Savannah River military plant.

The proceedings of the Symposium were published by the Oak Ridge Technical Information Service extension in three volumes totalling 1365 pages[19], and provided a practical summary of American knowledge of reprocessing: as R. B. Richards of Hanford[20] stated in his opening speech, "the series of papers in this session

---

[17] Communication from Rudolf Rometsch, 8 January 1992.

[18] The USAEC in March 1957 sent a memorandum – *Answers to OEEC Questions on Chemical Reprocessing*, SEN/CHEM(57)6 addendum, replying to the highly specific technical questions it had been asked in December.

[19] USAEC (1957), TID-7534. Volume 1 is in the following parts: Aqueous Processes, Review of the Redox, Purex and Thorex Processes; Ancillary Processes, Dissolving Zirconium and Stainless Steel, Plutonium; Storage of Gaseous, Liquid and Solid Wastes from the Plant. Volume 2 is devoted to non-aqueous processes and Volume 3 – Engineering and Economics – to the problems of plant and equipment, giving names and addresses of suppliers and estimates of costs.

[20] Vol 1 p. 3.

will treat an area of separations development capable of immediate application to the design and operation of facilities for reprocessing irradiated reactor fuels...We have attempted to organize an integrated presentation of aqueous reprocessing methods, rather than offer a collection of loosely-related technical papers". The Symposium papers were carefully studied by the Eurochemic engineers and researchers[21] and for some became a sort of bible. Personal contacts were made during the Symposium, which facilitated further co-operation. By participating wholeheartedly in the Brussels Symposium, the United States demonstrated in concrete terms that they were fully committed to transatlantic co-operation in the field of reprocessing, just at the time the Convention was being drawn up and technical options for the plant determined.

The initial technical choices[22]

The technical sub-group was able to use this information from unimpeachable sources for proposing options to the Study Group as regards both plant, equipment and scheduling, and to produce cost estimates.

The extraction process

The choice of process depended on the destination of the final product: the uranium and plutonium produced by the plant had to be capable of incorporation in new fuel pins and therefore of being handled without danger. This made a high decontamination factor (10^7) essential, together with adequately long storage, before or after reprocessing, to allow the worst of the radioactivity to decay.

Following a process of elimination, the PUREX method was selected. In 1957 available reprocessing techniques fell into two main groups:

- there were many non-aqueous processes (for example, pyro-metallurgy, scorification of oxides, and volatilisation of halides) but these were still at the stage of laboratory testing, and to adopt any of them would have incurred risks incompatible with the proposed industrial character of the undertaking;
- the sub-group therefore focused on the aqueous processes involving solvent extraction and, more particularly, those already used in large scale facilities. Three solvents appeared feasible: hexone, DBC and TBP, used respectively in the three processes known as REDOX, BUTEX and PUREX. The use of hexone was rejected because it made concentration and waste storage more complicated (a salt had to be added as a precipitant). This left DBC and TBP, to which only nitric acid had to be added. TBP had the advantage of a better decontamination factor than DBC, despite its higher radiation sensitivity and the PUREX process was accordingly chosen.

Thus a predominant factor in the Study Group's decision was the purity of the final product. Correlative advantages were that there was less industrial risk[23] and – a political plus – it would be possible to utilise two distinct sources of expertise, one internal – France – and one external – the United States – something which the other two processes could not provide. Of course the decision not to use BUTEX tended to sideline the British and thus contributed, quite apart from political factors, to making further co-operation less attractive for the United Kingdom.

Although TBP was selected as the solvent, some reprocessing aspects were not yet determined in 1957: the decladding method at the head-end of the process was not finally chosen, and the report stresses the advantages – in many circumstances – of mechanical decladding, especially for zirconium and stainless steel cladding. With regard to the other end of the process, the decision as to how the plutonium would be purified – using a solvent or by ion exchange – was postponed, since insufficient information was available about the processes used in industry. This stage too was closely bound up with military know-how, and relevant publications were scarce. The proceedings of the Brussels Symposium included only three papers on problems of the tail-end of the plutonium process, amounting to 50 pages from a total of 1365, or 3.6%[24]. Moreover these aspects were dealt with only as laboratory research topics in fundamental and applied chemistry, rather than from an industrial standpoint[25].

---

[21] Communication from Emile Detilleux, 11 December 1992.

[22] See final report of the Study Group in OEEC (1958), report by the Steering Committee for Nuclear Energy, pp. 31 *et seq.*

[23] The two most recent plants at Savannah River (opened in 1954) and Marcoule (being completed in 1957) were using or were about to use TBP.

[24] USAEC (1957), Vol. 1, pp. 296-348.

[25] It is possible that even in the contemporary military plants the plutonium was purified using the same methods. At Marcoule a building specially designated for working with plutonium was not built until 1963, some four years after production had started.

Plant and equipment

As regards plant and equipment for reprocessing, the Study Group opted for a plant with direct maintenance: "in other words maintenance was to be carried out directly on the plant, after decontamination". Thus the Study Group rejected the remotely-controlled maintenance used in the large American plants and that at Windscale, and chose the systems that already existed at ICPP and Marcoule. This decision was due as much to natural caution about something which was a novelty in continental Europe as to cost considerations. It also satisfied the requirements regarding research and flexible development. It was essential that staff should be able to work directly in the cells to check what was going on there and to make improvements and modifications as necessary. Direct maintenance also resulted in particularly reliable equipment being chosen in order to avoid too frequent stoppages, with the need to decontaminate the cells before people could enter. The fact that the incoming fuels contained varying amounts of fissile material inflicted special safety constraints aimed at ensuring that a critical mass was never approached. This meant a great deal of work on the geometrical safety of equipment had to be done at the design stage.

As far as the extraction systems were concerned, the Study Group showed itself more innovative in selecting pulse columns, making Eurochemic the first European plant to use this type of system in the main part of the process.

France used mixer-settlers[26] and the United Kingdom filled columns. Although the latter are very safe, their separation capacity is poor unless compensated by great height, so the columns were about 25 metres high[27]. Pulse columns are more efficient and modest in size: only about 10 metres high and requiring about 6 square metres of floor space. Mixer-settlers are efficient, but take up considerable floor space – six metres by three – for a fairly low height of about four metres[28], which has the result of making the concrete cells bigger and more costly to build. The shape of mixer-settlers also introduces the risk of accidental criticality when fissile materials are being processed. Finally the relatively long dwell time of the solvent in mixer-settlers raises special problems owing to the fact that it is fairly quickly degraded by the effect of radiation. This problem was less significant in a military plant where the irradiated fuel had low burn-up, the plutonium was rich in the 239 isotope and cost considerations were of little significance. The problem was liable to be more serious in civilian plants reprocessing fuel "burned" to the maximum for producing electricity, and where commercial operation required strict control of capital costs[29].

The highly innovative choice of pulse columns[30] was related to the problems that had been found in using European equipment and to the fact that Hanford had done development work on the application of pulse columns to the Purex process. The paper given by R. G. Geier to the Brussels Symposium[31] stressed the advantages in its conclusion: "The development work carried out at HAPO has indicated that the pulsed solvent extraction column would be a very attractive contactor for Purex process application. The columns comprising the battery would be of reasonable height, be easily capable of meeting the process requirements, and provide sufficient versatility to encompass flowsheet modifications if necessary".

Thus the arguments favouring pulse columns for Eurochemic were their smaller requirements for space in the concrete cells, their operational flexibility together with safety and efficiency. However the factor which decisively won the day was the shorter contact time between the aqueous and organic phases. This was particularly important since fuels were being irradiated to increasingly high burn-up levels, meaning more rapid degradation of the solvent. Pulse columns therefore seemed eminently suitable for a civilian plant, but considerable development was needed before they could be used on an industrial scale. Eurochemic was to achieve this transition in Europe. Generally speaking, the system constraints arising from the use of hot, acidic, radioactive solutions mean that tanks, pipes and columns must be made of fully austenitic stainless steel[32], of all-welded construction – using very sophisticated procedures for welding and weld inspection – and thus involving high quality materials and manpower.

The recommended instrumentation was of the type routinely used in the chemical industry, except for continuous monitoring of radioactivity, where new systems were needed.

---

26 Pulse columns had been tried out in the Châtillon pilot unit, but problems with controlling their operation led to mixer-settlers being preferred for Marcoule. See GALLEY R. (1958).

27 70 feet.

28 Floor space of 20 x 10 feet and a total height of 12 feet.

29 Application of the Marcoule approach to the UP2 plant – also equipped with mixer-settlers – led to operating problems when the La Hague plant had to deal with fuels of increasing burn-up at the end of the 1960s.

30 And, with hindsight, a clever choice, see Part V Chapter 1.

31 USAEC (1957), Vol. 1, pp. 107 *et seq.*, R. G. Geier, Application of the pulsed column to the PUREX process.

32 One of the three types of chromium-nickel stainless steel, possibly stabilized with niobium.

By September 1957 therefore, the main technical options had been adopted: TBP extraction in pulse columns, direct access to cells, the use of proven and safe equipment. The technical decisions resulted from a process of elimination, based not only on technical arguments but also on politically significant reasons while taking into account the requirements of cost, productivity imperatives and leaving open the possibility of applied research.

Initial outline of plant design

An initial approach to the plant design was sketched out: this involved about twenty buildings, eight for the processing installations, two for research facilities, six auxiliary buildings, two general service buildings and two service installations. From 1961 onwards the plant would employ about 450 people, including 250 in the processing facilities[33]. With a capacity of 100 tonnes a year, it would require an initial investment of 20 million dollars, including 12 million for construction, 7 million for operating costs during the first two years and 1 million dollars of working capital.

The process of choosing a site, which went on in parallel with that of the major technical options, was based on a number of factors involving technical, political and economic issues.

The geographical option: Mol

The process leading to the choice of the Mol site involved two overlapping arguments: the first one was technical, the other an intimate amalgam of political and economic considerations. General consideration of location criteria, related to the technical requirements of reprocessing, will reveal certain geographical constraints.

We shall then review – on the basis of these constraints – the sites member countries offered, reflecting their political and economic desire to have such a plant located on their territory, with the expected benefits for the local economy in terms of jobs and orders. In this way the number of possibilities is reduced.

However the final decision stemmed from an essentially political compromise, which developed along with the proposals for sites, and ran strongly against the technical reasoning involved in the first phase[34].

The choice of the site influenced the technical options adopted later for the construction of the plant, and goes some way to explaining the special emphasis at Eurochemic on safety problems, as well as the financial difficulties which very soon complicated the construction of the plant[35].

Stage 1. October-January 1956, technical reasoning predominates

On 27 October 1956 the Study Group sent member countries a questionnaire on "Needs for the Processing of Irradiated Fuel"[36]. The third of the eight questions put was: "If the Study Group decided to suggest that the chemical processing plant should be located in your country, would you be favourable in principle to this proposal and, if so, under what conditions?" A response was required by the 30 November. A summary of the replies sent in by ten countries[37] out of the thirteen members of the Study Group was presented in January 1957, excluding Sweden which was still unable to give a decision, showed that all were favourable to an international reprocessing plant being located in their country.

On 23 November 1956 the technical sub-group received a report from the United Kingdom, prepared by John Gillams, on the subject of location[38]. This report reviewed the factors affecting the choice of location for a chemical plant with a nominal capacity of 500 tonnes of uranium a year, based upon experience with the operation of Windscale. It examined four factors in turn: waste disposal, water supplies, transport costs and safety, and deduced their impact on location.

According to this review, water supplies and transport costs had little significance in the constraints governing location: water requirements of the order of 5000 to 10,000 m^3/day, including 1000 to 2000 m^3/day of drinking water, were regarded as relatively easy to meet. The impact of transport costs on the choice of location was regarded as low. With regard to low activity liquid wastes, routine disposal required the presence of water nearby: the sea or a river. At Windscale the shore location allowed daily disposal of 35 curies of beta activity, together with 16 curies of ruthenium-106, 7 curies of strontium-90 and 0.05 curies of plutonium-239.

---

33 Description in SCNE (1958), p. 40-41.

34 Quite apart from political interests, bitter criticism was expressed, for example by American scientists who did not miss the opportunity to make them known to members of the mission of November 1958, see Part II Chapter 1.

35 To the point of even threatening continued co-operation, see Part II Chapter 2.

36 SEN/CHEM(56)3 of 27/10/1956.

37 Austria, Switzerland and Turkey did not reply.

38 SEN/CHEM/WD.1 of 23/11/1956, annexed to SEN/CHEM(57)2 of 15/1/1957.

However the available choices were limited by the "consequences of unusual conditions", in other words safety conditions.

With normal gaseous discharges, a high stack would be required, and the effluent tested after scrubbing and filtering. In view of the possibility of an accidental release, it was best if the local population density was low.

In fact the constraints on location set out in the British report were governed primarily by considerations of safety: there were "special risks" needing preventive and surveillance measures, such as fire, criticality, and the danger of spillages or leaks. The seismic risk had to be limited, together with that of war. Sites situated below the normal water table were to be avoided, as well as those above a drinking water abstraction point.

The report concluded: "the main technical requirements for the site are the following: moderate water supplies; adequate communication facilities, preferably by rail; access to the sea or a river, preferably away from drinking water abstraction points. It would be better to choose a site in western Europe [i.e., not Mediterranean, according to the author], not in zones of high seismicity, remote from major population centres but nevertheless in a region with some degree of industrial tradition".

These conclusions were taken up in the later documents of the Study Group and were used in preparing the questionnaire sent to member countries seeking proposals for sites[39].

Proposed sites: January 1957 - June 1957

At its meeting on 12 January 1957, the technical sub-group invited interested countries to specify the site they proposed for a plant capable of reprocessing 50 tonnes of natural uranium or LEU a year, taking into account the constraints expressed in the British report for a larger plant, with a view to allowing subsequent expansion on the same site. The Study Group, at its third meeting on 22 January, accepted the sub-group's proposals and decided that "preliminary information on the situation and characteristics of proposed sites" should be provided by mid-February. The Secretariat accordingly wrote to interested governments on 29 January 1957 inviting them to reply before 20 February. The time allowed appeared too short, and it was not possible to provide governments with a summary of the replies sent in by Belgium, Denmark, Germany, Italy, Norway and Portugal, together with the United Kingdom and the Intergovernmental Conference for the Common Market and EURATOM until 25 March 1957[40].

Portugal declined in a letter of 16 February 1957, referring to problems of water supplies and seismic risks. The United Kingdom and Germany, although not formally declining, referred to problems in negotiations with local authorities.

Finally four countries made specific offers, between 15 March and 3 June 1957. In a ten-page letter dated 15 March, Belgium proposed three sites, with appropriate cost estimates, and five days later the Belgian delegate to the OEEC, A. De Baerdemaker, sent the Secretary of the Steering Committee for Nuclear Energy a 14-page booklet entitled "*Belgian Proposal for a Site for the Spent Fuel Reprocessing Plant*", prepared by Maurice D'Hondt and accompanied by five maps.

Denmark replied on 7 March but did not provide detailed proposals on three sites until 13 May 1957. Italy, having asked for more time, also submitted three sites. Finally Norway sent in two bids on 3 June.

The documents provided by the four countries were collected together on 4 June 1957 in a *"Report Prepared by the Technical Sub-group on the Choice of Site for the Chemical Processing Plant"* (Figures 18 to 20)[41].

Six of the eleven sites offered were coastal, those proposed by Belgium were not.

The first site proposed by Italy was at Garessio, in the foothills of the Alps, at an altitude of 600 metres and situated 20 kilometres from the Riviera di Ponente on the Ligurian sea, about 30 kilometres in a direct line from the French border. Disposal of wastes into the sea would have required a 6 kilometre channel to be excavated and the construction of 35 kilometres of railway to the plant.

---

39 See for example SEN/CHEM(57)4 of 15 March 1957.

40 SEN/CHEM(57)5. Since France was slow to respond, Bertrand Goldschmidt notified Pierre Huet on 27 March 1957 that it was "possible to find sites meeting the technical conditions required for the installation on the west coast of France"; since this would "represent a significant effort both in manpower and financial terms" France did not wish "to embark upon such a study unless the main countries involved in constructing this joint undertaking expressed the wish that it should be located in France". Source: Unreferenced letter attached to the SEN/CHEM(57) file.

41 SEN/CHEM(57)15, set of maps.

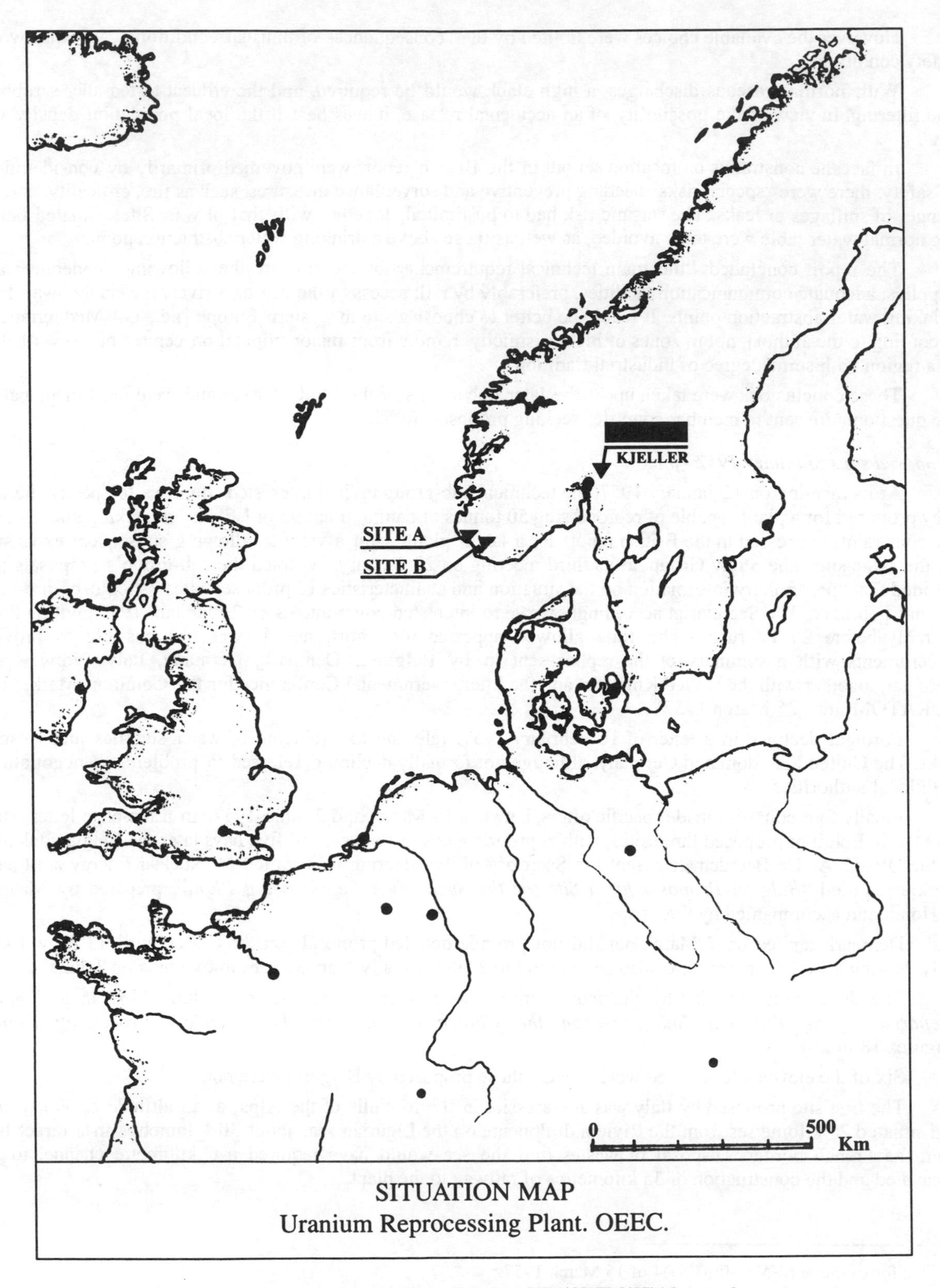

Figure 18. Norway's proposed sites (Source: SEN/CHEM(57)15, set of maps).

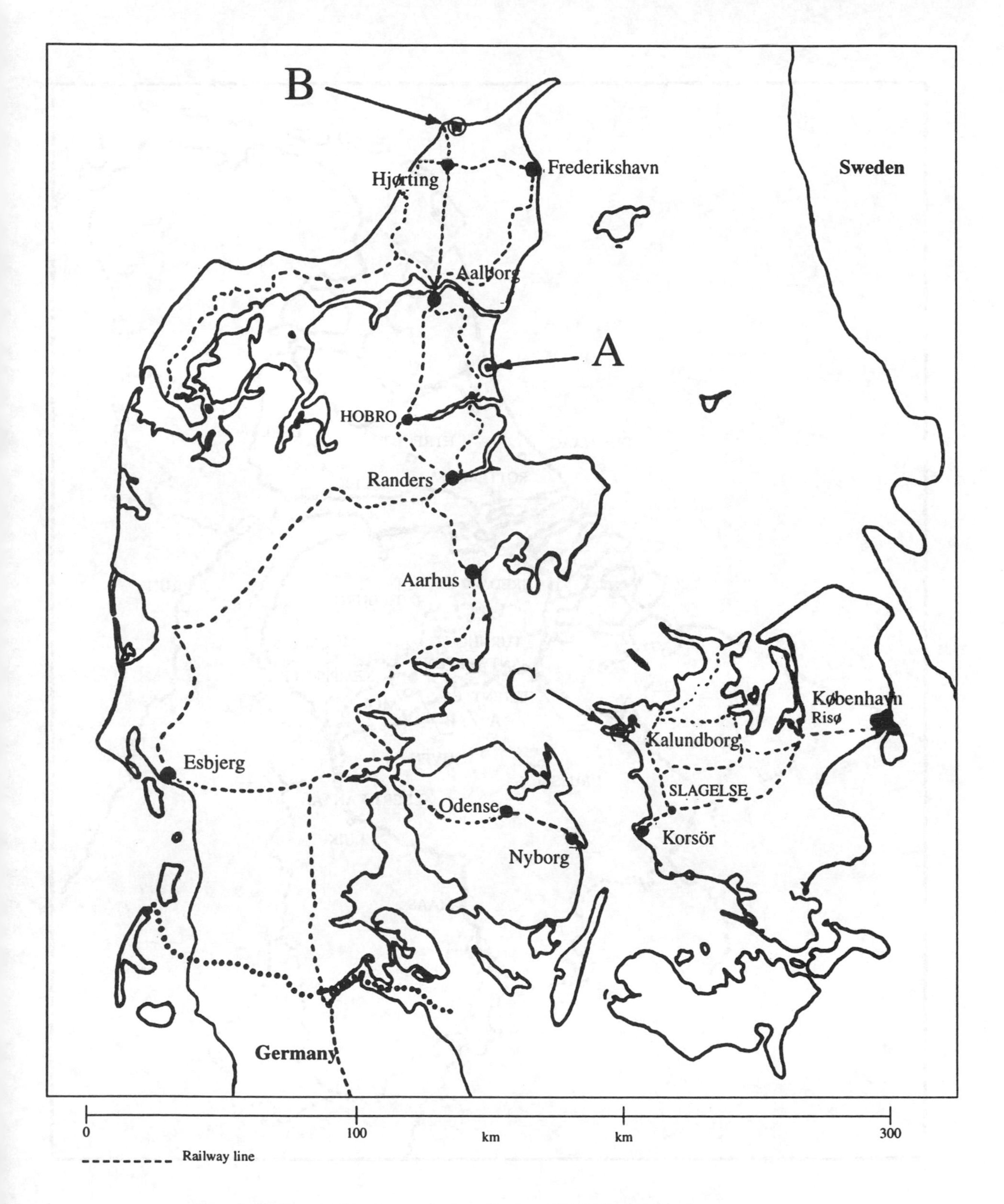

Figure 19. Denmark's proposed sites (Source: SEN/CHEM(57)15, set of maps).

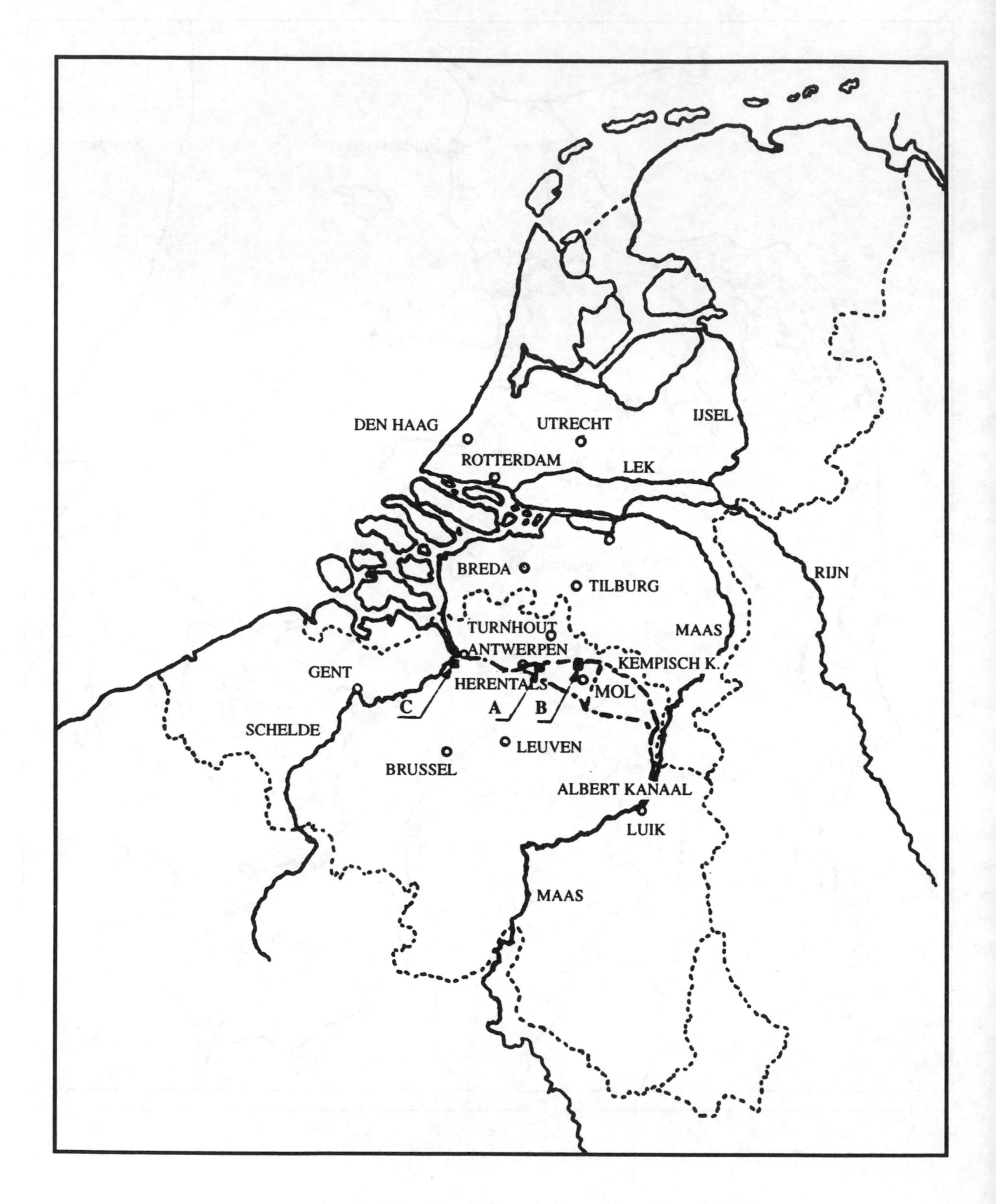

Figure 20. Belgium's proposed sites (Source: SEN/CHEM(57)15, set of maps).

The second site chosen by Italy was on the coast, on the plain of Maremme, 14 kilometres from Grossetto, opposite the Island of Monte Cristo, in a marshy area near the mouth of the Ombrone.

The final Italian site proposed was located on the coast of Sardinia, near the mouth of the Flumendosa, 65 kilometres to the north-east of Cagliari, in the hills near the village of Villaputzu on the right bank. "The deserted beaches are ideal for the chcap disposal of liquid waste in the sea". The reasons underlying the Italian choice of sites were the proximity of chemical and mining industries in the first two cases and a contribution to regional development policy in the third. At the time the hydroelectric development of the Flumendosa was in progress.

Norway proposed two sites on the south-west coast – Midbrödöen and Lista – where the sea currents, both the Gulf Stream and those of the Baltic, are strong and where the prevailing winds blow out to sea. The rocky and deserted island of Midbrödöen, which is 10 kilometres from Egersund and 75 kilometres from Stavanger, would however have required the construction of a 10 kilometre pipe to bring water supplies. Lista, 7 kilometres to the south of Farsund, had a natural harbour and concrete facilities.

Denmark suggested three sites: two were in the Jutland peninsula, one on the north coast near Hirtshals, to the north-west of Frederikshavn, the other on the east coast, near Lille Vildmose, 2 kilometres to the north-west of the small harbour of Oster Hurup, facing Sweden and 30 kilometres to the south-east of Ålborg. The third site was in the east of the island of Seeland, on the north coast of the Asnaes peninsula, 5 kilometres by road from the town of Kalundborg and 100 kilometres from Copenhagen.

The Danish authorities particularly recommended the third site owing to its good communications with the capital and with the Risø nuclear research centre which was then being built. The most important criterion was the proximity of the sea, but not for reasons of easy disposal of wastes. Denmark in fact found itself "completely in agreement with the American recommendations for avoiding radioactive disposal into the sea".

Belgium, supported by the Netherlands[42], also proposed three sites, all in the north of the country, none by the sea, but all three having links with the Hoboken company (SGMH). The first was at Olen, in Campine, 50 kilometres to the east of Antwerp, along the Meuse-Escaut canal, close to the radium and uranium factories of SGMH. The second was at Mol, near the Research Centre for the Applications of Nuclear Energy (CEAN), about 20 kilometres to the east of Olen, on land to the north of the CEAN and belonging to SGMH. Finally the third proposed site was in the southern suburbs of Antwerp, beside the Escaut, near the Hoboken factories, the cradle of the SGMH company.

The choice of Mol, or politics prevail: January-June 1957

On the basis of technical arguments only, the Belgian sites were excluded

In its report of 4 June 1957, the technical sub-group gave a comparative review of the different proposals received, focusing on those which were stated as being preferred by the candidate countries, i.e., Mol in Belgium, Asnaes in Denmark, Garessio in Italy and Lista in Norway. The result was that "following its original recommendation that the site chosen should be able to serve not only for a first pilot plant but also for a future large-scale plant, the technical sub-group found it could not recommend any of the Belgian proposals since they were inadequate from the point of view of general safety considerations and [...] do not offer sufficient advantages for processing and waste disposal in economic conditions". After reviewing the location factors, the sub-group identified the advantages and drawbacks of each country's offers. A perusal of this review shows that on technical grounds the sub-group leaned towards the Danish proposals, led by Asnaes: "the [Danish] sites are considered favourable from the point of view of transport. Communications and living conditions are good. The sites are all on the coast and the emergency risks are, generally speaking, less than in the case of Belgium. Denmark has the mechanical industries and labour needed for building the plant. The proposed sites do not appear to have any major disadvantage. The Danish Research Centre is still not in full operation, but it is expected to have at its disposal reactors and radiochemical laboratories soon"[43].

The choice of Mol was political and involved technical modifications to the project

As early as 15 March 1957[44] the Study Group recognised the limitations of the technical task it was accomplishing: "the Study Group considers that it will be very difficult to base the final choice of site on technical conditions alone, because several countries appear able to offer entirely satisfactory sites. In these circumstances the Study Group would therefore prefer merely to recommend a number of possible sites, leaving

42 Ibid p. 2: "The Benelux delegations indicated that if the Belgian proposals were not adopted by the Study Group, they might wish to look again at the suitability of proposing a site in the Netherlands on the North Sea coast."

43 Ibid. p. 8.

44 SEN/CHEM(57)4 § 72.

the final decision in principle to the governments concerned and to those who will ultimately participate in the undertaking".

The available archives say little about the political process that led to Mol[45]. There is some evidence, confirmed by interviews, suggesting that the decision was based upon the distribution of the sites for the future nuclear Joint Research Centre between the six EURATOM member countries. However the countries not involved in the Treaty of Rome still had to be convinced that the choice was valid.

In fact on 19 February 1957 Erich Pohland, as chairman of the Working Group for the Chemical Processing Plant of the Intergovernmental Conference for the Common Market and for EURATOM, notified the Secretary of the Steering Committee for Nuclear Energy that the Six had met on 16 February and that "the situation with regard to this subject (...) had been reviewed. Following this exchange of views it became clear that several of the six countries were offering the required conditions for siting such a plant. However certain technical points required additional, more detailed examination which the Study Group of the Six has been asked to carry out rapidly and the result of which will be notified to you in the near future".

It is likely that the Six made their choice between 15 March when the Belgian proposal was submitted, and 25 March when the Treaty of Rome was signed. However in order to get the decision made, Belgium had to fight hard and come up with highly attractive financial solutions. This is one way of interpreting the "D'Hondt counter-proposal" of 25 March 1957, which reduced the estimated cost of the project for a plant with a 50 tonne capacity prepared by Rudolf Rometsch on 10 February 1957 from 18.5 million dollars to 10.373 million dollars, by suggesting that the wastes could be processed in the Mol installations of the CEAN, the Belgian nuclear research centre adjoining the site proposed for the plant at that time[46].

Italy was excluded for technical reasons concerning manpower problems and the distance from the major European centres, and perhaps also because consideration was already being given to the Europeanisation of the Ispra centre[47]. Denmark and the other small countries not belonging to the EEC had more difficulty in accepting the decision that was emerging[48]. However the planned input of technical know-how and funding, as regards France and Germany – at the time consideration was being given to funding based upon GNP – caused the views of the "Six" to prevail[49] and, despite the impossibility of obtaining access to the archives, one may believe that they made this choice decisive.

The report of the technical sub-group says virtually nothing about the gulf between the Six and the others but does set out the positions of "some members of the group". While it does come out firmly in favour of the Mol site, it is at the expense of a shift in the project's technical premises, the abandonment of its industrial dimension and in fact the adoption of a "variant": it is difficult "to combine in one site needs typical of a research centre and pilot plant with the requirements of a large-scale industrial plant. A large-scale industrial plant is best located in a sparsely populated district which, by definition, is unfavourable for the plant's research activities". The same reasoning is adopted as regards wastes. As a result "the advantages originally envisaged in locating the two plants on the same site are no longer so clear". "Therefore the pilot plant should be built on the best possible site without regard to the possibility of ultimate construction of a large-scale plant on the same site".

In this particular case, obviously constructed to fit the circumstances, and shortly afterwards denoted "solution B"[50], Mol appears to be the best site.

---

45 The "Committee and Intergovernmental Conferences" archives, covering the period from April 1955 to January 1958, does not come under the 30-year rule decided by the EEC because the Conference took place before the Treaty of Rome was signed. Hence these are not Community archives and the 84 boxes are classified. However Michel Dumoulin was given special access to them in 1985 and gives a rapid description, see SERRA E. (1989), p. 195-210. However according to Jean-Marie Palayret, archivist in Florence (in a letter to the author responding to a request to consult these archives) the records are not in Florence and are not accessible to researchers at present.

46 Thus the plan was that the plant would be on the same side of the Meuse-Escaut canal as the CEAN, allowing it also to use the services of the centre for various analytical tasks.

47 However this raises a problem of timing. In fact the Ispra Centre was only Europeanised under the EURATOM-Italy agreement of 22 July 1959 and became the principal establishment of the EURATOM Joint Research Centre (JRC). The question is therefore to determine the date when this project was under consideration.

48 Letter from H. H. Koch, chairman of the Standing Committee of the Copenhagen Atomenergikommissionen to P. Huet on 16 April 1957: "I think any special preoccupations Austria, Switzerland and Sweden might have, should be taken into account".

49 In fact it was the "Five" since Luxembourg did not participate in the project.

50 Thus in the first report of the Steering Committee for Nuclear Energy, Annex A, *"Proposals to Build a Joint Spent Fuel Chemical Processing Plant"*, pp. 31 *et seq.*

The selection of Mol, with the abandonment of the possibility of building a major plant on the same site, meant reducing the project to that of a pilot plant. The characteristics of the site introduced substantial constraints in terms of safety and of monitoring liquid and gaseous discharges. No mistakes could be allowed, not only for the safety of the surrounding population but also because any accident would have been an "international scandal". These imperatives, which played a part in the pioneering nature of the undertaking as regards safety, were nevertheless regarded as an additional incentive to the enterprise culture, or a technical challenge the nature of which was well summed up by Rudolf Rometsch: "if it can be done at Mol, it can done anywhere"[51] (Figures 21 and 22).

The countries that were not members of EURATOM had to accept the choice of the Belgian site with its considerable cost advantages – and rely on hopes for the future. In fact the abandonment of plans for an extendable plant left open the choice of another site for a major plant to be built at a later date, Eurochemic II, which nobody, in the contemporary atmosphere of nuclear euphoria, doubted would be built. Norway was assured that the sites she had proposed would be regarded favourably in such circumstances.

Thus the major technical options had to be adapted to the constraints of a site which was far from being technically the best. The appropriate legal structures for an international research and production undertaking also had to be devised.

The legal option: an international shareholding company with its Statutes attached to an international Convention[52]

A desire for legal innovation and the Huet memorandum of 29 February 1956

On 29 February 1956 – the very day on which the OEEC Council set up the Special Committee for Nuclear Energy, Pierre Huet, at that time legal adviser to the Organisation, submitted to the OEEC ministerial Council a memorandum on "The Administrative and Financial Problems Relating to the Creation of Joint Undertakings"[53] proposed by the Nicolaïdis Report. This document reviewed "a number of questions that arise when joint undertakings are established and listed the approaches that have already been used in similar circumstances".

The memorandum covered five topics: administrative status, financial arrangements, privileges and immunities, governmental control and relations with non-participant countries. It gave examples of international institutions and undertakings involved in industrial or commercial activities in the areas of manufacturing, transport, public works, tourism and banking, set up between 1873 (the Compagnie internationale des wagons-lits) and 1955 (the Société européenne pour le financement de matériel ferroviaire – EUROFIMA). It also examined the advantages and drawbacks of the different institutional approaches as far as international co-operation is concerned.

From the information given in Part I – administrative status – the place of the Eurochemic project in the institutional genealogy of international co-operation can be determined.

It is possible, from Pierre Huet's 1956 presentation, to construct two decision-making charts, one institutional, the other financial:

One branch of the decision-making chart (Table 8a) is not respresented by an existing institution, that of international semi-public companies. These companies would combine private companies and government authorities on an equal basis, a different approach from French semi-public companies in which the government is predominant.

Similarly the chart showing the different approaches to funding (Table 8b) shows a joint financing formula which has so far been tried only in the financial sector. In this approach, governments initially contribute and subsequently transfer their holdings to private companies.

---

[51] Interview of 8 January 1992. The success of the undertaking in this area allowed a retrospective approach which emphasized how these constraints were stimulating ("technically sweet") to the engineer, but Rudolf Rometsch stressed the burden of responsibilities he felt when he was managing director of the undertaking.

[52] The text of the Convention and Statutes, incorporating the modifications made later is given in Annex 1.

For a contemporary analysis of this text, see in particular HUET P. (1958), pp. 519-523; see also the short presentation on the subject given by Erich Pohland to the Stresa Conference from 11 to 14 May 1959 in OEEC/ENEA (1960), pp. 103-106. A review of the legal problems raised by the first two years of operation is given in STROHL P. (1961). For a retrospective critical analysis, see BUSEKIST O. von (1980) and its translation (1982), notably pp. 9 to 12.

[53] The Huet memorandum is given in an annex to HUET P. (1958), pp. 524 *et seq.*

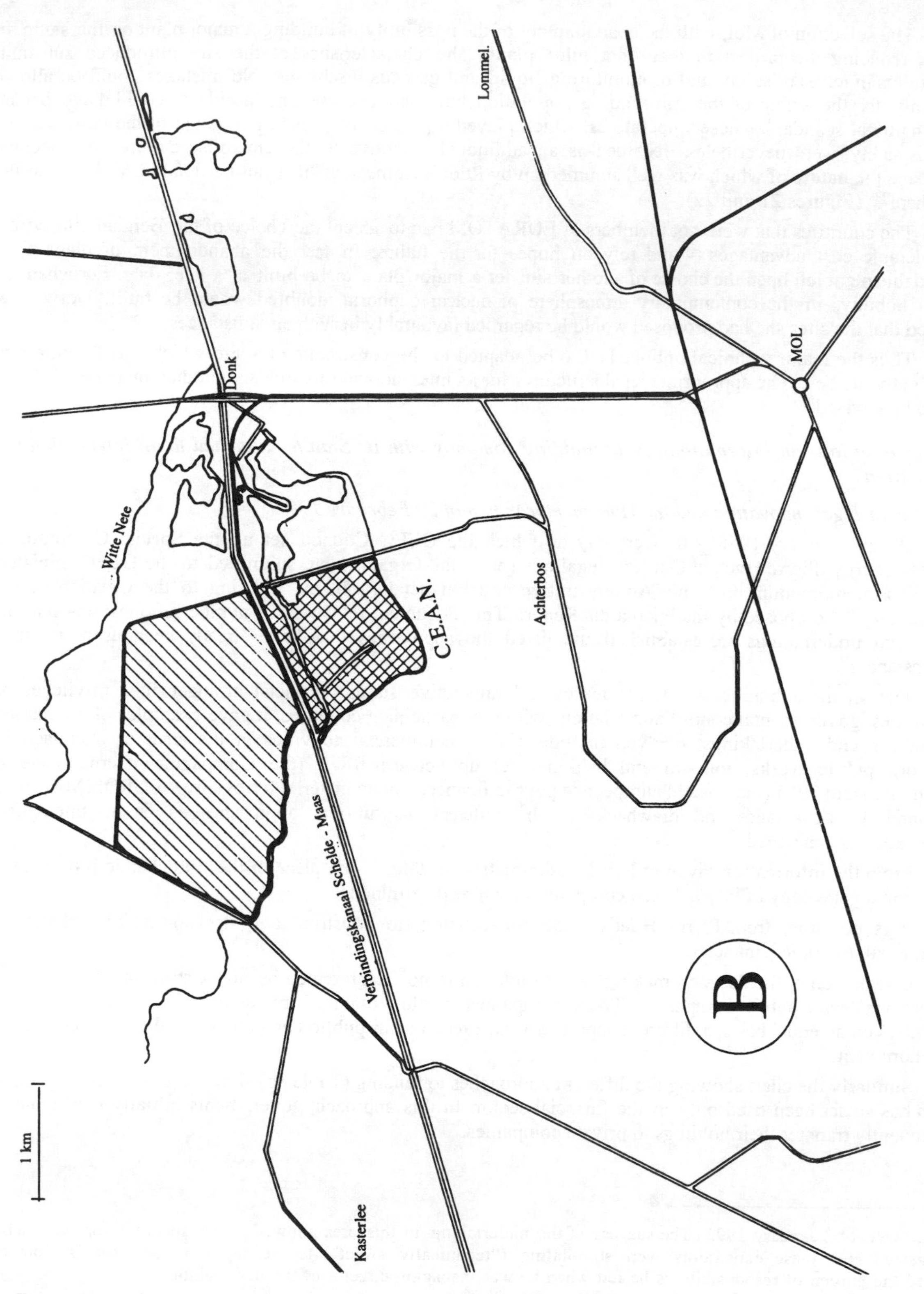

Figure 21. The site proposed by Belgium to the north of the CEAN (Source: SEN/CHEM(57)15, set of maps).

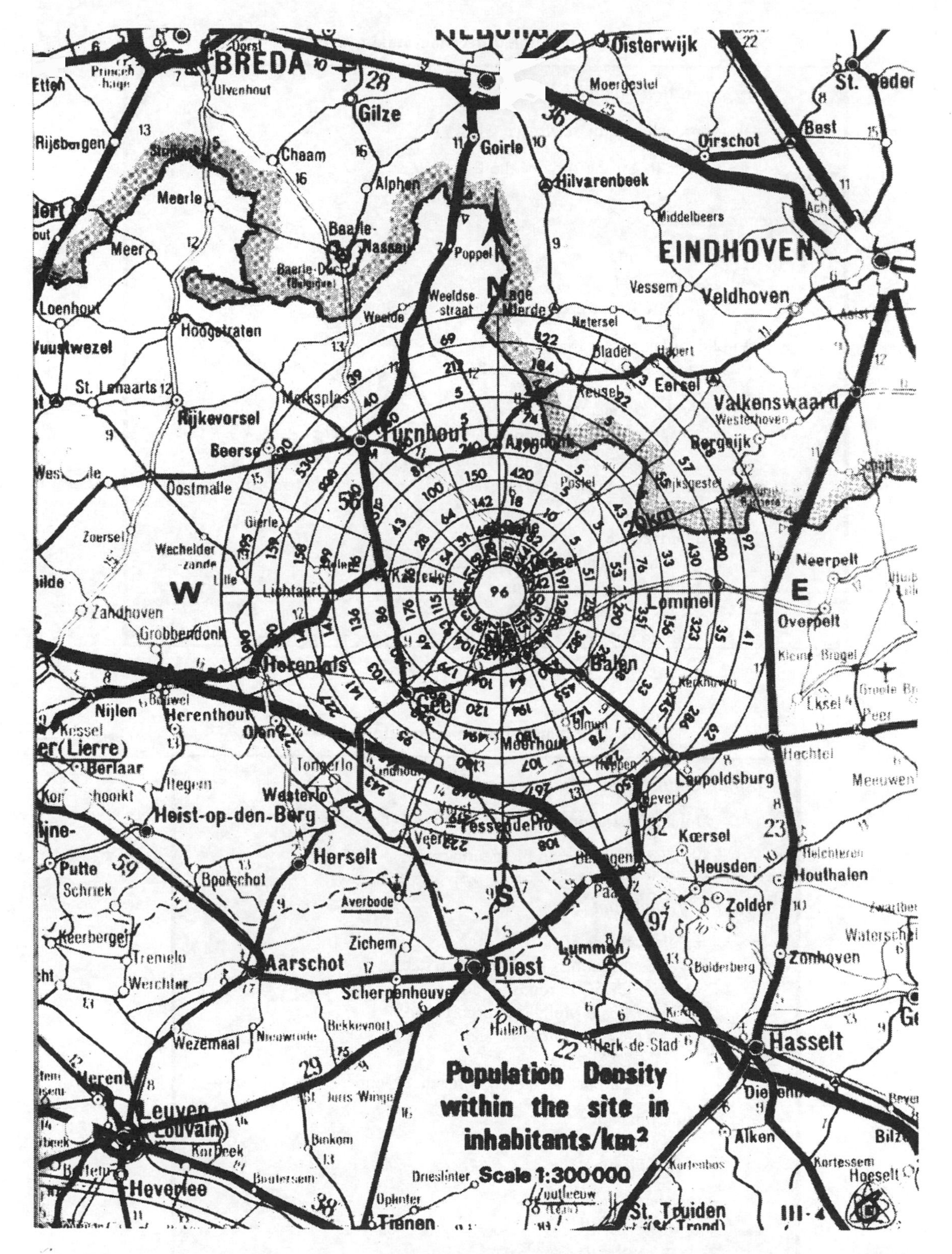

Figure 22. Population density within 20 kilometres of the site. The figures give the mean density per km^2 within the area bounded by two adjacent radii and two successive circles. The map also shows the proximity of major urban centres such as Eindhoven and Louvain (Source: *Safety Analysis* (1965), Volume of Figures, III-4).

Table 8a. **Institutional decision-making chart**

No new organisation created

- Iron production in blast furnace OEEC 1951
- Regularisation of the flow of the Rhine 1929

New organisation

- Intergovernmental organisation
  - – IBRD 1944
  - – Société financière internationale SFI 1955
- Public organisation
  - – Basle-Mulhouse airport 1949
- Commercial company
  - – No governmental control
    - - Compagnie internationale des wagons-lits 1873
    - - Société ferroviaire internationale de transports frigorifiques 1949
    - - Scandinavian Airlines System 1950
  - – intermediate
    - - International Joint Stock Company (IJSC)
  - – with governmental control
    - - BRI 1930
    - - Eurofima OEEE 1955

Table 8b. **Financial decision-making chart**

No joint capital created

- Joint infrastructure programme of NATO
- Production of iron in blast furnace 1951

Creation of joint capital

- publicly financed
  - – contributions from member states
    - - OEEC European Productivity Agency 1953
    - - Basle-Mulhouse airport 1949
  - – by levying duties or fees
    - - ECSC 1951
    - - European Comission for the Danube
- joint financing
  - – IBRD 1944
  - – SFI 1955
- privately funded
  - – BRI 1930
  - – EUROFIMA 1955

The Statutes of Eurochemic, which are those of an international semi-public company, meet the requirements that result from the objectives of the undertaking, notably that there should be no undue financial advantages for the host country and that no obstacles should be raised to the planned commercial operation.

The memorandum suggests that the joint undertakings should benefit from privileges, such as inviolability, legal immunity, extra-territoriality, and tax exemption. The allowance and extent of these exceptions to Eurochemic are a matter for debate.

The constitutional bases listed in the memorandum comprise private agreements, diplomatic Conventions between the founding countries and with the host country; consideration is also given to a hybrid procedure in the form of a governmental agreement to which a commercial agreement is attached.

The principles laid down in the Huet memorandum were adopted to a large extent by the Legal Working Party and by the Study Group. Indeed it was Pierre Huet who was to draw up the first version of the Convention and Statutes in April 1957. The Convention, to which the Statutes were attached, was signed by the governments after some small additions but no major modifications, on 20 December 1957.

The March 1957 report from the Legal Working Party[54]

The Legal Working Party, chaired by a member of the permanent Netherlands delegation, Adriaan Van der Willingen, was given the task in November 1956 of outlining the administrative and financial arrangements for the joint undertakings envisaged by the OEEC. It quite naturally focused its efforts on the project achievable in the shortest time, that of the reprocessing plant. It submitted its report in March 1957 and its work provided the basis for the constitutional basis of the undertaking the following month.

Three different approaches were considered for joint undertakings: the creation of international public bodies, a system of international operating contracts (concessions), and international companies. In all three the need to sign a diplomatic Convention – subject to ratification – was stressed. To avoid the delay introduced by such a procedure, the group suggested a "general treaty" to allow agreements for particular projects to be drawn up in simplified form later on.

With regard to the reprocessing company, the group had examined the advantages and drawbacks of the three possible legal approaches:

- an international public organisation, managed exclusively by governments or public institutions. An approach of this kind was rejected because it would have prevented private participation by placing all the financial and economic responsibilities on public bodies, whereas the objective of the undertaking was, as we have already seen, to develop a realistic mixed (public-private) formula;
- the operating contracts system (build and operate) approach was based directly on the American model, i.e., the links between the USAEC and the private companies which operated, on a concessional basis, most of the nuclear establishments it owned. Under this approach the governments of participating countries would have sub-contracted the construction and operation of the plant to an international industrial group, which would have had to comply with the instructions of the governments. However the concession contract would have placed the public and private organisations on an unequal footing. This too was far from creating the desired interplay between the public and private spheres;
- the approach involving an international shareholding company was therefore inevitable, insofar as, Erich Pohland put it, "we were seeking the closest possible co-operation between the public organisations and private industry within the undertaking itself"[55] and to ensure equality between the representatives of the public sector and those of the private sector. And then there was the hope that the predominance of the public organisations would "wither away" in the short term as the shares held by governments or nuclear agencies were sold to private companies.

However this predominance of the private approach was limited by the very objective of the undertaking: "a company established entirely under private law would not have been suitable to the present circumstances. International co-operation always raises special problems, not only political but also economic. Moreover, in many member countries, the State exercises a decisive influence over economic organisation in the nuclear

---

[54] NE(57)12. Its conclusions concerning Eurochemic were incorporated in the report of the Study Group dated 26 March 1957, NE(57)15, pp. 26-27.

[55] OEEC/ENEA (1960), p. 103.

energy sector. Nor must we forget that a plant processing a material of such high military importance as plutonium calls for particular surveillance on the part of participating governments"[56].

Hence the reason for choosing the form of an international shareholding company was the desire to involve "governments, public and semi-public institutions and private undertakings in the financing and management of the undertaking"[57]. Although the undertaking was commercial in form, it was to be hedged by a series of provisions which were both limitations and privileges by comparison with the ordinary status of a private firm.

In March 1957 therefore, the Legal Working Party came out in favour of adopting the form of an international shareholding company, although certain members – not specified in the documents – apparently expressed some hesitation about the suitability of this approach for an enterprise of public interest which would clearly have low viability in the first few years of operation.

The company's administrative bodies would therefore be those of an ordinary shareholding company, with an annual general meeting ("General Assembly"), a board of directors and a general management. However their decision-making powers were to be limited to day-to-day management. In fact overall policy was to be defined in liaison with a special intergovernmental committee, a kind of "international surveillance council" to use the expression of Otto von Busekist[58], in other words the future Eurochemic Special Group which reported to the Steering Committee of the future European Nuclear Energy Agency.

The Study Group also proposed that the undertaking should be constituted through the signature of a diplomatic Convention to which the Statutes of the company would be attached. Consideration might be given to negotiating a special agreement with the host country.

The Convention and Statutes of 20 December 1957

The first draft Convention prepared by the Secretariat was submitted to the Study Group for discussion at its fifth meeting held on 29 and 30 April 1957[59]. The final version of the Convention was submitted to the Steering Committee in September 1957[60] and signed by the twelve founding governments on 20 December 1957 in Paris.

An international shareholding company

The company's capital was 20 million EPU (European Payments Union) units of account, divided into 400 shares each of 50 000 EPU/UA. The initial distribution of shares was as follows, in decreasing order of value:

Table 9. **Initial distribution of shares**

Shareholder	Number of shares	% of 400 shares
Germany	68	17.0
CEA (France)	68	17.0
Belgium	44	11.0
CNRN (Italy)	44	11.0
AB Atomenergi (Sweden)	32	8.0
Netherlands	30	7.5
Switzerland	30	7.5
Denmark	22	5.5
Austria	20	5.0
Norway	20	5.0
Turkey	16	4.0
Portugal	6	1.5

This Convention came into force when it had been ratified by countries accounting for at least 80% of the capital, thus creating the Eurochemic company.

56 OEEC/ENEA(1960), p. 104.

57 NE(57)12 § 75.

58 BUSEKIST O. von (1982), p. 4.

59 SEN/CHEM(57)8 of 12 April 1957 for the Convention and SEN/CHEM(57)9 of 13 April 1957 for the Statutes.

60 C(57)204 Annex II B, reproduced in SCNE (1958), pp. 53 *et seq.*

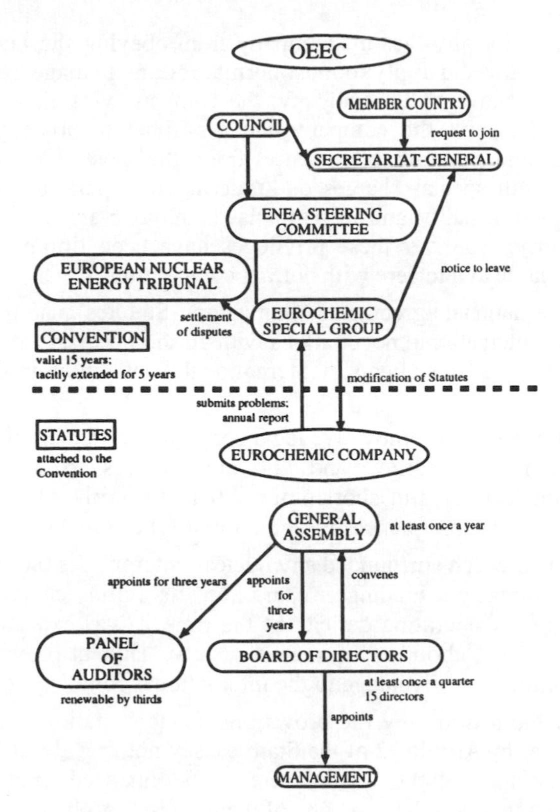

Figure 23. The different Eurochemic bodies[61].

The operational logic of the company

Eurochemic's existence was founded upon two legal texts of differing scope.

- The Statutes govern the relationships between the partners and establish a commercial enterprise ensuring equality of treatment[62] between the organisations participating in it or which may participate in it, whatever their nature: governments, government establishments, semi-public companies, private companies. The extent of representative power within the General Assembly and the Board of Directors depends upon the amount of capital invested and not on the privileges normally associated, in a national framework, with government authority.

  Similarly there is no discrimination between the participating nationalities. Thus nothing prevents a shareholder from selling his shares to any organisation of the same nationality, as this does not affect the international equilibrium within the undertaking.
- The Convention creates the company, gives it a legal personality under international law and controls the special nature of its activities; it makes the Special Group responsible for questions of overall policy and for relations between participating countries, and leaves total freedom of (economic) management, meaning scientific and industrial management. However the Special Group does dispose of a right to information and control, expressed in the submission of the annual report and in the need for the Board of Directors and the General Assembly to have certain decisions approved by the Special Group. This authority may also be reflected in the taking of decisions that are binding on the company. Finally the Special Group is entitled to make recommendations to governments and to the host government.

61 For the detail of the make-up and responsibilities of the different company bodies, see the text of the Convention and Statutes attached. This chart simply shows the general pattern of the company arrangements.

62 However this equality is not total: for example as regards the dissemination of information, the governmental participants can disseminate it, but the private companies may not.

To some extent the Convention absolves the company from obeying the law of the host country, in this case Belgium[63]. However Belgian law did apply to most normal routine management activities corresponding to the industrial and commercial objective of the company; the company was also subject to the requirements of Belgian law regarding safety[64]. However the company had a number of privileges which "were intended to guarantee its independence against possible interference from the host government, to prevent the said government from burdening it with special charges or hindering its operation, to place all the participating countries in a situation of complete equality and in particular to remove any monetary or customs obstacles to its operations within the country. However these privileges have been limited to ensure that they do not constitute financial advantages liable to interfere with normal competition"[65].

The dual nature of the international agreements (Convention-Statutes) also means that the Statutes can be adapted with no need for further international negotiations with all the necessary ratifications, which would have been the case if the whole process was based upon an international Convention. In this way a degree of flexibility was introduced.

The legal basis of the company is an innovative legal construction that applies the Keynesian principles of egalitarian co-operation between the "public" and "private" sectors to an experiment in international co-operation. However there are *lacunae* and shortcomings that are fairly easily explained by the historical background, but which had important consequences for the future of the company.

The contemporary optimism which surrounded any nuclear enterprise is the reason why the agreements are mute about the sources of the company's funding – apart from its initial capital – or the obligations of its shareholders, notably in covering any operating deficit. At the time it was expected that reprocessing activities would expand, and there was no doubt about its ultimate viability. This is proved by the emphasis placed on the need to avoid giving any competitive advantage to the international company, especially as regards contracts.

This enthusiasm is also the reason why the provisions for terminating the experiment in international co-operation, covered in particular by Article 32 of the Statutes, say nothing about how the decommissioning of the installations and the conditioning of the wastes were to be financed, merely envisaging "storage and surveillance". This lack of provision was to become one of the principal problems in the life of the company: it necessitated solutions being found outside the legal sphere defined by the Convention, particularly in negotiating a new legal basis and defining a new type of customary law, that of international nuclear obligations[66].

Thus the year 1957 saw the coming into being of the major technical, geographical and legal structures for European co-operation on reprocessing. The signature of the Convention was the project's first success, the different stages in which have been followed from the inside by examining internal sources.

Of course the above analysis could cause one to forget the role of external factors in the project's success. The success of the project must be placed in the western context of technical co-operation on fuel cycle activities, and linked to the failure of the European isotopic separation project.

Success of the Study Group for the OEEC reprocessing plant; failure of the Working Party on isotopic separation. An attempt to compare and explain[67]

The process that established Eurochemic is only meaningful when considered in parallel with the construction of EURATOM, by which it was largely determined. Conversely the emergence of Eurochemic played a significant role in reducing EURATOM's activities between the time of Messina in June 1955 and that of Rome in March 1957, marked by the abandonment of the joint project to build an isotopic separation plant, one of the fundamental factors behind the French commitment to EURATOM.

---

63 This "subsidiarity of the law of the host country" was to lead to the development of an internal "company law", analysed in STROHL P. (1961). This had definite effects on management-staff relationships within the undertaking, on the languages used and on the system of contracts, see below.

64 Here it must be emphasized that this legislation was established essentially at the same time as the under-taking and that it was formulated in conjunction with the managers of Eurochemic (see Part III Chapter 1).

65 HUET P. (1958), p. 521.

66 See in particular Part IV Chapters 1 and 2.

67 Since access to the archives of the Brussels Intergovernmental Conference is impossible, the following part is based essentially on the following: MELANDRI P. (1975), SCHEINMAN L. (1965) and (1967), WEILEMANN P. (1983), DAVIET J. P. (1993) and SEN(56)24, an OEEC document which reproduces a paper from the Conference.

France and the problem of the isotopic separation plant: 1952 to 1956

The military and civilian utility of an enrichment plant

Isotopic separation is used for enriching uranium in its fissile 235 isotope. At the end of the 1950s the only process used on an industrial scale, in the United States and the United Kingdom, was gaseous diffusion. Other approaches such as the ultracentrifuge and jet nozzles were being explored in the laboratory[68]. Consequently isotopic separation using gaseous diffusion had to be used for producing the uranium-235 bomb[69] as well as the enriched uranium fuel needed for operating the most compact reactors, materials testing reactors with highly enriched fuels or light water reactors, notably those used in submarines[70].

France and "forced" co-operation: the failure of bilateral attempts

At the beginning of the 1950s, France saw it as essential to master the technology of enrichment in order to cover all its own civilian and military needs. Thus by 1952 France was considering building a plant, in the first nuclear 5-year plan, similar to those the United States already had at Oak Ridge, Tennessee and Paducah, Kentucky and which the British were building at Capenhurst. Design work was entrusted to the army's explosives department.

However it soon became clear that such a plant would be extremely costly both to build and to run, and the alternative of international co-operation appeared unavoidable. When the "non-decision" to manufacture the bomb was taken in 1955[71] it appeared that France would have to get together with other countries to build a joint plant. In 1954 approaches had been made to the British, who had just commissioned Capenhurst. However in February 1955[72] the United Kingdom refused to take the project any further, referring to the 1943 Quebec agreements and American opposition.

The beginnings of a multilateral approach. From Messina to the Spaak interim report: June 1955-1956

With the "Messina relaunch" of June 1955, France was able to envisage European co-operation on a broader scale than it had originally considered. Following the British rejection, France had in fact sounded out Germany about possible co-operation. Jean Monnet had conceived the dual project of a European Atomic Energy Community and a European Economic Community in order to draw in the two leading powers on the Continent. Peter Weilemann[73] has clearly shown how France, at the beginning of the negotiations, was attracted more by nuclear co-operation and the possibility of building a joint isotopic separation plant (and also of co-operating on other nuclear projects) than in the creation of the Common Market. Germany's attitude was the reverse. In the "package deal" negotiations[74] leading to the Treaty of Rome, each party came to accept the aspect of greatest interest to its partner[75].

The intergovernmental Conference which met in Brussels as from 9 July 1955 set up a Study Group with the task of reviewing the problems involved in building this plant. The group submitted its first report on 14 January 1956[76]. The report estimated the cost of a 300 tonne/year plant to be $55 million, a 500 tonne plant at $80 million and a 1000 tonne plant at $160 million. The electrical power needed to meet the energy requirements of these plant was evaluated at 120, 200 and 400 MWe respectively. The incremental cost for the

---

68 Jet nozzle separation (Trenndüsenverfahren) or the Becker process had been developed in the laboratory by Erwin W. Becker at Marburg and then in Karlsruhe. In this process, a mixture of uranium hexafluoride and helium or hydrogen at high pressure is passed at high speed through a nozzle and the jet of gas then diverted through 180°. The process was never applied on an industrial scale. See WINNACKER K., WIRTZ K. (1977), p. 237-238, and MÜLLER W. D. (1990) p. 505-506.

69 It would also have been possible to build a bomb based upon uranium-233, which could be produced by separating it chemically from irradiated thorium. Reprocessing HEU can also produce uranium 235 but the first requirement is the HEU fuel, implying an enrichment plant or direct access to this fuel.

70 Enrichment to 3% is needed for PWR fuel, to 20% for ship propulsion and to 90% for the bomb.

71 COHEN S. (1988) supports the notions of a collective and diluted decision. MONGIN D. (1993) shows that secret decisions were in fact taken, notably by Pierre Mendès-France, which were subsequently denied by those responsible for political reasons.

72 According to BROMBERGER S. and M. (1968), p. 185.

73 Op. cit. Also see MONNET J. (1976), p. 619-620 and GUILLEN P. (1985).

74 SCHEINMAN (1967) on EURATOM.

75 However EURATOM lost importance in French eyes between the time the Treaty was formulated and the time it was signed, with the coming into power of Charles de Gaulle in the interim.

76 This report was to be used as a working document by the OEEC Working Group n° 1 and is reproduced under OEEC reference SEN(56)24.

European plant to produce a gramme of U-235, calculated on the basis of the enrichment rate compared with the natural rate of 0.7%, was: $30 to double the enrichment (1.4%), $35 to quadruple it (2.8%) and $50 to increase it by a factor of 30 (i.e., 21% of U-235). It would take at least 5 years to build the plant.

The conclusions of the Study Group were taken up in the interim Spaak report published on 11 February 1956, which came out in favour of the project and entrusted the preparatory work to a Study Group chaired by Bertrand Goldschmidt.

The collapse of the plans for a European isotopic separation plant. American commercial pressure: 1956-1957

The nuclear "Marshall Plan"

On 26 February 1956, only ten days after the publication of the Spaak report, the United States offered to sell 20 tonnes of enriched uranium to Europe.

Various interpretations were put on this offer: the British praised it as a "second Marshall Plan". It was seen as an expression of the American "open door" policy, a generous offer of co-operation to European nuclear development, continuing the spirit of Geneva[77].

No doubt it was also one way of finding markets for the output from the three enrichment plants then operating in the United States, with the opening of that of Portsmouth, or possibly an incentive for Europe to choose enriched uranium reactors of American technology for which there was no domestic market owing to the very low cost of fossil fuels in the United States.

The coincidence of dates at least suggests some form of pressure aimed at preventing the isotopic separation project from going any further. Similar pressure had already been exerted on a smaller scale with regard to Swiss heavy water in the summer of 1955[78].

The Reaktor AG group, led by BBC, had in fact signed a contract to build a heavy water production facility intended to supply the DIORIT reactor at a price of 1 SF a gramme, with two Swiss chemical firms associated with Sulzer and the Ateliers de construction Oerlikon. The aim was ultimately to provide the country with a national heavy water industry. During the summer of 1955 the United States offered to supply Reaktor AG with as much heavy water as it wanted at a price of 30 centimes a gramme. The acceptance of this offer led to cancellation of the contracts and, according to Peter Hug, made any further co-operation between BBC and Sulzer impossible. Sulzer was later to become the principal supporter of the national reactor at Lucens in the canton of Vaud, which was a failure[79], while BBC combined with Westinghouse and General Electric for the power plants at Beznau and Mühleberg.

According to Pierre Mélandri, the entire EURATOM project was opposed by the British until February 1956[80]. The main reason was Britain's hostility to the acquisition of atomic weapons by France. The opposition came to an end thereafter. These two facts suggest the conclusion that the United States may have promised the British that the plant would never be built. It is certainly true that the new security of supply and the immediate availability of enriched fuel meant that the proposed plant was of no further interest to the countries with civilian requirements. Moreover the proposed price for 20% enrichment – $25 per additional gramme – had a dissuasive effect on European projects which foresaw a price of $50 for enrichment to 21%, with a delay of at least five years.

The death throes of the European project

However the Working Party did not immediately abandon the project. It submitted its first report in September 1956, the very same month that the Eurochemic Study Group was established. One might wonder to what extent the latter was used as a counter-blast to the EURATOM project[81], which would explain first of all the support of the British, then their departure, because they participated in the Study Group without adhering to the Convention.

---

[77] That of the August 1955 Conference.

[78] HUG P. (1991) p. 343.

[79] And more precisely a dual failure: the Swiss reactor design using heavy water and pressure tubes was abandoned when the reactor began operation; the accident of January 1969 damaged the installation to such an extent that any plans to repair it were dropped.

[80] Op. cit. p. 82.

[81] Peter Weilemann believes that the United Kingdom gave strong support to the OEEC projects in order to squash those of the Six. Taking this view, and bearing in mind the fact that the United Kingdom never became a shareholder of the company, the country's initial participation may seem no more than tactical.

Nevertheless on 17 November 1956 came the final thrust: the United States now offered the Europeans any amount of enriched uranium at an incremental price of $16 a gramme in a 20% enriched fuel.

In May 1957 the "Report of the Three Wise Men"[82] no longer showed the construction of the isotopic separation plant as a matter of priority[83].

The Working Party chaired by Bertrand Goldschmidt submitted its report on 17 May 1957. It did not reach any conclusion, and revealed the technical gulf between the French and the Germans[84]. The former preferred the proven separation process involving membranes which would allow work to begin quickly, while the latter wanted joint development of the still experimental technique of Erwin Becker and were therefore under no pressure to move to an industrial scale, now that they were reassured by the possibility of commercial supplies of enriched uranium.

The final French attempt of the 1950s: military co-operation with Germany and Italy. 1957-1958

France had once again to fall back on a national solution. The second nuclear five-year plan of July 1957 released an initial tranche of 25 billion francs to build a national plant with a capacity of 300 tonnes. This was to be Pierrelatte[85]. The expected cost of the plant – calculated at $140 million in early 1958, which exceeded the total annual budget of the CEA for 1957 – again led to an attempt to find international funding, this time in a military and three-way framework[86].

On 28 November 1957 the French, Italian and German ministers of defence signed a tripartite agreement for co-operation on the military applications of atomic energy, which embraced the question of isotopic separation[87]. On 8 April 1958 the three Ministers of Defence, Jacques Chaban-Delmas (France), Franz-Joseph Strauss (Germany) and Paolo Emilio Taviani (Italy) adopted the "Rome protocol" the objective of which was to create a sort of strategic and nuclear community for the three countries and which set out, amongst other provisions, the basis for joint funding for the Pierrelatte plant. The expected capital cost of $140 million would be borne 45% each by France and Germany[88], and 10% by Italy.

However the fall of the Gaillard government on 15 April 1958 and the institution of the Vth Republic by Charles de Gaulle was to put an end to this project for military co-operation. Pierrelatte was therefore built using exclusively French military funding, and the French isotopic separation plant was to produce its first kilogramme of highly enriched uranium on 7 April 1967.

The plans for a civilian European isotopic separation plant were not to be taken up again until the early 1970s when it became clear that light water reactors with low enriched oxide fuel would predominate in Europe. In March 1970 the United Kingdom, the Netherlands and Germany (known as "the troïka") established the URENCO project. In February 1972 the EURODIF project brought together eight countries – including the members of the troïka, who abandoned it in 1973 – together with France, Belgium, Italy, Spain and Sweden. In fact three competing European plants were built: URENCO Capenhurst and Almelo by the troïka and then

---

82 Louis Armand, Frantz Etzel and Francesco Giordani.

83 According to Peter Weilemann, the downgrading of this priority by the Three Wise Men is apparently linked directly to their visit to the United States, particularly to Oak Ridge. It seems reasonable that the assurances of commercial supplies of enriched uranium also played a part.

84 For more details, see WEILEMANN P. (1983), p. 165.

85 With regard to Pierrelatte, see DAVIET J. P. (1993), pp. 181-207.

86 G. H. SOUTOU (1993), p. 13 and MONGIN D. (1993), p. 20. For the military aspects of European co-operation during this period, which are not central to the history of Eurochemic but which encompass it in a broader sphere, see the three articles produced for a GREPHAN colloquium managed by Maurice Vaïsse and published in the *Revue d'histoire diplomatique*, 1-2, 1990, BARBIER C. (1990), CONZE E. (1990), NUTI L. (1990); VAISSE M. (1990) for an introduction to the three articles.

87 Dominique Mongin, op. cit. p. 20, interestingly points out that although on 2 October 1954 Germany renounced the production of atomic weapons prior to the establishment of the WEU, it retained the possibility of possessing them so long as they were produced overseas. These circumstances can provide an explanation for the country's participation in a military isotopic separation plant. On the French side the enormous cost involved was a powerful incentive to finding partners.

88 Walter Kaiser, in BRAUN H. J., KAISER W. (1992), p. 294 gives another explanation for the failure seen from the German point of view. In his eyes the project was obviously directed against the United States and the Germans were afraid that their role would be no more than that of providing funding ("Die vermutete einseitige Rolle der BRD als Geldgeber").

EURODIF at Pierrelatte in 1979. They involved two competing processes: the ultracentrifuge for the former and gaseous diffusion for the latter[89].

The United States, EURATOM and Eurochemic post-1957: exemplary co-operation

The subsequently unfailing support of the United States for a European Atomic Energy Community now deprived of an isotopic separation plant[90], and for the Eurochemic reprocessing plant, can be seen in a new light.

The fact that the United States supported Eurochemic but not the reprocessing project of the Six deserves explanation. A simple answer might be that the Eurochemic project was supported both by the OEEC and by the Six and that it was in line with the American view regarding the control of activities with proliferation implications. In this context a 12-strong co-operative venture appeared more advantageous than one with only six participants.

To take a somewhat more complicated view, one might also think that the United States approved of European plans they found interesting as regards research and development but which were unlikely to generate commercial competition in the near future. On the contrary, they did in fact hold out potential for business[91]. Also, through the co-operative links that were formalized in 1958 through a USAEC-Eurochemic assistance programme, the Mol undertaking was to become one of the sources of information on reprocessing techniques for the Americans. Moreover the mutual control exercised in an international venture was such as to reassure them about the dangers of proliferation. Thus the work of the Study Group moved forward in parallel with the negotiations about the IAEA.

The success of the European reprocessing project was therefore more than merely the result of the failed plans for the European isotopic separation plant. However Eurochemic can also be seen as some compensation for the hostility to the isotopic project, in a framework extending beyond the active members of EURATOM although encompassing them all. This fact would explain why British co-operation was equally whole-hearted in the phase leading up to the Convention, but did not lead to active participation in the project. This British defection was not without its effects on the members of the Study Group, and left behind some bitterness towards the UKAEA.

In conclusion, it is obvious that Eurochemic came into being because those directly involved wished it to do so. Another reason is that the isotopic separation plant did not see the light of day.

Consequently European co-operation concerned with the head-end and tail-end of the fuel cycle was limited over a period of nearly fifteen years to reprocessing at Eurochemic until the advent first of URENCO and then of EURODIF.

Thus the signature of the Convention on 20 December 1957 placed the seal on the early co-operation of 12 European countries in an area which appeared to have a bright future; in the history of the company it marked the commencement of the "interim period".

---

89 For the details of European co-operation in URENCO, see Part V Chapter 2.

90 Precisely that in which the French were interested. Thus the extremely aggressive position of France in the early days of EURATOM may be seen not only as the expression of Gaullist attitudes but also as of a considerable disappointment at the loss of the original project. For Michel Debré, EURATOM was merely the "Trojan horse" of the United States.

91 The supply of fuel, subject to control, or even equipment.

Chronology of Part I

From the beginnings to 20 December 1957[1]

Italics: general milestones, overall nuclear history
Standard: Belgian nuclear history
Bold: OEEC or OECD and Eurochemic

1789-1914

24 September 1789: *in a paper given to the Prussian Academy of Science in Berlin, Martin Heinrich Klaproth announces the discovery of a new element which he has named uranium in honour of Herschel, the astronomer who discovered the planet Uranus in 1781. In fact it was an oxide (U_3O_8) contained in ore from silver mines at Sankt Joachimsthal, in Bohemia*

1841: *the Frenchman Eugène Melchior Peligot isolates uranium by reducing the tetrachloride with potassium*

in the nineteenth century uranium was used as a pigment for porcelain, glass and enamel

24 February 1896: *Antoine Henri Becquerel discovers natural radioactivity*

1898: *Pierre and Marie Curie discover radium*

10 April 1915: a prospector from the Union minière du Haut-Katanga (UMHK) discovers the Shinkolobwe uranium deposit in the Belgian Congo.

Between the wars

1921: mining of the Shinkolobwe deposit begins. The ore is sent to Belgium for extraction of the radium

1922: the Belgian company Le Radium, a subsidiary of UMHK, starts production at Olen. The uranium which is regarded as a waste by-product of radium extraction, is stored

1934: *Frédéric Joliot and Irène Curie discover artificial radioactivity*

January 1939: *Otto Frisch gives the physical proof of nuclear fission*

May 1939: *agreement between UMHK and the French CNRS to supply 5 tonnes of uranium dioxide for the construction of a reactor*

summer 1939: *suspension of scientific publications on nuclear questions*

Second World War

1939-1940

September 1939: creation of a subsidiary of UMHK in New York, African Metals, to which the 180 grammes of radium available in Belgium are transferred. In October 1940 the Congolese stocks of ore are transferred to the United States

April 1940: *creation of the Maud Committee in the United Kingdom*

winter 1940-41: *first production of plutonium-239 in the Berkeley laboratory*

1941

December 1941: *start of the "Manhattan Engineering District" project with the aim of producing a weapon using the properties of fission*

---

[1] Principal sources for the chronology: Eurochemic archives, AtW (1956 to 1994), EN (1957 to 1974), RGN (1975 to 1994), GOLDSCHMIDT B. (1962, 1967), LECLERCQ J. (1986), UKAEA (1984), VANDERLINDEN J. dir., 1994, WINNACKER K., WIRTZ K. (1975).

1942

1942:	*first commercial contracts between African Metals and the Manhattan project covering 3839 tonnes of uranium oxide (U_3O_8) contained in 30 000 tonnes of Belgian ore. Others followed until 1944*
April 1942:	*two objectives laid down for the "Metallurgical Project" of the University of Chicago as part of the Manhattan project. The first is to establish whether a chain reaction is possible in a natural uranium pile. The second is to develop the chemical method for extracting plutonium*
November 1942:	*decision to build a pile producing 100 grammes of plutonium a day (X10) at Clinton near Oak Ridge, and an adjoining extraction facility, the "Clinton pilot", the first facility for reprocessing irradiated fuel*
2 December 1942:	*first chain reaction in the pile built by Enrico Fermi in Chicago*

1943

19 August 1943:	*Quebec Agreements between the United States and the United Kingdom*
10 October 1943:	*construction of the three reprocessing plants designed for the industrial production of plutonium begins at Hanford*
end December 1943:	*the Clinton pilot extracts the first plutonium from fuel from the X10 pile*

1944

8 January 1944:	*final drawings for the first Hanford reprocessing plant*
26 September 1944:	a "Memorandum of Understanding" gives the British and Americans exclusive rights to the Belgian uranium
December 1944:	*first reprocessing campaign at Hanford*

1945

2 February 1945:	*first plutonium produced at Hanford delivered to Los Alamos*
April 1945:	*Britain decides to launch a major R & D programme in the nuclear field*
16 July 1945:	*first atomic bomb – a plutonium weapon – exploded near Alamogordo (New Mexico)*
6 August 1945:	*the first uranium-235 bomb dropped on Hiroshima*
8 August 1945:	*the second atomic bomb – a plutonium weapon – dropped on Nagasaki*
August 1945:	*publication of the Smyth Report*

After the second World War

18 October 1945:	*order establishing the French CEA*
5 November 1945:	*Switzerland sets up the Commission d'études de l'énergie atomique*
November 1945:	*construction of the Urals nuclear centre begins (Cheliabinsk 40)*
December 1945:	*Britain decides (kept secret until 1947) to build a plutonium-producing pile*

1946

March 1946:	*the Lilienthal-Acheson Report on the international control of atomic energy is published*
14 June 1946:	*the Baruch plan submitted to the UN provides for the creation of an international authority, one of whose tasks would be to operate reprocessing plants. The project is abandoned in 1948 as a result of the Cold War*
August 1946:	*the MacMahon Act comes into force (1946 US Atomic Energy Act)*

1947

9 June 1947:	a Belgian Royal Order levies a special tax on exports of uranium ore by UMHK, which is used to partly fund the colonial budget
June 1947:	creation of the IIPN (Inter-university Institute of Nuclear Physics) which on 6 September 1951 becomes the IISN (Inter-university Institute for Nuclear Science), an organisation for promoting and co-ordinating fundamental research
1947:	the Ateliers de construction électrique de Charleroi (ACEC) establish a nuclear design and research centre intended to provide the scientific and technical basis for the manufacture of laboratory equipment
5 June 1947:	**announcement of the Marshall Plan (Harvard speech)**

July 1947: **Paris Conference on Marshall aid**

July 1947: *Britain selects the Windscale (Cumbria) site for its plutonium-producing reactors and a reprocessing plant*

August 1947: *GLEEP, Europe's first experimental reactor goes critical at Harwell*

1948

16 April 1948: **the OEEC created by the Paris Convention. The sixteen member countries are Austria, Belgium, Denmark, France, Greece, Iceland, Ireland, Italy, Luxembourg, the Netherlands, Norway, Portugal, Sweden, Switzerland, Turkey and the United Kingdom**

7-10 May 1948: *"Congress of Europe" at the Hague*

28 July 1948: **the Paris Convention of 16 April comes into force**

1949

1949: *development of the PUREX process at Oak Ridge*

4 April 1949: *signature of the Washington Treaty establishing NATO*

29 August 1949: *the first Soviet atomic bomb – a plutonium weapon – exploded near Semipalatinsk (discovered by the United States in September)*

15 December 1949: **bilateral Treaty whereby the FRG accedes to the Marshall Plan (FRG participating in OEEC work since October 1949)**

1950

1950: *first instance of bilateral co-operation on nuclear reactors through a Norwegian-Dutch association at Kjeller (JENER) and at Petten*

25 June 1950: *start of the Korean war*

7 July 1950: *creation of the European Payments Union (EPU)*

27 October 1950: *Windscale pile n° 1 goes critical*

31 December 1950: A Belgian Royal Order creates the function of Commissioner for Atomic Energy; appointment of Pierre Ryckmans, former Governor-General of the Congo, who occupies the post until his death in February 1959. He is succeeded on 1 May 1959 by Jacques Errera and then by Paul De Groote on 15 December 1969. On 1 April 1971 the Commission is abolished and its tasks entrusted to the Ministry for Economic Affairs in a department known as the Atomic Energy Commission, soon to be integrated in the Energy Division.

1951

18 April 1951: *the Treaty of Paris creates the ECSC*

April 1951: *construction of the Windscale reprocessing plant completed*

July 1951: *the JENER reactor JEEP1 diverges at Kjeller*

summer 1951: tripartite Agreement between Belgium, the United Kingdom and the United States on delivery of uranium by UMHK up to 1956. A tax of 60 cents is levied on every pound of uranium oxide, the revenue being used to fund a Belgian research programme on nuclear energy

November 1951: *formulation of the first French nuclear five-year plan known as the "Gaillard plan". This provides for the construction of a French reprocessing plant*

1952

March 1952: *first plutonium production in the Windscale plant*

9 April 1952: the Belgian CEA creates the Research Centre for Nuclear Energy Applications (CEAN)

26 May 1952: *Italy sets up the Comitato Nazionale per le Ricerche Nucleari (CNRN)*

1952: *the United States becomes a net importer of petroleum products*

1952: *the French CEA begins to build a reprocessing pilot unit known as the "Châtillon pilot" under the auspices of the Fontenay-aux-Roses Nuclear Research Centre (CEN-FAR)*

3 October 1952: *explosion of the first British atomic bomb at Monte Bello in the Indian Ocean*

1953

February 1953: *commissioning of the so-called ICPP reprocessing plant at Idaho Falls*

5 March 1953: *death of Stalin*

May 1953: *creation of the European Productivity Agency (OEEC/EPA)*
1 July 1953: *signature of the treaty establishing CERN*
27 July 1953: *end of the Korean war (armistice)*
8 December 1953: *Eisenhower's "Atoms for Peace" speech*
14 December 1953: **memorandum by the OEEC Secretary-General to the Council on the rising cost of energy**
22 December 1953: the CEAN buys 184 hectares of land at Mol to build the Belgian nuclear centre
1953: *construction of the first British nuclear power plant begins at Calder Hall*

1954

1954: *the first Soviet nuclear power plant coupled to the grid at Obninsk*
1954: *the American "Nautilus" submarine reactor goes critical*
1954: *commissioning of the British gaseous diffusion enrichment plant at Capenhurst*
1954: establishment of the Study Group on Nuclear Energy (SEEN) involving industry and electricity suppliers on the initiative of UMHK, and the Study Group on Nuclear Power Plants (SYCA) made up of the electricity suppliers
1954: *Germany participates in the JENER at Kjeller*
1954: *the first (military) plant using the PUREX process comes into service at Savannah River, South Carolina*
17 April 1954: *construction of CERN begins at Meyrin, near Geneva*
15 June 1954: *creation of the Société européenne pour l'énergie atomique*
July-August 1954: *the UKAEA established*
August 1954: *amendment to the MacMahon Atomic Energy Act, the so-called Atomic Energy Act of 1954*
6 September 1954: *construction begins on the first American nuclear plant at Shippingport in Pennsylvania, a Westinghouse PWR which started generation on 2 December 1957*
7-8 October 1954: *first meeting of the CERN Council*
21 December 1954: **Louis Armand becomes an OEEC consultant**

1955

1955: *Electricité de France (EdF) decides to build its first reactor, the future Chinon 1 (or EdF1)*
1955: *Germany sets up the Federal Ministry for Atomic Affairs*
19 February 1955: *the United Kingdom publishes the White Book on electricity production giving considerable importance to nuclear power in its forecasts*
1 March 1955: *Switzerland sets up the Reaktor AG firm to build and operate a National Research Centre at Würenlingen*
24 May 1955: **the Armand Report submitted to the Council, published the following month**
1-2 June 1955: *the Messina Conference relaunches the European projects. The EEC and the ECSC are the result*
June 1955: *construction of the French military reprocessing plant begins at Marcoule (the future UP1)*
9 June 1955: **memorandum by the OEEC Secretary-General on the economic and financial aspects of the peaceful uses of nuclear energy**
10 June 1955: **resolution of the OEEC Council of Ministers setting up a Working Group to draw up the terms of reference for a nuclear energy committee**
29 June 1955: **resolution of the OEEC Council of Ministers for the setting up of an energy committee**
9 July 1955: *the Spaak Intergovernmental Committee preparing the Treaty of Rome starts work in Brussels*
8-20 August 1955: *first Geneva Conference on the Peaceful Uses of Atomic Energy. Presentation of the draft terms of reference for the IAEA*
October 1955: *Jean Monnet sets up the action committee for the United States of Europe*
1955: *during the negotiations concerning EURATOM, the CEA suggests that a joint uranium enrichment plant should be built*

1956

1956: *the United States sets up a system for inspecting their deliveries of enriched uranium, handed over the IAEA in 1957*
1956: *France sets up the Service central de protection contre les radiations ionisantes (SCPRI) attached to INSERM until 1983*
January 1956: **publication of the Nicolaïdis Report**

18 January 1956: *first meeting of the Committee for the United States of Europe, involving Jean Monnet and Louis Armand*

22 February 1956: *President Eisenhower announces that 20 tonnes of uranium-235 are available for overseas countries*

29 February 1956: **the OEEC sets up the Special Committee for Nuclear Energy (CSEN)**

10 April 1956: **the SCNE sets up Working Group n° 1 on joint undertakings, including the reprocessing plant**

21 April 1956: *the Spaak report, a summary of the work done since the Messina conference, proposes that a European Atomic Energy Community should be established*

23 May 1956: *the British Calder Hall power station, the first European nuclear power plant, starts generation (inaugurated in October)*

May 1956: **publication of the Hartley Report**

26 July 1956: *Egypt nationalises the Suez Canal*

July 1956: *a special group on reprocessing is set up as part of the negotiations on the EAEC, chaired by Erich Pohland*

8 June 1956: **the report of Working Group n° 1 is adopted by the SCNE**

18 July 1956: **the OEEC Council takes up the proposals of the SCNE in "Joint action by OEEC countries in the field of nuclear energy" in which it recommends the establishment of joint undertakings. On the same date it sets up the Steering Committee for Nuclear Energy with the task of preparing the statute for a nuclear energy agency, the future OEEC/ENEA**

19 July 1956: *establishment of the Karlsruhe Nuclear Research Centre*

28 September 1956: **a Study Group made up of representatives of 14 countries is set up to prepare for the joint construction of a plant for the chemical processing of irradiated fuels**

28 September 1956: *first nuclear electricity produced at Marcoule (G1)*

23 October 1956: *statutes of the IAEA adopted by 72 countries*

24 October 1956: **the Eurochemic Study Group holds its first meeting**

6 November 1956: *end of the Suez crisis*

17 November 1956: *the United States lowers the export price of enriched uranium and guarantees unlimited supplies*

1957

1957: *the United States adopts the Price Anderson Act, releasing the nuclear industry from its liability in case of accident for damages exceeding $100 million and setting the cover by the insurance companies to the same amount, the government undertaking to cover any difference with the real cost of damage*

1957: licensing agreement between Westinghouse and Belgian firms, under the auspices of the SGB, ACEC, Cockerill, SGMH and FN together with their subsidiary MMN

31 January 1957: creation of Belgonucléaire, which takes over the tasks of the SEEN

10 February 1957: **Rometsch plans and estimates for the reprocessing plant. 50 t/year, $18.5 million**

16 February 1957: **the delegation of the Six, the future signatories of the EURATOM Treaty, represented at the Intergovernmental Conference for the Common Market and EURATOM, reviews the proposals for the plant site: probable date of the decision to support the candidature of Belgium**

25 March 1957: **the Study Group receives the D'Hondt counterproject which reduces the capital cost of the project to $10.373 million by providing for wastes to be treated by the CEAN at Mol**

25 March 1957: *signature of the Treaty of Rome creating the Common Market and EURATOM*

30 April 1957: *the Swiss SAPHIR reactor (of American origin) restarted at Würenlingen*

May 1957: *the report of the Three Wise Men (Armand-Etzel-Giordani), "A Target for EURATOM" gives up the idea of a European enrichment plant*

20-25 May 1957: **Brussels symposium on the chemical processing of irradiated fuel**

27 May 1957: the CEAN becomes the CEN (Nuclear Research Centre), an establishment of public utility, on 570 hectares of land at Mol

29 July 1957: *the IAEA commences work*

1 August 1957: *commissioning of the CERN synchro-cyclotron*

September 1957: **the Study Group hands down proposals for the construction of a plant**

7 October 1957: *a fire at Windscale results in the two reactors being shut down*

September 1957: *an explosion – not revealed until 1976 – in a storage facility for highly radioactive liquid waste from the Cheliabinsk reprocessing plant*

17 December 1957: **adoption of the statutes of the ENEA and of the Convention on Security Control**

20 December 1957: **twelve countries sign the Convention setting up the EUROCHEMIC company.**

PART II

Completion of the project

January 1958 - November 1965

Chapter 1

The "interim period":

Early days of the company and the development of the detailed pre-project

January 1958 - June 1961

The company did not come into being when the Convention was signed on 20 December 1957, but only after it had been ratified by governments holding at least 80% of the shares. The parliamentary process of ratification takes a relatively long time, so to ensure continuity for a project which would take several years to complete, and avoid losing time, the Study Group and the SCNE Secretariat had given thought to organising an interim period starting in the summer of 1957[1].

Three days before the Eurochemic Convention was signed, the OEEC Council reached a decision "relative to the continuation of the work of the Study Group on the chemical processing of irradiated materials"[2] and took two actions for the purpose:

– it requested the Study Group to ensure continuity of the project and, to this end, amended its Statutes which, attached to the Decision, instructed it "to undertake [...] the preliminary work [...] with a view to construction of a plant by 1961"[3];

– it opened a line of credit amounting to 500 000 EPU/UA to allow funding through the OEEC budget for an interim period which was not expected to last more than one year. However this expenditure through the OEEC budget would be "funded only by contributions from the members of the Study Group and in proportion to the number of their votes"[4].

Thus began the interim period which was to continue until 27 July 1959 when the company was established. A feature of this period was the close co-operation that existed between the Organisation, ENEA and the Study Group. It was in fact on 1 February 1958 that the European Nuclear Energy Agency came into being[5] (see Figure 24). The Director-General of the Agency was Pierre Huet[6], with Einar Saeland (Norway) as Deputy Director-General. Pierre Strohl was to follow up the project on behalf of ENEA as Secretary of the Study Group and of the interim Board of Directors.

During the 18-month interim period, further important decisions were to be made by an organisation which was an extension of the first Study Group, in setting up the structures of the company.

This new organisation paved the way for co-operation with various external actors: these were non-member countries which possessed reprocessing technology – primarily the United States – and European firms that sought involvement in the construction of the installations so as to benefit from the contracts and gain technological skills in this new sector.

The project was progressively worked out by a Design and Research Bureau (BER), accommodated on the premises of the Belgian nuclear centre, working along with a restricted group of experts that met from time to time in the Management Committee and with an American expert.

---

1 See *"Proposed Measures for the Interim Period"*, report approved by the Study Group on 17 September 1957.

2 C(57)256, and *Acts,* XVII (1957) 171.

3 C(57)256 Annex, Article 4.

4 Ibid. Article 8. This was in fact an advance on the company's capital and was deducted when the shares were paid up.

5 It had been requested, for symbolic reasons, that ENEA should come into being one month after EURATOM, since the Treaty of Rome was not to come into force until 1 January 1958, hence after the Council Decision of 17 December which created the ENEA. See Figure 25.

6 Appointed by the Council on 24 January, *Acts* XVIII (1958) 8.

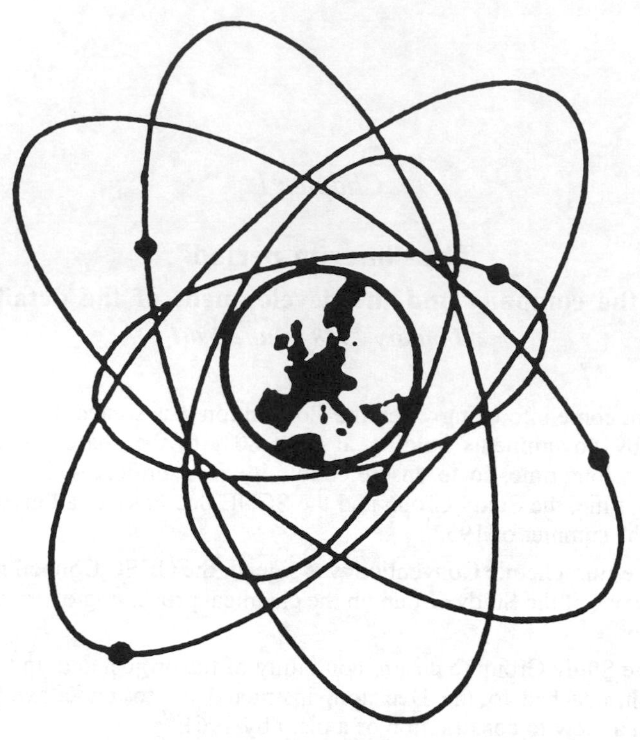

Figure 24. The OEEC/ENEA logo in 1959.

The entry into force of the Convention in July 1959 brought the various bodies of the company into being. The detailed pre-project needed before construction of the nuclear facilities could begin was drawn up by the Research Directorate which succeeded the BER, in conjunction with the principal architect engineer.

General organisation of the project

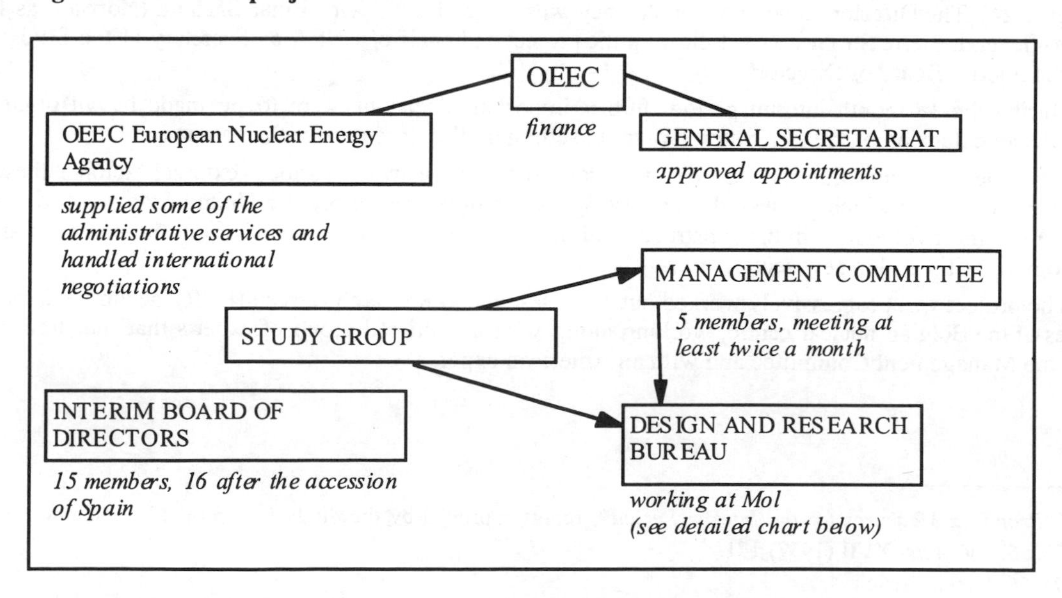

Figure 25. The organisational structure in the interim period[7]

[7] The names of the various bodies varied somewhat in the beginning; thus the Management Committee was known first as the Committee of Experts and then as the Technical Committee.

The modified Study Group: prefiguring the General Assembly[8]

The organisation and terms of reference of the Study Group were supposed to reflect those of the future company's General Assembly. Thus the non-governmental shareholders were involved in its work, with each government having a number of votes corresponding to its own shareholding in the company and that of its nationals.

The first meeting of the "Eurochemic Study Group" which took over from the "Study Group for the Establishment of a Joint Undertaking for the Chemical Processing of Irradiated Fuels" was held on 24 January 1958. Erich Pohland (Germany) was elected Chairman and Erik Svenke (Sweden) Vice-Chairman. In this way the continuity of the Study Group approach was assured[9].

The responsibilities of the Study Group included a large number of tasks which it could not handle alone. Under the modified Statute it was empowered to appoint an interim Board of Directors to direct the programme, and report to it. The Study Group appointed a Management Committee – the successor to the technical sub-group – to assist the interim Board of Directors in taking technical decisions and following up the work on site. The initial opening of the site and the first work done at Mol was carried out by a Research Group which soon became known as the Design and Research Bureau, whose Director was appointed by the Study Group.

The Study Group's special area of responsibility was that of financial decisions. It submitted forecasts of expenditure to the OEEC Council – these figures then providing a basis for the voting of credits – and approved detailed estimates so as to authorise expenditure.

The interim Board of Directors

Made up in the same way as the company's Board of Directors, it comprised fifteen (later sixteen) directors, the nationalities reflecting the holdings in the company's capital[10]. Belgium, France, Germany and Italy had two directors; Austria, Denmark, the Netherlands, Norway and Switzerland only one; Turkey and Portugal were represented by one director. Finally Spain, which joined the project at the end of 1958, had one seat. A representative of EURATOM took part in the meetings of the interim Board of Directors on an advisory basis.

The interim Board of Directors had general responsibilities for administrative and management questions; it supervised the activities of the Design and Research Bureau, and took technical decisions when these were such as to influence plant design, on the basis of information prepared and supplied by the Management Committee.

The Management Committee

The Committee was made up of five experts appointed by the Study Group; it assisted the interim Board of Directors, approved the appointments of experts and substantial contracts. Three of the five members of the Management Committee who were appointed on 3 March 1958 had already participated in the technical sub-group of the Study Group during the preparatory period: Yves Sousselier (CEA) as chairman, together with Teun J. Barendregt (JENER), Maurice D'Hondt (CEN) and Rudolf Rometsch (CIBA). Italy's seat was filled by representatives of the CNRN, first by M. Rollier who was quickly succeeded by Felice Ippolito, and then Alberto Cacciari. Maurizio Zifferero, a chemist who worked for a short time in the Design and Research Bureau before returning to the CNRN where he was responsible for reprocessing, was approached with a view to taking over at the very end of the interim period.

The Design and Research Bureau (BER)

The BER was the embryo of the establishment. During the interim period it worked in premises belonging to the Mol nuclear research centre. Its director had the final decision regarding technical work and could commit expenditure up to a certain level[11], but the staff was recruited by the Director-General of ENEA, acting on behalf of the OEEC Secretary-General. The appointment of members of the Bureau's staff required the agreement of the interim Board of Directors or of the Management Committee.

On 3 March 1958, Erik Haeffner (AB Atomenergi, Sweden) was appointed Director; he too had been a member of the technical sub-group prior to the signature of the Convention. The first six-person team was working at Mol as from May 1958.

---

8 Eurochemic(57)1 of 28 December 1957.

9 See Annex: Membership of Company Bodies.

10 The names of the directors are given in Annex 5.

11 1 million FF.

New shareholders: Spain[12] and private companies

Spanish accession

Spain had sought to be an associate member of the OEEC since May 1954, first on questions of domestic transport, then in the committees dealing with agricultural problems[13]. Spain had been able to take part in the Organisation's work on a general basis under an Agreement signed on 28 January 1958[14]. Its participation in ENEA followed, under a decision of the Council on 22 July 1958. On 11 October Spain notified the Secretary-General of its decision to accede to the Convention on Security Control and to the Eurochemic Convention. On 22 October 1958 the country asked to join the company. As a result the Junta de Energia Nuclear (JEN) declared its intention of acquiring thirty Eurochemic shares as soon as the company was established, thus placing Spain on an equal footing with the Netherlands and Switzerland. The members of the interim Board of Directors had decided, in an unofficial meeting during the second Geneva Conference in September, to issue thirty new shares. This increased the registered capital of the company from 20 to 21.5 million EPU/UA. Pending the constitution of the company, Spain was admitted to the Study Group and took part in the work carried out during the interim period, making a contribution of 37 500 EPU/UA. José Luis Gutierrez-Jodra, Director of Pilot Projects and Chemical-Metallurgical Industrial Projects in the Junta de Energia Nuclear, joined the Study Group at its fifth meeting held on 27 February 1959.

New non-governmental shareholders

The signatories of the Convention wanted to involve private firms in the management of the company; some even envisaged total privatisation on the assumption that the nuclear industry would rapidly develop. It was already possible for shares to be transferred freely between organisations in a given country, and the governments and nuclear agencies who held the overwhelming majority of the capital attempted to sell some of their shares to private companies. This trend, which was expected to intensify, began during the interim period.

The Belgian government, the French CEA and the Italian CNRN were the first to sell shares to companies.

As early as July 1958 the Belgian government announced its intention of selling 27 shares from its holding of 44 to the Société belge de chimie nucléaire (Belchim), an *ad hoc* group of companies interested in reprocessing formed only to acquire a holding in Eurochemic. Belchim had been set up on 17 June 1958 by twenty Belgian companies and was chaired by André Leroux. Its principal shareholder was the Société générale métallurgique de Hoboken (SGMH) which had put up a quarter of the capital, followed by the Belgonucléaire company which had subscribed 11%. Belgonucléaire itself had been established on 31 January 1957 to take over the activities of the SEEN and was formed of the Union minière du Haut-Katanga (UMHK) and the electricity suppliers. Staff was to be seconded to Belchim from the constituent companies[15].

In December 1958 the Compagnie Saint-Gobain was to receive 22 shares of the French holding of 68, the Societa Richerche Impianti Nucleari (SORIN) of Milan, which had built the private Saluggia nuclear centre and whose shareholders were Montecatini and Fiat, five shares. The Societa Elettronucleare Nazionale (SENN) of Rome was to obtain two of the forty-four Italian shares[16]. At the same time, Germany and Switzerland were considering selling some of their shares to private firms, but nothing was done before the company came into being[17].

The subsequent development of the company and the fact that the main private companies which had acquired capital were candidates for the function of architect engineer[18] suggests that this acquisition of shares was something of an "entry fee". Also the fact that SORIN and SENN bought only a few shares prefigured the high degree of fragmentation of the involvement of private German firms and was essentially aimed at the acquisition of information, since every shareholder received technical reports from the company.

---

12 C(59)5, Eurochemic(58)19.

13 See for example *Acts* (1954) 1195.

14 Report of the signature of the Agreement in *Acts* XVII (1957) 174.

15 Eurochemic(58)11 for the declaration of intent. The Belchim company managed the CEN wastes department. It was taken over by Belgonucléaire in 1964. See Jacques Leduc in VANDERLINDEN ed. (1994).

16 Eurochemic(58)21 of 10 December 1958.

17 Eurochemic(58)3, Annex. In the end, Swiss firms took part not through the transfer of shares but by buying bonds from the federal government equivalent to 10% of the total capital contributed by Switzerland (communication from Jean-Michel Pictet).

18 See below.

The framework for action: administrative and financial regulations

Administrative regulations for the interim period

Personnel status during the interim period was a delicate problem. Although the employer could only be the OEEC since the company did not yet exist, "the rules governing the different categories of staff employed by the OEEC – permanent civil servants, temporary officials, consultants – was not appropriate to the staff of the Design Bureau"[19]: these were skilled personnel, working away from the Organisation's headquarters, not under its authority, in a situation of temporary employment before being integrated into the new company under a private type of contract, and remote from the situation of international civil servants.

The OEEC Council therefore requested the Secretary-General to draw up special staff regulations. These were approved by the interim Board of Directors of 28 March 1958[20].

The principal provisions were the following:

- fixed-term contracts of up to one year, with salaries continuing to be paid for a maximum of six months for those members of staff not recruited by the Eurochemic company after its creation[21];
- a duty of discretion, a spirit of international independence – staff members not to accept any instructions from governments or from any authority outside the OEEC – all inventions to become the property of the OEEC;
- emoluments based upon Belgian salaries supplemented, for staff members not permanently residing in Belgium, by a monthly expatriation allowance of 30% for a head of family, 15% for others, to which a further allowance for special responsibilities or qualifications could be added; the administrative rules defined five categories of staff whose basic monthly salaries are shown in Table 10;

Table 10. **Basic monthly salaries for different staff categories**

Category	Type of post	Salary in EPU/UA
I	Director of Design and Research	925
II	Head of Research	700
III	Specialists and researchers	250 to 500
IV	Draughtsmen and skilled workers	180 to 220
V	Auxiliary staff	100 to 180

- subsidiary application of Belgian law as regards social security – with supplementary insurance similar to that applying to the staff of the Mol centre – and employment legislation;
- any disputes between the staff of the Design and Research Bureau and their employer to be settled by the OEEC Appeals Board in order to take into account the international nature of the employer.

Financial regulations and funding during the interim period[22]

Activities were funded by contributions from member countries paid to the OEEC. The OEEC administration ensured that proper procedures were followed as regards commitments, accounting and payments, in particular by seconding an official to Mol.

Under these arrangements, credits of 757 000 EPU/UA were opened during the interim period, in three phases. An initial credit of 500 000 EPU/UA was defined on 7 February 1958. Spain, as a new participant, contributed 37 000 EPU/UA on 9 February 1959, and the extension beyond the one-year period originally foreseen resulted in a new line of credit of 220 000 EPU/UA being opened on 14 April 1959. The money was spent at a slower rate than expected. Expenditure during the first year was as follows, shown in decreasing order: [23]

[19] Eurochemic(59)9 of 22 May 1959.

[20] Eurochemic(58)10 Final of 2 April 1958.

[21] The unexpected delay in ratification resulted in contracts being extended for six months in March 1959.

[22] Eurochemic(58)7 Final of 13 May 1958.

[23] See also the detailed budgets in the Annex. The archives show no overall state of expenditure for the entire interim period, since the 1959 balance sheet does not draw a distinction between the two phases between which the company was established.

Table 11. **Eurochemic committed expenditure during the interim period**

Budget chapter	First 3 quarters of 1958 (thousands of EPU/UA)	%	First quarter of 1959 (thousands of EPU/UA)	%	First year of operation (thous-ands of EPU/UA)	%
Personnel	98.456	39.4	48.080	38.2	146.536	39.0
Equipment and running costs	60.955	24.4	31.043	24.6	91.998	24.5
Start-up costs	49.474	19.8	30.523	24.2	79.997	21.3
Miscellaneous of which	41.115	16.4	16.338	13.0	57.453	15.3
- consultants	*7.039*	*2.8*	*1.124*	*0.9*	*8.163*	*2.2*
- design and works contracts	*15.680*	*6.3*	*7.967*	*6.3*	*23.647*	*6.3*
Total	250.000	100.0	125.984	100.0	375.984	100.0

Co-operation with external organisations

Co-operation between Eurochemic and the USAEC, links with the UKAEA and contacts with Canada

The United States; the establishment of technical co-operation; the USAEC-Eurochemic assistance programme

The contacts made with the United States during the preparatory period and which led in particular to the Symposium on Reprocessing in May 1957, were strengthened during the interim period and, to some extent, made formal. Oak Ridge National Laboratory (ORNL) and Floyd L. Culler, the head of the chemistry research division of the centre, formed the nucleus on the American side.

Official contacts with the USAEC

The Study Group decided, at its second meeting on 28 February 1958, to get in touch with Lewis L. Strauss, then President of the USAEC, to set up continuing links between Eurochemic and the main American laboratories and establishments involved in reprocessing[24]. In doing so, Eurochemic had three objectives: to make use of American consultants, to obtain advice from the USAEC and to send staff for training, particularly to Oak Ridge.

The interim Board of Directors decided to proceed in the same way with the British and Canadian authorities[25]. The Director-General of ENEA was asked to make these contacts at an official level, while the Management Committee arranged approaches to scientists. Two senior Americans, who had already taken part in the Brussels Symposium in May 1957, Floyd L. Culler, Director of the Chemical Engineering Department of ORNL and Myron B. Kratzer from the Division of International Affairs of the USAEC, met the Management Committee and the Design and Research Bureau on 22 July 1958[26]. The USAEC sent a Hanford expert, a specialist in liquid wastes, to spend a week at Mol during August. Another expert, E. L. Nicholson from ORNL, arrived at the end of September. His arrival marked the beginning of formal co-operation which was to continue until 1970.

The official letter from the USAEC, the principles governing USAEC-Eurochemic co-operation and the problem of patents.

On 29 August 1958 Allan J. Vander Weyden[27], interim Director of the USAEC Division of International Affairs, sent a letter to Pierre Huet, Director-General of ENEA, setting out the USAEC conditions governing the secondment of qualified United States personnel to the Eurochemic project: they would remain employed by their own laboratory but the OEEC would pay transport costs and expatriation allowances.

It was agreed in principle that American staff and members of the Eurochemic staff should make brief transatlantic visits for the purposes of consultation on specific problems.

---

[24] Eurochemic(58)11 of 3 March 1958.

[25] Eurochemic/CA(58)2.

[26] Management Committee: Minutes of the seventh meeting held on 2 July 1958.

[27] He had been liaison officer to the Canadian Chalk River centre but had returned to the Division of International Affairs of the USAEC in November 1955.

The USAEC declined any responsibility for the accuracy of information provided, limiting its experts' involvement to providing advice and making it clear that Americans visiting Eurochemic would not be concerned with detailed planning of installations[28].

In return for this assistance, which was agreed "because the development and design of this up-to-date multipurpose processing plant will yield useful and valuable information", the USAEC "shall receive, for whatever use and dissemination as the Commission may desire, all disclosable communicable information developed or acquired by Eurochemic. In return the AEC will provide any comparable, non-confidential information it acquires or obtains. Also, the AEC may at any time send representatives to the Eurochemic project for orientation and acquaintance with its program". Finally the letter implicitly suggests that the exchange might lead to a possible agreement for Eurochemic to process basic or special nuclear materials supplied by the United States.

Pierre Huet replied to this letter on 13 September, accepting the principles set out, but putting off the signature of the agreement until the next meeting of the interim Board of Directors. In the meantime he asked the USAEC to put the proposed co-operation into being immediately. In fact the proposals concerning the exchanges of information were incompatible with the relevant rules adopted by Eurochemic and required rewriting[29]. Also at the fourth meeting of the interim Board of Directors on 14 October 1958 the CEA and CNRN indicated their concern about communicating information and the problem of patents[30]. To calm their fears, the Chairman of the interim Board of Directors proposed that the Council should give its agreement to the USAEC only for the duration of the interim period. A draft agreement on the question of patents concerning inventions or discoveries made or designed by American personnel during assignment in Belgium was finally approved by the interim Board of Directors on 16 December 1958[31]. It did not meet the concerns of all members of the Board but was in fact valid only for the interim period. It was based on the relevant sections of the standard bilateral agreements concluded by the United States, but adapted to the international nature of the company. The USA acquired licensing rights in the United States; the OEEC, prefiguring Eurochemic, held such rights for the rest of the world.

The Eurochemic study visit to the United States: October-November 1958 (Figure 26)

The same meeting approved a study mission by the senior interim staff to the United States, including visits to American nuclear installations. The trip involved Erich Pohland, Chairman of the Study Group, Yves Sousselier, Chairman of the Management Committee, Erik Svenke, Vice-Chairman of the Study Group and Erik Haeffner, Director of the Design and Research Bureau. Pierre Regnaut, Head of the Plutonium Division in the CEA Chemistry Directorate, accompanied them as a consultant. A credit of 10 000 EPU/UA was opened for the visit.

Not all the centres ENEA had requested to be included in the visit – Oak Ridge, Savannah River, Hanford and Idaho – were opened for the visit which was organised by the USAEC Division of International Affairs. The industrial reprocessing sites were prohibited. Savannah River, where the most modern TBP plant was located, managed by Du Pont, was excluded from the programme. At Hanford all the visitors saw was the administrative building and the meeting room in the laboratories. However they were taken round centres which were of lesser interest to them, such as the Argonne National Laboratory and the Shippingport Power Plant, which they visited at the same time as Francis Perrin, the High Commissioner of the French CEA.

The visit lasted from Sunday 19 October to Thursday 6 November 1958 and followed an east-west route[32].

The Eurochemic representatives visited the new AEC buildings at Germantown (Maryland) near Washington on 20 and 21 October. They obtained an overall view of the reprocessing situation and the distribution of functions between the different American centres.

---

28 "They may not engage directly in the preparation of the plant design".

29 Eurochemic/CA(58)12.

30 Eurochemic/CA(58)13.

31 Eurochemic/CA(59)1 and Eurochemic/CA(59)17 for the text of the agreement.

32 Arrangements were made for a technical report to be prepared by Messrs Yves Sousselier and Pierre Regnaut, *Management Committee, minutes n° 12,* 20-21 November, and by Erik Haeffner. A report on the study visit written by Erik Svenke for Einar Saeland, prepared on 13 November and submitted on 1 December 1958 is attached (unreferenced) to the Management Committee documentation. The preliminary visit report and the final report are in file OEEC/ENEA 5.1 – Relations with the United States. It was stencilled by the Plutonium Division of the CEA Chemistry Department.

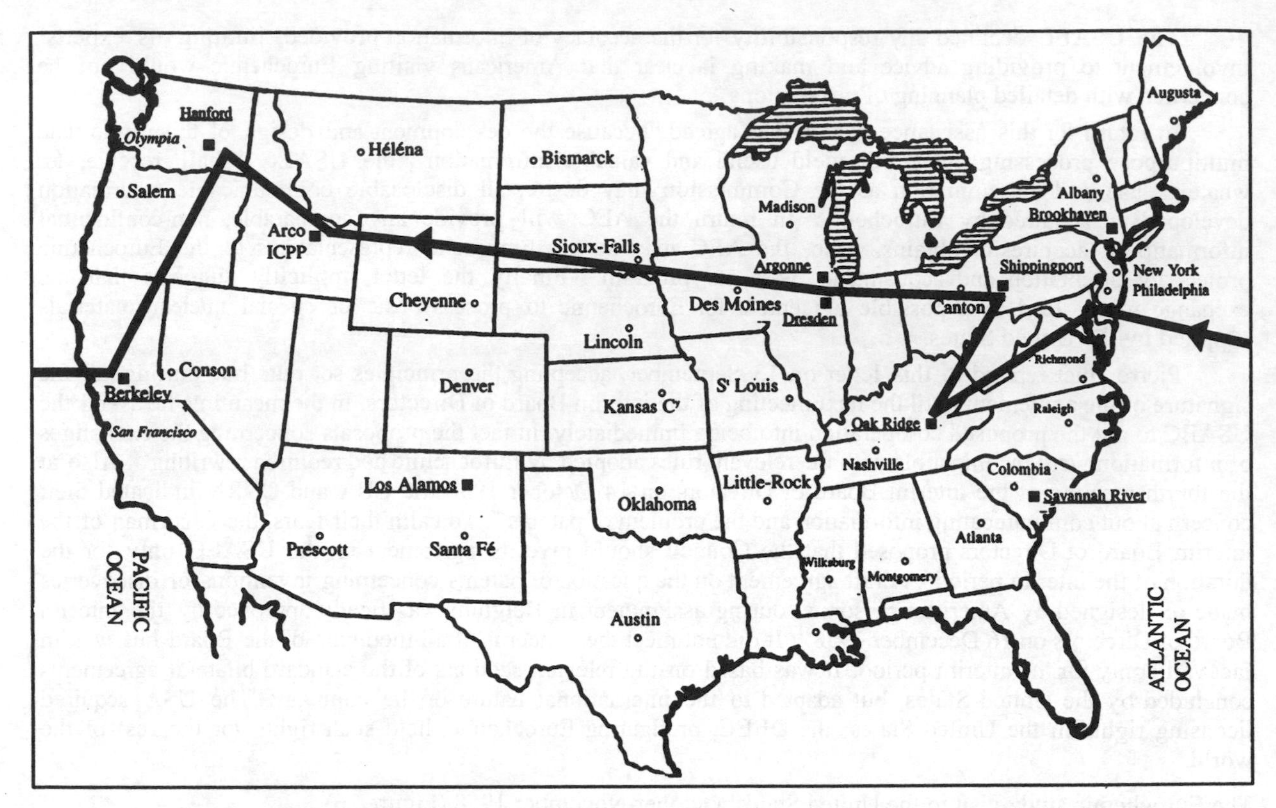

Figure 26. Route followed by the Eurochemic study visit: October-November 1958 (Source of data: SOUSSELIER Y. (1960), p. 384).

In 1959 the four American reprocessing plants of Hanford, Savannah River, Idaho and Oak Ridge appeared sufficient to meet the requirements arising from the power reactors being commissioned. The facilities were characterised by a flexible division of tasks: natural and low enriched uranium was reprocessed at Hanford for fuel clad in zirconium and stainless steel, and at Savannah River for fuel clad in aluminium and magnesium. Highly enriched fuel was processed at Idaho; Oak Ridge handled thorium and fuel from homogeneous reactors as well as 20% enriched uranium from naval reactors in the modified Metal Recovery Plant. The reprocessing system was thus extremely comprehensive (Table 12).

Table 12. **The US reprocessing system in 1958**

Site	Type of reprocessing
Hanford	Natural uranium clad in zirconium or stainless steel. REDOX process.
Savannah River	Fuel clad in aluminium or magnesium. PUREX process.
Idaho Falls	Highly enriched uranium clad in aluminium. Hexone process (a variant of REDOX). Fuel clad in zirconium or stainless steel. PUREX process.
Oak Ridge	Thorium fuel. THOREX process. Project for reprocessing fuel from naval reactors in the old pilot plant.

The transport of irradiated fuel appeared to be developing rapidly, although it raised the problems of standards as it did in Europe. The job of building flasks was put out to private contract.

It appeared most unlikely that it would be possible to send staff for training at reprocessing plants in the near future, except at Idaho.

Visits to the Oak Ridge installations then took place on 22 and 23 October. The visitors met Floyd L. Culler (he was already known to some of them) together with ORNL reprocessing specialists including Earl M. Shank.

The visit included the old processing pilot unit, the Metal Recovery Plant, at that time unused and decontaminated; the Thorex Plant, which was shut down owing to the lack of any material to be reprocessed; a pilot caesium and fission product production unit; and two facilities for processing low and intermediate activity liquid wastes. From their visit to ORNL the European group had the impression that the Americans themselves were doing little research on aqueous processes that appeared ready for industrial use.

The technical discussions covered, in particular, the problems of criticality, the experimental "pre-processing" methods developed in close conjunction with the PUREX process, notably ZIRFLEX for zirconium cladding and SULFEX for stainless steel cladding[33], and the testing of new solvents less sensitive than TBP to very high activity.

On 30 and 31 October the group visited the Idaho Chemical Processing Plant (ICPP) operated by Phillips Petroleum. This was the only American reprocessing facility to be visited. It was to provide substantial inspiration for the future European plant[34], with which it had many points in common[35]. The ICPP had a main building 80 metres long, 35 metres wide and a maximum height of 30 metres, housing the plant and its analytical laboratory side by side.

The plant had originally been built to process fuel rods from the MTR reactor[36], which were of 90% enriched uranium clad in aluminium, but was subsequently modified to cope with fuels clad in zirconium and stainless steel. It thus became multipurpose. The process head-end used either hydrofluoric acid for zirconium (which necessitated systems made of monel), or sulphuric acid for stainless steel, in which case the system was built of special stainless steel[37]. Two extraction processes were employed: hexone for MTR fuel and TBP for other types. The system used membrane pulse columns. The plant was maintained directly, so decontamination took a fairly long time. The cell floors were covered with stainless steel, extending halfway up the walls, and the working areas were coated with PVC. The radiation dose was held at 1 mSv per eight-hour period. Many access closures showed that accidental contamination was fairly frequent.

The laboratory included a routine testing section which received active samples by air lift directly into lead containers and two more conventional and very well equipped laboratories.

The visitors also inspected a liquid waste storage tank and a pilot unit for solidifying high activity liquid wastes by fluidized bed drying, both under construction.

The discussions were particularly concerned with costs. The ICPP plant had cost $38 million, including 9.5 million for the main building and equipment. The chemical engineering design and the costs of construction amounted to $5.7 and 10.5 million respectively.

The trip continued with a presentation on Hanford given in the meeting room of the centre's laboratories at Richland on 3 November. The Eurochemic representatives saw only a photograph of one of the two chemical plants that were in service at the time, supplied with fuel from the eight plutonium producing reactors on the site. However the discussions were fruitful and covered the plutonium recycling programmes and the project for reprocessing fuel from the power reactors[38].

It was planned that Hanford would modify the installations of the military REDOX plant so that by 1962 it could begin reprocessing about 100 tonnes of fuel from the dozen or so reactors expected to be commissioned in the United States between 1960 and 1963, using natural or low enriched (less than 5%) uranium, clad in zirconium or stainless steel. Contemporary planning envisaged chemical decladding using the ZIRFLEX[39], DAREX[40], NIFLEX or SULFEX processes, possibly preceded by mechanical decladding. Extraction would involve the REDOX process. Costs were estimated at $5 million for design work and $6.5 million for the necessary construction and modification.

The trip ended with a visit to General Electric's San José fuel fabrication plant, the related Vallecitos R & D laboratory[41], and the Berkeley laboratory.

---

[33] See UN(1958), 15/P/1930.

[34] See UN(1955), P/543 and USAEC (1955).

[35] See the comparative review at the end of Part III, Chapter 1.

[36] MTR: Materials Testing Reactor, a high flux reactor located in the centre of Idaho.

[37] Known as Carpenter 20 (29% nickel, 20% chromium and a trace of molybdenum).

[38] Programme known as the "non-production fuel reprocessing", clearly showing the predominantly military approach to reprocessing at Hanford, where production refers to plutonium and not electricity.

[39] Using ammonium fluoride and ammonium nitrate.

[40] Using *aqua regia*.

[41] The Svenke report terminates with the visit to ICPP.

Overall outcome of the mission. Eurochemic and the American reprocessing programme

The visitors retained the general impression that the USAEC was interested in the Eurochemic experiment. "In fact their current programme provides for no new reprocessing plants to be built [*sic*]". Despite the growing complexity of the materials used for cladding, research into preprocessing had not made much progress. Development work on aqueous processes appeared to be marking time in the United States, unlike the situation of dry processes (fluoridation, pyrometallurgy). However the latter were still only at the laboratory stage. In this connection there was a striking difference between Oak Ridge and Argonne.

"Thus (the Americans) believe that the Eurochemic plant, being more up-to-date, benefiting from current research and the results of experience in their own plants, could possibly be useful to them as a pilot unit in certain fields". However their interest did not go so far as to allow European trainees to be sent either to the plants or to the laboratories. "This clearly shows the existence of a substantial classified area...related to military questions". Classification appeared to the visitors to be "frequently a little ridiculous and exaggerated". To get round it, they recommend building up "personal contacts which have sometimes allowed us to penetrate the classification barrier which has always been strictly applied at an official level". "The presence of an American expert at Mol is highly beneficial because it means that information at the classification limit can be obtained unofficially".

As far as technical aspects were concerned, the visitors returned to Europe with the feeling that the $13 million envisaged as the cost of building the Eurochemic plant was liable to prove "clearly inadequate".

The European mission heard substantial and focused criticism about the chosen site, which rendered any discharge of activity quite impossible. The visitors concluded that "the greatest possible precautions must be taken as regards (the) storage (of liquid wastes) since the smallest incident could have major consequences". They were made particularly aware of the problems of the precautions to be taken against radiation and contamination.

"We noted once again in the various laboratories and chemical installations we visited that the precautions taken against the risk of radiation are simple and few in number. They are based upon the trust placed in the technicians and the discipline they put into their work. The results are generally very satisfactory.

However at ICPP we saw decontamination work in progress and many areas to which access was prohibited, clearly showing the difficulty that exists in preventing contamination in these plants".

The Eurochemic-USAEC assistance programme

On completion of the mission, the arrangements for United States-Eurochemic co-operation were put in place, and soon came to be known as the Eurochemic-USAEC assistance programme[42].

In November 1958 an ORNL reprocessing specialist, Earl M. Shank, whom the members of the Steering Committee had met during their visit, was appointed by Floyd L. Culler to head a secretariat "with the sole object to provide information to the Eurochemic Study and Research Office". Earl M. Shank[43] was a chemical engineer who, since his entry to ORNL after his demobilisation in 1946, had been involved in the preparatory work for four American reprocessing installations. Thus he had considerable practical experience of the problems of designing and building a reprocessing plant, and had retained personal contacts with senior staff on the production sites and with the researchers. This highly appropriate background shows the importance the United States was placing on scientific and technical co-operation with Eurochemic. Shank's job, which was supposed to be part-time, soon became full time. The Secretariat was in contact with Argonne, Idaho, Hanford and Savannah River. "In this way Eurochemic can receive information not only contained in declassified and published reports, but also technical know-how and information of an unclassified nature which can be extracted from classified reports".

At the other side of the Atlantic the American expert seconded to Mol, E. L. Nicholson, identified requirements and forwarded the technical reports – "Eurochemic's Technical Reports" (ETR) produced by the BER, and subsequently by the different Eurochemic divisions, for information and comment. These reports were recirculated to relevant American organisations by the National Technical Information Service in the form of USAEC reports[44]. This arrangement continued in being until the end of 1970.

---

42 Progress Report n° 4 of 1 December 1958, Eurochemic/CA(58)16. The classification and coding of the topics for co-operation are given in an Annex.

43 Interviews held on 19 April 1994.

44 For example the first ETR, signed by Erik HAEFFNER, *Eurochemic, Suggested Organisation and Work During the Interim Period*, was distributed in the United States under the reference USAEC report NP-7659.

E. L. Nicholson remained at Mol until July 1960. He was succeeded on 1 November 1960 by Robert J. Sloat from Hanford, who had worked on the REDOX plant and then in the Hot Semi-Works Pilot Plant and remained at Mol until 10 May 1962. Earl M. Shank himself then left Oak Ridge in August 1962 for Eurochemic where he remained until 1969, regularly transmitting Progress Reports in which he summarised the work carried out. In this way the United States was kept informed throughout the construction period and during the early plant operations. In return Eurochemic staff subsequently made numerous study visits and benefited from substantial scientific and technical assistance.

Other international links

Co-operation with the United Kingdom: limited outcome

The members of the interim Study Group had been disappointed that the British decided not to take part in the undertaking, even though the country had been supportive during the preparatory period. The reasons for this development were suggested in the conclusion to the previous chapter. According to the sources, the disappointment turned into a certain bitterness towards the British. Despite this the United Kingdom still had a substantial amount of experience and was a member of OEEC/ENEA. It was decided to make further contacts, and the UKAEA responded in June 1958.

The Management Committee obtained the impression that the British "appeared to want to be involved with Eurochemic primarily so that their industry take part in construction", but stressed "that real co-operation could be brought up in discussions about general projects"[45].

On 9 October 1958 the UKAEA Industrial Group at Risley (Warrington) sent the interim Board of Directors some proposals for co-operation[46], on a commercial basis, with the possibility of payment being delayed until the company commenced its activities. The Industrial Group offered its technical assistance during the interim period through written technical opinions, notably with regard to the process flow sheets. This assistance would be "based upon the Authority's own experience and the Authority would put no additional work in hand".

This assistance was to be continued during the construction period in the form of advice on the technical characteristics of equipment and the relevant inspection standards, safety problems, commissioning and operational schedules, and staff training.

The Industrial Group also supported the bids for services contracts made by three British contractors: W. S. Atkins and Partners, laboratory consultants, W. J. Fraser and Co. Ltd. and Nuclear Chemical Plant Ltd. for the processing plant.

Like the USAEC, the UKAEA declined any liability in connection with the use to which its information was put[47].

The interim Board of Directors approved the terms of this letter on 16 December 1958.

Canada: An agreement in principle, but no results

Canada had experienced problems with its own reprocessing system and at the time had virtually given it up. However the country was an associate member of the OEEC and in a letter dated 3 April 1958 Pierre Huet notified the delegate of Canada to NATO of the Study Group's wish to consider the possibilities of technical co-operation with Canadian experts. The Canadian authorities agreed in principle on 20 June 1958[48] but this led to no concrete results.

Beginnings of co-operation with European firms. The "Consultation Group" and the first choice of associate firms[49]

Principles governing relations with architect engineers and contractors

The problem of the most suitable form of co-operation with the private firms engaged to build the plant and the laboratory had been the subject of much discussion, particularly at the unofficial meeting of the interim Board of Directors held at the second UN Geneva Conference on the Peaceful Uses of Atomic Energy on

45 Eurochemic/CA(58)12.

46 Eurochemic/CA(58)14.

47 "The Authority can in no way guarantee that the use by Eurochemic of information provided by the Authority may not affect patent or other rights belonging to third parties but would grant a non-exclusive licence on any Belgian patents it holds in this field".

48 Unreferenced document, fourth meeting of the Study Group.

49 Eurochemic/CA(59)12 of 25 May 1959.

9 September 1958. Bertrand Goldschmidt had raised the question of whether there would be a project manager or general contractor, as at Marcoule, or whether construction would be managed directly by Eurochemic according to the British model[50]. The Director of the Design and Research Bureau, in conjunction with the Management Committee, was asked to consider the main options and produce proposals, bearing in mind "that the chosen formula should allow all member countries to participate to the greatest possible extent".

The interim Board of Directors, at its meeting held on 27 February 1959, set up a "Consultation Group" consisting of Erich Pohland, Chairman of the interim Board of Directors, Yves Sousselier, Chairman of the Management Committee, Pierre Huet, Director-General of ENEA, and Erik Haeffner, Director of the Design and Research Bureau. This group was asked to get in touch with companies the directors thought likely to be interested in contributing to the construction of the plant as architect engineers. This approach led to discussion in the Management Committee on 20 January 1959 about the works procedure, the distribution of jobs between Eurochemic and the contractors, and an initial outline of contract requirements[51]. It was on the basis of this document that initial contacts were made.

The six principles governing relations between Eurochemic and contractors were defined as follows[52]:

"i) Eurochemic shall exercise a permanent general control of the studies and work as a whole, at all stages;

ii) Eurochemic shall itself fully undertake certain studies and detailed projects;

iii) Eurochemic shall call on two Architect-Engineers, one for the plant and one for the laboratory. These persons shall be assigned, on the basis of projects prepared by Eurochemic, the detailed design of chemical engineering, the preparation of files concerning tenders, the co-ordination and supervision of contractors and suppliers;

iv) the construction of certain auxiliary or partial units (waste treatment station, ventilation of the main building, etc.) may be assigned to specialised Architect-Engineers or consulting engineers; the head Architect-Engineer shall be responsible for co-ordination of their work;

v) the actual construction and supply of equipment shall be assigned to contractors and suppliers selected by Eurochemic on proposals by the Architect-Engineers;

vi) all contracts with the Architect-Engineers, the consulting engineers, and orders and supply contracts with contractors, shall be concluded by Eurochemic."

Details of how work should be divided between Eurochemic, the architect engineers and the industrial consultants are set out for the various stages of the two projects, plant and laboratory, from the preliminary chemical engineering design through to construction itself, on the basis of these principles[53].

Accordingly the design and construction was planned around three focal points: first Eurochemic, designer of the project, as project manager, secondly the group of architect engineers led by the principal architect engineer charged with the detailed project and works supervision, acting as a design bureau and architect, and finally the contractors charged with executing the work with their suppliers and sub-contractors.

The choice of architect engineers

Sixteen firms in eleven countries[54] were consulted by members of the group between 31 April and 21 May 1959. As a result, the group drew up initial proposals for dividing the work between virtually all the contractors approached.

The German proposals involved four contractors: Friedrich Uhde GmbH, Dortmund; Leybold-Hochvakuum-Anlagen GmbH, Cologne; Lurgi-Gesellschaften, Frankfurt; and Siemens & Halske A.G., Karlsruhe.

Norway put forward the name of Noratom A/S, Oslo and Sweden that of Bofors, Stockholm. Denmark submitted the names of the four contractors then building the Risø Nuclear Research Centre: Preben Hansen and Paul Niepoort (architects), Steensen and Varming (civil engineers), Mogens Balslev, electrical engineering. Other countries' proposals were Belchim S.A. (Belgium), Saint-Gobain (France), Suter & Suter, Architekten B.S.A. (Switzerland), the C.A.L. Officina Técnica de Proyectos (Spain), Montecatini (Italy), Comprimo N.V. (the Netherlands). The size of these firms and the extent of their experience in nuclear construction was very

50 Eurochemic/CA(58)12. For the British organisation see GOWING M. (1974), Vol. 2, Chap. 21-22.

51 Eurochemic/CA(59)2 of 13 February 1959.

52 Ibid., p. 3.

53 Ibid. p. 5 to 11.

54 Neither Turkey nor Portugal were able to propose a contractor.

varied. The aim of the Eurochemic project was to get these firms to co-operate to enable them to acquire broader experience in the nuclear field.

From Table 13 giving details of the firms consulted, a number of conclusions can be drawn.

The major European chemical contractors are represented through their subsidiaries or joint ventures: this is the case of Montecatini and Saint-Gobain but also Farbwerke Hoechst, through Uhde and Leybold, and Royal Dutch-Shell, at the time the main shareholder of Comprimo. CIBA is not directly involved but did select the firm Suter & Suter. The metallurgical industries are also represented, Bofors directly, UMHK indirectly through Belchim, Metallgesellschaft through Lurgi. The same applies for electrical engineering through a subsidiary of Siemens.

The nuclear background of the different companies clearly reflects the state of progress of nuclear power in each country. There are few common points between a small Spanish company which had supplied non-nuclear services during the construction of an experimental reactor, and SGN which in 1959 had already built a pilot reprocessing unit and a full-scale military plant.

Behind the proposals for commercial groupings made by Belgium, Denmark and Norway respectively can be seen an option for organising the work of the architect engineer, which attempts to compensate for the country's lack of a firm specialising in the nuclear field by pooling skills for the project. Noratom, Belchim and the Danish Risø group put forward the idea of temporarily seconding specialists from the members of their group. However the Consultation Group decided not to use this type of formula for the principal architect engineer owing to the risk of compartmentalisation and lack of co-ordination.

Another possibility would have been to set up a new international company to carry out the work. However this kind of approach[55] had been rejected on the grounds that it would have delayed the start of the work and would have complicated the decision-making process by somewhat weakening the authority of Eurochemic.

An initial outline of the distribution of tasks

In May 1959 the first attempt to divide up the work between the different contractors consulted sought to tread the line between utilising existing experience and the primary objective of the European project, which was to enable the largest possible number of firms to take an active part in building the plant. The following breakdown emerged.

Saint-Gobain would be the principal architect engineer responsible for building the processing plant and for co-ordinating the work as a whole. The French firm fully met the requirements laid down by Eurochemic, being the only one with a specialist and well-tried division. Also France had made the choice of Saint-Gobain as architect engineer a condition for its participation. Marcoule had commenced operations in 1958 and the construction of a new reprocessing facility had not yet been decided. Yet working on the Mol plant project fitted in well with the SGN schedule[56]. René Grandgeorge, Managing Director of Saint-Gobain, a firm which had acquired 22 Eurochemic shares in 1958, attended nearly every meeting of the interim Study Group and was a member of the interim Board of Directors as the second French Director.

Bofors was made responsible for fuel element reception, mechanical handling and mechanical decladding (which was envisaged at the time). The reason for this choice was the firm's experience in designing mechanical systems, the remote handling of explosives, and its slight nuclear experience.

The treatment plant's ventilation and sampling system and the analytical laboratory was to be built jointly by Lurgi, Uhde and Leybold.

The control system for the processing operations, which bore a certain similarity to the advanced automatic facilities used in the petroleum industry, was to be entrusted to Comprimo.

Montecatini would be responsible for the liquid waste treatment unit.

Noratom was to be given the task of building the storage facilities for highly radioactive wastes.

The research laboratory would be built by the Danish group, CAL and Suter & Suter. These three firms formed a consortium known as SDS (Switzerland-Denmark-Spain). Hence there was no principal architect engineer for the research laboratory, contrary to the original intention. Siemens and Halske were to supply the laboratory equipment.

55 Proposed by one of the contractors contacted, not quoted in the source.

56 The La Hague site for the new plant was not selected until 1960. In fact the delay of the project compared with the original schedule was to result in the engineering design work for Eurochemic and UP2 taking place in parallel.

Table 13. **Firms contacted by the Eurochemic Consultation Group in 1959**

Name	Type of firm and business activities	Headquarters and establishments	Work-force	Remarks	Nuclear experience up to 1959
Belchim (Société belge de chimie nucléaire)	Ad hoc grouping of Belgian chemical and metallurgical companies. Limited company with registered capital of 7 million BF set up on 17 June 1958, shareholder in Eurochemic since 1 January 1959 (27 shares).	Brussels		Main shareholders: SGMH, UMHK, Société belge de l'Azote et des produits chimiques du Marly (biggest shareholder: SGB). Would act as central technical bureau, tasks to be shared between the different design bureaux and technical departments of the member companies.	Through Hoboken, industrial production of radium and uranium at Olen. Through one of the subsidiaries of Hoboken, MMN (Métallurgie et mécanique nucléaire), builder of a fuel fabrication plant at Dessel (near Mol) the future Belgian establishment of the FBFC.
Bofors	Explosives and artillery systems, military and civilian metallurgy (special steels), consulting engineering division.	Stockholm		For Eurochemic purposes, would work with the following Swedish firms: Uddeholm, Johnson Line (Avesta), ASEA, Stockholm Superphosphate Company, Reymersholm (later NOHAB).	Plans for the handling system for fuel from the Swedish R3 Adam
C.A.L. Officina tecnica de proyectos	Architect engineers	Madrid	n.a. (not available)		Building, civil engineering, electrical and ventilation work for the swimming pool reactor at the Nuclear Energy Centre of Moncloa (co-operation with American contractors).
Comprimo N.V.	Design and construction company working primarily for the petroleum and chemicals industry.	Amsterdam	300		Consulting engineer for the construction of the University of Delft swimming pool reactor, production of drawings for the KEMA homogeneous suspension reactor.
Friedrich Uhde GmbH	Design and construction company, subsidiary of Knapsack Griesheim A. G. Chemical business (initially ammonia, later diversified).	Headquarters and design bureau at Dortmund, construction at Farbwerke Hoechst.	Approximately 1000	Links with Farbwerke Hoechst	Design and operation of a pilot heavy water plant (Hoechst design) Plans and drawings for the laboratory at the Karlsruhe Transuranium Institute and the Julich high activity laboratory (Juliers).
Group of constructors at the Risø Centre	Preben Hansen, one of the official architects of the Danish government. Paul Neipoort, architect engineer. Steensen and Varming (civil engineering) and Mogens Balslev (electrical engineers).			Steensen and Varming, in conjunction with Mogens Balslev were supervising current work on the new NATO headquarters in Paris.	Paul Niepoort: construction of hot laboratories (experience in the United States) The Group built the Risø Centre in a very short time.
Leybold-Hochvakuum-Anlagen GmbH	Construction and service company (consulting engineers) High vacuum, particularly in chemical plants (distillation, melting)	Cologne	Approximately 200	Regular co-operation with Uhde.	Plans for the Institut für Radioaktive Körper at Düren, the Karlsruhe Transuranium Institute, isotopic treatment installation for the Berlin Hahn-Meitner Institut, design of fuel element unloading systems for German reactor projects.
Lurgi-Gesellschaften	Design and construction company, affiliated to the Metallgesellschaft group. High pressure technology in chemistry, metallurgy, purification of water, air and gases.	Frankfurt	1700, of whom 300 in the pilot plant		Uranium and thorium ore concentration and purification plants (for Société générale de Hoboken), methods for waste processing and liquid waste decontamination, heavy water purification in a research reactor.

(continuation of table)

Name	Type of firm and business activities	Headquarters and establishments	Workforce	Remarks	Nuclear experience up to 1959
Montecatini	Chemical industry, non-ferrous metals, substantial private electricity producer, large consulting engineering division	Milan 174 establishments	60000	Had a small nuclear section concentrating on the uses of radioisotopes in medicine, within its pharmaceuticals division.	Participation in CISE (Milan) since 1946. In conjunction with FIAT in SORIN, builder of the Saluggia Nuclear Research Centre.
NORATOM A/S	Group of contractors and research organisations for nuclear projects. Multiple skills.	Oslo	"Variable geometry" (25 000 in the different companies of the group)	Contractors wishing to be involved in Eurochemic through NORATOM: Norsk-Hydro-Elektrisk Kvaelstofaktiegeselskap, Elektrokemisk A/S, Borregaard A/S, Kvaerner Brug A/S, Norsk Spraengstofindustri A/S, Norconsultants. Research institutes: Institutt for Atomenergi, Chr. Michelsens Institut, Sentralinstitutt for Industriel Forskning.	Builder of the pilot reprocessing facility for the Kjeller Institutt for Atomenergi.
Saint-Gobain nucléaire (SGN)	Special division of the Saint-Gobain company (glassmaking and chemicals). Eurochemic shareholder (22 shares)	Paris 90 establishments for the parent company with the subsidiaries	20 000, of whom 250 in SGN	Nuclear activities since 1952, SGN set up in 1955	Pilot reprocessing plant at Fontenay-aux-Roses, reprocessing plant at Marcoule, uranium fabrication plant at Narbonne. Work in progress in 1959: Marcoule chemistry laboratories and hot cells, storage facility and mechanical decladding system for fuel elements from the G2 and G3 reactors.
Siemens & Halske A.G.	Company manufacturing electrical engineering equipment and instrumentation, subsidiary of the Siemens group.	Karlsruhe	n.a. (not available)		Builder of the Siemens nuclear laboratory at Erlangen and the Wuppertal Isotope Laboratory for Bayerwerke, involvement in planning of the Garching Radiochemistry Institute.
Suter & Suter Architekten B.S.A.	Architect engineers (multiple applications)	Basle	100 and over through "à la carte" arrangements	proposed by CIBA ?	None

Belchim, the main private shareholder in Eurochemic was to be made responsible for site works, general services, offices, and the supply of power and steam, etc. The Belgian group was selected owing to its proximity but it failed to obtain the role of principal architect engineer, which might have "raised the question of the equitable sharing of work between the participating countries", since a number of supply contracts were awarded to Belgian firms. In addition, the fact that it was an ad hoc group and particularly the lack of any permanent staff made it difficult to consider giving the group this highly important role.

Belgian disappointment was considerable, as made clear in a joint statement by Messrs. De Merre and Van der Meulen[57] to the interim Board of Directors, regretting "that it had not been suggested that Belchim should be given a task more directly related to nuclear technology, which it would greatly have preferred" and duly proposing liquid waste treatment.

At the first meeting of members of the Management Committee with the architect engineers, some adjustments were made to the distribution of tasks, in the light of the Belgian disappointment: Belchim was made responsible for liquid waste treatment, Montecatini was given the design of the head-end of the process instead, which reduced the amount of work for Bofors. The task entrusted to the three German contractors, who formed a consortium, the Ingenieurgemeinschaft Kernverfahrenstechnik (IGK), was specified: the analytical test laboratories and nitric acid recovery[58].

57 Eurochemic/CA(19)13 of 29 May 1959, p. 3.

58 Minutes of the twentieth meeting of the Management Committee, 8 July 1959.

At the same meeting, the architect engineers agreed to establish a joint engineering design office in Mol. Details of how the architect engineers should be paid, and of the type and detail of contracts, were still the subject of discussion.

The reason why the division of tasks could thus be outlined in July 1959 was that significant work had been done on the Mol site by the Design and Research Bureau under the direction of the Management Committee, and that an initial preliminary project had been established, prefiguring the still far-off detailed project of June 1961, which was the basis for construction.

The project takes shape. Preparation of the preliminary plans and the establishment of the international shareholding company

Installation at Mol and the work of the Design and Research Bureau (BER)

The Design and Research Bureau was accommodated in the nuclear research centre, with the planned Eurochemic site very close by. Design work began in May 1958 and research at the beginning of 1959, once the laboratories were ready.

The agreement with CEN: preparing to purchase the site

The Design and Research Bureau, guest of the CEN.

The CEN had been actively involved in drawing up the proposals for a Belgian site. The Director of the Chemistry Department, Maurice D'Hondt, was a member of the Management Committee. The CEN adopted a highly co-operative attitude in facilitating the beginnings of the undertaking in Belgium, even proposing to advance funds for the first few months (Figure 27).

The Design and Research Bureau did not have its own premises at Mol[59], but was accommodated by the CEN which, under an agreement concluded on 1 June 1958[60] between its Director and the Director-General of OEEC/ENEA, leased premises, offices and laboratories, provided the necessary services – furnishings, heating, lighting, telephone, ventilation – together with administrative facilities – assistance by the purchasing office, joint insurance policies and use of the Centre's canteen.

A Joint Committee, the permanent members of which were Messrs Goens and Hubert from the CEN and the Director of the BER, Erik Haeffner, met monthly as from December 1958 to deal with matters of common interest.

The active laboratories were not suitably equipped for the type of work envisaged by Eurochemic, so modifications had to be made to the existing infrastructure, which delayed the start of the experimental programme by several months. The programme began in January 1959 with the installation of the first few systems such as the dissolving and solvent extraction units[61].

Purchase of the site

In May 1958 neither the extent of the site which Eurochemic was to occupy nor the conditions governing its purchase had yet been precisely defined[62]. However the original idea of locating the plant on land adjoining the CEN had been given up owing to the Centre's own expansion plans, in favour of a site belonging to SGMH, nearby, but separated from the CEN by the Meuse-Escaut Canal and located in the commune of Dessel. The members of the Management Committee regretted this change because it meant that the plant's wastes had to be piped across the canal for treatment in the facilities of the Centre, which could represent a potential pollution hazard[63].

In June, the CEN notified the interim Board of Directors that it was prepared to build a bridge over the canal to provide a direct link between the Centre and the site intended for Eurochemic[64].

In July 1958, Mr. Van der Meulen officially notified the interim Board of Directors that SGMH was ready to sell its land to Eurochemic at cost and, in any event, at a price below 10 million BF. The interim Board was

---

[59] Construction of the first Eurochemic building, Administrative Building 11, did not begin until 1960.

[60] Eurochemic/CA(58)4. Some 336 m^2 of active laboratories, 80 m^2 of analytical laboratories and 138 m^2 of equipment stores were leased in this way.

[61] Progress Report n° 2, p. 2.

[62] Eurochemic/CA(58)4.

[63] Communication from Yves Sousselier.

[64] Eurochemic/CA(58)7.

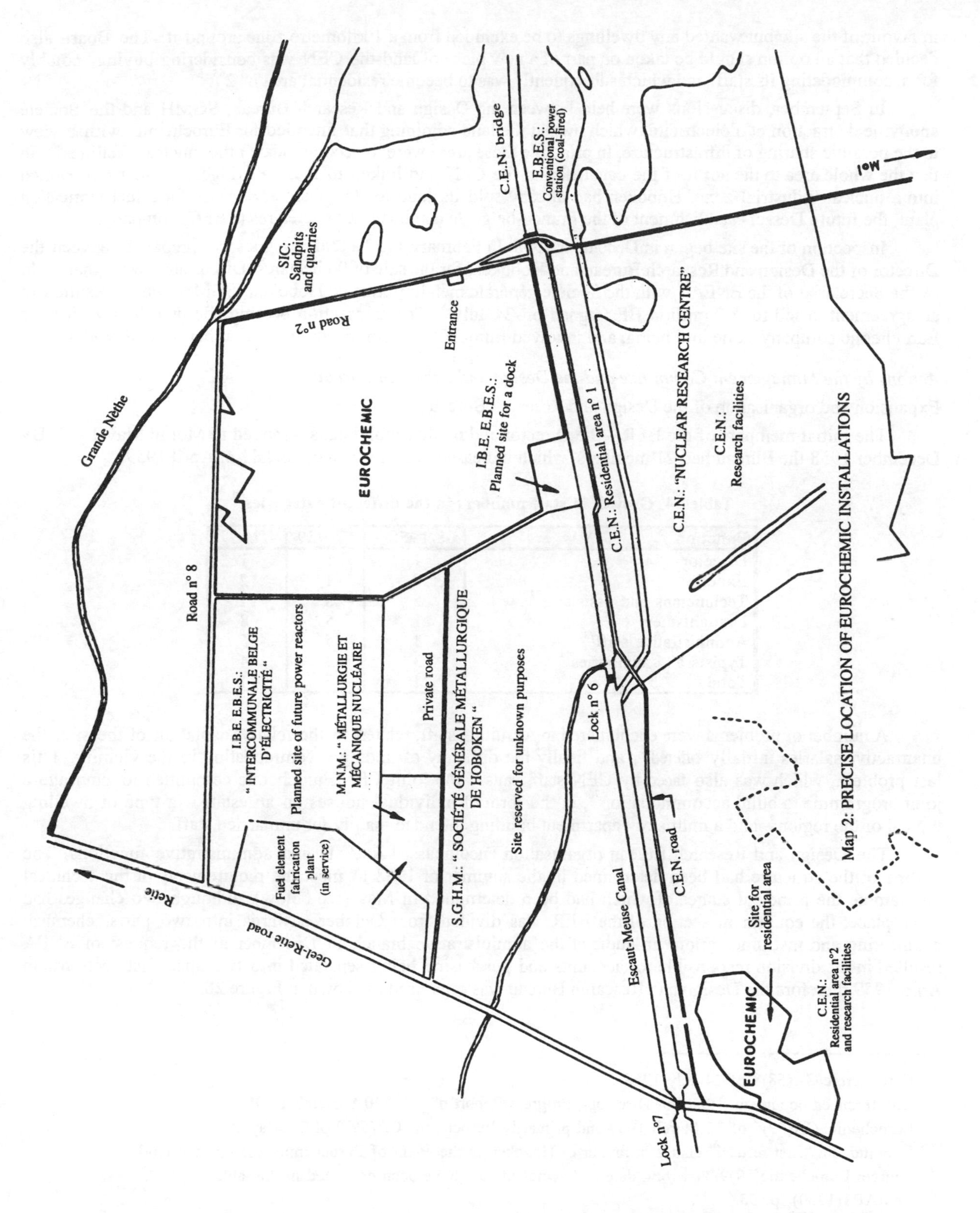

Figure 27. The Mol "nuclear complex". Eurochemic and its neighbours in 1961 (Source: RAE1, p. 54).

in favour of the site but wanted any dwellings to be excluded from a 1 kilometre zone around it. The Board also decided that an option should be taken on part of a new piece of land the CEN was considering buying, notably for accommodating its staff, and which subsequently was to become residential area n° 2[65].

In September, discussions were held between the Design and Research Bureau, SGMH and the Société anonyme de traction et d'électricité, which owned the land adjoining that intended for Eurochemic, with a view to the possible sharing of infrastructure. In principle these areas were to accommodate other nuclear facilities[66]. In fact the whole area to the north of the canal, facing the CEN and linked to it by a bridge, was to be converted into a nuclear industrial zone. Hoboken had already sold its land to its subsidiary MMN for a fuel fabrication plant, the future Dessel establishment of the Franco-belge de combustibles nucléaires (FBFC) company.

Inspection of the site began in December 1958. In February 1959 a draft contract was prepared between the Director of the Design and Research Bureau and Hoboken, for the sale of 96 hectares. Discussions were then held by the Secretariat of the ENEA[67] with the Société générale métallurgique de Hoboken (SGMH) which resulted in an agreement to sell for 8.7 million BF, signed on 24 July[68]. The transaction became effective shortly after the Eurochemic company came into being, and involved Eurochemic's contributing to a number of road works.

Actions by the Management Committee and the Design and Research Bureau

Expansion and organisation of the Design and Research Bureau

The initial members of the BER – the Director and 6 other individuals – moved to Mol in May 1958. By December 1958 the Bureau had 21 members, which increased to 33 in January and 41 in April 1959.

Table 14. **Growth in staff numbers in the different categories**[69]

Function	08/1958[70]	1/1959	4/1959
Director	1	1	1
Experts[71]	8	11	12
Technicians and assistants[72]	5	8	11
Draughtsmen	1	5	8
Administrative staff[73]	3	5	5
Typists and auxiliaries[74]	3	3	4
Total	21	33	41

A number of problems were encountered in recruiting staff, related to the relative isolation of the site, the unattractive salaries initially offered[75], and finally the difficulty of finding accommodation in the vicinity. This last problem, which was also faced by CEN staff, caused the joint CEN/Eurochemic committee to envisage a joint programme to build accommodation[76], in the form of individual houses on an estate – a type of dwelling typical of the region – and a multistory apartment building intended mainly for unmarried staff.

The Design and Research Bureau organisation encompassed R & D and administrative functions. The outline of the structure had been determined in the summer of 1958 to meet the requirements of the technical programme the principal stages of which had been determined in May (see below), although two changes had taken place: the equipment section of the BER was divided from October onwards into two parts, chemical engineering and instrumentation; an audit of the administrative branch in December at the request of ENEA resulted in the division responsible for accounts and purchasing being separated into two specialist sections. In April 1959 therefore the Design and Research Bureau was organised as shown in Figure 28.

---

65 Eurochemic/CA(58)8 of 24 July 1958.

66 Unreferenced document: Board of Directors, Progress Report n° 3 (of 30 September 1958).

67 Eurochemic/CA(59)6 of 25 March 1959 and primarily Eurochemic/CA(59)9 of 22 May 1959.

68 It included a "first refusal" clause in favour of Hoboken in the event of Eurochemic reselling the land.

69 Source: Eurochemic(59)9. A wrong date and switched data have been corrected in the table.

70 ENEAR1(1960), p. 58.

71 Plus one unestablished post in April 1959.

72 Plus two unestablished posts in April 1959.

73 Including an accounts officer seconded by OEEC to the BER.

74 Plus one unestablished post in April 1959.

75 Which soon had to be increased.

76 In April 1959 an architect prepared preliminary drawings for a hostel.

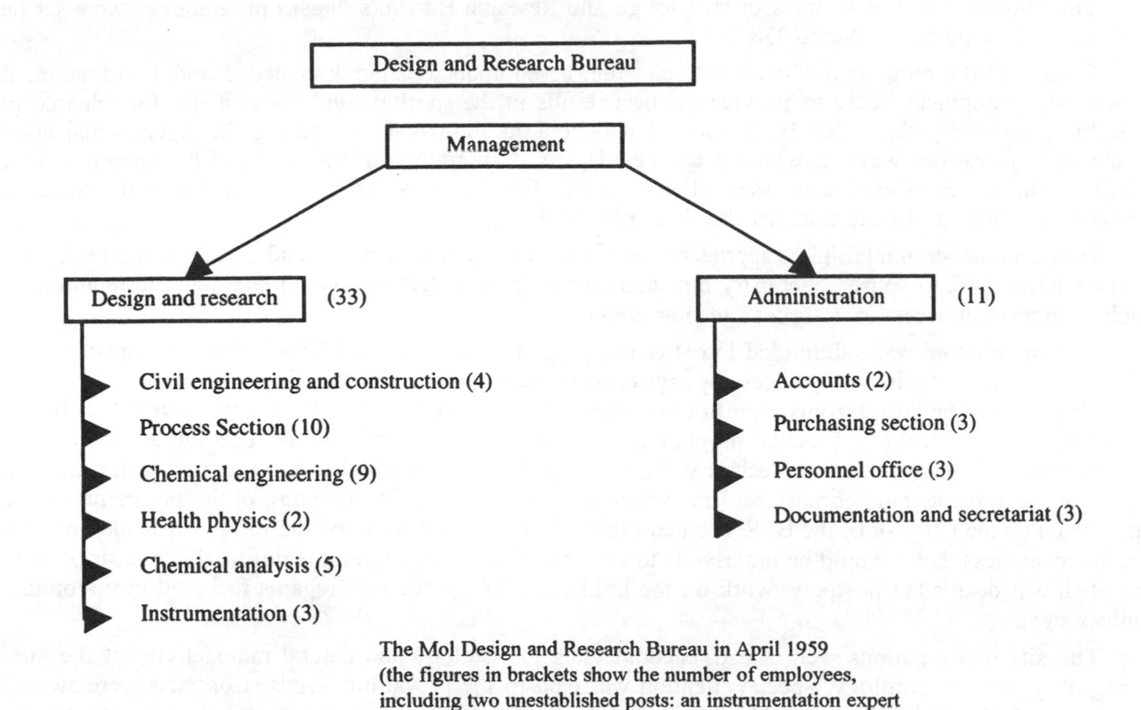

Figure 28. The Mol Design and Research Bureau in April 1959 (the figures in brackets show the number of employees, including two unestablished posts: an instrumentation expert and a documentation secretary).

The experts working in the different sections reflected the international structure of the undertaking.

The Civil Engineering and Construction Section was headed by a Swiss and a German, E. Lüscher and Walter Schüller; the Process Section by a Belgian and an Italian, Emile Detilleux and Maurizio Zifferero, soon succeeded by G. Calleri. The Chemical Engineering Section included an Austrian seconded from Lurgi, O. Jenne, a Frenchman and a Dane of Austrian origin, André Redon and Franz Marcus. A Spaniard, E. M. Lopez-Menchero, joined the Process Section once his country acceded to the project. The Public Health Section was directed by a Dutchman[77], H. H. Ph. Moeken, and Chemical Analysis by S. Ahrland and K. Samsahl, Swedish and Norwegian respectively. Finally the Instrumentation Section was led by a German, Rudolf Winkler. Most of these people were aged between 30 and 35 and an atmosphere of open co-operation soon emerged in this small group, most of whose members still retain nostalgic memories.

The technicians and draughtsmen were mostly Belgian, some of them coming from the CEN.

The team was completed by outside consultants, called in for specific tasks of variable duration. Some were involved in more than one particular job, including E. L. Nicholson, who was liaison officer with ORNL and worked essentially in co-operation with the experts of the Process Section, and Pierre Regnaut from the CEA Plutonium Division. Michel Lung, who had been permanently seconded to BER from SGN, remained at Mol until the detailed pre-project had been completed in 1961.

Principal activities of the Design and Research Bureau

The main task of the Design and Research Bureau was to draw up the pre-project for the plant and to start certain chemical research connected with its activities. It submitted a quarterly activity report to the interim Board of Directors, which enabled progress at Mol to be evaluated. It is impossible here to go into the technical details[78], but it is necessary to follow the institutional developments and to point out the main choices made.

---

[77] The presence of a Dutchman in this post reflected the special concerns of that country the border of which is about 10 kilometres from the plant. Mr. D'Hondt insisted on this arrangement at the meeting of the Management Committee held on 27 March 1958. The post came into being in November 1958.

[78] Almost all the technical reports are included in the Eurochemic archives.

The principles and main lines of the Design and Research Bureau's "basic programme" were set out in the first activity report of 17 May 1958[79].

In essence the programme "involved selecting, based upon existing knowledge and information, those methods and equipment likely to provide the best results in the specific conditions of the Eurochemic plant. Once this preliminary choice has been made, it will be a question of investigating the fundamental chemical process and the various ways in which it can be adapted. A thorough understanding of the chemical reactions underlying the processes used is an essential element. The final choice and design of the essential systems must be based upon industrial scale tests (as closely as possible)".

Hence a number of interlinked approaches were called for: a documentary and selective approach; research activities fairly close to "pure" chemistry but, and above all, activities of applied research, the requirements of which soon made it necessary to order a testing station.

The programme was subdivided into five sub-programmes of unequal length, covering installation, the fuel to be reprocessed, chemical processing, systems, instrumentation and buildings. As work progressed, the project expanded. The third activity report of September 1958 covered most of these sub-programmes but added two others: a pre-project experimental programme and a development programme. The fourth report issued in December focused on two items: a technical design programme, from which a sub-programme on the research laboratory was separated in February, and an experimental programme. The structure of the programme of work, as presented by the Director of the BER, seemed extremely flexible. It reflected the progresss made but also the growing awareness that it would be impossible to accomplish the full programme during the interim period. For example it was decided to postpone work on the head-end of the process as regards fuel clad in zirconium and stainless steel.

The site investigations went ahead, encompassing the geology and natural radioactivity of the subsoil, hydrogeology and meteorology. Special attention was paid to the prevailing winds. Contracts were awarded to the CEN and the Belgian Royal Meteorological Institute and, for measurements of radioactivity, to the Department of Health Physics of the Danish Risø Institute.

The Management Committee was particularly concerned with the type of fuel to be processed and the problem of its transport. These issues, "upstream" of the plant activities, involved a number of choices. As regards the amounts of fuel, the Committee conducted a survey which showed that the quantities of natural uranium and low enriched fuel to be processed would match the plant capacity up to 1966. A review of national reactor development programmes also led the Management Committee to consider the reprocessing of enriched uranium at Eurochemic.

The Management Committee also carried out a preliminary study of fuel transport, an area in which the main problem was not a technical one. Containers, also known as transport flasks, were being built not only in the United States but also in France and the United Kingdom, and the UK had issued technical standards as early as 1948. The main difficulty concerned legal liability in the event of accidents. At the time there were no standards governing liability as regards the transport of nuclear materials. Eurochemic was unable to settle the matter and pressed for an international solution, through its contacts with ENEA, EURATOM and IAEA.

As far as chemical processing and equipment were concerned, the Design and Research Bureau's programme of work covered the adaptation of the PUREX process to the different types of fuel to be processed, and the choice of system. It was a question of determining what process auxiliaries were necessary to enable the system to process the various types of fuel expected. This meant "adding to the chemical processing flow chart so as to include preparation of the elements to be processed, the treatment of wastes, the recovery of liquids to be recirculated (for example, solvents), and preparation of all the solutions involved in the process"[80].

The experimental study of the extraction process, which was essential "both to obtain the necessary information and to train staff in using it" was begun relatively quickly by using laboratory mixer-settlers which had been developed and used by the CEN.

Work on plutonium made a start thanks to the CEN[81]. The Belgian Centre had received from the USAEC a few fuel rods that had been irradiated in the ORNL graphite reactor and had reprocessed them in the laboratory.

[79] Eurochemic/CA(58)4, p. 4 *et seq.*

[80] Eurochemic/CA(58)4, p. 7. This concerns the treatment of solutions to be recycled in the process, such as solvents, and preparation of the necessary reagents.

[81] This was able to continue in 1960 as a result of a French donation. In the summer of 1960 Bertrand Goldschmidt in his Paris office handed over a small packet containing 10 grammes of plutonium oxide to Rudolf Rometsch, then Eurochemic's Director of Research.

It provided Eurochemic with the solutions already extracted and the fuel that had not yet been processed. This enabled Eurochemic to isolate its first plutonium.

The second Geneva Conference, which Eurochemic's chemists had awaited with some impatience, took place from 1 to 13 September 1958, but it provided them with little new information, although there was support for their choice of the PUREX process.

As work proceeded, the initial proposals were made for the process flow charts, operating schedules and equipment as well as locations. All this provided a basis for the negotiations that began with the industrial partners, some of whom had already been involved either because consultants had been employed by Eurochemic, like Michel Lund from SGN, or because they had done work under contract. This was the case of the Belchim company which came out top in Eurochemic's first call for tenders.

In fact it had been clear as early as May 1958 that "(since) determining the efficiency of the extraction columns by theoretical calculations often (gave) wrong results, (it would be) worth making provision for industrial scale – or thereabouts – tests on the extraction columns and other essential parts of the installations: absorption columns, evaporators, liquid transport systems, and so on. The Eurochemic senior staff believe it is of the highest importance to train the personnel to use the facilities in the future operating conditions"[82].

Priority was therefore given to the preliminary design of a test station, a pilot unit consisting of the full-scale system used in an extraction cycle. A comprehensive design, prepared by the BER and costed by André Redon at $151 000[83], was put out to tender. Five firms responded and presented their bids at the fourteenth meeting of the Management Committee held on 20 January 1959. They were Belchim, Bofors, Comprimo, Lurgi and Saint-Gobain. Montecatini had declared an interest at a late stage but was unable to bid within the deadline. The Belchim bid was the lowest and provided for the shortest completion time: 123 200 EPU/UA in seven months. Saint-Gobain was the highest bidder at 219 000 EPU/UA over nine months. Lurgi, which like SGN had experience in building nuclear facilities, bid 212 000 EPU/UA over nine months. Belchim finally built the test station which came into service in December.

The location of the plant which, as we have seen, was strongly criticised by American specialists, meant that priority had to be given to the study of safety problems. The nature of the ground and the shallow water table raised fears that any accidental spillage of active solutions could have serious consequences. The painstaking research carried out showed "movements of the water table to be very slow, so that it would be possible, in the event of an accident, to take the necessary emergency measures (pumping, injection of cement, etc.)"[84]. The problems of potential gaseous discharges, especially iodine, required provision to be made for retaining certain gaseous discharges for a given length of time to allow that element to decay sufficiently. The preliminary investigation of the economical and practical possibilities of treating and storing liquid radioactive residues was contracted to the British firm W. J. Fraser and Co. Ltd., which had been suggested by the Risley Industrial Group, and to the French firm Potasse et engrais chimiques[85].

A preliminary analysis of the accident potential was undertaken, stressing that "special care must...be given to designing the ventilation system, as that would be the weak point if an accident did occur"[86].

During the interim period the Design and Research Bureau drew up three pre-projects, as instructed by the Board of Directors and the Management Committee. The first of these[87], which provided for a plant throughput of 100 t/year, was based on an evaluation of quantities and characteristics provided by a new fuel survey carried out in June 1958. This paid little attention to enriched fuel. It provided for four PUREX cycles with a single dissolver, and final treatment using silica gel columns for the uranium and ion exchange for plutonium. This was abandoned in the spring of 1959 because the original estimates were out of date and the cost too high.

---

82 Eurochemic/CA(58)4, p. 12. This is a good illustration of the limitations of the theoretical or numerical approach to predicting the performance of complicated technical systems. This phenomenon of "non-computability" is one of the main reasons why full-scale experiments are necessary and makes it essential, in order to train truly operational specialists, for them to acquire "tacit knowledge" in the field, through human contact and shared experience. For this aspect in the production of the American nuclear bomb, see MACKENZIE D., SPINARDI G. (1994).

83 Progress Report n° 2 p. 10.

84 Eurochemic/CA(19)13 p. 2.

85 This firm was also an associate of Saint-Gobain and the CEA in the SRU (Uranium Refining Company), in operating the Malvési uranium fabrication plant the construction of which had been decided in 1957. CEA(1957) p. 28 and SOUSSELIER Y. (1960), p. 400.

86 Eurochemic/CA(19)13 p. 2.

87 For the characteristics of the first three pre-projects, see RAE1 p. 235-238.

The second pre-project focused on a pilot plant with a capacity of 350 kilogrammes of natural uranium a day, using three cycles, the low activity wastes being treated off-site by the CEN. This was a concise pre-project intended to show the feasibility of an inexpensive plant.

The third pre-project went into more detail. It was drawn up between May and September 1959 and presented in the form of an ETR with diagrams. Based on a capacity of 350 kg/day two dissolvers were proposed, one for fuel enriched up to 5%, capable of treating four cladding materials – aluminium, magnesium, stainless steel and zircaloy. It included two extraction cycles.

This pre-project was also considered too expensive. Erik Haeffner was asked to produce a new version at lower cost. A dispute about the feasibility of doing so then arose between the Director of the BER and the members of the Technical Committee which had succeeded the Management Committee[88]. The project was subsequently simplified by Rudolf Rometsch and turned into pre-project n° 4, which provided the basis for the detailed pre-project that SGN was asked to produce in March 1960[89].

The question of priority to industry or research, problems of time and cost

As work proceeded, the emerging details of costs and the delay in ratifying the Convention generated a degree of tension and resulted in a search for modifications to the original project.

It soon became clear that the date originally planned for starting up the plant – 1961 – could not be adhered to and that the ceiling of $20 million was not high enough. The capital cost of $12 million appeared too low for a 100 tonne capacity plant and a research laboratory with 100 staff. The higher capital cost resulted essentially from the underestimate of the costs of treating and storing the liquid and gaseous wastes.

The additional $8 million set aside for running costs in the start-up period and the beginning of the industrial phase would have been sufficient if commercial reprocessing contracts had appeared on time. However national nuclear plant development programmes did not keep up with predictions and it looked as though the plant would be unable to operate at full stretch before 1966.

Yves Sousselier, Chairman of the Management Committee, therefore proposed at its seventeenth meeting in April 1959 that steps be taken to prevent escalating costs and loss of time[90]. If a new large plant was to be built from 1966 onwards, as mentioned in paragraph 3 of Article 3 of the Statutes[91], this project would depend upon information about the operation of the first Eurochemic plant at Mol a few years earlier, a plant which would then have to be commissioned in 1962 at the latest.

This first plant could be converted at a later date to process highly enriched fuels. It was therefore necessary "not to lose any time in obtaining in the laboratory the additional information needed for building the plant", and it was essential to rapidly adopt "a priori" processes, ensuring that the facilities would be easily adaptable to new conditions. Yves Sousselier therefore proposed that the project should be directed more towards a pilot or demonstration plant than to a production plant[92]: by reducing the plant capacity to 350 kilogrammes of uranium a day compared with 100 tonnes a year; by no longer duplicating systems; by designing superstructure to last only about ten years compared with the usual 25 to 30 years in the chemical industry; by putting priority on the research laboratories.

Not all the proposals of the Technical Committee's chairman were accepted: the Board of Directors asked that pre-project planning for both a large and a small laboratory should go on in parallel. Costs would be about $3 million for the former and about half that for the latter.

In fact the shareholders were divided as to the main purpose of the plant, and it became clear that the three objectives originally envisaged[93] for the company would be difficult to accomplish together within the planned

---

[88] This was the transitional phase between the structures of the interim period and those of the company.

[89] See below.

[90] *"Considérations sur l'usine Eurochemic de Mol"*, April 1959, in the Management Committee documents.

[91] See Annex.

[92] While a pilot unit is used to test the technical systems, a demonstration plant can also be used to test its economic operation. A production plant performs its technical functions under economic conditions.

[93] These objectives are listed in a preamble to the "Considerations" of Yves Sousselier:

- to enable the different participating countries to familiarise themselves with the techniques of reprocessing irradiated fuels;
- to prepare the way for building a large capacity plant capable of reprocessing in acceptable economic conditions;
- to make it possible for irradiated fuel from the initial reactors of participating countries to be reprocessed in a European plant.

budget. Thus discussions about the future of the project already began during the interim period. They were to continue and intensify during the construction phase.

From the formation of the company to the adoption of the detailed SGN pre-project: 28 July 1959-June 1961

Formation of the company and establishment of its decision-making bodies

The international Eurochemic company comes into being

On 10 July 1959, Pierre Huet, Director-General of ENEA, convened the constituent General Assembly of Eurochemic to meet at 10 a.m. on 28 July 1959 at the OEEC headquarters at the Château de la Muette in Paris. The Convention had been ratified by sufficient countries for this to be possible. Switzerland had been the first country to lodge its instrument of ratification with the OEEC on 21 January 1959, followed by France on 23 February, then by Denmark on 23 May. The Convention was ratified by Norway and Belgium on 25 and 29 June respectively; in June it was the turn of Austria on the first and the Netherlands on the ninth. They were followed after 10 July by Germany, Turkey and Portugal.

Sweden however made its accession conditional upon the conclusion of the agreement between the OEEC and EURATOM on security control. Italy had not yet approved the Convention despite introducing a simplified procedure.

By 28 July eleven countries, holding 82.4% of the shares, had ratified the Convention. The international company had come into being and, as provided for in the Convention, was set up for at least fifteen years, until 27 July 1974.

Sweden lodged its instruments of ratification on 5 January 1960; Italy was the final country to join the company on 25 January[94].

The advent of new private shareholders

In October 1958 the Netherlands and German governments had sold a small number of shares. The Dutch company Neratoom obtained one share; Germany sold 18 of the 68 shares it held to 16 firms and one more to a seventeenth firm in June 1961. The two major chemical firms Hoechst and Bayer Leverkusen had two each, and the fourteen other companies obtained one share each. The German government had had some difficulty in accomplishing this process. The German shareholders included eight energy producers, two metallurgical firms (Metallgesellschaft and Degussa[95]), the engineering firms AEG and Brown, Boveri Mannheim, the other two being groups specialising in the construction of nuclear establishments – Interatom and the AVR Construction Association. In June 1961, Vereinigte Elektrizitätswerke Westfalen AG (Dortmund) joined the private shareholders. This represented the peak of private holdings in the firm at about one-sixth of its total capital (Table 15a)[96].

The holdings in the hands of private firms, at less than 18% of the company's capital, was far from the majority private sector holding wanted by the company's founders.

The list of shareholders remained virtually unchanged until the middle of the 1970s (Table 15b).

Table 15a. **Private shareholders**

Country	Firms	Number of shares	Capital holding (%)
Belgium	Belchim	27	6.28
France	SGN	22	5.12
Italy	SORIN	5	1.16
Italy	SENN	2	0.46
Germany	Hoechst	2	0.46
Germany	Bayer	2	0.46
Germany	15 firms	15	3.49
Netherlands	Neratoom	1	0.23
	Total	76	17.67

94 This ratification by Italy was accompanied by an interpretative statement regarding the use of bills of exchange, owing to problems of compatibility with Italian legislation.

95 On 22 December 1965 Degussa transferred its share to its subsidiary NUKEM, which comprised Rheinisch Westfälische Elektrizitätswerke (Essen), Metallgesellschaft (Frankfurt) and Rio Tinto Zinc European Holdings Ltd (London).

96 No private firms contributed to the later increases in share capital.

Table 15b. **List of shareholders as of 31 December 1961**

Country	Type of shareholder	Name of shareholder	Number of shares 50 000 EMA/UA held
GERMANY	Government	Government of the FRG	49
	Company	Farbenfabriken Bayer AG, Leverkusen	2
	"	Farbwerke Hoechst AG, Frankfurt am Main	2
	"	Vereinigte Elektrizitätswerke Westfalen AG, Dortmund	1
	"	Brown, Boveri & Cie AG. Mannheim	1
	"	Metallgesellschaft AG, Frankfurt am Main	1
	"	Degussa, Frankfurt am Main	1
	"	InteratomGmbH, Bensberg bei Köln	1
	"	Allgemeine Elektrizitätsgesellschaft (AEG), Frankfurt am Main	1
	"	Kraftübertragungswerke Rheinfelden, Rheinfelden-Baden	1
	"	Preussische Elektrizitäts AG (Preussen Elektra), Hanover	1
	"	Hamburgische Elektrizitätswerke AG, Hamburg	1
	"	Bayernwerk AG, Munich	1
	"	Badenwerk AG, Karlsruhe	1
	"	Steinkohlen Elektrizität AG (STEAG), Essen	1
	"	Grosskraftwerk Mannheim AG, Mannheim Neckarau	1
	"	[97]*	1
	"	Rheinisch-Westfälisches Elektrizitätswerk, Essen	1
AUSTRIA	Government	Government of the Republic of Austria	20
BELGIUM	Government	Government of the Kingdom of Belgium	17
	Company	Société belge de chimie nucléaire (BELCHIM), Brussels	27
DENMARK	Government	Government of the Kingdom of Denmark	22
SPAIN	Nuclear agency	Junta de Energia Nuclear, Madrid	30
FRANCE	Nuclear agency	Commissariat à l'énergie atomique (CEA)	46
	Company	Compagnie Saint-Gobain, Neuilly-sur-Seine	22
ITALY	Nuclear agency	Comitato Nazionale per l'Energia Nucleare (CNEN)	37
	Company	Societa Elettronucleare Nazionale (SENN), Rome	2
	"	Societa Richerche Impianti Nucleari (SORIN), Milan	5
NORWAY	Government	Government of the Kingdom of Norway	20
NETHERLANDS	Government	Government of the Kingdom of the Netherlands	29
NETHERLANDS	Company	S.A. Neratoom, The Hague	1
PORTUGAL	Nuclear agency	Junta de Energia Nuclear, Lisbon	6
SWEDEN	Company	Aktiebolaget Atomenergi, Stockholm	32
SWITZERLAND	Government	Government of the Swiss Confederation	30
TURKEY	Government	Government of the Republic of Turkey	16

[97] Arbeitsgemeinschaft deutscher EVU zur Vorbereitung der Errichtung eines Leistungs-Versuchs-Reaktors e.V. (AVR), Düsseldorf.

Establishment of the decision-making bodies and the Technical Committee

The management structures of the company were officially established at the meeting of the Board of Directors held on 22 October 1959[98]. The senior posts were distributed by nationality according to a points system. The German, Erich Pohland, became Managing Director at the beginning of 1960; Rudolf Rometsch, who had been Director of the BER since the departure of Erik Haeffner on 1 October 1959, was appointed Research Director[99], Pierre Strohl became both Secretary-General of the company and temporary Head of the Administrative Division, pending the appointment of Yves Leclerq-Aubreton as Director of Administration. Pierre Huet was appointed the company's adviser for international and legal questions[100]. At the beginning of 1960, Yves Leclerq-Aubreton was appointed Director of Administration and Teun Barendregt Technical Director.

The general management was established in Brussels together with the Secretariat of the Board of Directors and the Legal Section. The offices were a short distance from the headquarters of EURATOM in rue Belliard. The Brussels location[101] was chosen because of problems of communication between Mol and the different capitals and the need for direct negotiation with the Belgian government authorities. The different divisions remained at Mol, on the CEN premises. In February 1961 the administrative division was able to move into its own building, the foundation stone of which had been laid on 7 July 1960 by Prince Albert on behalf of King Baudouin.

As provided for by the management rules, a Scientific and Technical Committee[102] was appointed: this committee took over from the interim Management Committee and was chaired by Yves Sousselier, appointed Company Adviser for Technical Matters, and initially had 7 members: Einar Saeland, Teun Barendregt[103], Erik Haeffner[104] and Maurice D'Hondt, soon joined by Adelbert Orlicek[105].

Relations between Eurochemic and the architect engineers; the 1959-1960 negotiations and the first mechanism for co-ordination

All items still in dispute at the end of the interim period were the subject of negotiations between Eurochemic and the architect engineers, which led to the approval of a first overall specification[106]. This was based upon the standards used at Electricité de France (EdF). Remuneration was based upon the principles applied amongst Swiss engineers and architects[107].

In October 1959 guidelines were drawn up to govern relations between Eurochemic and the architect engineers; between SGN – the principal architect engineer (PAE) for the plant – and the other architect engineers, known as specialist architect engineers (SAE); finally between the specialist architect engineers themselves[108].

As far as the plant and its auxiliaries were concerned, Eurochemic, as project manager, delegated technical co-ordination and the planning of design and construction work to the PAE. The SAEs were subdivided into

---

98 CA/III/4 of 9 December 1959.

99 And the scientific side of the BER became the nucleus of the Research Division, its administrative side becoming the Administrative Division.

100 He retained this post after leaving the ENEA at the end of January 1964, when Einar Saeland became Director-General of the ENEA and Jerry Weinstein Deputy Director-General.

101 The retirement of Erich Pohland at the end of 1963 and Pierre Strohl's return to the ENEA at the end of 1964 led to the Brussels head office being given up.

102 It subsequently became the Technical Committee.

103 He left the committee early in 1960 to become Technical Director of the company.

104 He was subsequently to leave the company and the nuclear field.

105 Other members joined as follows: Maurizio Zifferero, C. F. Jacobsen, Luis Gutierrez-Jodra, Sven G. Terjesen (at 31 December 1961), then K. Giese (1 July 1964), J. Hoekstra (Netherlands), H. von Gunten (Switzerland) and A. Larson (Sweden) (on 31 December 1966). A EURATOM representative attended meetings on a consultative basis (C. Ramadier and then A. Baruffa). This increase in membership, which reached 13 in 1966, did raise some problems of co-ordination; the reason was the desire on the part of the countries to influence technical decisions, since its proposals were usually endorsed by the legal decision-making bodies, and to convert political and financial decisions into technical decisions. The membership of the Technical Committee is given in Annex 5.

106 The General Conditions of Contract were not in the archives owing to the patchy character of documentation for the period 1959-1960. However Pierre Strohl, who was responsible for its production, retained a copy in his personal papers.

107 More precisely the "Regulations and Fees of the Swiss Society of Engineers and Architects" (SIA).

108 Unreferenced document of the Board of Directors, 16 October 1959, *Proforma Contracts with the Architect Engineers.*

two groups: those involved in the construction of the processing plant, and the others. As regards the former, the PAE had both a co-ordinating and a management role. If it disagreed with the SAEs, it could require them to adopt a certain approach which then relieved them of any liability. The PAE also checked the relevant drawings, by a sampling process, the work actually being supervised by the SAEs. From the latter's point of view, the function of the PAE was merely co-ordination and advice.

Regular working meetings between the PAE and the SAEs had to be organised to examine co-ordination problems and review progress. Progress meetings involving the AEs and Eurochemic were supposed to facilitate Eurochemic's supervision and decision-making. The AEs received all the technical documents issued.

For the research laboratory, the Spanish, Danish and Swiss companies had formed SDS, a joint company established in Basle. The PAE had the task of co-ordinating the construction of the laboratory with that of the plant.

However it very soon became clear that this complex procedure, whose main purpose was to enable all the architect engineers to gain experience and information about the entire establishment, was costly to an extent difficult to reconcile with the increasingly stringent funding problems. As a result the Technical Committee in December 1959 suggested a more straightforward procedure that would eliminate duplication of effort in the detailed design work[109]. In the same way the series of checks applying to the SAEs' subcontractors was shortened. The industrial partners certainly lost information as a result, but the aim was to reduce costs.

These proposals were discussed on 22 January 1960 by the architect engineers meeting at the headquarters of Saint-Gobain. They decided in favour of overall remuneration equivalent to a maximum of 10% of the cost of the installations. A draft "General Conditions of Contract" for the plant, prepared by Eurochemic, was discussed with the AEs on 28 July 1960[110]. This led to an overall agreement except for three important points: the liability of the architect engineers, industrial property and remuneration. Discussions therefore continued on these three points for a further six months, before agreement was reached[111].

From pre-project n° 4 to the detailed SGN pre-project: October 1959-June 1961

The fourth pre-project[112]

Pre-project n° 4 was published in February 1960, based on the work the Research Division had done since October 1959. It was supposed to be a simpler version of pre-project n° 3 – the basic chemical flow chart of which was retained although considered too costly – to allow construction to be completed on schedule. Without going into the detail, it essentially involved reducing volumes and areas by about 25%, eliminating and simplifying equipment and reducing the size of the research laboratory (Figures 29 to 32).

As far as waste treatment was concerned, pre-project n° 4 provided for intermediate storage to cover at least the first ten years of operation. The construction of solidification units was postponed, since the relevant processes were not yet ready and needing further research, and also because this would reduce the immediate cost of construction.

It was pre-project n° 4 that adopted the principle for the main building of arranging galleries and corridors on only one side of the block of cells, which made it possible to reduce the length of pipework and simplified the "four contamination zone" system[113]. Of course another result was that one side of the active cells was an outside wall.

Rudolf Rometsch saw his restructuring work as a political not technical task, bringing the project within an acceptable budget. The total cost of construction in pre-project n°4 was reduced to \$15.93 million, a reduction of \$8 million compared with the first version of pre-project n° 3 and \$1.7 million less than its second version. It was approved by the Board of Directors in February 1960.

The detailed SGN pre-project

In a letter dated 5 March 1960, Erich Pohland requested SGN to prepare a detailed pre-project based upon pre-project n° 4 and additional documents published in the ETR and EIR series[114]. At the same time Belchim was asked to draw up the detailed pre-project for the effluent treatment plant, and Noratom prepared a proposal for the fission product storage unit.

---

109 CA/III/5 of 14 December 1959.

110 CA/VI/2.

111 See Part II, Chapter 3.

112 CA/IV/11. It comprises 23 pages of text and 27 pages of diagrams.

113 See Part III, Chapter 1.

114 EIR 39, ETR 44, 53, 62, 77, 79, 80, 81, 84, 86, 87, 89, 92, 94, 98.

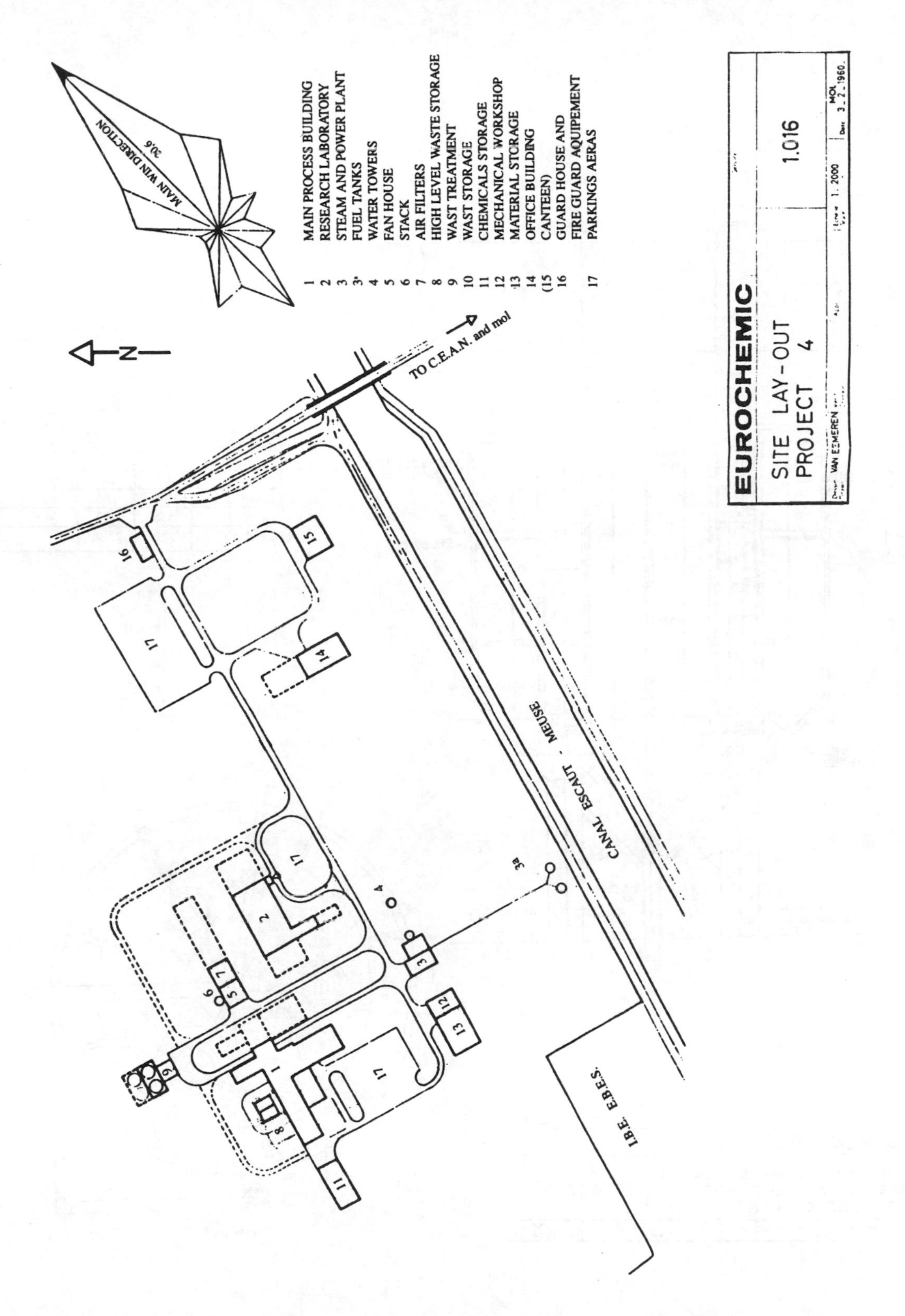

Figure 29. Planned overall layout of the company's buildings, from a figure in pre-project n° 4 dated 3 February 1960. The main facilities are shown but the form and arrangement of the buildings are not yet fixed (see Figure in Part III Chapter 1). The ventilation system is separate from the processing building, as at ICPP. Very little space is set aside for waste storage. The wind indicator at the top right of the drawing shows the prevailing south-west wind (Source: Pre-project n° 4, CA/IV/11, figures not numbered).

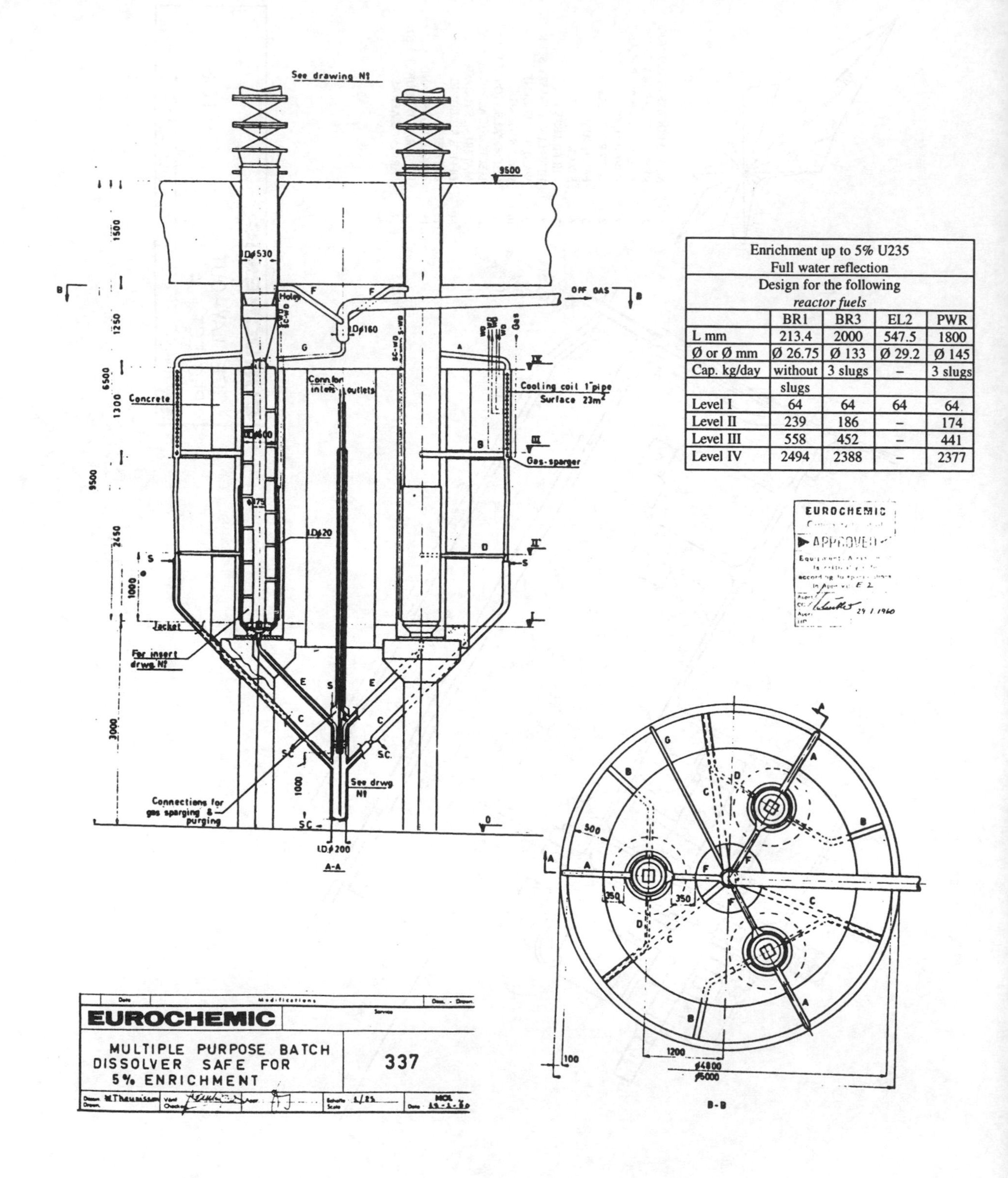

Enrichment up to 5% U235 Full water reflection Design for the following *reactor fuels*				
	BR1	BR3	EL2	PWR
L mm	213.4	2000	547.5	1800
Ø or Ø mm	Ø 26.75	Ø 133	Ø 29.2	Ø 145
Cap. kg/day	without slugs	3 slugs	–	3 slugs
Level I	64	64	64	64
Level II	239	186	–	174
Level III	558	452	–	441
Level IV	2494	2388	–	2377

Figure 30. Drawing of a multipurpose dissolver with three loading tubes capable of reprocessing all types of fuel up to 5% enrichment. To reach this level it was in fact necessary to build two separate dissolvers but the second one used the principle of three loading tubes (Source: Pre-project n° 4, CA/IV/11, figures not numbered, figure dated 29 January 1960).

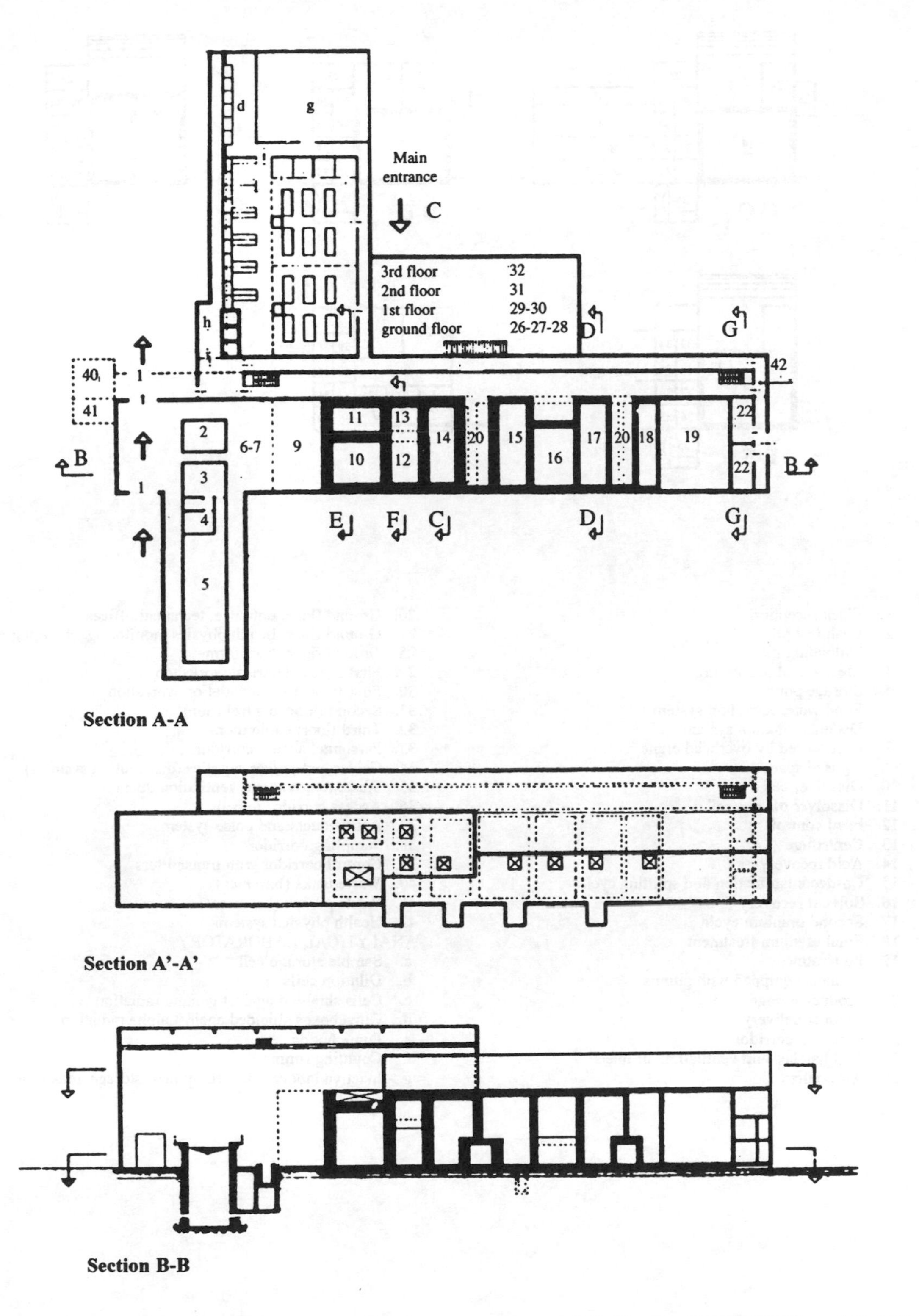

Figure 31. Layout of the main processing building, from pre-project n° 4 (I) (Source: RAE1 (1963), p. 252-253).

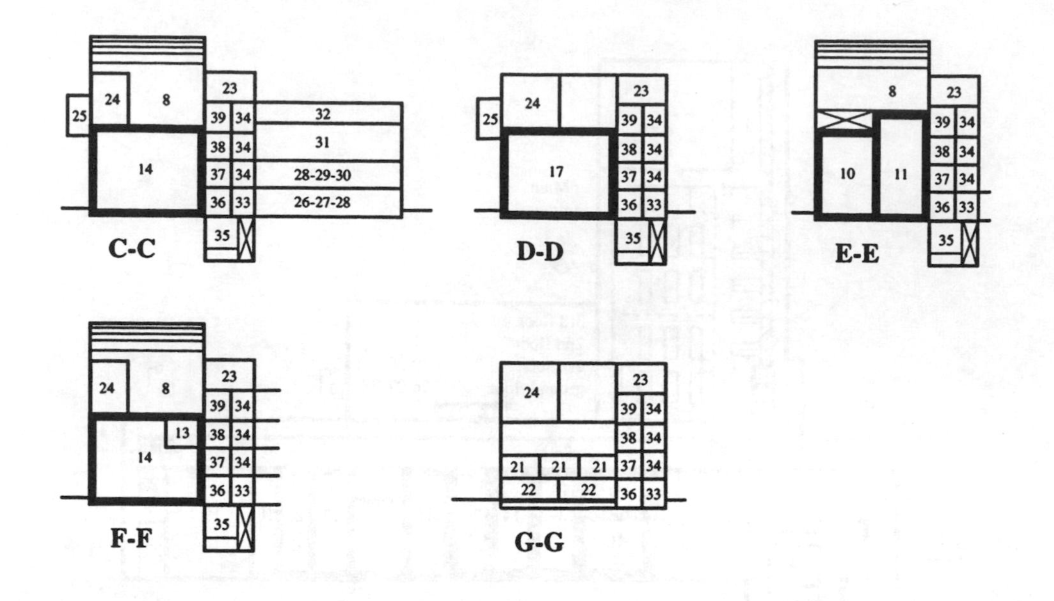

1. Flask reception
2. Leak test pit
3. Unloading pit
4. Mechanical processing
5. Storage pond
6. Pond water recycling system
7. Decontamination system
8. Hall served by overhead crane
9. Unused space
10. Dissolver cell
11. Dissolver off-gas cell
12. Feed control
13. Centrifuge
14. Acid recovery
15. Co-decontamination and splitting cycle
16. Solvent recovery
17. Second uranium cycle
18. Final uranium treatment
19. Pu treatment
20. Tunnels equipped with pumps
21. Product storage
22. Product delivery
23. Make-up corridor
24. Cell hatches and ventilation air inlets
25. Air coolers
26. Ground floor: entrance, technical offices
27. Ground floor: health physics monitoring, changing rooms
28. Ground floor: transformers
29. First floor: electrical workshop
30. First floor: instrumentation workshop
31. Second floor: control room
32. Third floor: cable room
33. Personnel access corridor
34. Cables and power supplies (distribution systems)
35. Waste removal and ventilation ducts
36. Access corridor to cells
37. Active ducts and pulse systems
38. Sampling corridor
39. Control corridor with transmitters
40. Waste tanks (basement)
41. Waste pumps (basement)
42. Health physics systems

ANALYTICAL LABORATORY

a. Sample storage cell
b. Dilution cells
c. Cells shielded against gamma radiation
d. Gloveboxes shielded against alpha radiation
e. Fume hoods
f. Counting rooms
g. Inactive laboratories, equipment storage, make-up

Figure 32. Layout of the main processing building according to pre-project n° 4 (II) (Source: RAE1 (1963), p. 252-253).

The SGN specification was as follows:

- capacity of 350 kilogrammes of uranium a day for fuels clad in aluminium or magnesium; 200 kilogrammes a day for fuels clad in stainless steel or zirconium; fuel should have a maximum enrichment of 5% and a maximum burn-up of 10 000 MWd/t. Cladding materials could be aluminium, magnesium, magnox, zirconium, zircaloy or stainless steel; fuel could be uranium metal, dioxide (UO_2) or a U-Mo alloy with up to 10% Mo; it had to be possible for the plant to be extended at a later date;
- maximum fuel element or bundle dimensions limited to a length of 4.2 metres and a diameter of 20 centimetres;
- basic chemical process to be of the 2-cycle PUREX type, with TBP solvent in a 30% solution in a hydrocarbon diluant;
- extraction systems to be essentially pulse columns.

Preliminary SGN-Eurochemic discussions on the detailed pre-project began on 2 May 1960. Agreement was soon reached on the extraction method and that of final uranium purification, but views differed about the fuel reception facility and the main processing building layout.

By the end of 1960, further work on the detailed pre-project suggested an estimated additional cost of $10 million over that of pre-project n° 4, clearly demonstrating its essentially political character. In fact a total of $26 million would have been required. It is true that this additional cost was partly linked to new architectural and technical choices. Thus SGN initially proposed that each important part of the system should be located in a separate cell, but this option, although considerably facilitating maintenance, would have substantially increased the volume[115] of the cells – from the 6300 m^3 of pre-project n° 4 to 15 000 m^3 – and hence their cost. The American consultant Robert J. Sloat suggested using the approach of the large United States plants with the processing systems installed in four large cells served by an overhead crane. The final choice was a compromise. Each major processing stage was to take place in a specialised cell which might contain several systems. However the final compromise reduced the estimated cost only by about $2 million.

The detailed SGN pre-project was completed in May 1961 and approved by the Board of Directors in June. It filled six large books, which formed the initial working tool for the architect engineer, the Technical Division and the contractors involved in building the plant, a process which then began. The expected capital cost of $24.01 million, together with a little over $6 million for running costs, largely exceeded the initial capital. It was therefore essential at a very early stage to begin looking for additional resources, just as foreseeable developments in the type of fuel raised the problem of further adaptation and hence of additional costs.

As the detailed pre-project was submitted, the company was facing a serious crisis concerning both its utility and its funding, and which was nearly fatal to European co-operation.

[115] RAE1, p. 248.

Chapter 2

The crisis of the early 1960s

What sort of plant, to do what, to be funded how?

As soon as the detailed pre-project had been adopted, the Eurochemic project was confronted by two considerable problems which threatened its very existence, particularly between the autumn of 1963 and the early summer of 1964. Although the company overcame these obstacles, it emerged from this difficult period somewhat transformed. The problems were of two kinds: technical and financial. Eurochemic owed its survival entirely to the fact that the senior staff of ENEA (backed by the OECD), representatives of certain countries and the company's own management team swung in favour of the project.

The technical problem stemmed from a disparity between the types of fuel selected for reprocessing by Eurochemic – as set out in the detailed pre-project – and member countries' reactor developments and changing needs for reprocessing. Nuclear power programmes were developing more slowly than expected, and increasing numbers of reprocessing plants were planned or actually being constructed in Europe and the United States.

The second, more substantial problem, which stemmed partly from the first one, concerned the project's growing deficit. The financial crisis, starting when the initial pre-projects were still at the design stage, had become obvious during the transition from pre-project n° 4 to the detailed SGN pre-project, dramatic when full-scale work began and was at its worst during the winter of 1963-1964. The financial problems profoundly strained the bonds between the signatory countries of the Convention which, by the time Eurochemic emerged from its troubles, had changed the character of the company.

Extension to encompass highly enriched uranium fuels; plans for a new plant adjoining Eurochemic for reprocessing 5% enriched fuels (1961-1963)

SGN had worked on adapting the PUREX process to Eurochemic so as to reprocess fuels clad with a variety of materials and made up of natural uranium, or uranium enriched up to 5% in the 235 isotope. In no case was the burn-up was not to exceed 10 000 MWd/t.

However, the fuel being planned for the new power plants, except in the United Kingdom and France, was no longer natural uranium, but uranium enriched to a little under 5% and utilised in the form of oxide, a fact liable to make some of the equipment obsolete fairly quickly. Also, the intention was that these low enriched fuels would be taken to high burn-up values, which was liable to complicate safety in the plant[1]. Finally Eurochemic was under considerable pressure to process fuels of even higher enrichment, for example from naval reactors, materials test reactors and the future fast breeders. This would have meant taking a fresh look at the planned multi-purpose nature of the plant – which referred only to the cladding materials – and modifying the installations, at a time when a financial problem was already emerging. However the first issue to be raised concerned fuels enriched to 20 and 90%.

Fuels from light water reactors and MTRs

In June 1961, when the detailed pre-project had just been adopted, it became clear that fuel enriched to over 5% would play a significant role in European nuclear programmes. The problem had already been raised in the Technical Committee during the interim period; however it was the EURATOM representative, E. R. von Geldern, who re-opened discussion of this topic in the Board of Directors on 13 June 1961[2]. At this time in fact, EURATOM was becoming involved in materials testing for BR2 and the Petten high flux reactor, HFR[3]. The Commission regretted that Eurochemic was not proposing to reprocess 20% enriched fuels, which

---

1. A high burn-up means more active fission products and necessitates more shielding in the plant.
2. CA/M(61)2 and a letter from E. R. von Geldern dated 20 June 1961 reproducing his statement.
3. For the EAEC projects, see the annual reports of the EAEC and NAU H.R. (1974).

were expected to develop substantially. EURATOM also deplored the fact that MTR fuel reprocessing was possible only in the United States at ICPP since 1953 or at Dounreay in the United Kingdom since 1958.

EURATOM therefore came out in favour of adding to the SGN project a unit for reprocessing highly enriched fuels or, as it was put at the time, of adopting "a new project for a limited capacity plant to reprocess only highly enriched fuel". There was a suggestion that in this case financial assistance from EURATOM might be possible.

France was also in favour of a project which would have made it possible to reprocess spent fuel from its nuclear submarine reactors the land-based prototype of which, known as PAT, was under development[4].

Between June and September 1961 the Technical Committee considered the possibility of reprocessing 20 and 90% HEU; it proposed[5] that consideration should not be given to extracting the plutonium from the 20% enriched fuel[6] and that only the uranium should be recovered. With this in view it recommended that the Board of Directors should take steps to allow the installations to be suitably modified at a later stage, at an additional estimated cost of $0.2 million. The American expert seconded to Eurochemic, Robert J. Sloat, was asked how HEU could be reprocessed in existing installations, and SGN was requested to produce a report on the reprocessing of irradiated HEU fuels from MTRs. The report was submitted in June 1962.

The reprocessing of 20% enriched fuels required an agreement with the Americans owing to the United States technological monopoly of civilian enrichment, and the American legislation governing the export of fissile materials.

On 5 December 1961 the American representative at the OECD informed the Steering Committee that Eurochemic might be able to reprocess spent fuel by virtue of an amendment to the 1959 EURATOM-United States Agreement. A joint ENEA-Eurochemic mission was accordingly sent to the United States in January 1962. It comprised the Deputy Director-General of ENEA, Einar Saeland, Jerry Weinstein, Teun Barendregt and Robert J. Sloat. Initially ENEA wanted a bilateral agreement between the international organisation and the United States[7]. However the USAEC Division of International Affairs raised problems of procedure, notably the need to involve a joint committee of Congress to have such an agreement adopted. Myron B. Kratzer explained in particular that it would be difficult to convince the Congress of the need to recognise the large number of European institutes involved in atomic co-operation. Accordingly he recommended a more flexible solution involving the agreements that already existed with EURATOM. The outcome of the mission was therefore an amendment to the EURATOM-United States Agreement signed on 21 and 22 May 1962[8]. The fuel was transferred legally to EURATOM through the procurement agency of the EAEC.

In this way the plant could widen its application to highly enriched fuel but modifications would be required.

Plans for reprocessing fast breeder fuels

The project for reprocessing new types of fuel was to take on a third dimension by being linked with the French[9] and German fast breeder reactors.

Accordingly contacts were made during the summer and autumn of 1962 between Eurochemic experts, the CEA – which had carried out the preliminary study[10] – and the German authorities, to consider how to graft onto the Eurochemic plant – or to build alongside – a unit for reprocessing fast breeder fuel. It is true that fast breeder fuels are not enriched in uranium 235 but contain about 20% of plutonium[11]. However it appeared reasonable to assume that HEU fuels containing 20% of uranium 235 and MOX containing 20% of Pu could be processed in the same installation owing to the similar criticality constraints, which were invariably very different from those applying to enrichments of 5% or less.

The Technical Committee submitted an appropriate report to the Board of Directors on 31 October 1962 which envisaged either modifications to the planned facilities or the construction of a new plant beside the first

---

4 Using enriched uranium supplied by the United States, pending domestic production at Pierrelatte.

5 CA(61)38.

6 In the case of fuels highly enriched in the uranium-235 isotope, recovering the remaining enriched uranium is much more attractive than recovering the plutonium, the formation of which depends on the uranium-238 content, which is of course low in a highly enriched fuel.

7 On this point and what follows, communication from Pierre Huet.

8 CA(63)13.

9 For the development of the French project, see FINON D. (1989). This was the era of RAPSODIE.

10 CA(62)15.

11 This mixture of uranium oxide and plutonium oxide is known as a mixed oxide fuel or MOX.

one. Certain services would be pooled. Five different scenarios were examined, the estimated additional cost varying from $3 to 8 million depending on the approach considered.

These options were discussed at the fourteenth meeting of the Board of Directors on 31 October 1962. The Board decided to set up a technical working group to examine the possibilities in more detail and proposed to make exploratory contacts with the Belgian authorities concerning the problems of health and safety that would result from an extension on the site. France and Germany were avid supporters of such enlargement, particularly for the reprocessing of fast reactor fuels. Spain, Switzerland, Portugal and Denmark showed no interest at the time. Italy declared that it preferred a second plant, separate from Eurochemic, because "too integrated a solution would be likely to draw the uninterested countries into building and operating the new project". In fact this proposed extension was in direct competition with a national project – EUREX – which was then being considered for joint funding with EURATOM[12] which Italy was proposing to carry out in parallel with the Eurochemic project. From Italy's point of view, it was essential that the two reprocessing projects – LEU at Mol and HEU at EUREX – should be completely separate. Thus any extension to the polyvalence of Eurochemic went against Italy's interests and hence contributed to its attitude to the financial problems encountered by Eurochemic[13].

The working group met on 16 November 1962 and submitted its report to the Board of Directors on 18 December 1962. It came out in favour of building two plants that would be entirely separate apart from final plutonium purification[14]. The Director of Research spoke of the concerns of the Belgian health authorities, but no firm answer was vouchsafed[15].

In fact the future of the new plant to be attached to the first one depended on funding from EURATOM, whose representative had pointed out the urgency of reprocessing MTR fuel and stressed that "the various options available for this process were being investigated". EURATOM had in fact made contact with the British and the Americans.

EURATOM hesitations, abandonment of plans for the second plant and the 1963 realignment

On 25 January 1963, in a letter to E. R. von Geldern, Pierre Huet defended the Mol extension project and asked EURATOM to state what financial support the Commission would be able to provide[16].

However the EAEC was slow to take a position, probably because it was being asked to choose between Eurochemic and EUREX and, when it did adopt a policy for reprocessing its irradiated fuel at the end of 1964, it "remained on the fence". Eurochemic would reprocess fuel from BR2 and HFR, but EUREX would be supported both by EURATOM contributions to its research programme, at the rate of 3 million UA for three years, and by the conclusion of reprocessing contracts at prices fixed in advance[17].

In the meantime, the project for building a new plant alongside the old one had been buried at the nineteenth meeting of the Board of Directors held on 2 October 1963[18]. The Board of Directors fell back on a solution submitted in August 1963 in an internal technical report[19]. This involved adapting the reprocessing installations to HEU fuels, seizing the opportunity that the final plutonium purification unit was still being built. This resulted in particular in consideration being given to building a third dissolver. The cost of these

---

[12] The Italian project had been revealed, to the great surprise of the other members of Eurochemic's Board of Directors, on 12 October 1961. See CA/M(61)3.

[13] See below.

[14] CA(62)30.

[15] The Italian delegate, in his "Comments on the Minutes of the Fifteenth Meeting of the Board of Directors" on 18 December 1962, gave a more negative version of the minutes written by R. Rometsch and stressed the weak support given to the Belgian solution: in his view only the Belgians, the Swedes and the private German shareholders had stated that they were in favour of Mol. At its eighteenth meeting the Board of Directors decided not to adopt his comments, since they referred to the attitudes of other directors, when it was up to them to make comments on their own remarks. A copy annotated by Pierre Strohl throws doubt on the accuracy of the transcription made by the Italian representative. In any event, Italy used all possible means to counter the plans for extensions at Mol.

[16] On 30 April 1963, a letter signed jointly by Erich Pohland and Teun Barendregt, who had learned that a decision was imminent, gave information that was likely to "torpedo" EUREX, stressing that in the view of Floyd L. Culler the cost of the Italian plant was apparently underestimated by a factor of three, that the process was apparently not developed and that the cost of reprocessing would be appreciably higher than that applied in the United States and the United Kingdom, as well as at Eurochemic. Unreferenced document, 1963 CA documentation.

[17] Annual report of the EAEC (1964-1965), volume of documentation, p. 59.

[18] CA/M(63)4 and CA(63)35.

[19] EIR 96.

adaptations – $500 000 – appeared minimal by comparison with the other options, and in any event $200 000 had been committed to modifications already decided upon.

However the financial crisis prevented the decision from being turned into action until June 1964[20], when the Board of Directors decided to adapt and add to the plant systems in order to process HEU so that the plant would be ready, as requested by EURATOM and Belgium, to reprocess fuel from BR2. It was planned to fund this work using what remained of a £2 million loan obtained in 1962 from the Belgian Caisse d'épargne (savings bank) for building 87 dwellings. Contacts were also made with respect to HFR.

This extension of Eurochemic's facilities to accommodate HEU fuel provoked the strongest protests from Italy, which regretted "a decision running directly against the interests of the Italian EUREX project", and increasing costs still further. Italy then threatened to review its undertaking to contribute to research costs and to covering the operating deficit, at a time when the company had just emerged from the most serious financial crisis of its existence.

The financial crisis and its resolution

The fact that the starting capital was used up well before the work was completed meant that new sources of funding had to be found against the background of accelerating inflation in prices and wages at the beginning of the 1960s. Estimates became more precise but the cost of the project also went up. Table 16a shows the evolution of the various project estimates, which are worth comparing with the final actual cost.

Although it was the beginning of 1960 before the first million dollars had been spent, the commencement of site works in July of the same year produced a sharp increase in expenditure. The initial capital of $21.5 million was used up by May 1963. A first increase in capital of $7.45 million already appeared necessary when the detailed SGN pre-project was approved in June 1961 but it was not provided until 18 June 1963. A second increase of $6.8 million was negotiated under very difficult conditions and raised the company's capital to $35.75 million on 1 July 1964. The sequence of capital payments ended in March 1966 (Table 16b).

As can be seen, the total capital payments made were $35.75 million while the total expenditure was $46.73 million. Securing the difference was the price paid to transform the company and complete the project.

Financial problems and fiscal privileges (May 1960-April 1962)

In May 1960[21], as a result of the overall estimate given in pre-project n° 4, which showed that the capital would be slightly insufficient for construction and the first few years of operation, a search was made for ways of limiting expenditure. With this in view the Special Group recommended that Belgium should exempt the company from indirect taxation for a period of 5 years, absolve it from customs duties on contractors' materials and equipment, and compensate for taxation on staff salaries through a special contribution to the company or by acquiring additional shares[22]. In fact the Special Group estimated that taxation on the company during the construction period amounted to $1.2 million, despite the exemptions provided for in the Statutes[23].

Eurochemic had originally agreed to pay these taxes to avoid interfering with the industrial competition which, in 1957, was expected to appear shortly. However by 1960 the outlook was different. The expansive industrial ambitions of the 1950s had disappeared or had been postponed. In fact the Special Group considered that Eurochemic now had a new vocation. The company now appeared "as a joint experimental centre or a European general purpose laboratory". Against this background "the idea of normal competition on which the 1957 Statutes were based appeared, at least in the immediate future, to be unfounded".

The Director-General of ENEA and the Belgian government therefore began negotiations on these fiscal issues. In April 1962, the Special Group recommended that Belgium – which supported the majority of its partners – should agree that the equipment used by the company should be imported free of duty, so long as it was used only by Eurochemic and that it was re-exported immediately upon completion of the work. This decision was applied with retroactive effect[24]. However the submission of the detailed SGN pre-project brought the financial problems into the light of day.

---

20 CA(64)22 and CA/M(64)3.

21 NE/EUR(60)3 of 20 May 1960; NE/EUR/M(60)2 of 29 June 1960.

22 Although the company benefited from fiscal privileges, it was required to pay indirect taxation and the staff had to pay tax on their income.

23 See Article 7.

24 NE/EUR/M(62)1 of 19 June 1962.

Table 16a. **"Financial drift" in the Eurochemic project 1958-1968, in million EPU/UA**

Date of forecast	Reference document	Capital investment	Operating costs	Total	Expected plant completion date	Planned plant capacity
01.1956	Nicolaïdis Report	40	n.a. (not applicable)	n.a.	1960	500 t/year
02.1957	Rometsch estimate[25]	18.5	n.a.	n.a.		50 t/year (500 staff)
03.1957	D'Hondt counter-proposal[26]	10.373				
09.1957	Steering Committee for Nuclear Energy (1958)	12	7 (+ 1)[27]	20	1961	100 t/year
09.1959	Pre-project n° 3, scenario 1			23.7		100 t/year
09.1959	Pre-project n° 3, scenario 2			17.7		350 kg/day (60-70 t/year)
02.1960	Pre-project n° 4	15.93	5.84	21.77	End 1963	350 kg/day (60-70 t/year)
03.1961	Detailed SGN pre-project[28]	24.664	6.036	30.7	End 1963	40-70 t/year
09.1963	CA(63)28	27.2	11.5[29]	38.7	End 1965	"
05.1968	BARENDREGT (1968)	29.6[30]	3.15 to 3.5[31]	n.a.	July 1966	"
Actual sums	Capital sums released from 1958 to 1966	n.a.	n.a.	46.73[32]	n.a.	n.a.

Table 16b. **Chronology of capital contributions to Eurochemic**

Date of release	Amount provided ($ million)	Proportion of initial capital (I), of first increase (II), of second increase (III)	Cumulative total ($ million)
07.1959	4.30	20% (I)	4.30
06.1960	6.45	30% (I)	10.75
06.1962	6.45	30% (I)	17.20
05.1963	4.30	20% (I)	21.50
06.1963	1.49	20% (II)	22.99
02.1964	3.73	50% (II)	26.72
03.06.1964	1.36	20% (III)	28.08
12.1964	2.23	30% (II)	30.31
23.06.1965	3.40	50% (III)	33.71
03.1966	2.04	30% (III)	35.75

25 RAE 1, p. 255. By shrinking pre-project n° 3, capital costs of $21 million.

26 Uses the Rometsch estimate with waste processing put out to the CEN.

27 Cumulative expenditure up to completion plus the first two years of operation. Plus $1 million reserve fund.

28 CA(61)4 and CA(61)14, the pre-project itself amounts to $24.01 million.

29 Excluding the running costs of the laboratory 1964-1965 ($2.9 million).

30 Real cost.

31 Annual estimate in NE/EUR(66)3.

32 Total of OECD contributions, calls for capital payments and loans from 1958 to 1966. See details below.

The first increase in capital (June 1961-June 1963)

The financial impact of the detailed pre-project

The company's initial capital which, with the accession of Spain, amounted to $21.5 million, was released in four tranches: 20% when the company was established, 30% in June 1960, 30% in June 1962 and 20% in May 1963. The adoption of the detailed pre-project for the plant drawn up by SGN and approved on 13 June 1961 called for the sum of $30.7 million up to the end of 1963. The surplus compared with pre-project n° 4 of the Design and Research Bureau was due "to modifications to the plant (+ $2.30 million) and the laboratories (+ $0.22 million) made necessary in the light of further research, the inclusion of new systems such as a second dissolver[33] (+ $1.45 million), and finally to the adoption of a contingency reserve much higher than in the previous estimate (+ $3.01 million)"[34].

In August 1961 Pierre Strohl send Pierre Huet a draft note on the additional funding of Eurochemic[35]. The mechanism was based on the issue of 184 new shares (amounting to $9.2 million). It was proposed that these should be divided between the members of the company on the basis, first, of the original distribution of the company's capital and, secondly, of national income. There was a provision for reducing the investment of Turkey and Portugal in view of their particular situation. Also Pierre Strohl had made sure of the support of Belgium: the country was in fact prepared to make "an additional capital contribution of about $2 million, on condition that the company's operating deficit subsequent to the period of construction, i.e., after 1 January 1964, be covered on the basis of national income, according to the usual practice of the international organisations". The aim of this condition was to decouple the covering of the deficit from contributions to capital increases.

The proposed breakdown was as shown in Table 17. This should be compared with the outcome of the first increase on 18 June 1963.

Three countries in fact rejected the proposed increase in capital, and Belgium purchased ten less shares than expected.

Table 17. **The first capital increase as proposed and adopted**

Country	August 1961 proposition	18 June 1963 outcome	Difference
Belgium	46	36	-10
Germany	32	32	0
France	32	32	0
Italy	20	0	-20
Sweden	10	10	0
Spain	10	8	-2
Netherlands	9	9	0
Switzerland	8	8	0
Austria	5	5	0
Denmark	5	5	0
Norway	4	4	0
Turkey	2	0	-2
Portugal	1	0	-1
TOTAL	184	149	-35

First breach in shareholder solidarity: the refusal of Italy, Portugal and Turkey

On 12 October 1961 Italy declared, through a statement by Achille Albonetti attached to the minutes of the tenth meeting of the Board of Directors, that she refused to contribute to any increase in capital[36]. In fact Italy had decided to minimise its contribution to the company. This position was defended firmly and constantly by

[33] Pre-project n° 4 provided for only one multipurpose dissolver, but later research showed it to be unsuitable for enrichment rates of 5%; it was decided to use two, the first one up to 1.6%, the second for higher rates. Later on, the second dissolver was regarded as safe only up to 4.6%; the decision to modify the plant for reprocessing HEU led to a third dissolver being built which was safe up to 93%.

[34] NE/EUR(61)1, p. 2.

[35] Draft note from the Director-General of ENEA to the Special Group on supplementary funding, 1 August 1961. Special Group documentation.

[36] CA/M(61)3. This point of view was developed in a letter of 4 December 1961 sent by Mr. Colombo, President of the CNEN and Ministry for Industry, to Pierre Huet. This letter is referred to in the archives but is missing, see NE/EUR/M(64)1 p. 9.

the Italian representative, which did not fail to cause considerable tension within the Special Group. In October 1961 the reasons given were financial: "the [Italian] committee does not in fact at present have any additional financial and manpower resources which could be placed at the disposal of Eurochemic". More precisely the CNEN considered that "if some directors desire to convert a pilot plant into an industrial plant, in the belief that this will meet their requirements, the CNEN does not for its part see any possibility of doing so and considers that such an aim changes the whole object of the company by entailing additional expenditure which is not not provided for in the Statute of the company and which moreover is not necessary to attain the research objective". The CNEN recommended that the original financial commitment should be strictly adhered to.

As from October 1961 Italy interpreted the Convention and the Treaty in a restrictive manner[37] and refused to consider Eurochemic as anything other than a research undertaking[38]. Subsequently this interpretation was linked to the demand that the budget should be balanced.

The position of Turkey was made clear on the same day in a statement by Mr. Erikan. This blamed the country's economic difficulties for its refusal to acquire further shares, and the principle of covering the operating deficit on the basis of national income, preferring an approach "on the basis of the technical benefits and interest to be derived by the shareholder". Application of this mechanism would have meant the country's being exonerated.

Portugal also refused further contributions and suggested that the deficit should be shared between countries in proportion to the quantities of fuel reprocessed[39].

Spain reduced its contribution to eight shares, which it believed corresponded more closely with its capacities.

These refusals and limitations led to the Belgian decision to reduce its own contribution.

The problem of covering the expected final deficit

At its fourth meeting on 13 April 1962[40] the Special Group proposed that "the operating costs of the company from 1964 onwards, so far as they were not balanced by receipts, should be covered by contributions from the governments of participating countries calculated on the basis of national income, and more specifically according to the scale of contributions to the OECD budget adjusted to the thirteen countries".

Participants hesitated to enter into commitments for the period remaining up to the expiry of the Convention, which had originally been signed for 15 years, and requested that a restricted committee be given the task of making precise forecasts.

The June 1963 increase in capital

The first increase in equity therefore amounted to 7.45 million EMA/UA and took the form of the issue of 149 new shares instead of the 184 originally planned. However this was not enough to allow site works to be completed. By November 1963 the Board of Directors were worried that funds would run out by the end of April 1964[41]. Predicted expenditure – capital and running costs – for the years 1964-1967 amounted to $21.4 million[42]. Moreover the General Assembly noted on 18 June that "the decision taken with regard to the

---

37 This attitude is illustrated in the quotation – abbreviated since it considers only the first paragraph – of Article 3 of the Statute with which Achille Albonetti began his statement:

"Article 3 of the Statute specifies that the objective is to build before 1961 and operate a plant and a laboratory for the processing of irradiated fuels 'for the development of techniques and the training of specialists in this field'. The CNEN believes that with the capital available to the company it is possible to build a plant capable of meeting the company's objectives".

It is worth recalling that the second paragraph of the article begins: "The company will carry out any research *or industrial* activity...", and that this expression is repeated in the fourth and final consideration of the Convention.

The planned nationalisation of the Italian electricity supply industry which led to its takeover by ENEL in 1963 was perhaps one reason for the CNEN position, the organisation being very concerned about its future. However the EUREX project was probably predominant.

38 A new statement was attached for example to the minutes of the fifth meeting of the Special Group NE/EUR/M(63)1. In June 1963 Italy threatened – since an amendment she had proposed to Article 4 ter. of the Statute had not been discussed – to apply the procedure of invoking the Tribunal provided for in Article 16 of the Convention.

39 Portugal had no fuel to be reprocessed.

40 NE/EUR/M(62)1.

41 NE/EUR(63)11.

42 CA(64)3.

increase in capital shall not imply any commitment on the part of shareholders with respect to the scale according to which any possible operating deficits of the company might be covered". At the time participants' attitudes to this problem were incompatible and may be summarised in Table 18[43]:

Table 18. **Position of shareholders regarding deficit coverage**

Basis for covering deficits	Countries adopting this formula
National income	Austria, Denmark, Norway, Spain, Sweden, Switzerland
Capital share	"Other shareholders"
Commercial management only (price of reprocessing, increases in capital, exceptional funding)	Italy
Use of plant by shareholders	Portugal

The Special Group held more meetings because it was essential to find common ground. It met on 27 June 1963, 26 November, 24 February 1964, 17 April and 1 July. In the meantime a number of diplomatic contacts were made. The winter of 1963-1964 was a period of considerable tension and the Board of Directors even considered the option of simply abandoning the project.

The financial crisis and its solution; second increase in capital and OECD contributions (July 1963-July 1964)

The Special Group's restricted committee and the November 1963 proposal

At its fifth meeting on 10 May 1963, the Special Group had decided to set up a restricted committee made up of representatives of four or five countries and of the company with the task of "estimating what the company's expenditure and income would be over the coming four years, to consider ways and means of reducing or eliminating any deficit" in order to "make proposals to the Special Group as to how they may be covered". The report was to be submitted before the General Assembly to be held on 18 June 1963 which was to endorse the first increase in capital.

In fact the committee was not set up until after that date, holding its first meeting on 19 July 1963. It consisted of representatives of Denmark, Germany, Austria, Belgium and Italy; representatives of EURATOM and Eurochemic also took part[44]. It held four meetings and its report was examined at the seventh meeting of the Special Group on 26 November 1963[45]. The report considered the company's financial situation, listing the expenditure up to 1967 and how it might be reduced. The report proposed a further increase in capital to cover capital and running costs[46] estimated at $9.3 million, to allow construction to be completed. The Board of Directors added a provision of $1.5 million, making a total of $10.8 million. The Director-General of ENEA therefore proposed[47] that 216 new shares should be issued, divided as shown in Table 19. Allowance was made for the "special situations" of Turkey and Portugal, who had made known their decision not to subscribe to this new increase in equity.

The gap between the proposed amounts and actual contributions was considerable and further solutions had to be found.

It was also proposed that the costs of the research programme and the net running costs for the years 1966 and 1967 – amounting to $10.6 million – should be funded by contributions from participating governments on the basis of the OECD scales, the amounts to be decided by the OECD Council in agreement with member countries and Eurochemic. The annual budgets of the company were therefore incorporated – like that of ENEA – in that of the OECD.

---

43 Source: NE/EUR/M(63)2, p.7.

44 H. von Bülow (Denmark – Chairman), B. Steinwender (Austria), R. Depovere (Belgium), C. Zelle (Germany), N. Catalano, then M. Zifferrero (Italy); C. Ramadier for EURATOM, T. Barendregt, Y. Leclercq-Aubreton, R. Rometsch, P. Strohl and Y. Sousselier for Eurochemic.

45 NE/EUR(63)10.

46 The delegate of Italy, in his statement of 26 November 1963, expressed a somewhat malicious pleasure that consideration was being given to funding running costs by increasing capital, an approach he had proposed in 1961 when he had received the reply that to do so would not be in line with sound commercial management.

47 NE/EUR(63)12 of 19 November 1963 "Funding the Company's Expenditure (Note by the Director-General)".

Table 19. **Second capital increase as proposed and adopted**

Country	November 1963 proposal	Position at 1 July 1964	Difference
Germany	40	28.5	-11.5
France	40	28.5	-11.5
Italy	22	0	-22
Belgium	22	22	0
Sweden	17	12	-5
Netherlands	15	10.5	-4.5
Switzerland	15	9	-6
Spain	15	8	-7
Denmark	11	5.5	-5.5
Austria	10	5	-5
Norway	9	4.5	-4.5
Portugal	0	0	0
Turkey	0	0	0
TOTAL	216	136	-80

The logjam: November 1963 - April 1964

An unfavourable international situation

In 1963-64 there was substantial uncertainty about the future of a project which appeared increasingly costly, just when the marketing prospects and even the technical importance of research into the reprocessing of natural and low-enriched uranium fuels appeared to be less than in 1957. In fact a number of participating countries were working on their own reprocessing projects. La Hague was under construction – UP2 was intended to reprocess fuel from the new French natural uranium graphite-gas power reactors[48]. Germany had decided in 1962 to build a small plant at Karlsruhe, the future WAK[49], for reprocessing fuel from the Mehrzweckforschungsreaktor (MZFR). Italy had started work in April 1963 on its EUREX[50] plant at Saluggia which was intended to reprocess highly enriched uranium fuel, in conjunction with the American firm Vitro, and was seeking financial support from EURATOM.

Moreover world capacity in the short and medium term was increasing with the United Kingdom project at Windscale and those of the United States at West Valley and Morris. At the beginning of 1964 the new Windscale 2 unit was being tested and the plant of Nuclear Fuel Services (NFS, a subsidiary of General Electric) had been under construction since January 1962 in New York State[51]. Also in 1964 General Electric was considering a new plant to utilise a new process on a large scale. This was to be the Midwest Fuel Recovery Plant (MFRP) at Morris in Illinois[52].

Finally there was still no sign of any upturn in the growth of nuclear power in Europe. Questions were therefore raised as to whether Eurochemic would ever operate at full capacity and would ever be viable[53].

---

48 The site had been selected in 1960; the plant was opened in 1966.

49 In the first half of December 1963, IGK (Uhde Leybold Lurgi) submitted to the federal government a pre-project for a plant with a capacity of 30 tonnes of natural U or LEU enriched up to 1.6%.

50 Capacity 150 to 300 kilogrammes of U-235 a year. The process foreseen in 1963 used tricaprylamine.

51 It was to be opened in 1966. Eurochemic had established links with NFS as part of the co-operative programme with the United States. An agreement on exchanges of information had even been signed.

52 The Aquafluor process had been successfully developed in the laboratory: its final product was uranium hexafluoride which could be directly recycled in the enrichment plants. However the move to the industrial scale was a disaster and the plant – which cost General Electric $60 million – was unable to process anything at all.

53 The NFS prices, for a capacity of 1 t/day, were the following in 1964: $30/kg for batches exceeding 20 tonnes, 40 for batches of European size, i.e., under 10 tonnes. Windscale 2 was reprocessing Magnox fuel at $15/kg. In early 1964 there was a possibility that it would reprocess natural uranium fuel from the Italian Latina reactor (SIMEA) although no contract had been signed.

Thoughts of ending the co-operative experiment

On 2 October 1963 the Board of Directors, which had been kept informed of the work of the Special Group's restricted group, decided to set up a working party of the Technical Committee with the task first of "finding ways of reducing the company's capital and running costs" by examining "the different possible options as regards the programme of work (research and construction of the plant)" and also "to consider an interim programme whereby the company could continue its activities until a definitive financial solution was found". The working party involved Yves Sousselier, Erik Haeffner and Maurizio Zifferero together with a German expert, Boettcher, subsequently replaced by Kurt Giese, together with the three Directors of the company. Their report was submitted to the Board of Directors on 12 November 1963. With regard to the first point the working party considered three scenarios.

The first involved halting all activities on 31 December 1963 and dissolving the company. To the \$16 million already committed there would be added \$5.5 to 7 million for redundancy payments to the staff, the cancellation of supply and works contracts, and paying for supplies and works already carried out. There would remain some \$5 to 7 million out of the capital sum of \$28.95 million together with buildings, furniture and fittings that would be difficult to realise.

The second scenario envisaged refocusing the plant exclusively on pilot activities, which would reduce staff numbers and running costs. Two "sub-scenarios" were considered: one involved maintaining the planned capital investment programme, leaving open the possibility of a subsequent restart for industrial purposes, the other proposed reducing the capital investment programme to what was strictly necessary for a pilot unit, and would therefore eliminate that possibility. The first option would require an extension of \$9.5 million, the second \$8 million. The working party gave preference to the former of these sub-solutions.

The third scenario, on the contrary, involved enhancing the production side of the operation in order to generate reprocessing income. The restricted group reviewed the various fuels available (from the Italian SIMEA reactor, LEU and HEU) and opted in favour of reprocessing low enriched fuels. The additional capital cost would be \$16.83 million, plus the costs of research – reduced but essential – evaluated at \$6 million. Reprocessing income would bring \$6 million and outside research contracts \$2 million.

The working party then proposed a series of economies: it believed the only effective measures would be to refocus research solely on the problems of starting up the plant and radioactive waste management, and move the general management from Brussels to Mol.

Finally it presented an interim programme based upon the proposals of the Eurochemic management: this involved continuing work at the present rate, which would be possible with the remaining capital up to 30 April 1964. The decision about reprocessing HEU was postponed.

The Board of Directors[54] decided against the pilot option on the grounds that it would produce only limited technical results – the main objective being to build an industrial pilot – would provide no economic information and would be able to reprocess only a small part of the available fuel.

To stop the company's activities altogether – a decision that would devolve upon the Special Group – was technically and financially feasible, but would have "serious psychological consequences" and would be "an international scandal".

It was therefore necessary to go on, but by reprocessing the largest possible amounts of fuel "in view of the high proportion of fixed costs in the plant's running expenditure" and by varying the types of fuel so as not to be dependent on a single supplier. In this way the proposal to specialise in reprocessing fuel from SIMEA (Latina) was implicitly abandoned, so it was necessary to intensify the search for contracts.

An interim programme for the first four months of 1964 was adopted and a provisional programme covering the period 1964-1967 recommended, this being a period "beyond which it was possible to envisage either viable operation of the plant or a halt to activities".

The crisis of the winter 1963-1964

The Special Group met on 26 November 1963 to review the proposals of the restricted group but was unable to reach a decision. The countries which had already declined the first increase in capital were no more disposed to subscribe to the second. While all the partners had accepted the reasons put forward by Portugal and Turkey, the bitterness with regard to Italy intensified. Through its statements[55] at each meeting of the Special Group, the General Assembly and the Board of Directors, Italy disputed the validity of the project and denounced, often in undiplomatic terms, what it regarded as departures from the original project.

---

54 CA(63)43 and CA/M(63)5.

55 Transcriptions of these are almost always annexed to minutes.

At the beginning of 1964 in particular, relations between the participants were somewhat strained. Thus France declared[56] that she "could not participate in the capital increase if Italy did not do so. It was not only a question of finance, but also of politics. In 1957 prospects had seemed more favourable and no procedure had been laid down for covering deficits. It was inadmissible that one of the partners should take advantage of this in order to get out of its obligations to the others, at a time when the company had to meet additional expenditure. In addition, the French authorities wished to avoid committing themselves beyond 1967".

Belgium also came in for criticism, having placed conditions on the effort she was prepared to make with regard to capital costs: this was to be conditional upon the running costs – and notably the coverage of the expected deficit – being settled using a mechanism related to national income. In the meantime she did not wish to take up the ten shares refused at the time of the 1963 increase in equity. "The French government had also gained a most unfavourable impression from the refusal of the Belgian government to subscribe the sums which it had been asked for, when the Belgian government was the only one whose capital subscriptions were offset by substantial revenues".

The Special Group decided to meet again in two months. In the meantime, several participating countries and the ENEA made diplomatic approaches to Rome, with limited success.

In a letter of 3 February 1964[57] the representative of Italy at the OEEC repeated his country's position: she refused to participate in any increase in capital, would take part in the research programme only and in funding any management deficit, and undertook to have Italian fuels reprocessed "to the greatest possible extent, on condition that all the other members of the company do the same and that the costs are competitive".

At its eighth meeting on 2 April 1964 the Special Group could do no more than note once again the Italian blocking operation; consideration was given to making up for the lack of Italy's subscription by taking out a commercial loan to be repaid by the income from reprocessing Italian fuel. However this solution was abandoned as too risky by the Board of Directors which recommended that a temporary reduced programme should be adopted[58].

Emerging from the crisis (17 April 1964 - 1 July 1964): the reduced programme and the second equity increase

The breakout

At the meeting of the Board of Directors held on 10 April, Belgium softened its position although continuing to refuse to take out the ten shares under dispute. It offered to grant the company a loan guaranteed by the government, for a maximum sum of $1 million, under the economic growth legislation in order to enable it to fund the initial programme of $21.4 million, on condition that all the countries accepted the proposed programme and that steps were taken to reprocess fuel from BR2, in other words HEU[59].

The Special Group met again on 17 April in the situation of crisis and took the following decisions.

It adopted, on a temporary basis, a reduced programme of work costing $14.8 million for the period 1964-1967, comprising completion of the installations in accordance with the plans adopted (for $6.8 million), the experimental operation of the plant with limited quantities of fuel – 8 to 20 tonnes a year ($3.8 million) – and a limited research programme ($4.2 million). It recommended that the costs of completing construction should be funded by a new increase in capital of $6.8 million involving the issue of 136 shares. The group noted the Italian refusal and reduced the programme accordingly; the group hoped that Belgium would purchase the ten shares set aside at the time of the first increase in capital on condition that an agreement could be found on ways and means of covering the expected operating deficit.

The Special Group also recommended, at the initiative of ENEA, that the costs of the research programme and the net operating cost – $8 million for the period 1964-1967 – should be covered by contributions through the OECD budget.

Italy stated that it was unable to accept these recommendations. In particular she rejected the idea that a deficit programmed in advance could be accepted, which she saw as turning the company into an intergovernmental organisation. Portugal reserved its position.

It had been planned to give the new shares issued since 1963 special status in terms of repayment; this scheme to amend Article 31 of the Statute was abandoned following protests by the representative of Belchim at the meeting of the Board of Directors held on 3 June 1964. This decision, which had been proposed to bring pressure on Italy, did in fact have an adverse effect on the interests of the private shareholders.

---

56 To the Board of Directors at the meeting held on 5 March 1964, CA/M(64)1.

57 Unreferenced copy in documentation for the eighth meeting of the Special Group.

58 NE/EUR(64)2.

59 CA(64)16.

The end of the financial impasse: June-July 1964

On 16 June 1964 the OECD Council adopted the Special Group's recommendations and opened a credit of 8 million EMA/UA for the period 1964-1967[60], to cover the costs of the research programme and the net cost of operating the plant. An initial credit[61] of 1.56 million EPU/UA was therefore provided for in the budget for the year 1964, rising to 2.33 million in 1965 and 3 million in 1966.

The second increase in capital was decided upon by the General Assembly on 1 July 1964. At the meeting of the Special Group on the same day Italy, which had abstained during the meeting of the Council on 16 June 1964 devoted to covering research expenditure and the net cost of operations, announced that she would contribute to the extent of 6% (i.e., $0.4 million out of $6.8 million), "a proportion (that is) substantially Italy's share in the current capital"[62].

This second increase in capital was the final one, but it still did not entirely settle the problem of funding the building work. In March 1966 the Special Group was informed that a further $1.335 million was needed to complete construction[63]. Although a further increase in capital was not excluded, the Board of Directors requested the Director-General to negotiate the conclusion of a loan as an interim solution. On 8 June 1966 the Special Group decided that repayment of the loan would be added to the operating deficit up to 1971 and would therefore be funded in the same way as the running costs.

Project funding from 1958 to 1966

Table 20. **Countries' contributions to capital, in decreasing order of amount (Value of shares purchased in the period 1959-1964 in million EMA/UA)**

Country	Value of shares acquired in 1959[64]	Value of shares acquired as of 18 June 1973	Value of shares acquired as of 1 July 1964	Total sums contributed	% of total capital subscribed
Germany	3.4	1.60	1.425	6.425	18. 0
France	3.4	1.60	1.425	6.425	18.0
Belgium	2.2	1.80	1.100	5.100	14.3
Sweden	1.6	0.50	0.600	2.700	7.5
Netherlands	1.5	0.45	0.525	2.475	7.3
Spain	1.5	0.40	0.525	2.425	7.3
Switzerland	1.5	0.40	0.450	2.350	7.0
Italy	2.2	0.00	0.000	2.200	6.1
Denmark	1.1	0.25	0.275	1.625	4.5
Austria	1.0	0.25	0.250	1.500	4.2
Norway	1.0	0.20	0.225	1.425	4.0
Turkey	0.8	0.00	0.000	0.800	2.2
Portugal	0.3	0.00	0.000	0.300	0.8
Total issued	21.5	7.45	6.800	35.750	100.0

Thus over half the capital had been provided by three of the thirteen participating countries: France, Germany and Belgium. Neither Italy, Turkey nor Portugal had subscribed to the further increases in capital of 1963 and 1964.

During the period in question the company was funded by a combination of the issue of tranches of capital (over three quarters of the total), contributions from the OECD (accounting for a little over 16%) and also by two loans following the type and sequence of financing shown in Table 21.

Once the capital was used up, the OECD contributions were to change the nature of Eurochemic. It became, as regards its financial operation, less and less a company and increasingly an international public body, a technical and scientific "subsidiary" of an international economic organisation (Tables 22 and 23).

60 CES/64.27; report by the Secretary-General to the Budget Committee SCB/64.25.

61 Figures taken from the balance sheets assuming 50 BF = 1 EPU/UA.

62 NE/EUR/M(64)4. However this was only half the amount Italy should have paid by applying the OECD scale.

63 NE/EUR/M(66)1.

64 As of 27 July 1959, but 29 October as regards Spain.

The funding problems resulted from cost overruns and work delays. These factors were not specific to Eurochemic: they appear in nearly every high technology project[65]. It was made worse by the rising inflation of costs and salaries from the beginning of the 1960s, which had not been foreseen in 1957. The figure of $20 million calculated and accepted in 1957 turned into $46.7 million. But above all, the date at which the company would become financially self-supporting at best receded into the distant future, at worst had become a chimera.

Further difficulties also arose from the lack of any compulsory provisions in the Convention about covering any deficit that might arise once the plant was in operation. In fact[66] when the Convention was written a deficit was inconceivable in the minds of those responsible. Later on[67] its authors were to regret two other *lacunae* in the company's Convention and Statute: the lack of "any provision which might have guaranteed Eurochemic exclusive rights to reprocessing contracts from shareholders", and of "any provision protecting Eurochemic against competition from its own shareholders". However it is likely that certain shareholders would not have accepted such provisions in the contemporary situation. Later events were to show that such an approach to international co-operation was not theirs.

In any event these shortcomings created a legal vacuum, which had to be filled by the good will of governments. Faced with the sheer extent of the resources needed, the solidarity of the partners was sorely tested. The tendency towards independent action did not only involve Italy which, with EUREX, found itself in direct competition with Eurochemic and where the internal problems of the CNEN only made things worse. Both France and Germany developed their own reprocessing projects in parallel, the former in an industrial plant, the latter in a pilot unit, making considerable use of knowledge and experience gained at Eurochemic[68]. Even a sensitive partner like Denmark did not hesitate in May 1965 to sign a five-year contract with Dounreay to reprocess fuel from the DR2 MTR, when Eurochemic could have done so within the desired timescale in view of its polyvalent capability.

After all what was built at Mol by the Eurochemic company, assisted by a large number of European contractors, was a reprocessing plant capable of accepting every kind of fuel that existed in Europe at the time.

Table 21. **Annual amounts and origin of Eurochemic funding from 1958 to 1966**

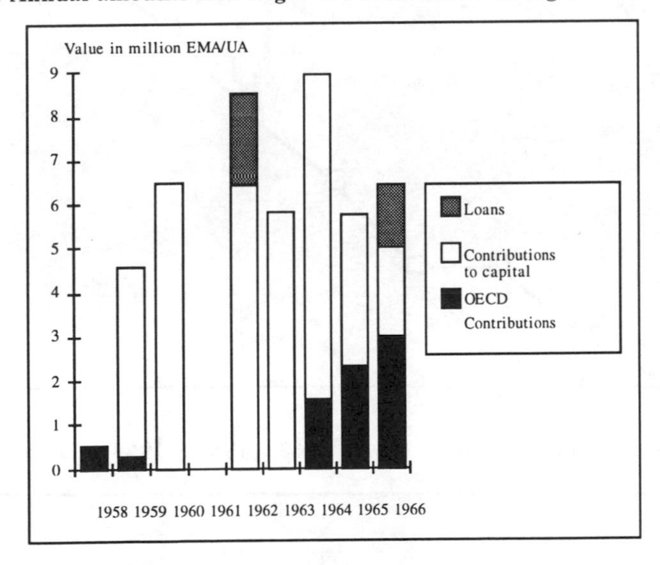

---

65 A joke circulating in nuclear circles is highly revealing: known to Pierre Huet and Pierre Strohl as "Gibrat's Theorem" after one of the Presidents of Indatom and then of the SFEN. It can be summarised as follows: "To determine the true cost of the project based on estimates from three experts, add the three together. If there is only one estimate, multiply it by π". In this "theoretical space" Eurochemic costs remained within reasonable limits...

66 Communications from Pierre Huet and Pierre Strohl.

67 STROHL P. (1983), in ETR 318, PP. 12-13.

68 Specific developments of these points are given in Part V, Chapter 1.

Table 22. **Sources of Eurochemic funding: Capital payments, loans and OECD contributions from 1958 to 1966 (in millions EPU/UA)**

	OECD contributions	Capital payments	Loans	Total	Running total
1958	0.500			0.500	0.500
1959	0.257	4.300		4.557	5.057
1960		6.450		6.450	11.507
1961					11.507
1962		6.450	2.000	8.450	19.957
1963		5.790		5.790	25.747
1964	1.560	7.320		8.880	34.627
1965	2.330	3.400		5.730	40.357
1966	3.000	2.040	1.335	6.375	46.732
Total	7.647	35.750	3.335	46.732	
% Total	16.4	76.5	7.1	100.0	

Table 23. **Relatively steady growth of resources produced by a combination of different types of funding**

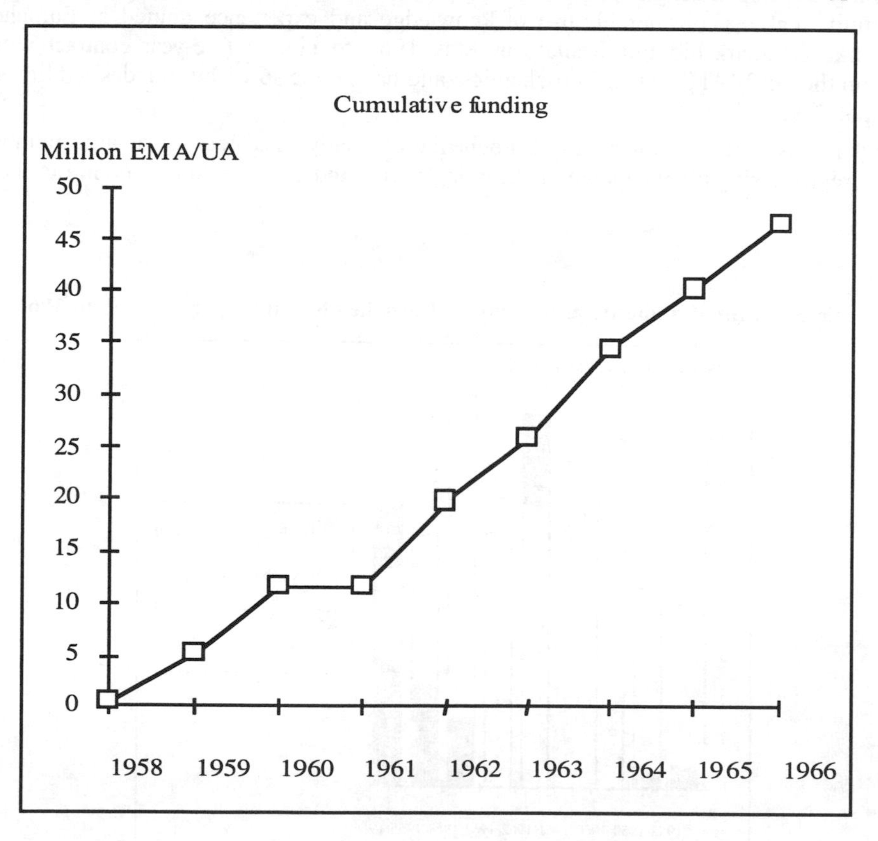

Chapter 3

European co-operation, applied research and industrial development in the construction of the plant and laboratories

1961 - 1965

The approval of the detailed SGN pre-project by the Board of Directors on 13 June 1961 marked the real beginning of the construction of the plant, and civil engineering works began in November 1961. The hot wing of the research laboratory was ready by February 1964 and the plant analytical laboratory commenced operations in August 1964. Construction of the plant was virtually complete by October 1965 and tests then began in preparation for industrial commissioning in 1966.

Work on the site between 1961 and 1965 was a European co-operative venture at several levels involving large numbers of participants.

The practical arrangements for construction were complex and articulated around two poles: Eurochemic on the one hand and outside contractors on the other as shown in Table 24.

Table 24. **Those involved in the construction of the plant**

EUROCHEMIC	OUTSIDE CONTRACTORS
General management	1. Engineering design services
	Principal architect engineer
Research Directorate	*Architect engineers*
Technical Directorate	2. Goods and physical services
	Building contractors
	Sub-contractors and suppliers

The external contractors worked under Eurochemic supervision, directed by the architect engineers who were responsible for a particular part of the site. A system of international invitations to tender enabled contractors from all over Europe to be involved in the construction. In reviewing the tasks and functions of the different firms involved, it is important to consider how their co-operative work was organised, and the practical problems that arose during site development. Indeed the international nature of the undertaking added further to the complexity of the arrangements for building the plant.

To begin with, European co-operation was internal to the company, whose structures developed in parallel with progress on the site. Staff numbers increased as people were recruited from all member countries. By the end of 1965 the Eurochemic workforce numbered 360, the composition of which reflected – albeit with some distortions – the aims of international co-operation on which the project was based.

A substantial R & D programme was co-ordinated by the Research Directorate. When site works began, priority was given to building the research laboratory and installing advanced equipment. As a result, research and development made rapid progress and produced significant results. However the company's financial problems soon forced a shift of priorities in favour of the plant itself, and the research side was therefore not as fruitful as had been hoped.

European co-operation was also active outside Eurochemic, *via* the close links formed between its Technical Directorate and the architect engineers, and the contacts made with hundreds of suppliers and sub-contractors in the tendering procedure, the awarding of contracts and their execution. However the large number of actors and the changes made to project planning as the work proceeded led to special problems of organisation which, although resolved, somewhat slowed down the process of construction.

European co-operation within the company

As an industrial firm and the creature of an international organisation, Eurochemic showed highly individual hybrid features which can be seen both in its organisation charts and in the changes that occurred in its workforce.

The company's structures and their development: an industrial firm bearing the international imprint

The overall organisation of the company and its development

The three organisation charts for 1961, 1964 and 1966 given in Figure 34 show the company's structure and its evolution.

The strong connection with the OECD is manifest in the company's General Secretariat which was closely linked to the ENEA. The firm's industrial character was reflected in its three functional directorates. The private model predominated but the international public character was expressed both in the structures and in the pattern of recruitment.

The company was organised on five hierarchical levels (Figure 33).

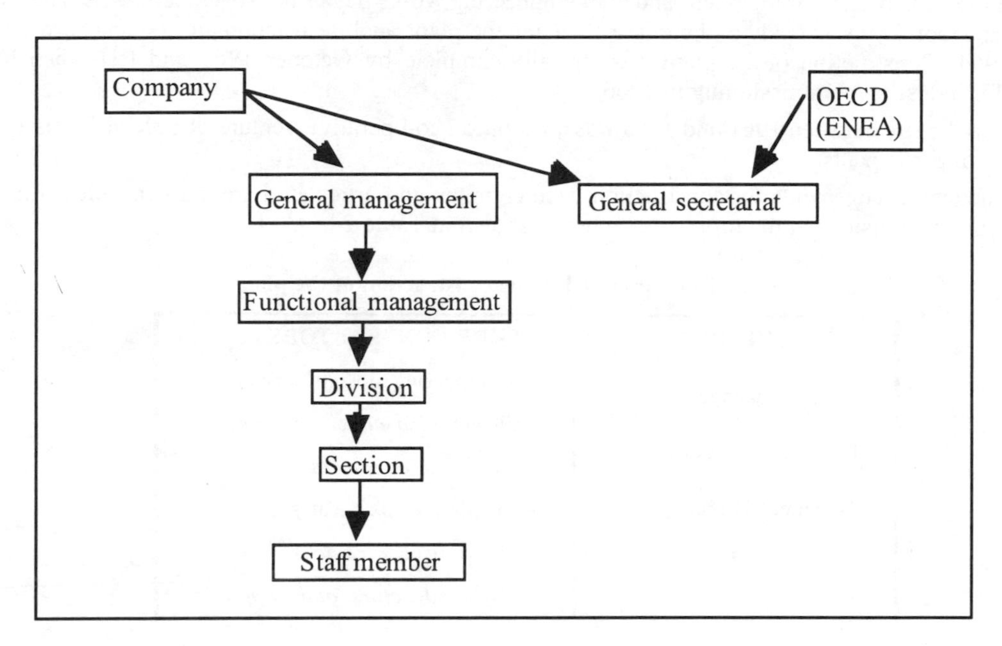

Figure 33. The five hierarchial levels[1]

Erich Pohland was Managing Director until his retirement on 31 December 1963. As the time for his departure approached, the search for a successor began, initially with a German in mind in order to adhere to the unofficial nationality quota system for posts[2]. However the German Ministry for Research proposed a person with experience neither of managing a company nor of international questions "which alone would enable him to take over the reins quickly at a time when there were important and urgent problems to be solved"[3]. Since no other German candidate could be proposed before the end of the year, it was decided to appoint Rudolf Rometsch, the Director of Research, to be Managing Director for an interim period of four months. In fact he was confirmed in the post by the Board of Directors on 3 June 1964 and was to remain in that position until 1969, a period covering all of the construction phase and the initial operations.

[1] Of which only the first three are shown on the general organisation charts.

[2] According to Yves Leclercq-Aubreton each function was initially allocated a number of points and the distribution of points was supposed to reflect the contribution of each shareholder. Thus a German Managing Director seemed essential in order to maintain the balance between the shareholders. Subsequently the points system was applied less strictly than in the beginning, because posts had to be filled.

[3] CA(63)35 of 14 October 1963.

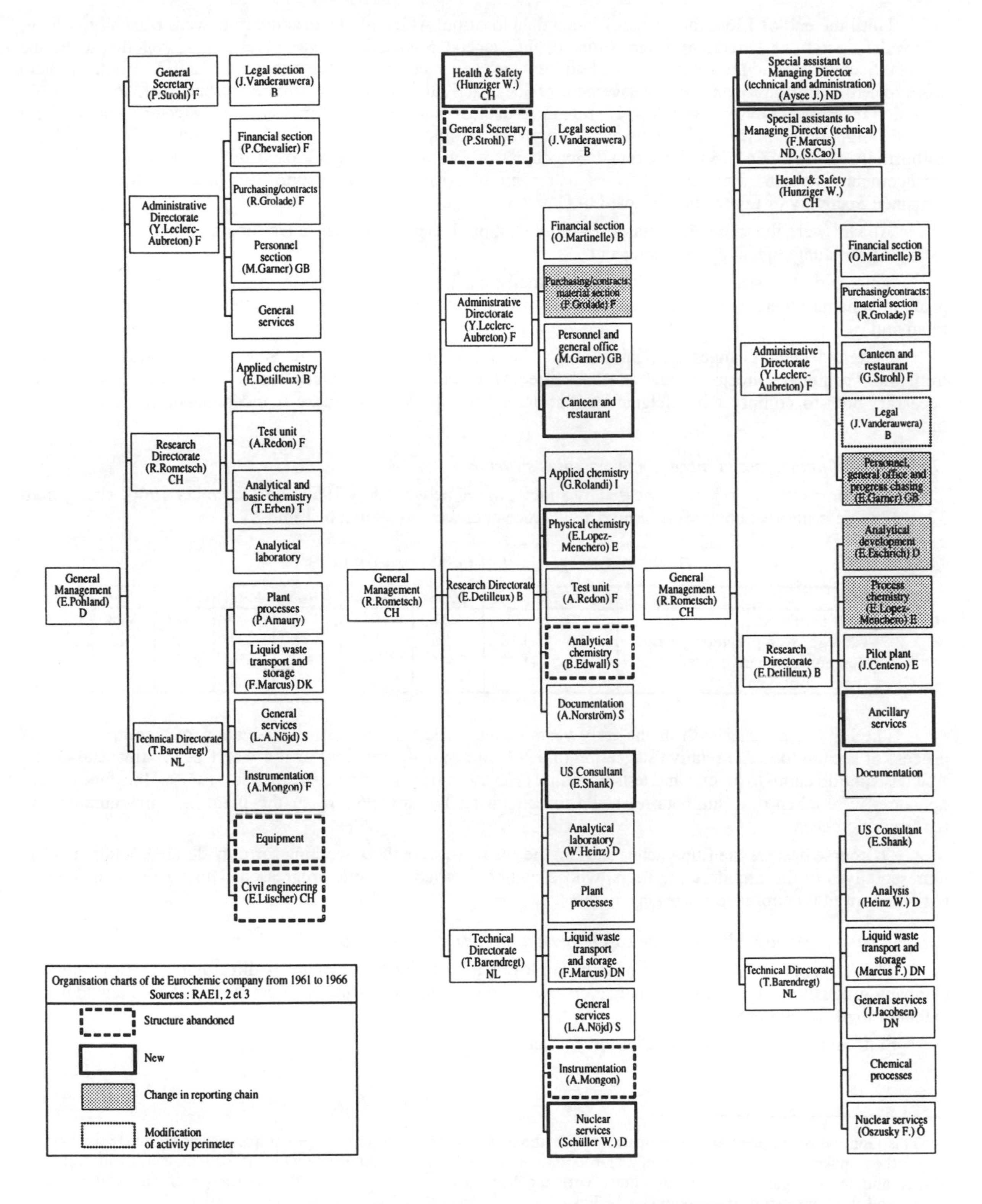

Figure 34. Organisation charts of the company at the end of 1961, 1963 and 1966.

Until the end of 1964, the company had a dual location. Although the headquarters were officially at Mol, where the operations Directorates were situated, the General Management was in Brussels, together with the General Secretariat[4] and the Legal Section, both of which reflected the supervisory role of ENEA. Erich Pohland went to Mol only rarely and his management of local activities was more symbolic than real. Pierre Strohl occupied the post of General Secretary. He remained in close liaison with ENEA and its Director-General Pierre Huet throughout the period. The Brussels office played an important role in the negotiations with the Belgian authorities. In October 1964 the General Secretariat of the company was closed down when negotiation of arrangements with the Belgian authorities had come to an end. Pierre Strohl then returned to ENEA but remained Secretary of Eurochemic's Board of Directors.

At Mol were the three directorates – Administration, Research and Technical, the presence of the last two reflecting the dual objective of the company.

This tripartite structure remained substantially as it was throughout the period of construction until profound changes were made in 1967 as the plant was started up and the four-year 1964-1967 programme came to an end[5].

Certain detailed changes are significant: in November 1962 the Health & Safety Section was taken away from the Technical Directorate and attached directly to the General Management. This change was made necessary both to comply with Belgian legislation and by the desire to make it independent of the operations units.

Growing staff numbers and increasingly complex structures

The company's workforce increased by a factor of ten between July 1959 and December 1966, rising from 37 to 378. Year-on-year changes measured at 31 December were as shown in Table 25.

Table 25. **Changes in staff numbers, 1959 to 1966**[6]

Year	1959	1960	1961	1962	1963	1964	1965	1966
Staff numbers	37	75	131	195	271	292	360	378
Change over previous year	+16	+38	+56	+64	+76	+21	+68	+18
Percentage change over previous year	76%	102%	75%	49%	39%	8%	23%	5%

The high rates of growth in the early years simply reflect the recruitment situation of a company in the process of starting up. The relative stagnation of 1964 corresponds precisely to the worst of the financial crisis. As construction came to an end and tests began in 1965 the staff numbers began to rise again as the future plant operators were taken on. Numbers reached a maximum at 378 in 1966 when the plant was inaugurated and reprocessing began[7].

Of course besides the Eurochemic staff on the site there were those working for outside contractors and the representatives of the architect engineers, whose numbers varied as work progressed. During the construction period about 600 people were working on the site.

The company's gradual shift from research towards preparations for production

The changes in the numbers of staff employed in the different functional directorates shows how the company gradually shifted away from research towards preparing for production, a development accelerated by the extent of the financial problems.

Staff numbers in the different directorates changed as follows[8]:

---

4 This General Secretariat is not comparable with those that exist in certain large companies, particularly in France, where they make up for the absence of a "Division of General Affairs". At Eurochemic it provided liaison between ENEA and the company in its negotiations with the host government and the other participants, as well as legal advice and the Secretariat of the company bodies.

5 These changes merged the Research Directorate and the Technical Directorate into a new Technical Directorate under the former Director of Research. See Part III Chapter 2.

6 Source: RAE 1, 2, 3.

7 The annual reports give staff numbers for 31 December each year. In fact the company had around 400 employees during 1966.

8 Table prepared from numerical data given in the Annex.

Table 26. **Distribution of staff amongst the directorates from 1959 to 1966**[9]

1. Absolute values	07-1959	07-1960	12-1961	12-1962	12-1963	12-1964	12-1965	12-1966
General management and administration	7	19	34	48	62	61	68	64
Research Directorate	18	31	59	84	84	79	79	81
Technical Directorate	12	25	38	63	114	139	197	204
Health & Safety Division					11	13	16	29
Total	37	75	131	195	271	292	360	378
2. Percentages	**07-1959**	**07-1960**	**12-1961**	**12-1962**	**12-1963**	**12-1964**	**12-1965**	**12-1966**
General management and administration	18.9	25.3	26.0	24.6	22.9	20.9	18.9	16.9
Research Directorate	48.6	41.3	45.0	43.1	31.0	27.1	21.9	21.4
Technical Directorate	32.4	33.3	29.0	32.3	42.1	47.6	54.7	54.0
Health & Safety Division					4.1	4.5	4.4	7.7
Total	100.0	100.0	100.0	100.0	100.0	100.0	100.0	100.0

These two tables show the role of the different directorates as the company grew, and how their relative influence changed. Until 1963 Eurochemic was a research company with a substantial administrative side; it subsequently became a technical undertaking.

Until the end of 1961, growth was fairly balanced between the three directorates, with a slight advantage for the Administrative Directorate, whose numbers stabilized at about 60 as from 1963. This directorate accounted for nearly 30% of the staff in July 1961, but only 16% in December 1966.

With a third of the staff in 1959 and 1960, the Technical Directorate fell below 30% in 1961 as the Research Directorate became relatively more active in recruitment. In 1963 the Technical Directorate caught up and overtook the Research Directorate both in numbers and percentages. From 1965 onwards, the Technical Directorate accounted for more than half the company's staff.

As regards changes in the structures of the Technical Directorate, the changes in the divisions paralleled the progress of site work. The Equipment and Civil Engineering Divisions worked closely with the contractors involved in construction and disappeared in 1964. The Instrumentation Division continued in being until work on installing the cells was complete. Progress with the licensing procedure resulted in the creation of the Nuclear Services Division. During this period the hard core of the directorate consisted of the analytical laboratory, the Process Division, the division responsible for operations preceding and following reprocessing (transport, storage, waste management) and general services.

In July 1959 nearly half the staff was working in the Research Directorate. In 1966, the proportion had fallen to 20%. During the first four years the numbers employed in this directorate went up to reach about 80 in 1962, after which they stabilized until construction came to an end. The percentage drop was linked to the rapid rise in recruitment by the Technical Directorate, reinforced to a small extent by the creation of the Health and Safety Division in November 1962.

In the Research Directorate, the "Analytical and Basic Chemistry" Division was split into two parts: the Physical Chemistry Division which took care of fundamental research related to the process[10] and the Analytical Chemistry Division, responsible for methods of measurement[11]. The structure of the directorate was modified in June 1965[12]. The Physical Chemistry Division was eliminated and research organised in a relatively symmetrical manner with the Technical Directorate: analytical development in the one matching the Analytical Division of the other, and process chemistry matching the Chemical Process Division. The Test Unit and Documentation Divisions were a permanent feature of the directorate throughout its history. The creation of an Ancillary Services Division in 1966 was intended to provide direct assistance to the reprocessing plant, which was then placed under the responsibility of the Technical Directorate.

---

9 And as of 1963 in the Health and Safety Division.

10 New solvents, preparation of a plutonium reducing agent based upon uranyl nitrate U $(NO_3)_4$.

11 In conjunction with the Instrumentation Section of the Technical Directorate.

12 CA (65)19.

A high technology company with international staff: wide-ranging qualifications

A substantial proportion of highly qualified staff

The staff hierarchy and the company's own qualification system

In 1961 the company defined five staff grades (Director, I, II, III and IV[13]), and five categories in 1963 (by incorporating the management into grade I and by creating a category II bis to which the draughtsmen of the former category III were added). In 1966 this system was replaced by one based upon four normalized salary ranges (above 480; between 265 and 470; between 145 and 260; and between 100 and 142).

The statistics given here group these different systems into four categories denoted A, B, C and D which are based upon the 1966 classification, with staff grouped as follows[14]:

A: Director + I;

B: II, II bis and draughtsmen from III;

C: former III except for draughtsmen;

D: IV.

In practice therefore, category A covers the directors and graduate or equivalent engineers, category B other engineers, senior technicians and draughtsmen, category C the operators, technicians and administrative staff, while category D covers operations assistants and unskilled workers.

Trends in the staff structure according to qualifications are shown in the following tables:

Table 27. **Staff structure according to qualifications from 1961 to 1966 (by number)**

Category	1961	1963	1966
A: directors and graduate and equivalent engineers	11	19	28
B: other engineers, senior technicians and draughtsmen	26	48	65
C: operators, technicians and administrative staff	62	98	193
D: assistant operators and unskilled workers	32	106	92
Total	131	271	378

Therefore as percentages of total staff:

Table 28. **Staff structure according to qualifications from 1961 to 1966 (as percentages)**

Category	%1961	%1963	%1966
A: directors and graduate and equivalent engineers	8.4	7.0	7.4
B: other engineers, senior technicians and draughtsmen	19.9	17.7	17.2
C: operators, technicians and administrative staff	47.3	36.2	51.1
D: assistant operators and unskilled workers	24.4	39.1	24.3
Total	100.0	100.0	100.0

The Eurochemic staff consisted of 7 to 8% of senior supervisory staff (A) and a quarter of highly skilled staff (A and B).

The proportion of total staff in categories A and B fell slightly from 28 to 24.6% but otherwise remain stable; most employees were in categories C and D. The reversal in the relative proportions in categories C and D between 1963 and 1966 stemmed from the recruitment policy and the advancement of staff in categories D, resulting from inflation and overall changes in wages and salaries in this first half of the 1960s in Belgium[15], but also from the increasing demand for operators – in category C – during the testing period. This followed the fall in the proportion of staff in category C and a temporary rise in category D, which reflected the company's financial difficulties.

---

[13] Source: RAE 1, p. 33. I: Head of Division, Technical Section or similar; II: Research engineer and Head of administrative office; III: Technician, draughtsman and administrative assistants; IV: Junior administrative and auxiliary staff.

[14] These are not arbitrary but emanate from the company itself.

[15] During this period the Managing Director periodically requested permission from the Board to raise salaries to keep pace with inflation and provided a whole range of data to support his requests.

A basically Belgian company: international at the top

Breakdown by nationality: overall approach

As an international co-operative undertaking, Eurochemic recruited staff from fifteen different countries, namely two more than those of the shareholders: eight Britons worked at Eurochemic in 1963 and an Irishman in 1966.

Taking the staff as a whole, the nationality structure over the period changed as follows (arranged in decreasing order for 1966[16]):

Table 29. **Staff structure according to nationality from 1961 to 1966 (by number)**

Country	1961	1963	1966
Total	131	271	378
Belgium	64	143	215
Netherlands	7	13	39
France	12	20	30
Germany	15	26	29
Spain	8	20	20
Italy	9	11	13
United Kingdom	3	8	8
Austria	1	3	5
Norway	2	6	4
Switzerland	3	5	4
Denmark	2	5	3
Portugal	0	3	3
Sweden	3	6	2
Turkey	2	2	2
Ireland	0	0	1

Belgian employees accounted for 48.9% of the staff in 1961 and 56.9% in December 1966. Throughout the period Dutch, French and German staff accounted for approximately one quarter of the workforce, although with a slight decline in the numbers of French and Germans. The high proportion of Dutch staff was due to the proximity of that country and the fact that an expatriation allowance was paid[17]. The other ten nationalities shared the remaining 25% in 1961. About half this category was accounted for by Italian and Spanish staff, although their share gradually declined (Table 30).

Nationality and qualifications: hierarchical distortions

A finer analysis is possible by examining the distribution of nationalities in each category of staff and comparing these with the overall statistics given above.

Tables 31 and 32 have been constructed by reducing the detailed statistics for the four categories of staff for the years 1961, 1963 and 1966 into two main categories AB and CD.

The tables show the relatively low Belgian presence in the upper A and B categories, particularly by comparison with the strong local recruitment for staff in categories C and D. However this under-representation of Belgium in the senior posts declined over time, indeed the trend began during this period. Conversely, there was an over-representation of the other member countries in the senior levels, albeit with fluctuations. Germany tended to be under-represented at the top at the beginning of the period, but provided the Managing Director. The proportion of Italian staff at the higher levels tended to fall, reflecting its relative financial disengagement. In general the distribution of staff by category and nationality reflects the concern for sharing out the senior posts between the main shareholders, although without introducing a formal system of national quotas. However this approach disappears lower down in the structure where staff did not need to be so qualified. Local Belgian and Dutch manpower was in the majority in grades C and D.

In 1966 the difference between the two main staff categories as regards nationality structure was still very marked, as shown in the double bar chart (Table 33) which shows the staff numbers with countries ranked according to their numbers in the CD category.

---

16 The basic data taken from RAE 1, 2, 3 are reproduced in Annex 3.

17 Thus a Dutchman living 12 kilometres from the plant received the same allowance as an Italian or Spaniard.

Table 30. **Staff structure according to nationality from 1961 to 1966 (as percentages)**

Pays	%1961	%1963	%1966
Total	100.0	100.0	100.0
Belgium	48.9	52.8	56.9
Netherlands	5.3	4.8	10.3
France	9.2	7.4	7.9
Germany	11.5	9.6	7.7
Spain	6.1	7.4	5.3
Italy	6.9	4.1	3.4
United Kingdom	2.3	3.0	2.1
Austria	0.8	1.1	1.3
Norway	1.5	2.2	1.1
Switzerland	2.3	1.8	1.1
Denmark	1.5	1.8	0.8
Portugal	0.0	1.1	0.8
Sweden	2.3	2.2	0.5
Turkey	1.5	0.7	0.5
Ireland	0.0	0.0	0.3

Table 31. **Staff structure according to nationality and category (by number)**

Country	AB61	AB63	AB66	CD61	CD63	CD66
Austria	1	3	5	0	0	0
Belgium	6	10	30	58	133	185
Denmark	2	4	3	0	1	0
France	7	8	12	5	12	18
Germany	4	9	13	11	17	16
Ireland	0	0	0	0	0	1
Italy	3	6	3	6	5	10
Netherlands	3	6	14	4	7	25
Norway	1	5	2	1	1	2
Portugal	0	1	0	0	2	3
Spain	3	3	4	5	17	16
Sweden	3	6	1	0	0	1
Switzerland	2	3	4	1	2	0
Turkey	2	1	0	0	1	2
United Kingdom	0	2	2	3	6	6
Total	37	67	93	94	204	285

Table 32. **Staff structure according to nationality and category (as percentages)**

Country	%AB61	%AB63	%AB66	%CD61	%CD63	%CD66
Austria	2.7	4.5	5.4	0.0	0.0	0.0
Belgium	16.2	14.9	32.3	61.7	65.2	64.9
Denmark	5.4	6.0	3.2	0.0	0.5	0.0
France	18.9	11.9	12.9	5.3	5.9	6.3
Germany	10.8	13.4	14.0	11.7	8.3	5.6
Ireland	0.0	0.0	0.0	0.0	0.0	0.3
Italy	8.1	9.0	3.2	6.4	2.4	3.5
Netherlands	8.1	9.0	15.0	4.3	3.4	8.8
Norway	2.7	7.5	2.1	1.1	0.5	0.7
Portugal	0.0	1.5	0.0	0.0	1.0	1.0
Spain	8.1	4.5	4.3	5.3	8.3	5.6
Sweden	8.1	9.0	1.1	0.0	0.00	0.3
Switzerland	5.4	4.5	4.3	1.1	1.0	0.0
Turkey	5.4	1.5	0.0	0.0	0.5	0.7
United Kingdom	0.0	3.0	2.1	3.2	2.9	2.1
Total	100.0	100.0	100.0	100.0	100.0	100.0

Table 33. **Difference between the two main staff categories as regards nationality structure**

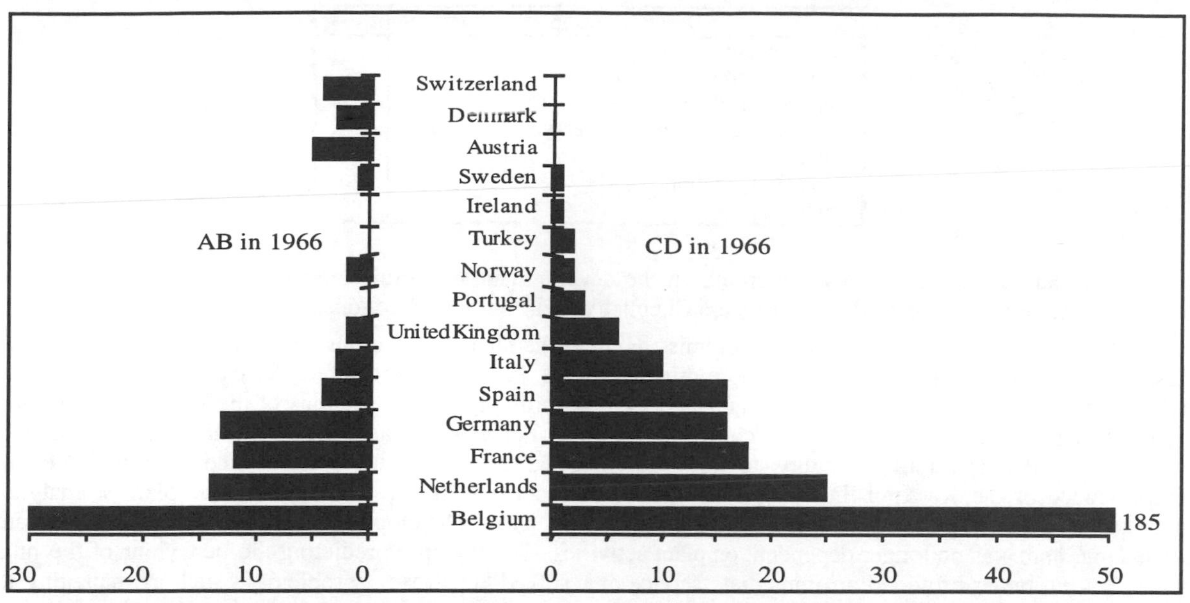

However it is important to look more closely at the make-up and functions of the different divisions of the company, beginning with the Research Directorate, for which these early years were the most important.

The Research Directorate and the role of R & D at Eurochemic during the period of construction[18]

The Research Directorate: organisation and objectives

The predominance of applied research

The Research Directorate had taken over from the Design and Research Bureau in October 1959. It was headed by Rudolf Rometsch until he was appointed Managing Director at the end of 1963. He was then succeeded by Emile Detilleux, previously head of the Applied Chemistry Division.

The Research Directorate employed 35 people at the end of 1960 and 59 at the end of 1961. At its peak in 1964 the laboratory had 84 employees[19].

The archives provide data for the breakdown of staff by qualifications for 1961 and 1963 and by sections for 1966.

Table 34. **Research laboratory staff by qualifications**

December 1961		December 1963	
Title or qualification		Title or qualification	
Director	1	Director	1
		Section heads	4
Section head, experts (engineers)	10	Engineers, researchers, office heads	17
Technicians, draughtsmen	41	Technicians, draughtsmen, administrative assistants	51
Secretaries and auxiliary staff	7	Junior administrative and auxiliary staff	11
Total	59	Total	84

The predominance of engineers and technicians in the qualification structure of the Research Directorate reflects that of applied research and industrial development in Eurochemic.

[18] Main sources: RAE 1, 2, 3, ETR 318, ESCHRICH, pp. 67 *et seq.*, DETILLEUX E. (1966) and a meeting held on 8 January 1992 between the author and Emile Detilleux, Rudolf Rometsch and Yves Sousselier to consider technical aspects.

[19] RAE 2, p. 60.

Table 35. **Research Directorate staff numbers by section in December 1966**

December 1966	
Management	4
Documentation Section	5
General Services Section	14
Analytical Development Section	14
Process Chemistry Section	24
Pilot Unit Section	21
Total	82

Over half the research staff was working on the development and refinement of the chemical methods and process operation methods in the two Process Chemistry and Pilot Unit Sections.

As from 1964, the acceptance and commissioning of the plant installations resulted in the members of the Research Directorate working more and more directly for the Technical Directorate; for example testing process columns following their installation in January 1965. Accordingly certain sections of the Research Directorate were gradually swallowed up by the Technical Directorate. For example as from 1964 the Analytical Section moved from the research laboratory to the analytical laboratory[20]. These trends led ultimately to the disappearance of the Research Directorate in 1967 whereupon the entire team joined the plant's analytical laboratory. The new group then became known as the Industrial Development Laboratory[21] and the research carried out there was no longer dependent on plant activities – for example predicting the behaviour of the pulse columns – but rather provided an immediate service or resolved "upstream" problems – such as managing the wastes actually produced by the plant. This change in the organisation marked the end of Eurochemic's peak period of reprocessing R & D.

Concentration on firm objectives

Eurochemic's research and development was exclusively of the applied type because its objectives were closely bound up with starting up the plant. As a result of the company's financial problems and as projects moved forward, activities were focused even more closely on increasingly precise goals, linked to the more immediate problems which faced the construction of the plant.

The R & D activities were closely linked to those of the Technical Directorate, which showed up in the similar way the sections were organised in the two directorates.

Research results were published in the form of ETRs and thus made available to shareholders. The company and ENEA also organised two scientific colloquia which reviewed different technical aspects: reprocessing[22] in April 1963 and instrumentation in November of the same year.

The symposium on reprocessing was held at the Brussels Palais des congrès from 23 to 26 April 1963 (Figure 35) and included a visit to Mol. The papers on aqueous processes, which were still the only ones suitable for industrial application, demonstrated the need for further work in areas such as cladding dissolution and new fuel materials, the improvement of solvents as regards chemical and radiation stability, and the development of improved methods for storing highly radioactive wastes, notably involving the solidification of fission product solutions.

The symposium on instrumentation[23] was held in Paris in November 1963 and was attended by 70 engineers. It compared procedures and experiments at EUREX, La Hague and Eurochemic with regard to automatic control, instruments for continuous analysis and data processing systems[24].

Contemporary R & D at Eurochemic covered five main fields[25], supported by four sections clearly separated in the 1964 organisation chart and by the Documentation Section which was attached to the Research Directorate[26]:

---

20 RAE3 p. 216.

21 Known as IDL.

22 ENEA/OECD-EUROCHEMIC (1963a). See also CA/M(62)2 and the report in CA(63)23 of 5 June 1963.

23 ENEA/OECD-EUROCHEMIC (1963b).

24 Problems of instrumentation are covered in a special section of Part V Chapter 1.

25 Hubert Eschrich indicates three main stages: "Preparation for active startup (1959-1966), plant assistance and waste treatment studies (1966-1974), decontamination and waste conditioning (1974-1983)".

26 The Fundamental Process Chemistry Section had previously covered physical and analytical chemistry. Commissioning of the plant in 1966 led to a further change to the organisation chart.

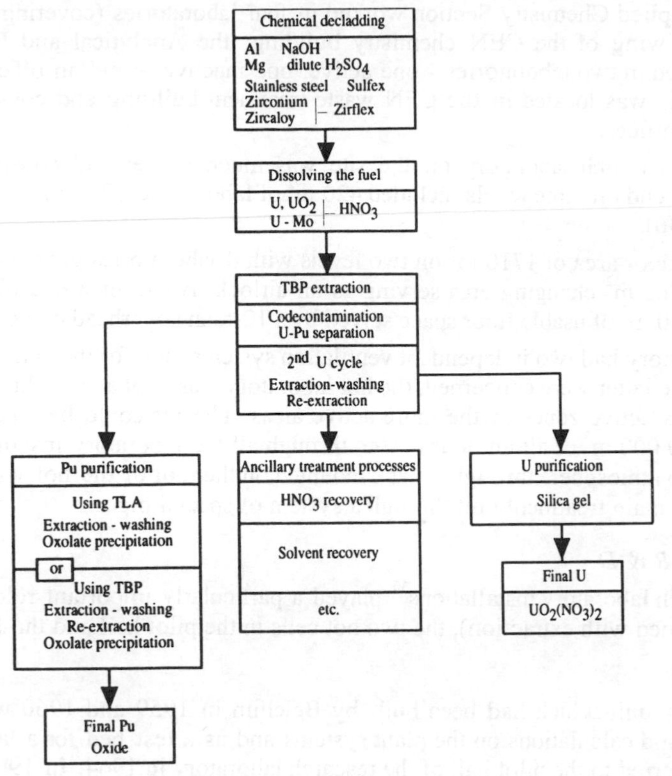

Figure 35. Chemical process flow sheet and variants envisaged in April 1963. Plutonium purification was not yet fixed and there was no provision for processing highly enriched uranium fuels (Source: EUROCHEMIC – ENEA/OECD (1963a), p. 221).

- physical chemistry, responsible for "laboratory investigation of the chemical reactions that might be applied to the project"[27];
- applied chemistry, charged with "developing processes on a larger scale, based upon fundamental research carried out by the Physical Chemistry Section";
- the test station, whose task was to "acquire the technological knowledge necessary for operating the installations, by testing the system at small or full scale";
- analytical chemistry, charged with "developing analytical methods for operating the plant" and with "carrying out special and routine analyses for the other sections";
- finally, the Documentation Section was given the task of collecting and disseminating information[28]. At this juncture it may be noted that of the 316 ETRs published during the life of the company, 184 were issued between 1958 and 1966, totalling 60% over 8 years.

Working conditions: premises and equipment

From the CEN to the research laboratory

Eurochemic began its R & D in rented accommodation at the CEN; this situation continued until the laboratory was built on the site in mid-1964; work was then transferred to the research laboratory in Building 10 and the premises at the CEN had to be decontaminated. One result was that research slowed down somewhat during 1963.

[27] This quotation and those which follow are taken from RAE2, p. 59-60.

[28] An inventory of the Documentation Section was carried out at the end of 1963, when it received the Sloat documentation comprising 2100 USAEC reports, 270 drawings and 300 miscellaneous reports and drawings; at the time the section held 2326 books, 2538 scientific reports and subscribed to 85 technical journals, 25 non-technical journals and nine newspapers.

At the CEN the Applied Chemistry Section worked in four laboratories (covering an area of 231 m^2) and five offices in the active wing of the CEN chemistry building; the Analytical and Fundamental Chemistry Section was accommodated in two laboratories – one active, one inactive – and an office in the same building. As for the test station, this was located in the CEN waste treatment building and consisted of a technical hall covering 140 m^2 and four offices.

Eurochemic's own research laboratory on the site was much bigger and consisted of three units[29]. A "cold" wing at the eastern end on three levels included 420 m^2 of laboratories, 380 m^2 of offices and libraries and 940 m^2 of stores (Figure 36).

A hot wing with a floor area of 1710 m^2 on two levels with 8 laboratories covering 560 m^2 was separated from the cold wing by a 263 m^2 changing area serving as an airlock. Adjacent was a pilot hall with 15 metres overhead clearance and 330 m^2 of usable floor space served by a 10-tonne overhead crane (Figure 37).

The research laboratory had two independent ventilation systems, one for the active areas, the other for the service zones. As far as the latter were concerned, the hot laboratory was kept at a slight variable vacuum so that air circulated from the less active zones to the more active areas. The air could be renewed up to 15 times an hour. In this way some 80 000 m^3 of air an hour passed through all the laboratory installations and were filtered before being discharged to atmosphere through a stack mounted on the roof of the hot wing. Liquid wastes were fed to the establishment's main treatment unit through a system of special pipes.

Systems for reprocessing R & D

Three of the research laboratory installations[30] played a particularly important role: the pilot installations (particularly those concerned with extraction), the two hot cells in the pilot hall and the Janus cell.

The pilot hall (Figure 38)

The extraction pilot unit which had been built by Belchim in 1959 and 1960 was used throughout the period for investigations and calculations on the plant systems and as a test bed for a large number of auxiliary installations. It was transferred to the pilot hall of the research laboratory in 1964. In 1966, following a series of modifications and improvements, it consisted of three air-operated pulse columns, of similar dimensions to the columns fitted to the first plant cycle[31], which could be used for full scale tests. It also included an evaporator similar to the inter-cycle evaporators of the plant, a mixer-settler for washing solvents, and another for washing aqueous phases[32]. It also had the necessary operational instrumentation and auxiliaries.

The pilot dissolving units played a very important role in dissolver development. Several generations of unit were developed in turn: a semi-pilot unit made of glass, a pilot dissolver for decladding and dissolving tests[33], a pilot dissolver unit at half the scale of the first dissolver, and a second dissolver at 1/20 scale of the same design as the second dissolver in the plant. The first two plant dissolvers were in fact tested in the pilot hall[34].

The two hot cells (Figure 39)

The two hot cells located in one corner of the pilot hall were used for chemistry and chemical engineering research using sources ranging up to 2000 curies gamma. Internally they measured 2.75 x 2 metres and were 2.45 metres high. They were equipped with two remote manipulators, a lead glass window at the front and a side window giving a perpendicular view. Access was normally from the back through a kind of porch closed off by a steel plate to give protection from alpha radiation and a 10-tonne plug system running on rails.

There was also an opening in the ceiling of each cell closed by a removable concrete plug. Thus in exceptional circumstances the overhead crane in the pilot hall could be used for transferring equipment between the cells. As a rule, there were two transfer systems, one for the inactive equipment at the front of the cells and another for the active systems at the back. The latter were usually transported by a special lead-lined flask.

---

29 Description in RAE1 p. 328; drawing on page 330 and 333.

30 DETILLEUX E. (1966).

31 Column HA for extraction-washing, IBX for uranium-plutonium separation and 1C for re-extraction of uranium. Initially there were only two columns, then a pulse column built of Pyrex glass was added.

32 Using a diluent in order to remove the TBP carried over from the foot of the first column.

33 RAE 3, pp. 62-64.

34 RAE 3, pp. 64-66.

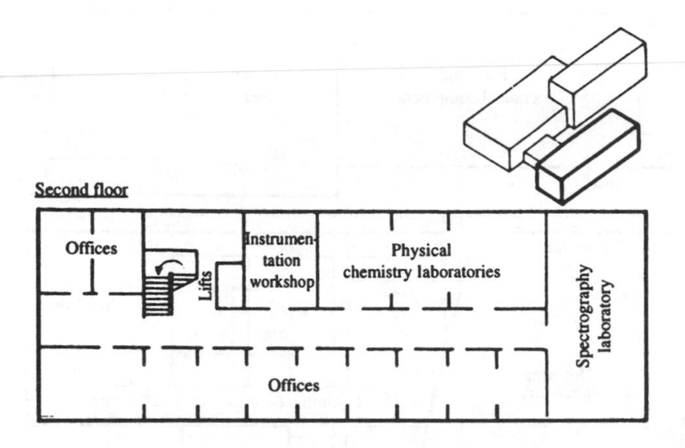

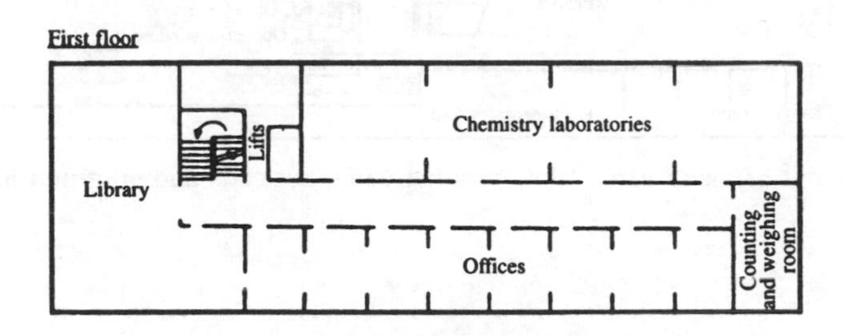

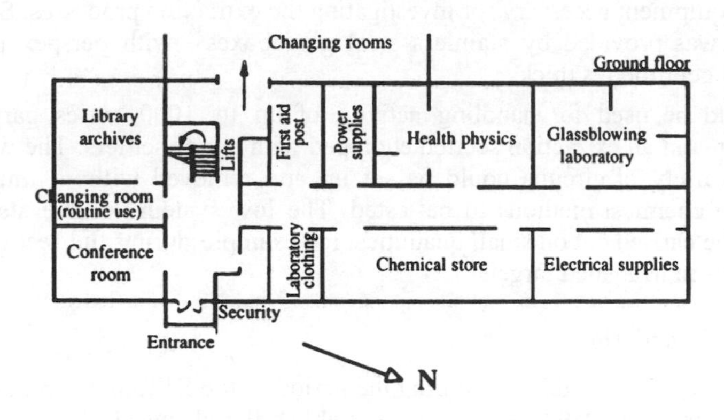

Figure 36. Layout of the research laboratory cold wing (Source: RAE1 (1963), p. 330).

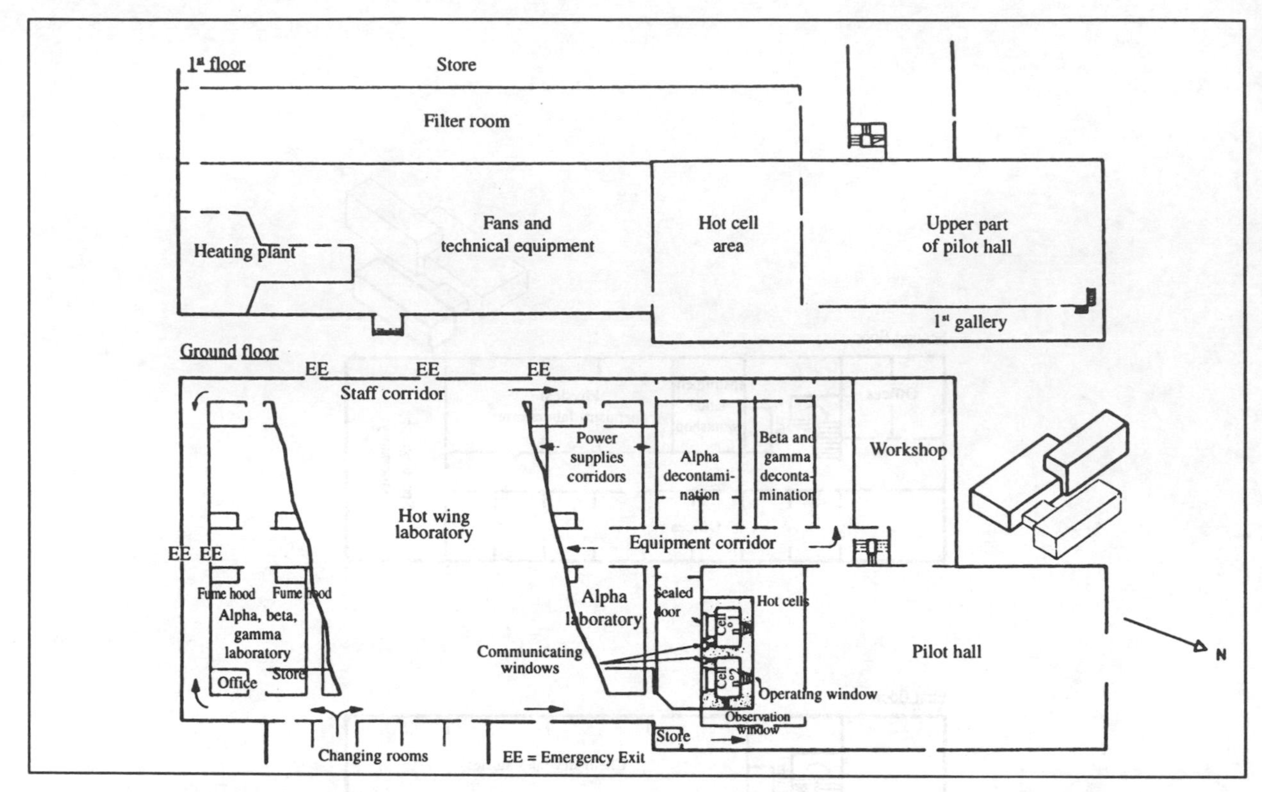

Figure 37. Layout of the research laboratory hot wing, hot cells and pilot unit hall (Source: RAE1 (1963), p. 333).

The "Janus" cell

The multipurpose "Janus" cell, a "miniature reprocessing cell", with a capacity of 3 to 6 kilogrammes of LEU a day and several hundred grammes of HEU, consisted of a series of 13 interconnected enclosures containing the laboratory equipment necessary for investigating the extraction processes. Shielding against alpha, beta and gamma radiation was provided by stainless steel gloveboxes[35] with perspex panels, located behind walls of lead bricks 5 to 15 centimetres thick.

The Janus cell could be used for handling activity of up to 1000 curies gamma. The cell had a dissolving-adjusting section and an extraction section equipped with mixer-settlers. The whole system had been designed so that a great variety of circuits could be set up and removed without much dismantling work, allowing a large number of chemical methods to be tested. The low system volume also meant that material recovery operations could be carried out on small quantities, for example during the recovery and purification of plutonium and neptunium from irradiated targets[36].

Research review[37] (Figures 40 and 41)

The first important task, which had begun during the period of the BER and continued in parallel with the formulation of the pre-projects from 1959 to 1961, was to establish the chemical flow sheets for the Eurochemic plant, on the basis of American publications. Its principal results were incorporated in the SGN project.

Later on, research went on along with the implementation of the detailed pre-project. However the financial problems and the reduced 1964 programme resulted in a slowing down of research, since the workforce stagnated at around 80 people and a significant number was involved in the plant startup as from 1965. Thus the research laboratory was relatively under-used once the construction period had come to an end.

---

[35] There was a total of 26 gloveboxes in the hot laboratory.

[36] However the financial problems seriously affected the work done with this cell, and it never progressed beyond the cold testing phase. Communication from Emile Detilleux.

[37] This section is based upon Eschrich, op. cit., Rometsch p. 45 *et seq.*

Figure 38. Top: General view of the installations in the pilot hall around 1966. Bottom: Detail of the pilot unit for the second dissolver (Source: Eurochemic photographs, undated).

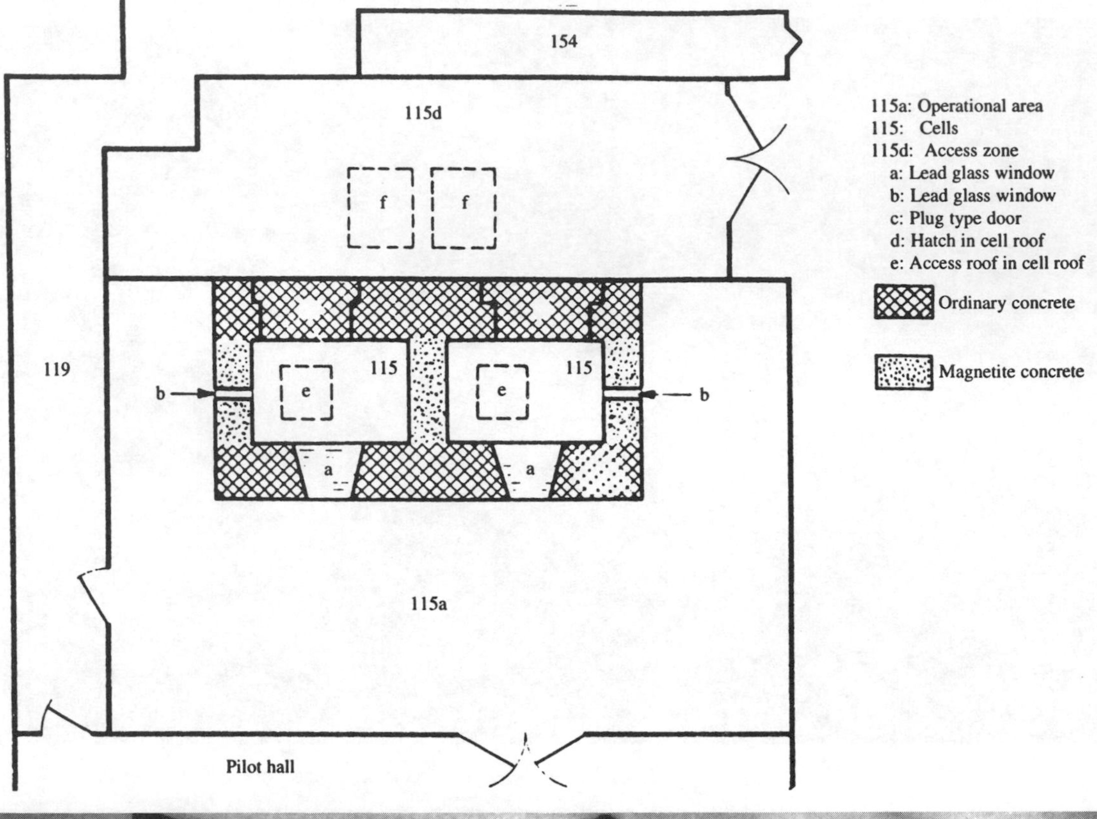

Figure 39. Top: layout of the twin alpha hot cells located at the pilot hall entrance (Source: RAE 3, p. 38). Bottom: Access zone 115d, with a plug-type door and the rails on which the door runs (Source: Eurochemic photograph, no date).

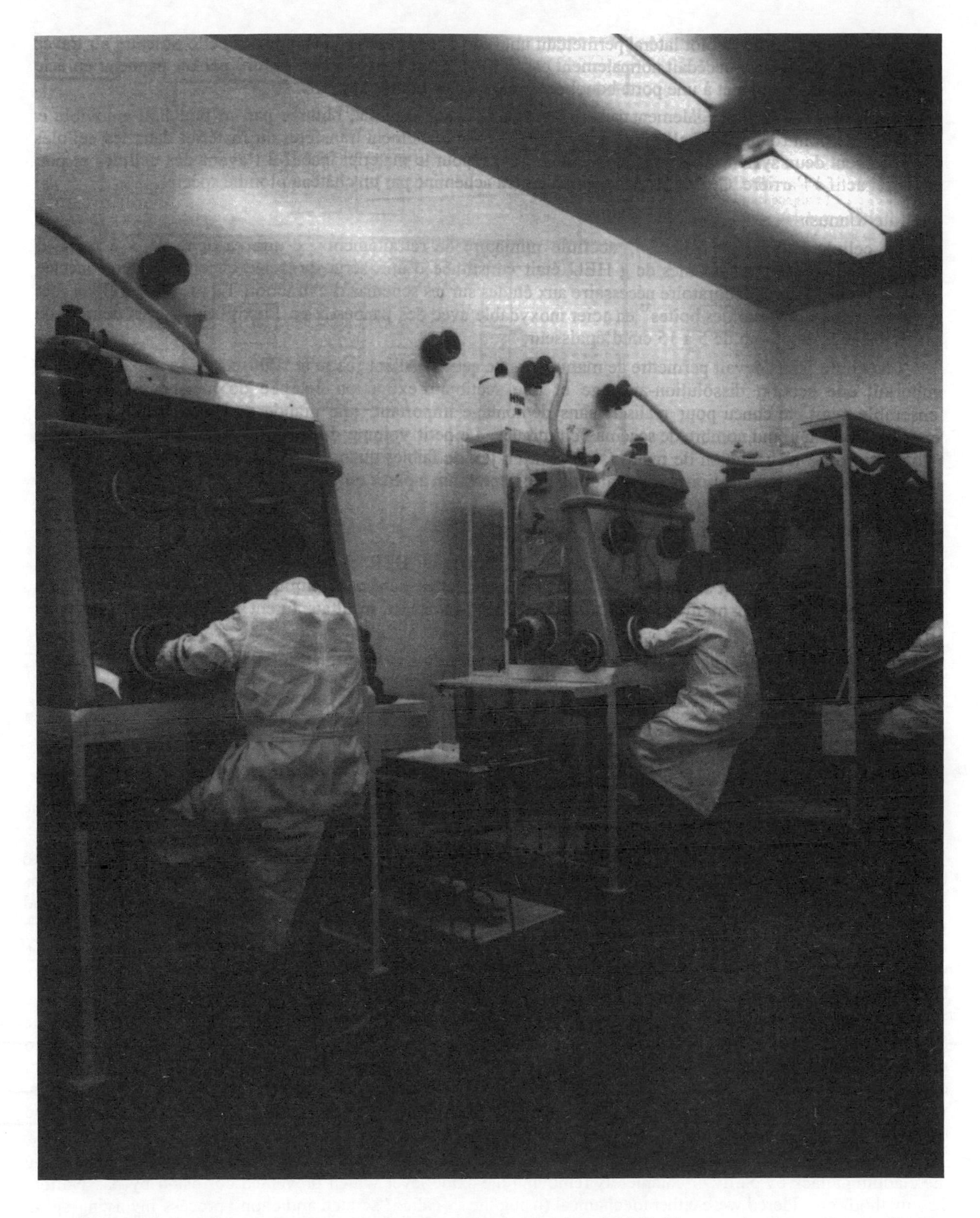

Figure 40. Set of three gloveboxes in the research laboratory around 1965. The pipes leading to the gas filtration system can be seen leaving the top of the gloveboxes (Source: Eurochemic photograph, no date).

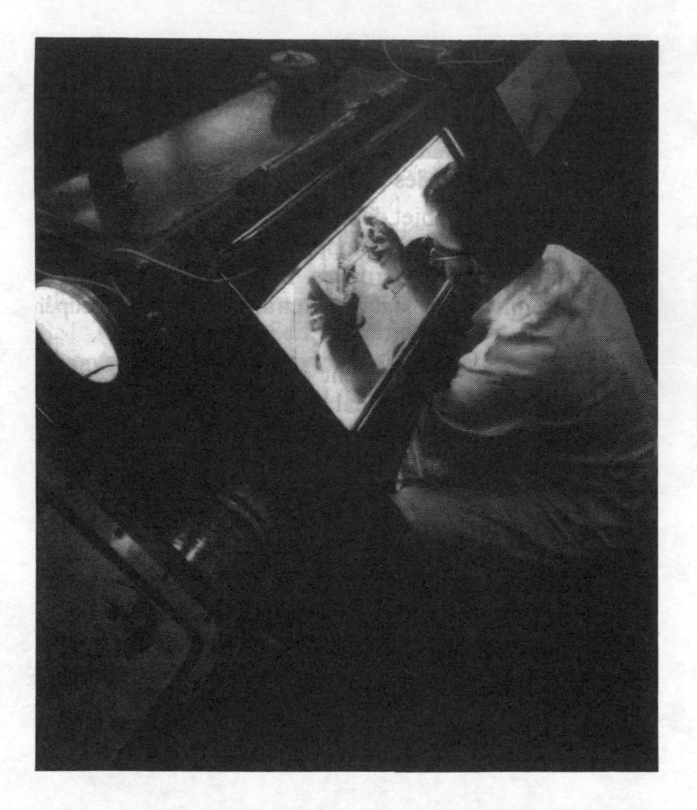

Figure 41. Working in the alpha glovebox in the research laboratory (Source: Eurochemic photograph, no date).

Eurochemic research covered three main areas: process improvements, the analytical methods needed for monitoring the reprocessing operations and, finally, waste management.

Process research

The main results of Eurochemic research during the construction period can be summarised for each of the different stages of the process.

As regards the head-end of the process, Eurochemic had finally opted for chemical decladding in order to be able to deal with the great variety of cladding materials: aluminium, magnesium, chromium-nickel stainless steel, zirconium and zircaloy, a zirconium-tin alloy. The flow sheets for all these had to be worked out and optimised. The same applied to the process of dissolving elements from the fuel itself: uranium metal, uranium dioxide, uranium-molybdenum alloys, uranium-aluminium and uranium-zirconium. It was essential to test the decladding and dissolving solutions in order to determine the best conditions as regards concentration, temperature and contact time.

As far as stainless steel and zirconium cladding was concerned, Eurochemic led the field in upsizing the SULFEX and ZIRFLEX processes (of American origin) to an industrial scale. The adaptation of these two processes required a very large amount of experimental effort. Eurochemic also developed and patented a variant of ZIRFLEX known as CITRIFLEX, in which the use of citric acid improved the solubility of the zirconium, with the oxide fuel being subject to less attack than in the ZIRFLEX process. However the process did not find any industrial application.

Improvements to the ZIRFLEX process were also investigated with a view to facilitating and accelerating the cladding attack by partly or completely removing the oxide layer which prevented corrosion by passivation. The methods considered were either mechanical (using the so-called "scratch and chop" process involving spiral notching, or electrochemical using the so-called "clean-up" process.

Extraction developments covered the determination of distribution data in the acid-solvent system[38], and formulating the extraction flow sheets which were then tested using low burn-up fuel in the shielded cell

38 Also called the distribution coefficient.

containing mixer-settlers. It was necessary at the time to determine the acid-solvent distribution data because the coefficients varied as a function of concentration. The tables of coefficients determined in military reprocessing experiments were still classified. In the civilian field, the PUREX process had so far hardly been used other than for fuel with a low plutonium content so the coefficients had to be determined beforehand in laboratory experiments.

The performance of the selected flow sheets was also verified in the three pulse columns of the test station, using uranium solutions. They were also tested using solutions obtained from real irradiated fuel under an outside contract signed with the pilot extraction unit at JENER in Kjeller.

Plutonium reduction in the partition cycle using uranyl nitrate instead of the conventional ferrous sulfamate was tested before being successfully transferred to the plant after only a few reprocessing campaigns[39]. Early tests had shown the unstable nature of uranyl nitrate, but Eurochemic succeeded in producing stable solutions by adding hydrazine during the electrolytic reduction of uranium to valency 4+. This modification was the main R & D contribution to improving the PUREX process during this period and paved the way for a practice still used today in reprocessing plants. The only process improvement made since is direct electrolysis in the columns. The behaviour of the pulse columns was investigated in detail in the test station. Original methods were devised notably as regards the use of compressed air for operating and controlling the pulse columns[40].

Research into final plutonium purification was conducted under two main headings: testing of the conventional process involving anion exchange and the development of a new extraction method using tertiary amines and quaternary ammonium salts. The process with trilaurylamine (TLA) was used exceptionally during the first ever reprocessing campaign for recovering the first 30 grammes of plutonium.

The Eurochemic laboratories also developed an electrolytic process for directly converting the uranyl nitrate – the normal final product of reprocessing – into uranium tetrafluoride, an intermediate compound subsequently allowing either the production of hexafluoride for enrichment or of oxide for the direct fabrication of recycled fuel. The laboratory preparation of tetrafluoride opened the way to the SFU project which was first applied on the site in 1969-1970[41].

Analytical research

Work in the Analytical Chemistry Section was under four main headings: the development and testing of analytical methods, equipment and continuous monitoring instruments, and the automation of analytical procedures[42].

More than 300 analytical methods[43] were tested or improved, as regards process control, analysis of liquid wastes and the specifications of the final products. Those giving the best results were included in the plant analysis handbook.

The research laboratory contributed to the development of equipment that could be installed in the plant, for example an X-ray and gamma absorption meter, a colorimeter, a neutron monitor and an alpha monitor. However only the alpha monitor was installed, then, irretrievably contaminated after a few days of plant operation, it was removed.

As regards the automation of analysis, laboratory systems for remotely analysing highly radioactive specimens were developed in one of the gloveboxes of the Janus cell.

A remotely-controlled sampling system using a pipette, known as a "calibrated stopcock"[44], more robust and reliable than those available on the market at the time, was developed and tested. Another of its advantages was a substantial reduction in manipulation time and hence in the risk of contamination or irradiation.

The limited extent to which laboratory research on hardware led to plant applications reflects a problem of co-operation between researchers and technicians. One of the researchers reflected, twenty years later, on the situation regarding in-line instruments:

---

[39] The use of this method is described in detail in Part III Chapter 2.

[40] Such as "bottom interface control using direct air purged dip tubes (located in the column decanter)", and an "air-pulsation system in which the compressed air inlet and outlet valves were controlled by an electrically driven camshaft".

[41] See Part III Chapter 2.

[42] These aspects are covered in more detail from a different standpoint in Part V Chapter 1.

[43] See for example the list of methods in RAE 3, pp. 124-128.

[44] Detailed in ETR 318, pp. 77-79.

"Today, I know there is nothing wrong with in-line instruments, but many things can go wrong between a plant operator and a research worker. They speak different languages and they are usually not too much inclined to learn to understand the language of the other. The application of in-line instruments in a reprocessing plant is not a technical problem, it is a communication problem or maybe even a problem of faith or confidence"[45].

The plant technicians preferred tried and tested methods and basically had little interest in testing out the products of their research colleagues on the plant. Also the contraction of the research programme owing to financial difficulties had prevented the researchers from developing their laboratory innovations to the stage of advancement required by the plant operators.

Initial research into waste conditioning

This research was carried out by the Process Chemistry Section and became one of the research priorities in the reduced 1964 programme.

Work began on the treatment of intermediate and high activity liquid wastes, as from 1963, with investigations of the existing processes, followed by tests and more advanced studies of the incorporation and solidification of intermediate activity wastes into bitumen or concrete[46]; as regards high activity waste, research was done on adapting a British vitrification process[47] to the characteristics of Eurochemic's irradiated fuel.

Finally the Eurochemic laboratory designed an original process for solidifying high and intermediate liquid wastes known as the Neutralisation Self Solidification Process (NSSP), but the programme was halted during active tests on intermediate activity liquid wastes following the failure of one of the essential components in the process[48].

Although the final outcome in terms of innovation appears fairly sparse, it must not be forgotten that the main R & D task was to prepare for starting up the plant in the best possible conditions. From this point of view its work was of excellent quality. However it was the Technical Directorate that was in direct contact with those involved outside the plant.

Co-operation between Eurochemic and outside contractors

The pivotal role of the company's Technical Directorate

Since the Technical Directorate was to be responsible for operations, its staff formed close links with the architect engineers and the Research Directorate in the preparation of the flow charts and kept a very close eye on construction. Hence by the time of startup, the Directorate had a detailed knowledge of this new type of plant.

The changing role of the Technical Directorate

The Technical Directorate was in fact the main link with outside contractors, and its functions and its staff evolved as work progressed. Initially a small group by comparison with those involved in research, it ultimately became the largest division of the company in terms of staff numbers.

From 1959 to 1961 the Technical Directorate had the task of reviewing and monitoring the pre-projects, matching them to the company's technical and financial requirements, and then co-operating with SGN in formulating the detailed pre-project.

Once this had been adopted and work begun, the Technical Directorate's job was to build the conventional auxiliary facilities, supervise the work of the architect engineers and co-ordinate construction along with SGN. Thus the directorate drew up invitations to tender, ordered equipment and supervised all work on the site.

When installation was complete, its task was to approve the acceptance tests and to conduct the commissioning tests on buildings and equipment.

The Technical Directorate was ultimately responsible for operating the installation. This involved it in three main tasks: first, and primarily, operation of the project reprocessing system and auxiliary services (power supplies, procurement of non-active products, compatibility of fissile materials, etc.); secondly evaluating the industrial feasibility of new methods; finally the collection of operating data for the purposes of technical and economic assessment.

---

45 Ibid. p. 80.

46 RAE 3 p. 109.

47 FINGAL.

48 This was a self-cleaning worm type heat exchanger manufactured by Lurgi. This process was the subject of three ETRs between November 1967 and February 1969, n° 219, 242 and 258.

The make-up and organisation of the Technical Directorate

Accordingly the number of staff working in the Technical Directorate rose rapidly: in December 1961 it employed 38 people including 16 engineers (civil, mechanical and chemical engineering) and 15 technicians and draughtsmen, directed by Teun Barendregt; they were grouped in seven specialist sections (see organisation chart in Figure 33), sub-divided into 11 detailed project supervision groups[49] corresponding to functions or buildings (for example ventilation-instrumentation, research laboratory, and so on). The Health and Safety Section was attached to the Technical Directorate until November 1962 after which it came under the General Management.

In January 1964 staff numbers in the directorate reached 114, including 10 graduate or equivalent engineers, 25 other engineers, 32 operators, technicians, draughtsmen and administrative staff, and 47 assistant operators and unskilled staff.

The Directorate consisted of six divisions.

By January 1966 it had a staff of 198 including 16 graduate or equivalent engineers, 35 other engineers, 91 operators, technicians, draughtsmen and administrative staff, and 56 assistant operators and unskilled staff. The main reason for the increase in numbers was the recruitment of plant operators who familiarized themselves with the systems during the period of acceptance and testing.

The need for large numbers of staff, particularly from the autumn of 1965, resulted in personnel being made available from the Research Directorate (see above). In fact Eurochemic took possession of the plant from the beginning of 1965 and as from October of that year assumed full responsibility, including the installation of the pipework in the final units. System testing began and work started on a four-shift basis.

In 1966 the Technical Directorate was restructured into specialist divisions as the start of reprocessing approached: the main division was responsible for the chemical process. A second division was responsible for receiving fuel, programming the plant and effluent treatment, a third for the analytical laboratory, a fourth for nuclear services – criticality prevention, radiation protection and nuclear accounting. The organisation also included general services and the American consultant, Earl Shank, who worked closely with the Technical Director.

Relations between the architect engineers and Eurochemic. The overall specification of January 1961 and the distribution of functions and tasks

The 1961 organisation

Contacts with the architect engineers were first made in 1959 and discussions continued in 1960[50]. The procedures for co-operation between Eurochemic and contractors were definitively laid down at the beginning of 1961[51].

These procedures were relatively simple for the research laboratory. The SDS group was responsible only for design work, invitations to tender and orders, work being supervised by the company itself. Payment was on a flat-rate basis.

Settling problems of liability, patents and payment

Procedures were more complicated however for the plant and its auxiliaries. Negotiations began in July 1960 with regard to the three items that were still problematical: liability, intellectual property and payment.

As regards liability, the agreement drew a distinction between "direct damage caused to Eurochemic, during its accomplishment of its own obligations, by negligence, contravention of normal practice or professional fault"[52] and "liability as regards injury and material damage caused to third parties, including Eurochemic staff and that of the other architect engineers [...] by a nuclear accident"[53]. In the first case, except for serious negligence, liability was limited to 20% of the amount of fees. In the second case, Eurochemic's liability supplanted that of the architect engineers.

---

49 RAE 1 p. 233.

50 See Part II Chapter 1.

51 RAE1, pp. 359-362 and POHLAND E., STROHL P. (1962).

52 General Conditions of Contract, p. 16.

53 Ibid.

The discussions on intellectual property were difficult, and resulted in highly complex provisions in the specification, covering five pages[54], attempting to deal with all possibilities, the starting point being joint ownership of inventions[55].

As regards payment, the overall principle was a percentage of the cost of the "works and supplies for which [the architect engineer] has prepared the documentation and contracts", these costs comprising "the duties, taxes and transport costs actually paid"[56]. The percentage figure for each type of work was determined with reference to the fees charged by Swiss engineers and architects.

The General Conditions of Contract of 16 January 1961

The General Conditions of Contract relating to the architect engineer contracts for the plant were finally published on 16 January 1961. This was a 24-page document consisting of 17 Articles, together with a six-page Annex. The principal tasks incumbent on the architect engineer were as follows:

"– preparation of the general project;
– preparation of invitations to tender;
– review of bids and formulation of orders;
– checking design work, construction methods, drawings and construction specifications;
– works supervision [tasks specified hereunder];
– commissioning and acceptance of the installations"[57].

Works supervision involved the following:

"– General organisation of site works;
– Co-ordination of contractors and suppliers;
– Checking that work conformed with the rules of practice and with the approved drawings and specifications;
– Acceptance of materials and equipment in the plant and on the site;
– Quantity and general surveying;
– Preparation of provisional and final accounts and proposals for payment [..];
– Monitoring committed expenditure, providing Eurochemic every two months, through the co-ordinating architect engineer (AIC), with a schedule of this expenditure"[58].

Co-operation between SGN, Eurochemic and the other architect engineers

Four different types of group were involved on the site: Eurochemic as employer, SGN as the co-ordinating architect engineer (AIC)[59], the architect engineers and, finally, the contractors. The fact that SGN was the only organisation on the site with knowledge and experience of reprocessing placed it in an unassailable position.

As architect engineer, SGN was responsible for all the civil engineering works for the plant and the extraction system, the heart of the process. It was also responsible, in conjunction with Montecatini, for the head-end and tail-end of the process. As the co-ordinating architect engineer SGN was Eurochemic's main adviser and co-ordinated the design and construction work in both time and space. It was also responsible for defining the essential joint standards. Thus Eurochemic and SGN worked very closely together as main contractor and project manager.

Eurochemic was the project manager and its Technical Directorate in particular was the liason with outside contractors. The work was carried out on the basis of the detailed pre-project (Figure 42)[60], which was of course modified as work proceeded, for example in the light of the results of laboratory experiments or decisions

54 Ibid. pp. 18 to 22.

55 These clauses were never applied.

56 Ibid. p. 22.

57 General Conditions of Contract, p. 5.

58 General Conditions of Contract, p. 12.

59 The term "principal architect engineer" (PAE) was abandoned; the same applied to the "specialist architect engineer" (SAE), which became simply AE.

60 Figure 42 shows the stages between the pre-project and actual installation. Note the "feedback" effect for pipework.

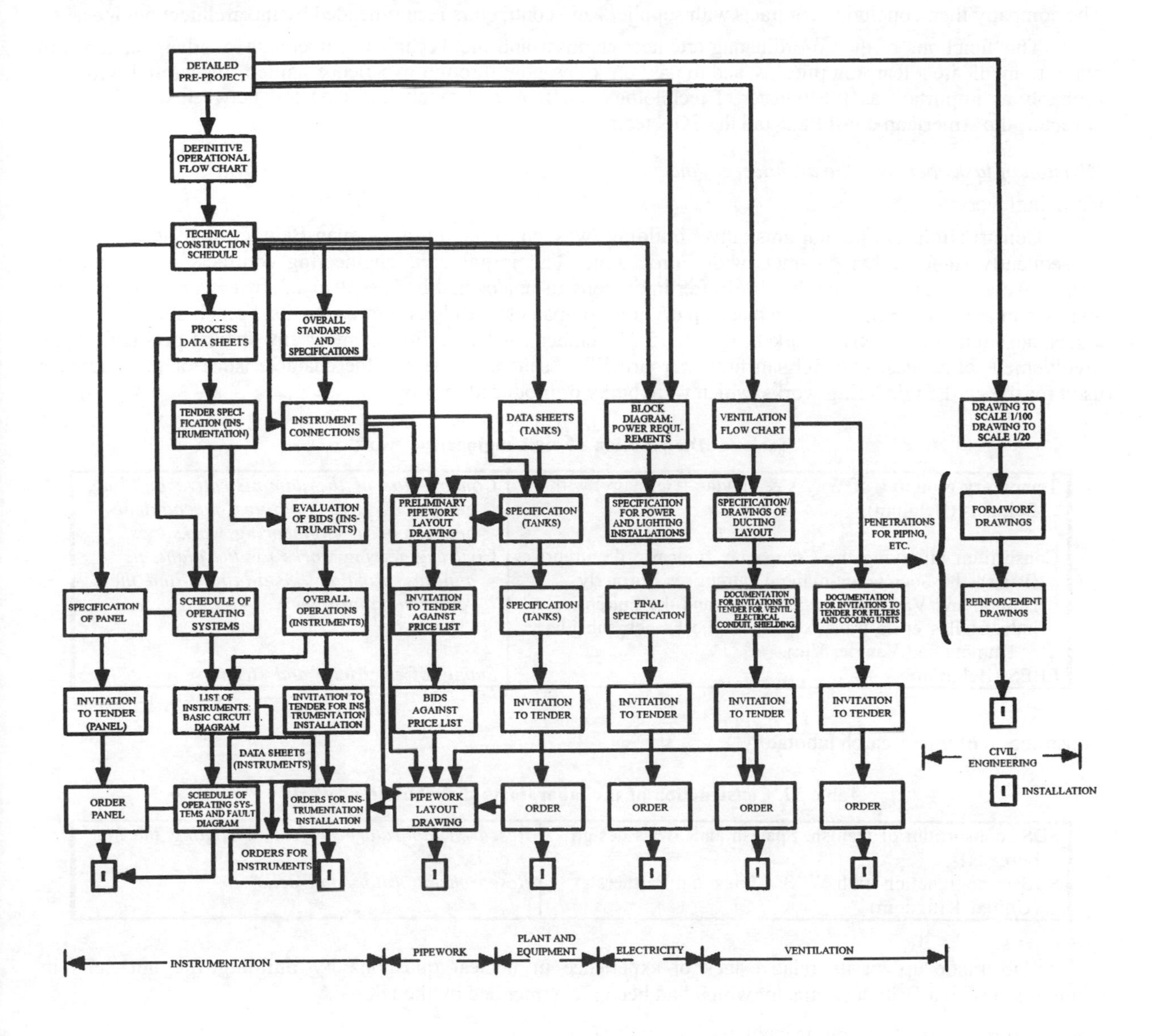

Figure 42. Diagram showing the main stages of project co-ordination, from the detailed pre-project to the tendering procedure for the plant and the final product storage building, and finally the installation (Source: RAE 2, p. 74).

to enlarge systems. At each stage of construction, the architect engineers proposed technical approaches from which Eurochemic then made a choice. Consequently all drawings were submitted to Eurochemic for approval. The company then concluded contracts with suppliers and contractors recommended by the architect engineer.

The functions of the co-ordinating architect engineer and the Technical Directorate overlapped, the aim being to inculcate a learning process, and to pass on know-how through experience gained, in parallel with and probably as important as the transfer of technology itself. A real symbiosis emerged between the Technical Director, the American consultant and the SGN team[61].

Division of tasks between the architect engineers

Civil engineering

Construction of the administrative building was entrusted to a German-Belgian consortium which subsequently signed other contracts with Eurochemic. The initial civil engineering contracts for the nuclear portion were concluded in October 1961 after invitations to tender had been sent to 23 firms. The contract was awarded to a Franco-Belgian group made up of three companies: a Belgian company, its French subsidiary, and a separate French company. Work began on 15 November 1961 and did not meet any special problems. The involvement of at least one Belgian firm was virtually essential in view of the characteristics of the materials used for the civil engineering works, which were bulky or produced locally.

Table 36. **Distribution of civil engineering work**

Temporary grouping of Wayss & Freytag (Germany) with Belmans (Belgium).	*Construction of the administrative building, general services buildings, mechanical and electrical engineering workshops.*
Consortium involving the Compagnie française d'entreprises (Paris), the Société anonyme d'entreprises, formerly Dumon and Van der Vin (Brussels), and the Entreprises industrielles et de travaux publics (the French subsidiary of Dumont and Van der Vin).	*Civil engineering works for the plant, its annexes, auxiliary installations and the research laboratory.*
EBES (Belgium).	*Supply of electricity and steam.*

Construction of the research laboratory

Table 37. **Construction of the laboratory: distribution of work**

SDS (consortium of Danish, Spanish and Swiss design bureaux)	*Research laboratory project (excluding the hot cells)*
SDS in conjunction with W. S. Atkins and Partners (United Kingdom)	*Research laboratory hot cells*[62]

SDS made up for its relative lack of experience in nuclear questions by building the hot cells in conjunction with a British contractor which had been recommended by the UKAEA.

Construction of the plant and its auxiliary buildings (Figure 43)

The distribution of work between the plant's architect engineers was finally settled when the detailed pre-project was submitted. For the activities planned at the time it was very close to the original breakdown[63]: Siemens and Halske had however lost their position as exclusive suppliers of laboratory equipment in favour of a broader tendering procedure (see below).

A German firm – a member of IGK – and the Norwegian architect engineer were made responsible for new activities.

On 28 March 1963, when agreement was reached on the method to be adopted, it was the firm Friedrich Uhde – a participant in IGK – which was appointed architect engineer for the plutonium purification unit[64].

[61] In his communication, Earl Shank lays considerable stress on his continuous co-operation with Teun Barendregt and on the value of his practical experience acquired from the construction of the American reprocessing plants.

[62] Decision of the Board of Directors of 14 March 1962.

[63] AG(62)7.

[64] CA(63)10.

Figure 43. Aerial view of the site, probably taken in May 1963 (Source: CEN/SCK photograph C/EC-60, print dated 4 June 1963).

Table 38. **Distribution of tasks amongst the architect engineers**

Firm	Tasks
Belchim (Belgium)	Final purification of uranium
Belchim	Effluent treatment and waste storage
Comprimo (Netherlands)	Instrumentation for the principal sections of the plant
IGK (Arbeitsgemeinschaft Leybold Lurgi, Uhde, Dortmund, est. in 1963) (Germany)	Solvent recovery; ventilation and analytical laboratory
Montecatini (Italy)	Preparation of organic and inorganic solvents; final product storage.
NOHAB (subsidiary of Bofors) (Sweden)	Fuel reception and handling
Noratom (Norway)	Storage of high activity products
SGN (France)	Uranium and plutonium extraction
SGN	Civil engineering for the main plant building
SGN and Montecatini (consortium)	Fission product concentration and recovery of nitric acid after concentration.
SGN and Montecatini (consortium)	Dissolving and nitric acid recovery thereafter

Subsequent to September 1964, Noratom was responsible for that part of the waste treatment project made necessary by the extension of reprocessing to highly enriched fuels[65].

Payment of architect engineers; the problem of cost evaluation

Payment was calculated as a percentage of the cost of the works plus miscellaneous costs. Payment at actual cost was planned for commissioning and works acceptance operations. Payments to the architect engineers were originally not expected to exceed 10% of the total cost of the works.

In March 1961 it had been estimated that the services of the architect engineers would cost \$1.8 million[66]. In November 1962 the architects indicated that unexpected problems and the unprecedented costs

[65] CA(64)26.

[66] CA/61/4.

of co-ordination had increased their expenditure and proposed that fees should be paid at a rate of 14%. On 16 September 1963 a revised estimate of $3.2 million was submitted, broken down as follows[67]:

Table 39. **Structure of architect engineer's cost estimate**

Pre-projects	$0.5 million
Fees (10% of costs)	$1.7 million
Miscellaneous costs	$0.5 million
Modifications and additional design work	$0.5 million
Total	**$3.2 million**

In December 1965, total expenditure under agreed contracts was estimated at $3.6 million, but it was to be expected that this figure would be exceeded following modifications and additional work[68].

The costs incurred by co-ordination in particular proved appreciably higher than expected. Owing to the company's financial difficulties, it was not possible to meet the request of the architect engineers which was repeated in 1963. By the end of October 1963[69] the position of certain specialist architect engineers became delicate. Pierre Danneaux, on behalf of Belchim, stressed the problem his firm was facing. Three months later, on 1 January 1964, Belchim's activities were transferred to Belgonucléaire.

Teun Barendregt acknowledged that losses averaged nearly 40% amongst the different architect engineers. The problem was raised again when the Board of Directors met on 23 June 1965. The Managing Director proposed a provision of $0.75 million to cover current contracts[70]. On 7 December 1965, a provision of $0.5 million was decided upon, but merely as a holding action.

The architect engineers' costs continued to inflate, doubling between 1961 and 1965, but they still appear to have made no money out of the operation. However the experience gained did allow them to make up on other projects[71].

To avoid further cost overruns, the management decided to limit the architect engineers' involvement in testing, for which it had originally been agreed that they would be paid. They were nevertheless allowed considerable unpaid access. By limiting the role of the architect engineers there was a commensurate increase in the need for Eurochemic staff, which accelerated the transfer of researchers to the Technical Directorate[72].

Eurochemic and its suppliers. Invitations to tender and contracts for works and supplies

During construction of the plant, several hundred European and non-European contractors supplied equipment and services[73]. It is possible to give a partial summary[74] in terms of the national distribution of contracts, with specific examples.

The distribution of supply and works contracts

Table 40 summarises the distribution (by nationality of contractor) of the cumulative supply and works contracts up to 31 December 1966, which total $25.89 million, arranged in decreasing order[75].

The main factor underlying the distribution appears to be geographical proximity. Over a third of the contracts were awarded to Belgian firms, with nearly 30% going to France. Germany comes far behind with 11%, which clearly expresses the difference between the two countries' nuclear industries at the time. The United Kingdom which, like France, had an advanced nuclear industry, was the company's seventh largest supplier.

67 CA(63)31.

68 CA(65)29.

69 CA/M(63)4 of 28 October.

70 CA(65)11.

71 Communications from M. Lung and W. Schüller.

72 To implement startup, 57 operations technicians had to be recruited. See CA(64)35 of 30 November 1964.

73 The quarterly Progress Reports list the contractors tendering for each contract, with the subject of the contract, its value and the reason why the particular firm was chosen. However the archives are incomplete, making any statistical analysis impossible. Therefore we must content ourselves here with an overall review for each country and significant examples.

74 A detailed review would require investigation of the files of calls for tender, which occupy several tens of metres of shelving in the Belgoprocess archives.

75 Table prepared from RAE3, p. 17, rounded figures, calculations from real data.

Table 40. **Distribution of supply and works contracts amongst the participants**

Country	Total (million $)	%	Through Belgian subsidiaries (%)[76]
Total	25.89	100.00	17.39
Belgium	9.14	35.31	17.79[77]
France	7.61	29.41	7.76
Germany	2.90	11.21	48.16
Netherlands	1.62	6.27	27.14
Italy	0.94	3.63	1.89
Norway	0.86	3.32	0.00
United Kingdom	0.71	2.73	21.47
Sweden	0.67	2.61	5.45
Austria	0.52	2.00	1.22
Switzerland	0.36	1.38	21.07
United States	0.20	0.79	72.95
Denmark	0.20	0.79	2.71
Portugal	0.11	0.45	0.00
Spain	0.02	0.08	0.00
Japan	0.001	0.00	100.00
Finland	0.0004	0.00	100.00

A significant example: plant and equipment

Equipment for the chemical plant alone consisted of nearly 200 main systems: dissolvers, evaporators, columns, mixer-settlers, large tanks; most involved equipment to transfer solutions – ejectors, air lifts and pumps – and valves of which there were 3000 in the plant. The process pipework totalled 70 kilometres in length, to say nothing of the steam, vacuum, water, and other pipes and those used for instrumentation. Other process buildings were less complex but still contained over 50 principal systems and 6 kilometres of piping.

The contractual process, particularly for special equipment[78] took a fairly long time owing to the political requirement to distribute orders as far as possible amongst the different member countries so that industry could gain experience.

For example the invitation to tender for the first dissolver and its Ni-O-Nel[79] piping was issued at the end of 1962 and the order issued in April 1963. Sixteen firms were invited: three German, three French, two Belgian, two Italian, two Dutch, two Swedish, one Swiss and one British. The German firm Essener Apparatenbau, whose equipment was the only one to meet the strict tolerances applied by Eurochemic, won this contract worth $132 013[80]. During the same quarter, an order for containers in special steel – silicon Uranus B6 – involved twenty-one firms of eight different nationalities. This contract worth $272 297 was won by the Italian firm proposed by the architect engineer, which submitted the lowest bid and was technically the most attractive.

A pilot dissolver unit in special stainless steel – also Uranus B6 – intended for preparing the specifications for the second dissolver, was the subject of another invitation to tender issued in 1962 to 15 firms in nine countries[81]. The contract was awarded to a French boilermaking contractor which already had nuclear experience, the Compagnie des ateliers et forges de la Loire (CAFL).

Most of the invitations to tender for special equipment were sent out during the summer of 1963 and the tendering process was completed in mid-April 1964. Standard equipment items (such as instruments, valves, steam ejectors and piping) were ordered in batches, still retaining the principle of distribution between industry in member countries. Deliveries of equipment began in the summer of 1964, although certain systems were installed up to a year later. Most difficulties were caused by the special steels: contractors were inexperienced in working with unusual alloys[82]. There were also problems with installation itself, especially as regards welding: for certain pipes the use of Uranus B6 had to be abandoned and suspect welds on the pulse columns and other systems remade.

---

76 The proportion passed through Belgian companies amounted to $4.5 million, including $1.4 million for Germany alone, usually through Belgian subsidiaries of foreign companies.

77 Through a French company.

78 Thus Portugal provided the overhead crane for the fuel reception hall.

79 Progress Report n° 23, Annex III, pages 14 and 15, and RAE3 p. 173.

80 The technical problems encountered and overcome in manufacturing the dissolvers are covered in Part V Chapter 1.

81 The relevant Progress Report is not in the archives.

82 For problems with materials, see Part V Chapter 1.

Stringent quality control

Owing to the nuclear character of the installation and its safety requirements, a very stringent quality control system was introduced which covered all stages of the project from the design and manufacture of systems through to their installation.

The decision in favour of direct maintenance made the requirements even stricter. Systems and piping had to be as tough as possible in order to minimise repairs and maintenance. In particular this decision had a considerable impact on connections between tanks and piping, and those between individual tubes. It was out of the question to use seals. Broken seals or mere leaks would have required several weeks or months of cell decontamination before they could be repaired. Consequently the whole system had to be welded and the quality of welds carefully inspected to avoid corrosion, notably as a result of oxide inclusions. Earl Shank's experience in building reprocessing plants in the United States was valuable, and Teun Barendregt used it as a basis for the quality control procedures he introduced[83]. Also the large number of people involved in the international undertaking made an advanced and effective inspection system even more necessary[84].

The specifications were drawn up by Eurochemic on the basis of standards proposed by SGN, with the architect engineers responsible for quality. Inspection was carried out by external approved national organisations: for example in Belgium by Controlatom.

In certain cases – for example with equipment important to nuclear safety – inspection was compulsory during manufacture. Particularly careful attention was paid to inside dimensions and the incorporation of neutron poisons. Staff also had access to the manufacturer's records and were able to comment on the test procedures used.

During installation, the inspection and approval process was systematic, conducted jointly by the architect engineer and Eurochemic, both of whom had access to all the results of radiographic and seal tests. Checks were made to ensure that the cells complied with the drawings. For certain special systems, a procedure specific to each was defined and applied.

Once each unit was complete, it was inspected by the Eurochemic process team; provisional acceptance was granted by the General Services Division, which then handed over the unit to the Chemical Process Division. On completion of each cell, a photographic survey was carried out to supplement the installation drawings in facilitating identification and the preparation of future repair and modification work. Also every pipe was identified by a numbered label.

The schedule of work: co-ordination problems

The work schedule

Civil engineering work on the research laboratory began on 15 November 1961 and on the plant building on 1 January 1962. The major works were carried out between 1962 and 1964[85]. Unfortunately there were two fatal accidents during this phase of site construction. The schedule was delayed by the hard winter of 1962-1963, a local shortage of labour in the spring of 1963[86], and finally by a strike over pay.

The sheer weight of the processing plant, built on sandy soil, required foundation piles to be driven to a depth of 8 metres. One of the trickiest problems encountered during the construction work was to determine where openings should be left in the cell walls for pipework to be installed when equipment was assembled later on (Figure 44). The precise location of these penetrations depended directly on where the equipment was to be installed, something not yet fully defined at the time the concrete walls were poured.

A further problem was the installation of the seven extraction cycle pulse columns, which came up at the end of 1964 and was finally resolved in January 1965. In fact the pulse columns in the plant were 12 metres high and very narrow. It was decided to install them through the top of the cells using a helicopter. A team stationed on the roof had to work with the pilot so the columns could be inserted smoothly into position (Figure 45).

While the architect engineers were completing the design work, preparing the final functional flow sheets as well as those for installation and execution, they were also preparing orders.

---

83 Communication from Earl Shank.

84 RAE3 p. 179 *et seq.*

85 RAE2 p. 169 *et seq*, and the chart showing progress with the construction of the three sections and galleries of building 1; section 1 (pretreatment), section 2 (extraction and evaporation), section 3 (final treatment and special treatments), each part being separated from the others by expansion joints, ibid. p. 202.

86 Workers had to be recruited from an area up to 50 kilometres from the site.

Figure 44. Photograph of the inside of a cell before the plant was commissioned, showing the extreme complexity of the piping (Source: Eurochemic photograph, undated).

Figure 45. A helicopter being used to install one of the pulse columns in the plant in January 1965 (Source: Slide taken by Earl Shank).

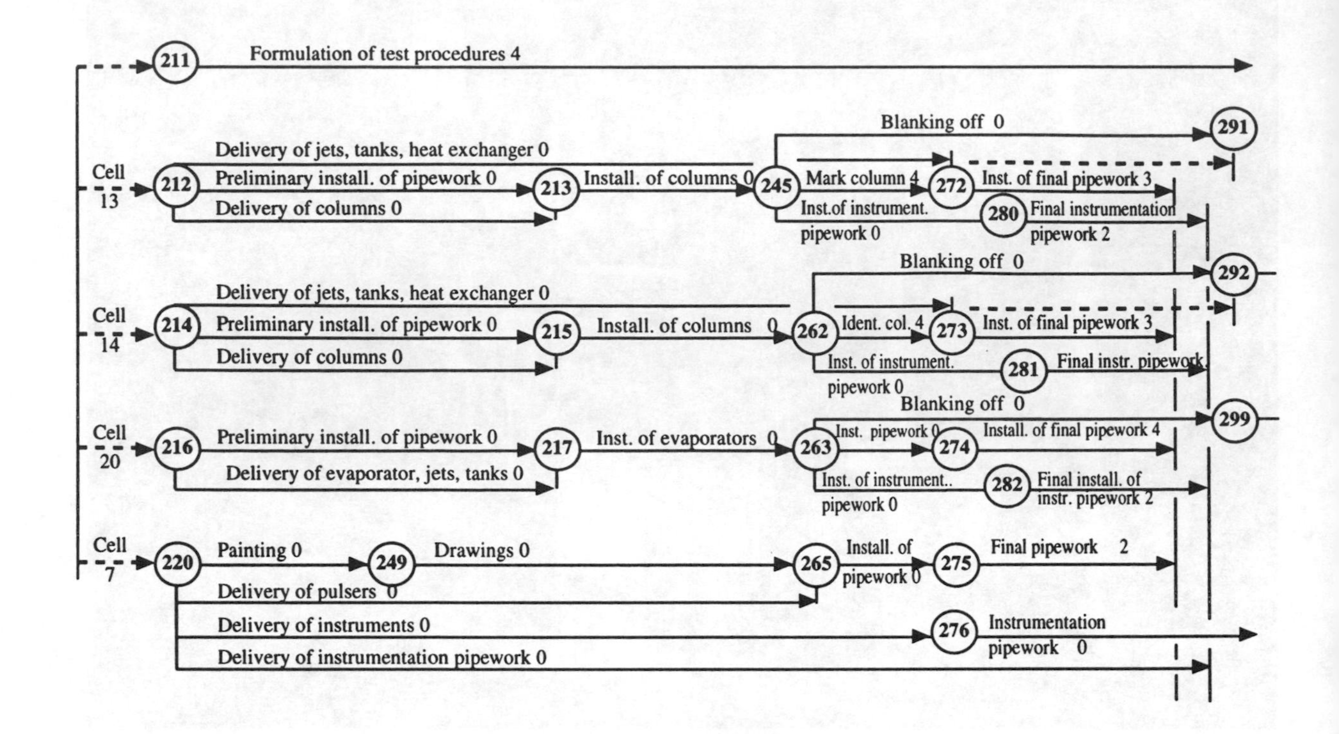

Figure 46. Part of the construction schedule for two cells in the treatment building prepared by Comprimo using the critical path method (Source: RAE3, p. 181).

Their task was complicated by the modifications made as construction proceeded, first by the decision taken in 1962 to make provision for reprocessing highly enriched fuel, and then by the decision taken during the second quarter of 1963 to reduce the size of the installations in order to reduce the total cost of the project.

As and when the buildings were completed, equipment was installed and accepted. The testing phase then followed, first on a non-active basis and then, as appropriate, using active material.

Problems in building the plant, co-ordination difficulties and the Comprimo charts

The most severe problem during construction of the plant was certainly the complexity of the pipework. Every tank and system had a large number of incoming and outgoing pipes – often at least nine connections[87] and sometimes even more than twenty – which made some cells look like an incomprehensible maze of pipes and tanks (Figure 44), despite the fact that pipes were theoretically arranged in layers.

Consequently as systems were installed they had to be inspected and tested before the installation of further piping made them inaccessible. Thus the order of events – installation and inspection – had to be carefully planned.

Piping installation began at the beginning of 1964 and very soon raised considerable problems of organisation, even to the point of halting the work owing to delays in orders and deliveries, combined with changes made to drawings as work progressed. It was necessary to call a temporary halt to the work.

A solution was found involving the use of computers. The task of preparing a co-ordination diagram was entrusted to the Dutch architect engineer Comprimo (Figure 46)[88], responsible for work on the electrical and instrumentation systems, which had it produced on a computer in Amsterdam. Comprimo prepared charts using the critical path method[89], first an experimental chart for the intermediate storage facility for intermediate activity

---

[87] These served many functions, for example feed pipes (by air lift, ejector or gravity), drains (idem), ventilation, instrumentation (for bubble tubes) sampling (outlet and return), stirring and rinsing.

[88] RAE3 p. 180-182. Example on page 181.

[89] Critical path planning is a technique for planning work which seeks to reduce the amount of time lost on a site. When two different specialists have to work at the same location this approach involves determining the schedule which minimises waiting time.

liquid wastes, and then a comprehensive chart for the plant at the end of 1964. The chart was updated and processed monthly to determine what jobs were outstanding, schedules worked out and distributed between the different trades, and the necessary equipment deliveries listed. Work then resumed normally and was completed satisfactorily.

By November 1965 the great majority of systems were installed. During the testing phase the plant was then progressively started up, commencing a new stage in the history of the company. Commercial preparations had already been made during construction.

Chronology of Part II

January 1958 - November 1965

1958

January 1958: *commissioning of the Marcoule reprocessing plant, UP1*
24 January 1958: **first meeting of interim Study Group**
February 1958: **OEEC/ENEA comes into being**
28 February 1958: **creation of a Design and Research Bureau (BER) at Mol. First team begins work in May**
29 March 1958: enactment of the Belgian law on the protection of the public and workers against the hazards of ionising radiations
17 April 1958: opening of the Brussels universal exhibition, symbolised by the atomium
11 June 1958: **agreement setting up the second joint undertaking of the OEEC/ENEA, Halden, a boiling heavy water reactor**
1-13 September 1958: *second Geneva Conference on the peaceful uses of atomic energy*
22 October 1958: **Spain seeks to adhere to the Eurochemic Convention, effective 27 July 1959**
October 1958: **Eurochemic mission to the United States**
8 November 1958: *EURATOM-United States Agreement (extended until 1995). Comes into force 18 February 1959*
December 1958: *launch of the Soviet nuclear icebreaker Lenin*
31 December 1958: establishment of Métallurgie et mécanique nucléaire (MMN), shareholders being SGMH and Fabrique nationale d'armement (FN), subsidiaries of SGB
31 December 1958: creation of Belchim (Société belge de chimie nucléaire)

1959

1959-1962: first Belgian three-year atomic programme
1959: **creation of OEEC/ENEA Security Control (transferred to the IAEA in 1976)**
1959: *France decides to build a new reprocessing plant for which the site at La Hague was selected in 1960*
2 February 1959: *a EURATOM directive lays down the first Community basic standards for radiation protection*
23 March 1959: **signature of the Dragon Agreement, the third joint undertaking. The high temperature reactor at Winfrith (UK) went critical in 1964 and co-operation continued until 1976**
June 1959: **the Halden reactor goes critical**
July 1959: *launch of the Savannah, the first civilian nuclear propelled ship, in New Jersey*
20 July 1959: **Spain becomes a member of the OEEC**
22 July 1959: *EURATOM-Italy agreement making the Ispra centre an establishment of the Joint Research Centre as from 1 March 1960*
10 August 1959: *signature of the United States-EURATOM co-operation agreement. In particular this provides for a loan of $135 million for building nuclear power plants in the EEC. The first application is from Italy for the future plant at Garigliano, north of Naples, which started up on 5 June 1963*
1959: **Belchim builds a test station for Eurochemic at Mol**
28 July 1959: **formation of the Eurochemic company. End of the interim period**
5 October 1959: **Germany sells 18 shares to 16 private companies**
16 October 1959: **the Netherlands sells one share to a private company**
November 1959: *IAEA Conference on radioactive waste management*
end 1959: **project expenditure exceeds the first million dollars**

1960

1960: *in Italy the CNRN becomes CNEN (Comitato Nazionale per l'Energia Nucleare)*
2 January 1960: **meeting of the architect engineers at Saint-Gobain concerning organisation and payment**

February 1960: **submission of pre-project n° 4. SGN requested to draw up the detailed pre-project**
13 February 1960: first French nuclear test in the Sahara
April 1960: closure of the Shinkolobwe uranium mine. Belgium stops producing uranium
7 July 1960: **work begins on the administrative building at Mol**
11 July 1960: creation of the European Atomic Forum, FORATOM
July 1960: **the architect engineers prepare the General Conditions of Contract**
September 1960: launch of the first American nuclear aircraft carrier, the Enterprise
30 September 1960: **co-operation and mutual consultation agreement between ENEA and IAEA**
December 1960: **Convention transforming the OEEC into the OECD following the adhesion of the United States and Canada. An additional protocol provides for the relations between ENEA and EURATOM which participates in ENEA and is represented on the Steering Committee**
1960-64: construction and commissioning of the Magnox reprocessing facility at Windscale

1961

1961: proposal of Vulcain as the focus of ENEA co-operation in the OECD/ENEA Working Group on Nuclear Ship Propulsion. Leads finally to an Anglo-Belgian agreement in May 1962 for BR3-Vulcain
1961: licence for a fuel fabrication plant at Mol-Dessel for a period of 30 years (renewed in 1991)
February 1961: **staff enter the administrative building**
March 1961: **initial contracts for the plant awarded to the architect engineers**
13 June 1961: **the Board of Directors adopts the detailed pre-project for the plant costing $30.7 million, including $24.01 million for capital expenditure and the remainder for operating costs up to the end of 1963**
June 1961: Convention between EURATOM and the CEN concerning the Bureau central de mesures nucléaires (BCMN) at Geel, not far from Mol
July 1961: agreement between the Netherlands and EURATOM concerning HFR (Petten)
September 1961: agreement between Cockerill, ACEC, EdF and Framatome establishing the Société électronucléaire des Ardennes (SENA) to build the Chooz power plant, the first joint undertaking of EURATOM and the first light water power reactor built in France
October 1961: Italy notifies the Eurochemic Board of Directors of its intention to develop an installation for reprocessing MTR fuel, the future EUREX
12 October 1961: **a first increase in capital is proposed**
5 December 1961: **in the Steering Committee the United States representative raises the possibility that Eurochemic might negotiate the reprocessing of irradiated American fuel under the EURATOM-United States Agreement. A group leaves for a visit to the United States in January 1962**
19 December 1961: **Italy, Portugal and Turkey indicate that they will not purchase new shares**

1962

1962: the FRG decides to build WAK
1962: in Japan, PNC decides to build an industrial reprocessing unit using imported foreign technology. First steps in the project for the Tokai Mura pilot unit
1962: construction of the Swiss Lucens power plant begins. Enters service in 1968
1 January 1962: adoption of the pre-project for the construction of the Transuranium Institute in Karlsruhe
January 1962: start of construction of the West Valley (New York) reprocessing plant designed to reprocess oxide fuels, by Nuclear Fuel Services (NFS), a subsidiary of General Electric. The plant was commissioned in 1966 and operated until 1972
January 1962: **Joint ENEA-Eurochemic mission to the United States**
February 1962: **SGN submits a pre-project for the reprocessing of highly enriched fuels**
22 May 1962: **an amendment to the EURATOM-USA agreement enables Eurochemic to reprocess fuel enriched in the United States**
June 1962: **first meeting of the "Committee for Contact with the Authorities" (CCA)**
September 1962: **completion of the water supply installation after a year's work**
October 1962: **Eurochemic, the German government, the CEA and EURATOM meet in Frankfurt to examine the possibility of creating a unit on the site for reprocessing fast reactor fuel**
November 1962: **general services building occupied after 15 months construction**

1963

1963: nationalisation of the electricity industry in Italy with the creation of Ente Nazionale per l'Energia Elettrica (ENEL) whereby nuclear power is separated from the private sector

1963: **the operations programme begins at the plant. Contacts with customers, model contracts, reprocessing pricing schemes**
28 February 1963: publication of the Belgian general rules for protecting the public and workers against the hazards of ionising radiations, in application of the law of 29 March 1958. Reprocessing plants are in category I. A nuclear safety committee known as the Special Commission, is charged *inter alia* with giving its views of plans for category I establishments is set up in December 1963
23-26 April 1963: **ENEA-Eurochemic Symposium on Reprocessing, Brussels**
10 April 1963: **Uhde appointed architect engineer for the plutonium purification unit**
April 1963: *work begins on the EUREX plant at Saluggia for reprocessing irradiated highly enriched uranium fuel. Italy-United States co-operation (Vitro C°) with the support of EURATOM*
18 June 1963: **first increase in capital**
July 1963: **opening of the laboratory cold wing after 18 months of construction**
2 August 1963: **enactment of the Belgian law on the use of languages for administrative purposes**
14 October 1963: **a working group is charged with devising an interim programme for the company**
22-23 November 1963: **Paris Colloquium on instrumentation in reprocessing plants**
Winter 1963-64: **serious financial crisis at Eurochemic**
31 December 1963: **Erich Pohland's contract terminates. Rudolf Rometsch temporarily appointed for 4 months as Managing Director of Eurochemic, appointment confirmed on 3 June 1964**

1964

1964: *General Electric plans to build the reprocessing plant known as the Midwest Fuel Recovery Plant (MFRP) at Morris (Illinois). Construction began in 1971; a new process known as Aquafluor was adopted. The project was abandoned in 1974 after two and a half years of fruitless tests*
1 January 1964: Belgonucléaire takes over the activities of Belchim
February 1964: *the United Kingdom stops producing plutonium in the Windscale 1 plant commissioned in 1952. A unit producing 2500 t/year is commissioned – Windscale 2*
23 January 1964: **the Board of Directors considers three scenarios for the future of the plant: shutdown, the pilot option and the production option**
February 1964: **opening of the laboratory hot wing**
March 1964: **co-operation agreement between Eurochemic and NFS**
20 April 1964: *the United States, United Kingdom and USSR sign a trilateral agreement on cutting the production of fissile materials for military purposes. Substantial increase in British civilian reprocessing capacity being offered*
27 April 1964: **following Italy's refusal to contribute to a further increase in capital, Belgium proposes a government-guaranteed loan if the plant's programme is modified to allow BR2 fuel to be reprocessed**
May 1964: **acceptance of the liquid waste treatment station and the transfer pipes to the CEN for low activity liquid wastes**
25 May 1964: **consideration is given to eliminating Italy's second seat on the Board of Directors**
3 June 1964: **plan to issue special shares**
12 June 1964: *launch of the German nuclear-propelled commercial ship Otto Hahn in Kiel*
21 June 1964: *startup of the Italian Enrico Fermi power plant at Trino Vercellese, in the plain of the Po*
1 July 1964: **second increase in capital, financial agreement for covering the operating deficit and a "reduced programme for 1964-1967". Commitment to cover deficits through the OECD budget for the period 1964-1967**
August 1964: **the Dragon reactor goes critical**
August 1964: **commissioning of the analytical laboratory**
31 August - 9 Sept. 1964: *third Geneva Conference on the peaceful uses of atomic energy*
16 October 1964: *first explosion of a Chinese nuclear device using uranium*
December 1964: **first reprocessing contracts signed**
end 1964: *signature of a EURATOM-EUREX Convention to build a pilot plant and conduct a 10-year research programme costing 3 million UA on the development of aqueous reprocessing processes for irradiated MTR and power reactor fuels, with provision for reprocessing MTR fuel at a price fixed beforehand.* **However Eurochemic is chosen to reprocess fuel from BR2 and HFR**

1965

1965-1967: second Belgian 3-year atomic programme
January 1965: **installation of the seven pulse columns in the main extraction cycle**

22 January 1965: *commissioning of the Indian reprocessing plant at Trombay, designed to take natural uranium aluminium-clad fuel from the Trombay CIR reactor. Plan to adapt the system to the Candu reactor fuel is in progress*
17 February 1965: *first delivery of spent fuel from HFR Petten to Eurochemic*
April 1965: *first American reactor in space*
1965: *Switzerland orders two reactors for Beznau (Westinghouse BBC) and Mühleberg (General Electric BBC)*
May 1965: *Denmark concludes a contract with Dounreay for reprocessing spent fuel from DR2 (Risø) and an MTR over a five-year period*
May 1965: **first non-active tests of the main plant, in the solvent recovery unit and the final uranium purification system. Tests accelerated from November onwards**
6 August 1965: **the application for a licence to operate the plant is submitted to the Belgian authorities**
September 1965: **at a press conference, R. Rometsch announces a reprocessing price of $20 to 30/kg of spent fuel up to an enrichment rate of 5%**
September 1965: *Pierre Huet becomes President of ATEN*
September 1965: **Turin Symposium on pilot reprocessing installations**
Autumn 1965: **completion of the two hot cells in the laboratory**
November 1965: **publication of the Safety Analysis for the plant**

PART III

The company's first life: reprocessing

December 1965 - January 1975

Introduction to Part III

The history of reprocessing at Eurochemic begins while construction was still going on, with the formulation of commercial policy, the search for contracts, initiation of the licensing procedure, and the tests carried out as soon as the plant was completed, which provided an opportunity to gain some practical experience in operating the new European plant. This involved technical contributions from France and the United States (Chapter 1).

Reprocessing as such took place in two stages, separated by two events, one of which was technical, the other political. The plant was shut down from 17 May 1970 to 15 February 1971. Between February 1970 and November 1971 the decision was taken to cease reprocessing.

This two-part history seems paradoxical. Between 1966 and 1970 in fact the political and financial problems appeared to be over, and the prospects for expansion gave rise to new plans. During this period, the undertaking learned about reprocessing, building up experience by overcoming the various problems that arose during the different campaigns (Chapter 2).

When the plant restarted operations, its closure was already being planned. The reason for this decision was a crisis in international co-operation against a background unfavourable to reprocessing. Between 1971 and 1974, Eurochemic was able not only to draw benefit from the technical experience gained during the preceding period but also to prepare for its shutdown (Chapter 3).

An overall review is essential in order to provide an assessment of the work accomplished during these years and to evaluate the importance of the period of reprocessing, insofar as this weighed heavily in the later history of the company (Chapter 4).

Chapter 1

The move towards reprocessing:
Contracts gained, licensing procedure and tests in the new plant

As work progressed on the site, preparations were also being made for commissioning the plant.

The financial problems gave impetus to the formulation of the company's commercial policy with the search for and signature of the first reprocessing contract, which in 1965 resulted in the first irradiated fuel being delivered and the first fuel elements from a number of European countries being stored in the *ad hoc* pond in the reception hall.

Before the plant could be commissioned a start-up licence was required from the Belgian authorities who were in fact drawing up their own legislation governing nuclear establishments while Eurochemic was being built. The builders paid special attention to the problems of nuclear safety and to worker safety. The licensing procedure also led to particular co-operation between Eurochemic and the Belgian authorities.

The plant was progressively put into service as from September 1965; acceptance of the facilities and testing then began, prior to the first reprocessing campaign on irradiated fuel which began in July 1966 with the inauguration by the King of the Belgians.

The reprocessing plant started up on that occasion was a hybrid design involving French and American technology, as can be seen from a comparison made between the three plants of Eurochemic, Marcoule and Idaho Falls.

Determination of contract policy; first deliveries of irradiated fuel

The standard contract and theoretical price formulation

Initial projects and general principles

From as early as 1962, the Board of Directors was working on the problem of reprocessing prices[1]. The plan was to adhere to American prices with a policy of evening out shipment costs to ensure equal treatment for all the countries involved in the company. However an exception was to be made for highly enriched fuels, the basis for which would be the costs actually incurred, higher than those offered by the United States. It was hoped that the difference in shipment costs would make the overall cost the same.

However it soon became clear[2] that if these principles were applied too strictly, the resulting prices were likely to be too high, even exceeding the value of the uranium and plutonium or the costs of storing unreprocessed irradiated fuel. An initial draft of a standard contract was drawn up in close liaison with the ENEA, notably with Jerry Weinstein[3]. It drew its inspiration from the existing contractual provisions in the United States[4] and was tabled at the meeting of the Board of Directors held on 28 March 1963. It was also submitted to the operators and to governments for comment. The original plan was to develop and use Eurochemic containers of an all-purpose design, which would be rented to the users[5]. This safe but expensive approach soon had to be abandoned owing to the overall contraction of the company's plans.

The contractual policy and the proposed reprocessing price were finally presented by the Managing Director and approved by the Board of Directors on 5 March 1964[6]: a summary of the standard contract was attached.

---

1 CA(62)25 of 9 October 1962.
2 CA/M(63)1 of 14 May 1963.
3 CA(63)12 of 6 March 1963.
4 Communication from Pierre Strohl.
5 CA/M(63)3.
6 CA(64)8 and CA(64)9.

Eurochemic undertook to return the uranium and plutonium to the customer, with the fission products and other wastes becoming the property of Eurochemic.

Deliveries were to be made in customers' containers to be approved by Eurochemic. The fuel was first to be dissolved and then an initial conformity analysis done, known as the "input determination"[7]. This was then to be compared with the information the operator was required to send with the fuel. In the same way an "output determination"[8] covering the composition of the final products could be carried out in the presence of a representative of the customer.

When fuels had been dissolved, Eurochemic was entitled to mix them together, since the return obligation did not pertain to identified materials but to products defined by their chemical and isotopic composition, within precise limits. This requirement meant that the plant had to resort to reprocessing fuels in batches having very similar characteristics, and thus contributed to breaking the plant's operating campaigns down to a very small size[9].

The company undertook to return 98% of the quantities of uranium and plutonium determined in the initial analysis. The enrichment of the uranium returned was not to vary by more than ± 0.3% from the value determined at entry. As far as the plutonium was concerned, its isotopic composition was to be the same as at entry; the chemical purity was defined in an addendum to the contract.

Any lapse by Eurochemic in meeting the return delivery characteristics resulted in a price adjustment. The same applied to late deliveries, except in cases of accident or *force majeure*.

The standard contract also included provisions for dealing with any difficulties in executing the contract, details of the insurance cover and sections relating to prices and financial conditions.

Eurochemic retained ownership of the wastes, to which a certain value was attributed at the time on the basis of the materials contained and which it was thought possible to extract and use[10].

The standard contract of June 1964

Table 41. **Calculation of reprocessing price in 1964**

Type of fuel	Price component n°1	Price component n°2	Price component n°3
Unalloyed uranium <1.6% initial enrichment and <1% after irradiation, Al or Mg cladding	Market price of natural uranium	Fraction of buy-back price of recovered plutonium	
As above, but Zr or stainless steel cladding	Market price of natural uranium	Fraction of buy-back price of recovered plutonium	Supplement for loss of capacity and increase in volume of effluent
Unalloyed uranium >1.6% initial enrichment and >1% after irradiation, Zr or stainless steel cladding	United States reprocessing price [i.e. charged by NFS], excluding shipment, including taxes	Fraction of buy-back price of recovered plutonium	
Other fuels	Case-by-case basis		

The final version of the standard contract was approved in June 1964 and sent to reactor operators.

For the price calculations, four formulae were defined, depending on the characteristics of the fuel, and applying to quantities ranging from 8 to 24 tonnes. One third was added for smaller quantities and a rebate of 20% subtracted for larger amounts.

To the price of the service provided are added overheads, the cost of insurance covering damage to fuel and containers, and the reimbursement of special charges.

---

7 "Détermination d'entrée".

8 "Détermination de sortie".

9 See Part III Chapters 2 and 3.

10 For example, the Head of the Process Chemistry Section in the Research Division, E. Lopez Menchero, considered recovering krypton and xenon for industrial purposes. One of the aims of the Janus cell was to recover neptunium, etc. However the objectives of the research programme were closely related to the plant's main activity with the result that these plans, the costs of which were never well defined, sank without trace.

The initial contracts and the problem of prices

The initial contracts

The first exploratory discussions took place in 1964, at the request of the operators of the Latina power plant (SIMEA) and of other Italian reactors, and with the Belgian authorities for reprocessing fuel from BR2.

Following the adoption of the standard contract, negotiations began with the French CEA for EL1, 2 and 3. Contacts were made with RWE for the VAK reactor in Kahl, BR3 and Diorit (Würenlingen) during the summer of 1964 and with ENEL (for SENN and SELNI), Halden, BR1 and SENA in the autumn[11].

Authorisation to sign the first contract was given at the meeting of the Board of Directors held on 22 September 1964; the contract covered 13 tonnes of uranium contained in fuel from the three French reactors EL1, EL2 and EL3, and amounted to a total of about $600 000. Delivery was to take place before the end of the first quarter of 1965. The contract was finally signed in early December 1964[12]. It covered 9.5 tonnes of natural uranium and 7.8 tonnes of slightly enriched uranium, the projected cost being $760 000.

Table 42. **First reprocessing contract**[13]

Reactor	Fuel characteristics	Weight of uranium contained (tonnes)	Weight of plutonium to be recovered (kg)	Price of reprocessing ($)
EL1	natural uranium metal, Al cladding	2	negligible	20$/kg of uranium contained
EL2	natural uranium metal, magnesium cladding	5	negligible	20$/kg of uranium contained
EL3	U-Mo alloy, slightly enriched (1.35 to 1.60% initially), American origin, Al cladding	6	6	47$/kg of uranium contained + 2.5$/g of recovered plutonium + a supplement for storing the wastes, related to the Mo present

On 17 March 1965, the Board of Directors authorised the signing of new contracts for 2.5 tonnes of natural uranium clad in magnesium from EDF1, under the same conditions as above; for 2.5 tonnes of enriched uranium initially in the range 3.7 to 4.4% from BR3, with 6 kilogrammes of plutonium to be recovered; finally for 4 to 5 tonnes of MTR type fuel from BR2 and HFR, reactors governed by a EURATOM co-operation agreement at Mol and Petten.

By the end of 1966, when the plant really began production, six organisations had signed eleven contracts with Eurochemic, covering a total of 54.5 tonnes of fuel, which did not even represent one year of reprocessing at full capacity. In fact reactor development had been slower than expected.

Table 43. **Status of contracts in 1966**

Organisation	Reactors	Type of fuel	Tonnage
CEA (France)	EL1, 2, 3, EDF1, Triton, Mélusine, Siloe, Cabri, Pégase, Osiris	natural uranium, LEU and HEU	27 t
CEN Mol (Belgium)	BR3	LEU	2.5 t
EIR, Würenlingen (Switzerland)	Diorit	natural uranium	6 t
EURATOM	HFR, BR2	HEU	10 t
Nukem (Germany)	FRJ1and other research reacotrs	HEU	2 t
GfK (Germany)	FR2	natural uranium	7 t

The considerable influence of the French CEA in the early contractual policy of Eurochemic must be stressed; one reason is the fact that France was at that time the leading nuclear power in the group of shareholders, another being that the CEA's reprocessing policy was to use the capacity of Eurochemic for reprocessing "exotic" irradiated fuels, with normal irradiated fuel being sent to the installations at Marcoule or La Hague[14].

---

[11] Further information on the reactors for which reprocessing contracts were concluded are given in the Annex at the end of Part III.

[12] Between 7 and 11 December 1964.

[13] CA(64)29.

[14] UP2 was then coming into production.

The problem of prices

However it was impossible to adhere to the price policy defined in 1964 owing to the competition then emerging amongst the providers of reprocessing services. In September 1965, the Managing Director of the company, Rudolf Rometsch, announced publicly that the price for reprocessing low enriched uranium would be between 20 and 30 dollars per kilogramme of uranium contained and referred to the dumping policy followed by the United Kingdom.

By 1966 it was realised that the prices charged for the early contracts had been below the optimal theoretical cost, i.e., calculated on the assumption of operation at full capacity[15]. Capacity would have had to be doubled to achieve financial viability.

It was clear therefore, even before the company began its processing programme, that the operating deficit would be structural.

In the meantime the plant's raw material – irradiated fuel – began to accumulate in the storage pond, but the plant could not be started up without a licence from the Belgian authorities. This was a long procedure, involving careful study and analysis of the safety and security aspects of the undertaking.

The licensing procedure: the Contact Committee with the Belgian Authorities (CCA) and the *Safety Analysis*[16]

The Contact Committee with the Authorities (CCA)

According to the third paragraph of Article 6 of the Eurochemic Convention, the execution by the competent authorities in the host country "of judicial decisions or regulations for the protection of public health and the prevention of accidents" was given as an exception to the inviolability of the installations and records. In Belgium these rules stemmed from the *"General Rules for the Protection of Workers"* approved by the Royal Orders of 11 February 1946 and 22 September 1947 and many times amended.

In particular the specific features of nuclear installations led to changes being made in 1965. The regulations for the "Protection of the Public and Workers Against the Hazards of Ionising Radiations" were formulated while Eurochemic was under construction and promulgated by the Royal Order of 28 February 1963. The fact that the construction of the plant and the institution of the legal framework took place at the same time resulted in close co-operation between the government authorities concerned and the management of the company, supported by the lawyers of the OECD European Nuclear Energy Agency (ENEA). The Belgian contact was Samuel Halter, Director General and Head of the Private Office of the Minister for Public Health and the Family[17], with whom contacts were made as early as 1961.

In 1962 an informal committee, known as the Contact Committee with the Authorities (CCA) was set up under the chairmanship of Samuel Halter, and met once a month until 1967[18]. The committee had three to five members from the Ministry of Public Health, three members from the Ministry of Employment, and representatives of Eurochemic, depending on the questions at issue. Representatives of EURATOM and the Netherlands – the border is close by – also took part when necessary. In this way the government authorities were kept abreast of progress with the work and co-operated on setting up the safety system of the undertaking well before the plant actually started up, which facilitated the licensing procedure. We may give two significant examples.

In the Belgian system, although responsibility is placed on the "Head of the Enterprise" – here the Managing Director – the safety structure is headed by a "Head of Safety" who can intervene directly in cases of extreme emergency. Eurochemic modified its organisation accordingly in 1962, by bringing the Head of Safety directly under the Managing Director while previously he was part of the Technical Division.

The CCA also helped the management to produce a clear plan of the chain of authority in emergencies; although legally the Managing Director was the responsible person, authority was held by the Technical Director and, in his absence, by the "plant superintendent" during the day and the shift manager at night.

When construction came to an end the company began to prepare the safety review necessary to obtain the start-up licence. The CCA provided an interface between the two partners in preparing the document and

---

15 RAE3 p. 22.

16 Sources: *Safety Analysis (1965)*, communication to the author by R. Rometsch and W. Hunzinger, archives of the CA and RAE.

17 See VANDERLINDEN J. dir (1994).

18 It continued in existence until September 1983, meeting as required. The archives of the CCA are preserved by the Société Belgoprocess.

ensuring that nothing was missing. This enabled the Belgian authorities, for whom this was their first experience of such a process, to focus their requirements.

The accident analysis was prepared jointly. A report[19] summarising the measures included in the *Safety Analysis* for monitoring and protecting the environment was also prepared for the Belgian, Netherlands and EURATOM authorities in 1965.

The November 1965 Safety Analysis and the language problem

The preparation of the Safety Analysis

The Head of Safety, W. Hunzinger (Switzerland), had the task of drawing up the report needed to obtain the start-up licence. He made use of the safety reports prepared since April 1962 by the different divisions, the project staff and the operating personnel, in co-operation with the Health and Safety Section[20].

Speedy presentation of the safety report

The *Safety Analysis*, published in November 1965, consisted of three volumes of text and one volume of figures, divided into 11 sections with individual page numbering, and a total thickness of 6 inches.

Volume one contains six sections: "Legal Background, Basic Fuel Data and Plant Capacity, Site Description, Description of the Plant Buildings, Description of the Low Enriched Uranium Process (LEU), Description of the Highly Enriched Uranium Process (HEU)".

Volume two is in three sections: "Description of the Main Equipment, Auxiliary Installations of the Process Buildings, Plant Operation".

The descriptions of the installations which are a feature of the first two volumes naturally focus on the safety aspects; thus Section VII dwells on the way in which the problems of corrosion and transfer techniques are envisaged. Section IX deals with staff organisation and training, the monitoring of radioactive materials, wastes, accidents and incidents during operations[21].

Volume three is entirely given over to the general problems of safety, in two sections, "Safety Aspects of the Plant" (X) and "Accident Analysis" (XI).

Section X is concerned primarily with radiation protection and monitoring, and with the systems providing protection against criticality accidents. There are also indications of the expected amounts of liquid and gaseous effluents and the relevant organisation in the event of an accident.

Section XI, the last one, reviews 20 accident scenarios and their consequences, criticality accidents, chemical explosions and losses of radioactive materials. The accidents described were selected on the assumption that up to two problems might occur simultaneously.

Each section of the report is illustrated by diagrams in the Volume of Figures. Thus the report gives a complete description of the plant as well as the expectations of its builders.

The Safety Analysis as an expression of the linguistic status quo

With the agreement of the Belgian government, the *Safety Analysis* was produced in English, with a covering letter in Dutch. This device overcame the language problem facing Eurochemic as a result of the Belgian law of 1963[22].

As a product of the OECD, Eurochemic worked, like that organisation, in both French and English, and the company's archives reflect this fact. In practice, the management used French in its relations with the government authorities; the supervisory staff also mainly used French; English was more frequently used in relations with staff at a lower level, mostly recruited locally and therefore Dutch speaking; hence a sort of "Eurochemic English" was developed as a language of communication.

This linguistic apparatus, which was one of the consequences of the international character of the company[23], was upset by the Belgian law of 2 August 1963 on the use of languages for administrative purposes: very soon this law was strictly applied by the provincial authorities, leading to obstacles and delays. The fact

---

19 Special Report on the Environmental Monitoring Programme.

20 RAE 2 p.51 *et seq* lists the reports produced between April 1962 and May 1964.

21 Recording and reporting.

22 AG(64)8.

23 Another consequence was initially the lack of any staff representation system, until the adoption in 1965 of a system of the company's own making, which however had no links to the Belgian trade union structures. This arrangement remained in force until the early 1970s.

was that this law required that relations with government authorities should be conducted in the language of the region in which the head office of the enterprise was located, in this case, Dutch.

It seemed likely that considerable additional costs for translating documents would be added to the normal overheads at a particularly difficult time. The Managing Director therefore requested the annual General Meeting held on 3 June 1964 to add a further paragraph to Article 33 of the Statutes relating to the company's communications and publications in order to affirm the bilingualism of Eurochemic and thus abrogate the effects of the Belgian law as far as the company was concerned; the Board of Directors considered this modification to be inappropriate[24]; the Belgian representative indicated that it would be ineffective because the law, in any event, applied to the Belgian authorities, and that a parliamentary amendment would be necessary, which would certainly raise political difficulties at a time when the problem of contributions for the second increase in Eurochemic's capital was coming up. Pierre Huet, who had been in touch with the Belgian Ministry for Foreign Affairs, while stressing the importance of the principle, which went against the international character of the company, emphasized that the primary requirement was "to find, with the Belgian authorities, the best practical way forward". This common sense approach was effective and the trouble ended.

However neither party ever tackled the problem head on and issues were dealt with on a case-by-case basis. Thus the *Safety Analysis*, at the heart of the licensing procedure, was written and presented in English, with a simple covering letter in Dutch, and the authorities of the province of Antwerp passed on the licence issued by the Government without difficulties.

The safety system at Eurochemic

Nuclear safety depended on the safety of equipment and on the checks of the amounts and composition of the fissile material to be processed, carried out during operations[25]. Every fissile material container was designed in such a way as to limit the probability of criticality; this was done either through the geometrical design of the unit, or by means of several detection, locking and alarm devices. Needless to say, the requirement for safety was also a powerful factor in the development of quality control.

As regards the safety of personnel, the plant was subdivided into four contamination zones[26], ranked according to the probable level of radioactivity (Figure 47)[27]. Each of these zones, symbolised by colours, had its own inspection procedures and special working rules. The ventilation system was arranged in stages of increasing vacuum, the air always flowing from green zones to red zones.

The green zone included inactive areas and was normally free of any contamination.

The yellow zone contained equipment that was normally inactive but which could become contaminated, such as the valve corridor and the transmission system corridor.

The orange zone encompassed normally inactive areas where there was a permanent risk of contamination. Operations in that zone were subjected to continuous radiological monitoring, such as the access to active cells and the sampling corridor.

The red zones were those in which the radiation level normally exceeded the danger threshold and which might also be contaminated. Access to those zones was impossible without prior decontamination. In the Eurochemic plant these zones consisted of the processing cells.

All workers at the plant wore dosemeters and were monitored at the end of the working day. Health checks were reinforced for people working in the green and yellow zones; in orange zones special clothing was worn and no-one was allowed to work there except as part of a group of at least two people. The red zone could be accessed only through the orange zone. Entry required the specific permission of the Health and Safety Section; tasks were required to be strictly defined and analysed in a Hazardous Work Permit (HWP) before access to this zone could be accorded.

In the event of an emergency a special organisation came into being, directed from an emergency control centre located in the entry building, thus fairly remote from the plant (Figures 49 to 51)[28].

---

24 CA/M(64)3.

25 For aspects of fissile material accounting and its consequences in terms of international security monitoring, see Part III Chapter 4.

26 RAE 1, p. 278-290.

27 Figure 47 shows the arrangement of these zones on the first and second floors of the plant. Figure 48 shows the location of the various radiation detectors.

28 The evaluation system was supplemented by an external surveillance system for gaseous discharges, see Figure 48.

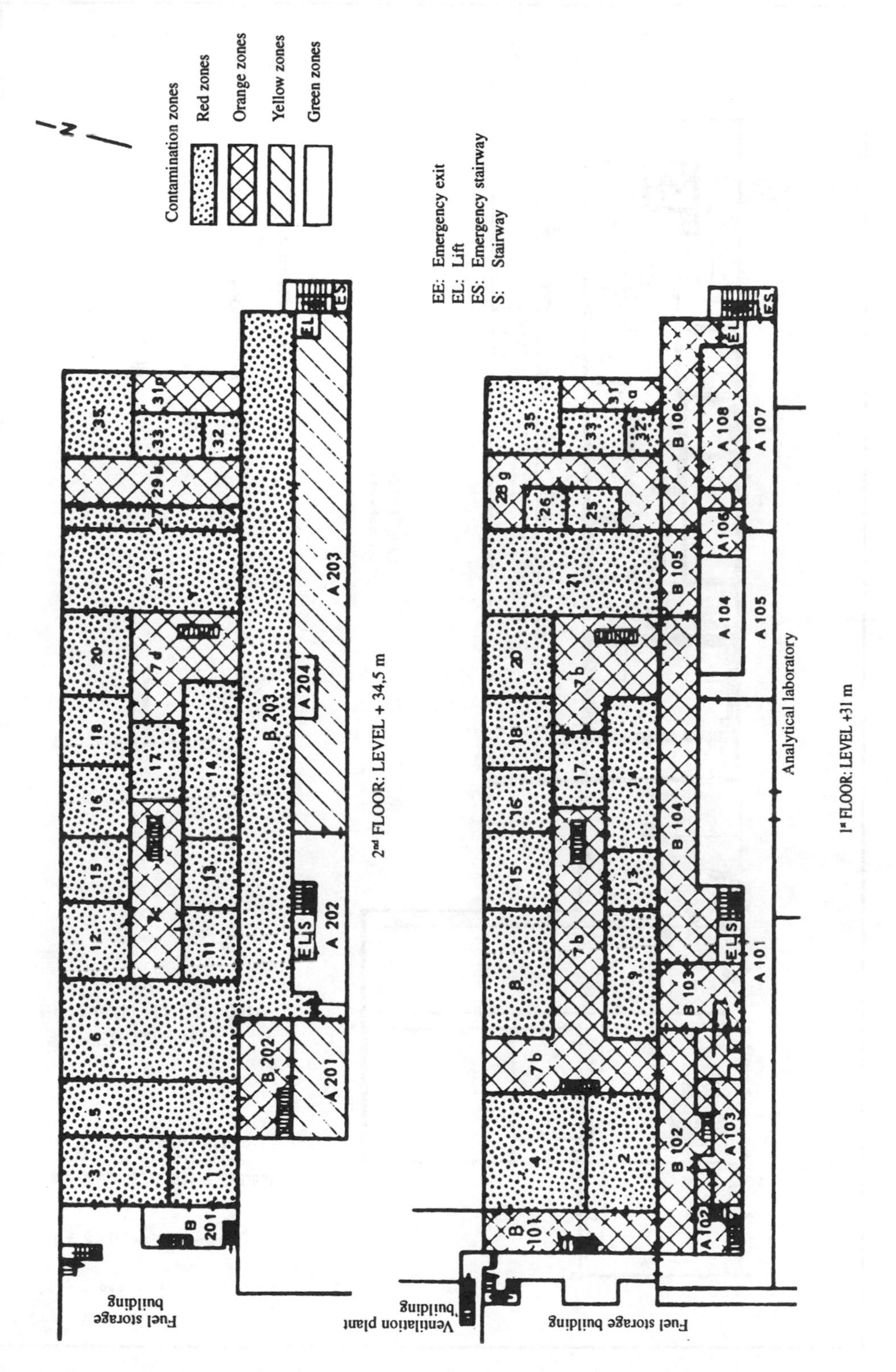

Figure 47. Layout of the four "contamination zones" on the first and second floors of the treatment building (Source: RAE 1, 1963, p. 283).

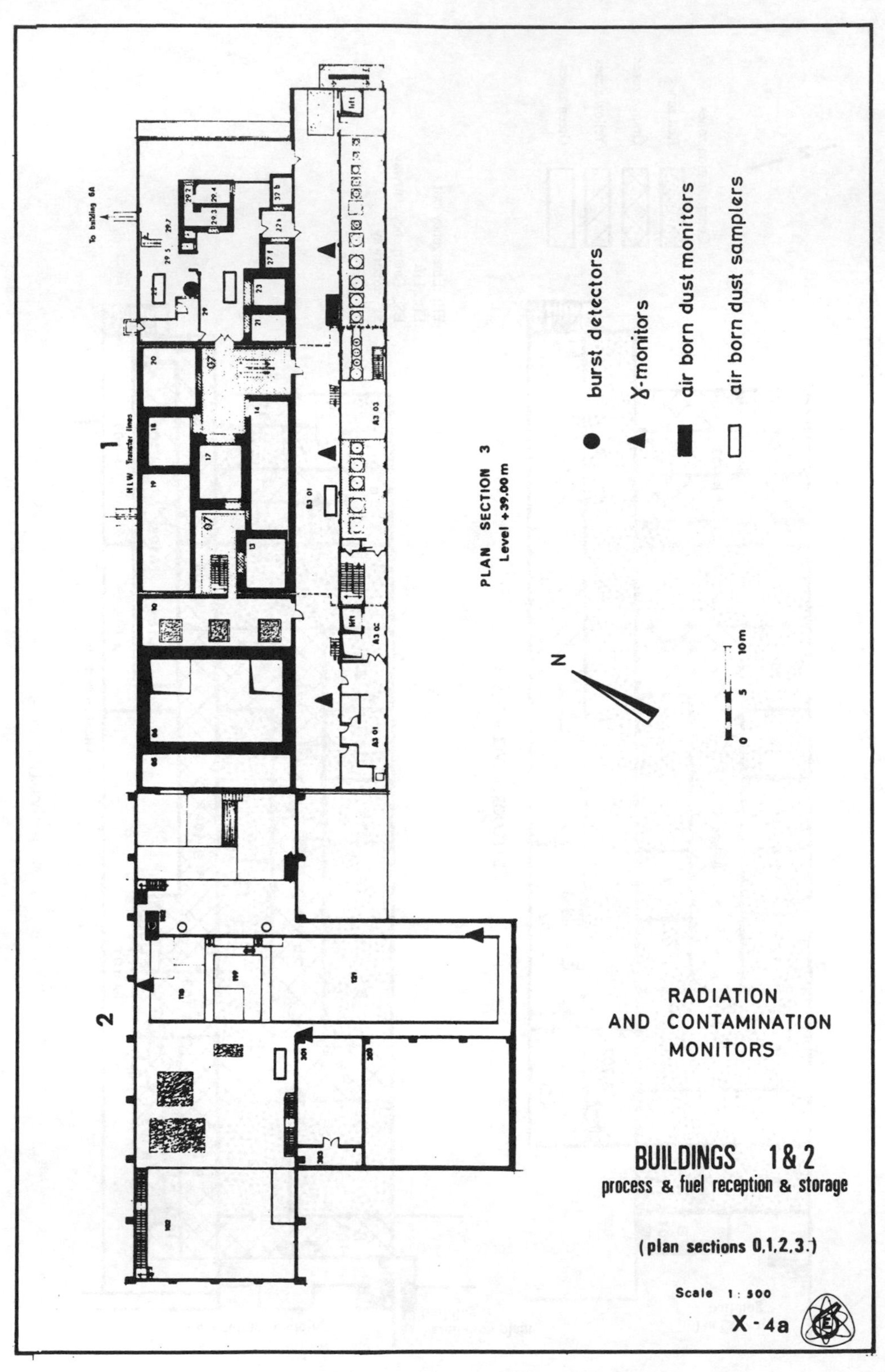

Figure 48. The safety system inside the buildings: example of the location of explosion, radiation and contamination detectors at the 39 metre level in Buildings 1 and 2 (Source: *Safety Analysis* (1965), Volume of Figures, X-4a).

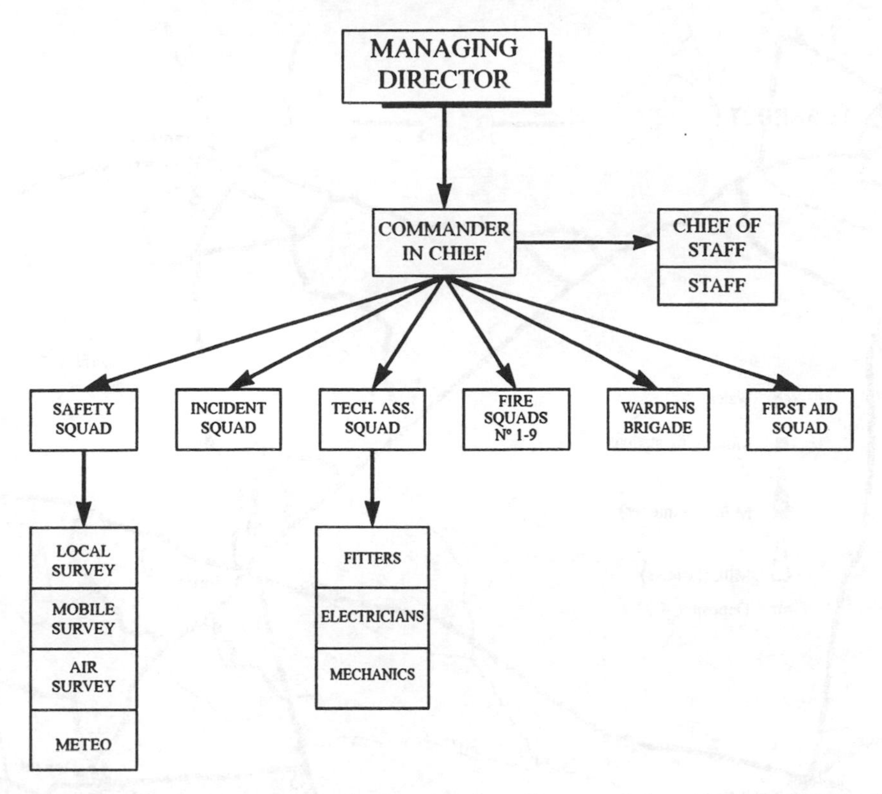

Figure 49. The emergency organisation in the plant (Source: *Safety Analysis* (1965), Volume of Figures, X-9).

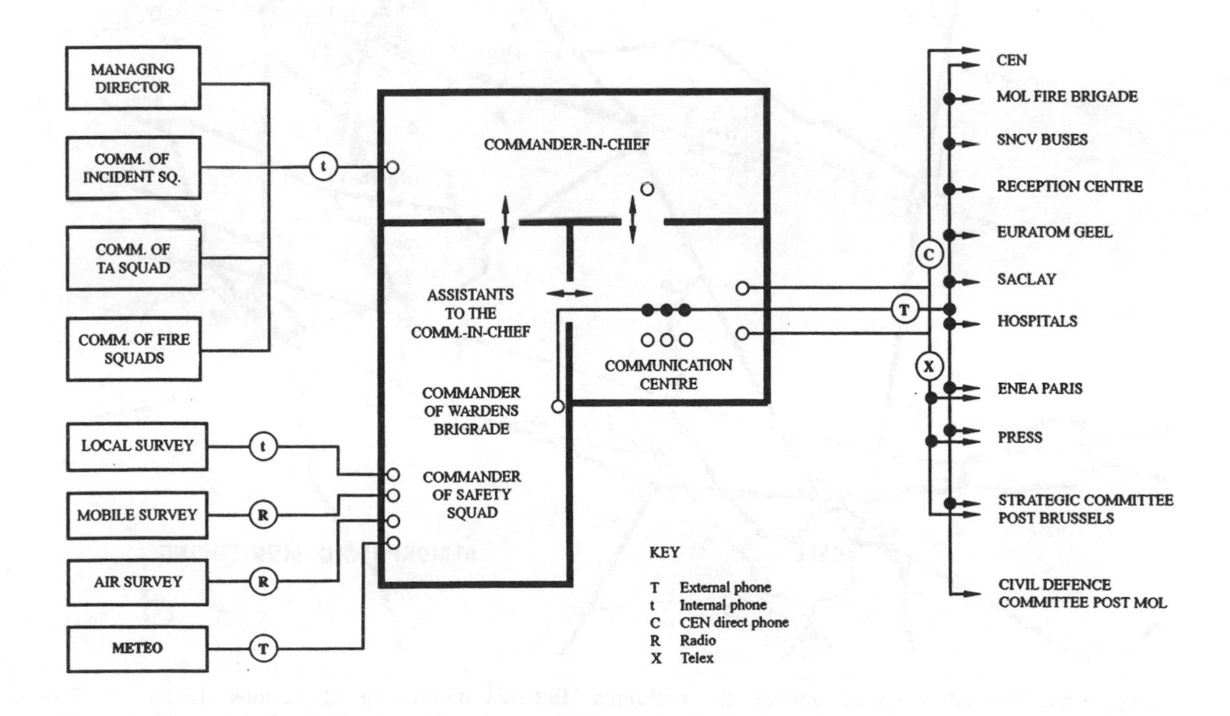

Figure 50. The communications system in the emergency control centre in Building 16 (Source: *Safety Analysis* (1965), Volume of Figures, X-11).

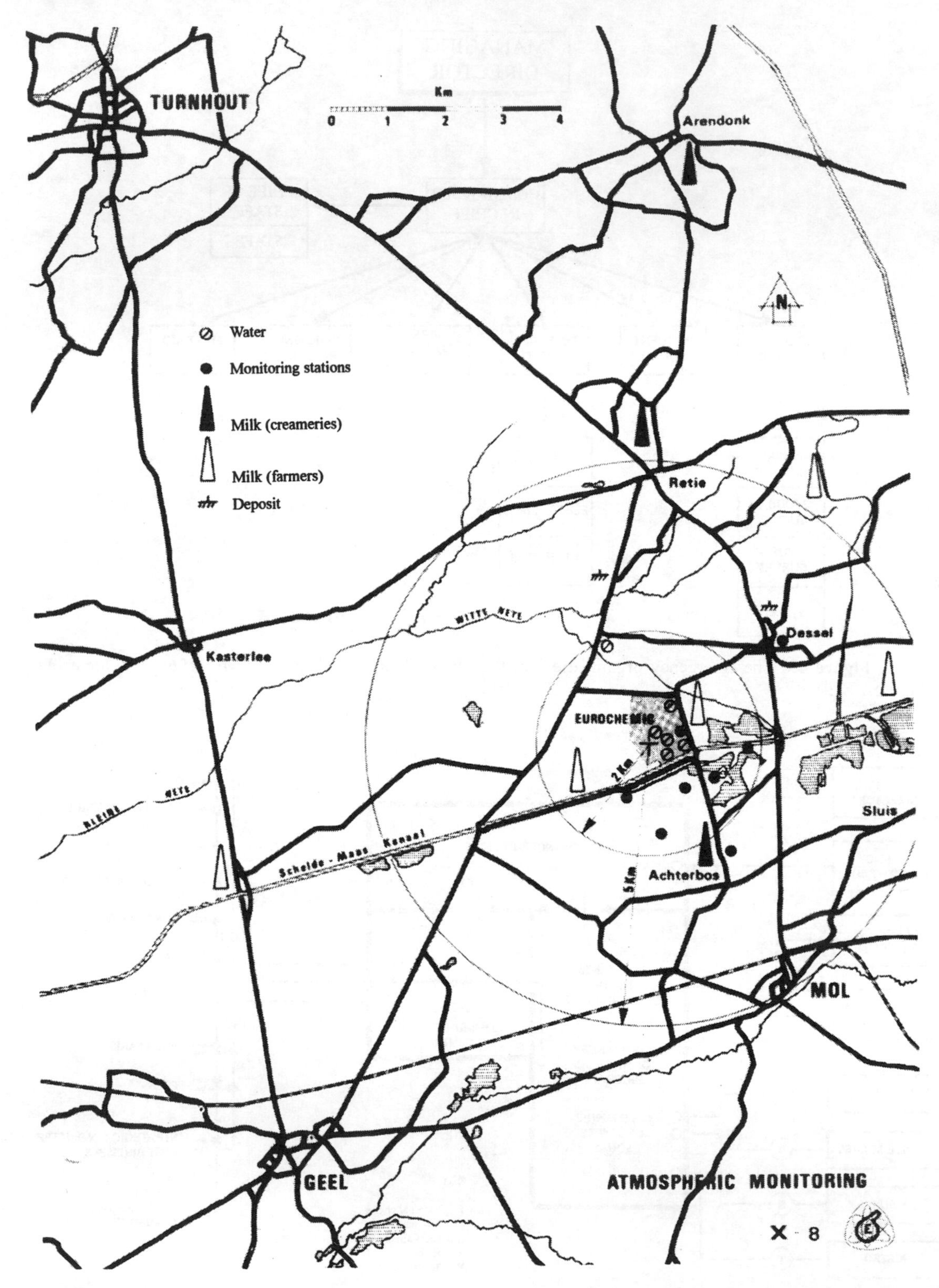

Figure 51. The safety system outside the buildings. External monitoring of gaseous discharges. Sample locations for water and milk, location of monitoring stations (Source: *Safety Analysis* (1965), Volume of Figures, X-8).

The dual licensing procedure

The licence application was lodged by Eurochemic on 15 July 1965. Its evaluation by the Belgian authorities took time. Indeed this was the first application the authorities had had to deal with and at that time neither the Ministry for Public Health and the Family nor the Ministry of Employment had any nuclear specialists.

The evaluation procedure leading to the licence was twofold and involved experts at national and international level.

National evaluation by CORAPRO

According to the procedure laid down in Belgian law, Eurochemic chose from the two organisations existing in Belgium an inspection company made up of former employees of the CEN with experience in health and safety problems, Contrôle Radio-protection (CORAPRO), which was set up on 8 December 1965[29], and instructed it to evaluate the report, the results of which were forwarded directly to the ministry.

The international evaluation by EURATOM

Secondly, at the request of Samuel Halter, the EURATOM Commission set up a consultative committee of seven high level experts, the "Advisory Panel on the Safety of Eurochemic", which met under the chairmanship of two American experts, E. D. Clayton and J. C. McBride[30].

Also taking part in the work of the Panel were seven members of EURATOM, four representatives of the Belgian government, and those responsible for safety at Eurochemic, i.e., the Managing Director, the Technical Director and the Head of Safety. During the discussions the criticality experts of the company's nuclear services division were also invited to give their views. Additional information was provided in writing and design calculations were verified[31].

Tardy licensing and a start-up without formal authorisation

The Panel drew up recommendations which, together with the supplementary information requested by the Ministry of Health in November 1966 and February 1967 following the CORAPRO inspection, resulted in the licence being issued by Royal Order on 25 September 1968 and forwarded by the provincial authorities on 20 November 1968. Article 2 of the order listed 19 conditions to be fulfilled for the operating licence to be valid.

In the meantime the plant had been inaugurated by the King at a time when only the operating licence for the research laboratory and that for the pumping of effluent to the CEN had been issued[32].

Thus from July 1966 to November 1968, the Managing Director assumed the responsibility for running an undertaking whose licence was still being reviewed. This caused him a certain amount of anguish since he would have been personally responsible in the event of a difficulty. However Rudolf Rometsch, who had been involved in design and construction from the outset, was confident in the company's technical capability for ensuring safety and security both inside and out. Indeed since 1965 the start-up procedure had demonstrated that the plant could operate as planned.

The commissioning of the plant[33]

The basic principles

The organisation

The plan was that the plant would be commissioned unit by unit by a series of specialist teams, according to a "vertical organisation" reflecting the different aspects of the process stages. Thus at the beginning of this phase groups were set up for the operations of decladding, dissolution, extraction, final purification, acid recovery, sampling, and so on. The tests were standardized.

---

[29] More precisely, l'Association sans but lucratif (ASBL). Radioprotection, set up on 8 December, took the name of CORAPRO on 24 January 1966.

[30] The other members were F. Baumgärtner and A. Birkhofer (Germany), P. Giuliano and Simoneta (Italy) and W. K. A. Walrave (Netherlands).

[31] Spot checks on calculations.

[32] Lodged in December 1963, they were granted by royal decree on 28 January 1966.

[33] RAE3, pp. 202 et seq.

However, as the initial checks progressed, it appeared that a horizontal organisation for each type of inspection – piping, ventilation, etc. – would be more effective in preventing interference between the groups. Teams were therefore set up under group leaders with technical assistance from the staff of the research division, both for the tests and for commissioning the equipment.

The start-up was scheduled by a Comprimo-owned computer using the graphical method which had been employed during construction, with priority systems added to the program as required. The commissioning of one unit was in fact not supposed to be delayed by problems occurring in the others. Spare time created in the schedule was used for secondary repairs.

The main technical stages of plant commissioning

The commissioning procedure necessitated the following stages: post-construction clean-up, a series of system and operational tests progressively approaching real operating conditions, and finally finishing-off jobs.

Following acceptance of the plant, the different cells were cleaned up and then tested, notably piping inspection, during which the operators affixed labels, thus familiarizing themselves with the installation. Tanks were then calibrated by being filled with water and weighed.

The plant was then cleaned from top to bottom using water and, where necessary, steam or compressed air. There followed the first functional tests which identified open ducts, blocked pipes and unexpected siphon effects. The instrumentation was then checked and additional instruments installed wherever this appeared necessary. Heating and cooling tests were then carried out on the dissolvers using steam, hot water and cold water, the systems being boiled and then rapidly cooled.

The next stage involved inactive chemical tests using acid and/or solvents, which completed the cleaning of the pipework, followed by corrosion inspection with determination of temperatures, pressure and levels. Overall operations were simulated and incidents deliberately caused to evaluate the consequences of any mistakes.

When these tests had been successfully completed, the first nuclear materials were introduced, in this case unirradiated solutions of uranium. The first active tests were carried out in conditions as close as possible to those of the operational flow chart, with system efficiencies and mass balances being determined. Blockage tests were carried out, the nuclear material accounting system tested as well as the radiological safety precautions.

Next came flushing tests using acid or water which made it possible to determine outage times and to obtain a first idea of any decontamination problems.

Finally came the reference tests using unirradiated fuel elements which demonstrated the efficiency of the process for recovering uranium. These tests employed 2.5 tonnes of unirradiated EL2 fuel following the LEU flow chart. Commissioning was concluded with the finishing-off jobs, such as welding covers and tanks and completing paintwork. The cells were then closed and ventilation circuits balanced.

The commissioning schedule: difficulties encountered

Commissioning encountered no major problems as regards the conventional parts of the plant, but the situation was more complicated for certain special systems.

The final schedule is shown in Figure 52. The last unit to be completed was the third dissolver which was specially designed for MTR fuel assemblies of very high enrichment, the installation of which had been decided upon in June 1964[34]. Most of the preliminary tests using natural uranium took place between September 1965 and February 1966.

These tests brought the future operators face to face with radioactivity for the first time, and led to a number of minor incidents[35]. During December 1965 there were five incidents during plant testing; three overflows of solution containing 60 g/litre of uranium; a mistake in a density calculation led to the overflow of a 400 g/litre solution; a leak occurred in a plastic connecting tube.

These problems did not cause any human contamination, but resulted in the inspection procedures being reviewed and the overflow problems re-examined. Most of the reference tests took place in June 1966.

Despite the clean-up, construction debris still remained, blocking ejectors, fouling valves and pumps; in some cases this necessitated systems being unblocked with acid, and certain pipes had to be cut; waste material

---

[34] See Part II Chapter 2.

[35] VIEFERS W. (1986), pp. 45-47.

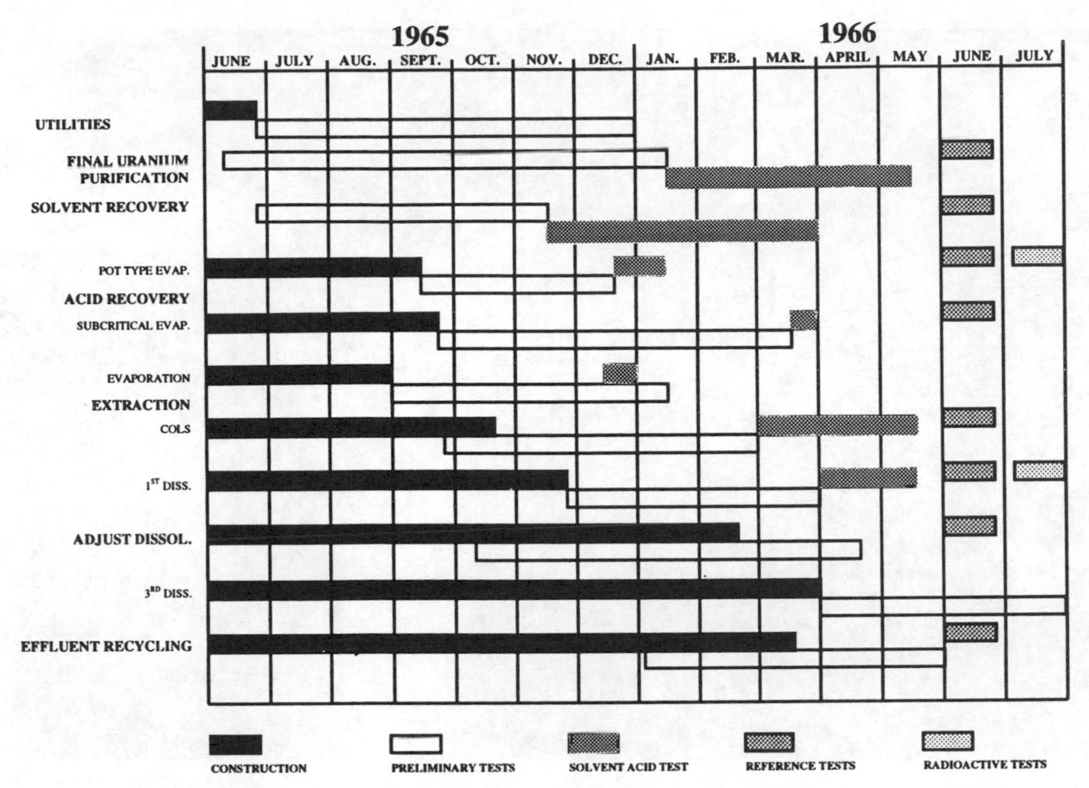

Figure 52. Acceptance timetable for the main plant units, showing the series of tests between the completion of construction and the commencement of active operations. When the plant was inaugurated in July 1966, only the first dissolver and an evaporator were active (Source: RAE3, 1968, p. 206).

built up in the reconditioning unit, blocking a manifold. Piping installed with an incorrect curvature, diameter or gradient resulted in air being trapped, which had to be dealt with. Welding defects were found and had to be put right sometimes under difficult conditions.

The tests on the extraction columns were a good example of these initial difficulties[36]. These tests revealed problems of interface stability, which were corrected by installing an automatic control system. Pulse tests showed that certain stacks of perforated plates were vibrating abnormally and substantially, so they had to be welded to their seats after the columns had been opened. Foam formed in the "air lifts" and the system had to be modified. However all the other components of the pulsed columns were found to be satisfactory. Incident simulations were run in order to determine the time necessary to shut the system down and restart extraction, which turned out to be three or four hours.

The process of learning about reprocessing and its techniques thus began well before the plant was started up and the staff had already acquired some experience when King Baudouin officially opened the plant (Figure 53) on 7 July 1966 when he fed the first charge of irradiated fuel into the plant's first dissolver. His action symbolically marked the commencement of the period of reprocessing at Eurochemic.

The technical characteristics of the reprocessing plant inaugurated in 1966 reveal its originality, which can be assessed by comparing it with the two plants commissioned earlier from which it drew most inspiration. From this standpoint, Eurochemic can be seen both as a hybrid of the European approach and the American design and as a unique installation owing to its polyvalent capability. A description of the original features of the process will give a better idea of its uniqueness and show the conditions in which subsequent reprocessing took place.

[36] RAE 3, p. 204-210 detail the commissioning of the process systems. The columns are dealt with on p. 207.

Figure 53. Photograph taken at the inauguration of the plant on 7 July 1966. From right to left: King Baudouin, Rudolf Rometsch, Managing Director of the company, Einar Saeland, Director-General of ENEA and Walter Schulte-Meermann, Chairman of the Board of Directors of Eurochemic (Source: RAEEN n° 8, 1966, p. 25).

Portrait of the Eurochemic reprocessing installation (Figures 54 to 56)

To grasp the originality of the reprocessing plant, it is useful to try to compare it with the two plants with which it is "generically" most similar: the one which was regarded by the Eurochemic engineers as the "big sister" of the Mol plant, the Idaho Centre Processing Plant (ICPP – Figure 57), and the Marcoule plutonium plant (Figure 58), built by SGN for the CEA and using processes developed by the CEA.

When the Marcoule[37] plant was commissioned, its primary purpose was military, since it produced metallic plutonium. The first five-year plan in 1952 defined its capacity in terms of plutonium production – 50 kg/year – modifications having been made to raise the production target to 100 kg/year in 1955. The plant resulted from a very rapid expansion, dictated by the urgency of the military programme, of the small Châtillon pilot plant into a veritable factory[38].

---

[37] On the origins of the Marcoule plutonium plant project, see Part I Chapter I. The comparison with Marcoule is founded upon documents published between 1956 and 1965 and is based on the processes used between the first campaign of 1958-1959 and the completion of the Mol plant in 1961-1963. The Marcoule plant has undergone a number of transformations and extensions since its first years in operation. Sources: TARANGER P. (1956), GALLEY R. (1958), the 1963 issue of *Energie nucléaire* devoted to Marcoule with articles by ROUVILLE M. (1963), ALLES A. (1963), JOUANNAUD C. (1963), CURILLON R., COEURE M. (1963), FERNANDEZ N. (1963), RODIER J., ESTOURNEL R., BOUZIGUES H., CHASSANY J. (1963), as well as information revealed over the course of the third Geneva Conference in 1964 in ONU (1965), JOUANNAUD C. (1965), FAUGERAS P., CHESNE A. (1965).

[38] Mr. de Rouville, Director of the Marcoule Centre in 1963, describes the feverish atmosphere of that period as "the Marcoule mystique", that emerged "when a large team, driven by a common enthusiasm, came together to carry out a major project in a new field which, faced with severe international competition, raises a country to a higher level of civilisation, marked by a victory over nature, one of whose new secrets allowed man to tame a new form of power", ROUVILLE M. (1963), p. 216.

1. Main processing building
2. Fuel reception and storage building
3. Ventilation building
4. Fission product store
5. Final product store
6. Inactive liquids store
7. Liquid waste treatment
8. General services building
9. Research laboratory
10. Administration building
11. Water treatment plant
12. Inactive solid product store
13. Intermediate activity liquid waste store

Figure 54. Aerial view of the site from the south-east in 1965. The EBES power station can be seen beyond the bridge over the Meuse-Escaut canal. The BR3 reactor is in the distance on the right, on the CEN site. The photo was published in the seventh AEEN report (Source: RAEEN n°7, December 1965, p. 13).

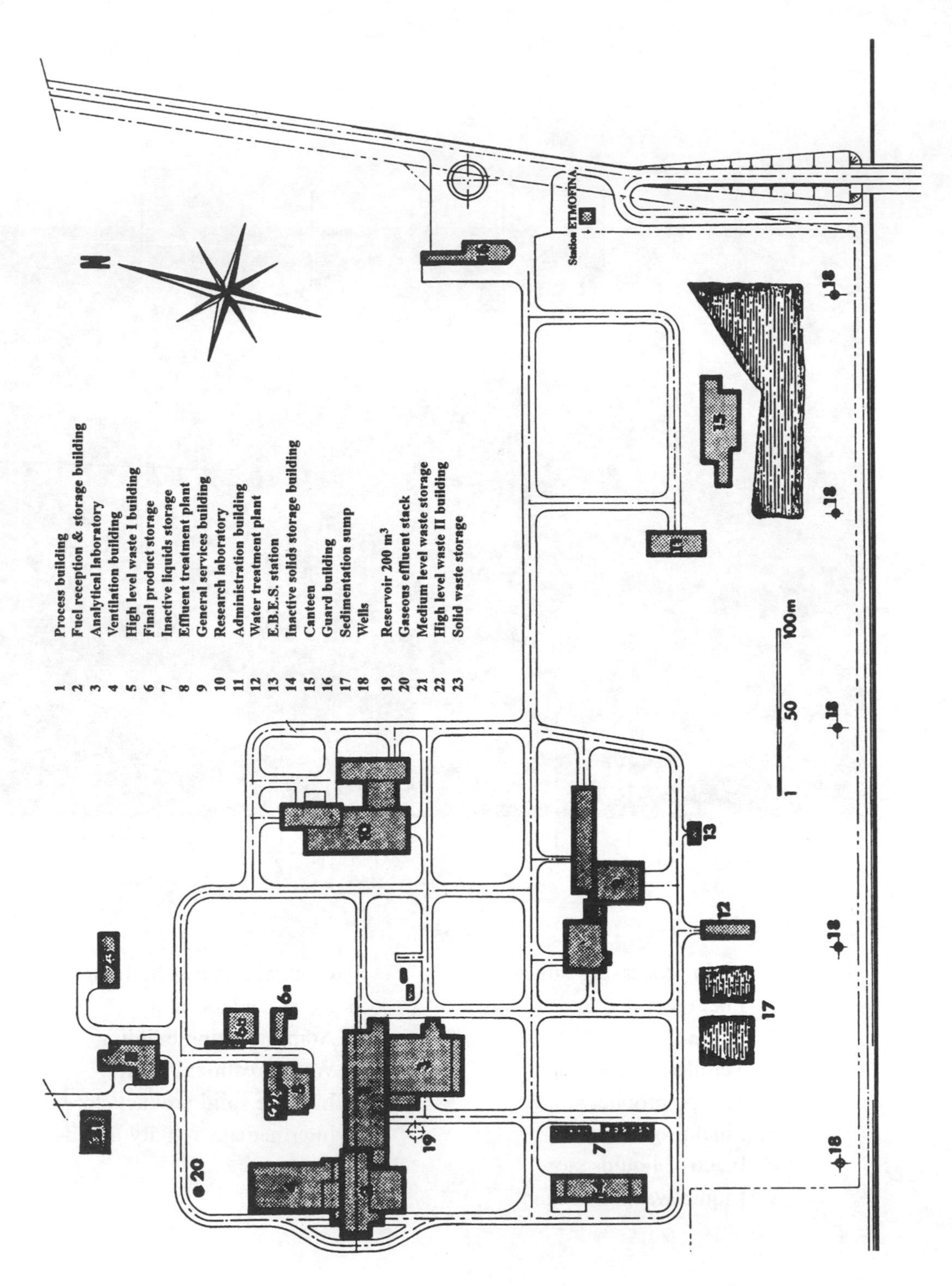

Figure 55. General site plan, November 1965 (Source: *Safety Analysis* (1965), Volume of Figures, IV-1).

Figure 56. General view of the plant towards the end of 1964, taken with a telephoto lens from the roof of the administrative building (n° 11) (Source: Eurochemic photo, no date).

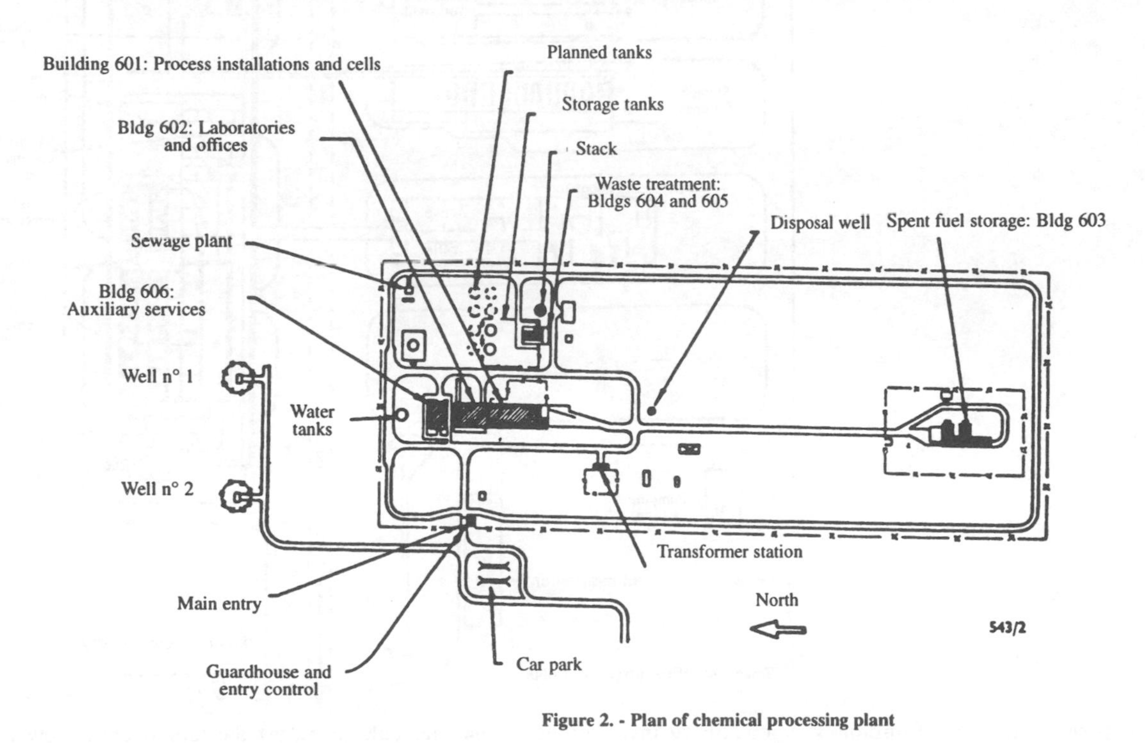

Figure 57. Layout of the Idaho Centre Processing Plant (ICPP) in 1955, two years after it was commissioned (Source: USAEC (1955) p. 14).

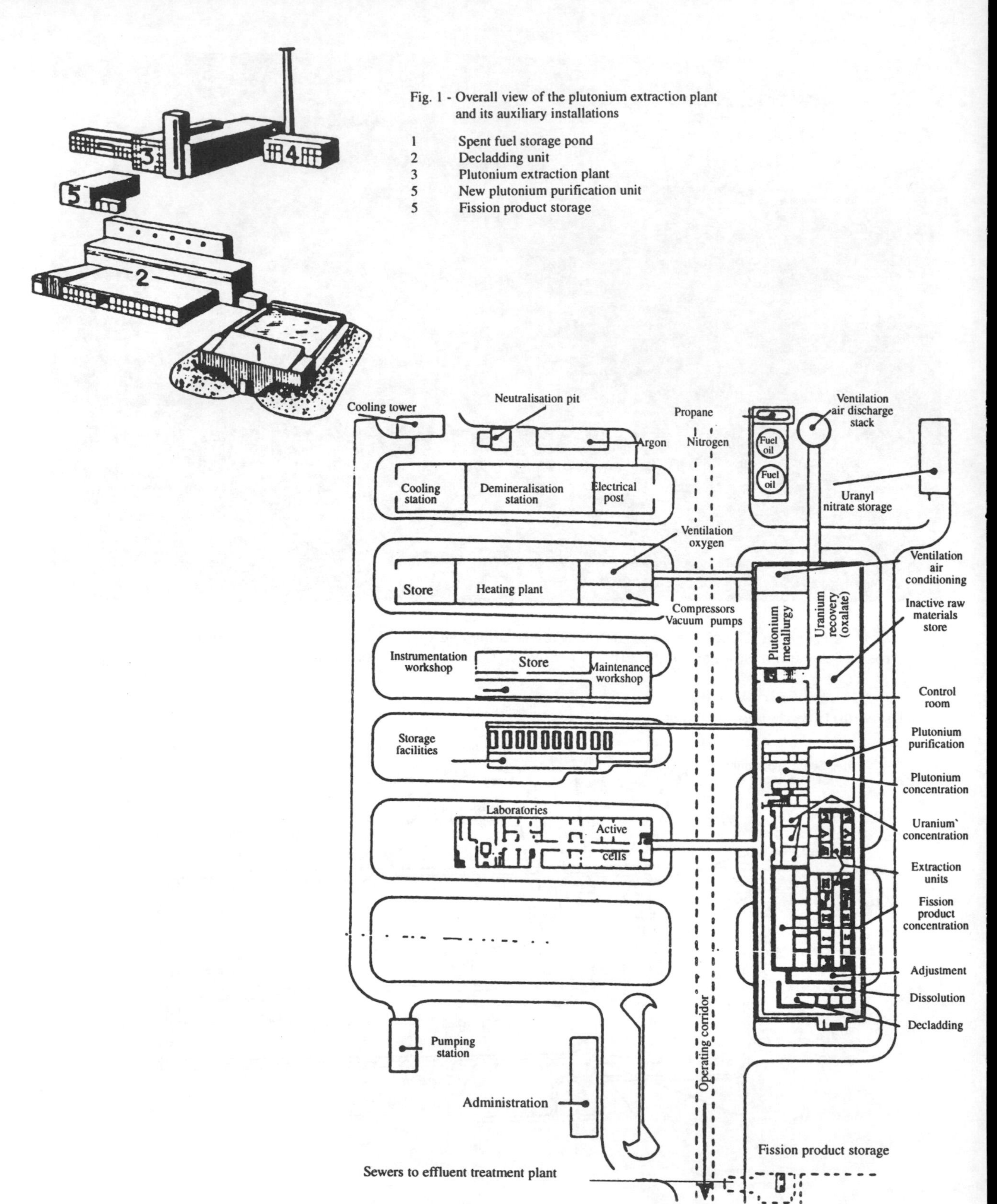

Figure 58. Layout of buildings and auxiliary installations at the Marcoule plant. At the top: overall view in 1963, a drawing accompanying a photograph of the site in CURILLON R., COEURE M. (1963), p. 271. At the bottom: plan showing the layout of the plutonium extraction plant in 1958 (Source: GALLEY R. (1958), p. 11).

Although it was able to produce enough plutonium to manufacture the weapon which was exploded at Reggane on 13 February 1960, at that time not all the auxiliary installations, such as those for fission product storage and waste management, were completed. Moreover the system for purifying the plutonium and converting it into metal was largely derived from laboratory methods and processes. A new building, the so-called Workshop 100, intended for continuous production, was due to come into service in April 1963 and a continuous dissolution building, known as MAR 200, was under construction.

The similarities between Marcoule and Eurochemic resulted from the fact that the two plants had been designed by the same architect engineers. However their objectives differed considerably.

By contrast the Idaho Falls ICPP plant[39] was older – it began operations in February 1953 – but like Eurochemic it was an "initial production plant"[40], in other words an industrial pilot. Initially the REDOX process, used in the new Hanford[41] plutonium plant, was applied to the highly enriched uranium fuel elements used in the Materials Testing Reactor (MTR) at the Idaho Falls National Reactor Testing Station. The plant was intended to recover – owing to its high value both in energy and money terms – the very highly enriched uranium which was alloyed with aluminium, in a cladding of the same metal. The declared capacity of the plant was one tonne of uranium a day.

The application of the process had been developed at ORNL, which had also drawn up plans jointly with Foster Wheeler. The plant was built by Bechtel, and the USAEC awarded contracts for commissioning and initial operation to the New York chemical firm American Cyanamid Cy. The operating contract was subsequently transferred to the Atomic Energy Division of Phillips Petroleum, which was running the site in 1955. During the Eurochemic mission's visit to the site[42], the plant had been diversified to encompass the processing of other types of fuel by the PUREX process, and was also testing the processing of zirconium and steel cladding. Spare cells had been built at the outset with a view to this possibility.

Eurochemic had a number of points in common with its two cousins, but it did have highly original features.

Three plants using solvent extraction and operating with direct maintenance

Solvent extraction

The first point the three plants had in common was their use of solvent extraction methods with nitric solutions. The solvent used was TBP at all three plants although ICPP initially used hexone. This type of extraction required facilities for storing the basic products such as acids, solvent and diluent. These facilities were grouped together in the open air and the chemicals prepared in *ad hoc* areas of the plants. The solvent used was recycled and any "used solvent" that could not be recovered had to be stored as a waste product.

Direct maintenance and its consequences

The three plants also had the common feature of using direct maintenance. Any repairs and modifications were carried out by the operators who had to enter the plant and, in particular, the areas containing radioactive material. This feature, unlike the remotely operated maintenance used in the first major plants at Hanford, had a number of consequences.

Cells of modest size, containing a large quantity of relatively common commercial equipment.

First of all it is possible to use equipment similar in its function to that used in the ordinary chemical industry. However emphasis was placed on minimising the number of mechanical parts and those exposed to wear. These systems could be of small size, unlike the situation at Hanford where remote manipulation, in the state of the art at the time, was possible only using parts of fairly large size. The advantage was that it was possible to reduce the surface area and total volume of the concrete cells containing the equipment exposed to radiation.

---

[39] The sources used for the ICPP are USAEC (1955), LEMON R. B., REID D. G. (1956), SCHWENNESEN J. L. (1958), and the Eurochemic mission report of 1958. The ICPP was later known as the Idaho Centre Reprocessing Plant (ICRP).

[40] SCHWENNESEN J. L. (1958), p. 332 draws a distinction in the United States between the pilot plants, THOREX at Oak Ridge and PUREX at Hanford, the initial production plants such as ICPP, and the production plants such as REDOX at Hanford and PUREX and Savannah River.

[41] See Part I Chapter 1.

[42] See Part II Chapter 1.

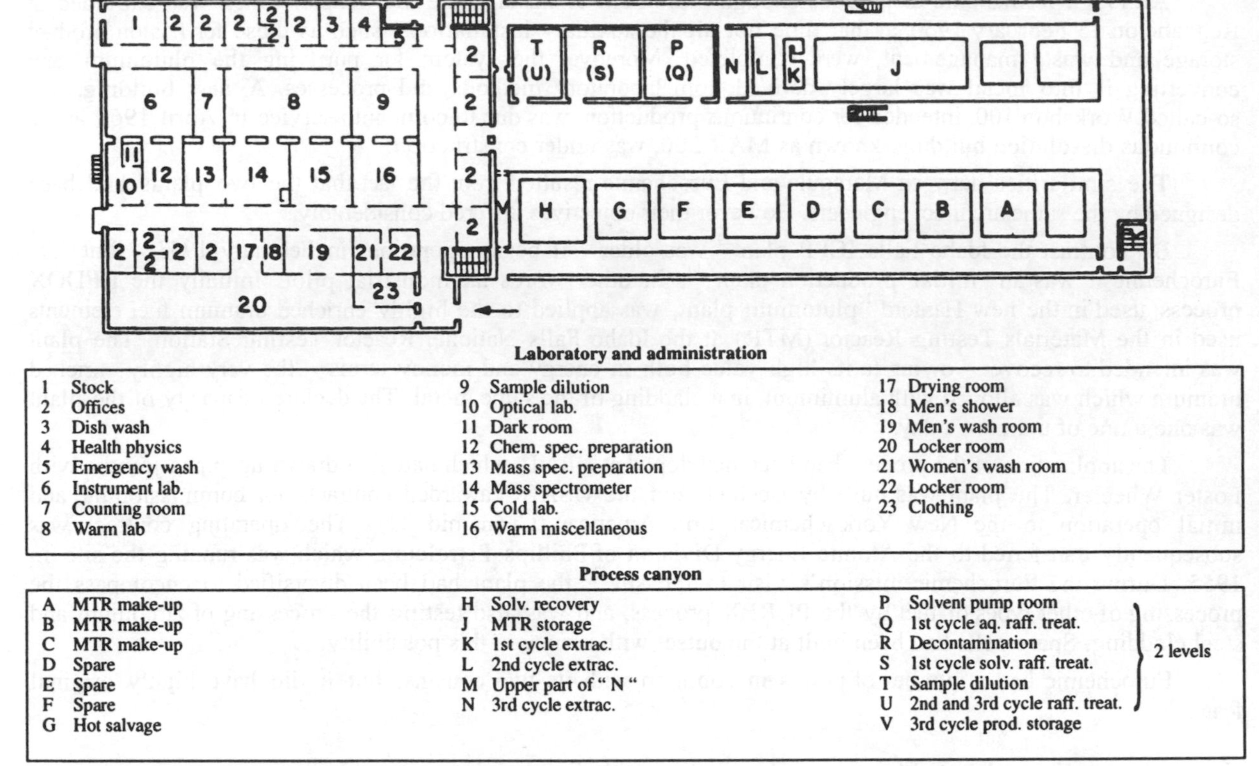

Figure 59. Drawing of the ICPP treatment building and laboratory in 1955 (Source: LEMON R.B., REID D.G. (1955), p. 615).

The need to be able to decontaminate the cells so that equipment could be maintained or modified argued in favour of a large number of cells all differentiated[43] according to their intended purpose. The fact of having direct access also made it possible to fit more equipment into the cells. It was merely necessary for men to be able to pass between the piping, the cells and the equipment to carry out repairs or modifications following decontamination. The ICPP had a system of platforms and ladders inside the cells. The Eurochemic processing building had 29 cells of different sizes, the largest containing the extraction equipment, which were pulsed columns 12 metres high.

The presence of radioactive areas alongside areas routinely frequented by personnel calls for a sophisticated ventilation system.

If the cells were to be accessible a highly elaborate building ventilation system was also required. The rule adopted in all three plants was to create a pressure differential between the areas routinely frequented by the operators and those exposed to radiation, and to ensure that the pressure in the process cells was lower.

Historically, the degree of sophistication has varied. In 1955 at the ICPP, there were two levels (Figure 59), the controlled areas being at a vacuum of 0.2 inch of water compared with the areas where staff were present. At Marcoule (Figure 60) the system in 1958 was organised at three levels, the active areas (zone I), the semi-active areas (zone II) and the inactive areas (zone III). Zone III was at an overpressure of 5 milimetres of water with respect to the outside of the plant, zone II 10 milimetres less compared with zone III and zone I at a further 10 milimetres less compared with zone II[44].

At Eurochemic, as we have seen, the system involved four levels symbolised by colours (Figures 61 to 63). The green zone was at an overpressure, the yellow zone 5 milimetres less, the orange zone 7 milimetres less and the red zone 25 milimetres less.

---

[43] The initial Hanford plants had standardized cells which were constructed before the necessary equipment had been completely designed. Remote maintenance was carried out through overhead hatches, requiring the equipment to be dispersed horizontally. Thus substantial ground space was necessary.

[44] The pressures were soon modified: compare GALLEY R. (1958), p. 14 and CURILLON R., COEURE M. (1965), p. 315.

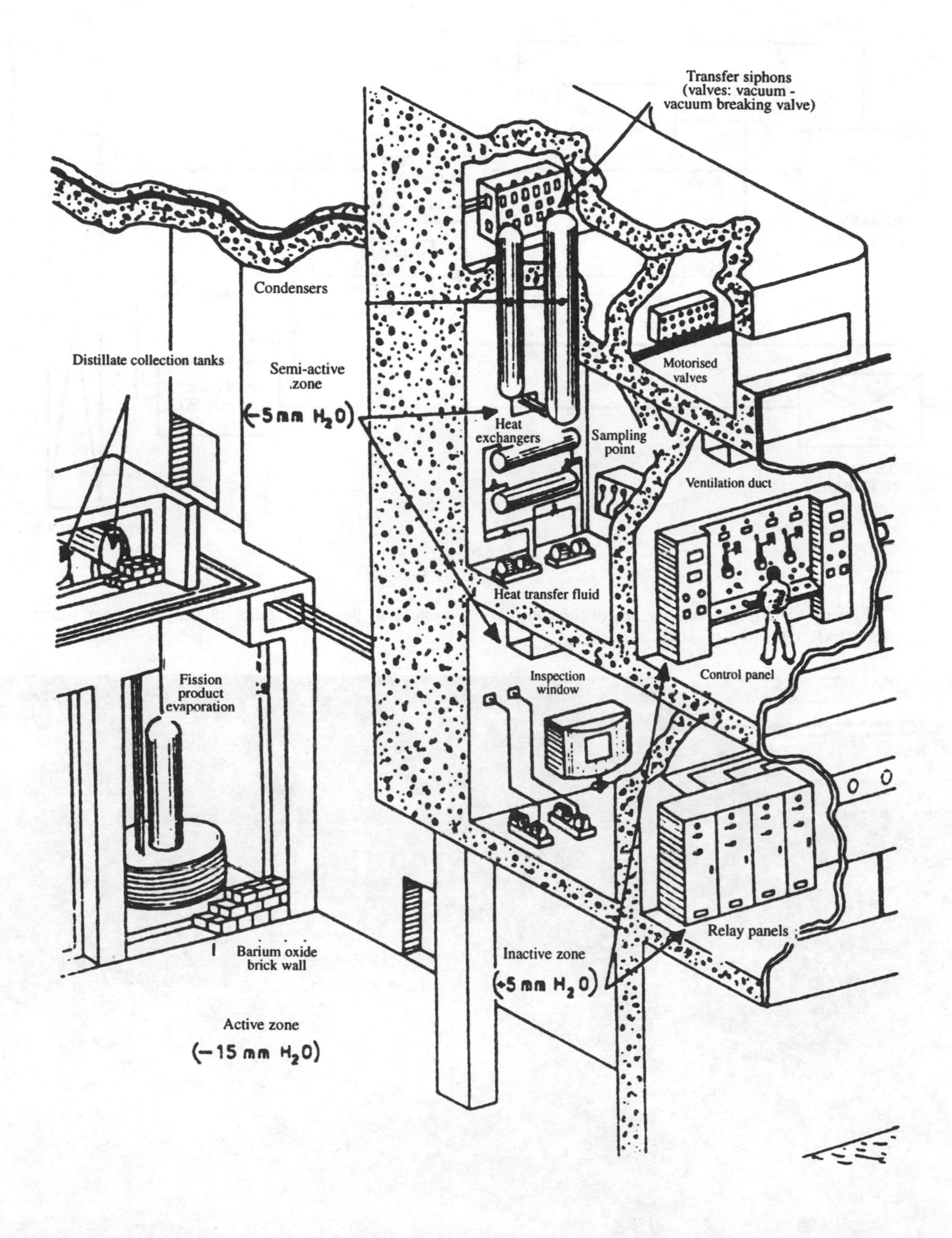

Figure 60. Sectional drawing showing the layout of the three plant zones at the fission product evaporator (Source: GALLEY R. (1958), p. 14).

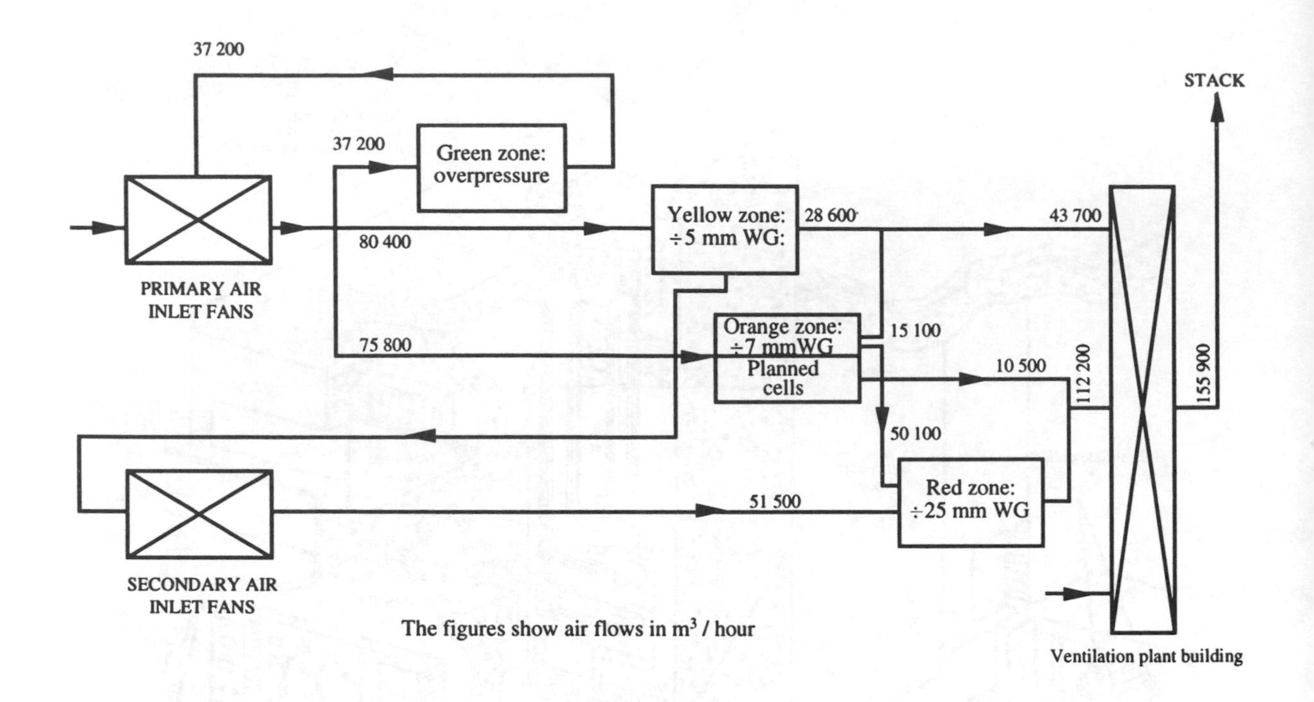

Figure 61. Simplified diagram of the ventilation system in the four zones of the treatment plant (Source: RAE 1 (1963), p. 315).

Figure 62. View of the battery of absolute filters in the ventilation building (Source: Eurochemic photos, no date).

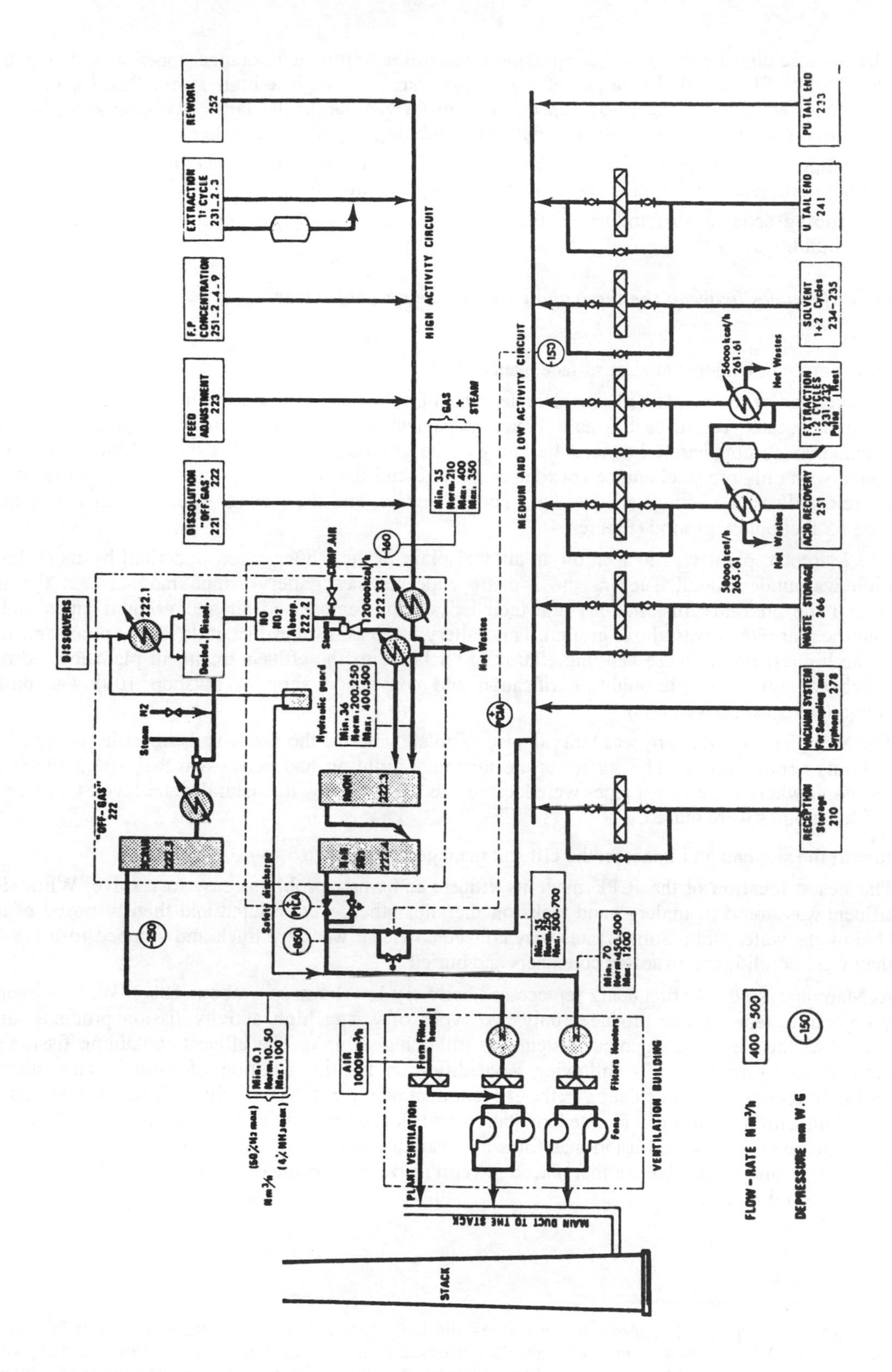

Figure 63. Diagram of the ventilation system for the treatment plant tanks (Source: *Safety Analysis* (1965), Volume of Figures, VIII-4).

The air reaching the red zone after passing through the yellow and orange zones was drawn through a battery of absolute filters and discharged after being checked[45] through a high stack. Besides the ventilation system for the different zones, there was a special system for ventilating the tanks, working at a higher vacuum compared with the active cells and provided with gas scrubbing and filtering systems.

The importance of ventilation can be seen in the plant design by the presence of ventilation air conditioning buildings, filter batteries and a high stack. It also called for substantial electrical supplies.

The zoning scheme also influenced the architecture, particularly as regards the movement of people through the buildings.

The main differences between the three plants: sites, size and objectives

The sites

Nature of the ground. Underground or surface plants

The location of the plants had considerable impact on their architecture and layout. The ICPP was built in a desert, in a research centre as big as a French *département* or English county. In practice, no particular environmental constraints had to be met. The nature of the ground was such that the buildings could be placed underground with only the steel and perspex superstructure and the inactive systems visible at the surface. The thick concrete walls of the active cells had been poured *in situ*, and the leakage drain system led to the lowest point some 55 feet below ground (Figure 64).

The Marcoule plant was located on an alluvial plain of the Rhône river, underlaid by marl. Part of the installation was underground, such as the 3 metre wide services gallery which ran between the treatment building and the auxiliary installations and included steam, compressed air and vacuum pipes and electric cables, but the plant itself was above ground. This gallery had a rainwater duct at its lowest point and the active drains at the highest point, these carrying effluent of all kinds to the effluent treatment plant located nearer the Rhône. Subsequently, the plutonium purification and conversion shop (Workshop 100) was built partly underground (the most active parts).

The Mol plant, for its part, was built almost completely above the sandy soil, since the water table lay at a depth of only about 1 metre. The surface at the treatment building had been excavated to 1.5 metres, so that the lowest point where the effluent pipes were located was hardly below the natural water level. Only certain low activity effluent pipes were buried.

The nature of the site had an impact on the effluent management system

The desert location of the ICPP made its effluent and waste problems easy to resolve. While the most active effluent was stored in underground tanks on the site, others were diluted and then disposed of in a well located below the water table. Solid wastes were collected twice a week by truck and shipped to a USAEC site where they were conditioned in sealed containers and buried[46].

At Marcoule, since the fuel being reprocessed had very low burn-up – about 100 MWd/t – owing to its military objective, reprocessing produced only two types of waste: high activity fission products and liquid effluent. In 1963 the waste management system was still being set up[47]. The effluent containing fission products was stored in underground tanks following de-acidification by the addition of formol. The other effluent underwent sedimentation or flocculation in the effluent treatment plant. The resulting sludge was stored in metal drums, pending bituminisation[48]. The sedimentation and flocculation fluids were diluted and then discharged into the Rhône river through diffusion manifolds at a variable rate depending on the flow of the river and the residual activity. Solid wastes were either placed in ventilated concrete-lined pits – this included the fuel element cladding – or stored in drums.

---

[45] With regard to the type of checks and the quantities discharged, see Part III Chapter 4. At the ICPP, the gaseous effluent was discharged after passing through glass fibre filters and after being diluted in the inactive zone ventilation air, without being checked. At Marcoule, paper filters were used in the active and semi-active zones and, although there were qualitative checks on gaseous discharges in 1958, the quantities discharged were not measured.

[46] The AEC Burial Ground, see UN (1955), p. 43.

[47] Pre-1963 documents consulted do not mention the problem of waste management. We therefore had recourse to oral evidence.

[48] A pilot bituminisation plant was tested in 1963. It appears that in the early years at Marcoule this type of effluent was disposed of at sea and not bituminised.

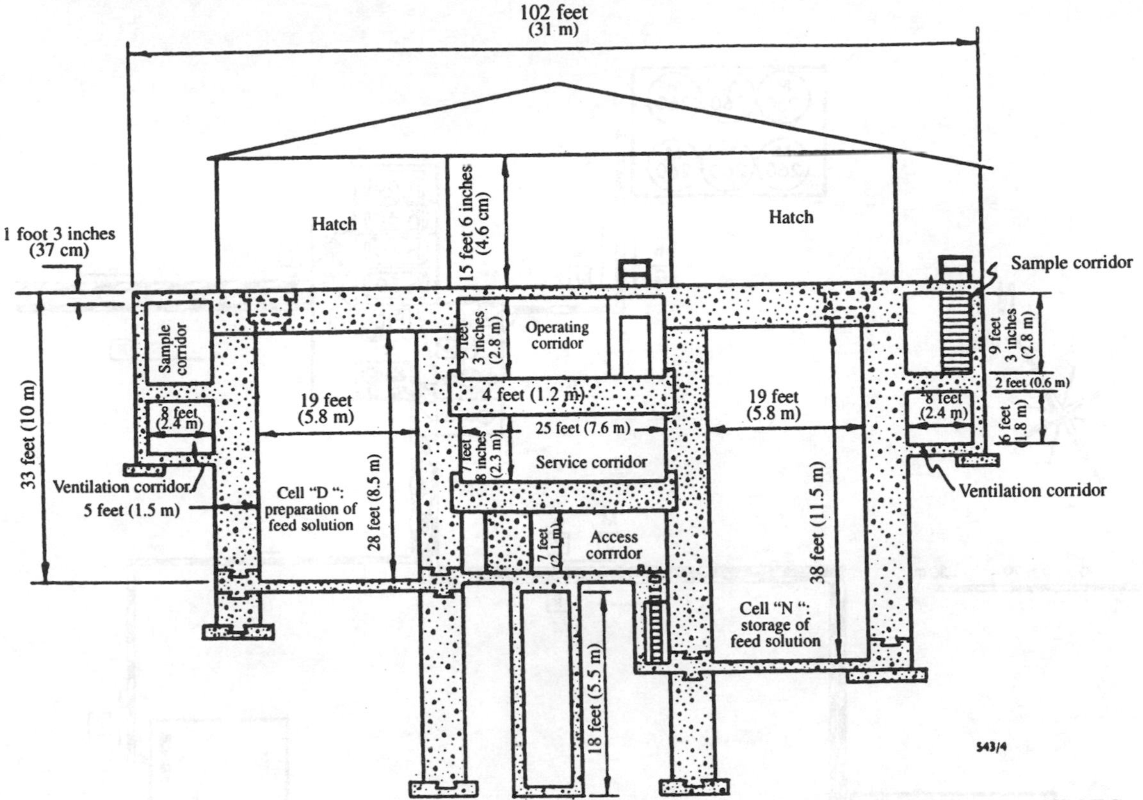

Figure 64. Sectional view of the ICPP treatment building, mostly underground (Source: USAEC (1955), p. 19).

In the early days of Eurochemic, the waste treatment system (Figure 65) followed the same logic of separation and storage. However the constraints on environmental disposal were more stringent owing to the high local population density and the lack of any substantial watercourse. High activity effluent was stored in a specially prepared building – Building 5 – which contained two 40 m^3 tanks and two 210 m^3 tanks. All the tanks were cooled. The medium and low activity effluents were pumped to an effluent treatment plant in Building 8, where they were separated or concentrated in two evaporators. Medium activity effluent was sent to a storage building – Building 21 – containing six 260 m^3 tanks. Low activity effluent was pumped through a double pipeline to the CEN where it was treated before being discharged into the Molse Nethe some 12 kilometres to the south, where the Belgian centre had a discharge authorisation.

The nature of the process appeared to limit solid wastes to those of low activity resulting from decontamination operations. These had to be transferred to the CEN for drumming and storage, until it was possible for them to be dumped at sea, under a European programme being planned by the OECD European Nuclear Energy Agency from 1964 onwards[49].

Eurochemic: a small, compact and highly integrated plant

One explanation of the relatively compact design of the Mol installations could be the modest size of the plant compared with that at Marcoule. In fact one series of buildings (numbers 1 to 4) combine the treatment plant, the reception, flask decontamination and fuel storage block, the analytical laboratory and the ventilation plant[50].

---

49 For the waste management policy, see Part III Chapter 4.

50 The ICPP also included the laboratory and the treatment building. However fuel reception took place more than one kilometre to the south of the treatment plant and the fuel was brought in by truck. At Marcoule, the auxiliary treatment buildings are laid out perpendicularly to the plant, at the other side of the underground services gallery. Thus the analytical laboratory is separate from the plant and connected to it by a simple walkway over which the samples were carried. The fuel storage installation is located near the G1 reactor, and not near the plant.

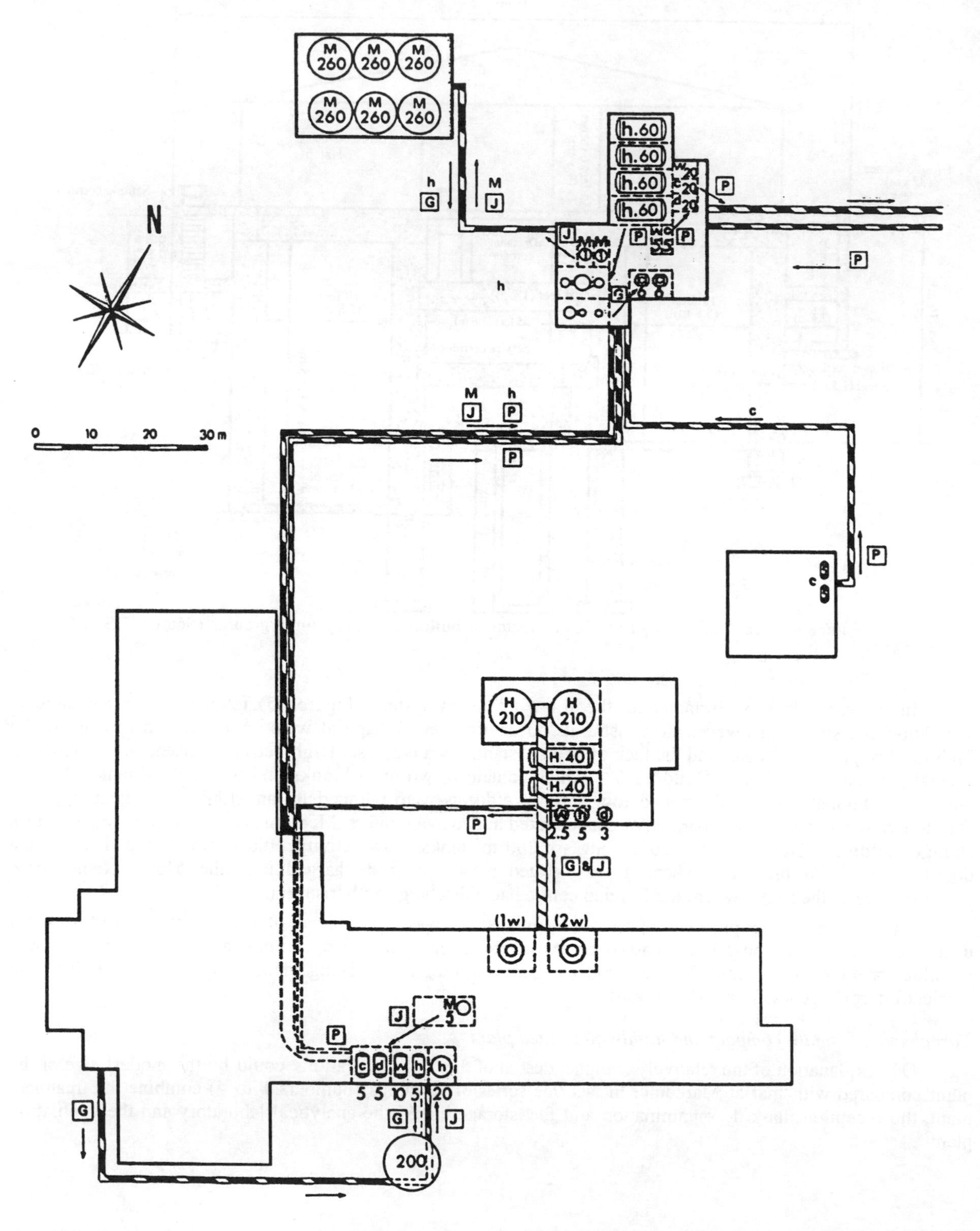

Figure 65. Diagram of the plant effluent transfer system. H: high activity effluent; M: medium activity effluent. Lower case letters concern low activity effluent: hot (h), warm (w), cold (c) and inactive condensate (d). The numbers show tank volumes in m^3 (Source: *Safety Analysis* (1965), Volume of Figures, V-16).

The treatment building is approximately the same size as that at ICPP (90 x 30 metres), but it is only about half as long as that at Marcoule (160 x 37 metres). However it is higher – 30 metres compared with 19 metres – disregarding the Marcoule siphon tower[51]. The difference in surface area results from the high extraction equipment and the lack of underground plant.

The high degree of integration of the process buildings avoids long transfer distances and thus reduces the hazards related to transfer operations in moving radioactive materials.

The compactness of the plant can also be seen from the internal layout of the treatment plant. At the ICPP the cells are served by a central corridor, which saves space but is risky in terms of safety. At Marcoule, the non active zones surround the active zones, which is the best solution but also the most expensive. At Mol (Figures 66 to 68) the inactive areas cover seven floors giving on the southern facade of the treatment building, the active zones being grouped along the northern facade. The overall layout of the buildings also corresponds to this internal layout of the treatment building: non-active buildings to the south, near the Meuse-Escaut canal, active buildings to the north.

However the fundamental differences between the three plants stem mainly from their objectives: Marcoule given over to military production[52], ICPP being an industrial pilot[53], and Eurochemic a combination of research laboratory and industrial pilot plant.

A synergy between laboratory and plant

The presence of a research laboratory specialising in reprocessing and a multipurpose plant on the same site was to generate new synergies at Eurochemic.

Because of its twofold objective, Eurochemic combined a pilot plant and a research laboratory.

The role of the R & D laboratory in process development has already been described. It was a question of developing the systems to be installed in the plant and then testing them at full scale in the pilot hall. It was expected to have the same function in the future, since the plant was supposed to be able to adapt to the characteristics of new fuels, with what was perhaps the plant's most original feature, its multiple capability.

The polyvalent nature of Eurochemic

The multiple capability of the plant called for a highly unusual organisation of operations[54] prior to the extraction process, for receiving fuel at the head of the process, including decladding and dissolution.

Reception and storage of spent fuel (Figure 69)

At Marcoule and ICPP, spent fuel came from a small number of reactors located near the reprocessing site and stored in a particular plant building. At Eurochemic, spent fuel could come from any member state and was brought to site in transport casks of various designs carried on truck trailers. The reception zone in Building 2 could accommodate transport casks to a maximum size of 2.5 x 3 x 5 metres, with a truck-trailer assembly up to 15 metres in length. This area was served by an 80-tonne overhead crane capable of lifting the cask and placing it on an inspection and decontamination platform measuring 13 x 13 x 8 metres. Checks were carried out to ensure that the transport casks had not been contaminated by the fuel they contained. If necessary, they were decontaminated before being moved into a reception pond deep enough to ensure at least 3 metres of waterbetween the top of the open cask and the operators whose job was to remove the fuel using the overhead crane and long gaffs and hooks. The fuel discharged under water was then moved through a communicating passage to the 25 x 9 metre storage pond capable of holding 80 tonnes of natural uranium or slightly irradiated fuel, where it was placed in racks so arranged as to avoid any risk of criticality (Figure 70)[55].

---

51 Most of the liquid transfers in the French plant made use of siphons, while ICPP and Mol used airlifts. Thus Marcoule needed a "water tower" system while ICPP and Mol used compressors.

52 The military production requirement shows up for example in the duplication or even triplication of systems in the plant, in order that any breakdowns should not interrupt the plutonium production programme.

53 R & D was done at Oak Ridge.

54 Simplified chemistry flow sheets for the plant are given in an Annex.

55 The capacity for highly enriched uranium fuel was lower and required the installation of neutron poisons and special racks.

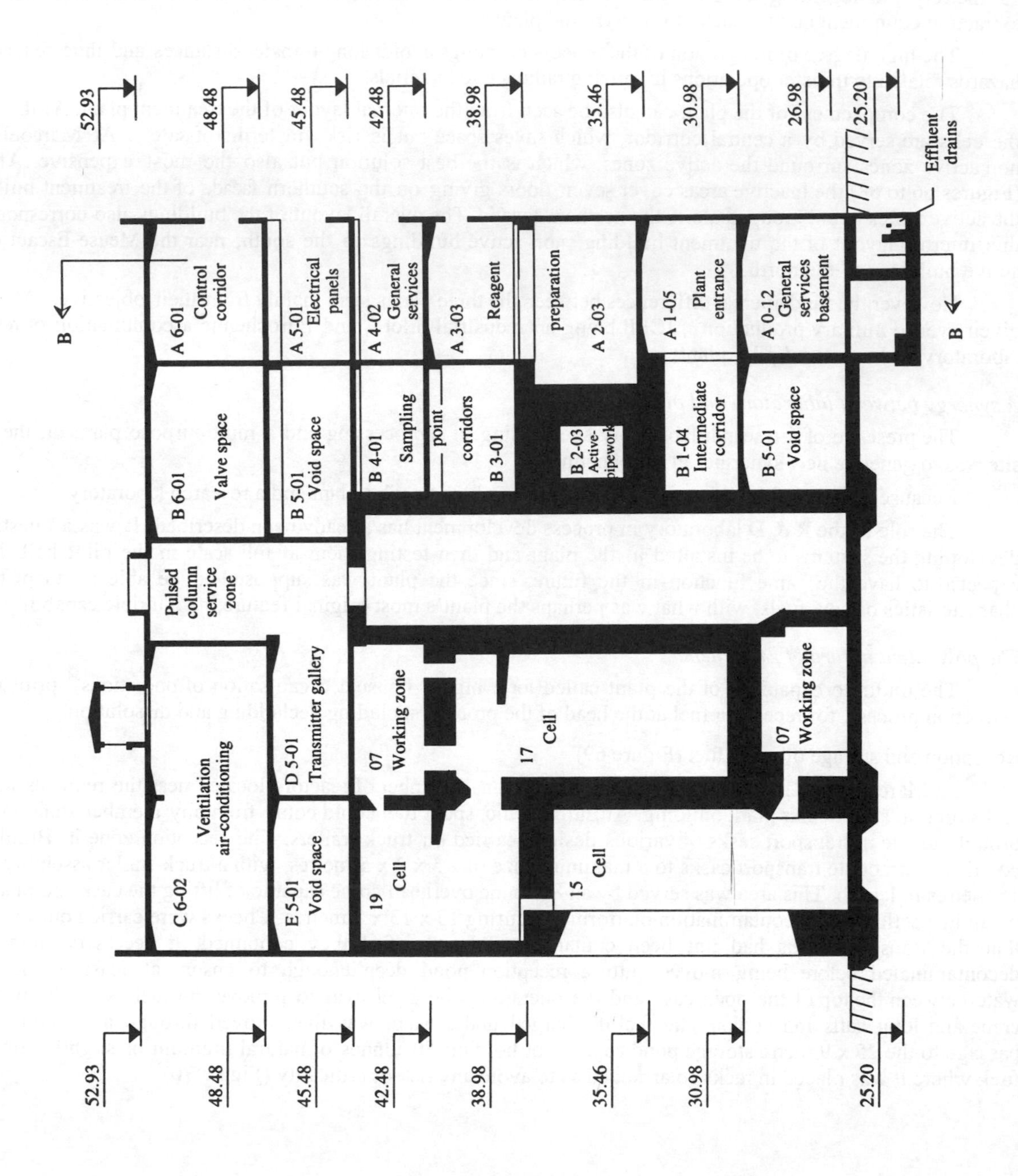

Figure 66. Elevation cross-section of the main building of the chemical plant, showing the asymmetrical layout of the active cells and the service corridors. The southern facade is on the right of the figure (Source: ENB n° 13, March 1965, Figure 2).

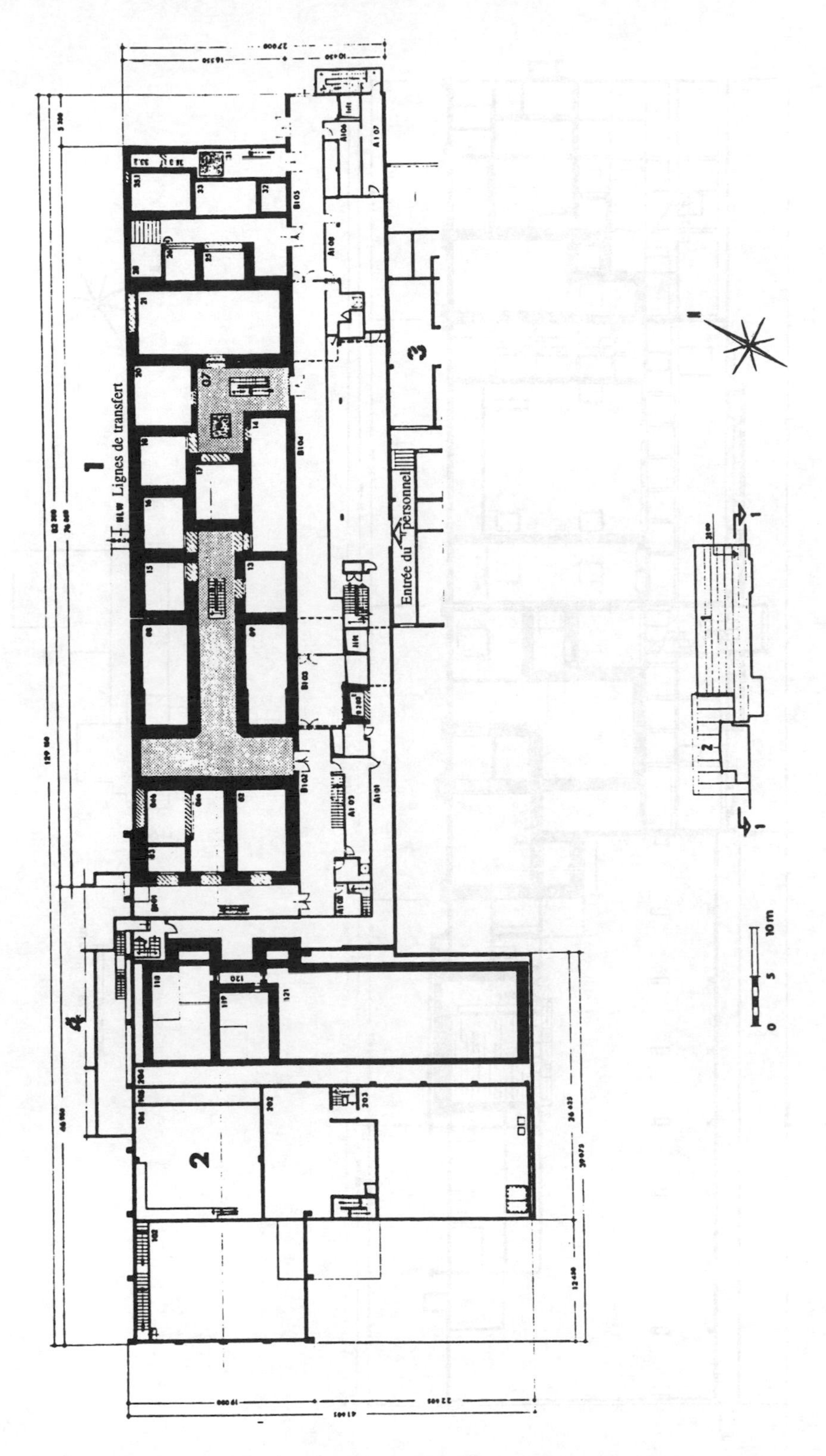

Figure 67. Plan of the fuel treatment, reception and storage buildings, showing the layout of the green (A), yellow (B), orange (shaded) and red (within the concrete-lined cells) zones (Source: *Safety Analysis* (1965), Volume of Figures, IV-4).

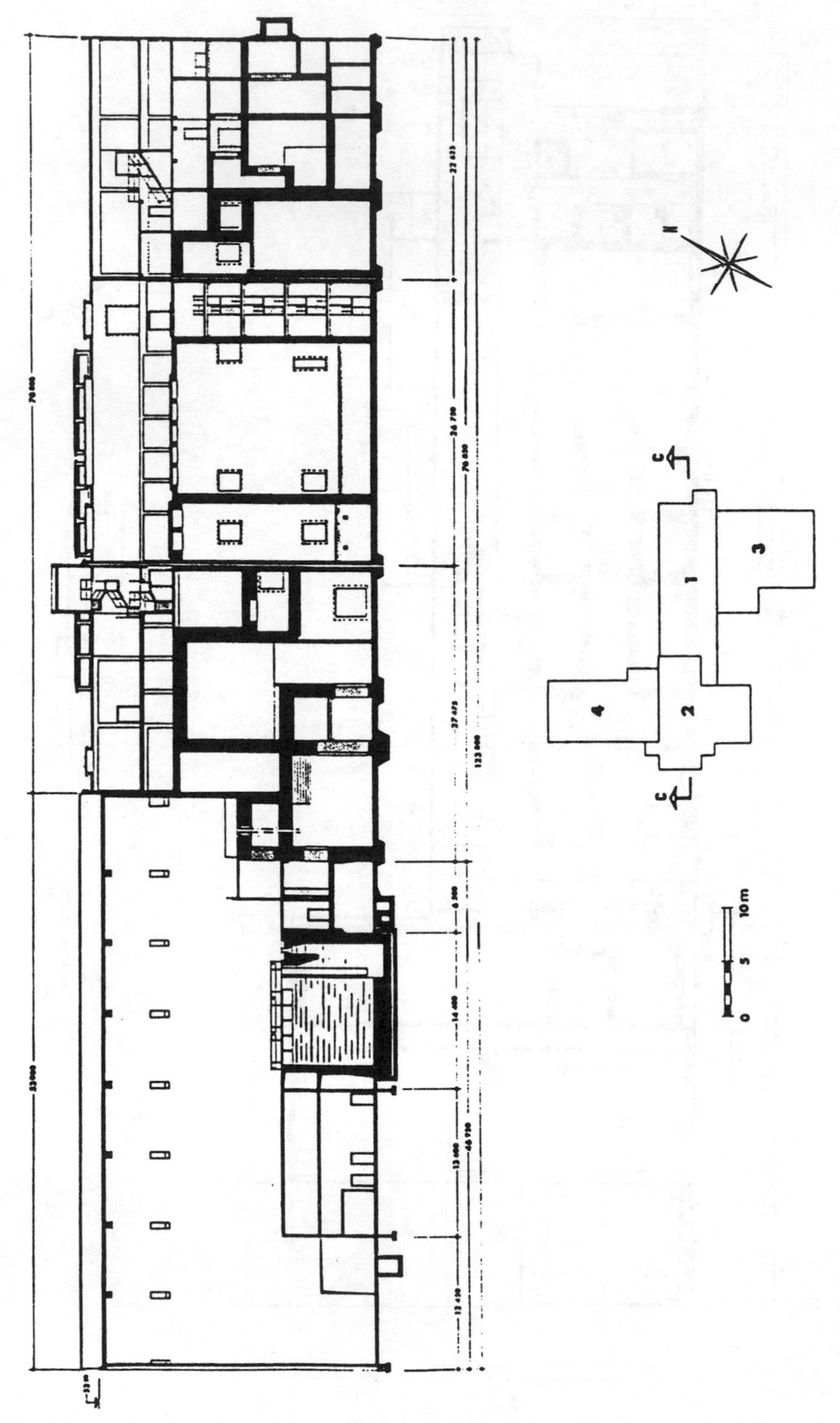

Figure 68. Elevation longitudinal section (median plane) of the main building of the chemical plant and of the fuel reception and storage building. The process commences in cells 1 to 6, 9, 11 and 12. Extraction takes place in the large cells 13 and 14. The reconditioning unit is in cells 21 and 23. Final uranium purification is done in cells 30, 32 and 33 and plutonium purification in 29, 35 and 36 (Source: *Safety Analysis* (1965), Volume of Figures, IV-12).

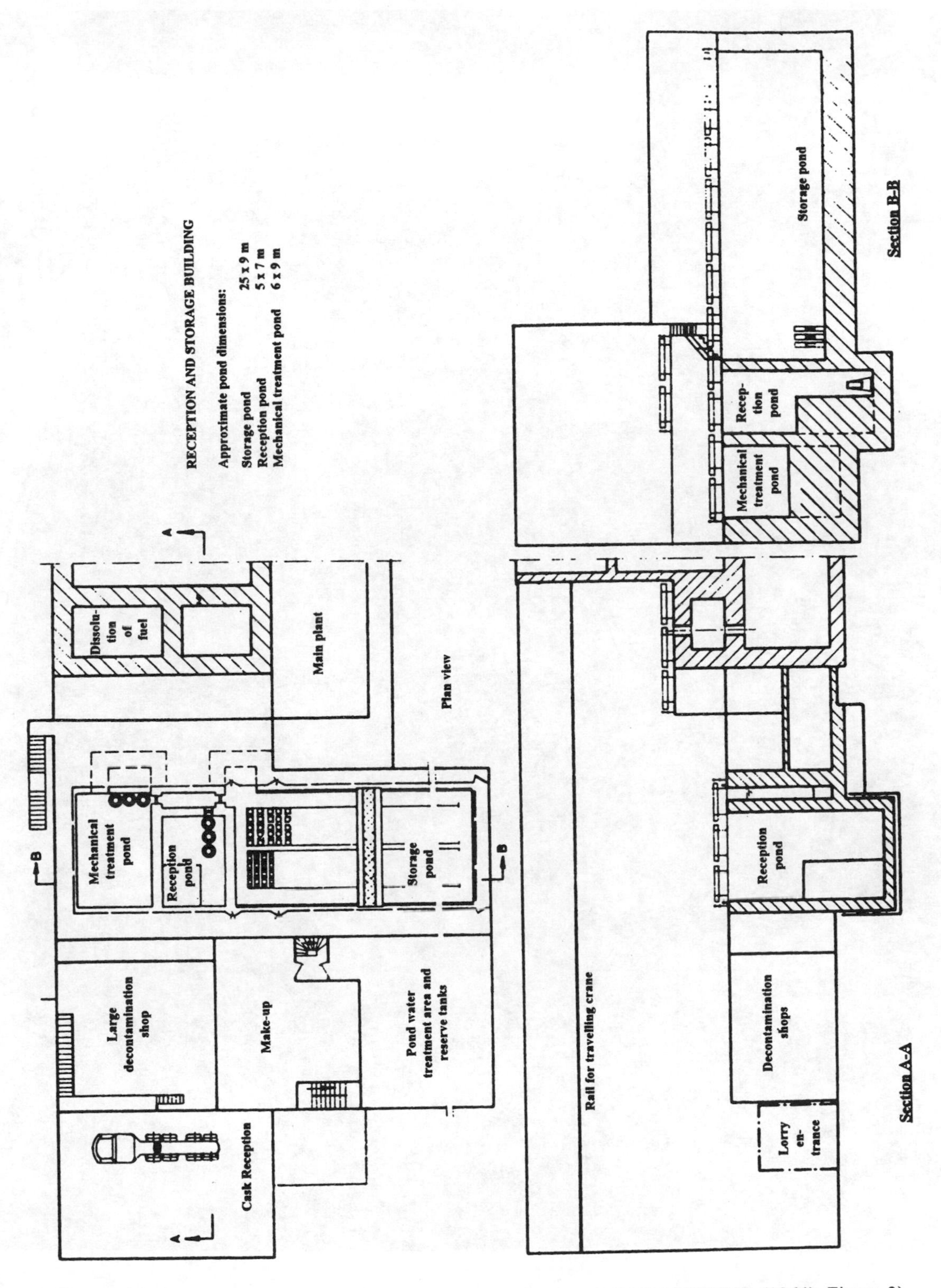

Figure 69. Plan of the fuel reception and storage building (Source: BARENDREGT T. (1964), Figure 3).

Figure 70. General view of the storage pond, with the spent fuel storage racks in the foreground (Source: Eurochemic photo, no date).

Figure 71. View of the charging machine for natural uranium or slightly enriched fuel, located beside the underwater transfer duct for moving fuel from the storage pond to the mechanical treatment pond (Source: Eurochemic photo, no date).

A third pond included an underwater saw for cutting up any structural components to facilitate loading the fuel into the dissolvers. Because of the variety of fuel shapes, related to the plant's multiple capability, chemical decladding was used, rather than a mechanical process which would have required a special installation for each type of fuel[56].

The decladding process and the chemical dissolving of fuel elements of various compositions required the use of three different dissolvers

The fuel to be processed was lifted from the pond by the overhead crane and transferred to charging machines (Figures 71 and 72), which were then sealed on the top valves of the dissolver cells. The fuel elements were then lowered into the cell and placed above the dissolver loading positions by remote manipulators. Three dissolvers had been built to accommodate fuel elements containing proportions of fissile uranium-235 ranging from 0 to 93%.

The first unit, which could accept a maximum enrichment of 1.6%, had a capacity of 1 tonne of fuel and was 7 metres high.

The second unit could accommodate fuel containing up to 4.6% of uranium-235. In order to guard against criticality it had a flat configuration and had three charging positions (Figures 73 to 77).

The third unit was designed to deal only with highly enriched fuels. It consisted essentially of a narrow tube (Figure 78).

---

[56] Thus at Marcoule the original dry installation for decladding fuel from the G1 reactor was never used, probably because it had been hoped that the non-active pilot plant stage would be unnecessary; the system bent the rods and jammed. It was very quickly decided to build a new decladding unit, this time under water, and this operated satisfactorily for fuel from G1, G2 and G3. However Marcoule used a chemical "peeling" system after mechanical decladding in order to free the fuel of any dust or residual pieces of cladding. ICPP, like Eurochemic, used chemical decladding.

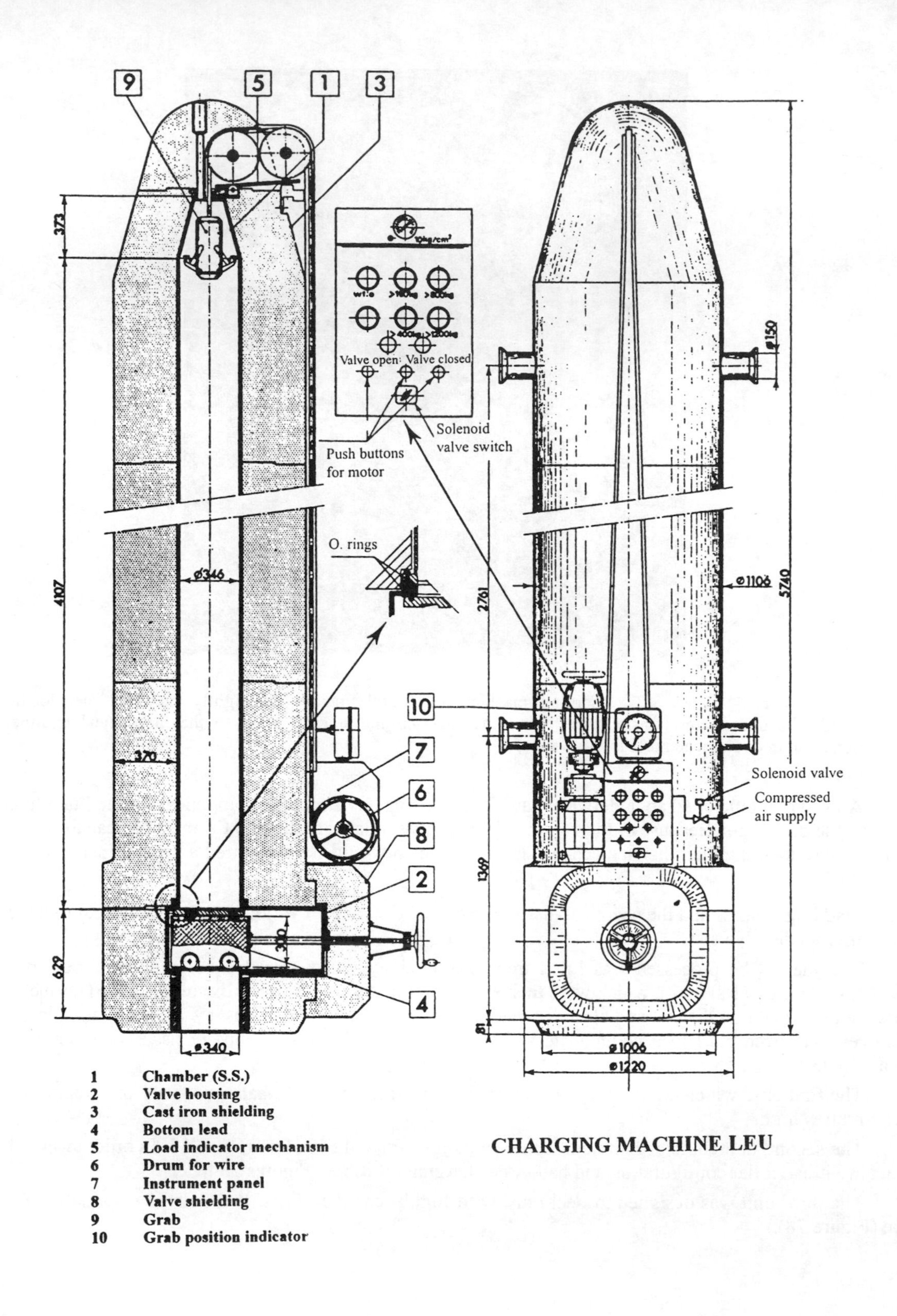

Figure 72. Diagram of the LEU charging machine (Source: *Safety Analysis* (1965), Volume of Figures, VII-1).

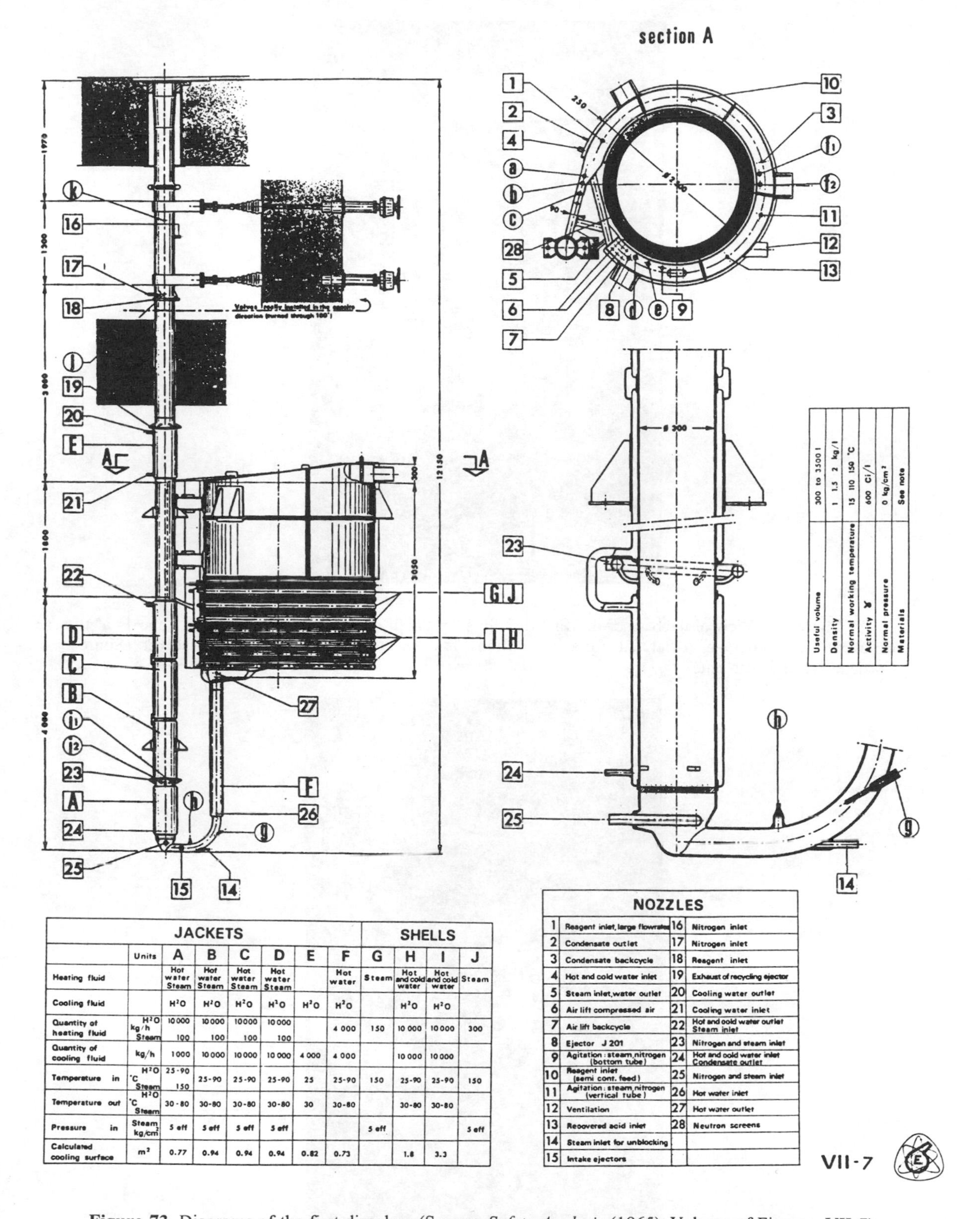

	Units	A	B	C	D	E	F	G	H	I	J
		JACKETS						SHELLS			
Heating fluid		Hot water Steam	Hot water Steam	Hot water Steam	Hot water Steam		Hot water	Steam	Hot and cold water	Hot and cold water	Steam
Cooling fluid		H^2O	H^2O	H^2O	H^2O	H^2O	H^2O		H^2O	H^2O	
Quantity of heating fluid	kg/h H^2O / Steam	10 000 / 100	10 000 / 100	10000 / 100	10 000 / 100		4 000	150	10 000	10 000	300
Quantity of cooling fluid	kg/h	1000	10 000	10 000	10 000	4 000	4 000		10 000	10 000	
Temperature in	°C H^2O / Steam	25-90 / 150	25-90	25-90	25-90	25	25-90	150	25-90	25-90	150
Temperature out	°C H^2O / Steam	30-80	30-80	30-80	30-80	30	30-80		30-80	30-80	
Pressure in	Steam kg/cm²	5 eff	5 eff	5 eff	5 eff			5 eff			5 eff
Calculated cooling surface	m²	0.77	0.94	0.94	0.94	0.82	0.73		1.8	3.3	

Useful volume	300 to 3500 l
Density	1 1.5 2 kg/l
Normal working temperature	15 110 150 °C
Activity γ	600 Ci/l
Normal pressure	0 kg/cm²
Materials	See note

NOZZLES			
1	Reagent inlet, large flowrates	16	Nitrogen inlet
2	Condensate outlet	17	Nitrogen inlet
3	Condensate backcycle	18	Reagent inlet
4	Hot and cold water inlet	19	Exhaust of recycling ejector
5	Steam inlet, water outlet	20	Cooling water outlet
6	Air lift compressed air	21	Cooling water inlet
7	Air lift backcycle	22	Hot and cold water outlet Steam inlet
8	Ejector J 201	23	Nitrogen and steam inlet
9	Agitation: steam, nitrogen (bottom tube)	24	Hot and cold water inlet Condensate outlet
10	Reagent inlet (semi cont. feed)	25	Nitrogen and steam inlet
11	Agitation: steam, nitrogen (vertical tube)	26	Hot water inlet
12	Ventilation	27	Hot water outlet
13	Recovered acid inlet	28	Neutron screens
14	Steam inlet for unblocking		
15	Intake ejectors		

Figure 73. Diagrams of the first dissolver (Source: *Safety Analysis* (1965), Volume of Figures, VII-7).

Figure 74. View of the base of the first dissolver installed in its cell. The dissolver tube is at the centre of the photograph. A flat tank for acid solutions is connected to the top of the dissolver (Source: Eurochemic slide, no date).

Figure 75. Photograph of the second dissolver before its installation on the site (Source: Eurochemic photo, no date).

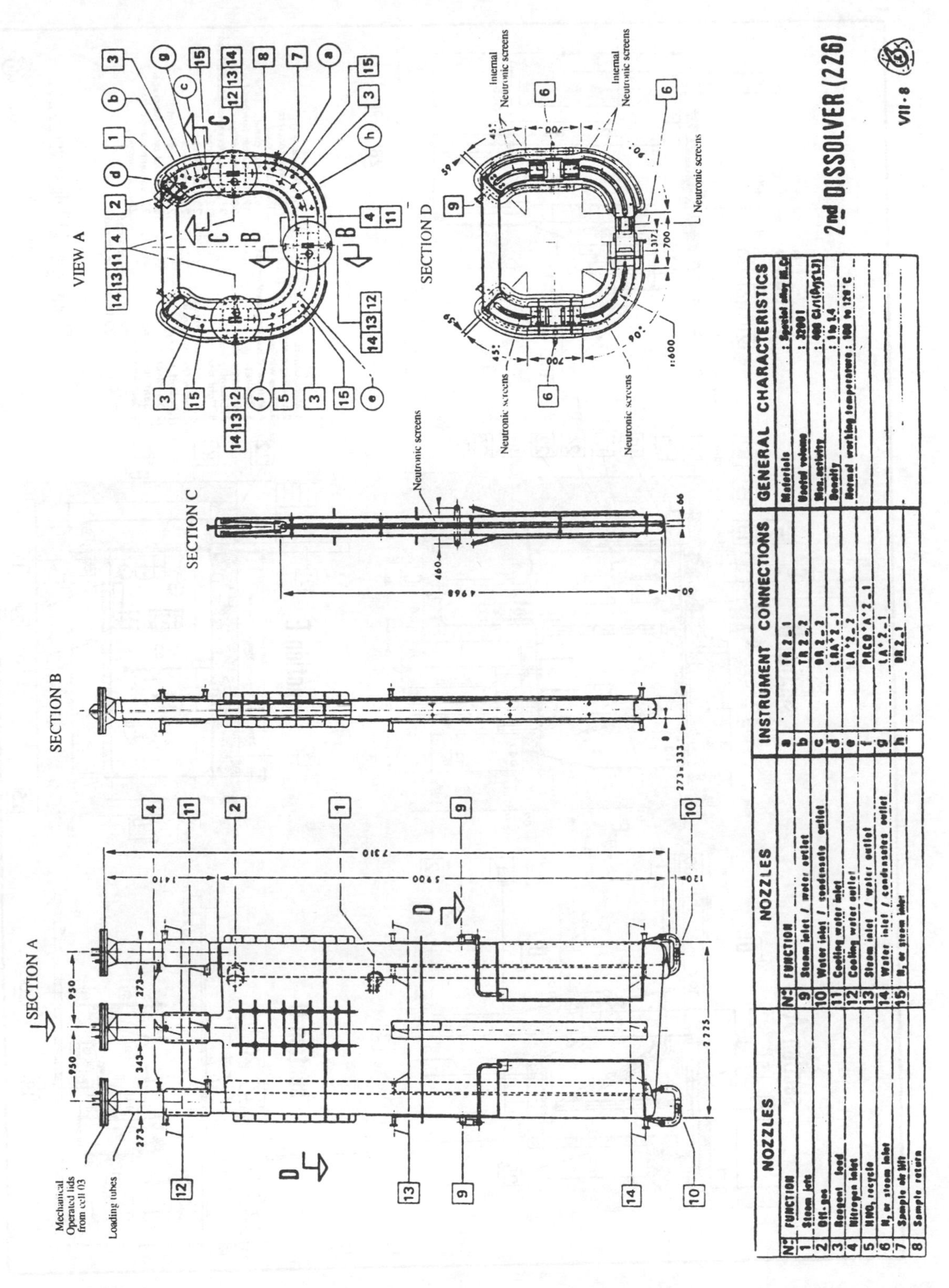

Figure 76. Diagrams of the plant's second dissolver (Source: *Safety Analysis* (1965), Volume of Figures, VII-8).

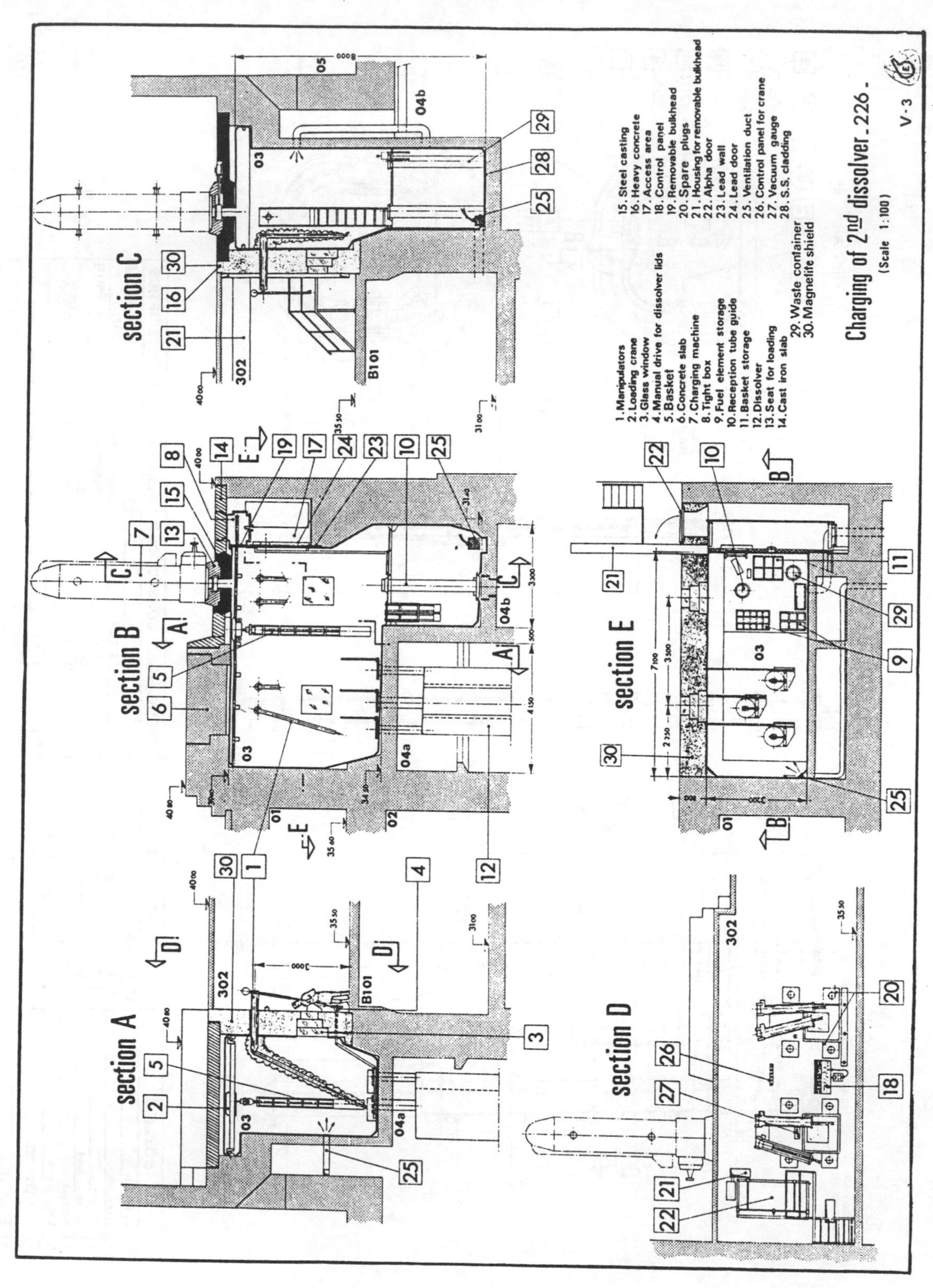

Figure 77. Diagrams of the charging systems on the plant's second dissolver (Source: *Safety Analysis* (1965), Volume of Figures V-3).

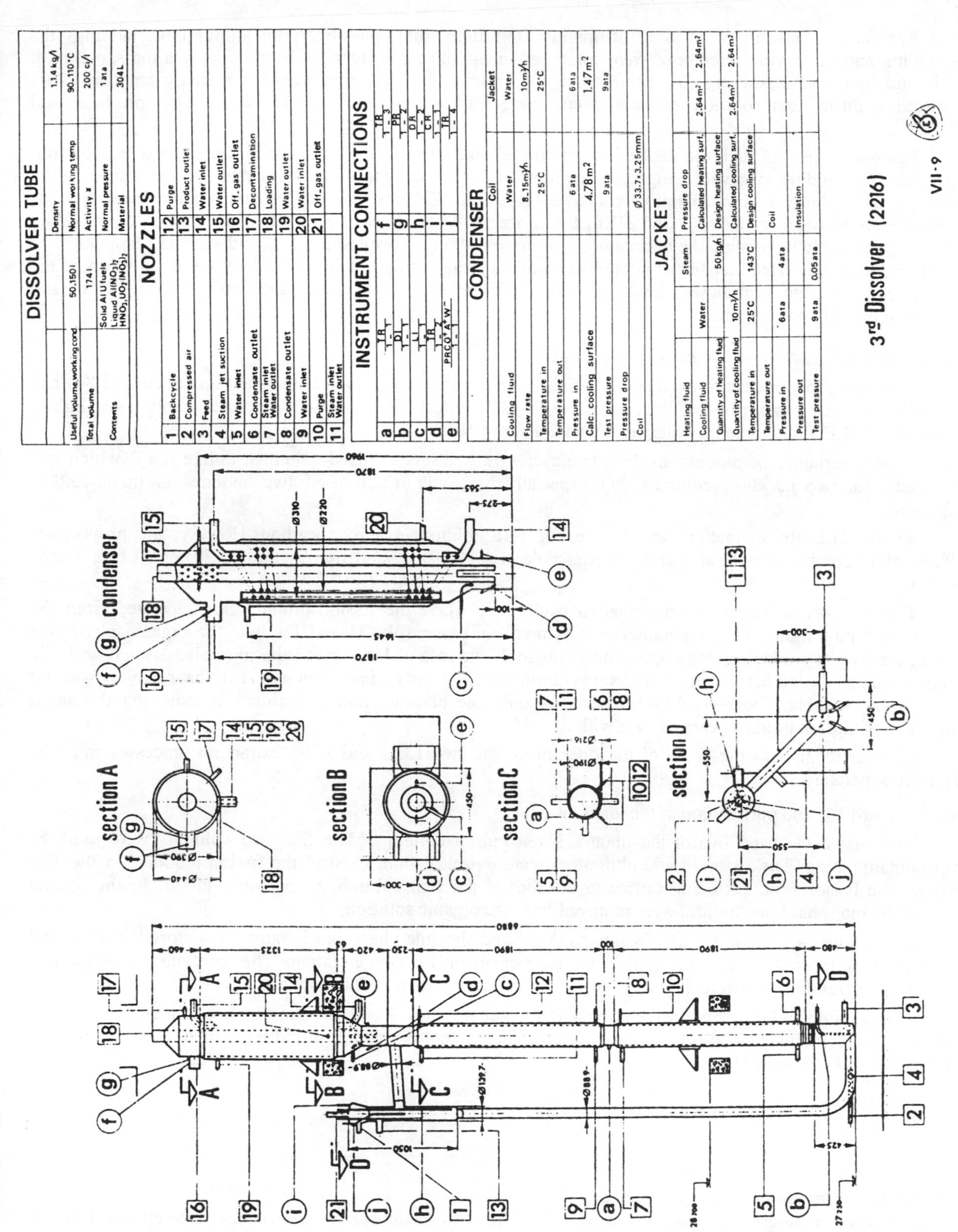

Figure 78. Diagrams of the plant's third dissolver (Source: *Safety Analysis* (1965), Volume of Figures VII-9).

For fuels consisting of an alloy of uranium and aluminium, possibly with aluminium cladding, the decladding and dissolving operations were combined in the third dissolver[57]. The process was the same as at ICPP and involved dissolving the elements in hot nitric acid in the presence of a mercury catalyst. This produced a nitric acid solution essentially containing different isotopes of uranium, fission products and mercury[58].

For other types of fuel, the decladding and dissolving operations took place in the same dissolver, but separately. Decladding involved using acids appropriate to the cladding materials, using processes invented at ORNL, with sulphuric acid for magnesium and steel (the SULFEX process), and essentially ammonium fluoride for zirconium or zircaloy cladding (the ZIRFLEX process). When it was decided that the chemical decladding had continued for long enough, the decladding solution was drained away, the solid, undissolved elements were rinsed, and then they were transferred to the fuel dissolving stage using hot nitric acid. When dissolution of the fuel was complete, the solution was drawn off from the dissolver tank, its concentration and acidity adjusted and then fed to the cells for extraction.

The two original features of the extraction process

The extraction process showed two original features: the use of pulsed columns (Figures 79 and 80) and the use in the same system of two variants of the PUREX process, one adapted to natural uranium or slightly enriched fuel (LEU), the other to highly enriched uranium (HEU).

In both variants, the process involved two cycles. It used seven pulsed columns, unlike ICPP which was equipped with two backflow columns, but especially Marcoule which used five batteries of mixer-settlers (Figure 81)[59].

As at Marcoule, extraction was done using TBP diluted in a hydrocarbon. However the proportions differed at the commencement of operations: Marcoule used 40-60%[60], and Eurochemic 30-70% for LEU, 5-95% for HEU.

The first phase was a decontamination cycle, i.e., where the fission products are separated from the uranium and plutonium. This separation process involved passing the U and Pu into the organic phase and leaving the fission products in the original nitric phase. In the case of HEU fuels, the plutonium is left with the fission products during the decontamination cycle, since its isotopic composition was such that recovery was not economically feasible. The second cycle concerned only the uranium and was aimed at removing the small proportion of fission products carried over with it.

In order to go into the detail of the operations, the two LEU and HEU extraction processes must be described separately.

Extraction of LEU and purification of the uranium

The first cycle used five of the plant's seven pulsed columns. The first two columns were used for co-decontamination. The uranium and plutonium were separated from most of the fission products in the first column, the fission products being carried to the foot of the first column in the nitric phase. In the second column, the interphase precipitates were removed from the organic solution.

The solutions containing the fission products and the interphase precipitates were stored, settled and analysed in buffer tanks before being fed into an evaporator for concentration. The concentrates were then transferred to Building 5, where they were kept in 40 m^3 tanks with artificial cooling.

The uranium and plutonium were separated in columns 3 and 4 with the addition of a reducing agent. The nitric solution containing the plutonium was removed at the foot of column 4 and transferred to the plutonium purification unit (see below).

---

57 Sometimes there was no cladding at all.

58 The presence of the catalytic mercury was one of the most important specific characteristics of effluent from the processing of highly enriched uranium fuel.

59 Marcoule used three batteries for extraction, and two for uranium purification. Each battery consisted of between eight and thirteen mixer-settlers. The system was duplicated so that production could continue even if one battery broke down, the estimated MTBF of the mixer-settlers being about a year. Pulsed columns had been tested at Fontenay at the pilot stage, but the mixer-settlers, a more familiar system and having advantages in terms of control in the event of a shutdown, were preferred.

60 These were the proportions expected in 1958. As from 1961 they were the same as for Eurochemic.

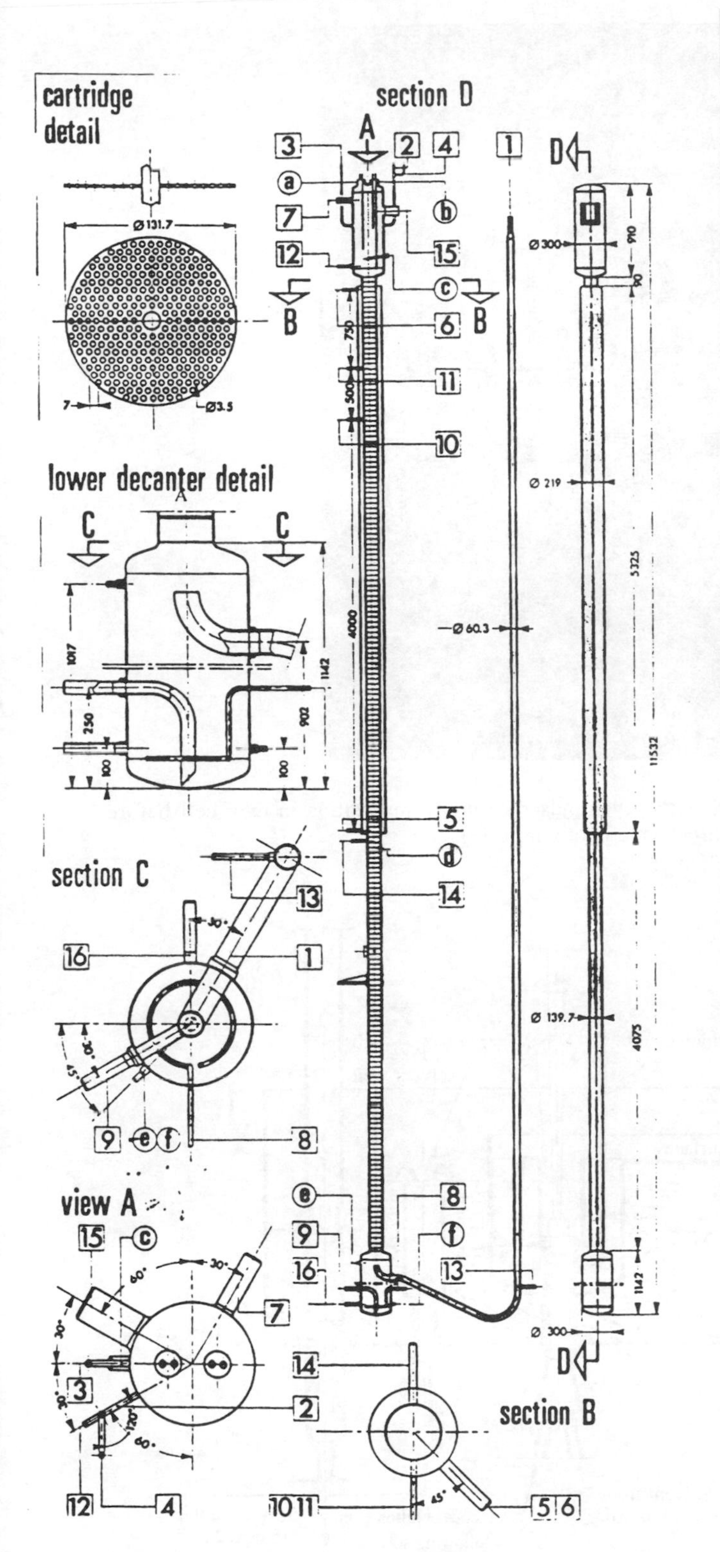

NOZZLES			
N°	FUNCTION	9	Emptying
1	Air pulse	10	2DS
2	Sampling outlet	11	2DS
3	Backcycle sampling	12	2DS
4	Sampling air	13	2DX
5	Heating water inlet	14	2DP
6	Heating water outlet	15	2DU
7	Ventilation	16	2DW
8	Agitation		

INSTRUMENT CONNECTIONS			
b	DRAW± / 2-1a	d	T-R / 2-2
		e+f	LiRCW± / 2-1a
a+b	LIW+ / 2-1		
c	T-R / 2-1		

PROCESS REQUIREMENTS

		INLET FLOWS				OUTLET FLOWS	
	fluid	2D1S	2DS	2DF	2Dp	2DU	2DW
U. enriched 5%	flows l/h	12	22	32.8	163	163	66.8
	spec.gra. kg/l	1.00	1.10	1.59	0.82	0.943	1.03
U. enriched 20%							
U. enriched 93%							
Temperature °C		30	30	30to55	30	30to54	30

FLOODING CONDITIONS

	continuous phase	cartridge	cycle / mn	mm	flood cap. l/h.dm²
extraction	organic	B	60	25	320

NORMAL OPERATING CONDITIONS

	% flooding cap.	oper. flow-rate	theoretical stages N.T.S.	theoretical stages H.E.T.S.
extraction	53	169	5	0.7

DECANTING CHAMBERS

	cross section		designed Ø		design height
extraction	136 cm²		131.7mm		4 m
			diameter	height	
top			300 mm	910 mm	
bottom			300 mm	1142 mm	

CARTRIDGE CHARACTERISTICS

TYPE	free area.	holes Ø	plate thick.	spacing	nozzle depth.	material
B with nozzles downwards	22.5 %	3.5 mm	1.2 mm	50 mm	2.5 mm	AISI 304 L
pulse	frequency	30 to 100 cycles/min		amplitude extraction	10 to 40 mm	

Pulse Column
232.2

VII-21 

Figure 79. Diagrams of one of the plant's seven pulsed columns, with its lower and upper settling tanks (Source: *Safety Analysis* (1965), Volume of Figures, VII-21).

Figure 80. Partial view of three of the plant's seven pulsed columns, in position in their cell before the commencement of reprocessing (Source: Eurochemic photo, no date).

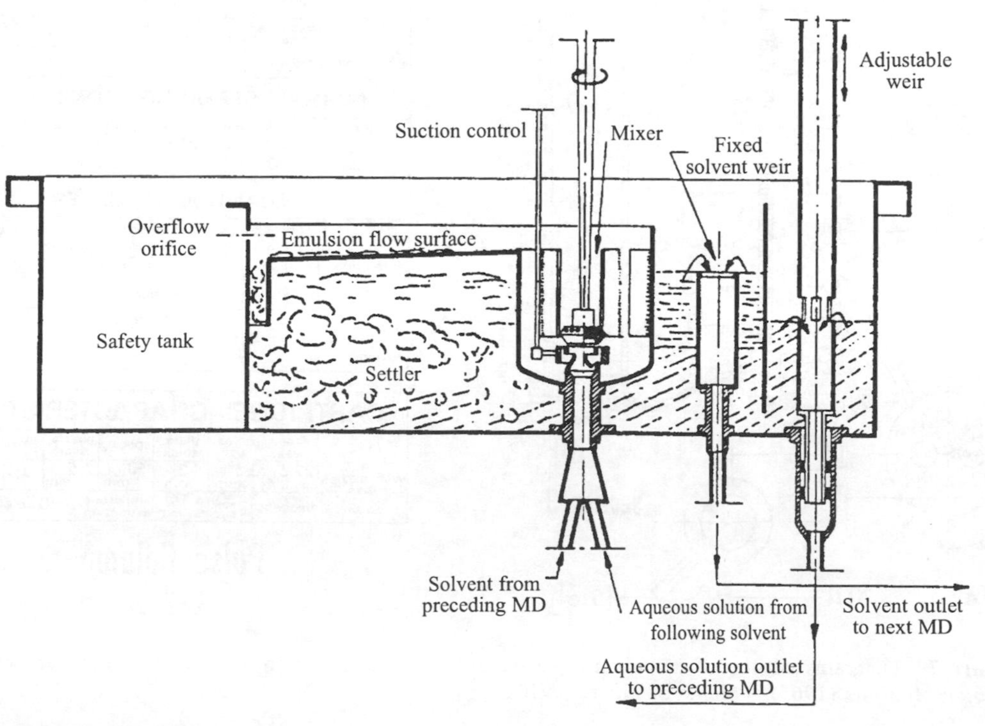

Figure 81. Section through a mixer-settler (MD) used in the Marcoule plant in 1963 (Source: JOUANNAUD C. (1963), p. 264).

The uranium in turn was then passed into the nitric phase, removed at the bottom of the fifth column, concentrated by evaporation and fed to the second extraction cycle consisting of two columns. It was then decontaminated once again. Finally, the solution was concentrated in an evaporator and purified on silica gel in an absorption column.

The final product of the operation was uranyl nitrate which was transferred to a storage building and delivered to customers in truck-mounted tanks.

Extraction and purification of HEU

The first cycle employed four of the five columns of the first LEU cycle, whereupon the decontaminated solution was concentrated in a special evaporator.

The second cycle involved the plutonium purification system (see below) prior to purification on silica gel in the *ad hoc* LEU column[61].

The fact that the same equipment was used for different types of fuel illustrates the compact design of the plant, but also reflects the financial problems encountered by the company. This dual use meant that painstaking rinsing operations were necessary between operations involving major categories of fuels.

The fission product solutions which also contained aluminium and mercury were stored in two 212 m^3 tanks located in Building 22.

Purification of the plutonium: mixer-settlers and a laboratory style process

The process used for plutonium purification was based directly on that used at Marcoule and involved mixer-settlers for re-extraction.

However, whereas Marcoule – a military plant – continued the plutonium nitrate conversion through to the production of metal, Eurochemic – a civilian plant – went only as far as producing the dioxide by calcination of plutonium oxalate obtained by precipitation from the purified solution of plutonium nitrate[62].

As at Marcoule, during early operations the final phase used installations that were virtually on a laboratory scale (Figure 82) involving a series of gloveboxes as from the oxalate drying stage.

However, while Marcoule concentrated the separated plutonium by precipitating it in caustic soda and then redissolving it in a small amount of nitric acid, Eurochemic used a special evaporator.

Eurochemic was a thus compact and multipurpose plant, employing and mixing together certain aspects of Marcoule and ICPP, but also having certain original features, the most important being chemical decladding and the use of pulsed columns.

The first few years of reprocessing constituted a learning period for the staff of the joint undertaking; it was a period that demonstrated the overall efficiency of this hybrid design but also revealed its imperfections.

---

[61] Before the enriched uranium left Europe to be returned to the United States for re-enrichment, it was transferred to the Pierrelatte plant where it was converted into hexafluoride after a further decontamination.

[62] At Marcoule, the plutonium oxide was next heated in a current of hydrofluoric acid to produce plutonium fluoride, which was then converted into plutonium metal by heating over lime in an argon atmosphere.

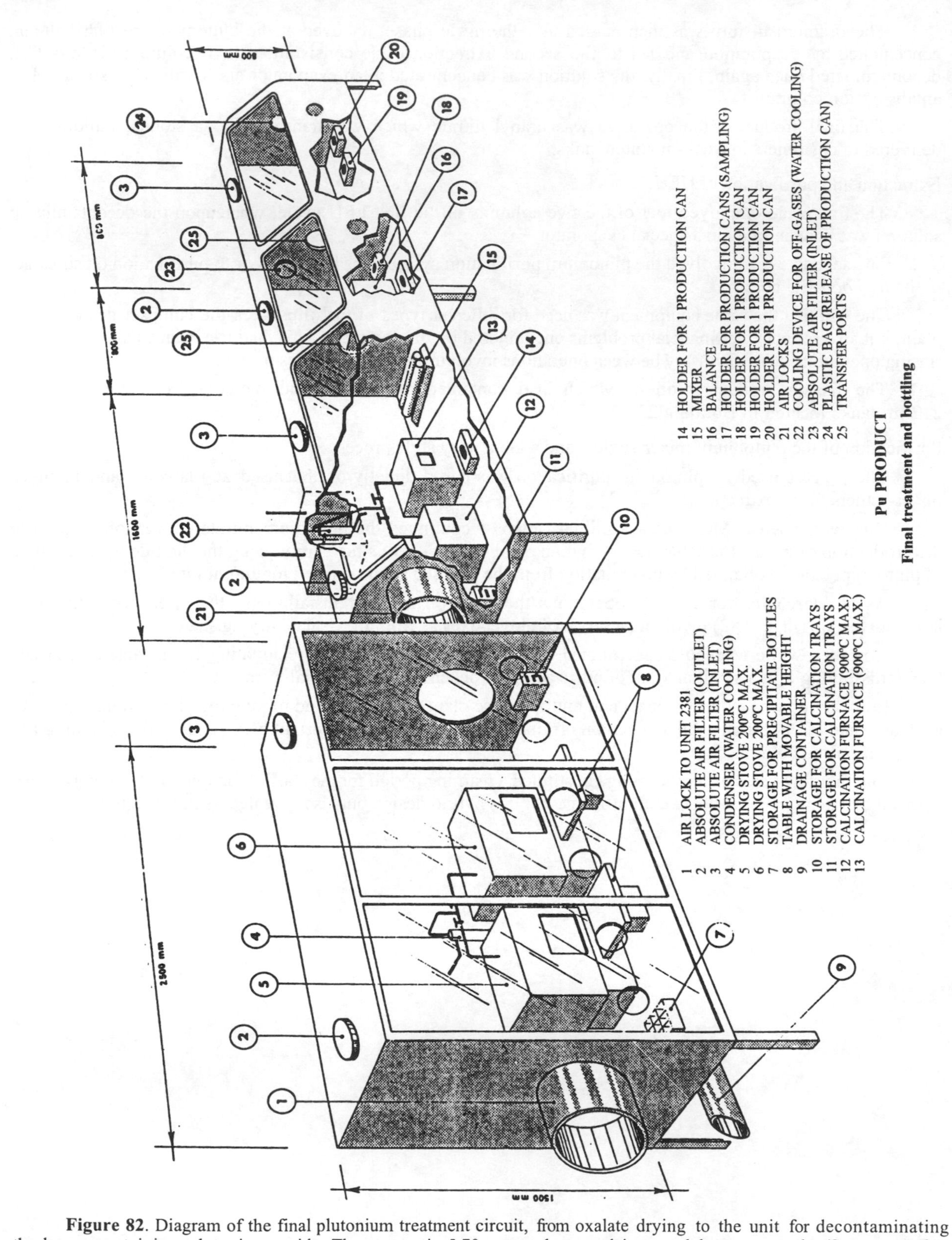

Figure 82. Diagram of the final plutonium treatment circuit, from oxalate drying to the unit for decontaminating the boxes containing plutonium oxide. The system is 5.70 metres long and is on a laboratory scale (Source: *Safety Analysis* (1965), Volume of Figures, VII-38, addendum of November 1966).

Chapter 2

The learning years

July 1966 - February 1971

During the first few years of reprocessing, the funding problems appeared to have been settled, at least in the medium term. This made it possible to define a five-year programme of operations and to deploy an ambitious commercial and industrial policy which suggested that the undertaking might become viable by 1974.

With this in view, the structures of the company underwent a far-reaching reorganisation and were even more tightly focused on reprocessing than before. This is clear from the changes in the activities of the Industrial Development Laboratory and the Analytical Laboratory.

During this initial period of learning, the plant teams tried out the reprocessing of different types of fuel, encountering unexpected difficulties which they overcome, experiencing incidents which gave them food for thought, and evaluating the strengths and weaknesses of the system. When considerable experience had been built up, the plant was shut down for a long period during which the changes considered essential were made so that the company, which had hitherto operated as a pilot plant, could go on to reprocess fuel at full scale.

Secured medium-term funding permitted an ambitious commercial and industrial policy

The short and medium-term settlement of the funding problems: from the Transitional Programme for 1968-1969 to the Five-year Programme for 1970-1974[1]

The construction phase had been marked by continuous financial problems, but these difficulties were resolved relatively easily by the responsible authorities for, it was thought, a period which would run up to 1974.

Up to the end of 1967, funding of the company had been provided under the Temporary Programme from April to July 1964. In March 1967 a Transitional Programme for the years 1968 and 1969 was accepted by the Board of Directors. In May a procedure was initiated which resulted in the adoption of a programme to cover the years 1970-1974, the Future Programme of the company. This was subsequently named the Five-year Programme. In July 1969 the Council of the OECD adopted the conclusions regarding funding. Thus it appeared that operation of the undertaking was assured beyond the 15-year period provided for by the Convention, 27 July 1974.

The Transitional Programme for the years 1968 and 1969

Proposals of the Working Group

A Working Group on Work Progress and Funding was set up on 24 March 1966 and submitted its report on 20 May[2]. Its proposals served as a basis for discussion which led to the definition of a transitional programme for the years 1968 and 1969. The report mainly dealt with two points.

As regards funding the completion of the works, the principle of a further increase in capital, which had originally been envisaged, was rejected. The increase, estimated at 1.335 million dollars, was considered to be inappropriate for "psychological and procedural" reasons. The two previous increases in capital had, as we have seen, been secured with great difficulty. For this reason it was decided that funding should be in the form of a loan backed by the Belgian government, to be reimbursed up to 1971.

As regards covering the operating deficit for the years 1968 to 1971, it was recommended that the system of government contributions paid through the OECD budget, set up in 1964, should be continued. The deficit was in the range 7.77 to 10 million dollars. However, different views were again expressed about how the scale of contributions should be calculated. France and Germany were in favour of contributions being proportional to

1 The Report is attached in the file of the Special Group meeting held on 1 September 1966.

2 NE/EUR(66)7.

countries' shareholdings in the company; the other countries, except for Italy which once again showed its overall opposition in view of the changing situation, were favourable to contributions being proportional to GNP. Switzerland suggested a compromise, and a basis for calculation was agreed[3]. These proposals were carried over into the Transitional Programme.

The Programme and its financial arrangements

The Transitional Programme for the years 1968 and 1969, drawn up by the general management in conjunction with the Technical Committee, was proposed to the Board of Directors which approved it on 2 March 1967[4].

Three main decisions were taken:

- the structures of the company were recast. This resulted in changes to the organisation chart with, in particular, four divisions being merged into two, covering operations and development, as well as technical and industrial development[5];
- the development programme was cut and a maximum programme of operations adopted[6];
- a programme of progressively reducing staff numbers was implemented. The company was to employ about 320 people in 1970 compared with 378 at the end of 1966.

On 17 May 1967, the Special Group adopted the funding measures of the Transitional Programme for 1968-1969[7]. The scale that was finally adopted for calculating government contributions applied the two relevant variables in a complex manner[8], with special arrangements for certain countries[9].

Switzerland expressed its disappointment at the scale applied but agreed to the arrangement after recalling that its interest was primarily industrial.

France, for its part, gave its agreement subject to three conditions, one being a *sine qua non*. She demanded first that changes in the expected deficit should be closely monitored so as to keep it as low as possible – Germany also expressed the same request. France then asked that a study of the suitability of closing the plant should be started before the summer of 1968. This attitude clearly showed the relative lack of French interest in the undertaking. Above all France made it clear that the exclusive application of EURATOM security control to Eurochemic was a condition for its participation[10]. In fact security control could legally be exercised by three authorities: EURATOM, for two reasons – the location of the plant and the fact that five of its six member states were participating; ENEA which had its own system; the IAEA since the intention was that fuel of American and Canadian origin would be reprocessed at Eurochemic. In fact France was resolutely opposed to any control by the IAEA, which she regarded as "the eye of Washington"[11] on her reprocessing activities, even minor operations such as those at Mol.

Finally, the future of the company should be reviewed by an ad hoc working group.

Two months after the adoption of the Transitional Programme, a programme covering the period 1970-1974 was prepared.

The Future Programme of the company: 1970-1974

Preparation of the Five-year Programme

In May 1967 a working group was established to consider the company's objectives and structure. In November 1968 it submitted an interim report to the Special Group, stressing in particular the need to take conservative measures to increase the effluent storage capacity[12]. Results from the initial reprocessing had shown

---

3 One-third according to holding, two-thirds according to GNP.

4 CA(67)1 of 31 January 1967. Approval CA/M(67)1.

5 See the description of the structures for 1966-1971 below, and the figures.

6 CA(67)1 Annex I.

7 NE/EUR/M(67)1.

8 NE/EUR(67) gives a detailed description.

9 Belgium, Portugal and Turkey, the Italian contribution being reduced to the R & D costs.

10 See the section on security control in Part III Chapter 4.

11 Mention must be made here of the contemporary crisis in Franco-American relations and its effects on NATO, which came to a head in March 1966.

12 NE/EUR(68)3 of 7 November 1968.

in fact that the amount of effluent produced was greater than expected. The final report was submitted on 21 March 1969[13] and approved by the Special Group at its sixteenth meeting on 23 April 1969[14].

The Council of the OECD approved the conclusions as regards funding on 22 July 1969[15]. The OECD released an amount of 15.09 million EMA units of account, with the CNEN committing itself to 1.480 million. The outstanding deficit for the period 1964-1969, amounting to 0.575 million EMA units of account, was also covered by the OECD[16]. The existing scale of contributions was applied to the funding arrangements up to the end of reprocessing.

A medium-term deficit now accepted

The Programme, after reviewing the situation of European reprocessing[17] and its prospects, went on to consider future options and incorporated these into a programme of operations and development; it also looked into the problems of funding and suggested modifications to the Statutes.

Although it was unlikely that the company would be viable in the medium term, it was nevertheless planned to continue its activities. It is important to try to understand the reasons.

According to predictions of the amounts of fuel that would become available for reprocessing, it was impossible for the company to be viable in the medium term, even if it succeeded in attracting 50% of LEU fuels[18] and, as was already the case, virtually all HEU fuel. The breakeven point between operating costs – even disregarding depreciation on the investment – and operating income would be reached only with 200 t/year of LEU, at the European prices charged by the UKAEA, which were much lower than the stated American prices[19]. The trend in the amounts of irradiated fuel being unloaded from reactors in member countries of Eurochemic were as shown in the following table (these figures are very optimistic). If these quantities are compared with those actually reprocessed by Eurochemic and shown in the same document, the commercial optimism appears even greater[20].

Table 44. **Predicted amounts of spent fuel discharged for the period 1970 to 1974**

	1970	1971	1972	1973	1974	Total
Fuel reprocessed at Eurochemic (predicted in tonnes)	63.7	63.7	73.7	123.7	138.7	463.5
Fuel unloaded in Europe (predicted in tonnes)	103	94	117	179	302	795
Eurochemic share of European market (%)	*62.0*	*67.7*	*63.0*	*69.1*	*46.0*	*58.3*

Thus Eurochemic should have acquired nearly 60% of the European market: even had the company succeeded in achieving this goal, it would still not have been viable.

According to contemporary estimates, a reprocessing undertaking could be commercially viable at a capacity of 1500 t/year, assuming operation at two-thirds capacity, or 1000 tonnes a year. In fact predictions at the time suggested that the amount of spent fuel unloaded would not reach this level until 1979-1980.

Thus an operating deficit was inevitable, and the question of terminating reprocessing was then raised. However the Special Group judged that the undertaking should be kept in being[21] for at least three reasons, the first of which – technical in nature – may seem hardly convincing, the other two clearly showing the predominant political importance of continuing activities:

- a reprocessing plant in the heart of Europe was a geographical advantage, and certainly minimised problems of transport;
- countries without their own reprocessing capability could obtain information about the true costs of reprocessing and thus "exercise a degree of control on the level of prices";
- finally Eurochemic provided a "unique centre for the industrial development of reprocessing techniques", with immediate possibilities; this final proposal attempted to turn the under-use

---

13 NE/EUR(69)1 of 28 March 1969.

14 NE/EUR/M(69)1.

15 C(69)96 of 30 June 1969.

16 NE/EUR(72)1 p. 4 refers to C(69)96 final, an OECD document missing from the Eurochemic archives.

17 At the same time the FORATOM report was being prepared, see Part III Chapter 3.

18 This was the lower hypothesis of the programme.

19 They were 35 $/kg of uranium for NFS and 28 $/kg for the future Allied Chemicals Plant.

20 Source of raw data: NE/EUR(69)1.

21 Following a poll of member countries.

of the laboratory – estimated at 30% of capacity – to Eurochemic's advantage. By continuing operations it was hoped to attract other co-operative experiments to the site.

Funding would therefore be made up of the reprocessing income estimated at 9.8 million dollars for the period 1970-1974 – accounting for 37% – and by the annual credits from the OECD budget, up to 16.5 million dollars or 63%. Hence there was no longer any question of achieving financial independence, even in the medium term[22].

Founders' shares

The changes made to the pattern of funding was to result in changes to the Statutes, with the creation of founders' shares, "in order to bring the rights and obligations of shareholders into line with their contribution to the funding of the company"[23].

The allocation of founders' shares was decided by the General Assembly held on 26 June 1969 and approved by the Special Group on 16 December of the same year. It was made proportional to the contributions "that most [of the countries] have made in addition to their contribution to the company's capital in order to cover its operating deficit"[24]. One share was allocated for every 5000 EMA units of account of contribution, the aim being to give entitlement to a portion of the sums recovered following liquidation of the company. The result of this allocation was to change the distribution of powers amongst the shareholders[25] of the company as and when contributions were paid[26], minimising the share of Turkey, Portugal and, above all, Italy, which contributed solely through voluntary contributions to research and development costs alone. On the other hand this process increased the shares of Germany, France and Belgium.

Programme of operations and development

A "possible programme" for operating the plant foresaw the reprocessing of 445 tonnes of LEU and 18.5 tonnes of HEU between 1970 and 1974, with the LEU tonnages increasing according to the timetable given in the summary table below. The accomplishment of the programme, to the extent that the initial reprocessing campaigns had demonstrated the need for it, required additional investment. This was necessary in particular to increase the effluent storage capacity, modify the extraction cycles, and construct installations for mechanical decladding and effluent solidification. This additional investment was estimated at 4.1 million dollars with 1.7 million dollars for renewal of equipment.

A separate development scheme for the operational programme was also decided upon, amounting to 4.8 million dollars, including 2.5 million dollars for staff costs[27].

The declared objectives went beyond the technical area of reprocessing alone[28], in that they included analytical and experimental work likely to satisfy all participants, whether current or future "reprocessors" as future customers. The ten points listed covered three topics: technical development, economic research and problems of safety and security (Table 45).

It was also planned to expand co-operation with national research centres, particularly as regards waste solidification processes.

Five years of future assured

The Five-year Programme approved on 23 April 1969 appeared to ensure that the company – and therefore international co-operation – would be kept going until the end of 1974. Its principles provided the basis for the company's industrial and commercial policy (Table 46).

---

22 For a comparison with actual income, see the balance sheet in Part III Chapter 4.

23 NE/EUR(69)1, p. 14.

24 For the founders' shares see NE/EUR(69)2 of 1 December 1969.

25 Statutes, Article 14. With regard to the powers of the General Assembly (of shareholders), the "higher authority of the company", see Article 10.

26 Each allocation of founders' shares gave the right to one additional vote in the General Assembly, each original share being worth 10 votes.

27 Consideration had been given to breaking down the research activities into specific research contracts. However this idea was dropped, mainly because the countries who were not interested in sending fuel to be reprocessed would then have received no benefit from participating in Eurochemic. By maintaining a development programme they enhanced their research potential.

28 As regards the technical aspects it was planned for example to look into the possibility of adapting a system of mechanical decladding – a "chopping machine" – to the plant, and to consider the possibilities of modifying the extraction cycles.

Table 45. **The "development programme"**

Topic (s)	Title
Technique	Development of special decladding methods
	Rationalisation of analytical services
	Effluent processing and solidification
Technical-economic	Re-use of final products
	Evaluation of yields and specifications of final products
Economic	Analysis of plant operating costs
	Continuous evaluation of markets for by-products
Safety-security	Fissile material accounting
	Evaluation of security measures and security control
	Publication of manuals on security and criticality

Table 46. **Financial predictions for the Future Programme[29] (in millions EMA u/a)**

	1970	1971	1972	1973	1974	Total
Expenditure						
Plant running costs	3.1	3.0	3.0	3.0	3.0	**15.1**
Plant investment	2.0	1.6	1	0.6	0.6	**5.8**
Total plant	*5.1*	*4.6*	*4*	*3.6*	*3.6*	***20.9***
Laboratory	0.6	0.6	1.2	1.2	1.2	**4.8**
Repayment of loans	0.0	0.1	0.1	0.2	0.2	**0.6**
Total	*5.7*	*5.3*	*5.3*	*5.0*	*5.0*	***26.3***
Income						
Income from reprocessing	1.8	1.8	1.8	2.2	2.2	**9.8**
OECD contributions	3.9	3.5	3.5	2.8	2.8	**16.5**

An ambitious commercial policy

The race for contracts

Organisation

Although the conclusion of contracts required the authorisation of the Board of Directors, the tasks of finding and negotiating them were the responsibility of the Managing Director assisted, following the 1967 restructuring, by a small group attached directly to his office, linked to the Administrative Division. When Yves Leclerq-Aubreton was appointed as Managing Director, this small group disappeared since he assumed this role on a more personal basis.

Rate of signing of contracts

Twenty-eight contracts were signed between the end of 1964 and 1970[30], whereas only two were signed after 1970[31].

Table 47. **Chronology of contracts signed, or of authorisation by the Board of Directors**

1964	1965	1966	1967	1968	1969	1970
1	3	4	7	5	3	5

The number of contracts negotiated during the period was higher because some of them were not carried through. Thus in 1967 a contract was authorised to deal with fuel from the Argentinian MTRs. In 1968 there were negotiations for reprocessing fuel from the Spanish Zorita reactor and with the CEA for reprocessing fuel from the land prototype of a naval reactor (PAT).

Thus Eurochemic concluded contracts for fuel from 38 European and Canadian reactors[32].

The contracts fell into two types: conventional contracts and those which involved a new kind of service.

---

29 NE/EUR(69)1 p. 12.

30 It was possible for any particular contract to concern several reactors, and for several contracts to relate to deliveries of fuel from one and the same reactor. A complete list of contracts in chronological order is given in Part III Chapter 4.

31 One of these, signed in 1971, concerned the German Stade reactor but was not carried through since the reactor did not come into service until 1972. The other contract, signed in 1972, concerned the reactor in the German nuclear-propelled ship, Otto Hahn. Approaches were made through the British for reprocessing fuel from the Japanese MTR, but these did not come to fruition.

32 See Part III Chapter 4 and its Annex for the main characteristics of the reactors whose fuel was reprocessed by Eurochemic.

"Conventional" reprocessing contracts

There were two types of such contracts: individual contracts covering the reprocessing of one charge or one reactor, such as the 1966 contract concluded with the Würenlingen Centre for reprocessing 6 tonnes of fuel from Diorit, and framing contracts providing for the reprocessing of several charges from different reactors; these were in the majority.

For example the framing contract with the CEA signed in 1966 concerned six research reactors: Triton, Mélusine, Siloe, Cabri, Pégase and Osiris. Annual extensions were signed and Eurochemic regularly reprocessed fuel for the CEA.

New contracts for long-term services: the Mühleberg contract

The 1969 Future Programme set out guidelines for general commercial policy, requiring it to follow the emerging trend in the world reprocessing industry with the new concept of so-called fuel cycle services: "in addition to reprocessing, there is an increasing demand on the part of reactor operators for additional services such as the transport of irradiated fuel, conversion of the end products, return of the recovered uranium to the USAEC or the re-use of this uranium in initial enrichment for fabricating new fuel elements. Another request on the part of operators includes the re-purchase of irradiated fuel elements. In such a package deal the reprocessing undertaking [..] must find a buyer for the recovered fissile material. The facilities they offer thus cover the entire fuel cycle, therefore allowing undertakings to attain an overall financial equilibrium"[33].

Hence this process relieved the electricity producers of all fuel cycle operations preceding and following power reactor operations.

For example, taking into account the changing demand, the 1969 Programme offered services involving "taking in charge irradiated fuel elements in the electricity producers' installations and replacing them with new elements ready to use"[34].

This integrated approach involved more collaborative and sub-contracting agreements, notably with the European spent fuel transport industry – Transnucléaire and its Italian and German subsidiaries – the uranium convertors, Ugine-Kuhlmann, the fuel element manufacturers, the nearby MMN in Belgium, CERCA in France and Nukem in Germany. There was also a possibility that it could stimulate Eurochemic into developing certain new activities.

The company's management tried to adapt to this changing market. Yves Leclerq-Aubreton, as Director of Administration, had already considered facilitating the conclusion of such agreements through a project – which was never carried through – for setting up a financial organisation to act as a clearing house for the movement of fissile material between the different organisations involved: FINUCLEAIRE[35]. He was to pursue this policy vigorously when be became Managing Director.

An advertising brochure issued between 1969 and 1972, the first page and extracts from the third page of which are reproduced in Figure 83, clearly shows Eurochemic's desire to project itself as a many-faceted service provider, or as the hub of nuclear sub-contracting in Europe.

This new contract policy resulted in the negotiation and signature of a long-term contract concerning the Swiss power station at Mühleberg, under construction at the time. This was a good illustration of the new strategy which involved establishing long-term commercial relationships by getting in touch with reactor operators as soon as construction began, which at the time meant some six years before the first fuel change, in order to conclude service contracts, which were also on a long-term basis[36].

The Mühleberg contract[37] was signed with the Swiss operator, Bernische Kraftwerke AG, and the German fuel supplier and carrier STEAG (Steinkohlen Elektrizitäts AG) on 31 December 1969. This was a long-term contract which was to commence in 1975 at a price fixed in advance (65 million DM). It included a whole series of services other than reprocessing, such as the shipment of the reprocessed uranium and plutonium. It had

---

33 NE/EUR(69)1 p. 3.

34 Ibid. p. 7.

35 LECLERQ-AUBRETON Y., ASYEE J., NORAZ M. (1968), conclusion.

36 This was a gamble on reprocessing continuation beyond the period originally provided for under the Convention, and contradicted the decision taken in the Five-year Programme not to enter into commitments beyond 1974. However this strategy of the Managing Director was followed with no criticism from his Board of Directors at the time. This suggests that some members of the Board were in favour of continuing reprocessing after that date.

37 AtW (1970), p. 57. The contract provided for the reprocessing and shipment of uranium and plutonium "sowie die Versorgung des Reaktors mit Natururan als Ausgangsmaterial für die Herstellung von angereichertem Uran".

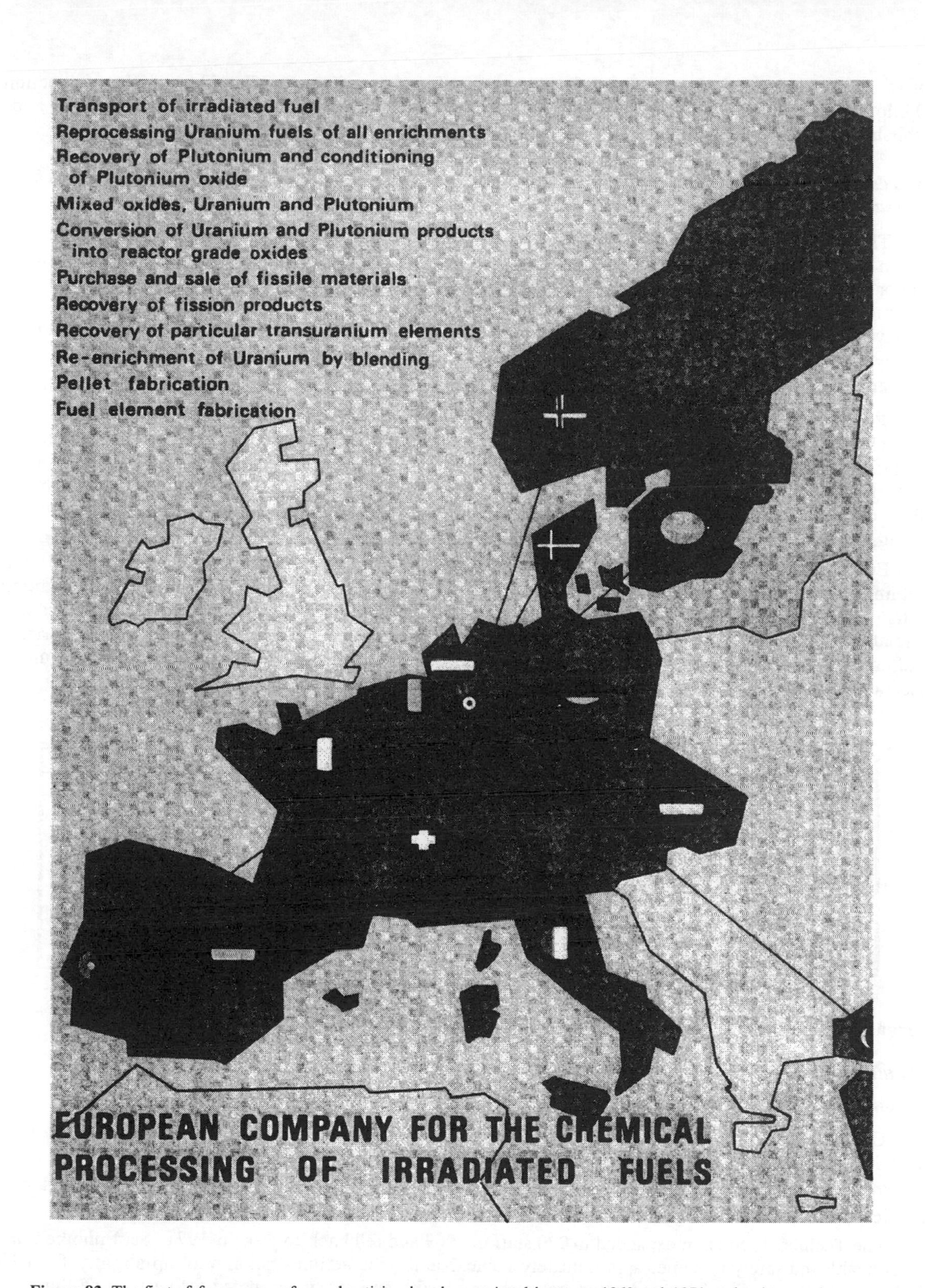

Figure 83. The first of four pages of an advertising brochure printed between 1969 and 1971, enlarging on the many services offered by Eurochemic, as regards both reprocessing and "nuclear services". Page 3 states (the bold characters are in the original): "**Eurochemic** forms the necessary industrial link in the **European fuel cycle** between reactor operators and the various national and international firms dealing with irradiated fuel transport, uranium and plutonium product conversion, fuel element fabrication, etc. [...] **Eurochemic**, in the peaceful application of nuclear energy, promotes fuel cycle activity in **Europe** and co-operates with **nuclear industry** in providing tailor-made fuel cycle contracts for reactor operators. **Eurochemic** in close cooperation with nuclear research centers continually develops new techniques in the field of aqueous reprocessing and the recovery and supply of **by-products**. **Eurochemic**'s ability to reprocess large quantities of all different types of fuel elements **guarantees** a future for **European nuclear industry** and consequently helps to reduce the cost of nuclear power in Europe."

clauses, *inter alia*, on the conversion of 100 tonnes of uranium into hexafluoride and the production of 600 kilogrammes of plutonium. In fact the contract was linked with the plans to extend activities, and the establishment of the Société de fluoration de l'uranium (SFU) was to be the first stage.

Towards industrial diversification on the site. Integration of a service at the back end of the fuel cycle with the creation of the Société de fluoration de l'uranium (SFU)

The policy on developing reprocessing services found its first application in September 1969 with the creation of SFU on the Eurochemic site[38]. Eurochemic was both the consulting engineer and the industrial operator for this company. Its purpose was initially to convert solutions of uranium nitrate up to a maximum enrichment of 3.5% (the final product of the reprocessing plant) into uranium tetrafluoride UF_4. Later on, a unit for converting tetrafluoride into hexafluoride UF_6 – the basic product of enrichment plants – was to be added. In this way Eurochemic was indirectly extending its activities into the recycling of reprocessing products.

SFU was set up by nuclear undertakings and agencies from seven member countries of Eurochemic, Italy taking part through the CNEN but without taking a shareholding.

The Belgian share exceeded one-third of the capital; this expressed the continuing wish of Belgium to see installations encompassing the entire fuel cycle built on its territory. The German energy producers accounted for 21.125%. The contribution of Ugine-Kuhlmann was not surprising because this company had a strong presence in the front end of the cycle, notably the production of enriched uranium[39]. The other shareholders had minority holdings.

Eurochemic produced a design for the installation. Work was completed at the end of May 1970 and the plant immediately entered service while the reprocessing plant had just been shut down for decontamination and repairs. The first tetrafluoride produced was converted to hexafluoride in a pilot plant located beside the Pierrelatte enrichment plant and shipped to the United States. Plans for a unit to convert material into hexafluoride at Mol were then initiated but no work was even started. On the contrary, the tetrafluoride unit was closed at the end of 1970. Eurochemic subsequently gave up operating SFU.

Table 48. **SFU shareholders in decreasing order of holding**

Company or agency	Capital share	Nationality
SGMH (Hoboken)	21.125	Belgian
Société technique d'entreprises chimiques	15.625	Belgian
Ugine-Kuhlmann	15.625	French
STEAG (Steinkohlen-Elektrizität AG)	15.000	German
RCN (Reactor Centrum Nederland)	9.625	Dutch
Gesellschaft für Stromerzeugung und Energieversorgung mbH	6.125	German
Atomic Energy Commission Denmark	5.625	Danish
Institutt for Atomenergi (IFA)	5.625	Norwegian
AB Atomenergi	5.625	Swedish

Reorganisation of the company; the limited role of the laboratories

Structures of the company from 1966 to 1971

Declining manpower: divisional variations

Under the 1967 Programme, the staff numbers at the company slowly fell until 1971. These changes are shown in detail in Table 49, the data being for 31 December.

The reduced staff numbers primarily affected research and development and, to a lesser extent, the Technical Division. With 81 staff in 1966, R & D had only 28 in 1971.

The Technical Division expanded to 220 staff in 1969 and fell back to 176 in 1971. Staff numbers in the area of health and safety remained approximately stable. Numbers in administration went up slightly from 1970. In 1971, staff not involved in production, research and health and safety accounted for nearly one-third of the total. This growth in administrative and similar staff numbers was linked to the importance placed on finding contracts and the development of new products.

---

38 RAE 1969, p. 4 and 5. This source gives only the proportional breakdown of the capital of this non-Eurochemic company, and not the amount of the capital, which at the time was probably fairly modest.

39 For the nuclear activities of Ugine-Kuhlmann at the time, see RENAUD J. (1970).

Table 49. **Changing staff numbers, 1967 to 1971**

Year	1967	1968	1969	1970	1971
Staff numbers	364	354	326	310	301
Change on year	-14	-10	-28	-16	-9
% change compared with end of previous year	-3.7	-2.7	-7.9	-4.9	-2.9

Table 50. **Change in staff numbers by category, according to the 1966 pattern**[40]
Combined statistics to permit comparison with the period 1961-1966

1. Raw data	1966	1969
Health and safety	29	33
Technical division	204	220
General management, administration and others	64	58
R & D	81	34
2. As percentages	**1966**	**1969**
Health and safety	*7.7*	*9.6*
Technical division	*54.0*	*63.8*
General management, administration and others	*16.9*	*16.8*
R & D	*21.4*	*9.9*

Table 51. **Distribution by major function**

1. Raw data	1969	1970	1971
HSD (health and safety)	33	33	30
POD (plant)	179	171	148
ALD (analytical laboratory)	41	39	28
IDL (research laboratory)	34	40	28
Administration (excluding management staff)	25	23	35
Other	33	39	32
Total	*345*	*345*	*301*
2. In percentage terms	**1969**	**1970**	**1971**
HSD (health and safety)	*9.6*	*9.6*	*10.0*
POD (plant)	*51.9*	*49.6*	*49.2*
ALD (analytical laboratory)	*11.9*	*11.3*	*9.3*
IDL (research laboratory)	*9.9*	*11.6*	*9.3*
Administration (excluding management staff)	*7.2*	*6.7*	*11.6*
Other	*9.6*	*11.3*	*10.6*

Changes in the nationality of the staff also confirmed the trend to increasing the number of Belgians which began in the first half of the 1960s.

An organisation in flux (Figures 84 to 87)

The general organisation of the company underwent a number of modifications as a result of the succession of programmes up to 1969, and of changes in personnel.

At the end of 1966 the structure was still based upon three divisions: Administration, Research and Technical. The departure of Teun Barendregt in June 1967[41] and the introduction of the Transitional Plan resulted in the elimination of the Research Division, whose activities were transferred to the Technical Division, to be led by the former Director of Research, Emile Detilleux. As from 1967 the management structure was split in two. The departure of Managing Director Rudolf Rometsch to the IAEA at the end of August 1969[42] together with the introduction of the Five-year Programme led to a reorganisation.

40 No individual data for 1967 and 1968.

41 Teun Barendregt left the company while the plant was still fully operational. He became Managing Director of one of Eurochemic's industrial architects, Comprimo, where he worked on instrumentation and engineering design in both the nuclear and petroleum fields. In the middle of the 1970s he went to Brazil as technical adviser to Nuclebras during the negotiation of the "nuclear contract of the century" with Germany. See Part IV Chapter 1.

42 Rudolf Rometsch became Inspector-General of the IAEA which gave him the rank of Deputy Director-General. He was involved in setting up the system of safeguards under the Nonproliferation Treaty. He remained at the IAEA until 1978, when he was appointed Chairman of the Swiss Organisation for Radioactive Waste Management, CEDRA/NAGRA.

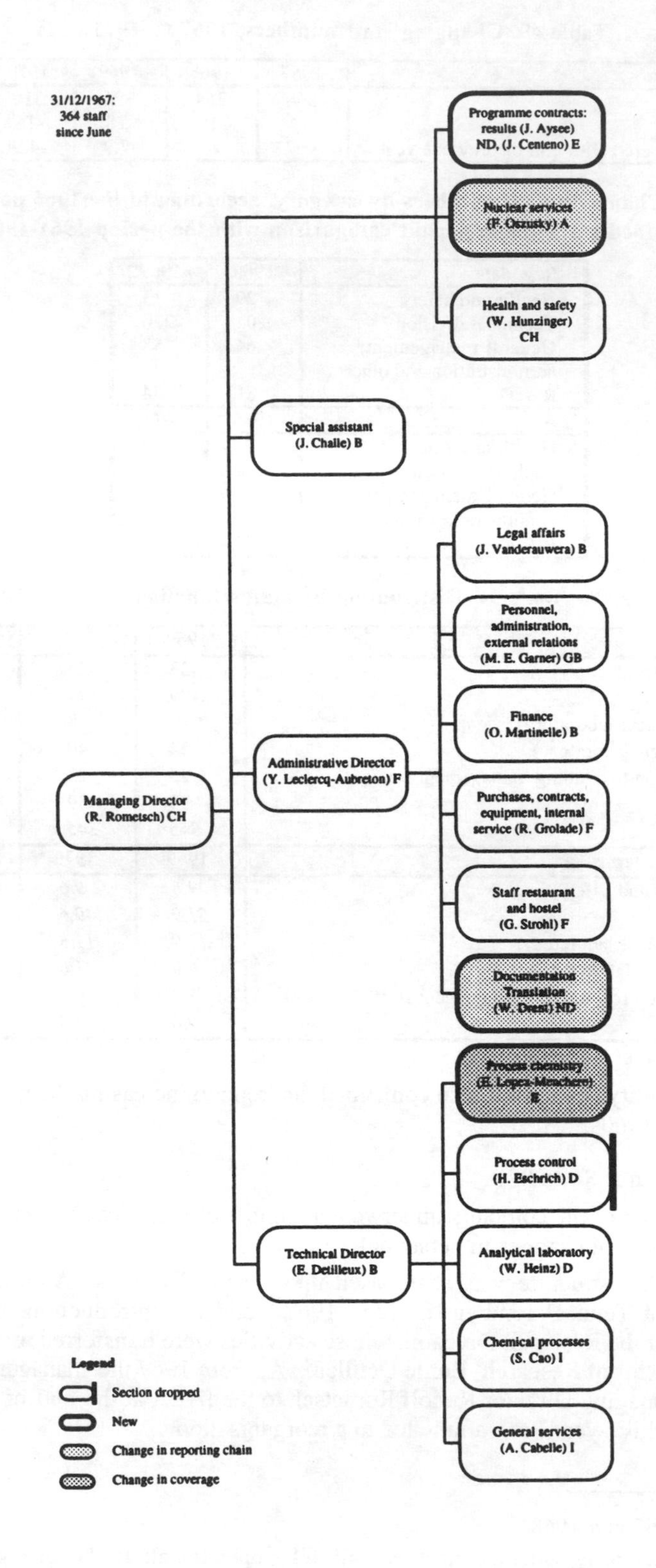
31/12/1967:
364 staff
since June
Programme contracts: results (J. Aysee) ND, (J. Centeno) E
Nuclear services (F. Ozrusky) A
Health and safety (W. Hunzinger) CH
Special assistant (J. Challe) B
Legal affairs (J. Vanderauwera) B
Personnel, administration, external relations (M. E. Garner) GB
Finance (O. Martinelle) B
Administrative Director (Y. Leclercq-Aubreton) F
Managing Director (R. Rometsch) CH
Purchases, contracts, equipment, internal service (R. Grolade) F
Staff restaurant and hostel (G. Strohl) F
Documentation Translation (W. Drest) ND
Process chemistry (E. Lopez-Menchero) E
Process control (H. Eschrich) D
Technical Director (E. Detilleux) B
Analytical laboratory (W. Heinz) D
Chemical processes (S. Cao) I
General services (A. Cabelle) I
Legend
Section dropped
New
Change in reporting chain
Change in coverage

Figure 84. Organisation chart of the company as from June 1967 (Source: RAE 1967).

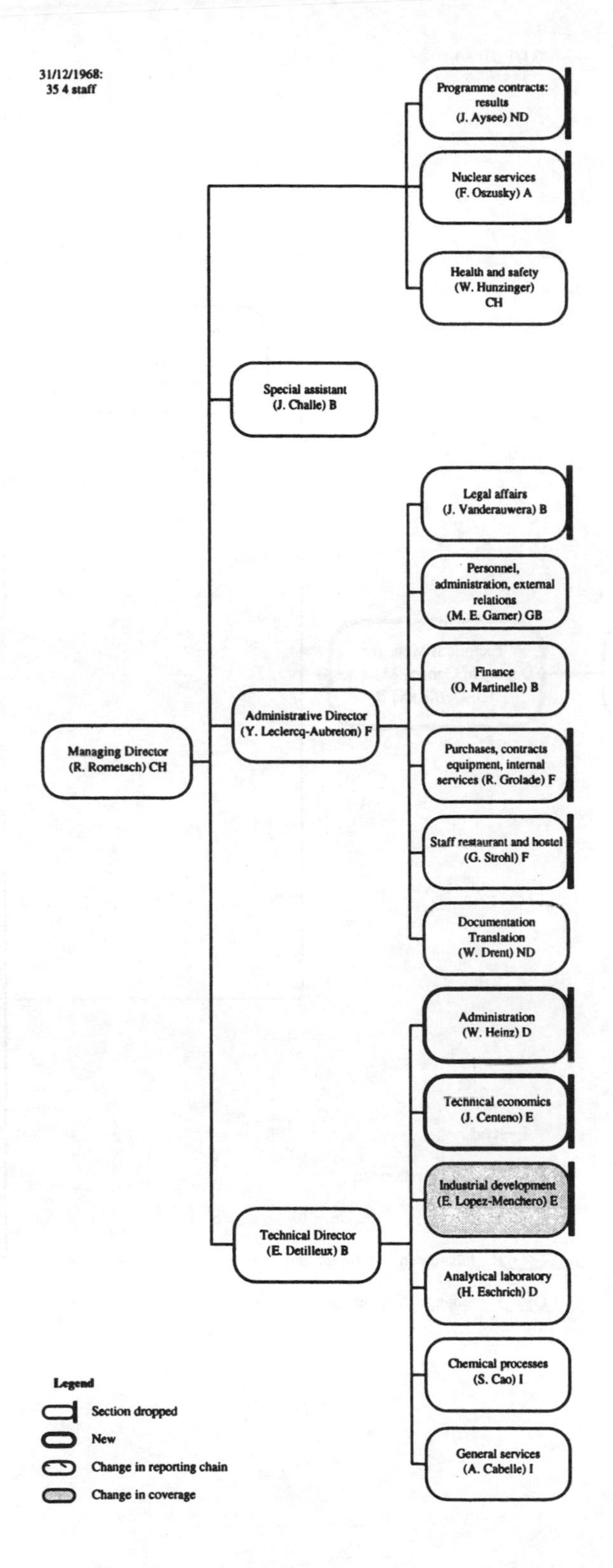

Figure 85. Organisation chart of the company at the end of 1968 (Source: RAE 1968).

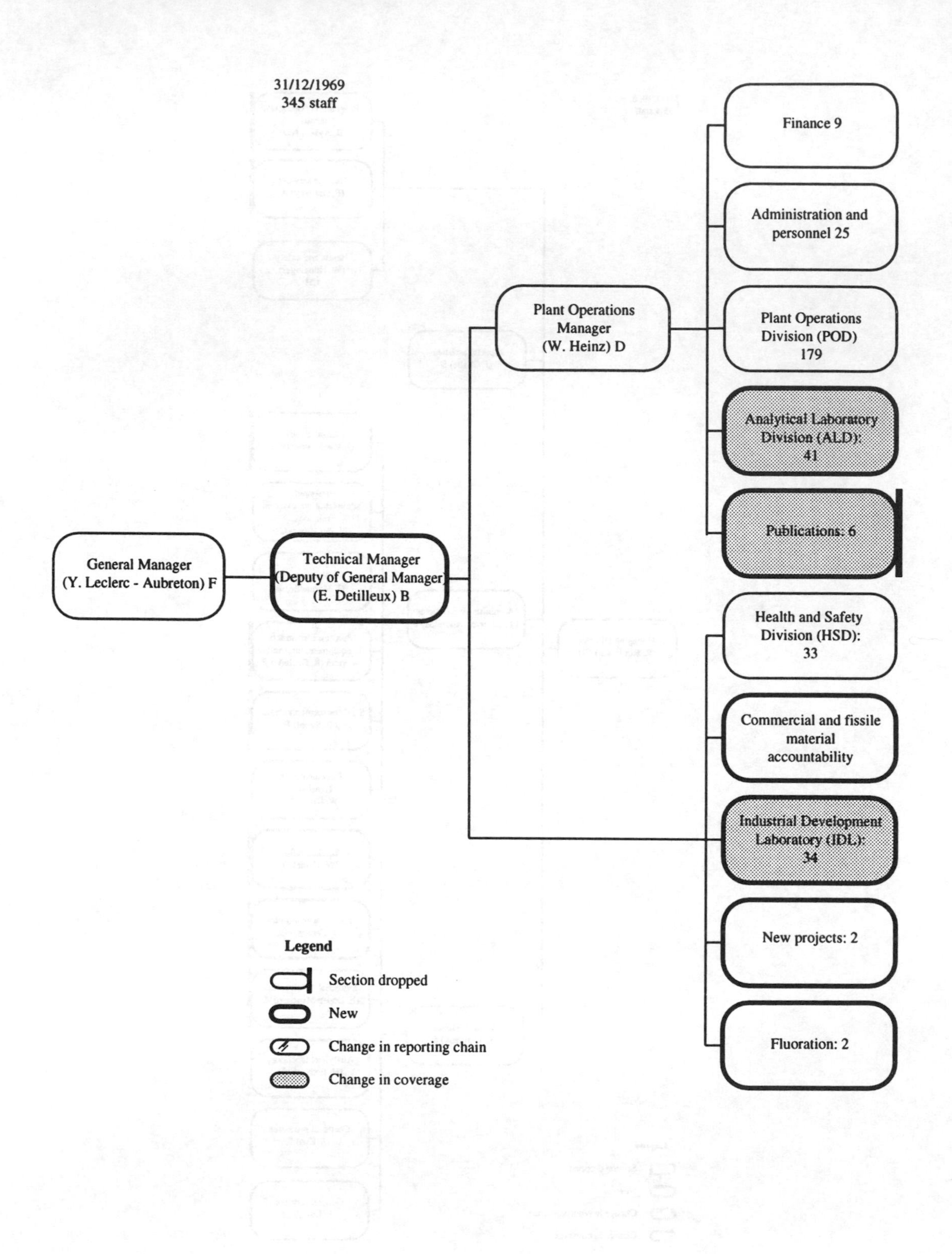

Figure 86. Organisation chart of the company at the end of 1969 (Source: RAE 1969).

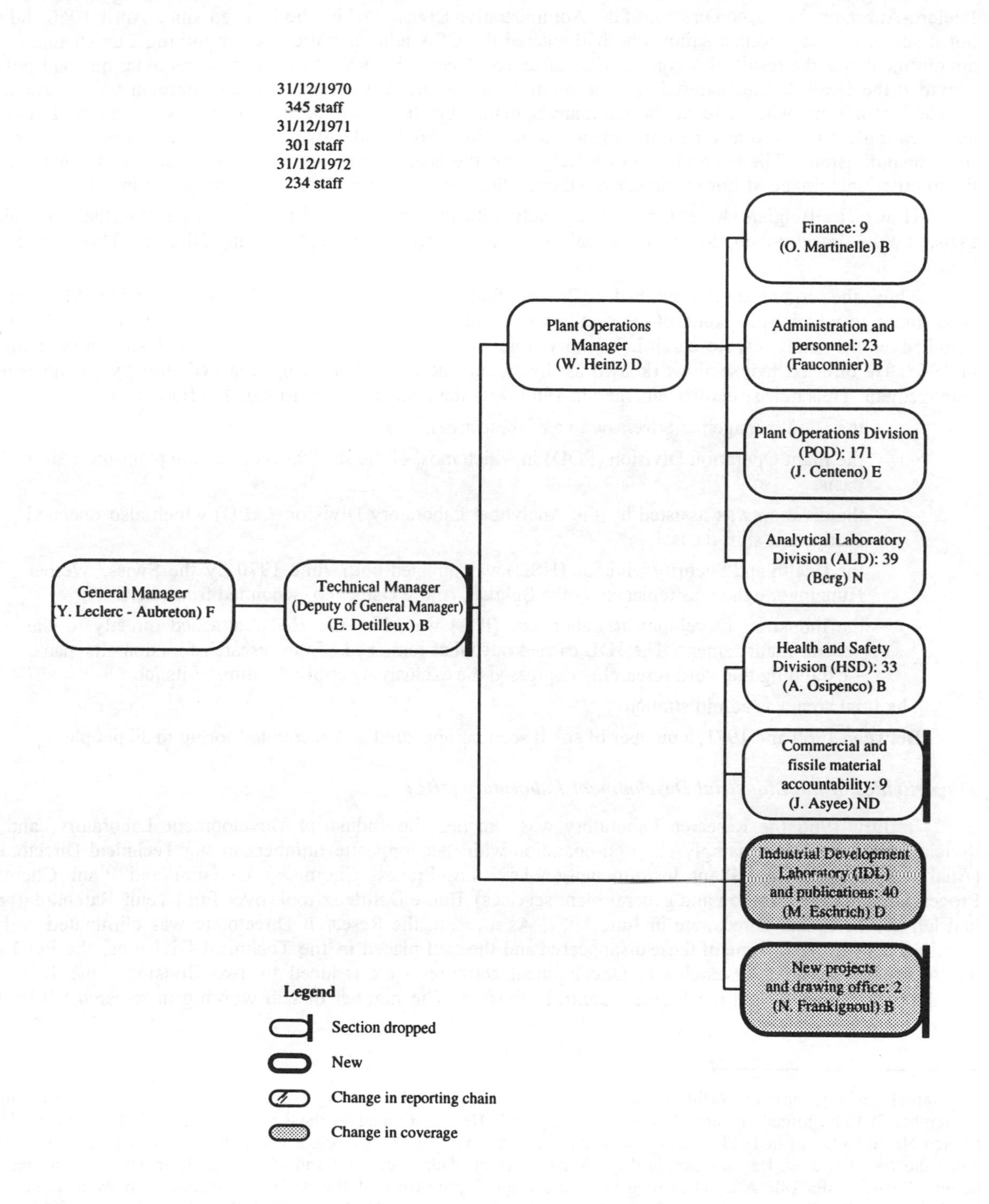

Figure 87. Organisation chart of the company: 1970 to the end of 1972 (Sources: RAE 1970, 1971, 1972).

The general management now became a two-headed organisation. The new Managing Director, Yves Leclerq-Aubreton, had been Director of the Administrative Division which he had led since April 1960; he was not a scientist, but a former sailor who had entered the CEA administration before joining Eurochemic[43]. His appointment was the result of a combination of several factors. France was a main shareholder but had not yet provided the Head of the undertaking; the administrative talents of Yves Leclerq-Aubreton were particularly welcome at a time when one of the company's main objectives was to negotiate new contracts. Finally it appeared logical that a director from a functional division should take over from another former director of a functional division[44]. The appointment of a Belgian to the head of the company would have somewhat limited the international image of Eurochemic, even though the "Belgianisation" of the company continued.

However a Belgian charged more specifically with the scientific and technical aspects of the undertaking assisted Yves Leclerq-Aubreton as Technical Director and Deputy to the Managing Director. This was Emile Detilleux[45].

Thus the organisation now had only one functional directorate, the Plant Operation Management Directorate, to which only some of the sections reported[46]. The Plant Director was Wilhelm Heinz, a German who had led the Analytical Research Laboratory from 1964 to 1967 and the Technical Division administration in 1968. He directed five sections (known as divisions), the other four being attached directly to the general management. Thus new structures emerged in 1969, with the company now arranged in five functional areas:

- the plant management oversaw two of these areas;
- the Plant Operation Division (POD) in which most of the staff worked ran the plant on a shift basis;
- the division was assisted by the Analytical Laboratory Division (ALD) which also operated partly on a shift basis;
- the Health and Security Division (HSD) was directed until June 1970 by the Swiss, Werner Hunzinger, who was replaced by the Belgian, André Osipenco, seconded from the CEN;
- the Industrial Development Laboratory (IDL) was, like the HSD, attached directly to the general management. The IDL carried out what remained of the research function. Its name – excluding the word research – expressed the exclusively applied nature of its job.

The final area was administration.

Between 1969 and 1971, a number of small sections appeared and accounted for up to 30 people.

The activities of the Industrial Development Laboratory (IDL)

In July 1966 the Research Laboratory was renamed the Industrial Development Laboratory, and its divisions worked in increasingly close co-operation with their opposite numbers in the Technical Directorate[47] (Analytical Division and Plant Instrumentation Division; Process Chemistry Division and Plant Chemical Process Division; test station and general plant services). Emile Detilleux took over from Teun Barendregt who had left the Technical Directorate in June 1967. As a result, the Research Directorate was eliminated and its divisions restructured. Some of these disappeared and the staff placed in the Technical Divisions, the POD and the ALD; the Applied Research and Development activities were reduced to two divisions, the Industrial Development Division and the Process Control Division. The number of staff working in research fell by half

---

[43] Marcel Leclerq, born in northern France in 1920, had embraced a military career as a volunteer in 1938 and in September 1940 he joined General de Gaulle in England. He was trained at the Ecole navale de la France libre [Free French Naval College] in 1941. During the war he took the name Yves Aubreton to which he attached his family name when the war was over. He took part in the colonial wars of Madagascar and Indo-China as Commander of the navy's bombardment flotilla 28F. After Dien Bien Phu, he became Deputy Head of the Aviation Section at navy headquarters, but left in 1957 following differences of opinion with his superiors. He then joined the administration of the CEA, before going to Eurochemic in 1960. Although he was not technically qualified in chemistry or nuclear questions he was not unaware of the problems of administering technical matters as a result of his training and, as a former student at the naval college, he had the title of engineer.

[44] Rudolf Rometsch had been Director of Research in the time of Erich Pohland.

[45] This two-headed structure may recall the organisation of the CEA with its General Administrator and Scientific High Commissioner.

[46] As from 1969, as a money-saving exercise, the activity reports were no longer bilingual but published in English only.

[47] DETILLEUX, E. (1966).

between the end of 1966 and the end of 1969[48]. In 1969 these two divisions were combined under the 1967 Transitional Programme. The Industrial Development Laboratory (IDL) now referred not only to the building but to the division.

The staff changes are a quantitative illustration of Eurochemic's reduced R & D ambitions and the refocusing solely on activities directly related to plant operations, which can be grouped under three headings[49]:

- assistance with plant operations;
- process and control improvements;
- waste management studies.

Assistance with plant operations: extraction of plutonium from ZOE fuel

Assistance with plant operations was particularly important during the initial phase. One significant episode was the co-operation between the plant and the laboratory for the reprocessing of fuel elements from ZOE[50] (EL1). These consisted of two tonnes of a mixture of uranium oxide and uranium metal, containing between 30 and 35 grammes of Pu. The concentration of the fissile 239 isotope was exceptionally high at 99.88%, owing to the low burn-up of the fuel, 24.5 MWd/t. This fact meant that the material was of great importance for physics research. Moreover, since the Eurochemic installation was still free of any contamination, it was possible to recover this plutonium without its isotopic composition being affected by the residues of earlier campaigns. It was for this reason that the CEA urgently requested that attempts be made to recover at least one gramme.

The fuel was processed in the plant using special extraction conditions, notably very high acidity, with the solution being recycled several times during the first cycle in order to concentrate the plutonium in the refined extract. The first extraction cycle was repeated several times over a period of weeks. The plutonium was then reduced using a high dose of ferrous sulphamate[51]. This process produced 2000 litres of a highly acid aqueous solution containing, apart from 1 kilogramme of iron and another kilogramme of sludge of uncertain composition[52], 30 grammes of plutonium. This solution which could not be processed by the plant was handed over to the laboratory. Maurizio Zifferero, a member of the Technical Committee, had reported the effectiveness of trilaurylamine (TLA), developed in Italy as a particularly efficient extraction agent for plutonium. Experiments with TLA had already been done at Eurochemic during the construction period.

Mr. Lopez-Menchero therefore ordered 200 litres of TLA and twenty 20-litre separation containers[53]. Technicians were borrowed from the plant for a period of a week to carry out the extraction. Thus for extracting the plutonium, an unusual method was adapted to an unusual problem. Some 26 grammes of plutonium dioxide containing 99.88% of Pu 239 were thus extracted and supplied to the CEA, which paid an additional 6000 dollars for this service which had not been originally included in the contract.

Improvements to the plutonium production process. The replacement of ferrous sulphamate by uranium nitrate

Another example of co-operation was the development of certain process phases which were not fully defined when operations started, such as the problem of reducing plutonium using reagents other than ferrous sulphamate, which had the known disadvantage of leaving iron in the effluent.

At the time, consideration was being given to replacing it by uranium sulphate-4 as a reducing agent. However the use of this salt increased the risk of corrosion to the system. The Technical Committee then considered using tetravalent uranium nitrate. Research carried out in the laboratory at Eurochemic showed that although it was impossible to prepare stable crystals[54] of pure uranium nitrate $U(NO_3)_4$, it was however possible to prepare a double salt of uranium nitrate-4 and hydrazine dinitrate.

Transferring this process to the plant involved close co-operation between the members of the IDL and POD staff and finally resulted in the adoption of another method, the electrolytic reduction of the uranium nitrate

---

48 The annual reports significantly say nothing about the numbers of staff working in research during the years of restructuring, 1967 and 1968.

49 Hubert Eschrich, ETR 318, p. 68 and 80-82.

50 Rudolf Rometsch, ETR 318, pp. 45-46. The first charge had been reprocessed in the Châtillon pilot plant. This had produced about 100 grammes of plutonium, handed over to the CEA metallurgists for testing with a view to the production of the bomb.

51 "The tenfold amount of reducing agent".

52 "Indefinite crud".

53 Separation funnels.

54 In fact this meant combining a reducing cation and an oxidising anion.

in a very pure solution, to give a solution of uranium nitrate stabilized by hydrazine[55]. Thus uranium nitrate solutions were used in the plant for reducing the plutonium[56]. On other occasions, IDL helped find solutions to problems encountered on the plant, working for example on mechanical decladding or contributing to the construction of the first continuous plutonium oxalate precipitator.

When the plant was fully operational, IDL dealt with problems related to reprocessing, such as the dissolving of graphite fuels[57] or the recovery of secondary products of reprocessing, neptunium-237, krypton-85, or the conversion of uranium nitrate into uranium tetrafluoride – in conjunction with the projects for setting up the SFU – or into dioxide. Chromatography separation was developed for recovering strontium and caesium from wastes. Extraction methods using chromatography were tried out for both uranium and plutonium in fuels from homogeneous reactors and for the wastes arising from the production of MOX fuels.

Developments related to wastes

As the end of the operational programme approached, the IDL became increasingly involved in testing processes for stabilising the different effluents from the plant, i.e., the solidification of high activity effluents, the bituminisation of medium activity effluent and the treatment of used solvents. In this way it contributed to the choices that were made for the radioactive waste management period[58].

However most of the company's staff contributed directly or indirectly to the daily running of the reprocessing plant. This was the case of the engineers and technicians of the Analytical Laboratory.

The role of the Analytical Laboratory (ALD): 1966 to 1971

The plant's Analytical Laboratory (Figures 88 to 91) had a staff of 41 in 1969 and 28 in 1971. During the period of construction, an analytical manual was drawn up which set out the most efficient methods and indicated how the analytical work should be carried out. Over the years, other methods were added.

When reprocessing commenced, the Analytical Laboratory was submerged in requests for analysis and in 1967 the gap between the demand and the amount of work the technicians could handle resulted in delays and then omissions in the accomplishment of the analytical programme. A fall in demand then followed.

In certain cases the installed equipment proved defective, particularly certain active gloveboxes; there were some contamination incidents and the staff received non-negligible exposure[59] owing to the combination of their relative inexperience and the heavy workload.

Until 1970, the ALD was divided into two groups: the Process Control Laboratory (this term was used in the reports until a late stage) operated in parallel with the plant, i.e., in three shifts: it was responsible for routine analyses carried out to ensure that the reprocessing operations were being carried out satisfactorily; in 1968 for example it carried out 20 111 determinations, or 387 a week, with a maximum of 850 in one week.

The second group was concerned with special analyses, using mass spectrometry, and spectrometric techniques for input analyses. It worked closely with the analytical accounting service[60].

During the 1970-1971 period of plant decontamination, verification and improvement, routine work was reduced to the analysis of rinsing solutions. The ALD took the opportunity to improve certain analytical methods, clean and decontaminate the gloveboxes, install new systems and become familiar with the use of a computer.

The fall in staff numbers in 1971 was due primarily to voluntary departures which did raise problems. In fact the prospect that the plant was to close resulted in a number of specialists looking for work elsewhere. They could not be replaced.

---

55 In fact in the laboratory a double salt of uranium nitrate-4 and hydrazine dinitrate was crystallised [$U(NO_3)_{4.4}(N_2H_{4.2}HNO_3)_{.2}H_2O$].

56 Subsequently another variant was developed: direct electrolytic reduction in the separation column, used at La Hague and intended for Barnwell.

57 According to B. Lenail, Director of External Relations at La Hague, it was a CEA "glue" which had caused the worst problems with this type of "exotic" fuel. In fact Eurochemic did no better than the French.

58 See Part IV Chapter 2.

59 See analysis of incidents in Part III Chapter 4.

60 See the section on security control in Part III Chapter 4.

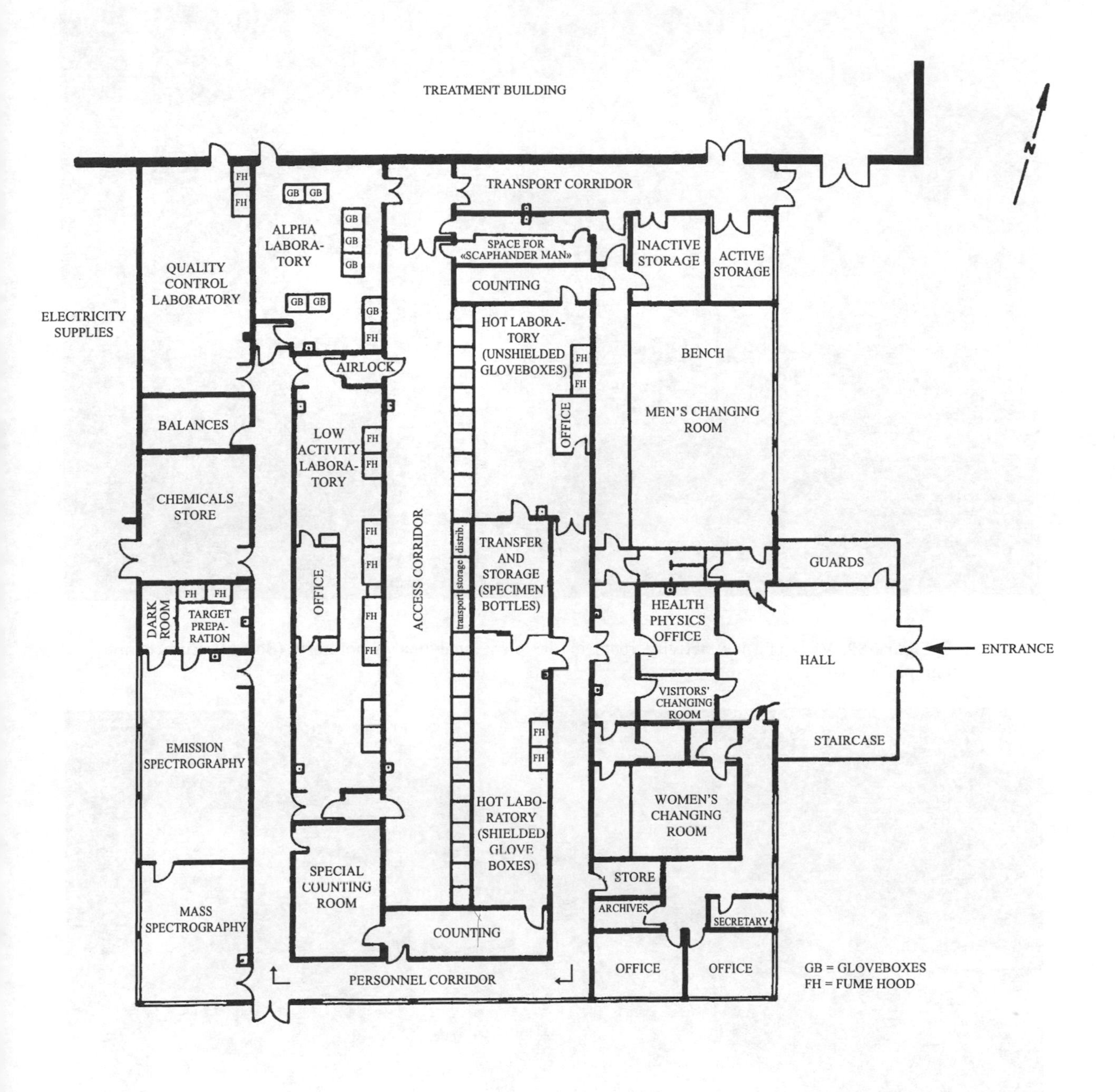

Figure 88. Simplified plan of the analytical laboratory (Source: RAE2 (1965), p. 211).

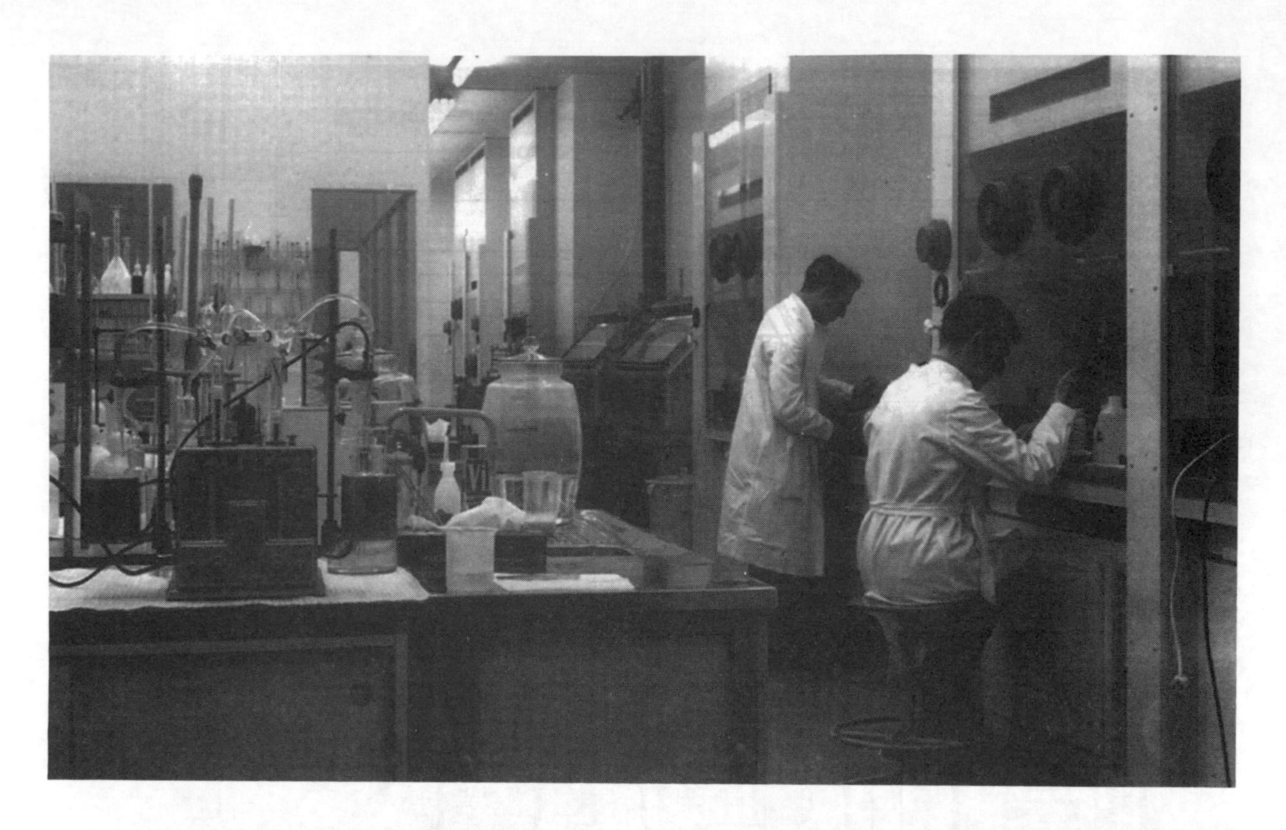

Figure 89. View of a low activity room in the plant analytical laboratory (Source: Eurochemic photograph, no date).

Figure 90. View of the row of shielded gloveboxes in the high activity analytical laboratory (Source: Eurochemic photograph, no date).

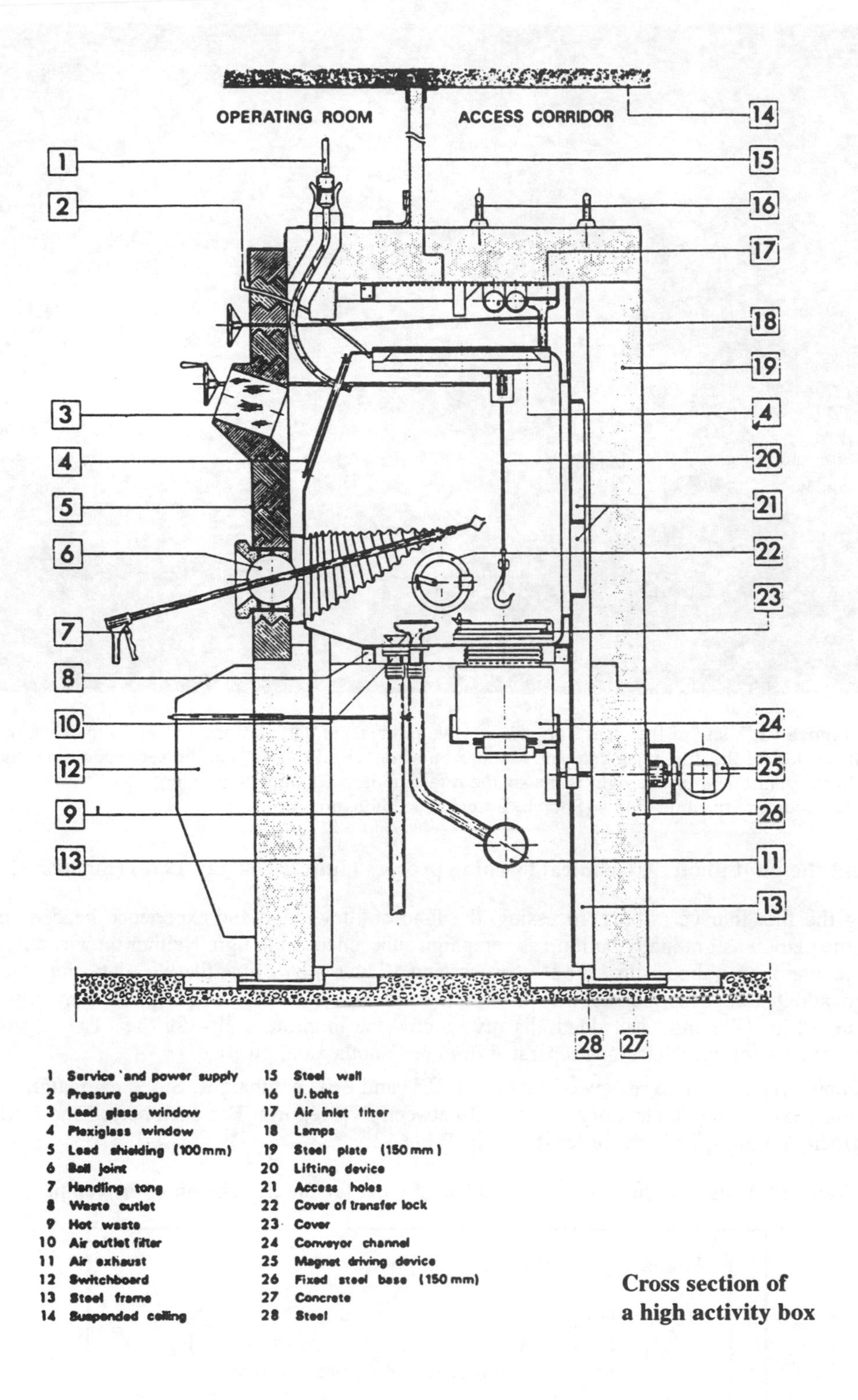

Figure 91. Sectional drawing of a high activity glovebox in the analytical laboratory (Source: *Safety Analysis* (1965), Volume of Figures, VIII-14).

Figure 92. View of the plant from the inactive product storage platform (n° 7). In the background on the left the top of the fuel reception and storage building (Building 2) can be seen, and also the plant stack. In the centre and on the right is the southern facade of the treatment building, with the windows of the galleries on the top four floors (Source: Eurochemic slide, no date).

The POD and the pilot plant. A technical learning process (July 1966-May 1970) (Figures 92 and 93)

During the first four years of reprocessing, the fund of knowledge and experience needed for satisfactory operation of the plant was gradually built up as campaign succeeded campaign. Neither the written sources or the oral reports contain any theorising about reprocessing. They give quantitative data for each campaign, sub-campaign and batch, describe the way in which each fuel element or group of elements was in turn discharged, stored in the pond, decladded, dissolved, how the uranium and – for the LEU – plutonium were separated from the fission products, then separated from one another and purified.

Each annual report gives a review of the year's HEU and LEU campaigns. Since campaigns ran over from one year to the next the complete story is divided between two reports. For each period the reviews for each technical segment are grouped in the order shown in Table 52.

Table 52. **Phases of the project used as background for the description of the campaigns**

- Unloading and storage in the pond
- Reprocessing operations
  - Decladding and dissolving (head of process)
  - Extraction and separation (core of process)
  - Treatment of final products (tail of process)
- Treatment of waste

The highly narrative approach adopted in the pages that follow – a kind of topical "chronicle" – is not only the one that best reflects the sources, but is also most suitable for demonstrating how the reprocessors acquired professional experience in a highly progressive and cumulative manner: a technical learning process in action. It divides the first four years into three phases.

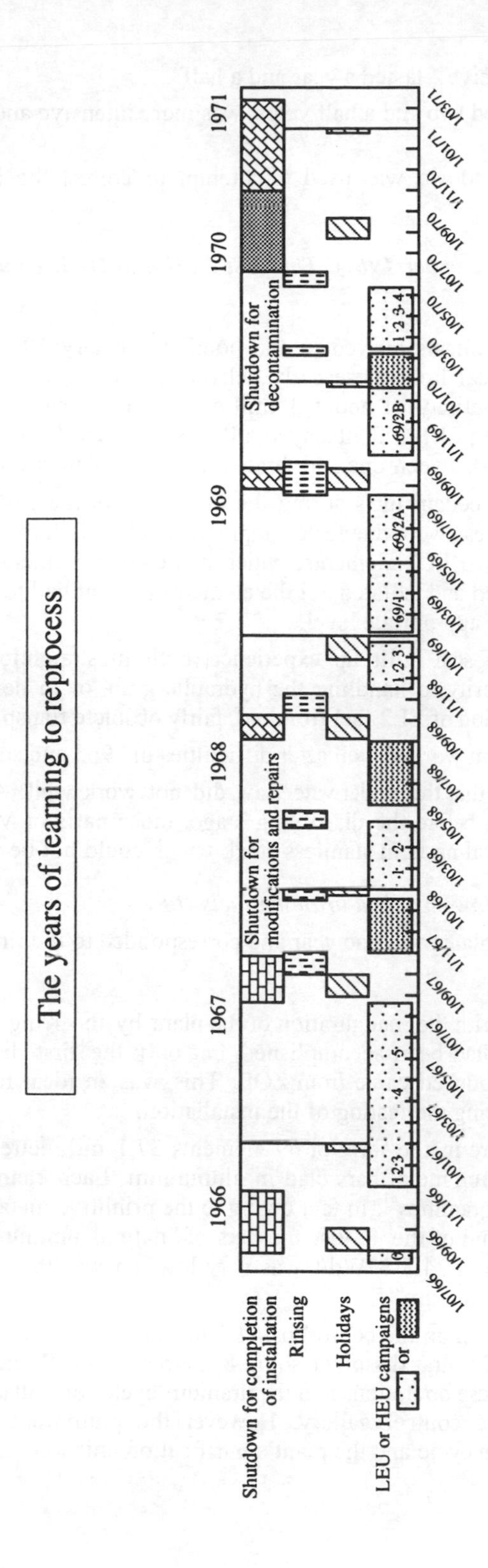

Figure 93. The operating pattern of the reprocessing plant in the period from July 1966 to February 1971. The period was marked by many interruptions and the campaigns followed one another fairly quickly. The lower chronological strip shows the series of reprocessing campaigns, numbered by year and sub-campaign for the reprocessing of natural or slightly enriched uranium fuels (LEU). The three upper strips indicate closures for holidays and periods of shutdown with their reasons, rinsing between two campaigns, installation of new equipment, repairs, modifications or decontamination (Source: RAE 3, RAE 1967 to 1971).

The start-up – very progressive – lasted a year and a half.

The next phase, which lasted two and a half years, was more intensive and revealed a number of operating problems.

A planned nine-month shutdown was used to attempt to correct the imperfections discovered in this learning period.

A gentle start-up (July 1966 - December 1967). The first LEU and HEU campaigns

Fuel reception and storage

The first irradiated fuel elements arrived in the pond in January 1965 (Figures 94 to 96). The fuel handling team learned a great deal from these early deliveries. During December 1966, 52 shipments were received, amounting to a total activity of about 3 million Ci. They came from France, Belgium and the Netherlands and included fuel from EL1 at Fontenay aux Roses, EL2 and EL3 at Saclay, EDF1 at Chinon, BR3 at Mol and HFR at Petten. It took between one and three days to completely unload a shipment.

There were problems with certain fuels during the unloading of the casks, during transfer to the storage ponds or actually in the ponds. These were due for example to cladding failures during transport, deformation, or corrosion. For instance in 1966 two EL3 elements, which had cladding cracks, contaminated the storage pond water; before they could be located and isolated, all the elements present had to be tested, and it took six months for the water purity to return to an appropriate level.

In the period 1966-67 the staff built up experience, both in straightforward unloading operations, for example by improving their dexterity in handling the hydraulic grab, or in dealing with difficulties that arose, such as that of the delicate extraction of EL2 fuel from old, fairly obsolete transport casks.

The pond purification system worked well after difficulties in 1966 and subsequent modifications.

Work in the pond revealed that the underwater saw did not work well. A saw had to be leased from the Petten centre in order to separate, before the dissolving stage, those parts of MTR fuels that were in uranium-aluminium alloy from the structural parts in stainless steel, which could not be dissolved.

The reprocessing of natural and low enriched uranium fuels (LEU)

The commissioning of the plant took one year and corresponded to the first LEU campaign.

King Baudouin and ZOE

The first campaign began with the inauguration of the plant by the King of the Belgians on 7 July 1966, when most of the reference tests had been accomplished, but only the first dissolver was operating on active material. The first fuel to be introduced came from ZOE. This was an ideal fuel for a gentle start or, and this comes to the same thing, completing the testing of the installation.

The two tonnes of fuel were in the form of 69 elements 37.1 milimetres in diameter and 1.5 metres in length, consisting of natural uranium metal bars clad in aluminium. Each element, once the ends had been cut off, weighed a little under 29 kilogrammes[61]. In fact owing to the primitive metallurgical methods used when the fuel was made in France at the end of the 1940s, the bars of "natural uranium metal" included about 30% of oxide (U_3O_8)[62]; since the burn-up of 24.5 MWd/t was very low, so was the residual fission product activity, about 0.05 Ci per kilogramme of uranium[63].

The King pressed a red button on the control panel. This initiated the lowering of the 700 kilogrammes of fuel in the charging machine to the first dissolver some 8 metres lower. Some 300 guests gathered in a stand erected near the ponds watched these operations and the uranium cycle took place without incident, Eurochemic staff having taken over from the control gallery. However the plutonium was recovered in the research laboratory[64] because the plutonium cycle and the plant's purification unit were not yet operational.

---

61 It consisted of 28.5 kilogrammes of uranium and 375 grammes of aluminium.

62 In 1966 it was thought that the presence of the oxide was due to the oxidation of the uranium. However the latter, being contained in aluminium cladding, could not have been oxidised. Rudolf Rometsch concluded that the oxide had arisen during fabrication of the fuel.

63 Of the fission product activity, 44% came from Sr/Y-90, 44% from Cs-137, 10% from Pm-147, 1% from Sm-151 and under 1% from Pr, Eu-155, Ru/Rh-106 and Kr-85. No Zr-Nb remained owing to the long cooling period.

64 See above.

Figure 94. View of the transport cask unloading area, showing the great variety of shapes. From left to right, a Diorit cask, a small unidentified cask, the AEG cask for VAK and BR3, and two EL3 casks, one of which has been lifted from its transport base by two chains attached to the overhead crane. Photograph taken from the large decontamination shop (Source: Eurochemic photograph, no date).

Figure 95. A cask arrives in the treatment area of the large decontamination shop (Source: Eurochemic photograph, no date).

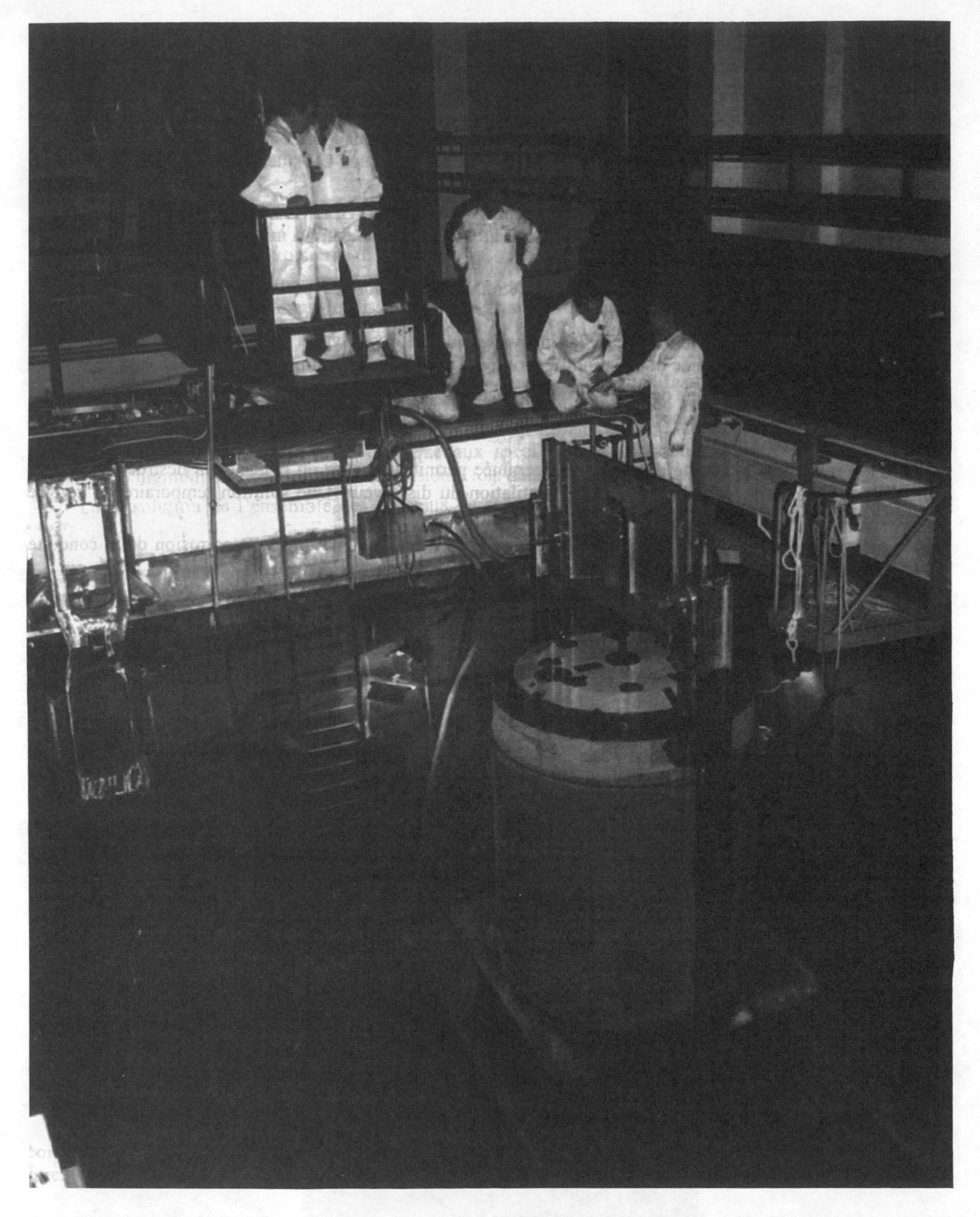

Figure 96. Inspection of a cask resting submerged on the platform of the reception pond before being opened for the fuel to be unloaded (Source: Eurochemic photo, no date).

A slow increase in throughput

Some thirty-three tonnes of low enriched natural uranium were then reprocessed between October 1966 and July 1967, coming from the EL2[65], EL3, Diorit, FR2 and BR1 reactors.

This was natural or very slightly enriched uranium (1.6% for EL3), with relatively low burn-up between 200 and 1250 MWd/t; apart from the EL2 fuel, which was clad in magnesium, the fuel was clad in aluminium. The cladding was dissolved in dilute nitric acid and after rinsing, the fuel was dissolved in concentrated nitric acid. All these operations took place in the first dissolver.

For the EL2 fuel, dissolution and extraction took place in sequence, to allow the staff to build up the necessary experience. The separated plutonium was stored pending the start-up of the second plutonium cycle.

This first LEU campaign revealed a number of problems and was the opportunity for a number of tests to be carried out.

The blockage of a column caused by poor solvent quality interrupted production, since the system had to be rinsed out and the solvent treatment procedure modified.

During the third EL2 sub-campaign, in early January 1967, the use of uranium-IV instead of iron-II was tried in the second cycle, but this first experiment did not produce any significant improvement.

After slight activity had been detected in the stack, it was found that the filters in the dissolver ventilation circuit had been destroyed by nitric acid vapour. A temporary solution was found but the incident called for more substantial modifications.

Also the fact that the manufacturer had used an unsuitable material resulted in considerable corrosion of the pipe feeding nitric acid to the dissolver.

The fuel from EL3, whose core consisted of an uranium-molybdenum alloy clad in aluminium, had to be dissolved at low temperature (50°C) and in batches of gradually increasing size.

The input compatibility was unsatisfactory and it was found difficult to resolve the differences[66].

The final sub-campaign on fuel from Diorit, FR2 and BR1 (uranium-aluminium alloys clad in aluminium) proceeded satisfactorily, apart from a few operating errors[67], one of which caused the blockage of a column.

The sampling units suffered from continuous problems which had to be dealt with urgently by modification and simplification. The ALD was soon submerged in requests for analyses and these had to be restricted, since it was not possible to take on more staff.

During testing of the plutonium purification unit, some imperfections were found; modifications had to be made during the second half of 1966 and the unit did not come into active service until 6 May 1967. It involved two successive stages: a cycle of purification by extraction in mixer-settlers of safe geometry, known as the "wet process", followed by a plutonium oxide production line using precipitation and calcination of oxalate, known as the "dry process". Although the results were acceptable in 1967, they did not get any better, indeed rather the contrary, and it was necessary entirely to rethink this phase of operation by the beginning of the 1970[68].

At this time and during the first HEU campaign, tests were started on the second dissolver. Installed in March with its loading cell and its auxiliary equipment, it came into service in December 1967 during the second LEU campaign with VAK fuel.

In the meantime there was the first HEU campaign, which demonstrated the multiple capability of the plant.

The first campaign of reprocessing highly enriched uranium fuels (HEU): October-December 1967

The first HEU campaign began in October 1967; the summer holiday period and the month of September were utilised for completing the various units and for putting right the various faults that had appeared during the first LEU campaign. The entire plant was flushed out, and more particularly the extraction cycle in the plutonium purification shop, which was to serve as the second decontamination cycle for the highly enriched

65 The reprocessing of the EL2 fuel began on 24 October 1966. It was carried out in three sub-campaigns up to 20 January 1967.

66 The problems of compatibility of fissile materials are dealt with in more detail in Part III Chapter 4.

67 RAE 1967, p. 39.

68 See below. It appears that plutonium purification caused problems in every contemporary reprocessing plant. Thus Eurochemic did not have a monopoly of the difficulties, but the opening of the archives means that they can be revealed here without restriction. The chemical behaviour of plutonium, together with its radiological characteristics, make it a very delicate material to handle. The learning process was a long one.

uranium. This rinsing process generated large quantities of effluent. The campaign was preceded by a non-active test of the HEU circuit.

The campaign covered the processing of fuel from the CEA research reactors (Siloe, Triton, Cabri, Pégase and Osiris), and from Astra, FRJ1, R2, HFR and BR2. Elements from Astra were the first to go into the third dissolver. Chemical problems quickly appeared in the dissolving process: the aluminium was dissolved first. Then an amount of uranium – difficult to evaluate – accumulated in the dissolver, with an obvious criticality risk. It was therefore decided after a few days to modify the process by making it more acid. Adjustments had to be made to the extraction stage, particularly as a result of the changes to the dissolving process, but it did function well, apart from an interruption necessary for rinsing the first column as a result of the accumulation of impurities in the interphase. It appeared that a third extraction cycle would be necessary if the final product was to be in line with the contractual specifications, which were the same as those of the USAEC. The uranium was therefore recycled a third time in the extraction units of the second cycle.

All the elements were dissolved by 27 November and extractions completed in December 1967.

Management of waste and effluent[69]

Production

The reprocessing operations generated a larger quantity of effluent than expected, particularly during decladding, but also because of the routine rinsing operations between each type of campaign or others made necessary by the different incidents. It very soon became clear that consideration should be given to building additional storage facilities and an attempt made to minimise the generation of effluent through process modifications. However it appeared equally quickly that the problem of solid waste had been underestimated[70]: the output of dissolution hulls was greater than expected and a suitable storage facility had to be designed after they had been stored for a few months in Building 14.

The effluent treatment plant (*station de traitement des effluents* – STE) and the disposal system

The progressive nature of the plant start-up was an opportunity for running in the effluent treatment installation. No major problem was found in managing medium and high activity effluent, apart from that of sheer volume mentioned above. The storage system for these effluents was put into service in July 1967, following a few modifications that tests showed to be necessary. The main problems encountered concerned the checking of solution activities, solids, temperatures, acidity and organic substances. "These problems, which appeared during construction, were gradually resolved through better understanding of the operation of the plant and improved discipline"[71].

The weak point appeared to be the pipes transferring warm non-active effluent and non-active condensate to the CEN, which became blocked by the bitumen lining, which was of poor quality; clearing these pipes required six months of intermittent effort and resulted in the complete replacement of 60 metres of piping and the construction of a new monitoring pit. This was only the beginning of a long series of difficulties[72].

Increasing technical problems (December 1967 - May 1970)

The second LEU campaign was the commencement of a more intensive phase of operation. The values of burn-up and enrichment were higher, and the first zirconium and stainless steel cladding was dissolved.

The higher burn-up values, ranging up to more than 20 000 MWd/t, caused problems in a plant which had originally been designed for burn-ups of up to only 10 000 Mwd/t[73]. Wherever possible, it was necessary to enhance protection and shielding and modify procedures in order to limit the length of time staff spent in certain areas, particularly certain operating galleries. These improvements took up a large part of the shutdown times.

Reception and storage

The operations of receiving and storing irradiated fuel were purely routine. However in 1968 a new pond for solid process wastes was built in Building 2.

---

69 For the overall policy on waste, and the amounts produced, see Part III Chapter 4. Only the technical aspects arising during the learning process are described here.

70 In designing the mechanical processing pond, it was planned only to store small amounts of sawn-up structural parts in a corner of the pond set aside for the purpose.

71 RAE67, p. 49.

72 See the analysis of incidents in Part III Chapter 4.

73 See the 1961 specification. In 1958, Marcoule was receiving fuel with burn-up values lower by a factor 5 to 7.

Dissolution and extraction

The LEU 1968-1 campaign

The first LEU campaign of 1968 began on 15 December 1967 with part of the VAK fuel, which inaugurated the second dissolver. This was completed on 9 April 1968.

LEU 68-1 marked a new phase in the operation of the plant, because it involved reprocessing oxide fuels with enrichment between 1.48 and 4.43% and burn-ups between 2300 and 17 000 MWd/t. But above all, this was the first time that irradiated fuel clad in zirconium and stainless steel had been dissolved using the ZIRFLEX and SULFEX processes outside the laboratory.

The SULFEX process was satisfactory from the outset, but ZIRFLEX led to a number of disappointments.

The decladding of VAK fuel using the ZIRFLEX process was slower than laboratory work had suggested, and its incomplete success meant that the operation had to be repeated with further dissolution of the solid residues. But above all ZIRFLEX caused problems related to the release of ammonia in the dissolver. The ammonia combined with the nitric acid vapour in the ventilation system to form crystals of ammonium nitrate. These blocked the absolute filters, which frequently had to be replaced[74].

It was during this campaign that uranium nitrate 4 was tried as a replacement for ferrous sulphamate as a plutonium reducer.

During the LEU 68-1 campaign, the ferrous sulphamate [$Fe(NH_2SO_3)_2$] was entirely replaced by uranium nitrate [$U(NO_3)_4$] for reducing the plutonium in the first and second uranium cycles. As regards efficiency of decontamination this procedure had no significant effect compared with the sulphamate. However the installation, during the summer, of a buffer tank between the two columns so that they could be supplied independently did bring about a significant improvement in the productivity as from LEU 68-2.

A one-month flush of the plutonium unit was necessary before it could be used for the HEU campaign.

The HEU 1968 campaign

The HEU 68 campaign ran without problems from 30 April to 26 July 1968, but showed the need for a third cycle to ensure adequate decontamination.

The LEU 1968-2 campaign: attempts to reduce effluent volume

Following the annual shutdown, the system was given a further rinse to allow the second campaign, LEU 68-2, to be carried out: this was supposed to begin on 5 October. However start-up was delayed by the problems encountered in clearing the feed tube to the first extraction column. LEU 68-2 was completed on 23 December.

The fuels were of the oxide type enriched to over 4.45% and natural uranium. They were clad in aluminium and a magnesium-zirconium alloy. For the decladding of these two types of fuel element, the processes were modified in order to minimise the amount of effluent produced.

In 1967 and 1968 the sampling systems were modified since they required frequent, difficult and costly maintenance.

In 1968, to avoid filter blockage problems caused by the ZIRFLEX decladding system, the gaseous effluent scrubber was connected to the high activity ventilation system.

The LEU 1969 campaign: problems with evaporators and piping

In 1969 only LEU campaigns were run. The first took place between January and April 1969, the second began immediately thereafter and continued to mid-July; however a leak due to the corrosion of an evaporator in the plutonium purification cycle had the result of preventing the processing of 11.8 kilogrammes of plutonium. The plant was shut down for inspection, repair and modification, and the LEU 69/2 campaign restarted only in the autumn. In its turn, this campaign was delayed by a pipe blockage.

The fuel came from three sources: the Swedish R3 reactor (an unenriched oxide, clad in zircaloy), EDF2 and EDF3 (a natural uranium alloy clad in Mg-Zr), and E. Fermi at Trino (slightly enriched oxide – 2.72 to 3.9% – clad in stainless steel).

Decladding

The improvements made in 1968 reduced the time taken by the ZIRFLEX decladding process, but the problems of blocked filters continued. The decladding of EDF2 and EDF3 with dilute sulphuric acid and of the Trino fuel using the SULFEX process encountered no problems.

---

74 This was particularly troublesome because these operations are the source of considerable doses to the maintenance staff.

The uranium cycle and the tail-end of the process (Figure 97)

The uranium extraction cycles and the tail-end of the uranium process in 1969 confirmed the value of using uranium nitrate as a plutonium reducing agent; the optimal operating conditions were determined[75]. At the beginning and end of the LEU 69-2 campaign, two columns were unstable because the neighbouring column was too small, despite the installation of a buffer tank in 1968.

On 5 October 1969, when LEU 69-2 was restarted, it was found that a connecting pipe between the first column and the solution adjusting unit was blocked. All the usual remote action measures were tried in vain: water, steam, acid, vibration and so on. It was necessary to effect entry into the cells, which meant first decontaminating them and then installing a shield to give protection against very high radiation, of the order of 10 Sv/h, in contact, emitted from a corner of the cell. The pipe was sectioned and part of the blockage was found, consisting of a highly active black material (2 Sv/h contact). The task of unblocking the pipe was given to a specialist German firm, WOMA Rheinhausen, who began to feed in very high pressure water through a metal nozzle on 11 November 1969; two days later the incident was over. Analysis showed that two blockages had formed, probably due to zirconium decladding residues.

The lost time meant that it was impossible to carry out an HEU campaign during the year.

Plutonium cycles

In 1969, the second and third plutonium cycles experienced problems. Some four weeks of rinsing between the LEU-1 and LEU-2 campaigns were necessary before a satisfactory plutonium balance could be achieved, owing to the quantities of plutonium adhering to the walls of the equipment and piping. The leak on the evaporator in May 1969 interrupted the campaign at a time when 11.8 kilogrammes of plutonium out of a total of 115 kilogrammes remained to be purified.

The 1968 improvements made to the Pu precipitation and calcination units had borne fruit. However the extraction process experienced new problems owing to difficulties with an evaporator and the mixer-settlers (geometrically subcritical) of the plutonium purification unit. These difficulties arose from the presence of a heavy organic solution containing plutonium, due to the very high plutonium concentration in the solutions treated, the unstable operation of the evaporator and to abnormally high degradation of the TBP. It was demonstrated on this occasion that the mixer-settlers, owing to their very special geometry – a very flat parallelepiped – were more sensitive to the presence of a heavy phase than the pulsed columns. This demonstration influenced subsequent technical choices.

As far as the dry process treatment unit was concerned, the problems essentially involved the system transferring the wet precipitate of plutonium oxalate to the drier. Improvements were planned for the 1970 shutdown period.

The HEU 1970 campaign: only two extraction cycles

The HEU 1970 campaign began on 29 December 1969 and ended on 2 February 1970, since contractual requirements for plutonium delivery required the plant to be shut down at that time and an LEU campaign had to be carried out before the planned shutdown for repairs and maintenance.

A feature of this campaign was the low capacity of the front-end of the process, owing to a failure in a steam injector in the dissolver unit. However the extraction cycles were operated satisfactorily resulting in a product that met customers' specifications (Figure 98).

The LEU 1970 campaign: mediocre productivity

The HEU 1970 campaign was followed by a system rinse in preparation for the LEU campaign. This began on 18 February and continued until 17 May, consisting of four sub-campaigns involving the reprocessing of fuel from DP, VAK and Trino, together with Canadian "Cristal de Neige" (CdN) fuel which had been used in EL3. These fuels were oxides between nil and 4.5% enrichment, the first two being clad in zircaloy-2 and the last two in stainless steel and aluminium.

As usual, the SULFEX decladding system caused no problem. The operation of the ZIRFLEX decladding system had been improved. However the rate of attack was higher for the DP fuel than for the VAK owing to its lower burn-up (6000 MWd/t) compared with 16 000 MWd/t for VAK, which had thus resulted in a thinner

[75] For the LEU 69-2 campaign, they were fixed at a Pu/U4 stoichiometric ratio of 10 with 20% of the reducing solution introduced at the top of the column, the rest being injected at the mid-point. It was found that the distribution at the bottom of the column led to considerable losses of uranium.

Figure 97. Loading a "SAFRAP" container of low enriched uranyl nitrate, one of the end products of the LEU campaigns, on a truck trailer (Source: Eurochemic photo, no date).

Figure 98. Working in a glovebox, an operator bottles highly enriched uranyl nitrate, an end product of the HEU campaigns (Source: Eurochemic photo, no date).

oxide film being produced and owing to the prior mechanical scratching of the surface of the cladding[76]. Blockages of the ventilation filters by ammonium nitrate were substantially reduced by adding nitric acids to the tanks collecting the ammoniacal condensate. The solutions produced by dissolving fuel which had first been decladded by the ZIRFLEX process were centrifuged in order to prevent any recurrence of the blockages of October-November 1969. The fuels were also dissolved without problems. Exceptionally, the 4.5% enriched CdN fuels[77] were dissolved – cladding and fuel together – in the third dissolver normally used for highly enriched fuels, but which was able to accommodate them owing to the small size of the elements.

Uranium extraction was less satisfactory. Many problems arose as soon as the first cycle columns were started, and they had to be shut down for two days: there was a fault in the solvent feed pump, the reducing solution was wrongly made up, its feed supply failed, there were problems in controlling the pulsed columns, and so on. As a result of these difficulties, the uranium solutions had to be recycled several times towards the uranium purification cycle.

Because of the contractual requirement to return the recovered products within narrow margins of isotopic composition, fuels from certain reactors could not be mixed with others. This meant more and more sub-campaigns, which reduced productivity still further. Between each sub-campaign, the system had to be rinsed with nitric acid in order to minimise the likelihood of any batch being "contaminated" by another.

The plutonium from LEU 70-1 and LEU 70-2 was reprocessed without difficulty, however the 12 kilogrammes from the 1969 Trino campaign was more complicated owing to the presence of $S0_4$ ions resulting from the sulphuric acid decladding; this took 35 days of treatment out of the 119 days of the second cycle to complete the process. The precipitation and calcination stages were almost uneventful.

Reprocessing was halted on 17 May 1970 and the plant was thoroughly rinsed to prepare it for the modification and other work based on the lessons learned from four years of experimental reprocessing on a pilot scale.

Experience had also been built up at the back end of the cycle.

Recurrent problems in the Pu purification unit

During the summer of 1968, modifications were made to the plutonium purification shop to correct the defects that appeared shortly after it was put into service. It was now able to treat more concentrated solutions up to 100 g/l. A new evaporator using steam was installed to replace the previous effective electrically-operated unit. The connections between the tanks were modified particularly so that recycling did not affect the remainder of the plant. Also a new plutonium oxide production unit was installed in order to reduce the amount of manual work to be done in gloveboxes[78]. The systems in this new unit were divided between six gloveboxes to improve safety and reduce the risk of contamination during repairs and maintenance: one glovebox for precipitation, one for calcination at 400ø, one at 800ø, one for decontamination, one for weighing, one for mixing and sampling the final product. The cost of the unit was $54 000 excluding Eurochemic manpower and design work. However despite all this, the modifications did not result in completely satisfactory operation of the purification unit.

Effluent: substantial production and pipework problems (December 1967 - May 1970)

The reprocessing campaigns during this second phase confirmed that predictions had underestimated the amounts of liquid effluent and solid waste that would be generated and revealed difficulties in the transfer of low activity effluent to the CEN.

It was decided in 1968, with a view to reducing the volumes of wastes, to separate acid and basic effluents prior to concentration. The sheer quantities of medium activity effluent necessitated the construction in 1969 of a 2000 m^3 storage unit to accommodate the medium activity effluent up to the beginning of 1974, on the basis of 500 m^3 a year.

Piping problems continued to affect plant operations to differing degrees. In 1970 a leak in the feed and monitoring system of an evaporator for low activity solutions in the processing plant led to contamination of the ground and of a pipe intended for carrying very low activity effluent.

---

76 Using the "Scratch and Chop" process.

77 The reprocessing of Canadian fuel by Eurochemic needs explanation. France had bought the fuel because of its plutonium content. Its reprocessing for France by Eurochemic raised difficult problems of security control.

78 For the radiation protection problems arising from work in gloveboxes, see the radiological review in Part III Chapter 4.

Inspection problems on the pipeline leading to the CEN were continuous from 1967 to 1969. A new leak was discovered in September 1969 which took two weeks to repair. In 1968 storage tanks for used solvent were installed in Building 6 since the CEN was unable to accept this material.

A pond for high activity solid waste

The production of solid waste had been largely underestimated in the plant design owing to the complete lack of practical experience with chemical decladding. In particular it had been wrongly believed that virtually all the cladding and the structural components of fuel assemblies, together with the soluble baskets used for charging them into the dissolver, would be dissolved during the chemical decladding process. But the first few campaigns showed that this was far from being the case. Ways and means had to be found for storing these wastes that were highly contaminated with fission products as well as alpha emitters – essentially plutonium. It was therefore decided to build a pond where they could be stored under water.

The new Solid Waste Pond (SWP) was built in the south-west corner of the hall of Building 2 in 1968. With a surface area of 120 m^2 and a depth of 5.4 metres, this pond was intended to store highly contaminated solid waste, i.e., essentially undissolved parts of fuel elements, end fittings that were sawn off underwater, and equipment from cells no longer in use. The pond was built above the floor of the hall, its walls being lined with 2 milimetres of stainless steel and the bottom with a 3 milimetre sheet of steel and a 3 milimetre sheet of lead between two layers of bituminous felt. Surrounded by a walkway, the pond was fitted with a loading cage and an overhead crane. It was connected to a ventilation system. It cost $53 000 excluding Eurochemic manpower and design costs.

In 1969 this pond experienced a few problems connected with the unloading of hulls arising from the decladding of zirconium. Air contamination was noted whenever the baskets were emptied into the pond. Contaminated particles rose to the surface and gave a contact dose of 10 mSv/h. Consequently during the summer of 1969 the pond was covered by a tent with its own ventilation system which renewed the air contained 60 times an hour. The pond water circuit was connected to the purification system used for storing spent fuel prior to reprocessing.

The build-up of experience and a review of operating problems led to a decision to temporarily halt reprocessing after the LEU 1970 campaign for decontamination of the installations prior to verification and modifications. This shutdown was to last nearly 9 months.

The plant shutdown (17 May 1970 - 15 February 1971). Four months of rinsing and decontamination

The installation was first thoroughly rinsed, in order to recover as much fissile material as possible. Nitric acid solutions and solutions containing uranium 4 were used. It had been found in fact that the plutonium deposited on the stainless steel walls in the form of the "polyoxide" in which it took a valency IV, was more easily detached when it was reduced to valency III. In this way 9.5 kilogrammes of uranium and 1.7 kilogrammes of plutonium were recovered.

On completion of rinsing, a start was made on the decontamination of the system which was planned to make repair and maintenance work possible. Over a short period this phase provided the first experience of the type of work which was to follow the closure of the plant in 1975.

Various types of decontaminant were used with differing corrosive power according to requirements[79]. The decontamination effluents, of which there were large amounts, were taken, according to their composition, to the appropriate treatment and storage units. In this way some 13 000 Ci were removed from the installation by the final rinse. The levels of radioactivity which reached 10 Sv/h at certain points were reduced almost everywhere to levels of about 0.1 to 0.25 mSv/h.

The reduced ambient radioactivity made it possible to start work on the plant in October. This involved maintenance and inspection operations, modifications and new installations.

---

[79] In increasing order of corrosive power:

- non-corrosive agents: in turn 2-10 M HNO_3 and 0.5-5 M NaOH, sodium citrate, and nitrate of Vantoc (lauryl-dimethyl-benzyl-ammonium);
- slightly corrosive: 5% sodium tartrate in 5 M of NaOH, 0.1 M $KMnO_4$ in 0.1 M NaOH; a 7% solution of $H_2C_2O_4$;
- corrosive, with a maximum contact time of 5 hours; 3% NaF in 5 M HNO_3, and 0.1 M $Al(NO_3)_3$, added for complexing the F ions.

The decontamination programme and a schedule are given in detail in CA(71)2, pp. 39-42.

Far-reaching maintenance and inspection work was carried out: overhaul of fans and pulsed column drives, replacement of defective and worn equipment, such as 120 metres of steam pipe and the sampling system on the tanks receiving hot effluent before it was transferred to the treatment unit.

Four types of modification were carried out:

– the second LEU uranium cycle was modified in order to serve as the second cycle for the HEU campaigns;
– the new installations for plutonium purification and neptunium recovery were connected up;
– the entire system was modified in order to improve the technical and safety aspects of reprocessing. Attempts were made to apply the lessons learned from the incidents which had occurred;
– a new building – n° 24 – was built to store the medium activity effluent.

In February 1971 the plant was ready to start up again. However the future prospects were not bright and the situation had entirely changed since 1969. In fact the joint undertaking was experiencing a new economic and political crisis.

Chapter 3

The final years of reprocessing at Eurochemic:
Co-operation under political and economic threat and technical maturity
The early 1970s to January 1975

By the end of the sixties, the growing number of projects for new reprocessing plants was giving rise to fears about overcapacity; at the same time prices were depressed by the slower than predicted growth of nuclear power and the need to keep existing plants in operation. In February 1970 the crisis in reprocessing was acknowledged and reviewed in a report by an expert group appointed by the European Atomic Forum. The report drew up recommendations as to how the crisis might be overcome.

These recommendations might have led to a strengthening of European co-operation in reprocessing. In fact they blighted it within two years. The period 1970 to 1972 saw a process which led to plans for European reprocessing to be terminated in 1974, to be replaced by a cartel involving the two main shareholders in Eurochemic, France and Germany, together with the United Kingdom. However this decision did not put an end to co-operation in Eurochemic: the company still had to discharge a number of reprocessing contracts and, moreover, it is impossible simply to close down a nuclear establishment without first resolving a large number of problems. But it did create a deep and lasting gulf between the member countries, within NEA and in the company itself.

The closure decision meant that the company had not only to prepare to halt its reprocessing activities but also to complete those outstanding as quickly as possible. Following the 1970-1971 shutdown therefore, reprocessing resumed at a rate that was closer to "industrial" practices than a pilot plant approach. The final years of operation continued to provide substantial technical experience.

A crisis in European reprocessing at the end of the 1960s. The FORATOM report of February 1970

The origin and objectives of the report

- In 1968 the Board of Management of the European Atomic Forum, known as FORATOM[1], representing the nuclear interests of European industry, decided to set up an expert group with the task of reviewing the future of reprocessing in Europe.
- FORATOM's aim was to co-ordinate European efforts at a time when there was an increasing number of national schemes despite the real medium-term surplus in reprocessing capacity: "In 1968 the total quantity reprocessed in Europe was about 1000 tonnes, while the available reprocessing capacity was three to four times as big"[2]. The aim of the report was to review the situation and devise proposals to "make reprocessing into a sound industry"[3].

The expert group had 16 members, and was chaired by Mr. Seynave, from the Trade Federation of the Belgian Nuclear Industry; the rapporteur was Mr. Marcus, of the Nordic Nuclear Co-ordination Committee, which had co-financed the report[4]. It met on 10 December 1968 and then in June and October 1969. Its report

---

1 FORATOM had been set up on 11 July 1960. Its first Chairman was Henri Ziegler, a leading figure in the French aviation industry. AtW, 1962, p. 410.

2 P. 1.

3 Abstract, p. I: "The European reprocessing specialists have found a joint forum in which they can express their point of view about how their activities should evolve. They believe that if their ideas are adopted, reprocessing will become a healthy industry, but they are not unaware of the obstacles that will have to be overcome".

4 The other members were: Y. Leclerq-Aubreton and J. Asyee (Eurochemic), J. Couture and Y. Sousselier (CEA), Mr. Mamelle (SGN), Messrs Avery and Kemp (UKAEA), W. Schüller and P. Zühlke (GWK), Messrs Calleri and Zifferero (CNEN), E. Svenke (AB Atomenergi), with Belgium represented by Messrs de Beukelaer (Belgonucléaire) and Spaepen (CEN).

was approved on 10 December 1969 at a final meeting in Karlsruhe. The 62-page report was written in English with abstracts in English, French, German and Italian.

After reviewing the significance of reprocessing to the nuclear fuel cycle, and describing the techniques used and the plants operating in Europe and the United States, the experts examined the market in detail and the costs of reprocessing fuels from ordinary water reactors and other types, from the technical, economic and political points of view. They drew up conclusions and recommendations.

Reprocessing in Europe[5]

The fear of overcapacity

The reprocessing situation was in fact giving cause for concern. The deemed overcapacity[6] caused prices to fall to levels considerably below the operating costs of the purely civilian reprocessors. The pursuance of national development projects could only accentuate the problems in the future, since the development of nuclear electricity generation was still slower than expected. However it was estimated that in the long term, by the 1990s, some 10 000 to 40 000 tonnes of spent fuel a year would be available for reprocessing[7], equivalent to a market of 900 million dollars (1970 values).

The problem did affect fuels discharged from materials testing reactors and fast breeder reactors, but primarily concerned those from light water reactors which, at the end of the 1960s, were gaining ground everywhere in commercial power stations.

For reprocessing fuel from materials testing reactors there was not only the Dounreay plant – 5 to 10 tonnes declared capacity – and that of Eurochemic – 1.5 tonnes – but also EUREX – 0.3 tonne. Even for the fast breeder reactors there were increasing numbers of facilities. In addition to Dounreay, AT1 was introduced at La Hague in 1968 and WAK was also considering reprocessing these fuels.

As regards natural or slightly enriched uranium fuels, the market was already dominated by two large capacity plants which were being converted from graphite-gas reactor fuels to other types.

At Windscale, which at 2500 tonnes had the biggest capacity in Europe for reprocessing graphite-gas fuels, it was planned to convert the plant in the early 1970s by adding an oxide head-end to the plant first commissioned in 1964. This was not the result of a change to reactor systems, but reflected the desire to earn maximum commercial return on a major system of military origin for which the problems of depreciation clearly did not arise in the same terms as for the commercial plants. The prospects for reprocessing oxide fuels led to a search for overseas reprocessing contracts. Discussions had been held in France about changing from the indigenous graphite-gas system to the American light water type of reactor, and had resulted, a few months prior to the submission of the FORATOM report, to the government decision in November 1969 to abandon

---

5 Note on sources: the FORATOM report is often somewhat allusive but is developed by the review in AtW.

6 The calculations of overcapacity made at the time must be considered with care. In fact they were based on capacity figures given by the builders and the assumption was made that plants would be fully operational by their planned commissioning date. In fact quite apart from the commercial schedule, technical problems during operation reduced performance to the extent that actual capacity was below the declared figures. As Pierre Strohl has put it, there was in fact no real overcapacity but an overcapacity "on paper" or "in peoples' minds".

7 Report p. 1. It is difficult to relate these projections to reality. They appear highly optimistic, even if not unrealistic. Indeed if it is assumed, as the FORATOM report suggests, that an installed generating capacity of 1000 MWe produces 35 tonnes of [spent] fuel a year for light water reactors, this was equivalent, in the lowest hypothesis of the FORATOM report estimate, to an installed capacity of 285 GWe in 1990. In fact the installed generating capacity in the European countries of the OECD in 1992 was 274 GWe, according to OECD/NEA (1994), *Data on Nuclear Energy*, p. 12. It should also be noted that the estimated ratio between installed generating capacity and the amount of irradiated fuel produced was reduced to 30 tonnes per 1000 MWe of installed generating capacity, according to HOSTE A., JAUMOTTE A., dir. (1982), section VIII, p. 3.

Figure 99 shows how nuclear power generation in the Eurochemic countries actually developed in the period 1960 to 1990, with growth falling short of prediction. The trend expressed as a percentage of total electricity generation is even more striking, while showing the "speed" of growth that begins at a very low level. From 1968 to 1970 the share of nuclear power in the Eurochemic countries rose from 1.1% to 2.1%, compared with 12.3% in 1980 and 35.02% in 1990. These lumped statistics must be examined in the light of national trends: in 1990 five member countries were not generating any nuclear electricity (or had stopped doing so). Developments in particular countries are detailed in the statistical annex on nuclear electricity generation.

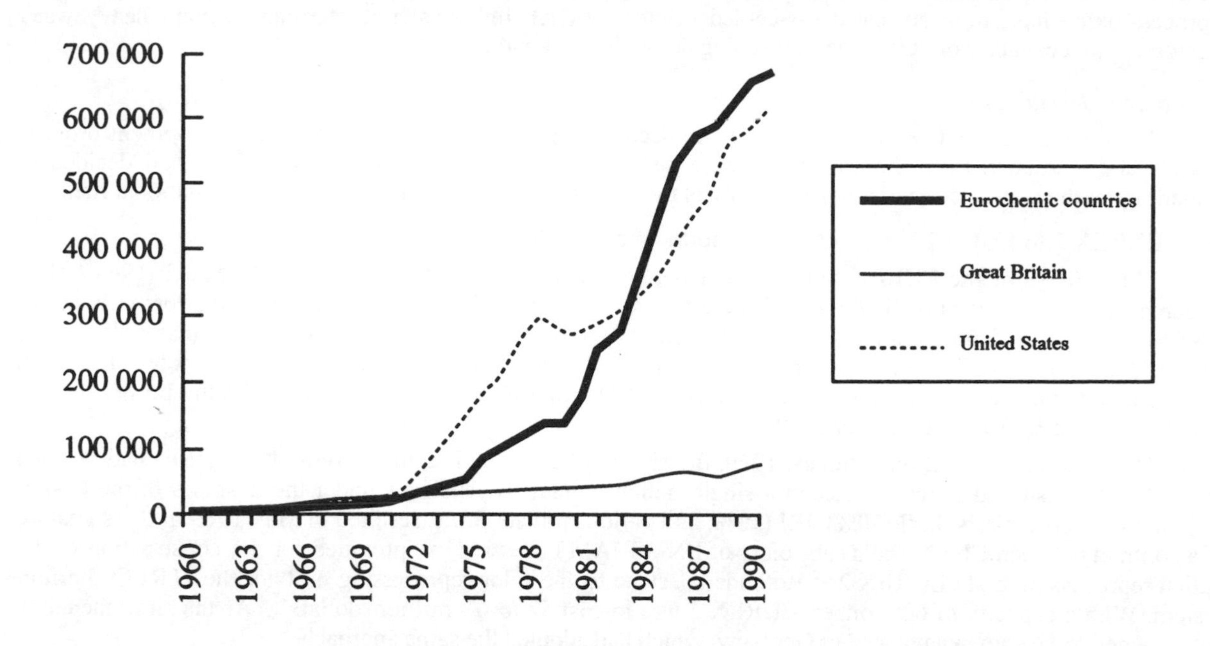

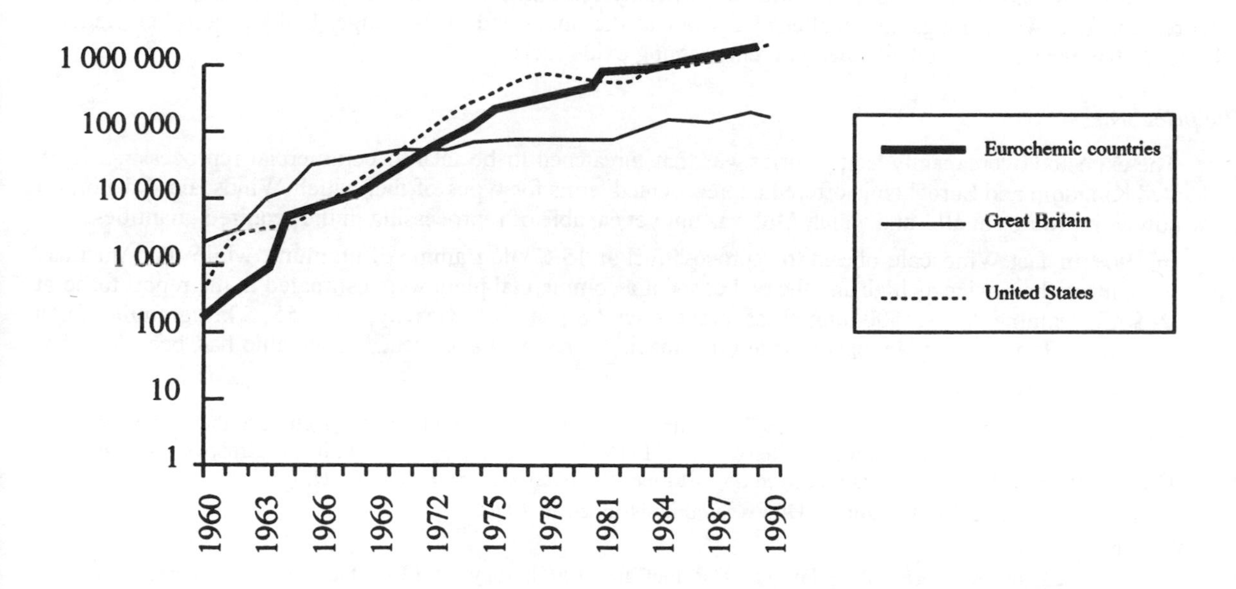

Figure 99. Development of nuclear electricity generation in Eurochemic member countries, compared with that in the United Kingdom and the United States from 1960 to 1990, in Gwh. Upper graph: arithmetic scale. Lower graph: logarithmic scale. The two graphs show, first, how continental Europe caught up with the United States and the United Kingdom and, secondly, the very modest level of nuclear electricity generation in the Eurochemic countries up to the end of the 1970s (Source of data: OECD/IEA (1990, 1991, 1992)).

the system developed at the CEA[8]. It was then entirely logical for La Hague to consider converting UP2 to reprocess oxide fuels from advanced gas-cooled reactors (AGR), light water reactors and certain heavy water reactors, with a capacity of 900 tonnes, by adding an oxide head-end[9].

Italian and German plans

The overcapacity situation was ultimately to become even worse owing to domestic development projects in Italy and Germany. Both these countries were just completing their pilot plants, and Italy had decided to embark upon the construction of a large capacity plant intended mainly for reprocessing its own national fuel.

From EUREX 1 to EUREX 2: reprocessing in northern and southern Italy

The builders of the EUREX pilot plant at Saluggia handed it over to CNEN on 26 March 1968[10]. The declared capacity of this pilot plant was low: 25 tonnes of low enriched fuel from light water reactors (LEU from LWRs). The original programme provided only for reprocessing highly enriched fuel from materials testing reactors (HEU from MTRs) but in a move contrasting with that adopted at Eurochemic the project had been refocused on the mixed reprocessing of LEU and HEU[11]. The first irradiated fuel, from the Latina power station, was delivered to EUREX in October 1969[12].

Three months earlier, on 2 August 1969, the national Economic Planning Committee (CIPE) had decided that Italian fuels should be reprocessed in a single national plant. Discussions under the auspices of the CNEN and the expected backers, ENI (70%), IRI (10%) and various private firms, resulted in May 1969 in the signature of a contract between CNEN and a subsidiary of ENI, SNAM Projetti. This provided for the construction of the Italian reprocessing plant EUREX 2 at Rotondella, close to the pilot reprocessing unit for the ITREC thorium system. With a capacity of 600 tonnes, EUREX 2 was to cost 52 to 61 million dollars[13]. At the time therefore, Italy seemed to be somewhat ahead of Germany, which had adopted the same approach.

In 1967 in fact, construction work had begun on the WAK pilot plant at Karlsruhe, a plant with a capacity of 35 to 50 tonnes of LEU[14]. Germany wanted ultimately to built its own major plant[15], but its plans were less advanced than those of the Italians. Against a background of European overcapacity, the Italian situation could hardly fail to cause concern to Germany.

In terms of capacity, Eurochemic, with its officially estimated 100-tonne capacity for reprocessing light water reactor fuels was amongst the smaller plants but at the upper end of the range. Unlike the other plants it did already have experience, albeit recent, in reprocessing oxide fuels.

The price war

The expected overcapacity led to a price war that threatened to be fatal to commercial reprocessing. Both the United Kingdom and Eurochemic offered unprecedented terms for types of fuel which Windscale did not yet know how to reprocess at all[16] and which Mol was not yet capable of reprocessing in the required quantities.

In 1969 in fact Windscale offered to reprocess fuel at 15 $/kilogramme of uranium, while the American prices were more than twice as high and the real costs of a commercial plant were estimated in the report to be at least 22 $/kilogramme for a 1500-tonne/year plant operating at full capacity, and 55 $/kilogramme for a 300-tonne/year (1 tonne/day) plant operating at full capacity[17]. As we have seen, Eurochemic had been forced at

8 For the "argument about reactor systems" in France and the choice of the light water system, see the highly relevant views of the chairman of EdF, Marcel Boiteux,: BOITEUX M. (1969), together with a retrospective view in BOITEUX M. (1993). This situation is placed in an historical perspective in SOUTOU G. H. (1991).

9 The high activity oxide (HAO) unit on UP2 was commissioned in 1976.

10 AtW, 1967, p. 228.

11 The declared capacities were 30 kg/day of MTR fuel and 100 kg/day of LEU fuel, or three times less than Eurochemic.

12 EUREX was to be put into service in October 1970.

13 AtW 1969, p. 227. The source indicates 192 to 224 million DM. EUREX 2 was never built.

14 The WAK plant, from the standpoint of its relationship with Eurochemic, is reviewed in detail in Part V Chapter 1.

15 This was to be the Wiederaufarbeitungsanlage project (WA). This term is of later origin and was used to designate the plant planned for Gorleben and then Wackersdorf.

16 However the United Kingdom could discharge its contracts without actually having to reprocess oxide fuels by supplying its customers with low enriched uranium from its enrichment plant at Capenhurst, and plutonium arising from the reprocessing at Windscale of its own spent fuel from its natural uranium reactors. Competition from the British was therefore real in *commercial* if not in technical terms.

17 Abstract, p.III.

the beginning of its commercial programme to abandon its initial system of determining prices and was now proposing prices of between 20 and 30 $/kilogramme for reprocessing natural or slightly enriched uranium fuels[18].

This price war was accompanied by a contract policy whereby a wider range of services was offered, the effects of which on Eurochemic's commercial policy have already been analysed.

The growing overcapacity and the price war were regarded as suicidal by European reprocessors. FORATOM drew the appropriate consequences and formulated a number of recommendations aimed at ensuring the survival of the largest possible number of reprocessors. The conclusions drawn were not always as respectful of the *status quo*.

The recommendations of the FORATOM report and their consequences

The main conclusions of the report[19]

The fact that Europe had three major plants of military origin and three smaller, exclusively civilian plants[20] was leading to distortion of competition, which was likely to maintain European prices at a level that would not encourage the industrial development of civilian reprocessing[21]. This meant that any new project had to be on a sufficiently large scale to be economically viable.

Nuclear power development prospects were such that no new reprocessing plant would be needed in Europe before 1978-1980. Such a plant should essentially reprocess fuel from light water reactors. It should be of large size, with a capacity of the order of 3 to 5 tonnes a day, and would thus be capable of servicing 70 reactors.

The increasing number of small projects could only delay the advent of a sound market[22]. However these small projects reflected national industrial and political interests which had to be taken into account. The answer therefore had to involve international co-operation.

If an industry was to be built "which is viable, not requiring subsidies, in a competitive system but with some degree of co-ordination", then attempts should be made "through commercial agreements between all interested parties [...] towards making optimal use – without discrimination – of existing reprocessing capacity in Europe. The aim of such agreements would be to avoid pointless financial loss and to achieve a sound level of reprocessing prices. Similarly decisions about the construction of additional capacity should depend on actual demand. Consequently, planned extensions and conversions [...] would be postponed. Similarly, new plants would not be put into service until they were justified by the market situation [...]. Finally agreements should be concluded to ensure that the existing small plants were operated as pilot systems, for special fuels, for processing peak tonnages, etc. [...]. Development work should be co-ordinated to avoid duplication of effort and accompanied by commercial agreements with a view to its application in new plants".

"In order to counter the political arguments in favour of the early construction of small plants, despite their higher cost, new major plants should be distributed geographically in such a way that each region that has a high density of nuclear power plants is served by a reprocessing plant"[23].

There should be a degree of harmonization in the controls and safety standards "to avoid any danger that competition is at the expense of safety standards".

The group concluded by proposing to review the changing situation on a regular basis and offering its services, notably in helping "to formulate regulations and standards for the management of waste, including its final storage". In fact the report stresses the need for improvements in these areas.

---

[18] The archives give no information about the prices charged by Marcoule and La Hague, probably because the relevant markets were captive and non-commercial, internal to the CEA or between public bodies such as the CEA and EdF.

[19] P. 55-59. The report is politically neutral and written from a purely industrial standpoint: it was consequently open to different political interpretations, as described below.

[20] Windscale, Marcoule and La Hague; Eurochemic, EUREX and WAK.

[21] The report considers the European market, although it does mention the existence of competition outside Europe. These limitations are recalled in the conclusions.

[22] P. 58: "... the overall reprocessing situation will further deteriorate by the building of more, smaller plants than warranted by the market".

[23] This is the first reference to the scheme for what the IAEA was to call in 1977 "regional fuel cycle centres", intended to reprocess fuels under international control in order to avoid any risk of proliferation, see Part IV Chapter 1.

The issues raised by the FORATOM report: What plant for the 1970s – a major international plant or a major national plant?

Most members of the FORATOM expert group – who were involved in European or regional (between the Nordic countries) co-operation – took the view that the resolution of the reprocessing situation called for international co-operation and hence the abandonment of further national projects. In this context, Eurochemic may seem to have a dual identity, being both a fully-fledged plant and an international structure. The report, in a concern for consensus, stresses the limited but positive role of pilot plants. It implicitly suggests that Eurochemic could be the embryo of a future major European plant.

However certain protagonists read the situation differently. UNIREP pointed out the consequences of the situation for three countries, the two European giants of reprocessing and Germany. Because of the risk of overcapacity, France and the United Kingdom called for other national projects to be stopped. Germany, which might have been sidelined, managed to retain a foothold amongst the reprocessors, at least in what it hoped would be the relatively near future[24]. In fact in March 1970 difficult negotiations began between the UKAEA, the CEA and the German firms involved in reprocessing – Bayer, Hoechst, Gelsenberg and Nukem – with a view to setting up a marketing pool, the future UNIREP company.

The fact that the two principal shareholders in Eurochemic were involved in the parallel UNIREP negotiations was bound to have an impact on the future of the European reprocessing undertaking.

The closure decision and the crisis in European co-operation[25]

The Future Programme of March 1969 provided for a review of the Eurochemic situation by 1972 at the latest, particularly in order to update the funding arrangements. This review was in fact undertaken at the beginning of 1971 and resulted in the plans to halt reprocessing.

On 19 November 1971, the Board of Directors[26] adopted the following conclusion:

"After thorough study of the problems raised by the future of the company, a majority of the Board was in favour of a progressive slowing-down of the operation of the plant in order to prepare the closing down of the installations within the most favourable time-limits. This should be undertaken at the minimum cost compatible with the safety requirements and the observance of commitments of the company, in particular towards its personnel. The stages and schedule for the entire set of operations intended to meet these objectives [...] will be determined by the Restricted Group with the assistance of the Group of Experts. In parallel, negotiations will be broached with the Belgian Government concerning the conditions for taking over the site".

This decision led to the 1969 Five-year Programme being revised on 16 March 1972. The termination of operations was planned for 1 July 1974 and a draft agreement covering staff redundancies was adopted.

It is necessary to explain the process which led to the decision to abandon reprocessing. Most of those who were involved at the time put the "beginning of the end" of Eurochemic with the publication of the FORATOM report in February 1970. They relate the decision of 19 November 1971 described above to the creation, a little over a month earlier, on 13 October, of United Reprocessors Gesellschaft (UNIREP)[27] by the French, British and German reprocessors. According to this view, UNIREP "killed off" Eurochemic. This view has been fleshed out by reading the archives and conducting interviews. The closure decision was taken before the negotiations that led to the establishment of the company were completed. However the advent of UNIREP put an end to any hope of extending European technical co-operation in reprocessing beyond the Eurochemic experiment. What then emerged, with the ever-present fear of overcapacity, was trilateral commercial co-operation in the form of an oligarchical cartel[28].

The Future Programme adopted in 1969 held out an ambitious future for Eurochemic, taking up an idea declared at the outset in one possible interpretation of a clause in the Statutes[29], reaffirmed on 12 December 1968

---

24 In the absence of any sources presently open for research, it is impossible to know how. However its financial capabilities and its prospects for nuclear power development made it one of the main customers of the reprocessors.

25 NE/EUR(72)1 – Part I – reviews the principal stages and gives references to the original documents.

26 CA/M(71)3, p. 6.

27 The sources sometimes use the abbreviation URG.

28 Technical co-operation became bilateral or trilateral and was dominated by commercial arguments: the same applied to the BNFL-SGN merger of December 1972.

29 Article 3, paragraph 3: "When the quantity of irradiated fuels which Member countries of the Organisation for European Economic Co-operation wish to send for processing in a joint installation seems likely to exceed the capacity of this plant, the Company should examine means to meet the demand of these countries under economic conditions".

by the Managing Director, Rudolf Rometsch, in agreement with the Board of Directors at the Symposium on Nuclear Energy organised in Milan by the FORATOM Scientific and Technical Federation[30]. The Mol plant was to be the prototype of a "large European reprocessing plant", built by "an international industrial consortium formed by companies carrying out important nuclear activities in the fuel cycle"[31]. There was no other possible explanation for the signature of long-term contracts at a time when the future of the Mol plant was not absolutely certain beyond the lifetime of the company.

Yves Leclerq-Aubreton, Managing Director since 1969, defended this concept with conviction and force[32]. His dream was a new European plant, Eurochemic 2, which he wanted to be comparable with La Hague or Windscale. This view of things led him to try, through personal determination, to save European reprocessing. His initiatives, at a particularly difficult time, helped damage his relations with some members of the Board of Directors and exacerbated the dissension and schisms that were appearing amongst the shareholders of the company, disagreements which were also a factor in the process leading to the closure decision. The schedule of plant operations had similar effects. The halt in reprocessing from May 1970 to February 1971 delayed the completion of contracts and was a source of discouragement to those who had put their industrial hopes in the plant. However although UNIREP was clearly not the only cause of the termination of reprocessing, the establishment of the European cartel was nonetheless a powerful catalyst.

The process leading to the closure decision

Realisation of the impossibility of the Five-year Programme, the closure decision in principle and the conditions of closure. January 1970-May 1971

The decline in reprocessing income and the 1970 funding problems

The technical problems described in the last chapter led to delays in the completion of contracts with a consequent impact on income.

Table 53. **Predicted and actual income from reprocessing, 1969 to 1971**

Year	Reprocessing income according to the Five-year Programme (million BF)[33]	Actual income (million BF)[34]
1969	n.a.	58.28
1970	89.56	28.45
1971	89.56	104.74

These technical problems also incurred extra costs, which exceeded the provisions made for operational contingencies.

Growing inflation subsequent to 1968 raised prices and put salaries under pressure, with the result that the salaries bill remained between 122 and 124 million BF from 1967 to 1970, despite the reduced workforce. Once again, no provision had been made. To face its cash flow problems, the company had to borrow from the Staff Provident Fund and minimise overheads[35].

It was urgent to find new income, and this was done by speeding up the process of obtaining and signing contracts in 1970[36], necessitating more administrative staff. Attempts were made in the Special Group to obtain

30 Reported by Rudolf Rometsch. This paper declared that if current needs were met, any new plant should have a capacity of between 1000 and 1500 tonnes and that the reprocessing price should be set at such a level as to enable recovery of capital and operating costs over a ten-year period. In practice this was tantamount to multiplying the price by a factor of 10. This was precisely what was done by UNIREP later on.

31 NE/EUR(69)1, p. 8.

32 During his farewell speech on 11 December 1972, he was to say: "Personally I have always had a particular idea of...Europe in general and international co-operation in particular. I have tried to put this into practice since I came to Eurochemic".

33 Figure obtained by converting from 1969 dollars.

34 Source: balance sheets.

35 The cash flow problems continued in 1971. Together with the personnel difficulties, they had the disadvantage – for the historian – of reducing the quantity and quality of paper records, thus reducing the material available for this chapter. For example the annual reports, which since 1967 had included only rare drawings and no photographs, were issued from 1970 to 1974 in the same format as Board of Directors documents.

36 The result was a sharp rise in reprocessing income for 1971, which was the best year of all from this standpoint; see the financial results in Part III Chapter 4.

preferential contracts with member countries, but without any tangible results. It soon became clear that the Five-year Programme was untenable.

Threats to the Five-year Programme and the adoption of the principle of closure in January 1971

On 20 January 1971, a restricted group of the Special Group which had been set up in 1970 sent the Special Group a paper on the future of the company. It evaluated the increasing deficit that would be encountered if the 1969 programme were to be applied in full, pointing out that the deficit would continue to rise well beyond 1974. It was therefore essential to review the programme and "consider how the operation of the installations could be terminated". The Special Group asked that a more detailed study setting out procedures for the review should be carried out by an expert group and that closure scenarios should be explored, that Belgium should be particularly involved with these studies and, finally, that an approach should be made to the UNIREP negotiators in the hope of finding a place for Eurochemic in the agreement that was then under negotiation.

The decision of 26 May 1971 on the conditions for closure

The Expert Group submitted its study[37] on 26 May 1971 to the fifty-first meeting of the Board of Directors, held in Mol; it set out scenarios for closure in the short, medium and long-term – April 1972, December 1974, the end of 1977 and the end of 1980 – together with scenarios for dealing with outstanding contracts, involving cancelling either the majority or only those covering longer periods. The Board of Directors requested the group to examine further the scenarios for rapid closure and for maintaining a pilot plant operating at a low level[38]. It also decided to consider an installation for the bituminisation of intermediate activity wastes, which was necessary whatever programme was adopted. This began the company's mutation towards waste management.

A restricted group of the Board of Directors was set up at the same time to monitor the work and issue the necessary directives.

The Eurochemic staff had heard rumours that the plant might be closed[39]. They tried to get more detailed information from the management, but no clear reply was forthcoming. The Staff Committee then demanded a meeting with the Board of Directors in order "to obtain clear decisions and guarantees for the future"[40]. In fact the Board of Directors was held captive by Eurochemic staff for seven hours. The Board informed the staff representatives that the decision to close the plant had been taken, that the position of the staff would be carefully examined and that the staff would be involved in the work of the restricted group of the Board of Directors. Negotiations about redundancy conditions then commenced, and a draft agreement was produced at the end of the year.

Eurochemic, UNIREP and the German strategy for entering the reprocessing industry. May-October 1971

With regard to the approaches made to the UNIREP negotiators, José Luis Gutierrez-Jodra, Chairman of the Board of Directors, sent a letter on 29 April 1971 to the UKAEA Production Group, the French CEA and the German Ministry for Education and Science. While the reply from Pierre Taranger, on behalf of the CEA, referred to discussions with the French and German representatives on the Board of Directors, that from Wolf J. Schmidt-Küster, head of the International Affairs Division in the Bundesministerium für Bildung und Wissenschaft and member of the Eurochemic Board of Directors, was more explicit, setting out clearly the principles which Germany believed should now govern reprocessing activities:

"Certainly the German Ministry does not have any objections against contacts between Eurochemic and the group of European industries [which was preparing the way for UNIREP]. On the other hand, I doubt very much that such contacts will be helpful in bringing about the decisions to be taken by the competent Eurochemic bodies on the long-range future of the Eurochemic plant at Mol. I might point out in this context that in our opinion reprocessing of oxide power reactor fuel ought now to be a *commercial responsibility of interested industries* [it is for this reason that the Ministry] does not intend to...*subsidise* reprocessing of such fuel except to the extent covered by the present Eurochemic programme"[41].

---

37 CA(71)11. The study itself is missing from the archives.

38 This option envisaged that reprocessing would no longer be carried out on a commercial basis but only for R & D purposes.

39 The concern of the staff accelerated the move towards normalising the organisation of the company's staff, with the growing involvement of Belgian trades unions – most of the staff were Belgian – the prelude to institutional normalisation in 1972.

40 AtW (1971)(1971), p. 325.

41 CA/M(71)2, p. 10-11.

In fact Germany's negotiating aim at this time was to "join the bigger players". In return for sending its fuels for reprocessing in British and French plants, Germany wanted Frankfurt to be selected as the location of the planned company for selling reprocessing services, but most of all a guarantee that the major plant to be built at the end of the 1970s would be German. With this in view, German interests were grouped into the KEWA on 23 August 1971[42].

On 13 October 1971 therefore, United Reprocessors Gesellschaft mbH was set up and its head office was in fact in Frankfurt. The company was owned equally by KEWA, the CEA and British Nuclear Fuels Ltd. (BNFL), which had taken over from the UKAEA for the fuel cycle in Britain[43] (Figure 100). The company was managed by P. Zühlke (Germany). The chairmen of the CEA and BNFL, André Giraud and C. Allday, between them chaired the different company bodies[44]. The immediate aim of UNIREP was to divide the European reprocessing market between the British and French plants until their capacity was saturated. At this point, a large German plant would take over, for which a capacity of 1500 tonnes was being envisaged[45] (Figure 101).

Eurochemic was not involved in this agreement in any way. The major reprocessors considered that the joint undertaking's task was complete and that in any event it was unrelated to their alliance – which concerned fuel from power reactors – and was not commercial in nature. This was a clear expression of the original disagreements about the purpose of the Mol plant. Those members of Eurochemic who were in favour of an industrial approach considered that they had been deceived, and reacted angrily at the next meeting of the Board of Directors[46].

At the meeting of the Board of Directors held on 19 November 1971 it appeared quite clear that the future of European reprocessing had been decided elsewhere than by those responsible for European co-operation. The meeting examined the two detailed scenarios "most likely to be envisaged in practice" for putting an end to co-operation. The first option, involving closure at the end of 1972, implied that the company might be dissolved in 1980. The alternative approach, prescribing a gradual reduction in reprocessing activity, could be accompanied by a new research and development programme focused on the problem of discharges and on the conditioning and final storage of radioactive waste. If this approach were to be adopted, closure of the plant would be in 1984 at the latest.

The choice of closure date and the revised programme for 1973-1974. November 1971 - November 1972

The Board of Directors chose the closure date and revised the programme on 16 March 1972, these decisions being approved by the Special Group when it met on 8 November 1972. The decisions represented a compromise between the immediate cessation of activities and the continuation of a long-term industrial programme on radioactive wastes.

The cessation of operations and the redundancy programme.

It was decided that operations would continue up to 30 June 1974, since "it was not possible to decide a brutal change in orientation in a short time without creating a serious crisis"[47] for technical and social reasons. In fact the cancellation of all contracts which a 1972 closure would have made necessary would have raised many commercial problems, particularly as regards the irradiated fuel that had already been delivered. There were also technical reasons: certain fuels could only be reprocessed at Eurochemic. Accordingly the principle adopted was that contracts already entered into should be executed to the greatest possible extent. Apart from the technical and commercial reasons for the choice of closure date there was also a social requirement, which was to arrange for the end of reprocessing to coincide with the opening of the bituminisation unit the construction of which had been decided in May 1971. This schedule had the result of "the experienced engineers and operators required by the company would still be there at the start of radioactive waste solidification operations".

Finally, from the purely social standpoint, the Board of Directors "had to take account of staff reactions...who were working in the perspective of the Five-year Programme and safe employment during that period".

---

[42] Kernbrennstoffwiederaufarbeitungs-Gesellschaft. Its capital was held equally by Bayer, Hoechst, Gelsenberg and Nukem.

[43] As did COGEMA at the CEA in 1976.

[44] ATW (1971) 1971 p. 557 and 565. In fact P. Zühlke was "Geschäftsführer", André Giraud "Präsident", C. Allday, "Präsident des Beirates".

[45] This occurred at a later period than the events being described here; it does, however, help place things in their proper context.

[46] Notably the Scandinavians. See the table of reactions given below.

[47] NE/EUR(72)1 p. 10.

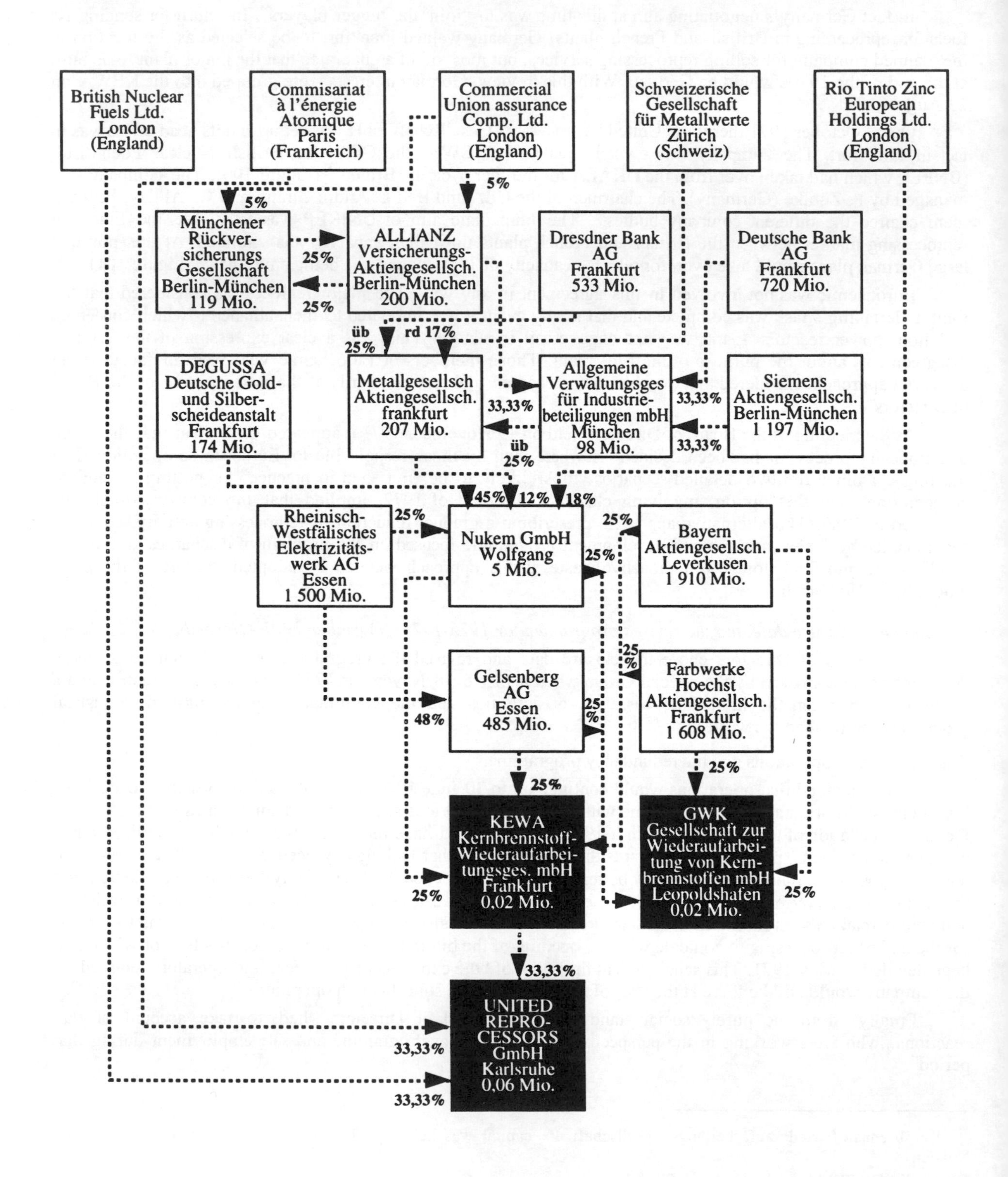

Figure 100. UNIREP and the structure of reprocessing in the Federal Republic of Germany in June 1974 (Source: SOLFRIAN W. (1974), p. 314).

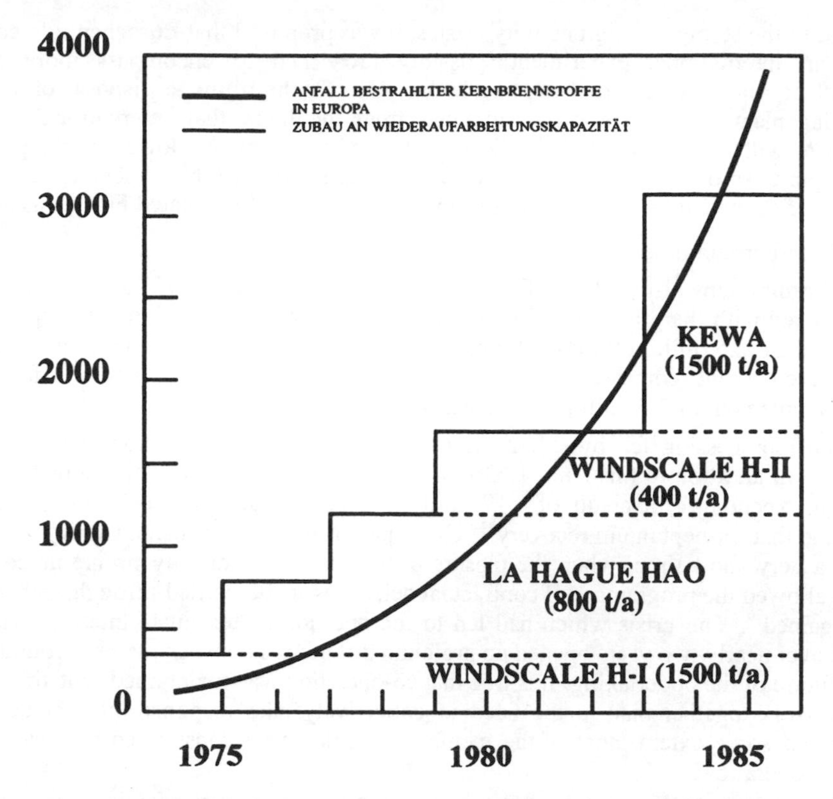

Figure 101. The future of reprocessing in Europe seen by UNIREP in 1974. Construction of a very large German plant by KEWA was planned for 1983, when the La Hague and Windscale plants would reach saturation. Bold curve: predicted availability of irradiated fuel in Europe. Block diagram: increases in reprocessing capacity (Source: ZÜHLKE P. (1974), p. 349).

A draft agreement on the redundancy conditions had in fact been negotiated with the Staff Committee[48] and signed on 30 December 1971. This made provision for the dismissal of 64 members of the staff as from 1 February, and laid down the conditions and period of notice[49] both for employees dismissed in 1972 and for other staff. In order to avoid resignations prior to dismissal and to keep staff in place, a lump sum payment calculated on the basic salary was made available to staff remaining, this payment being progressive as the end of the period approached[50]. Total redundancy costs were estimated at 2.960 million EMA units of account [EMA/UA] (1972 values).

Radioactive waste management[51]

The second part of the revised programme covered the conditioning and storage of radioactive waste. The Board of Directors did not propose an overall project but merely the initial steps that would lead to the completion of conditioning by the end of the 1970s.

The Board of Directors adopted the principle that intermediate activity wastes should be coated in bitumen, and began a research programme into the conditioning of high activity wastes, covering exploration of four possibilities, involving three research centres: bituminisation for concentrations above 1 Ci/litre in conjunction with Karlsruhe; vitrification, with twofold co-operation proposed with the Marcoule centre (the PIVER process) and the Karlsruhe centre (VERA); fluidised bed calcination, tested at Idaho; the final possibility was an "in-house" process of low temperature solidification, LOTES, the industrial value of which was to be demonstrated before the end of the year.

---

[48] CA(72)3.

[49] This involved a combination of length of service, age and salary (in increasing order of the resultant coefficient), and offered the possibility of payment in lieu of notice, in which case the lump sum payment was halved.

[50] 15% of basic salary up to June 1972, then 20% and finally 25% from 1 July 1973 to 30 June 1974.

[51] Part IV Chapter 2 includes a detailed study of radioactive waste management from the beginning of the 1970s. Consequently the subject is only treated in outline here.

With regard to the storage of high activity wastes, it was proposed that Eurochemic's contribution should be regarded as an international experiment, possibly broadened to encompass more than its original shareholders[52]. In fact, this was "the first large-scale operation for the ultimate disposal of highly-active wastes from a reprocessing plant"[53]. This proposal stemmed from the hope that international co-operation would continue on the site with a new purpose – radioactive waste management. However this proposal, which the NEA took up in the form of organising a symposium in Paris from 27 November to 1 December 1972 on "The Management of Radioactive Waste Arising from the Processing of Irradiated Fuel" was not followed up[54].

The cost of the revised programme

The revised programme also included financial estimates for the years 1973 and 1974. These amounted to 11.8 million compared with the provision of 10 million in the Five-year Programme. Expenditure was set out under various headings: 2.5 million for conditioning wastes, 2 million for staff redundancy costs, 0.950 million for research and development (compared with 2.4 million in the Five-year Programme for the same period). There was also a supplement of 2.6 million to counter the effects of inflation from 1969 on.

Income from reprocessing fell by 2.8 million, so government contributions had to be increased from 5.6 to 10.2 million. In addition to this sum, 1.536 million was needed to cover the deficit from 1969 to 1971. As regards capital expenditure, a credit of 0.17 million was included for completing the final plutonium purification unit and that for neptunium recovery[55]. The apparent paradox of manufacturing systems that were to be used for only a very short time, when the closure of the plant was already programmed, was justified by safety, but it also allowed the programme of contractual deliveries to be carried through and some new practical experience to be gained[56]. The crisis which had led to the decision to terminate international reprocessing and which had spread over nearly two years opened up profound differences between member countries. These had an influence on the future of the undertaking. It is true that co-operation was maintained, but this was due less to a positive desire to work together than to the need to collectively take responsibility for decontamination and waste processing. To some extent most of the member countries were constrained by past events to consider continuing their co-operation.

To understand these differences, it is necessary to carefully review the position of the different countries in the decision-making process.

Breakdown in co-operation. An analysis of the national and international positions – and those within the company – regarding the decision to close the plant

National differences

Although the closure decision was taken by a majority vote, it did not reflect a consensus on the part of the member countries. It was in fact the only way out of the conflict in which they were involved. Although UNIREP crystallized the criticisms of a number of small countries, the continuing deficit of the company also played a role. Only Belgium – with shaky support, it is true, from the partisans of European industrial reprocessing somewhat disappointed by the technical difficulties encountered – came out as entirely hostile to the principle of closure.

We have used the two most explicit documents to illustrate the positions of the different countries: the minutes of the meeting of the Board of Directors held on 19 November 1971[57], at which the principle of rapid closure was decided upon, and those of the meeting of the Special Group held on 8 November 1972[58], at which the proposals of the Board of Directors on the revised programme were approved and the recommendation of the OECD Council on funding was adopted.

These two meetings also provided opportunities for the representatives of member countries to clearly express their points of view on the circumstances which had led to this decision and to indicate their positions for the future. It should be noted that the meeting of the Board of Directors was very shortly after the UNIREP

---

[52] In April 1972 the OECD European Nuclear Energy Agency became the OECD Nuclear Energy Agency following the accession of Japan to membership.

[53] NE/EUR(72)1 p. 14.

[54] OECD/NEA-IAEA (1973). The proceedings of this symposium, published as a volume of 1266 pages, gave a good review of the state of the art around the world at the beginning of the 1970s.

[55] The neptunium recovery unit, which used some components of the plutonium circuit, had originally been funded by certain customers outside the Eurochemic budget.

[56] NE/EUR (72)1, Addendum p. 30 gives some details of the actual problems of operating this weak link in the plant.

[57] CA/M(71)3.

[58] NE/EUR/M(72)1.

announcement and that the extreme views of Switzerland for example changed over the period of one year. Indeed Belgium sought to exploit in 1972 the fact that "views have matured"[59] in order to get certain countries to go back on their decision to abandon reprocessing and to propose new forms of co-operation to them. In fact a review of the positions shows that the countries could be grouped according to five different "psychological" attitudes to the closure decision.

A first group of countries recalled that their wish had been that Eurochemic should itself embark on international industrial reprocessing, or mark a stage towards that goal. However even amongst these countries, there were differences. Some countries were resigned to the fact that the Franco-German alliance sealed within UNIREP made it impossible to continue Eurochemic's activities. This applied particularly to the Scandinavian countries and the Netherlands. Others considered that they had been let down and therefore wished to bring the earliest possible end to the co-operation in which they believed they had been abused. The most explicit complaints came from Spain and Switzerland in 1971[60].

Belgium, for its part, was ready to do everything to maintain a reprocessing activity at Mol.

Finally there was France, somewhat ahead of Germany which chaired the Board of Directors.

Italy reserved its position, because CNEN was in the middle of a reorganisation and because its contribution to expenditure was on a voluntary basis. Turkey was not directly represented in the decision-making bodies since the director's post it shared with Portugal was occupied by latter country at the time.

Table 54 is an attempt to summarise countries' attitudes. The terms used are not quotations, because the style of the statements made was more diplomatic. However they illustrate the general state of mind as it appears from the documents. The analysis that follows reveals three sharply differing attitudes – positions 1, 3 and 4 – which were the subject of closely argued and sometimes detailed annexes to the minutes contributed by the French, Spanish, Swiss and Belgian representatives.

Table 54. **Summary of countries' attitudes to the shutdown of reprocessing**

Attitude	Countries
1. Eurochemic has run its course. What is needed now is purely commercial reprocessing in very large national plants, working for all European countries.	France (analytical arguments) Germany (arguments of principle)
2. Disappointment and resignation. If it had only been up to us, Eurochemic could have become the centre for international co-operation in reprocessing. For reasons over which we have no control, this co-operation is now impossible. The plant must therefore be closed.	Denmark, Sweden (and probably Norway, which remained silent), Netherlands (will decide to leave)
3. Irritation at the betrayal by the French and the Germans. No point in continuing co-operation destroyed by the selfishness of a few countries.	Spain, Switzerland (Austria and Portugal held a position somewhere between 2 and 3)
4. Eurochemic must absolutely continue its reprocessing activities.	Belgium
5. No position	Italy: in the middle of reorganising its nuclear structures; already largely disengaged. Turkey: not represented (Portugal took its place at the time).

France and Germany: from Eurochemic to UNIREP

Germany remained somewhat in the background, repeating to the Board of Directors[61] what had already been written during the consultations of UNIREP during May:

"The Government of the FRG was not prepared to subsidise this sector of the fuel cycle which had now entered the commercial phase".

The French position in favour of closing the plant and the reasons for French participation in UNIREP were set out in more detail at the meeting of the Board of Directors held on 19 November 1971 by Jacques Sornein[62], Director in the Production Division of the CEA and occupying the second French seat on the Board of Directors following the death of Jacques Mabile in January 1971.

59 Preamble of his declaration to the Special Group.

60 However the Swiss position became more moderate in 1972.

61 CA/M(71)3 p. 4.

62 The text of his statement is attached to CA/M(71)3, p. 14-16.

France had, recalled Jacques Sornein, "while the English declined, 'contributed' at the outset to this undertaking [i.e., Eurochemic] all the experience and know-how it had acquired through the creation of the Marcoule plant". France had thus passed on to its partners the necessary experience for building larger plants.

When the Italians decided to build EUREX and the Germans WAK, the French had asked "whether Eurochemic had not completed its task and whether consideration should not be given to terminating its activities". France "however did not wish to withdraw before the end of the agreed 15-year period" and had therefore approved the Five-year Programme, while nevertheless seeking a review. In fact since 1969 three new factors had emerged:

- the commercial viability of Eurochemic had been adversely affected by "the decision of certain plant operators in member countries not to have their fuel reprocessed at Eurochemic". This referred to Italy and Denmark who had signed reprocessing agreements with the United Kingdom. Thus France indicated that she was not alone in lacking solidarity;
- technical and economic progress in reprocessing had substantially increased the optimum size of a reprocessing plant, which was now ten times the capacity of Eurochemic. "Now the OECD member countries had not stated that they intended to reprocess, in a joint installation, quantities which would justify construction of a plant with a greater capacity";
- finally the fact that France had adopted a new reactor system meant that UP2 had to be adapted to reprocess oxide fuels. "It is easily understood that the CEA could not possibly contemplate giving up this adaptation in order to participate in the possible joint creation of a plant with a large capacity at a cost ten times greater".

Since keeping the Mol plant in service would not be possible beyond 1974 without "a subsidy without security of at least two million dollars per year" [....] "the reprocessing needs of users in the Community could only be met at satisfactory economic conditions by using existing large capacity reprocessing plants in priority, and also by a harmonized development policy to be adopted by the main industrial groups with vast technical possibilities, having already decided to enter this field".

This reasoning was more specific than that used by France in replying to the Belgian declaration on the revised programme, which opened the meeting of the Special Group. In fact according to Bertrand Goldschmidt, it appeared that the situation in which Eurochemic found itself, with a structural deficit, would not allow the plant to continue in operation. Indeed in the heat of the discussion, Bertrand Goldschmidt said this situation was "the reason why certain shareholders have had to embark themselves on reprocessing irradiated fuel on a commercial basis"[63].

He believed therefore that it was essential to evaluate the costs of closure and thus produce a technical specification for liquidation as quickly as possible. He suggested that this should be used as a basis for an international call for tenders and that an outside contract should carry out the work under NEA supervision, which would allow the shareholders to profit from the technical and economic experience that would be acquired.[64]

As far as funding was concerned, France – which approved the programme – again made its position very clear following the rejection by Spain.

Eurochemic had run its course and fulfilled its task – incidentally being limited to its R & D objective – and European solidarity should now be expressed in joint funding of the shutdown[65]. "Eurochemic had broadly fulfilled its role by considerably improving the knowledge of reprocessing techniques in Europe". The continuation of industrial activities would require substantial investment, which "would have been paid by taxpayers in the participating countries, to the benefit of the electricity producers". Therefore it was now appropriate to face together, in a spirit of international co-operation, "the financial obligations resulting from the termination of activities and the closure of the plant".

If this solidarity could not be found "owing to the defection of certain countries", "the contribution of France would be limited to the most urgent operations, i.e., to the completion of the reprocessing services to which the company was already committed. However the capital expenditure and running costs relative to the processing of radioactive wastes should be postponed until a definitive agreement as to how this expenditure should be covered could be found within the OECD framework".

---

63 Ibidem p. 5. Claiming that Eurochemic's budgetary deficit was the main cause of the signature of UNIREP is at the very least exaggerated.

64 One may ask who, if not the French or the British, would have been able to assemble the necessary technical capabilities.

65 A/M(71)3 p. 7.

This threat with regard to funding made by one of the two main shareholders and sponsors of the undertaking provoked responses from the Chairman, the Belgian delegate and the Director-General of NEA.

Violent reactions from Switzerland and Spain

The Swiss position had been made clear in a letter of 3 December 1971 sent by the representative of the Swiss delegation to the OECD to the Deputy Director-General of ENEA, giving the reasons why the Swiss Director was unable to assume the chairmanship of the Board of Directors as had been proposed at the end of the Board's meeting on 19 November 1971: "the Swiss government has done its utmost to support the development of Eurochemic by providing, amongst other things, financial support greater than that commensurate with its real interests as a country not considering the construction of a reprocessing plant on its own territory. The direction taken by Eurochemic activities from March 1970 on has departed even further from the objectives needed to make the plant as viable as possible. In particular it became clear that *Eurochemic would be unable to obtain the equitable position it should normally have secured on the European reprocessing market.* This situation is fairly remote from the spirit of solidarity that is a feature of other international undertakings in the field of research and development. The Swiss government, which feels in no way responsible for this state of affairs, believes it must therefore adopt a strictly commercial standpoint...It therefore takes the present opportunity to announce its wish to withdraw from the company as quickly as possible".

The declaration of Switzerland's wish to withdraw was not translated into fact but does reflect the state of mind of the supporters of international reprocessing with regard to what they considered to be betrayal, even if the diplomatic language attenuated the strength of feeling.

Spain did not beat about the bush in explaining why it rejected both the programme and the funding arrangements. In the Special Group Spain "considered that the co-operation begun in Eurochemic had failed because certain special interests had predominated over the common interest"[66]. At the meeting of the Board of Directors held on 19 November 1971, Spain had been even more outspoken[67]: "Eurochemic has provided the information sought on many aspects of reprocessing and the training of specialised personnel, in accordance with the purposes for which it was created.

The recent commercial reprocessing agreement concluded between Germany, France and the United Kingdom practically eliminates consideration of Eurochemic as the basis of a European system of scientific and technical co-operation in this field, and totally eliminates its use at industrial and commercial levels [...] As a result it seems appropriate to take an acceptable position in respect of future liabilities in case of approval of the proposed programme. This position takes into account the efforts made in erecting and building the Eurochemic facilities as well as in covering the deficit produced in the past years, in addition to the interruption of the international co-operation system once the reprocessing services enter a commercial stage".

At the 1972 Special Group meeting, the delegate for Spain concluded that "in these circumstances, he had asked that the company's activities should be terminated as soon as possible and that no new commitments be entered into. Moreover he believed that the R & D programme had no value since it had been decided to terminate co-operation in the field of reprocessing".

Support from Belgium for continuing reprocessing on its territory

In a statement attached to the minutes of the Special Group, Belgium recalled the opposition to closure she had expressed at the meeting of the Board of Directors held on 19 November 1971[68]. She believed the decision had been taken because a number of shareholders "had in fact let it be understood that they were not interested in co-operation which was essentially of benefit to the two main shareholders and their new partner, since these bodies had declared their intention of organising the market and co-ordinating both their commercial policy and their projects"[69].

Belgium recalled that she was not persuaded "that the solution adopted...was the best one, particularly as there were not very many installations effectively processing oxide fuels in Europe or even in the western world". Belgium therefore recommended that the period up to the planned closure should be used in order "to find a more constructive solution"[70]. She believed that all the costs of closure would be a matter for the company and doubted whether the company had the necessary reserves.

66 Ibid, p. 6.

67 CA/M(71)3, Annex I, p. 12.

68 CA/M(71)3 p. 3.

69 NE/EUR/M(72)1, p. 9.

70 Ibid.

She thought the precise costs that would result from the closure decision should be worked out and expressed the wish "that the costs of liquidation in case activity should be continued on a new basis should be evaluated". She intended to carry out a "broad information programme [...] aimed at defining a possible programme and at finding new co-operative arrangements to carry it out", possibly involving the establishment of a new takeover company.

Belgium thus set out the broad lines of its position in the negotiations in which, if the company were to be liquidated, it would be involved in as host country, by virtue of Article 32 of the Statutes. She would examine "any proposal that would result in the continuation or extension of the installations' activities". If it appeared impossible to continue operations, "the Belgian authorities would take their share of responsibility as a shareholder and would consider the proposals made for disposing of wastes and decommissioning the main installations. [They] would consider the possibility of taking over certain auxiliary installations".

Finally Belgium pointed out that the sale of shares by members not wishing to remain in the company after 27 July 1975 should mean, since "the value of the latter (i.e., the shares) was negative as things stood", purchasing an "authority to leave" rather than by "receiving the value of its shares"[71].

However the Belgian position set out at that time evolved considerably up to the signature of the 1978 Agreement, the first negotiations for which began the same month[72].

The prospect of closure meant that Belgium would lose a reprocessing plant on its territory. It also deprived Denmark, Italy, the Netherlands, Norway, Portugal, Spain, Sweden and Switzerland of a plant and these countries, in November 1972, "expressed their interest in the study proposed by Belgium[73], for the continuation of reprocessing activities, while expressing some doubts about the existence of new factors likely to change the situation". Portugal recalled "that it had continued its co-operation in Eurochemic with a view to embarking upon the industrial reprocessing of irradiated fuel in a European context"[74]. With this co-operation coming to an end, Portugal decided to reduce its contribution for 1973 and 1974 to a lump sum of 10 000 EMA/UA. Germany and the Netherlands believed "that in any event these activities should be continued in a new legal framework". Denmark, Sweden and Switzerland recalled their support for the continuation of international co-operation up to the industrial phase, but "having regard to the current situation, they were ready to adhere to the proposals...concerning the shutdown of the plant"[75].

NEA embarrassment and divergences within the undertaking[76]

The NEA

Because of the differences between its member countries, the NEA played a very discreet role in the crisis which ended Eurochemic's first life.

There was no broad front of support within the Agency for the undertaking to continue its reprocessing activities. Neither Einar Saeland, the Norwegian who had succeeded Pierre Huet as Director-General, nor his then British Deputy, Ian Williams, were favourable. The former found it logical that the plant should close, whereas the latter was frankly hostile to the whole enterprise[77].

However Pierre Strohl obtained carte blanche to make final approaches with Pierre Huet, notably towards the UKAEA. These were unsuccessful.

Dissension within the company

The threat of closure which had been present since 1970 caused the Managing Director to take some personal initiatives to save his company, sometimes against the interests of the main shareholders on the Board of Directors.

---

71 Ibid, p. 11.

72 The main reason for the change was the wish to restart reprocessing in Belgium, a position maintained against all the odds. The concessions made by Belgium for this purpose were substantial.

73 For the study carried out by the ERSA group, headed by Pechiney, see Part IV Chapter 1.

74 NE/EUR/M(72)1, p. 6.

75 Ibid, p. 5.

76 Communications from Pierre Strohl and Pierre Huet.

77 This hostility was of long date. Rudolf Rometsch had crossed swords about reprocessing prices. According to him, this situation got worse following the publication of an article in *Atomwirtschaft* in which, answering complaints about subsidised prices at Eurochemic, Rudolf Rometsch stated that BNFL received as much if not more subsidy, simply in a different way, through long-term contracts with the military authorities.

Yves Leclerq-Aubreton[78] took as his starting point the fact that Eurochemic was at that time the only European plant capable of reprocessing oxide fuels[79]. He believed that this technological advance would allow the company to steal a march on its competitors and he wanted to play the international co-operation card for all it was worth. At the end of 1970, his strategy was threefold.

First of all it was a matter of finishing off the long-term contracts concluded when the reactors were built and which were supposed to provide the company, far into the future, with abundant markets[80]. The Mühleberg contract was typical of these, and Yves Leclerq-Aubreton made similar contacts with other enterprises.

The next step was to internationalise the La Hague plant[81] which, in his view, could ultimately have become Eurochemic 2. A first step would have been for La Hague to reprocess fuel covered by contracts signed by Eurochemic for substantial tonnages which the Mol plant was unable to process. This would have turned Eurochemic into a kind of services broker for La Hague. With this in mind he made contact with the CEA Director of Production, Jacques Mabile, who at that time was Vice-Chairman of the Board of Directors. According to Yves Leclerq-Aubreton, Jacques Mabile did not consider a subsequent internationalisation of UP2 to be totally out of the question. However Jacques Mabile was killed in an aircraft accident in January 1971, and this possibility was never taken any further.

Finally agreement had to be reached with the British. On 15 October 1970, Yves Leclerq-Aubreton signed a "memorandum of understanding" with the Director of UKAEA providing for the European market for oxide fuels not yet covered by reprocessing contracts in the years 1971-1974 to be shared. The memorandum allocated a tonnage of 124 tonnes over four years to Eurochemic compared with 126 tonnes over three years, as from 1972, for the UKAEA. Provision was also made that in the event of the association then being negotiated between the French, the Germans and the British actually being formed, "a more formal arrangement would take over."[82] This document appeared to indicate that the British would not oppose the survival of Eurochemic if the UNIREP agreement were to proceed.

However this strategy of the Managing Director somewhat overestimated his margin for manoeuvre with respect to the main shareholders of his company. In fact it short-circuited the national projects which had been running in Italy and particularly Germany for a number of years[83]. This became a problem once it was clear that the European market would shortly be divided up between the existing French and British plants, and that it was liable to put off the time when a decision would have to be taken about building a German plant. The German representatives considered the actions of the Managing Director as a threat to the construction of the large national plant they wanted. The growing dislike of the Chairman of the Board of Directors for the Managing Director was known to all[84]. The French, for their part, had also decided to continue reprocessing on a national basis[85] and under a trilateral agreement with the United Kingdom and Germany. In the reorganisation of the undertaking following the decision to shut down, the post of Managing Director was therefore suppressed[86].

To sum up, the closure decision resulted mainly from the predominance of national industrial aspirations which were clearly leading to a reprocessing overcapacity if the French, Italian, German and British plans were to

---

78 Interview on 3 June 1994.

79 A visit he made to Windscale on 15 October 1970 had convinced him that the British would not be capable of reprocessing oxide fuels in the immediate future, despite the announcement made in 1969 that their oxide head-end unit had been put into service. In fact until the accident of 26 September 1973 which was to render the plant's oxide head-end unserviceable for five years, before a fire finally closed the building down, the Windscale plant reprocessed only 130 tonnes of oxide fuel although it had declared a capacity of 400 tonnes a year. In the United States, the West Valley plant was also capable of reprocessing oxide fuels, but under conditions that were highly dangerous for its staff, because the average dose had exceeded the legal limit of 50 mSv/year since 1970.

80 See Part III Chapter 2.

81 The managers of UP2, which experienced operational problems during its first few years owing to the rapid increase in the burn-up of the fuel it had to reprocess, took a great interest in the work going on at Eurochemic, unlike their colleagues from Marcoule. There were excellent relations between UP2 and Eurochemic.

82 MEMORANDUM OF UNDERSTANDING (1970), p. 1, § 3: "this understanding may be overtaken by a more formal arrangement if and when a European Reprocessing Organisation of the type presently under discussion between the CEA, the German Ministry of Science and the UKAEA is formed."

83 The WAK pilot plant came into service on 7 September 1971.

84 The written sources naturally say nothing on this point, but the oral reports are many and unanimous.

85 Highly revealing, although post-dated, is the analysis done by Jacques Couture on the "reprocessing crisis", for which Eurochemic is held jointly responsible with the British. The decline in prices compared with the actual processing cost estimated by him at 90 dollars/kg, was due to "the unbridled competition between Windscale and Eurochemic", COUTURE J. (1975), p. 32, and to "the ruinous competition between pilot installations", ibid. p. 34.

86 See below.

be completed on schedule. In this context, the continuation of reprocessing at Eurochemic could only postpone the time at which this overcapacity would be absorbed and in the meantime the three main European powers agreed to share the market between them.

Eurochemic might still have saved its bacon by adopting a low profile, as envisaged by certain members of the Board of Directors and Technical Committee. The company might then have obtained support from EURATOM[87] and the IAEA[88]. However the aggressive strategy of Eurochemic's Managing Director ruled out this moderate approach, which in fact gives rise to doubt as to whether it would have been financially viable.

The planned shutdown of international reprocessing, together with the accident of 26 September 1973 which put the Windscale oxide head-end unit (installed in 1970) out of service for a long period, the delayed and slow commissioning of the oxide head-end unit grafted onto UP2 as from 1976[89], and the slow start-up of the Italian and German pilot plants[90], had a paradoxical outcome. This was the creation during the 1970s[91] of a real reprocessing undercapacity, in other words a situation diametrically opposite to that feared in the FORATOM report, which had led to the formation of UNIREP.

In any event the decision by the Board of Directors in November 1971 which led to the March 1972 closure schedule, itself endorsed by the Special Group in November, led to substantial changes in the undertaking as regards structures, commercial policy and R & D. It also speeded up execution of the contracts that were still in hand.

The shutdown decision: impact on the undertaking from 1971 to 1974

Scheduling the end of reprocessing

Structural reforms (Figure 102)

Meeting on 16 March 1972, the Board of Directors decided to adopt the principle of a far-reaching reorganisation of the company, since "the progressive cessation of commercial activities no longer justified a structure with a Managing Director, a Technical Director and an Operations Director". In fact a tighter management structure was organised. The post of Managing Director was eliminated on 31 December 1972. Yves Leclerq-Aubreton, dismissed with compensation, then left the company[92]. The resignation of the Operations Director Wilhelm Heinz was accepted[93]. Emile Detilleux became Director of the company and André Redon came back from Marcoule to be his Deputy, to help implement the remaining reprocessing programme and to carry through the programme of redundancies[94]. The restricted group was kept in being and appointed a permanent representative to the management, P. Dejonghe[95]. The Technical Committee "on which all the countries were represented" received more powers. There was an initial wave of redundancies during 1972, affecting 67 people or a little over one fifth of the staff.

A much simplified organisation appropriate to the reduced staff numbers was established on 1 January 1973 and lasted until 1 July 1974.

---

87 Communication from Jules Horowitz.

88 And Eurochemic had an ardent defender in the Vienna Agency in the person of Rudolf Rometsch.

89 Its true capacity went up only slowly.

90 EUREX was commissioned in January 1971 and WAK in September.

91 Only small quantities were reprocessed in WAK and EUREX.

92 He also left the nuclear field.

93 He then returned to WAK.

94 Following his first departure from Eurochemic, André Redon worked on the extraction of tritium produced by the Célestin reactor and on the Marcoule tritium plant project, for which he became the manager in the programme for producing the French thermonuclear weapon. The extraction of tritium is technically very different from the extraction of plutonium and uranium because it is done from gas and not from liquids, but André Redon believes that the major principles, and particularly the approaches built up through his long experience with PUREX reprocessing as regards safety and decontamination, remained valid and were useful to him.

He had the opportunity to make use of the experience he had gained in the labour relations field at Eurochemic during his second spell when he was called to UP2 as a "visiting fireman" to help overcome the serious crisis in industrial relations which shook La Hague when the establishment came under the management of COGEMA in 1976, an event which crystallized the unrest related to the worsening operating conditions in the plant since the beginning of the 1970s.

95 He was then the Deputy Managing Director of the CEN, and also the Deputy and Technical Adviser to the Belgian government's representative on the Board of Directors and the Special Group.

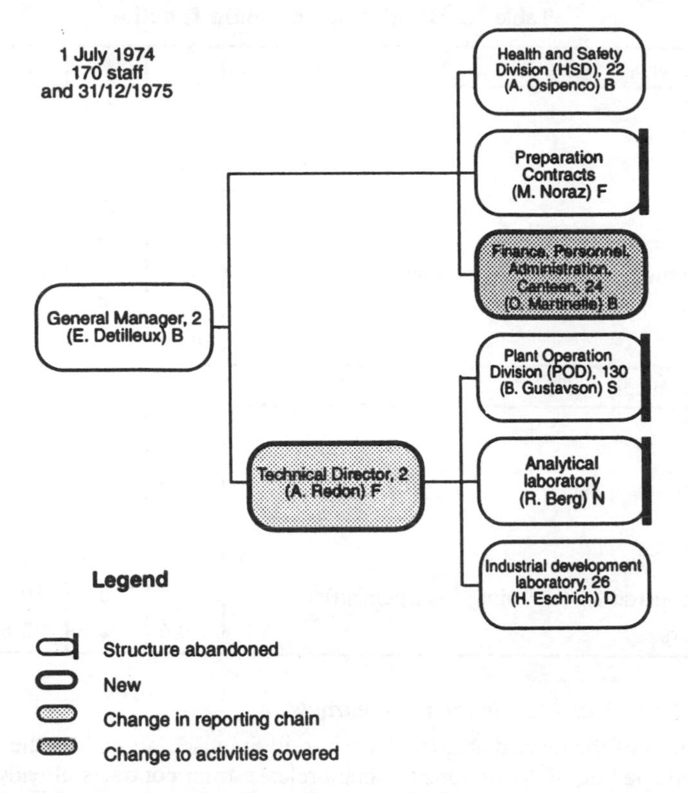

Figure 102. Organisation chart of the company from the beginning of 1973 to the end of June 1974 (Source: RAE 1973 and 1974).

Staff numbers remained steady throughout 1973 and the planned second wave of redundancies took place in 1974, bringing staff numbers to less than half of what it had been when reprocessing commenced. By July 1974 the size of the company was reduced by a quarter and it employed 170 people, while reprocessing was continued, causing some problems of organisation.

The company organisation shown in Figure 102 was focused on five basic functions described in Part III Chapter 2 and shown below in tables. There was also a small team monitoring contracts.

Analysis of the changes shows relative gains for the Health and Safety Division (HSD) and the Operations Division (POD), a depletion in the laboratories (ALD, IDL), fewer administrative staff, and primarily the sudden elimination in 1972 of parallel structures which had increased in number and which concerned 30 of the 67 departures (Table 56).

One of the main problems then encountered by the management was how to retain skilled staff in the company. Despite the system of rising premium payments as the end of reprocessing approached, plenty of staff were looking for work elsewhere. The example of the ALD provides a good illustration of the problems of running a department which is destined to disappear.

A PDP-8 computer had been used by the ALD from 1970. Automation of analytical and inspection procedures became a major concern as staff numbers fell. However simply replacing men by machines was not a panacea. Thus in 1973 it was found impossible to replace the experienced Mass Spectrometry Manager. With the end of reprocessing in view, the ALD was often called upon to help IDL in its work. The ALD disappeared in the restructuring of July 1974; 12 out of 26 staff left the company and what was left of the Process Control Laboratory was incorporated in the POD, the technicians in the Special Analysis Section being integrated in the IDL, now charged exclusively with jobs related to waste management.

Table 55. **Staff numbers at Eurochemic from 1972 to 1975 (as of 31 December)**

Year	1972	1973	1974	1975
Staff numbers	234	234	170	170
Change	-67	0	-64	0
Percentage change over previous year	-22.2	0	-27.3	0

Table 56. **Breakdown by main function**

1. Numbers concerned				
Year	*1971*	*1972*	*1973*	*1974*
HSD	30	22	22	21.5
POD	148	130	130	104
ALD	28	26	26	13
IDL	28	26	26	12.5
Administration (excluding management)	35	24	24	13
Others	32	6	6	6
Total	301	234	234	170
2. Percentage figures				
Year	*1971*	*1972*	*1973*	*1974*
HSD	10.0	9.4	9.4	12.6
POD	49.2	55.6	55.6	61.2
ALD	9.3	11.1	11.1	7.6
IDL	9.3	11.1	11.1	7.3
Administration (excluding management)	11.6	10.3	10.3	7.6
Others	10.6	2.6	2.6	3.5

Changes in commercial policy and termination of contracts

The implementation of the revised programme naturally imposed an end to the negotiation and signature of new contracts and the opening of discussions to obtain release from contracts already signed. Thus the Board decided not to reprocess fuel from the German Stade reactor, a contract signed in 1971, made no bids for the EURATOM ESSOR reactor, and suspended the negotiations in progress since 1971 with BNFL for reprocessing fuel from a Japanese test reactor on a sub-contracting basis[96]. All that was signed were annual agreements pertaining to the Swedish R3 reactor and the Swiss SAPHIR reactor, and a new contract in 1972 for a fuel charge from the nuclear ship Otto Hahn.

The management had the task of negotiating the transfer of contracts. For most of those in hand this did not lead to any particular problems. However the transfer of the long-term Mühleberg reprocessing contract to another firm gave rise to a dispute the answer to which was found after reprocessing had ended[97]. UNIREP certainly showed itself prepared to take over the reprocessing, but at much higher cost than the terms in the contract signed with Eurochemic. A clause in the contract provided that if Eurochemic abandoned the agreement, it would assume the difference in cost.

The planned cessation of reprocessing led to another dispute, when SFU was informed that Eurochemic was no longer able to operate the UF4 unit on its behalf, owing to the reduction in staff numbers.

However because of its dual nature, the company was forced not only to terminate the contractual obligations related to its commercial activities, but also to deal with the execution and funding of the legal obligations arising from its being both an international and a nuclear undertaking, particularly as regards ensuring the safety of the installations once they were shut down.

Financial arrangements for the cessation of activities

The Special Group was given an initial estimate of the total cost of terminating activities in the covering note of 29 August 1972[98]. For the post-1973 period, the estimated costs ranged from 27.5 to 33 million EMA/UA (1971), including 5 million for 1973-1974, this sum covering the solidification of wastes, staff redundancy costs and the repayment of loans. This assessment was subsequently proven to be much too low.

96 This fuel came from the JMTR at Oarai, managed by JAERI.

97 Under a contract signed with UNIREP on 21 February 1975. Eurochemic finally signed a compensation agreement for 100 million BF with Mühleberg on 26 May 1975 and the second tranche of 75 million BF was paid on 1 July 1976.

98 NE/EUR(72)1, p.20-21.

Initial moves towards waste treatment

Start-up of the EUROBITUM project[99]

Pursuant to the 1971 decision, work was started on the site for a bituminisation installation for intermediate activity wastes, the future EUROBITUM project. A design contract was signed with Belgonucléaire and Comprimo who formed a temporary association.

Refocusing of R & D policy of wastes

In conjunction with the closure operations, the research programme was refocused on intermediate and high activity wastes: as from 1972, the IDL conducted research into rendering fission products insoluble and on the operating conditions of the bituminisation installation.

However the greater part of the company's efforts were devoted to executing current contracts since it was essential that all should be discharged before the planned completion of activities on 1 July 1974.

Execution of outstanding contracts. Eurochemic as a reprocessing plant: February 1971 - January 1975 (Figure 103)

On 15 February 1971 the plant was restarted after an eight-month shutdown. In order to discharge the contracts and as a result of technical improvements made during the 1970-1971 shutdown, a virtually industrial rate of working was then adopted. The pilot plant was run like a factory.

The 1971 HEU campaign: the benefit and limitations of the shutdown (Figure 104)

The plant restarted with an HEU campaign which lasted until mid-July, reprocessing 12 tonnes of a fuel which had cooled for only 136 days and which therefore contained large quantities of fission products. The campaign went smoothly. Productivity was up sharply by 73% compared with the previous HEU campaign. For the first time, the installations of the second LEU cycle were used for the second HEU cycle. However an accidental alpha contamination incident meant that 30% of the final product had to be recycled. However prior to the refurbishment of the plant, 100% had to be recycled. Following the summer break, preparations were started for the LEU campaign. While system rinsing to allow use of the second LEU cycle lasted only one week, decontamination of the second plutonium cycle took a great deal of time and the plutonium unit could not be used for the LEU campaigns until 30 October, the initial tests having shown the unsuitability of the dry process. Subsequently its productivity was so low that planned operating times were extended by 50%. The second dissolver was not ready to accept low enriched fuel until the LEU-71-2 campaign.

The 1971-1972 LEU campaign: a full year's reprocessing (Figure 105)

There was a long LEU campaign lasting from 9 July 1971 to 12 July 1972, interrupted only by the summer and Christmas holidays and by the effects of an incident in the solvent processing unit. The programme included four sub-campaigns in 1971, involving the reprocessing of fuel from Lucens, Trino, NPD and SEP, and three sub-campaigns in 1972, with SENA[100], DP and a further Trino sub-campaign.

This campaign, the longest in the history of the company, generally went well despite a few problems which demonstrated – if this was still necessary – the inherent difficulties in reprocessing zirconium-clad fuel and the problems related to changing from one kind of campaign to another, rather than any technical shortcomings of the plant; it is true that the only purpose served by demonstrating the technical capability of the plant at this time was to feed the Belgian desire to continue operations. But it did point up the weak point of the system: the plutonium cycle, in particular the dry process.

Reception and storage

As regards reception and storage, cracks were found in certain SENA fuel assemblies delivered in 1972, so it was decided that these should be dissolved at once without intermediate storage.

Early in 1972, pond water purity gave rise to problems linked to unsatisfactory operation of the treatment systems, mechanical prefilters and resin beds; the water activity rose to 0.01 Ci/m^3 and the pond became fouled; the system was repaired and the pond cleaned out. The situation returned to normal in June.

---

99 EUROBITUM is the subject of a special review in Part IV Chapter 2.

100 Reprocessing of the SENA fuel led to occasionally animated discussions about the compatibility of the U-235 and Pu it contained. The views of experts from the CEA Cadarache and the BCMN at Geel had to be obtained, which revealed measurement problems which in fact were not confined to Eurochemic. The contamination of flasks also led to problems with Transnucléaire. On completion of reprocessing, the fuel could not be returned until 1977 owing to congestion both at La Hague and at Chooz. Source: SENA documentation.

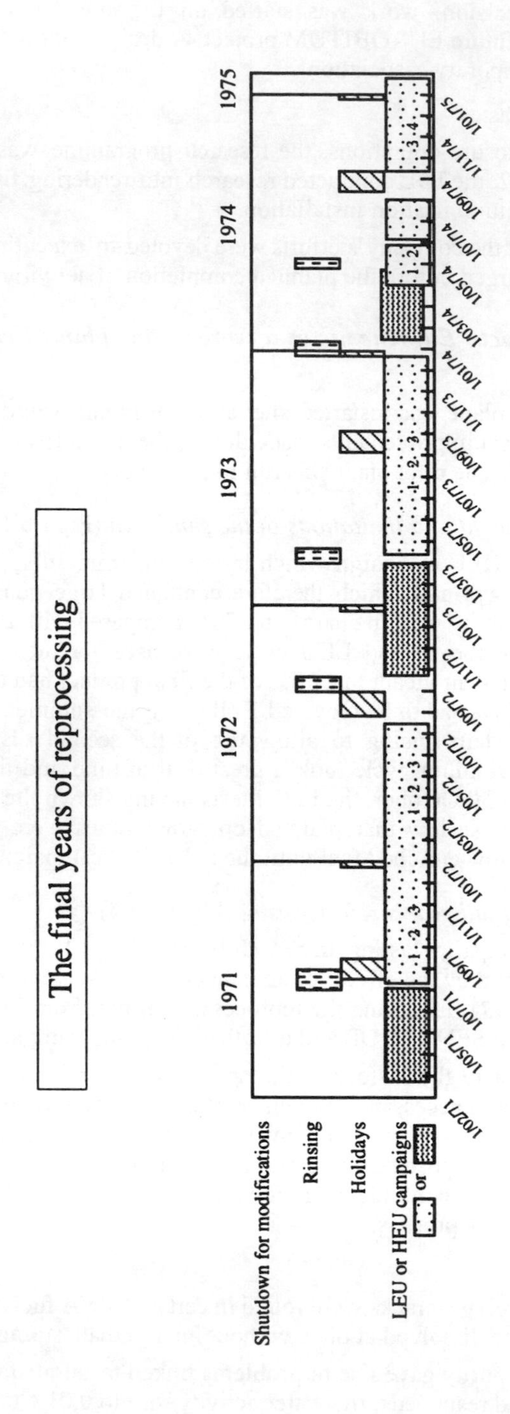

Figure 103. Load factor of the reprocessing plant from February 1971 to January 1975. The chronological chart below shows the series of reprocessing campaigns, numbered by year and sub-campaign, for reprocessing natural and slightly enriched uranium fuel (LEU). The three upper strips show plant closures for holidays and the shutdown periods with their causes, which were inter-campaign rinsing operations. Work was much more continuous than during the early learning years. There were no extended reprocessing shutdowns for reasons other than rinsing operations between campaigns (Source: RAE 1971-1975).

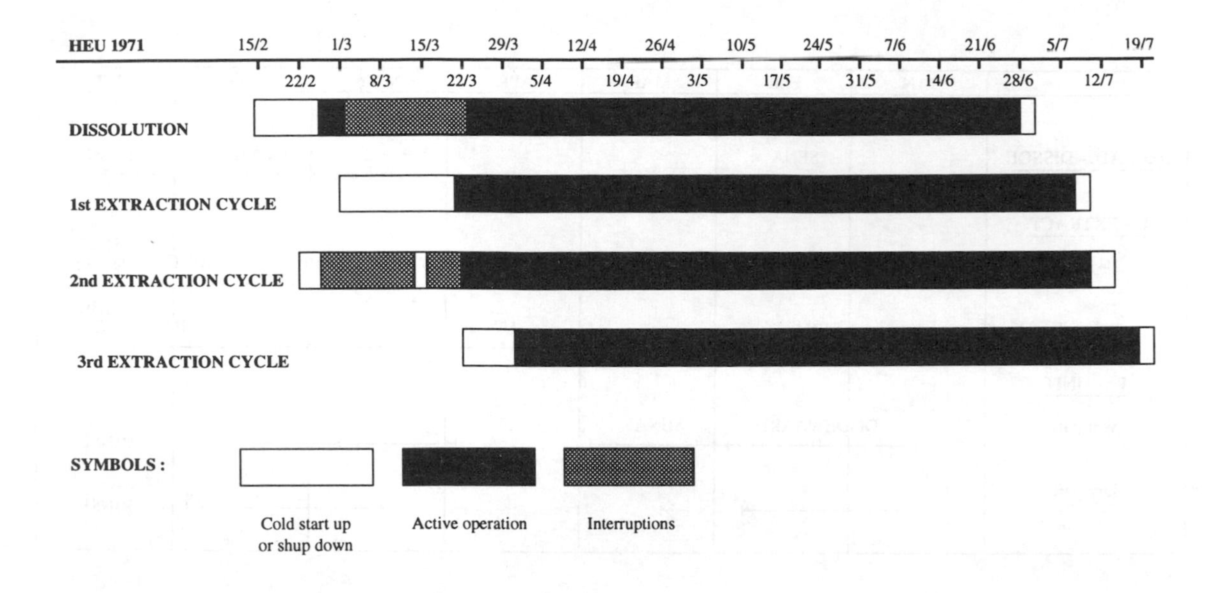

Figure 104. Chronological chart showing the series of operations in the different plant units during the reprocessing campaign for highly enriched uranium fuel in 1971 (Source: CA(72)1, Figure between pages 50 and 51).

Decladding and dissolving

At the head-end of the process, dilute sulphuric acid was used in 1971-1972 for the Magnox fuel from the Lucens reactor, the SULFEX process for Trino and SENA, and the ZIRFLEX process for NPD, SEP and DP. Decladding of the Lucens fuel was slow because the second dissolver was not available; the SEP fuel caused problems owing to its high Zr/U content and as a result of difficulties with the crane and the manipulators in the loading cell of the second dissolver, which caused a 20-day delay in dissolution. However the Trino and NPD fuels were dissolved satisfactorily. In 1972 the Trino sub-campaign was delayed by a further manipulator fault in the cell of the second dissolver, and by operating problems with the overhead crane. These problems were linked to the heavy use of the equipment for removing solid wastes that had remained in the baskets. Dealing with them was costly in terms of doses to workers' hands. It was decided that these systems should be duplicated.

The first and second uranium extraction cycles

During the 1971 restart, some difficulties arose from the rinsing and decontamination operations: the pipe transferring solutions from the third dissolver to the reception tank was blocked, and the HA column was unstable owing to the presence of Vantoc, one of the decontamination agents used. It took three weeks before the column again worked properly.

The SEP fuel caused problems during extraction, as it had done during dissolution, owing to its high zirconium content, and made two columns (HA and BX) unstable. The make-up of the solutions was quickly modified, whereupon the cycle again worked satisfactorily.

In February and March 1972, the output and quality of uranium extraction fell, owing to increased activity in the solvent and the TBP – its beta activity reached 2500 mCi/litre – necessitating systematic recycling which was done using the first cycle system. The cause of the problem was found following the blockage on 2 March of a pipe between two mixer-settlers in the solvent treatment unit. Attempts to unblock the pipe from outside lasted two weeks, but were in vain. Direct action had to be taken. The situation then returned to normal and the solvent activity fell to 10 mCi/litre. This blockage delayed the reprocessing programme by seven weeks.

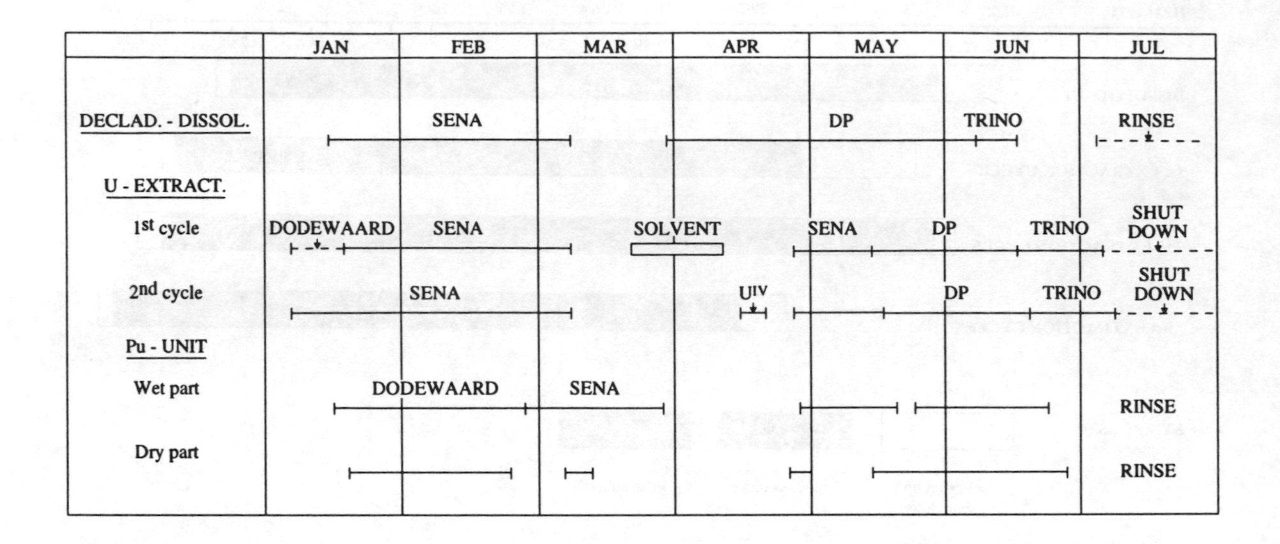

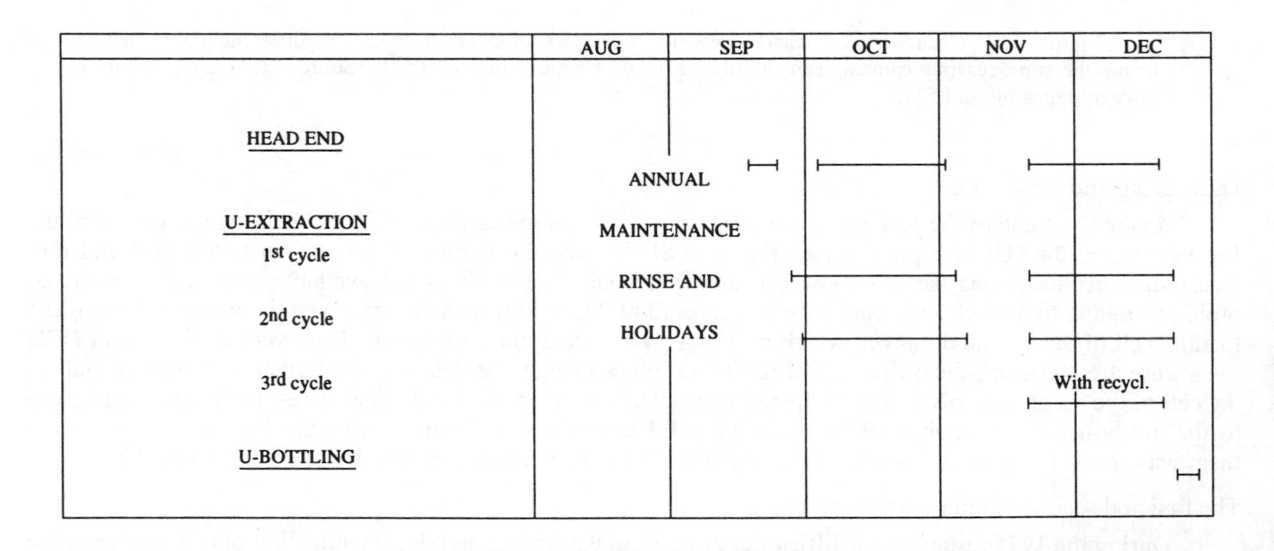

Figure 105. Operational programme for 1972. At the top: the reprocessing campaign for slightly enriched uranium fuel; At the bottom: the reprocessing campaign for highly enriched uranium fuel (Source: CA(73)1, Figure between pages 25 and 26, 34 and 35).

The problems of plutonium treatment

Plutonium treatment, which was possible only in the last two months of 1972, was still unsatisfactory.

As from January 1972, frequent recycling operations were necessary to lower the fission product content during the second plutonium cycle; in May 1972 a further leak was discovered in an evaporator, which could only be repaired during the plant shutdown for the summer holidays.

The dry cycle continued to cause problems. The system of valves between the precipitator and the calcination unit did not work properly, allowing the entry of material that was too wet, leading to breakdowns in the calcination unit and other problems throughout the circuit. The decision was taken to completely re-organise the dry process.

A leak in the low activity liquid waste pipe

The only incident in the operating history of Eurochemic which had any effects outside the plant occurred in 1972 with the leak from the line transferring low activity liquid waste to the Belgonucléaire installations at the CEN[101].

The 1972 - 1973 HEU campaign: improvements to the nuclear accounting methods

Following the summer holidays, a long HEU campaign began on 19 September, interrupted for the Christmas holidays on 19 December, restarting in January and ending on 10 March 1973. During this campaign 5.8 tonnes (1263 elements) of MTR fuel was reprocessed, containing 259 kilogrammes of highly enriched uranium.

The novelty lay in the application of a separate accounting system for the material for each reactor, which made it possible to refine security control by the method of isotopic correlation, which had been tried out in respect of a contract with the IAEA[102].

In general the campaign was satisfactory and its productivity was fully comparable with that of the HEU 71 campaign. However it showed once again that the series of LEU and HEU campaigns led to a certain number of problems, and that it would be therefore best to change the type of fuel as infrequently as possible.

In the start-up phase, aluminium elements containing no fuel were dissolved, to supply the initial solutions for the first cycle. This cycle began on 25 September, but was rapidly held up by the effects on the head-end of the process of problems encountered in the third cycle; a satisfactory work rate was not reached until 19 November, despite the reduction in yield of a column resulting from an air leak in one of the pulsation units, which was responsible for three-fourths of the uranium losses in the cycle.

The start of the third cycle was delayed, because the decontamination process at the end of the previous LEU cycle had been inadequate; the third cycle had to be rinsed once again, an operation which generated 46 m^3 of additional liquid waste; in spite of this the HEU produced proved to be to contaminated by plutonium. It therefore had to be decontaminated in a new unit using a new plutonium solvent[103], which was put into service at the end of 1972. The plutonium quality remained fairly poor despite the use of this new unit which was completed on 10 March 1973.

LEU 73: commissioning of the new plutonium purification unit (Figure 106)

A smooth campaign

The remainder of 1973 was taken up by an LEU campaign which saw the commissioning of the new plutonium purification unit.

Reception of spent fuel from SENA still caused problems owing to the fact that there were cracks in the cladding of most of the fuel elements. Only two could be placed in the pond; others were cut up in the pond for inspection under a joint CEA-CEN non-destructive inspection programme of irradiated pins.

Decladding and dissolving were satisfactory, despite the large amounts of solid residue from ZIRFLEX treatment of KRB fuel.

The uranium processing stage went smoothly. The operating conditions chosen for the new plutonium purification cycle meant that the plutonium could be concentrated without passing through an evaporator. However since the Trino and SENA fuels contained a high proportion of plutonium at the outset, the alpha contamination of the uranium was nevertheless fairly high.

---

101 This incident is analysed in detail in the incident review in Part III Chapter 4.

102 See Part III Chapter 4.

103 See RAE 72, p. 39. This was thenoyltrifluoroacetone, TTA.

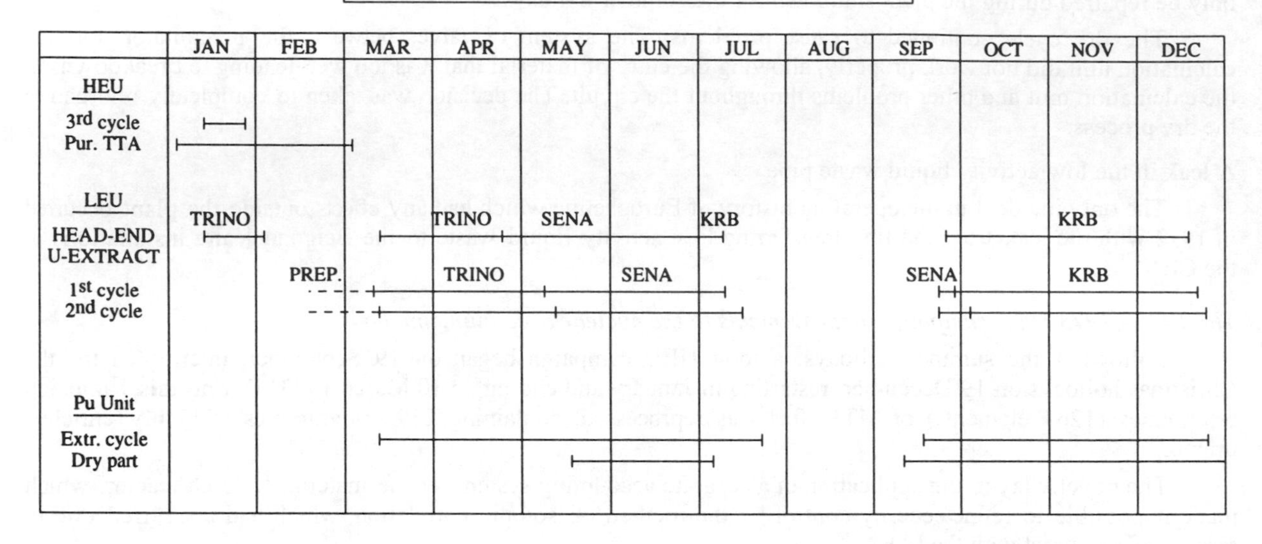

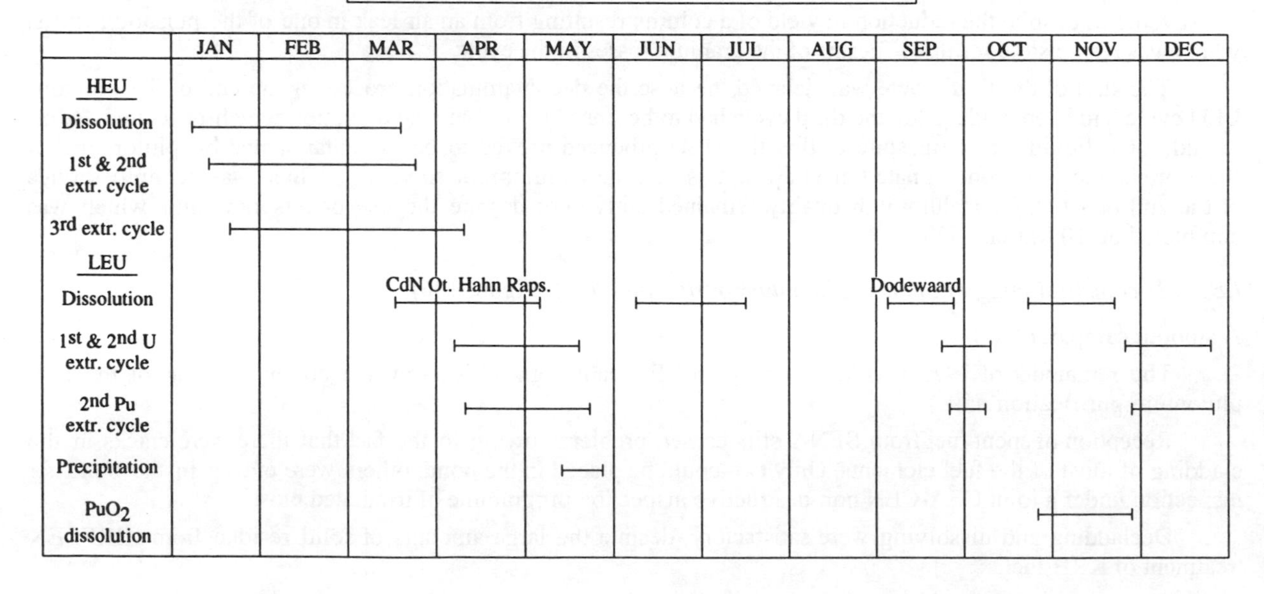

Figure 106. Operational programme for 1973 and 1974 (Source: CA(74)2, p.24 and CA(75)2 p. 23).

The campaign continued with no further incident and was completed in the first half of January 1974. The plant appeared to have reached its "cruising speed".

The major event of 1973 was therefore the commissioning of the new plutonium treatment installation.

Construction principles

The aim was to eliminate the problems that had been encountered: the third HEU cycle was now quite separate from the plutonium cycle, thus avoiding the rinsing operations and cross-contamination which had so delayed and perturbed the campaigns; the evaporator problems were a thing of the past because the solution was no longer concentrated; the mixer-settlers had been replaced by pulsed columns which were easy to use. In order to prevent problems in the dry process from interfering with operations in the wet line, an intermediate storage facility for 15 kilogrammes of plutonium had been planned. Prior to the precipitation and calcination of the plutonium oxalate, an anion exchange cycle was used to remove americium and, if necessary, carry out a further decontamination of fission products and uranium.

Two months behind

The installation started up two months behind schedule, owing to the time necessary for the adjustments that had to be made in the different component units.

One of the extraction columns was difficult to control. The operation of adjusting the plutonium valency using nitrous vapour did not give the expected results which laboratory studies had predicted. Instead a washing solution containing uranous nitrate and hydrazine had to be used; the solvent separator had been incorrectly designed; the reason for the delay was bubbles in the pipes and pump failures. However start-up, at reduced speed in mid-May 1973, went smoothly.

The anion exchange unit intended for purification purposes was started up without any problems on 12 May. However after a little over a month's operation, particles of resin and a plutonium-rich white precipitate appeared in the solution leaving the unit. It was shut down and taken out of circuit.

For precipitation and calcination, a rotary SGN filter similar to that used at La Hague in UP2 was installed between the purification unit and precipitator. This system was started up on 14 May; the filter was blocked by resin from the ion exchange system at the end of June, but the shutdown of the latter improved the situation.

Wastes

As far as wastes were concerned, there were two significant events. A new pipe was opened to transfer low activity liquid effluent to the CEN; this put an end to the system used since the 1973 incident involving transporting this liquid between the two sites in a tank mounted on a semi-trailer. Solid particles from KRB waste, some of which was tubing hardly attacked by the acid and which was stored in the pond, blocked the overflow pipe; water spilled over and contaminated the floor of the reception hall.

The final year of reprocessing: technical maturity for Eurochemic?

HEU 1974: a routine campaign

The final year of reprocessing at Eurochemic began with an HEU campaign which started on 14 January and lasted until 12 April. During this campaign 1338 elements containing 266 kilogrammes of uranium and 5.37 tonnes of aluminium were reprocessed. The consequences of the third cycle system having been used by the plutonium cycle until 1972 were still being felt and meant that additional rinsing operations had to be carried out before it was used. Otherwise the reprocessing campaign went off smoothly.

LEU 1974. The effect of staff reductions. A piecewise campaign

Shortly before the end of the HEU campaign, the final LEU campaign began: contracts required 11.8 tonnes of uranium to be reprocessed before the plant was closed. The current redundancy programme was affecting the POD and necessitated changes in the work plan; although fuel from EL3 (CdN), Otto Hahn and Rapsodie were reprocessed on a continuous basis, after the holiday period only two teams remained and SEP fuel was therefore reprocessed discontinuously, the staff operating one unit after the other in two cycles: decladding and dissolving from 2 to 20 September; temporary storage of the solution; commencement of uranium and plutonium extraction from 23 September to 11 October; storage of the plutonium nitrate; reopening of the decladding unit and dissolving on 21 October; and finally a new extraction cycle which was completed on 13 December.

Technically speaking, the head-end of the process showed the following characteristics: Rapsodie fuel was dissolved using the SULFEX process without difficulty, but for the fuel elements from the Otto Hahn ship reactor, which were dismantled first because of their unusual design, the dissolving process was slow and had to be repeated, leading to significant losses in the effluent; ZIRFLEX was used for SEP and, as usual, generated substantial amounts of solid wastes; for CdN an attempt was made to dissolve the aluminium and the uranium oxide together using nitrate of mercury.

Uranium extraction went well, as did that of plutonium. During October and November, 9 kilogrammes of plutonium in oxide form, which had accumulated since 1966 since it was outside contractual specifications, were dissolved in a small dissolver installed in a glovebox, before being recycled in the plant.

All the plutonium nitrate produced in this way was converted into oxide during January 1975, by which time the other units at the plant had been closed down; on 24 January 1975 all reprocessing activity at Eurochemic came to an end.

Chapter 4

Assessment of reprocessing

It is important to conduct an overall assessment of the eight years of Eurochemic reprocessing to see whether the company did in fact meet its original objectives. The evaluation should also provide a measure of the "inheritance of the years of reprocessing", which those involved were soon to call the "nuclear liability", which had a considerable influence on the company's situation during the fifteen years between the end of reprocessing and its liquidation.

The assessment therefore covers three aspects: material production, the financial outturn, and technical results.

While material production was modest, the financial outcome is frankly disappointing.

The technical aspect on the other hand is extremely positive. Despite the small quantities of material reprocessed, Eurochemic acquired multi-faceted experience unrivalled in the reprocessing field.

The international undertaking also built up a fund of practical knowledge through the feedback of experience with incidents, radiation protection problems and fissile material accounting methods.

However in 1975 all these issues were in the past. It is necessary to conduct a separate review of the waste produced during the reprocessing operations. In fact while some of the wastes were disposed of as they were produced, the greater part, in terms of the quantity of radioactivity contained rather than their volume, was accumulated on the site and represented the company's inheritance of its years of reprocessing.

Assessment of reprocessing: production and financial aspects. Small quantities, limited income

Modest material output[1]

The four histograms given below (Figure 107)[2] and Figure 108 show how the company's production of nuclear materials developed.

During the period of reprocessing, the following quantities of fissile material[3] were returned to customers[4].

- about 185 tonnes of natural or slightly enriched (up to 5%) uranium (LEU);
- about 1360 kilogrammes of highly enriched uranium, containing between 65 and 75% of uranium 235 (HEU)[5];
- finally, around 650 kilogrammes of plutonium (Pu).

The plant was operated for 1366 days for the LEU campaigns and 583 days for the HEU campaigns, making a total of 1949 days over a period lasting 3124 days[6].

The "load factor" was therefore 62.4%, the remaining 36.7% of the time being taken up by holiday periods and outages related to plant breakdowns, repairs and modifications.

---

1 HUMBLET L. (1987).

2 Sources: RAE 3, RAE 1967-1975; ETR HUMBLET L. (1987). The results of LEU 71-72 have been broken down according to data from RAE 1971. However the results of the HEU 72-73 campaign, which overlapped the two years, are listed under the year 1972.

3 The estimates of overall quantities vary according to the sources, although not by very much. Also it is sometimes difficult to determine whether the quantities mentioned are those entering or leaving the plant. The maximum ranges are as follows: 181.3 to 188.43 tonnes of LEU, 644.4 to 671 kilogrammes of Pu, 1363 to 1364 kilogrammes of HEU.

4 In a few cases the products were sold to an addressee designated by the customer, such as Alkem for R3.

5 The original content at manufacture was 90 to 93%.

6 From 6 July 1966 to 24 January 1975.

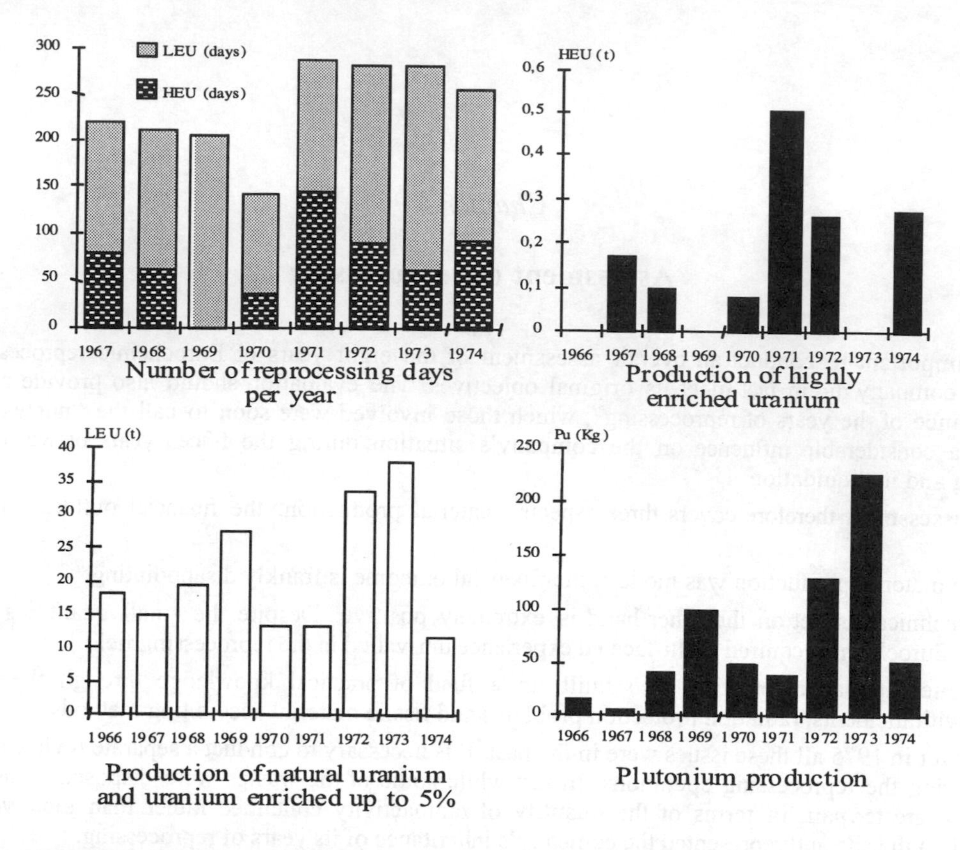

Figure 107. Development of the company's production of nuclear materials.

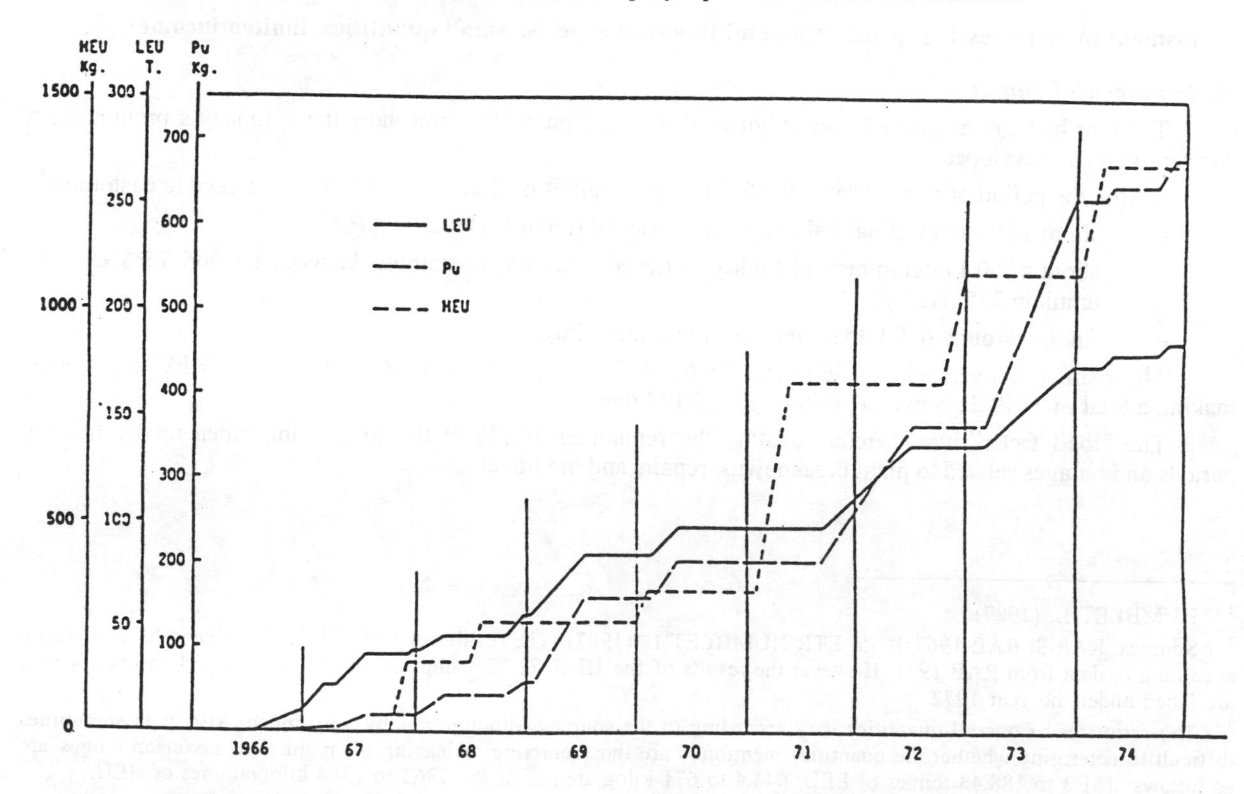

Figure 108. Cumulative production of fissile materials during the reprocessing period. Over half the production came after the interruption of 1970-1971 (Source: HUMBLET L. (1987), p. 13).

Daily output figures obtained by dividing the quantities produced by the total length of the campaigns[7] are 137.94 kg/day of LEU, 472 g/day of Pu, and 2.34 kg/day of HEU, in other words well below the capacities declared at the start of the project, which were 350 kg/day for LEU and 5 kg/day for HEU[8].

One of the main reasons for the poorer than expected performance is the proportion of campaign time that was non-productive, being taken up by rinsing operations between sub-campaigns, and by the work needed to switch from an HEU campaign to an LEU campaign.

To some extent therefore, low overall productivity was the inevitable price of having a multiple-capability plant and clearly shows the contradiction between experiment and production that was a permanent feature of the undertaking, defined as an "industrial pilot plant". In the specialised production plants, the specific constraints of a multiple HEU/LEU capability are by definition absent and those due to the sub-campaigns are much less significant owing to the processing batch sizes and to the relative homogeneity of their make-up in terms of composition, burn-up, structure, and so on.

However the maximum values achieved do give an idea of the potential performance of the plant: LEU 1967 and LEU 1972 achieved 180 and 170 kilogrammes/day respectively, cover wide variations in enrichment which was close to "natural" in 1967 and 5% in 1972, with the burn-up higher by a factor of almost 10.

Clearly therefore the productivity levels of the best campaigns do indicate undeniable progress in the use made of the plant capacity, but even the maximum values are well below the predicted values, which certainly erred on the side of optimism[9].

As far as plutonium is concerned, the calculation of the total duration of the LEU campaigns is a fairly arbitrary process. If instead we take account of the period during which the plutonium cycle was used, we obtain values ranging from 200 grammes in 1967 to 1300 grammes in 1973. These figures reflect the difficulties encountered in the plutonium cycle and the tail-end of the process, but also the efficiency of the installations that were commissioned *in extremis* in 1973. As regards HEU, the values fluctuated from one campaign to another, from 1500 grammes/day in HEU 1973-1974, an output value related to alpha contamination which required long processing, and 3100 grammes/day in the HEU 1974 campaign, which is still well below the declared 5000 grammes.

While the results in terms of productivity are average, the financial outturn is disappointing.

A disappointing financial outcome

Inspection of the company's annual income shown under the "nuclear services" heading on the balance sheet, of which the largest item was reprocessing income[10], clearly shows that the undertaking was not a commercial venture, contrary to what was hoped when the project began (Table 57).

Income for nuclear services totalled half a billion Belgian francs, accounting for only one-fifth of total income during the period 1966-1974[11]. Grants from the OECD covered 80% of expenditure. It was only very rarely, for example in 1971, that reprocessing income and grants for operating costs contributed equally to the company's income.

From the financial point of view, the industrial target which Eurochemic had been given – if not by the Convention then at least by some of its signatories who wanted their fuels reprocessed by Eurochemic on attractive terms[12] – had clearly not been achieved. Thus Eurochemic's failure as an international industrial

---

7 The figures used in the calculations are those of HUMBLET L. (1987), i.e., 188.3 tonnes of LEU, 1362 kilogrammes of HEU and 644.4 kilogrammes of Pu.

8 The output of plutonium could not be predicted because it depended on the characteristics of the LEU.

9 It appears that until recently there was a tradition of overestimating reprocessing plant capacity, or at least underestimating the difficulties in the way of achieving nominal output. For example at UP2 it took some 15 years to achieve nominal output. With a declared capacity of 400 tonnes a year of oxide fuel in 1976, its productivity was very slow to increase. It was 1980 before 100 tonnes a year were reached, and 1985 before the 300 tonnes a year were exceeded. In 1987 and 1989 however, annual output was above the declared capacity. The rise in output of UP3 was much quicker. Commissioned in 1989 with a declared capacity of 800 tonnes a year, the plant reprocessed a little under 200 tonnes in 1990 and nearly 450 tonnes in 1992. Source: BETIS J. (1993), p. 271 (graph). This kind of underestimate is widespread in chemistry for a new process not yet on an industrial scale, but in reprocessing it was slower than the sector norm.

10 It also included the costs of protecting nuclear materials, initial storage and the management of final wastes.

11 Under the Five-year Programme these services were supposed to contribute a third of total income.

12 This was the case for example of Switzerland.

Table 57. **Structure of Eurochemic from 1966 to 1976 (in million BF)**[13]

Year	Income for "nuclear services" (million BF)	%	OECD[14] contributions (million BF)	%
1966	16.38	9.8	150	90.2
1967	69.1	47.7	75.7	52.3
1968	39.4	20.3	155	79.7
1969	58.28	32.7	120	67.3
1970	28.45	11.3	223.75	88.7
1971	104.74	49.9	105	50.1
1972	80.61	35.7	145	64.3
1973	71.26	31.5	155.16	68.5
1974	32.84	10.4	282.23	89.6
1975	1.29	0.3	446	99.7
1976	2.14	0.6	345	99.4
Total 67-74	*484.68*	*22.1*	*1707.84*	*77.9*
Total reprocessing income	504.49			

reprocessing unit is obvious, but the reasons were not only internal. There is no doubt that it was speeded up by the British "dumping" policy which dragged reprocessing prices downwards and by the predominance of national approaches, extended through the cartel, over the international co-operative approach, which certainly deprived the company of markets. The slow development of power reactors with its consequent lowering of reprocessing demand also played a part[15].

However the reason for the industrial failure was to some extent present at the birth of the company. The primary objective of the plant was never to be an industrial unit, but to provide, on the scale of a large pilot unit, the experience that would be needed to build an industrial production plant. This particular technical objective was certainly achieved.

A highly positive technical outcome

In order to make a technical assessment of the company it is necessary to tackle two subjects: first the acquisition of know-how and, secondly, the accumulation of experience.

Qualitatively speaking, Eurochemic's success is undeniable. The plant reprocessed virtually every type of fuel used in the different types of reactor in member countries, and tested new ones[16]. In this way it built up considerable know-how. The technical outcome is also significant in quantitative terms, when it is expressed not in terms of tonnages but as a proportion of the reactors existing at the time.

Eurochemic reprocessed a wide range of fuels from the principal reactor types in member countries

The major types of reactor in Europe during the 1960s

To begin with it is necessary to recall some of the main characteristics of the different reactor types developed in Europe in the 1950s and 1960s, highlighting their main features from the reprocessing standpoint[17].

---

13 Source: company balance sheets.

14 This only covers contributions to operating costs. Grants for equipment amounted to 70 million BF in 1971, 100 million in 1972, 63 million in 1973 and 120 million in 1974.

15 European nuclear power programmes finally took off after the first oil shock, when Eurochemic was in its last year of reprocessing. For the effects of this nuclear power "awakening" on reprocessing programmes, see Part IV Chapter 1.

16 Such as the "Cristal de Neige" elements, in uranium-molybdenum alloys containing up to 10% of Mo.

17 The best review of reactor development policy in Europe during the 1960s is in NAU H. R. (1974), despite its political and non-technical approach.

First of all it is appropriate to consider the research reactors, which varied widely in detail depending on their research objectives, for example the testing of materials, new fuels, reactor geometry, and so on. The quantities of fissile material involved were generally low, but the amount of enriched uranium could be very high.

As regards power reactors[18], the modern success story of the light water design in Europe and elsewhere in the world should not obscure the fact that in the 1960s the question of reactor design was not yet settled[19] and that various combinations of reactor characteristics were still being explored[20].

Thus very different reactor systems co-existed in the research programmes or power stations of the different countries of the OEEC, all with their national and international champions, and all at different stages: experimental, pilot or prototype.

The main families were then the following:

- Research reactors. These were of two types, either tank reactors or pool reactors.
- Power reactors, including prototypes and commercial reactors. Within this group there were five families:
  - reactors using heavy water as moderator (HWR) on which there were many experiments in Europe: pressurised heavy water (PHWR), high temperature (HTHWR), with varied coolants, heavy water (HWCHWR), gas (GCHWR), organic liquids (OLCHWR), and so on. In Canada a specific system was developed known as CANDU (Canadian Deuterium Uranium);
  - graphite-moderated, gas-cooled reactors (GCGR), with two main types, the French natural uranium-graphite-gas reactors (UNGG) and the British version known as GGR or Magnox;
  - advanced gas-cooled reactors (AGR), derivatives of the GGR. A high temperature version was developed at Winfrith, England (the Dragon project), known as a High Temperature Gas-Cooled Reactor (HTGCR);
  - light water reactors (LWR), of two types, pressurised water reactors (PWR) and boiling water reactors (BWR);
  - finally, work was started on fast neutron reactors often known as fast breeders or FBR.

---

[18] The military aspect of the plutonium-generating reactors excluded them from the range of reactors whose fuel was available for reprocessing, but it must not be forgotten that the first power reactors were also military or derived from military types.

[19] However there was a significant event at the end of the 1960s: the conversion of France to the light water system in November 1969.

[20] Technically speaking there is an infinity of possible combinations for a reactor design, the variables being as follows:

- fuel composition: apart from thorium or a uranium-thorium mixture, this may be natural uranium (containing 0.71% of the 235 isotope) or enriched uranium (with more than the natural content of uranium 235, possibly up to 100%), or a mixture of uranium and plutonium (such as the mixed-oxide fuel or MOX);
- the chemical form of the fuel: metal, oxide, alloy (for example U-Al for the MTR);
- fuel cladding material: this can be magnesium, aluminium, an aluminium-magnesium alloy, stainless steel, zirconium (with niobium), zirconium alloys known as zircaloy, and pyrolitic carbon. Alternatively there may be no cladding at all;
- the physical arrangement of fuel and cladding: homogeneous in the form of balls or granules, heterogeneous in the form of bars, pins, rods or pellets, assembled in tubes or in bundles of elements;
- the moderator, which often gives the name to the reactor type: light or heavy water, graphite, beryllium. Also there may be no moderator at all;
- finally the coolant: boiling or pressurised light or heavy water, gases (carbon dioxide, helium or quite simply air), and sodium or potassium.

Of course some combinations are unacceptable for technical reasons but there are finally several tens of thousands of possibilities. Although power reactor designers have opted for the PWR system with slightly enriched uranium pellets clad in zircaloy or stainless steel and arranged in bundles, the variety of designs and materials was considerably greater and more varied in Europe during the 1960s.

The reasons why certain reactor types were adopted in preference to others are a combination of the technical, economic and political factors. At the beginning of the 1960s, when Eurochemic had to decide which type of fuels it would process, no decisions had yet been made and Eurochemic therefore reprocessed virtually all of them.

Qualitatively speaking, Eurochemic was able to reprocess fuel from virtually all types of European reactor

The table given in Figure 109 shows how the reactors whose fuel was reprocessed by Eurochemic fit into the major families of reactors existing in Europe (and Canada).

The pattern shows that Eurochemic reprocessed fuel from most existing reactor types, except for heavy water reactors cooled by an organic liquid, an approach explored by EURATOM at Ispra, and the British reactor designs, which is easily understood since the United Kingdom was not involved in the company.

The experience built up in this way prepared the company for all technical developments foreseeable at the time; in particular Eurochemic was the first European plant[21] to reprocess significant quantities of fuel from light water reactors, both PWRs and BWRs, i.e., those which were finally adopted in Europe.

Quantitatively speaking, Eurochemic essentially reprocessed fuels from light water reactors

From the standpoint of tonnages reprocessed, Table 58[22] shows the breakdown of materials recovered as a function of the types of fuel reprocessed, ranked in order of the amounts of uranium recovered.

Table 58. **Breakdown of materials recovered as a function of the types of fuel reprocessed**

Type	U (tonnes)	Burn-up (Mwd/tU)	Pu recovered (kg)	kg Pu/tU recovered[23]
I. LEU				
Pressurised light water reactor	71.4	3200 - 21 000	402.6	6.54
Heavy water reactor	69.4	50 - 15 000	104.7	1.51
Boiling light water reactor	29.5	3800 - 17 300	157.1	5.33
Graphite-gas reactor	7.9	900 - 1500	5.1	0.65
Fast breeder (blanket)	3.1	1300	8.2	2.65
Total natural or low enriched uranium	181.3	50 - 21 000	677.7	3.74
II. HEU				
Materials testing reactors	1.4 (90-95% U235)[24]	n.a.	n.a.	n.a.

The quantitative predominance of light water reactors, including pressurised water reactors, is clear. Most of the plutonium recovered came from these fuels.

It must also be pointed out that Eurochemic was the first plant to reprocess fuels of very high burn-up, up to 21 000 MWd/t for SENA[25].

Eurochemic reprocessed fuel from over half the reactors existing in member countries

The company's commercial policy was to obtain contracts from most of the major operators of reactors, whether these were of the research or power type, as shown in Table 59 which lists the contracts in chronological order[26].

Table 60 shows existing research and power reactors in each country with the reactors whose fuels were reprocessed by Eurochemic shown in italics.

---

21 The world's first plant was NFS at West Valley, but in conditions that were hazardous to workers' health.

22 VIEFERS, W., Wastes, Table 1. Total figures that are different from those of other sources.

23 Plutonium was not recovered in the HEU campaigns.

24 Alloyed with 29 tonnes of aluminium.

25 Fuels currently reprocessed at La Hague have burn-up values of the order of 33 000 MWd/t, expected to go up to about 40 000 MWd/t since the power generators, for obvious economic reasons, are tending to lengthen the time fuels spend in the reactors.

26 Taken from the RAE, together with information from the archives of the Board of Directors.

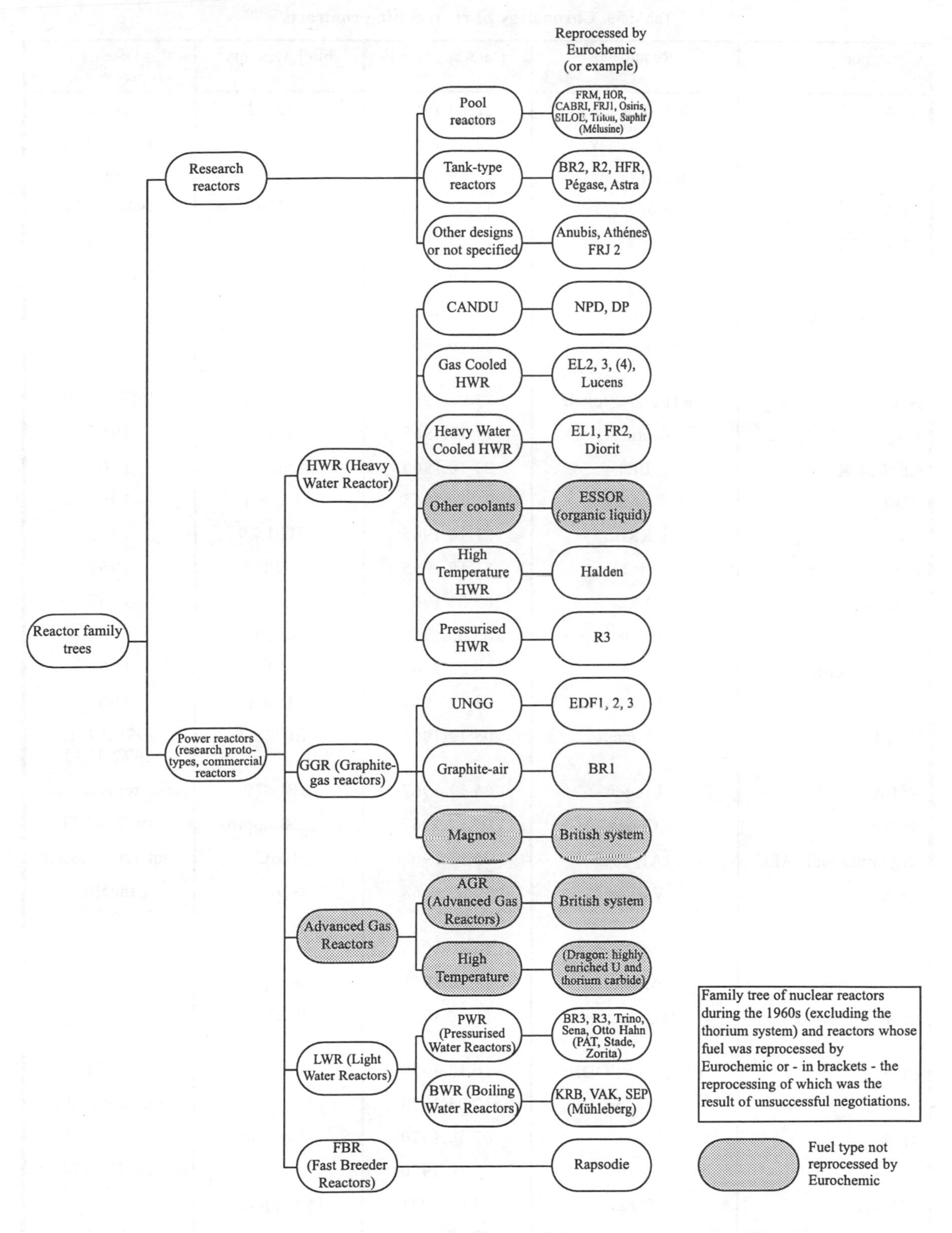

Figure 109. Existing reactor types and those from which fuel was reprocessed by Eurochemic.

Table 59. **Chronology of reprocessing contracts**

Customer	Reactor(s)	Date approved	Fuel type/qty (tonnes)	Date of reprocessing
CEA	EL1 (Fontenay)	22.09.1964	U nat./2	1966
CEA	EL2 (Saclay)	22.09.1964	U nat./9.5	10.1966 - 1.1967
CEA	EL3 (Saclay)	22.09.1964	LEU/8.5	1967
CEA	EDF1/2/3	17.03.1965	LEU/6	1968-1969
CEN/SCK	BR3	17.03.1965	LEU/2.8	1968
	BR2	23.06.1965	HEU/0.475	1967, 1971
EURATOM (Petten)	HFR	23.06.1965	HEU/1.7	1967,1971
CEA	Pégase, Triton, Mélusine, Siloe, Osiris	23.06.1965	HEU/3.8	1967
Nukem	FRJ1 and 2 Jülich	23.06.1965	HEU/0.5	1967, 1970
EIR	Diorit (1)	21.10.1965	U nat./6	1967
CEN/SCK	BR1	21.10.1965	U nat./1.9	1967
RWE	VAK	21.10.1965	LEU/6.1	1967
RWE	KRB	21.10.1965	LEU/14.9	1973
AB Atomenergi	R2	21.10.1965	HEU/0.6	1967
GfK	FR2	02.03.1967	U nat./6.7	1967
IFA	Halden	06.06.1967	U nat./1.3	1968
AB Atomenergi	R3	06.06.1967	LEU	1969
Öst. St.	ASTRA	05.12.1967	HEU/0.11	1967
ENEL	Trino	05.12.1967	LEU/28.4	1970,1971, 1972,1973
STEAG	Lingen	05.12.1967	LEU/70	not reprocessed
SENA	Chooz	05.12.1967	LEU/50 approx.	1972, 1973
Argentina Nat. AEC	RA1, RA2	05.12.1967	HEU	not reprocessed
CEA	Various	29.02.1968		cancelled (no penalty)
EIR	Diorit (2)	15.10.1968	U nat./7	
Belgonucléaire	Waste	15.10.1968	"scraps"	
BKW (FMB)	Mühleberg	16.10.1969	LEU/87.5 (1975-1982)	not reprocessed
CEA	EL4, RAPSODIE	16.12.1969		1974
	Lingen	17.03.1970		not reprocesed
SNA	Lucens	07.10.1970	U nat./6t	1971
	SEP	07.10.1970	8 t	1971, 1974
STEAG	Stade	24.03.1971	15 t approx.	
GKSS	Otto Hahn	27.06.1972		1974
EIR	Diorit (3)	27.06.1972		CEA subcontract

Table 60. **Existing research and power reactors in Europe and reactors from which fuel was reprocessed by Eurochemic (the latter in italics)**

Country	Research reactors[27]	Power reactors[28]
Germany	*FRM*	Ohrigheim
	FRG1	*Otto Hahn*
	FRG2	*VAK*
	FR2	*KRB*
	FRJ1	Lingen
	FRJ2	MZFR
Austria	*Astra*	
Belgium	BR1	*BR3*[29]
	BR2	
Denmark	DR2	
	DR3	
Spain	JEN1	José Cabrera
		Santa Maria de Garona
France	*Cabri*	*SENA*[30]
	EL1	G1
	EL2	G2
	EL3	G3[31]
	Mélusine	*EDF1*
	Néreïde	*EDF2*
	Osiris	*EDF3*
	PAT	St-Laurent 1
	Pégase	EL4
	Rapsodie	
	Siloe	
	Triton 1	
Italy	Ispra1	*E.Fermi/Trino*
	Essor[32]	Garigliano
	Avogadro	Latina
	RTS1	
Norway	JEEP2	
	Halden[33]	
Netherlands	*Athene*	*SEP*
	HFR[34]	
	HOR	
Portugal	RPI	
Sweden	R1	Oskarsham
	R2	*R3*
Switzerland	*Saphir*	Beznau
	Diorit	*Lucens*
Turkey	TR1	

27 Reference is made only to research reactors rated at 1 MWe or above, except for EL1, Cabri and HOR.

28 In service in 1971 except for Lucens, which was finally closed down in 1969 after an accident.

29 EURATOM.

30 Joint Franco-Belgian undertaking on French territory.

31 Military reactors supplying UP1 at Marcoule.

32 EURATOM.

33 A joint undertaking of the OECD.

34 EURATOM.

To produce this table it was necessary to construct a database of the research and power reactors that were in service between 1960 and 1971 in the countries participating in Eurochemic[35].

To consider the overall results[36]: out of 39 research reactors in thirteen countries, 24 had their fuel reprocessed at Eurochemic, 61.5% of the total. As regards power reactors, 12 of the 26 reactors in service – 46% – had at least one of their fuel charges reprocessed at the plant. Thus Eurochemic reprocessed fuel from a total of 36 European reactors from a total of 65, a little over 55%, to which must be added the fuel channelled by France from two Canadian reactors, NPD and DP.

During the 1950s and 1960s therefore, Eurochemic reprocessing was highly representative of the many technical approaches explored in reactor technology.

The second problem arising in drawing up a technical balance sheet is to determine the extent to which Eurochemic was able to reprocess satisfactorily. One answer to this question can be found in analysing incidents and in a review of the feedback of experience. Eurochemic experienced technical problems, like any undertaking trying out new processes on fuels of different composition and shape[37]. These must be considered in more detail and an attempt made to determine the extent to which they were resolved.

Incidents, incident analyses and Eurochemic's system for experience feedback

Incidents at Eurochemic and at other reprocessing plants. Some definitions and fixed points

The "in-house" definition of incidents

At Eurochemic, an incident was subjected to a special review when there was "an unusually high risk of irradiation, contamination or damage to workers or property"[38]. In 1967 there were 66 cases of external contamination affecting 79 employees.

Two of these were regarded as incidents. The definition given leaves room for a degree of subjective assessment[39].

Eurochemic and the International Nuclear Event Scale (INES)

The meaning attributed by Eurochemic to the term "incident" is entirely different from that used in the international scale of seriousness of accidents and incidents applicable to all nuclear installations, known as the International Nuclear Event Scale (INES), the adoption of which was recommended jointly by the IAEA and the OECD Nuclear Energy Agency in April 1992. This scale has eight levels in increasing order of seriousness, numbered from zero to seven, and also named as deviation, anomaly, incident, serious incident, accident without substantial off-site risks, accident with off-site risks, and major accident, the definitions of which, together with examples, are given in Table 61. Applying the INES scale, all the incidents at Eurochemic during the period of reprocessing were at level zero, one or two. From this standpoint, the plant was safer than many other reprocessing plants. It is true of course that it was not in service for a long period.

---

35 This database was constructed from data prepared by the OECD European Nuclear Energy Agency, ENEA Annual Report [ENEAAR] n° 3 (1961), pp. 217-246, ENEAAR n° 8 (1966), pp. 111-125 [1966 was the last year in which research reactors were included in the lists; the lists and statistics therefore do not include research reactors commissioned between 1967 and 1971], ENEAAR n° 12 (1970), no page numbers, final page of the report [this list includes only power reactors and prototype reactors], ENEAAR n° 13 (1971), no page numbers, final page of report. Later reports from the NEA did not include a list of reactors.

Only reactors in service between 1960 and 1971 have been referred to. The type categories used are those in the sources.

For more information on the reactors from which fuel was reprocessed by Eurochemic, see the Annex to the chapter on reactor characteristics, which was based upon other sources, notably the RAE and two lists published in the journal *Die Atomwirtschaft*, which gave additional information on certain reactors: AtW(1962), pp. 435-436 which, for the member countries of FORATOM, indicated the builders and suppliers of reactors, and AtW (1969), p. 83 which indicates fuel types and manufacturers for 1969.

36 Detailed consideration of the extent to which countries made use of Eurochemic's reprocessing potential would reveal a wide variety of attitudes.

37 The first and only industrial experience with ZIRFLEX and SULFEX.

38 RAE 1967, p. 22.

39 As an illustration (but not for comparative purposes owing to differences in formulation), the German WAK plant published, between 1986 and 1989 in the journal *Atomwirtschaft*, an annual list of "particular events" ("besondere Vorkommnisse") occurring in the plant, classified using a four-level scale specific to German legislation, based upon clearly measurable criteria. The results for 1989 for example are given in AtW (1991), p. 251-252. During that year the plant experienced 45 events, of which 39 were classified at level IV (the lowest), and 6 at level III.

Table 61. **The eight levels of the INES scale**

Level	Name	Criterion	Example
0	Deviation	No importance to safety	
1	Anomaly	Departure from authorised operating conditions	
2	Incident	A substantial fault but no suggestion of further failures; leading to a dose above the permitted limit or the presence of significant amounts of radioactivity inside the establishment	
3	Serious incident	Discharge outside the establishment exceeding authorised limits and, on the site, doses high enough to affect the health of workers or cause serious contamination; can – in the event of a further fault – lead to accidents	
4	Accident not generating substantial hazards off site	Limited discharge outside the establishment not requiring protective measures apart from the control of local foodstuffs. Substantial damage to the nuclear installation. Irradiation of one or more workers making early death probable	Windscale 1973 (reprocessing), Saint-Laurent 1980 (reactor)
5	Accident generating hazards off site	Discharge of radioactive material outside the establishment necessitating countermeasures to limit the serious effects to health. Serious damage to the nuclear installation with large quantities of radioactivity being released in the installation	Windscale 7 October 1957 (reactor), Three Mile Island 28 March 1979 (reactor)
6	Serious accident	Discharge of radioactive material outside the establishment, requiring evacuation of the population to limit serious effects to health	Kychtym 29 September 1957 (explosion of wastes)
7	Major accident	Discharge of large quantities of radioactive materials, acute effects to the health of the neighbouring population and contamination of a vast area (several countries), with long-term consequences for the environment	Chernobyl 26 April 1986 (reactor)

A few "events" occurring in other reprocessing plants

As examples, we may mention a few "events" affecting various reprocessing plants which are analysed in a remarkable book published by the OECD Nuclear Energy Agency on the safety of the fuel cycle[40]. These events are classified into four categories: criticality incidents (I), fires, explosions or exothermic reactions (II), leaks of radioactive material and instances of contamination (III), and finally the loss of electricity supplies (IV).

Table 62 is both a selection limited to reprocessing installations and a summary, setting out the major types classified in chronological order.

By comparison with the types of incident described above, Eurochemic experienced no criticality incidents nor losses of electricity supplies. It was affected by a few fires and limited exothermic reactions. It had many problems in the form of leaks and contamination incidents, although these were all in categories zero to two of the INES.

Incident recording at Eurochemic: the Health and Safety Division (HSD), the Incident Reporting System (IRS) and the Internal Project Applications (IPA)[41]

The roles of the Health and Safety Division and the Health and Safety Sub-committee.

The system for incident monitoring, prevention and control at Eurochemic during the period of reprocessing was operated by the Health and Safety Division.

The very nature of the company's activities and its experimental character meant that there was a strong possibility of incidents or accidents occurring, and special measures were adopted to minimise such events and to manage them in the best possible way if they did occur. This concern for prevention and management was reflected in the special position of the HSD and by the establishment, as soon as reprocessing commenced, of a Health and Safety Sub-committee. Chaired by the General Manager, this sub-committee met every Friday afternoon. Members discussed the events of the week, decided what improvements should be made, reviewed IPAs and decided what action should be taken, drew up incident reports and determined what corrective action should be taken.

40 OECD/NEA (1993a), pp. 214-235.

41 Written communication from R. Rometsch and W. Hunzinger referring to the Belgoprocess archives. Interview with Maurits Demonie.

Table 62. **Major types of incidents affecting reprocessing plants from 1959 onwards**

Type	Plant and location	Date	Analysis of the event
Criticality incident (I)	ICRP[42], waste tanks	16 October 1959	Critical excursion in uranium while solutions were being transferred from subcritical tanks to liquid waste tanks. Two persons were seriously irradiated. Operational error (unexpected siphon effect).
I	ICRP, reprocessing cells	25 January 1961	A pulsed air system used to unblock a pipe moved 40 litres of uranium 235 solution from a subcritical evaporator to a vapour discharge tube which was not geometrically safe. No victims since operation took place in a shielded cell. Operational error, poor appreciation of its consequences.
I	RECUPLEX plant (auxiliary installation for Pu recovery), Hanford	7 April 1962	Overflow of liquid from a pit, sucked out by a temporary pipe installed in a tank during decontamination operations. Excursion due to plutonium. No victims. Liquid not identified, unexpected consequence of an operation.
I	Windscale: installation for recovering Pu from waste	24 August 1970	Excursion due to plutonium during the transfer of a solution from a treatment tank to a dosing pot via a transfer pot (incident in the transfer pot). Related to the simultaneous presence of an aqueous phase and a solvent phase (the latter was unexpected and remains unexplained).
I	ICRP, first solvent extraction cycle	17 October 1978	During recovery of U from solutions containing highly enriched uranium, the fall in the concentration of the aluminium nitrate solution raised the concentration of U at the base of a column. No victim because the operation took place in a shielded cell. Failure in plant operations (management error, lapses in administrative checks and alarm instrumentation in poor condition).
Fire and exothermic reactions (II)	Kychtym: tank storing high level liquid waste	29 September 1957	Accidental breakdown in the cooling system to a tank containing a solution of HNO_3 and highly radioactive acetate, produced by military reprocessing since 1949. Explosion resulting from the rise in temperature leading to the discharge of two million Ci [pro mem: Chernobyl discharged a total of 20 M Ci] over an area 100 to 300 km long and 8 to 9 km wide. The central contaminated zone covered 1120 km^2. Cause: mechanical failure, faulty (or absent?) malfunction alarms, special composition of the solution (no PUREX equivalent).
II	Hanford, REDOX plant Pu treatment installation	6 November 1963	Fire in the sulphur building. Cause: burning of bichromate ion exchange resins containing Pu at a contactor. The resins had degraded and oxidised very quickly, which made them easier to burn. The incident was made worse by an error of judgement regarding firefighting methods (strict application of the instructions forbidding the use of water, even though water would have put out the fire in five minutes). No injuries.
II	Windscale, head-end building B-204 for oxide fuels (commissioned in 1969)	26 September 1973	Release of radioactive gas into the working area at the start of a campaign. 35 people were contaminated, essentially by ruthenium 106. Cause: an exothermic reaction in a feed tank between fission products containing large amounts of Ru zirconium fines and acidified BUTEX. This caused the decomposition of the BUTEX and perhaps the burning of the Zr. Emission of radioactive gases into the working area and, to a lesser extent, through the stack to the outside (10 Ci of Ru-106). Causes: poor understanding of the interactions between BUTEX, fission products and dissolution fines, inadequacy of the seals (for example at instrumentation penetrations) and of the system for ventilating and filtering the gases, inadequacy of the emergency arrangements (unexpected incident). Definitive closure of the head-end unit.
II	Savannah River, unit for converting uranyl nitrate into solid uranium oxide	12 February 1975	Thermal decomposition of a TBP-uranyl nitrate mixture added accidentally into a nitrate removal unit. Emission of NO_2 and combustible gases from the decomposition of the TBP. Explosion ejecting a major part of the content of the unit and fire. Two persons slightly injured. Repairs made and operations resumed in August. Causes: human error. Heating procedure slower than usual, probably leading to abundant and rapid discharge of gas. Inadequate analysis of components.

(see following page)

42 ICRP: Idaho Chemical Reprocessing Plant, the new name of ICPP.

(continued from previous page)

Type	Plant and location	Date	Analysis of the event
II	Savannah River, unit concentrating uranyl nitrate solutions	12 January 1983 or 1993?[43]	Explosion of an intermediate evaporator during the concentration of the fourth and final batch of uranyl nitrate solution being reduced. Two persons slightly injured. While the three batches totalled 1900 litres, the final batch was a "remainder" of 265 litres. It was necessary to add 600 litres of materials already evaporated in solution into the water to give the minimum evaporator charge. The explosion resulted from a combination of several factors: the equipment was in poor condition; an operating error; a temperature recorder was broken, hence the system was operated by guesswork by analogy with the three previous batches, resulting in an excessive concentration of the solution; partial blockage of a pulsing plate causing a rise in pressure; abnormal composition of the solution; presence of TBP in the aqueous solution of uranium nitrate.
	Tomsk[44] tank for adjusting solutions of U, Pu and PF	6 April 1993, 12:48 pm	Explosion of a 35 m^3 tank containing 25 m^3 of a mixture of 8.773 tonnes of U, 320 g of Pu together with caesium 137, niobium 95, ruthenium 103 and 106, strontium 90 and zirconium 95 (totalling 22.4 Ci of alpha emitters and 536.9 Ci of beta emitters). During adjustment of the solution by addition of nitric acid, which was not done according to the rules – the acid was too concentrated, the flow too high, and the tank stirring system was not operating – the pressure rose suddenly and caused first one tank and then a second to explode. Release of 7.5% of the tank contents. Discharge of 1.5 Ci of alpha radiation and 40 Ci of beta emitters outside the building, contaminating an area of the order of 200 km^2.
Leaks of radioactive materials and contamination (III)	La Hague, pond water treatment unit	2 September 1977	Overflow of a tank during an operation to unclog a filter, due to a valve not closing fully. Substantial contamination of the ground.
III	La Hague, Pu conditioning and storage unit	26 November 1977	Contamination of the air in the unit following an increase in pressure in and even the temporary overpressurisation of two twin gloveboxes used for conditioning plutonium dioxide (normally kept at a slight vacuum). Cause: equipment incident, when the bellows of a manipulator grab became detached from the manipulator penetration ring on one of the two gloveboxes, and unduly slow reaction of the pressure control system.
III	La Hague, HAO building, sampling system on a high activity process tank	3 February 1978	An air lift system was accidentally left running while its lower end was no longer in the liquid at the bottom of the tank, and transferred radioactive aerosols into the gas circuit. Error of design and/or construction (there should always have been a residual volume of liquid at the bottom of a tank to immerse the end of the air lift system), together with a lack of co-ordination between the laboratory teams and the operating staff.
III	La Hague, HAO building, pipe carrying liquid waste from the chemical decladding shop	10 May 1978	Leak at shielded valves whose flanges were no longer tight. Failure of the alarm connected to the liquid waste probe. Some 2 m^3 of liquid spread over the floor of the cell, which triggered the alarm.
III	Sellafield, Silo B 38 and Building B 701	15 March 1979	Discovery of fission products in the basement during a geological search following the discovery of a leak from a tank in a silo. Proof of other, old leaks which had contaminated the ground under the building (100 000 Ci over several years), but at some depth.

(see following page)

[43] The source indicates 1953. This is wrong for two reasons: in 1953 Savannah River was not yet in service, and the incident is placed chronologically between the previous Savannah incident and that at Tomsk. Two dates are possible because both lie in the interval between February 1975 and April 1993.

[44] See also *Le Monde*, 13 May 1993, p. 10, which refers to a study by the French Institute for Radiation Protection and Nuclear Safety submitted to the CSSIN.

(continued from previous page)

Type	Plant and location	Date	Analysis of the event
III	Sellafield, accidental discharge to sea of radioactive rinsing and decontamination solutions	10 November 1983	Some of the contents of a tank holding washing solutions resulting from the annual maintenance campaign on the B205 reprocessing plant was accidentally discharged to the sea through the pipeline for three and a half hours, until an alarm was triggered by the gamma radiation detectors on the pipeline. Some of the discharge containing solvents was washed up on the beach (1600 Ci discharged to sea, including 1214 Ci of Ru-106). The dry organic material which built up on the beach had a maximum activity of 270 mSv/h and the beaches were closed to the public. Cause: poor communications between the teams responsible for analysing the tank contents – who knew that they were mixed – and those responsible for the discharge, who assumed that the contents were homogeneous, since a sample taken from the bottom of the tank had shown a composition suitable for discharge, and who therefore did not consider leaving in the tank the supernatant solvent or the highly radioactive interface sludges.
Loss of electrical supplies	La Hague	15 April 1980	A fire caused by a short-circuit in an electrical cable destroyed the site's electricity distribution control room, two transformers and the electrical panels for the plant's four emergency diesel generators. The plant had no electricity for half an hour. The internal emergency plan provided for spare mobile diesel generators to be connected to the most sensitive parts of the installation, i.e., those where fission products were stored and where plutonium oxide was conditioned. When the ventilation system stopped, the controlled areas were evacuated as a preventive measure (the alarm was given through the internal battery-powered telephone network). The fire was brought under control in two hours and a special connection to the EdF electricity grid was made the same day. Operations restarted on 22 April.

The Incident Reporting System (IRS)

The experimental nature of the activities meant that incidents were likely, and called for close monitoring to ensure that incidents were understood and that appropriate corrective action could be taken. As far as off-site discharges were concerned, it was of primary importance to notify the Belgian authorities, as required by the law and the operating licence.

Thus the company held records of incidents which in 1986 were collated (but not published) for the period 1965-1979[45], upon which the following review relies heavily.

Internal Project Applications (IPA)

The Internal Project Applications (IPAs) were the mechanism whereby the analysis of incidents was translated into modifications to the procedures contained in the Operations Manual. There were two sources for the IPAs: operational experience leading to suggestions for modifications, and technical changes which were made when problems arose.

Thus all modifications had to be explained in an IPA which, before being appended to the Operations Manual, was verified at a number of levels; its application became possible only after it had been approved or modified first by the senior staff of the plant or the laboratory, and then by the HSD. The large numbers of IPAs in 1971 and 1972 were a direct result of feedback of information following the improvement works carried out during the major shutdown in 1970-1971 and of the accelerated pace of the campaigns.

Table 63. **Numbers of IPAs issued from 1969 to 1974**[46]

Year	1969	1970	1971	1972	1973	1974
Number of IPAs	43	37	56	44	21	16

45 VIEFERS W. (1986a) and (sd). The two studies were approved by the Deputy Director of Eurochemic (Eschrich) and were intended for publication in ETR. However they were not included in the publication programme which ran until 1990.

46 Source: RAE 1966-1974. This data was not recorded until 1969.

The system in action: the example of the evaporator leak in the second plutonium cycle in May 1968 and the prevention of a criticality accident

At the eleventh meeting of the Health and Safety Sub-Committee in May 1968, priority was given to the question of monitoring the concentration in the plutonium evaporator.

The Technical Director, Emile Detilleux, explained that the results of the weekly calculations of the amounts of plutonium entering and leaving the cell showed an inexplicable deficit of nearly 3 kilogrammes which was much greater than the normal proportion of losses. The IPA suggested a further check. An examination by an inspector wearing a protective suit (a "scaphander man") was decided upon and carried out immediately. The inspector, J. N. C. Van Geel[47], notified the Sub-committee that the plutonium evaporator was leaking and that a solid crust of plutonium nitrate had formed "like a wasps' nest" along one of the legs of the evaporator and that a further quantity of plutonium nitrate had accumulated in a sort of flat cake on the floor of the cell. Sent once again into the cell, J. N. C. Van Geel made sure that it was impossible for the two masses of nitrate to come together by separating them with a metal drip tray. The meeting was stopped and restarted at 10 p.m.

In the meantime, the criticality specialists met to consider the probability of an accident in view of the amounts of plutonium involved. Their conclusion was that no criticality accident and no additional contamination was to be feared since all operations in the cell had ceased and the cell was hermetically sealed. The Sub-committee therefore ended its meeting.

The following week the clean-up operation was started in close co-operation with the HSD. The specialist team at the Louvain University Hospital had been alerted. A technician wearing a protective suit, Maurits Demonie, recovered the nitrate in metal canisters, which were decontaminated outside the cell and then hermetically sealed by operators working in a temporary sealed airlock. Biological tests were carried out frequently and did not detect any plutonium ingestion by members of the team[48].

The incident reporting system makes it possible to establish a geography, typology and a causal analysis of the statistics of the incidents.

Incidents at Eurochemic

A total of 99 incidents were recorded during the period of reprocessing at Eurochemic. Table 64 shows the annual breakdown and compares the incidents with conventional accidents (i.e., those having no nuclear aspects) occurring on the site[49]. The so-called "learning" period saw most incidents: 51 out of 99 from 1966 to 1969 compared with 38 during the 1971-1974 "industrial" period. During the plant shutdown between May 1970 and February 1971 there was a marked fall in the number of incidents but a sharp increase in conventional accidents.

Table 64. **Statistics on accidents and incidents occurring at Eurochemic during the period of reprocessing**

Year	1965	1966	1967	1968	1969	1970	1971	1972	1973	1974	Total
Incidents	4	2	12	14	23	6	7	17	12	2	99
Conventional accidents	n.d.	n.d.	27	20	19	32	17	11	8	6	140
Total	n.d.	n.d.	39	34	42	38	24	28	20	8	239
% incidents/accidents	n.a.	n.a.	44.4	70.0	121	18.7	41.1	154	150	33.3	66.4

Geography of incidents[50]

The distribution by location covers the 99 incidents recorded between 1965 and January 1975. They highlight the hazardous areas of the plant (Table 65).

[47] The term "Scaphander man" was used for an operator equipped to enter an active cell and who therefore had to be completely isolated from the radioactive environment. J. N. C. Van Geel had just submitted his thesis in January 1968 to the University of Delft on *Recovery of Pure Plutonium*, based upon his work at Eurochemic. Today he is director of the Karlsruhe Transuranium Institute.

[48] Similar procedures and precautions were adopted during the replacement of the evaporator which was done after thorough decontamination of the cell and the equipment it contained.

[49] Source: RAE 1967-1974. The incidents during the period 1965-1966 are included in VIEFERS W. (1986), but RAE 3 gives no information about conventional accidents.

[50] VIEFERS W. (1986), p. 16 gives the raw data.

Thus over a quarter of the incidents were due to malfunctions in the plutonium purification unit, whose regular difficulties have been described in the reprocessing history. Two other areas account for a little under a quarter of the events: the treatment of low activity liquid waste, subject to numerous leaks, and the operations of handling irradiated fuel on arrival, due to cask contamination and mistakes in handling the fuel when it was extracted from the casks. These three sectors therefore saw nearly half the incidents.

By grouping these data according to the major stages of the process, it is possible to obtain an idea of which parts of the Eurochemic process were particularly hazardous (Table 66).

The tail-end of the process accounts for nearly 40% of the incidents and liquid waste treatment for one sixth. Thus over half the incidents affect the phase subsequent to extraction. It appears that problems in handling plutonium and uranium at the tail-end of the plant and liquid waste leakages were the plant's main weak points from the risk standpoint.

Table 65. **Location of incidents at Eurochemic**

Location	Number	%
Final Pu purification	25	25.3
Treatment of low level liquid wastes	12	12.1
Fuel reception and storage	11	11.1
First extraction cycle	8	8.1
R & D	8	8.1
Second U cycle	5	5.1
Final U purification	5	5.1
Concentration of fission products and acid recovery	5	5.1
Ventilation system (vessel and room)	4	4.0
Charging dissolvers	4	4.0
Analytical process control	4	4.0
Intermediate storage of low level liquid wastes	3	3.0
Decladding and dissolving	2	2.0
Final Pu storage	2	2.0
Treatment of off-gas	1	1.0
Final U storage	1	1.0
Rework	1	1.0
Treatment of intermediate level liquid wastes	1	1.0
Clarification	0	0.0
Solvent recovery	0	0.0
Storage of high level liquid wastes	0	0.0
Storage of solid wastes	0	0.0
Other[51]	-3	-3.0
Total incidents	99	100.0

Table 66. **Particularly risky stages of the process**

Process stage	Number	%
Pu and U purification and final storage	33	33.3
Liquid waste treatment and storage	16	16.2
Head-end of process	14	14.1
Extractions	13	13.1
Other (ventilation, process control)	12	12.1
R & D	8	8.1
Other aspects of the tail-end of the process	6	6.1
Not included	-3	-3.0

[51] The tables include a margin of error, expressed in this line, due to inaccuracies in the source: VIEFERS W. (1986). In fact VIEFERS covers all the incidents from 1965 to 1979. The statistics presented here refer only to the period of reprocessing and so necessitated a chronological selection and cross-referencing of the locations, pp. 16-17 with the description of incidents on pp. 20 to 70, together with an attached chronological list. In three cases it was impossible to relate the location to the date. The margin of error on locations is therefore a maximum of three cases, but the probability that all three affected the same location is low.

Cause of incidents[52]

Table 67. **Causes of incidents**

Nature	Number	Percentage
Human error	85	76.5
of which:		
Operational error	37	33.3
Mistakes during repairs	6	5.4
Other[53]	42	37.8
Construction defect	15	13.5
Non-classifiable	11	9.9

Incidents were rarely caused by poor quality equipment. Some three quarters were due to the "human factor", the reason being that Eurochemic reprocessing was only very slightly automated.

One-third of the incidents were related to procedural errors during operations. It is possible that language problems contributed to this high figure. Operating instructions were in fact prepared in English although the operators were largely Dutch-speaking and mostly had no English. It is always a delicate matter to fully understand instructions and it is in any case slower when they are not expressed in one's language, despite the fact that all the staff spoke a common basic English known as "Eurochemic English". Thus this high rate of incidents was the price paid for internationalisation.

Another factor was also the complexity of the procedures in a multi-purpose plant, where the detail of the process very often depended on the type of fuel.

However the relatively high number of errors during operations contrasts with the scarcity of incidents during repair and maintenance work. This is clearly a result of the great care taken in drawing up the HWP and PREPs.

Typology of incidents and case studies[54]

The predominance of discharge problems and leaks, some leading to contamination, is highlighted in the following table:

Table 68. **Types of incident and annual frequency**

Year	1965	1966	1967	1968	1969	1970	1971	1972	1973	1974	Total	% total
Discharges or emissions of radioactive material	3		1	5	3	1	1	5	5		24	24.2
Leaks			3	1	4	3	1	3	1		16	16.2
External contamination of persons		1	1	1	1	1	4	1	3	1	14	14.1
Contamination of objects and surfaces		1	2	2	3			1			9	9.1
Operating errors, mixtures of solutions			1	1	3		1	2	1		9	9.1
Technical incidents, operational defects					4	1		1	1	1	8	8.1
Blockages	1				3			3			7	7.1
Burns or injuries without contamination			2	2	1			1			6	6.1
Internal contamination of persons				2	1						3	3.0
Explosion			1						1		2	2.0
Fire			1								1	1.0
Total	4	2	12	14	23	6	7	17	12	2	99	100.0

52 According to VIEFERS W. (1986). Covers all incidents from 1965 to 1979, a total of 111. A single type of cause is attributed to each incident.

53 Other than in the plant: mistakes in laboratory handling operations, in the liquid waste treatment plant, and so on.

54 Typology prepared from the topical list of incidents recorded, VIEFERS W. (1986), non-paginated Annex, the topics having been refined from a reading of the description of incidents, ibid. pp. 22-70. Only one aspect of each incident is given. The table is ranked in decreasing order of totals.

Case studies

A few characteristic incidents are described and analysed below: evaporator leaks, pipe breaks, glovebox explosions, and internal contamination at the head-end of the process.

Leaks from evaporators: a cause of campaign delays

Leaks from evaporators were the cause of seven incidents. Two evaporator leaks were discovered in 1968 in the first and second uranium extraction cycles. In both cases, the solution leaking out evaporated completely, leaving only a solid deposit on the outside wall of the system. As a result, no liquid reached the collecting pot at the bottom of the cell so the alarm provided to indicate any unusual presence of liquid was not triggered. The units were considered repairable. The opportunity was also taken during the repair work to replace the alarms on the HEU intercycle evaporator – which had not detected the blockage of the system by uranium nitrate crystals – by a more reliable system.

A further leak due to local corrosion occurred during the LEU-1969 campaign on the uranium cycle evaporator; the evaporator could not be used during LEU 69-2.

Three leaks occurred on the evaporators in the uranium cycles and the plutonium purification cycle; the first occurred in November 1967, the second, dating from May 1968, has already been described in detail above. Two further incidents arose in May and June 1969, contributing to the decision to shut the plant down for repairs and modifications in the summer of 1969. One of these affected the uranium intercycle evaporator, the other the evaporator in the first plutonium cycle.

In the two 1969 cases, the cause was localised corrosion to parts which had been made from the wrong material, a connecting pipe and a heating tube penetration respectively. The fact that this material was not the required type 304 L stainless steel had escaped the quality control system.

Problems with liquid waste pipes led to the only incident having off-site effects

Low activity liquid wastes[55], including "Warm Waste" (WW) and "Cold Waste" (CW), and condensate – water produced by condensing the steam used for heating the process solutions which in normal operation was not contaminated – were transferred to Eurochemic's liquid waste treatment plant through an internal system and thereafter, through underground pipes, to the liquid waste treatment plant of the CEN, a distance of 1600 metres away, operated at that time by Belgonucléaire, where they were decontaminated and monitored before being discharged into the nearby river. The CW and WW pipes were of identical design so that either could be used in the event of a problem. They were made up of two concentric pipes made of ordinary steel, with the liquid flowing in the inner pipe. The outer pipe, which served as a protective barrier against leaks, was fitted at its lower side with sampling tubes at regular intervals to allow any leaks from the central pipe to be detected and located. The interspace between the two concentric pipes was filled with air kept at a pressure higher than the maximum pressure expected during transfer of liquid; any substantial leak from either pipe should reduce the pressure, trigger an alarm and stop the transfer.

Six incidents affected the system of low activity piping, one each in 1965 and in 1969 and two in both 1967 and 1972.

When the CW pipe within the site was commissioned in 1965 it became blocked as the internal protective layer of bitumen began to flake off. This lining had to be removed. However the principal difficulties were experienced with the WW pipe.

The problems began in August 1967 when corrosion caused a leak. Then on Friday 1 December 1967 there was a break in the pipeline connecting the industrial development laboratory to the Eurochemic liquid waste treatment plant. This pipeline was used to take away the WW generated by the laboratory in batches of 5 m^3, amounting to a total volume of about 1000 m^3 a year. The transfer was monitored by level detectors in the 5 m^3 reception tank. A liquid spout was observed indicating that one of the pipes had broken, so the transfer was immediately stopped. Some 5 m^3 of soil were very slightly contaminated. The total activity of the substances released in this way was under 100 μCi.

In September 1969 another leak occurred on the pipeline running to the CEN, and two further leaks occurred on 2 February and 1 May 1972. The first concerned the warm waste pipe: repairs were made and pipeline operations reversed so that the warm liquid waste was pumped through the pipe originally assigned to cold liquid waste. This modification accelerated degradation of the replacement pipe and resulted in the May leakage, which affected a portion of the CEN site located between the road and the canal.

This incident involved the release of under 15 m^3 of warm liquid waste containing 0.5 Ci of mixed fission products. The layer of sandy soil and some of the concrete from the road which were contaminated were removed

55 The classification of low activity liquid wastes is described in Figures 111 and 112.

and placed in six hundred 200-litre drums. An area of contamination estimated to represent 0.1 Ci was located on the canal bank, leading to restrictions being placed on fishing and boating and to the installation of a system for continuously monitoring the canal water, near the point where the pipelines crossed[56].

The two pipelines were then taken out of service and liquid waste transferred using a semi-trailer truck fitted with a 24 m^3 tank providing a twice-daily shuttle service between the Eurochemic site and that of the CEN. Complete reconstruction of the pipeline began in November 1972 and the new pipe was commissioned in 1973. This pipe, made of stainless steel, was inside a protective envelope consisting of an ordinary steel tube protected externally by a layer of bitumen. It was fitted with an improved detection system. There were no further problems with pipeline leaks; those which did occur had been due to the poor quality of a section of the installation regarded as of secondary importance.

A glovebox explosion: luckily a non-serious incident

Of the incidents that affected the laboratories it is worth mentioning that of 12 January 1973 involving the explosion of a shielded glovebox in the high activity laboratory located in the research laboratory. Caused by a technical failure, it fortunately caused no injury.

During a routine task – determination of a standard solution – the operator pressed the centrifuge brake button whereupon there was an explosion and fire which partially shattered the front upper perspex window and made large breaches on the side of the access corridor. The extraction box was destroyed together with the equipment in the adjacent boxes; dust from inside the box caused heavy contamination to the access corridor. However contamination in the laboratory itself was light and the building as a whole was not affected. By a fortunate chance no personnel were injured or contaminated. Analysis of the causes of the incident showed that the ventilation filters had become blocked, preventing the evacuation of organic vapours of hexone and acetone which had built up during the four earlier uranium extractions carried out in the glovebox.

Contamination on the dissolving platform: a human error of judgement following a mechanical problem

On 25 January 1968 an incident during dissolving affected the dissolver service platform adjoining the pond area. It resulted in personnel inhaling fission product dust while working on the basket of the second dissolver.

The removable bottom on the insoluble basket fitted to the second dissolver, intended for fuel enrichments of up to 4.6%, was opened at the end of the operation to allow the dissolution hulls to be emptied into a waste container, before a new batch of fuel was loaded for reprocessing. On 25 January the mechanism for opening the bottom of the basket jammed owing to a bent shaft. The basket was washed down with water in the loading cell of the second dissolver, reassembled and covered with plastic sheets.

The blockage problem appeared a simple matter to resolve and it was considered that containment was adequate since the basket was wet. The work team wore plastic clothing, boots, hoods and gloves, but not masks. The method initially used proved ineffective and the foreman, wearing only a laboratory coat, used a hammer to strike the basket which opened. The basket was then lowered again into the loading cell of the dissolver which was closed again.

It was then that the dust detector in the pond area raised the alarm and the area was evacuated. The vibration of the basket caused by the hammer blows had caused the dispersion of highly contaminated dust into the air. The dose was estimated to be 72 mSv/h at 5 centimetres and 0.6 mSv/h at 2.5 metres. Nose swabs were taken, indicating possible take-up; and all the personnel present were sent to be decontaminated. Urine tests proved negative. Whole body counts were taken and for three persons these proved to be higher than previous readings by less than 1%. The two people who were most exposed received 2.6 and 1.5 mSv.

This incident initiated Eurochemic's radiation protection arrangements.

It is important to review the exposure of persons to radioactivity. It can certainly be affirmed that the incident rate at Eurochemic tended to be modest, bearing in mind the fact that the very special activities involved handling highly radioactive substances dissolved in boiling acids, organic solvents, hydrocarbons and other inflammable and corrosive materials. However the radiological outcome, although remaining within legal limits, was not totally negligible, owing to the decision made to carry out maintenance directly. This situation has resulted in improved working conditions in modern plants[57].

---

[56] It must be remembered that the site originally envisaged for the plant was on the same side of the canal as the CEN. The fears expressed by the Technical Committee when it learned that the site was on the other side of the canal, necessitating construction of a bridge, proved justified on this occasion.

[57] See Part V Chapter 1.

Radiation protection and doses received during operation of the plant

The legal framework and the organisation of radiation protection, Hazardous Work Permits (HWP) and Planned Radiation Exposure Permits (PREP)[58]

Radiation protection arrangements at Eurochemic were in line with the Belgian regulations set out in the Royal Order of 28 February 1963, which took up the 1959 recommendations of the International Commission for Radiological Protection (ICRP) concerning the exposure of workers to radiation: since 1950 the maximum permissible dose had been set at 50 mSv/year, but the ICRP had added a quarterly maximum of 30 mSv/quarter, with a maximum cumulative dose of 50(N-18) mSv, N being the age of the person. These standards remained unchanged throughout the operation of the Eurochemic plant[59].

Besides the Belgian regulations, Eurochemic had its own special rules. Any work involving an additional risk of external radiation, contamination or conventional hazard required a Hazardous Work Permit (HWP) to be issued. In such cases the work was painstakingly prepared by the technical management in conjunction with those responsible for safety.

As from 1970, along with the decontamination of the system and the modifications made during the shutdown in reprocessing, the permit procedures for the most dangerous tasks were made even more stringent with the creation of a Planned Radiation Exposure Permit (PREP), covering any task in an area where there was a risk of receiving a dose exceeding 0.75 mSv/hour or 2.3 mSv/week.

Large numbers of HWPs and PREPs were issued.

Table 69. **Numbers of HWPs and PREPs issued from 1969 to 1974**

	HWP	PREP (>2.3 mSv/week)	of which >5mSv/week
1969	1380	n.a.	
1970	1556	180	
1971	1593	375	121
1972	1325	425	133
1973	1555	225	23
1974	1187	187	9

Doses were monitored by personal detectors, usually two thermoluminescent dosemeters (TLD). One of these was read monthly or whenever an unusual dose was suspected. The other was read every quarter and totalled at the end of the year. Special detectors were used in addition whenever there was a risk of neutron irradiation. Also workers involved in areas where beta or gamma radioactivity might be present wore a film badge dosemeter which was read monthly. Daily dose readings involved a direct-reading pocket ionisation chamber. Those working in the cells wore a dosemeter fitted with an audible alarm which could be programmed in advance.

Doses to the extremities were measured by "fluoride-teflon discs" (FTD) inspected monthly or immediately in cases of uncertainty.

As a general rule, the results from measurements of external irradiation connected with the plant were overestimated, since the background radioactivity was not subtracted. If the dosemeters disagreed, the higher value was used. Calculated doses were rounded upwards.

Bodily contamination was evaluated from urine samples taken over a 24-hour period, this being done every three to six months according to the hazards or immediately if ingestion or inhalation was suspected; the evaluation looked for uranium, plutonium and strontium.

These arrangements gave the following overall radiological picture.

---

58 OSIPENCO A., IAEA (1980), pp. 442 *et seq.*

59 In 1977 it was recommended that the average dose should be reduced by adopting the "as low as reasonably achievable" (ALARA) principle. In 1990 the ICRP recommended that the dose should not exceed 20 mSv/year as a moving average calculated over 5 years.

General results during the period of reprocessing

Table 70. **Annual collective doses, 1969 to 1974**[60]

Year	Doses in mSv/year
1969	3765
1970	3396
1971	3152
1972	4270
1973	3715
1974	2672

It seems difficult to establish any direct link between the length of the reprocessing campaigns and the collective doses received. However the connection is more obvious if we consider the quantities of fissile material leaving the plant, at least as regards slightly enriched uranium between 1971 and 1972. However the year 1973, which was the most productive as regards this type of fuel, was marked by a fall in the collective dose received.

Table 71. **Maximum individual doses, 1969 to 1975**

Year	Dose in mSv	In excess of the legal maximum (50 mSv/year)
1969	56	6
1970	50.3	0.3
1971	78	28
1972	76	26
1973	105.6	55.6
1974	76.5	26.5
1975	50.3	0.3

Doses above the legal maximum were related essentially to plant incidents, as can be seen from the distribution of mean dose by category of personnel during operations.

Dose distribution by personnel category (Figure 110)

The distribution of the average annual doses received by those working in the different departments of the technical management of the company shows the relatively high doses received by those involved in operations, radiation protection, health and safety, and maintenance, who were working directly in the plant. On the average, the doses exceeding the permitted annual maximum involved 4.1% of staff, but 9% of the operating personnel. The R & D team is the least affected, probably because of their experience in handling fissile materials and the fact that the IDL's objectives were redirected towards basic research on waste, which initially did not require the handling of active solutions.

Average doses and comparison with other contemporary reprocessing plants

Table 72. **Average doses received by staff working in radiation areas, 1969 to 1974**

Year	Average dose in mSv[61] (maximum legal dose: 50 mSv)
1969	14.6
1970	12.7
1971	11.8
1972	18.2
1973	16.5
1974	14.3

60 Source: RAE.

61 Data prepared from the RAEs: DETILLEUX E. (1978), p. 185 gives 18.6 mSv for 1972 and 14.5 mSv for 1974.

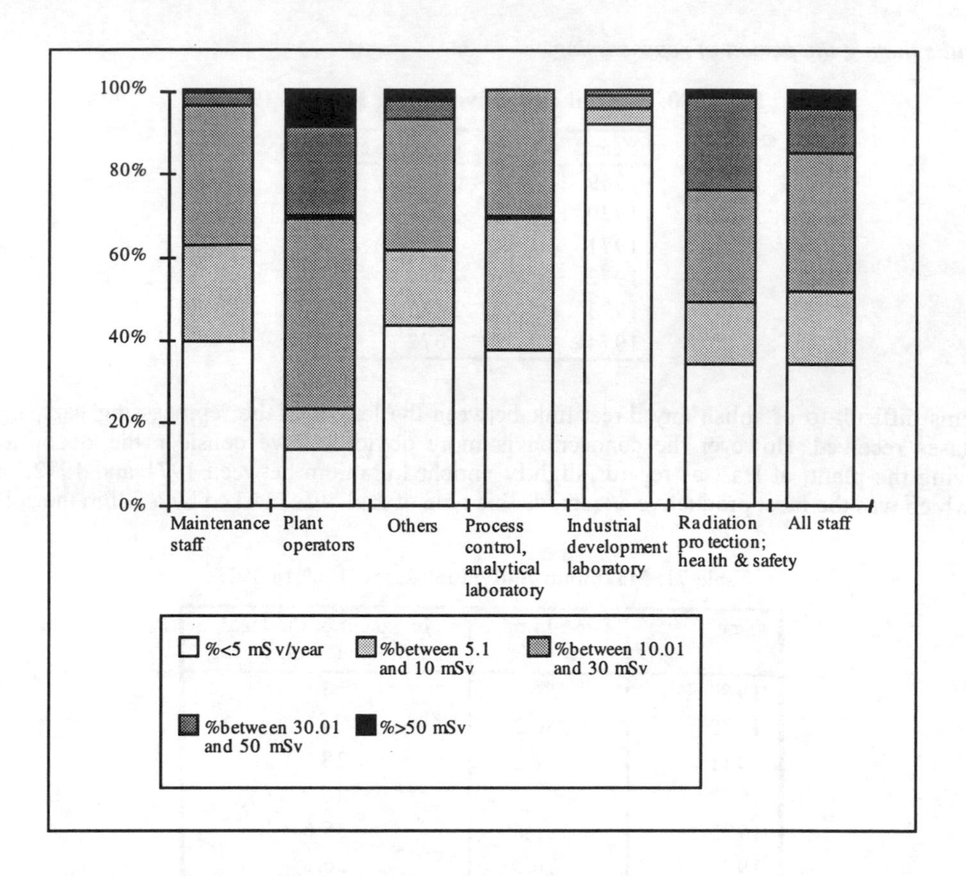

Figure 110. Average distribution of annual dose by personnel category, 1969 to 1974

It is important to determine how the doses received at Eurochemic compare with those at other reprocessing plants. This is not easy, because documentation is incomplete and the basis on which average doses are established is not always clearly specified. Table 73 attempts a summary, but the footnotes show the uncertainties.

Table 73 shows how the average doses developed in a few civilian reprocessing plants[62] between 1967 and 1992: these are Eurochemic[63], WAK[64], Tokai Mura (TRP)[65], UP2 and UP3 at La Hague[66], Windscale[67] and West Valley (NFS)[68]. The dates given in brackets indicate the periods of reprocessing. The term "n.d." means that no data are available. These data, except those for NFS, are plotted graphically in Part V Chapter 1.

---

[62] A request for information was sent to the Marcoule establishment but no reply was received. The available documents on Marcoule do not indicate how the statistics were prepared. It appears that the data pertain to the entire establishment and perhaps to all staff and not only to reprocessing alone and the people exposed. Consequently these data were not used.

[63] Sources: Eurochemic archives.

[64] Source: AtW, February 1989, p. 100 and OECD (1993), p. 207 for 1988 and 1989. The calculations concern only persons working in controlled areas.

[65] Tokai Reprocessing Plant, the pilot Japanese plant. Source: TAKASAKI K. (1991), p. 223 Fig. 2 (quantitative data taken from the graph). The data do not make it clear whether the calculations are done for all staff or for only those exposed to radiation. The difference may be considerable because this plant, of a size comparable to that of Eurochemic, employed between 889 and 2407 staff with considerable variations from year to year.

[66] Source: Documents provided by COGEMA. The statistics give zero doses for UP2 until 1973, but excludes them thereafter. Since 1989 the results include the performance of UP3.

[67] Source: CFDT (1980), p. 197, which uses a COGEMA document, forwarded by La Hague through the kindness of B. Lenail.

[68] Ibid. These two series are given as an illustration although they are not confirmed by other sources.

Table 73. **Trends in average doses in a few civilian reprocessing plants, 1967 to 1992**

	EUROCHEMIC (1966-1974)	WAK (1971-1990)	TRP (as from 1978)	La Hague (as from 1966)	Windscale (as from 1952)	NFS (1966-1972)
1967	n.d.			1.3	n.d.	
1968	n.d.			1.4	n.d.	29
1969	14.59			1.56	n.d.	45
1970	12.67			3.1	n.d.	65
1971	11.76	n.d.		4.75	13	75
1972	18.17	11.61		3.24	15	
1973	16.51	9.29		6.53	14	
1974	14.29	6.02		6.96	13	
1975		5.11		7.51	12.5	
1976		2.93		6.68	12	
1977		3.78		6.34	10	
1978		4.07	0.29	5.35	9	
1979		4.50	0.61	4.59	7.5	
1980		3.35	0.55	4.74	8.1	
1981		2.94	0.47	4.21	7	
1982		2.66	0.44	3.47	6.7	
1983		1.75	0.90	2.96	7	
1984		2.29	0.69	3.83	6.8	
1985		2.04	0.29	3.76	5.8	
1986		2.53	0.35	2.97	5.7	
1987		2.84	0.75	3.2	4.7	
1988		2.04	1.78	3.18	4	
1989		1.53	0.87	3.66	3.1	
1990		n.d.	0.32	2.08	3	
1991			n.d.	1.88	n.d.	
1992			n.d.	1.56	n.d.	

Considering all the plants analysed, mean annual doses at Eurochemic were rather high. However they were much lower than those recorded at NFS (West Valley) which can be seen to exceed the legal limit between 1968 and 1971[69]. They are more comparable with those of Windscale and fairly close to those of WAK during its early operation.

Doses at La Hague appear to have been much lower, moving against the trend of the other European plants examined up to 1975. The increased tonnages and higher burn-ups of the fuels reprocessed in UP2 led to an increase in average doses from 1968 to 1975, when they were temporarily higher than WAK[70]. The figures for the Windscale staff very probably pertain to all the personnel at the nuclear centre and not just to those working in radiation areas on reprocessing operations. The dose levels appear to have been comparable with those of Eurochemic.

A specific factor in the way work was organised also contributed to placing the doses at a comparatively high level. Unlike many nuclear undertakings, Eurochemic made relatively little use of contract workers during the period of reprocessing[71].

The proportion of contract staff working in controlled areas at Eurochemic never exceeded 11%. Although in 1969 and 1970 the average dose received by contract workers is appreciably higher than that of the company's own employees, the opposite is true thereafter. In any event, recourse to external workers, which can sometimes "dilute doses" in case of problems, was limited by the company's financial capability.

The reason for the rather high doses received at Eurochemic can be found in the lack of advanced systems for remote operation and control[72], and in the experimental nature of a large part of Eurochemic's work, rather than in any shortcomings of the safety systems.

---

[69] According to various oral communications from persons external to the plant, commercial pressures appeared to have diverted attention away from the requirements of staff safety and probably played a part in the decision to shut it down.

[70] With regard to these difficulties, see for example COUTURE J. (1975).

[71] However Eurochemic did have recourse to contract workers at the end of the 1970s for waste management purposes, see Part IV Chapter 1.

[72] This point is examined in detail in Part V Chapter 1.

Table 74. **Variation in numbers of contract workers for radiation work at Eurochemic and the average distribution of doses, 1969 to 1975**

Year	Eurochemic staff working in radiation areas (number)	Contract staff working in radiation areas (number)	Total working in radiation areas (number)	Average dose received by Eurochemic staff (mSv/year)	Average dose received by contract workers (mSv/year)	Reminder: average dose to all staff working in radiation areas (mSv/year)
1969	252	6	258	14.4	22.0	14.6
1970	255	13	268	12.4	18.7	12.7
1971	238	30	268	11.9	10.8	11.8
1972	213	22	235	18.6	14.0	18.2
1973	206	19	225	16.6	16.0	16.5
1974	171	16	187	14.4	12.8	14.3
1975	152	17	169	9.6	6.4	9.3

The causes of exposure[73]

The main sources of exposure to the operating staff were the following:

- the gloveboxes in the plutonium purification installation with their filtration, sampling and weighing systems;
- the cask decontamination area, owing mainly to the heavy contamination of the inside of the casks;
- the filters and resins used for purifying the pond water; first the filters on the tank ventilation system, which had to be replaced more often than planned in the event of abnormal conditions and, secondly, the absolute filters in the high vacuum system when these were replaced.

The main sources for maintenance staff were repair and maintenance work on valves and pumps, repairs to the sampling enclosures, and to remote manipulators and the handling equipment in the dissolver cells.

Radiation doses to control laboratory analysts resulted essentially from the handling difficulties with high activity specimens because of the design adopted for the working enclosures.

For both the maintenance staff and the analysts, doses to the hands were often substantial, above the permitted quarterly limit of 140 mSv. In these cases the exposed workers were moved away from this type of work so that the maximum permitted annual dose to the hands – 600 mSv – was exceeded in only three accidental cases.

The health and safety staff were exposed primarily during the detailed inspection of working areas during preparations for special tasks for which HWPs or PREPs were to be issued, or during the repair and maintenance of systems that had contained highly active solutions.

Security control at Eurochemic: the search for "material unaccounted for"

Security control at Eurochemic: objectives and organisation

The purpose of security control was to guarantee that the fissile materials handled in a nuclear undertaking supposed to be civilian could not be diverted for military purposes. Eurochemic was the first European reprocessing unit to be placed under this control[74] and as such it was a test bed for the control and fissile material accounting techniques applied. Internal control was managed by the team responsible for fissile material accounting. External control was conducted by inspectors from the specialist international organisations.

The problem of the external inspector was fairly complicated. There were three international control organisations: the ENEA had written a Security Control Convention in 1960 and Eurochemic, as a joint undertaking of the Agency, was naturally governed by it. Owing to its geographical situation as much as to the presence of five of the six members of EURATOM, it also came under the security control of the European Atomic Energy Community (EAEC). Finally the reprocessing of fuels of American origin from countries not belonging to EURATOM – for which there was a delegation of control which was established in principle as early as 1962 – came under the control of the IAEA.

France was opposed to any IAEA control and this raised a problem the solution of which – delegating the inspection powers of the IAEA to EURATOM – was initiated by the visit made by Glenn Seaborg, then

73 Principal source: OSIPENCO A. (1980).

74 The first plant inspected by the IAEA was that of NFS at West Valley in August 1967.

chairman of the USAEC, to Mol on 11 March 1966. There was another problem: that of the meaning to be placed upon the planned "continuous inspection of reprocessing plants". Told about this in a conversation that took place in the car of Rudolf Rometsch in the presence of Einar Saeland and Pierre Strohl, Glenn Seaborg requested that discussions should be started between EURATOM and the American experts, to deal in particular with the number of inspectors needed. EURATOM suggested one, the United States demanded twenty. Finally a compromise was reached on three and, on certain occasions, five. In 1969 the USAEC and the IAEA agreed jointly to delegate control to EURATOM, but representatives of the IAEA were present on the site, being involved in research contracts on materials accounting. In fact control was exercised by representatives of the OECD/ENEA on particular occasions and continuously by those of EURATOM. The two groups of inspectors were empowered to make unplanned measurements at any point in the plant.

Initially the internal inspection procedures focused on verifying fissile material accounting. This was done in conjunction with the plant's technical staff by an Accounting Section which until 1967 was integrated with the Nuclear Services Division as part of the Technical Directorate, attached to the general management until 1969, when it was integrated in the Commercial Department.

The Accounting Section checked fuel reception, the determination of input volumes, the measurements of output weights and the relevant sampling operations. In this way any losses or gains of fissile material during operations could be monitored. Percentages of "material unaccounted for" (MUF) were worked out from the differences between the accounting inventory and the actual inventory. This accounting system was also used to verify tank calibrations and to add to or modify the plant's *"Manual of Standard Procedures and Specification"* accordingly. In the accounting process there was a good case for computerising the inventory system and the calculation of inputs and outputs, and this was done in 1969 using a computer shared with the plant's analytical laboratory.

The Accounting Section compared these figures with data from actual measurements made during plant shutdowns. The problem was that these shutdowns, although initially frequent, became increasingly rare as operations proceeded; real measurements were liable to be too infrequent.

Eurochemic therefore looked into the possibility of daily monitoring under a contractual arrangement with the IAEA[75]. Trials were carried out during the HEU-1970 campaign and published in a technical report[76]. This system was computerised in 1971 and applied once again during the HEU 1971 campaign.

Another method of checking known as "isotopic correlation" was tried out for the HEU campaigns, again under a contract with the IAEA. This related the values given by the fuel manufacturer, the values measured during reprocessing and those obtained from the checks carried out by the inspectors. These results were also published as a technical report[77].

In this way Eurochemic contributed to the development of methods on which the effectiveness of fissile material monitoring depended closely, at a time when the IAEA was setting up the system of guarantees[78] necessary for the application of the Non-Proliferation Treaty which, signed on 1 July 1968, had come into force on 5 March 1970[79] for 43 countries.

Results: materials unaccounted for (MUF)

The problem in accounting for fissile materials in a reprocessing plant comes from the fact that not all the quantities entering show up at the exit, and that sometimes the amount of material leaving exceeds that entering. The MUF loss indicator does not account for process losses but for the quantities that do not appear in the input-output balance. Materials present in liquid waste form part of the outputs[80].

The main cause of "disappearances" usually turned out to be the accumulation of material in the plant as it adhered to pipe walls or was spread in the cells when leaks occurred[81]. Occasionally "disappearances" were purely an accounting matter, arising from measurement errors. In fact the accounting process was based upon measurements which had variable margins for error, these being more significant when involving small quantities, as for plutonium, and occasionally analyses which were difficult to carry out. It is possible to imagine

---

75 RAE 70 p. 60.

76 ETR 266.

77 ETR 277.

78 Foreseen by the NPT in its Article III. See CANTELON P. L., HEWLETT R. G., WILLIAMS R. C. (1991), p. 239.

79 However no member country of Eurochemic was a signatory at that time. As regards the NPT, American policy and European reprocessing, see Part IV Chapter 1.

80 Although there are problems of evaluation: see below.

81 See the case of the evaporators considered above.

a situation where the MUF is due entirely to accounting errors, for example when the plutonium input to a sub-campaign is overestimated and the plutonium present in the waste from the same sub-campaign is under-evaluated.

The aim of security control is to ensure that the MUF is not due to fissile material diversion. In these circumstances, only continuous monitoring allowed changes in MUF to be determined and their causes to be found. Conversely, when the system was rinsed or during subsequent campaigns, fissile material could turn up again, generating a positive balance.

The process of introducing fissile material accounting was long and delicate, and the statistics given in the annual reports were sometimes corrected from year to year. The reasons for this were the complexity of the operations, their lack of correspondence with the reprocessing schedule and the problems of measurement encountered; Table 75 is a compilation of cumulative MUF from year to year.

Table 75. **Cumulative year-on-year MUF, 1968 to 1975**

Date	LEU		Pu		HEU	
	kg	% of inputs	kg	% of inputs	kg	% of inputs
End 1968		+0.62		+ 5.10		- 0.17
End 1969		- 0.43		- 3.25		- 0.17
End 1970		- 0.25		- 2.78		- 0.48
End 1971[82]		- 0.19		- 1.79		+ 0.01
End 1972	+ 107.0	+0.39	- 13.26	- 3.30[83]		+ 0.01
End 1973		- 0.30	- 7.18	- 1.90	- 3.810	- 0.40
Beginning 1975[84]	-706.2	n.d.	- 16.025	n.d.	- 7.499	n.d.
End 1975[85]	-407.4	- 0.22	- 12.728	- 1.88	- 6.959	- 0.51
Total reprocessed[86]	185 000		677		1364	

On completion of the reprocessing operations at the beginning of 1975, MUF accounted for over 700 kilogrammes of natural or slightly enriched uranium, about 7.5 kilogrammes of highly enriched uranium and over 16 kilogrammes of plutonium[87].

The operations that followed reprocessing[88], notably rinsing and decontamination in the period 1975-1978, led to the recovery of nearly 300 kilogrammes of natural and slightly enriched uranium, 500 grammes of highly enriched uranium and 3.5 kilogrammes of plutonium. Later operations carried out up to 1984 led to the MUF being reduced a little further. At the end of 1984 the cumulative MUF was as shown in Table 76.

Table 76. **Cumulative MUF as of the end of 1984**

Date	LEU		Pu		HEU	
	kg	% of inputs	kg	% of inputs	kg	% of inputs
End 1984[89]	-344.5	-0.19	-11.863[90]	-1.75	-6.959	-0.51

[82] Source: RAE 1972. In RAE 1971 the percentages are - 0.25, - 2.01 and + 0.01 respectively. The difference for the plutonium MUF is related to the plutonium recovered during a rinsing operation.

[83] Taking account of the plutonium entering the waste.

[84] On completion of the last reprocessing campaign.

[85] Following rinsing of the installation.

[86] However there remains a problem as to the difference between the calculated quantities and other production data: 181.3 tonnes LEU, 644 kilogrammes of Pu, 1.36 tonnes HEU. Other source: 188.43 tonnes LEU, 644.4 kilogrammes of Pu, 1362 kilogrammes HEU. Third source – VIEFERS W., wastes: LEU 191.3; Pu: 677.7; HEU: 1.4 tonnes.

[87] By comparison, a recent study of MUF at Tokai Mura, quoted in the economic supplement to the newspaper *Le Monde* of 21 June 1994, p. I, note 1, gave a cumulative MUF of 70 kilogrammes of plutonium for the Japanese pilot plant which had operated since 1978. This quantity justified the introduction of a joint IAEA-Japan programme, announced on 10 June 1994, to recover 80 to 85% of this material over three years.

[88] See Part IV Chapter 1.

[89] RAE 1984.

[90] The small amount recovered is due to the 1975 overestimate of the quantity of plutonium present in the liquid waste, an overestimate revealed by subsequent analysis.

Of course it will be impossible to prove that this material unaccounted for is actually in the plant until the buildings are finally dismantled.

For a number of reasons, related to the conditions in which the installation was operated and the way in which the MUF was evaluated, any external causes, such as a cloak-and-dagger diversion, are highly unlikely.

In fact the losses in the decladding solutions were very difficult to evaluate, owing to the sometimes painstaking nature of experimental operations which left occasionally significant fragments of fuel undissolved, especially during ZIRFLEX operations.

The difficulties encountered in the plutonium cycles and in the various purification units, and the problems of establishing the accounting system, can also partly explain the accounting losses of plutonium. Also around 2% of the accounting loss can be accounted for by the fact that a relatively low mass of plutonium was subjected to a very large number of operations. Finally, an unknown but significant quantity of plutonium was and is still today inside the pipes and machines of the plant, as could be seen[91] from the changes in the concentrations of fissile materials – notably plutonium – in the rinsing liquids after termination of operations and the balance sheets drawn up in the dismantling operations in the years that followed. This residue of fissile material is liable to complicate the decommissioning operations and must be added to the overall balance sheet of the wastes produced given below.

Balance sheet of reprocessing wastes at Eurochemic[92]

Waste management and classification principles at Eurochemic during the reprocessing period

The principles of waste management at Eurochemic

The principles of waste management at Eurochemic initially depended mainly on three factors which it was hoped would minimise the quantities involved: the first two were aimed at disposing of the largest possible volume of wastes off site, the third at avoiding the management of high level solid wastes.

The presence nearby of the CEN and the participation of Belgium in the international sea dumping operations under the auspices of the OECD/ENEA did in fact facilitate the disposal of low level wastes and of some intermediate level wastes. It will be remembered that this externalisation of the problem had played a significant role in the choice of the site since Belgium had proposed, in the D'Hondt counter-proposal in March 1957, that the CEN should manage the plant's wastes.

Ideally, the choice of chemical decladding should have meant that no high level solid wastes were produced at all. Consequently no arrangements were made at the outset for managing these wastes, apart from a corner of the pond in which the fuel elements were sawn up. As we have seen, this initial "minimal" approach made wastes into one of the reasons for the financial failings of the undertaking. Calculations done after the detailed preliminary project had been adopted had led to provision being made for additional storage tanks[93]; the initial campaigns soon revealed the need for additional capacity and for modifying procedures, especially in order to reduce the volumes of high and intermediate level liquid wastes. The termination of reprocessing was to focus attention on the problem of what was to be done with the types of waste which had hitherto simply been stored on the site.

The classification of wastes at Eurochemic

Reprocessing generates wastes of all kinds – gaseous, solid (waste proper) and liquid (effluent) – and of all degrees of radioactivity. The fact that the cost of treating waste increases with the amount of radioactivity contained, combined with safety requirements, imposes in nuclear establishments what is today somewhat pleonastically called "selective sorting".

A very precise nomenclature was drawn up for the liquid wastes and gaseous emissions well before the plant was put into service. However it was not really applied to solid wastes until 1968, until which time they had all been stored on the site. There was a major campaign of preconditioning that year to allow disposal at CEN.

---

91 See Part IV Chapter 1.

92 The term "wastes" is used here in a broad sense and covers solid, liquid and gaseous wastes.

93 RAE3, p. 187. Two 210 m^3 tanks were installed in Building 22 for storing fission products.

Figure 111 sets out the waste nomenclature used at Eurochemic. It differs from the nomenclature subsequently defined by the IAEA. The correspondences are set out in the table of Figure 112. For each type of waste there was a particular management procedure. Generally speaking, low level wastes and certain intermediate level wastes were taken off site for disposal. Most of the intermediate level waste and all the high level waste was stored on site. Thus when reprocessing came to an end, there remained the highly radioactive waste which had to be dealt with before the plant could definitively be closed.

The production of waste during reprocessing

Table 77 summarises waste production in cubic metres during the period of reprocessing[94].

Table 77. **Waste production in m^3, 1967 to 1974**

Year Volume of wastes in m^3	1967	1968	1969	1970	1971	1972	1973	1974
High level liquid waste (HLLW)	92	70	13	41	214.7	149	18.3	118.1
of which from LEU campaigns	92	2	13	13	8.7	12	18.3	11.1
Intermediate level liquid waste (ILLW)	283	255	140	335	210	302	346	119
of which decladding liquid waste	143	99	70	6	130	247	n.d.	n.d.
Low level liquid waste (LLLW)	35000	27300	22600	18205	16956	15785	16440	14352
"Inactive" condensate	22000	22056	28000	34995	32524	25161	29412	22188
Low level solid waste – beta and gamma (LLSW beta-gamma)	127	341	54	243	376.4	631.7	500.8	355.9
of which combustible waste	107.8	309	n.d.	207.9	335.5	519.2	432.3	302.5
Alpha contaminated solid waste (LLSW alpha)	16.8	59.9	1655 g (of Pu contained)	33	14.5	21.6	0	20.9

The only pertinent comment on the table concerns the rate of waste generation and the annual – not cumulative – quantities, owing first to the fact that low level waste was disposed of as it was produced, and secondly to the decay in the activity of the high and intermediate level waste with time, which made further concentration operations possible. The sum of the volumes produced is considerably higher than the amounts present on the site when reprocessing came to an end[95].

Three factors contributed to increasing the volumes of liquid wastes. The reprocessing campaigns on highly enriched fuels generated large quantities of high level liquid wastes containing a great deal of aluminium, an element difficult to concentrate.

The reprocessing campaigns on natural and slightly enriched uranium fuels on the other hand produced appreciably smaller quantities of high level liquid wastes, but their chemical decladding did generate large amounts of intermediate level liquid waste.

Liquid wastes were also generated during the many rinsing operations between LEU and HEU campaigns, and those required between batches of a given campaign. For example during the LEU campaigns the objective was to avoid mixing depleted uranium – arising from natural uranium fuels – with uranium that was still slightly enriched – arising from enriched uranium burned in light water reactors. The impact of these rinsing operations on the quantity of liquid wastes was particularly marked since the campaigns and the batches were of limited extent.

94 Considerable work had to be done on the data to obtain the figures shown, since the basis for waste compatibility varied prior to 1970. Thus for example in 1967 and 1968 the stocks of intermediate level liquid wastes were presented in terms of each tank. As from 1970, ETR 328 gives continuous, retrospective data, since the annual reports produced between 1970 and 1974 did not include an annual review. With regard to solid wastes showing beta and gamma activity, the number of 200-litre drums is shown for 1967 and 1968, a distinction being drawn between combustible and non-combustible waste, but with no indication of volume. The volume has been deduced from the number of drums. In 1969, the presentation was based upon the radioactivity at the surface of the drums. In 1970 the data refer to the pre-compacting volumes in 220-litre drums or 30-litre canisters. It should be noted that the compacting process reduced volumes by a factor average 2.51. As from 1971 the data refer to waste after compacting. For alpha-active waste, drums and canisters are given from 1967 to 1971. After 1971 these wastes were conditioned only in canisters. The data for 1969 give the Pu contained (1655 grammes) in the packages, without the nature of the latter being specified.

95 This amount is evaluated at the end of the chapter.

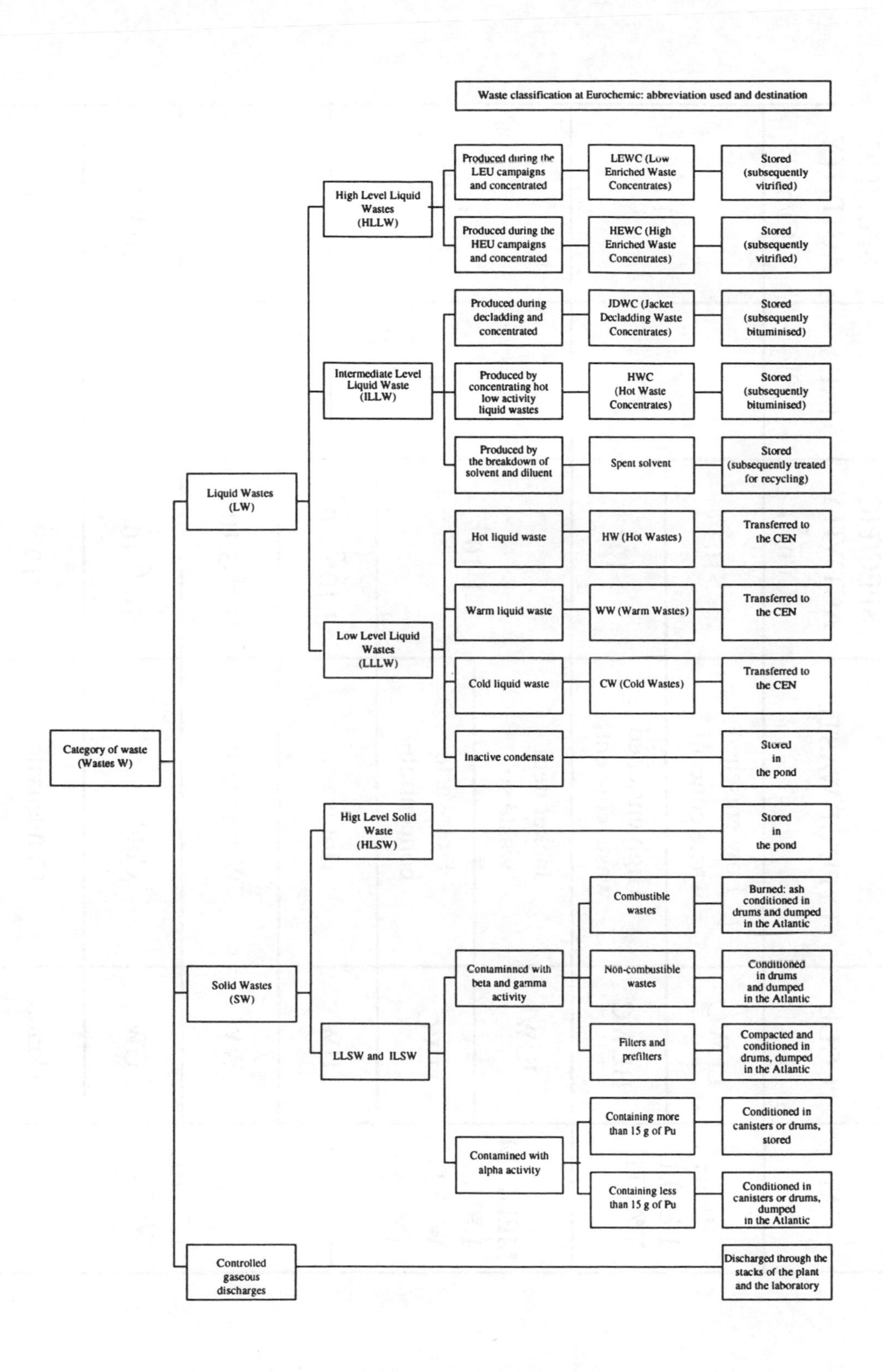

Figure 111. The system of waste classification at Eurochemic: a trial summary.

CAT.	CODE	TYPE OF WASTE	SPECIFIC ACTIVITY (Ci/m^3)	IAEA Cat.	SPECIFIC ACTIVITY (Ci/m^3)
HIGH LEVEL WASTE	LEWC	Low enriched waste concent.	$>10^4$	5	$>10^4$
	HEWC	High enriched waste concent.	$>10^4$		
MEDIUM LEVEL WASTE (MLW)	JDW	Jacket decl. waste sol.	$1\text{-}10^4$	4	$10^{-1}\text{-}10^4$
	HWC	Hot waste concentrate			
	HW	Hot waste	$3.10^{-2}\text{-}10^3$		
LOW LEVEL WASTE (LLW)	WW	Warm waste	$10^{-4}\text{-}3.10^{-2}$	3	$10^{-3}\text{-}10^{-1}$
	CW	Cold waste	$10^{-6}\text{-}10^{-4}$	2	$10^{-6}\text{-}10^{-3}$
	Cond.	Condensate	$<10^{-6}$	1	$<10^{-6}$

Figure 112. Table of correspondence between the Eurochemic liquid waste classification system and that of the IAEA (Source: HUMBLET L. (1987) p. 9. Simplified document).

Present-day industrial reprocessing plants specialise in reprocessing fuels with a relatively low enrichment belonging to the LEU category. They all use mechanical decladding. For these reasons, these plants produce much smaller quantities of medium and high level liquid wastes than did Eurochemic.

Having considered the rates of production, it is important to review the treatment processes that followed.

The destination of wastes: those which left the site

The wastes that were disposed of as they were produced were gases after filtration, liquid wastes of activity below 0.03 Ci/m^3, the low activity solid wastes and some of the intermediate level solid wastes. The gases were discharged to the atmosphere, the other wastes were transferred to the CEN and treated before being discharged into the river (this was authorised for certain liquid wastes[96]) or into the sea as regards other liquid wastes and the final solid wastes.

Low activity liquid wastes (Table 78)

The Eurochemic plant had no direct access to river or sea and so had no authorisation to dispose of wastes into water but had to utilise that from which the CEN benefited[97]. The liquid wastes transferred to the CEN represented 98% of the total volume of liquid wastes, but only an infinitesimal proportion of total waste activity[98].

The rate of transfer varied from year to year, as did the activity discharged. During the first two years of reprocessing the low activity liquid wastes accounted for a larger proportion of the liquid waste transferred to the CEN than did the condensate. The methods of filtration and separation were then improved. The rinsing and decontamination operations of 1970-1971 resulted in a peak in the production of low activity liquid waste. The leaks in the low activity liquid waste transfer pipelines also made a contribution to reducing the transfers to the CEN. On the average every tonne of LEU or HEU generated 150 m^3 of WW, 150 m^3 of CW and 250 m^3 of condensate[99].

Gaseous waste and gaseous emissions

Certain reprocessing operations produce significant amounts of radioactive gaseous waste. These were filtered, sometimes after scrubbing, in a complex system of prefilters. Also the plant, the laboratories and some of the storage buildings were kept at a slight vacuum.

The air was regularly renewed then passed through the filters and discharged through the stack. The ventilation systems terminated at the absolute filters which were not supposed to allow anything to reach the exterior. The filters and prefilters were regularly changed and treated as solid wastes. Thus the stacks from the laboratory and the plant discharged air and gases the nature and quantities of which were monitored by alarm and measurement systems.

---

[96] At the beginning of the 1970s, the CEN had permission to discharge 0.9 Ci/year alpha and 4.5 Ci/year of beta and gamma activity. According to the results of a study by the European Community (EUR 6088) the main results of which were published in AtW (1979), pp. 612-613, the establishment substantially and continuously exceeded these limits in the period 1972-1976, except for tritium.

[97] The activity discharged is out of all proportion with that from Windscale and La Hague, even when expressed in terms of the quantities reprocessed, and even though the French plant rapidly reduced its discharges to the sea. For example, the permitted alpha discharge for La Hague in the 1980s was 45 Ci for alpha discharges and 45 000 Ci (excluding tritium) for beta and gamma discharges. The actual maximum discharges during the decade were about 19.5 Ci in 1985 for alpha activity and a little over 34 000 Ci in 1982 for beta and gamma activity. Discharges during the 1990s are much lower owing to the company's implementation of a much stricter policy. In 1993 they were 2760 Ci of beta and gamma activity and 2.7 Ci of alpha activity. Source: 1990 tests, COGEMA La Hague, p. 14 and COGEMA documents provided by Yves Sousselier.

At Windscale at the beginning of the 1970s, the limit was fixed at 6000 Ci alpha activity and 300 000 Ci of beta and gamma activity (excluding tritium) for the site as a whole. Between 1972 and 1976, the discharges fluctuated between 1614 Ci and 4896 Ci of alpha activity and between 245 000 and 127 000 Ci of beta and gamma activity excluding tritium.

[98] The rough estimate given below produces discharge figures of 1380 curies over 96 months or 14.37 curies a month on the average. However it takes no account of the tritium activity, which was measured only after 1972. Thus 2700 Ci of tritium were discharged by the CEN in 1972, 5000 Ci in 1973 and 1325 Ci in 1974: RAE 1972, p. 12, RAE 1973, p. 12, RAE 1974 p. 13. The shutdown of the plant led to a sharp fall in tritium discharges by the CEN from 2510 Ci in 1974 to 523 Ci in 1975: AtW (1979), p. 613.

[99] VIEFERS W. (not dated), Wastes, Table 4.

Table 78. **Types of low activity liquid wastes; quantities disposed of**

Type of waste	Composition and source	Mean activity in Ci/m^3	Average quantity	Total radioactivity discharged[100] (estimated over 96 months)
Low activity liquid waste discharged by the CEN				1388.75
Warm Waste	Liquid waste from the plant and the laboratory, distillate from the treatment of Hot Waste	10^{-4} to 3.10^{-2}	950 m^3/month	1386 (14.4/month)
Cold Waste	Liquid waste from the laboratory and the plant	10^{-6} to 10^{-4}	500 m^3/month	2.5 (0.026/month)
Condensate		Inactive (below 10^{-6})	2300 m^3/month	0.25

The maximum rates of discharge into the atmosphere upon which the operating licence was based were 0.5 mSv/week to a person standing at the point of maximum concentration downwind[101]. Discharges from the stack were governed by standards which required four types of measurement to be made until September 1971 (iodine 131, plutonium 239, strontium 90, krypton 85) and five thereafter, when tritium measurements were added. The annual discharges were as listed in Table 79 and are worth comparing with the official standards in Table 80[102].

Table 79. **Annual discharges through the stack, 1967 to 1974**

Emissions measured at the stack (in Ci)	Alpha aerosols (Pu 239) in Ci	Beta aerosols (Sr 90) in Ci	Iodine 131 in Ci	Krypton 85 in Ci	Tritium in Ci
Standards 1967-1971	1.3	1000	130	6 300 000	-
Standards 1971-1974	1.25	220	12.5	6 100 000	4 400 000
1967	$4\ 10^{-4}$	$4\ 10^{-3}$	$<1.6\ 10^{-2}$	37 000	n.a
1968	$1.8\ 10^{-4}$	$3.1\ 10^{-2}$	$<1.6\ 10^{-2}$	n.d.[103]	n.a
1969	$2\ 10^{-2}$	$4\ 10^{-2}$	$<1.6\ 10^{-2}$	60 000	n.a.
1970	neglibible	neglibible	neglibible	50 000	n.a.
1971	neglibible	neglibible	neglibible	126 440	532
1972	$<1.2\ 10^{-2}$	<0.12	$1.5\ 10^{-2}$ [104]	199 600	706
1973	neglibible	neglibible	neglibible	216 730	1 893
1974	$<5\ 10^{-4}$	$<8.7\ 10^{-3}$	"small discharge"	100 232	1 579

100 The activity was calculated from the data. The total volume of low activity liquid waste is worked out according to the number of months between January 1967 and December 1974, i.e., 96 months. The activity was estimated as the arithmetic mean of the elements in the range for WW and CW (WW: 0.0152 Ci/m^3; CW: 0.0000505 Ci/m^3). The condensate was assumed to have the maximum permitted value.

101 Downwind of the stack.

102 Sources: RAE 1967 to 1969; ETR 70-74; VIEFERS W. (not dated), Wastes; AtW (1979), p. 612-613 for alpha and beta-gamma aerosols in 1972-1974.

103 A small emission of Sb-108 was noted for this year, linked to an incident which occurred during a Zy decladding operation.

104 In June.

For purposes of comparison:

Table 80. **Standards and gaseous discharges from European reprocessing plants for 1974**

Plant	Standard in force in 1974 (Ci/year) Alpha	1974 emission (Ci/year) Alpha	Standard in force in 1974 (Ci/year) Beta-gamma	1974 emission (Ci/year) Beta-gamma	Standard in force in 1974 (Ci/year) Krypton 85	1974 emission (Ci/year) Krypton 85	Standard in force in 1974 (Ci/year) Tritium	1974 emission (Ci/year) Tritium
Eurochemic	1.25	<5 10^{-4}	220	<8.7 10^{-3}	6 100 000	100 232	4 400 000	1579
WAK	1 10^{-2}	1.5 10^{-4}	20	1.4 10^{-2}	350 000	850		<30
La Hague		6.8 10^{-6}		1.3 10^{-2}		720 000		190
Marcoule		3.8 10^{-5}		5.9 10^{-4}		110 000		340
Dounreay	as low as possible	1.2 10^{-2}		<6.2				
Windscale	as low as possible	0.18		2.8		800 000		8000

Thus the gaseous emissions remained well below the legally allowed limits.

Low and intermediate level solid wastes: Eurochemic and the "International Dumping Operations in the Atlantic Ocean"

Initial conditioning and transfer to the CEN

As solid wastes[105] were produced they were sorted according to type and activity level. In 1967, the solid wastes were divided into three categories, conditioned in 200-litre drums: combustible, non-combustible and alpha wastes. These drums were stored in Building 14. As from 1968, for the first transfer to the CEN and in preparation for the 1969 sea dumping operation, the wastes were divided into those which could be disposed of and those which had to remain on the site[106]. The latter were stored in the controlled area (Building 22). The alpha wastes in 28-litre canisters were added to the drums. Hence the system was fully operational in 1968-1969. In the early 1970s, the drum size was increased to 220 litres as containers were standardized.

– *beta-gamma wastes:*

  Fuel wastes were collected into 50-litre plastic bags, then precompacted in groups of 10 inch drums which from 1968 to 1973 were transported to the CEN for treatment – i.e., incineration – the ash then being embedded in concrete and placed in drums. Non-combustible waste was also collected in plastic bags or wrapped in plastic sheeting and separated into that which could be compacted and that which could not, placed in drums and transferred to the CEN for final conditioning before disposal at sea.

– *alpha wastes:*

  Alpha wastes were sealed in plastic bags and collected in 28-litre metal canisters or 200-litre (subsequently 220-litre) drums. If the total amount of plutonium contained did not exceed 15 grammes – which at the time was the maximum permitted amount in a 220-litre drum for sea disposal – the canisters or drums were transferred to the CEN, otherwise they were stored on the site[107].

– *ventilation filters:*

  The prefilters and absolute filters from the ventilation system were separately sealed into plastic bags, grouped in cardboard boxes and transferred to the CEN where they were compacted and embedded in concrete.

---

[105] VIEFERS W. (not dated), Wastes, Table 3b and p. 9 *et seq*. The information is not dated, but it would appear that this procedure was followed at least from 1968.

[106] Each of the three categories was dealt with according to two criteria based on surface dose: up to 200 mrem/h they were combined; those giving a surface dose exceeding 200 mrem/h and up to 2 rem/h were separated and their level of activity clearly indicated on the containers.

[107] A number of these were subsequently treated in the plutonium installations, before being specially processed: see Part IV Chapter 2.

As from 1973, Eurochemic had its own system for conditioning low level wastes, which were then disposed of together with Belgian wastes.

The intermediate level solid wastes, which during the period of reprocessing consisted essentially of off-gas filters from the container and head-end cell ventilation system, were stored on the site in concrete-lined drums and concrete containers. Subsequently some of these were conditioned for disposal at sea.

The drums prepared for sea disposal were stored at the CEN awaiting the regular disposal operations in the Atlantic co-ordinated by the OECD/ENEA.

Transfer to Zeebrugge, loading on the *Topaz* and disposal in the Atlantic.

Solid wastes were transferred from the CEN under Belgian responsibility to the port of Zeebrugge where they were loaded on board the British ship *Topaz* which since 1967 had been regularly chartered for the International Dumping Operation in the Atlantic, under the auspices of the OECD/ENEA[108].

The ENEA had in fact been investigating the problem of the accumulation of solid wastes on nuclear sites since 1964, and had initiated a study into the feasibility of disposal in the North Sea or the Atlantic. An experimental disposal operation had been planned. The preliminary studies showed that the dumping of 10 000 Ci a year was feasible without danger at a depth of about 5000 metres. The search for a dumping site had resulted in the definition of a perimeter of 70 nautical miles about the point 46°15'N - 17°25'W, about 900 kilometres south-east of Land's End and 450 kilometres from the edge of the continental shelf. This area was remote from shipping lanes and was not frequented by fishermen. The water depth was about 4500 metres. The minimum requirements for the containers had been determined and a vessel sought.

The *Topaz*, a 2250-tonne ship belonging to William Robertson Ltd. of Glasgow, was chartered from May to August 1967 to carry the first containers. Between 17 May and 5 June it collected British, German, Dutch, Belgian and French drums from Newhaven, Emden, Ijmuiden, Zeebrugge and Cherbourg before leaving for the dumping site. The operations were supervised by an international team from the participating countries.

The operation was evaluated in a report published in September 1968, *Radioactive Waste Dumping Operation in the Atlantic, 1967,* which concluded that this approach was feasible.

The first wastes from Eurochemic were dumped during the first of the three voyages in the second operation between 23 June and 26 August 1969[109]. A further operation took place in July 1971. The second and final voyage of the operation was devoted entirely to Belgian wastes. Thereafter the operations became annual[110]. However as from 1972, there was growing criticism in the media. The London Convention of 29 December 1972 on the "prevention of sea pollution resulting from the dumping of wastes and other materials", which was to come into force on 30 August 1975, led the NEA to reconsider its programme and to conduct a further evaluation, which confirmed the "soundness of the arrangements adopted" subject to a few minor modifications and, primarily, the standardisation of containers.

Eurochemic's low activity solid wastes and some of the intermediate level solid wastes were therefore disposed of using this procedure in 1969, 1971, 1972 and 1973. That year, 52.6 m^3 representing 236 tonnes of a mixture of low activity Eurochemic wastes in 220-litre drums, were dumped in a total of 2200 tonnes of Belgian wastes.

Table 81 shows the main features of the dumping operations in which solid reprocessing wastes from Eurochemic were dumped, except for the first operation and that of 1974.

However most of the intermediate level wastes and all the high level wastes were stored on the site.

Wastes remaining on the site: intermediate and high level liquid and solid wastes

Intermediate and high level solid wastes

The intermediate level wastes which could not be disposed of were conditioned in concrete containers and stored in Building 22.

The high level solid wastes produced mainly in the decladding operations, had been stored as they were produced, under water in special baskets in the storage pond built in 1968 specifically to accommodate them. At the end of the period of reprocessing, they amounted to about 20 m^3 in volume.

---

108 The following passages are based upon 1964-1974 reports from the ENEA (subsequently NEA).

109 Wastes from seven countries were dumped. On its first voyage the ship carried containers from a variety of sources in Belgium, Italy, the Netherlands, Sweden and Switzerland. The second voyage carried only British wastes and the third only French wastes, which once again reflects the different progress made in nuclear programmes.

110 They continued until 1982.

Table 81. **Main features of waste dumping operations in the Atlantic (1967 to 1974)**

Year (number of voyages)	Participating countries	Tonnage of wastes[111] (number of drums)	Alpha activity in Ci	Beta-gamma activity in Ci (tritiated wastes not included except for 1973)	Beta activity of tritiated wastes[112]	Belgian data/remarks
1967 (5)	G, B, F, N, UK	10 893 (35 800)	273	7635	-	no Eurochemic wastes
1969 (3)	B, F, I, N, S, CH	9000	500[113]	22 100	-	first Eurochemic voyage
1971 (2)	B, N, CH, UK	4000	630	11 100	-	second exclusively Belgian voyage (1800 t, 2100 Ci beta gamma, 300 Ci alpha)
1972 (2)	B, N, UK	4130	685	6180	15 400	First Belgo-Dutch voyage[114]
1973 (2)	B, N, UK	4500	750	12 660	included in left-hand column	First exclusively Belgian voyage[115]
1974	N, CH, UK	2300 (4000)	416	6000	95 000[116]	no Belgian wastes

G: Germany; B: Belgium; CH: Switwerland; F: France; I: Italy; N: Netherlands; S: Sweden; UK: United Kingdom

Intermediate and high level liquid wastes

However most of the wastes produced were in liquid form and the high and intermediate level liquid waste was accumulated on the site in tanks, those containing high level wastes being fitted with cooling systems. They were stored separately. They consisted of three types, in decreasing order of activity: high level liquid wastes; concentrates of low activity liquid wastes and decladding liquid waste, forming the category of intermediate level liquid wastes; and finally, used organic solvents.

High level liquid wastes

These wastes had been produced during the combined decontamination cycle in the first extraction column. The solutions containing 99% of the fission products were concentrated as far as possible and then stored. In addition to the fission products whose activity generated heat, the liquid wastes contained acid from the head-end of the process, solvent extraction and the uranium and plutonium cycles. In terms of volume, most of the high level liquid wastes came from the reprocessing of HEU.

The LEU campaigns initially produced large amounts of liquid wastes, placing the storage capacity under threat. For example in 1969 it was believed that the storage system would be saturated by the following year[117]. This problem was dealt with in two ways: concentration and storage.

Various methods for reducing volumes by concentrating high level liquid wastes were successfully introduced. The rapid reduction in activity made it possible to use non-cooled tanks for storing the less concentrated solutions, which were the most abundant by volume. In fact the possibility of reducing the volumes of solutions varied enormously depending on whether they were of LEU or HEU origin.

111 Data up to and including 1974 give the overall tonnage, thereafter the tonnage of radioactive wastes, disregarding the weight of concrete and bitumen.

112 Tritiated wastes, regarded as only slightly radioactive, were dumped in the sea in large amounts as from 1972. NEAR 1972, p. 39: "the best available method for disposing of this radio nuclide is (...) to dilute it in large quantities of water".

113 Source: Retrospective NEA data, 1982 report, p. 36.

114 Data for the first voyage are as follows: 2250 tonnes, 2580 Ci beta-gamma, under 10 Ci alpha.

115 Data for Belgian waste dumping voyage: embarkation on 26 June in Zeebrugge of 2200 tonnes totalling 950 Ci beta-gamma.

116 Dumping of tritium-contaminated wastes from the Amersham Radiochemical Centre.

117 For liquid wastes from the HEU campaigns: RAE 1969, p. 31.

It was possible to achieve high concentrations with the wastes produced from LEU reprocessing. The limiting factors were the salt concentration (100 g/l) and the amount of fission product heat, which had to be removed by the tank cooling systems. In this particular case the upper limit was 6 kcal/h/l, which had not to be exceeded so that even in the hottest part of the summer the temperature did not exceed 60°C. These characteristics are the reason for the relatively moderate volume of high level liquid waste produced by the LEU campaigns, which at the end of reprocessing amounted to 65 m^3[118]. They were stored in two 40 m^3 tanks.

However it was impossible to achieve these concentrations with the HEU since these wastes contained, apart from fission products and plutonium, all the aluminium which had been alloyed with uranium in the fuel in a proportion of about 95% by weight. Concentration was not possible by more than a factor of about two. In this way 800 m^3 were stored at the end of the reprocessing period, in two forms. As and when they were produced, they were fed to two 210 m^3 tanks capable of removing up to 1.5 kcal/h/l, for cooling. When their activity had fallen, they were transferred to the uncooled 260 m^3 storage tanks in the same building as the intermediate level liquid wastes.

The chemical composition of these liquid wastes was reviewed in July 1974 just as reprocessing was approaching its end[119].

Table 82. **Chemical composition of the two categories of high activity liquid wastes in 1974**

Type	LEWC	HEWC	Total
Volume (m^3)	61	789	850
Beta activity of fission products (Ci)	19 500 000	17 270 000	36 770 000
Beta activity per m^3	320	22	n.a.
Sr 90 activity (Ci)	2 660 000	1 660 000	4 320 000 (12%)
Cs-137 activity (Ci)	3 800 000	1 580 000	5 380 000 (15%)
U (kg)	55.0	2.6	57.6
Pu (kg)	1.1	3.5	4.6
Al (tonnes)	0	37	37
Na (kg)	333	12	345
Hg (kg)	0	269	269
Fe (kg)	14	56	70
SO_4 (m)	480	2000	2480
Mo (kg)	141	53	194
Zr (kg)	176	78	254
Ru (kg)	79	25	104
HNO_3 (m)	7	about 2	n.a.

The "cocktail" of elements varied according to the source of the wastes. The LEWC took up less volume than the HEWC but contained more salts.

Each tonne of LEU had produced 0.5 m^3 of high level concentrates, compared with 21 m^3 per tonne of HEU. The nitric solutions in the LEWC were more acid than in the HEWC (7 metres compared with about 2) being more highly concentrated.

Intermediate level liquid wastes: concentrates and decladding solutions

The intermediate level liquid wastes were stored in two buildings, one containing six 260 m^3 tanks, the other four 500 m^3 tanks. The liquid wastes came from two sources: the hot waste concentrates (HWC) and the decladding solutions (JDWC).

The former resulted from low activity liquid wastes which could not be transferred to the CEN. They were concentrated[120] in an evaporator down to the salt (essentially $NaNO_3$) saturation limit. It was possible to achieve very high reduction coefficients: for example in 1969[121] some 2500 m^3 of HW were produced, which were

---

118 This figure included 4 m^3 of liquid waste produced in the 1970 decontamination operation, with an activity of 435 W/m^3. This relatively small volume was achieved by a number of concentration operations.

119 Table produced from VIEFERS W. (not dated), Wastes, Table 6.

120 VIEFERS W. (not dated) on page 8 quotes 3 x 19^{-2} Ci/m^3 as a limit.

121 RAE 1969 p. 31.

reduced to 73 m^3 of HWC, the extent of concentration that year varying between 18 and 113 depending on the batches of material. At the end of reprocessing their average activity was 1000 Ci/m^3.

Every tonne of LEU and HEU had generated 1.5 m^3 of concentrate.

The second category of intermediate level liquid waste consisted of decladding solutions arising from LEU operations[122], transferred from the plant through a special pipeline. These solutions were stored in appropriate tanks ranging from mild steel for solutions containing aluminium to special stainless steels[123], and by categories wherever their mixing led to precipitates, except for magnesium and stainless steel. The activity of these solutions came from the cladding materials made radioactive by irradiation in the reactor and from fission products entrained in the solutions.

The amounts of liquid decladding waste produced varied a great deal depending on the campaign, the difficulties encountered during certain operations and the type of cladding. With regard to the last point, the situation is summarised in Table 83.

Table 83. **Amounts of intermediate activity liquid waste produced according to the type of cladding**

Cladding type	m^3 ILW produced per tonne of uranium
Zirconium ou zircaloy	6
Stainless steel	6
Magnesium	3
Aluminium	2.4

Because of problems with zirconium cladding, some 400 m^3 of related liquid waste remained bound up with this cladding at the end of the reprocessing period. Some 260 m^3 came from the mixture of magnesium and stainless steel decladding solutions, while 140 m^3 contained only stainless steel solutions. The rest of the stock was 150 m^3 from aluminium-clad LEU.

Organic liquid waste from used solvents and diluents[124]

These consisted essentially of used solvent, TBP and its hydrocarbon diluent, Shell-Sol-T (SST), partially decomposed by radioactivity and containing, in addition to zirconium and ruthenium, alpha emitters and some fission products.

Stocks of liquid and other wastes on the site at the beginning of 1975

When reprocessing at Mol came to an end, the stock of liquid waste present on the site can be evaluated as shown in Table 84[125].

Virtually all the radioactivity in the liquid waste (over 20 million curies) can be regarded as having remained on the site. The high level liquid wastes accounted for 91.4% and the intermediate level liquid wastes 8.6%.

In addition to this there was the activity in the high and intermediate level solid wastes.

Thus the stored wastes constituted the inheritance of the years of reprocessing. The liquid wastes had to be stabilized and the solid wastes treated before the international company could consider bringing its activities to an end.

However the cessation of reprocessing did not mean the end of waste production. The experience acquired in 1970-1971 showed that the planned decontamination procedures would generate significant quantities, quite apart from the solid wastes that could be produced by the partial dismantling of certain units.

The changes initiated by the company following the shutdown decision in 1972 were to accelerate as from 1975, the start of the fifteen-year period between the end of reprocessing and the end of the company: the "years of waste management".

---

[122] The HEU campaigns did not produce specific decladding wastes because the U-Al alloy and the Al cladding were dissolved together, so everything entered the high level liquid waste. Rinsing solutions from the LEU 70 to 74 campaigns, containing NaOH and sodium tartrate, were stored with the Al decladding solution.

[123] VIEFERS W. (not dated) p.7.

[124] The solvent and diluent were recycled. Some of the solvents and diluents were also sent to the CEN as low level wastes. This was done in 1968 for 1400 litres of slightly contaminated solvent.

[125] HUMBLET L. (1987) p. 31, and VIEFERS W. (not dated) Table 2 for the liquids, with a few differences indicated in the note.

Table 84. **Stock of liquid wastes in early 1975**

Type of waste	Main components	Mean beta and gamma activity (Ci/m^3)	Volume (m^3)	Total calculated activity (Ci)
I. High level liquid waste (HLLW)	Fission products and...			
LEU liquid waste	HNO_3 (2 m)	178 000	65	11 570 000
HEU liquid waste	HNO_3 (0.5 m), Al^{+++}(1.8 m)	8 400	800	6 720 000
Total high level liquid waste			***865***	***18 290 000***
II. Intermediate level liquid waste				
A. Liquid decladding waste				
Aluminium cladding	NaOH (2 m), Al (2 m) and decontamination solution (NaOH and sodium tartrate)	40	150	6 000
Magnesium cladding and stainless steel cladding	SSO (6 m), MgO (2 m); H_2SO_4 (1.8 m)	800	260	208 000
Stainless steel cladding	SSO (8 m); H_2SO_4 (2 m)	1 000	140	140 000
Zirconium cladding	ZrO (4 m), F (2.7 m), NH_4^+ (1.1 m); NO_3 (0.1 m)	800	400	320 000
Total decladding solutions			***950***	***674 000***
B. Concentrate from the treatment of hot liquid waste of low activity	$NaNO_3$ (5 m) HNO_3 (2 m)	1 000	1 050	1 050 000
Total intermediate level liquid waste			***2000***[126]	***1 724 000***
III. Used solvent	9 to 30% of TBP in kerosene, zirconium, ruthenium, alpha emitters and some fission products	8.5	60[127]	510
Liquid wastes stored on site			***3 875***	***20 014 000***

[126] Strangely enough, the two sources add up to 2100 m^3.

[127] 25 m^3 according to VIEFERS W. (not dated).

Annex to Chapter 4

Alphabetical list and characteristics of the reactors from which fuel was reprocessed by Eurochemic

RR: Reactor used for research, materials testing and irradiation purposes.
PRR: Research reactor in the field of power reactors.
PPR: Pilot or prototype power reactor.
Burn-up in Mwd/tonne of uranium

ASTRA
Site and country: Seibersdorf, Austria
Owner: Österreichische Studiengesellschaft für Atomenergie GmbH
Builder: AMF Atomics (USA)
Date of commissioning: 1960
Power (MW): 5
Reactor category: RR
Family: MTR
Fuel type: U-Al
Cladding type: Al
Dates of reprocessing at Eurochemic: 1967, 71, 72, 74

ATHENE
Site and country: Eindhoven, Netherlands
Owner: Technische Hogeschool Eindhoven
Date of commissioning: 1967
Reactor category: RR
Family: Argonaut
Fuel type: U-Al
Cladding type: Al
Dates of reprocessing at Eurochemic:
1967, 68, 70, 71, 72, 74

BR1
Site and country: Mol, Belgium
Owner: Centre d'étude de l'énergie nucléaire (CEN)
Builder: ACEC
Date of commissioning: 1956
Power (MW): 4
Reactor category: PRR
Family: GGR
Remarks: Graphite-air
Purpose: Physics research, radioisotope production
Dates of reprocessing at Eurochemic: 1967
Burn-up: 900

BR2
Site and country: Mol, Belgium
Owner: Centre d'étude de l'énergie nucléaire (CEN)
Builder: CEN, NDA (Canada)
Date of commissioning: 1961
Power (MW): 57
Reactor category: RR
Family: MTR
Remarks: Tank, H_2O
Purpose: Materials testing
Fuel manufacturer in 1969: MMN-Nukem
Fuel type: U-Al
Cladding type: Al
Dates of reprocessing at Eurochemic: 1967, 68, 72, 74

BR3
Site and country: Mol, Belgium
Owner: CEN
Builder: Westinghouse
Date of commissioning: 1966
Power (MW): 11.5
Reactor category: PPR
Family: PWR
Fuel manufacturer in 1969: Belgonucléaire-MMN-UKAEA
Fuel type: UO_2 (4.4 % enr)
Cladding type: Stainless steel
Dates of reprocessing at Eurochemic: 1966, 68
Burn-up: 3200-7800

CABRI
Site and country: Cadarache, France
Owner: CEA
Builder: CEA, GAAA
Date of commissioning: 1963
Reactor category: PRR
Family: Swimming pool
Purpose: Research reactor for French fast reactor programme
Fuel manufacturer in 1969: CERCA
Dates of reprocessing at Eurochemic:
1967, 68, 70, 71, 72, 74

DIORIT
Site and country: Würenlingen, Switzerland
Owner: Reaktor AG (1960), then the Swiss Federal Reactor Research Institute (EIR) in 1962
Builder: BBC, Escher-Wyss, Sulzer Elektrowatt, Motor Columbus
Date of commissioning: 1960
Power (MW): 30
Reactor category: PRR
Family: HWR
Type: HWCHWR
Purpose: Materials testing, isotope production
Fuel type: U-Al
Cladding type: Al
Dates of reprocessing at Eurochemic: 1967, 68
Burn-up: 50-160

DP
Site and country: Douglas Point, Canada
Owner: Ontario Hydro
Builder: AECL (Atomic Energy of Canada Limited)
Date of commissioning: 1966
Power (MW): 208
Reactor category: PR
Family: HWR
Type: CANDU

Purpose: 1st Canadian power reactor
Fuel type: UO_2 (natural ;0.71% enr)
Cladding type: Zy-2
Dates of reprocessing at Eurochemic: 1970, 72
Burn-up: 6000
Remarks: Contract via the CEA

EDF1
Site and country: Chinon, France
Owner: EdF
Builder: EdF, CEA
Date of commissioning: 1964
Power (MW): 70
Reactor category: PR
Family: GGR
Type: NUGG
Fuel manufacturer in 1969: SICN
Fuel type: Natural U metal
Cladding type: Al (Mg-Zr in 1968)
Dates of reprocessing at Eurochemic: 1968-69
Burn-up: 1200-1500

EDF2
Site and country: Chinon, France
Owner: EdF
Builder: EdF, CEA
Date of commissioning: 1966
Power (MW): 200
Reactor category: PR
Family: GGR
Type: NUGG
Fuel manufacturer in 1969: SICN
Dates of reprocessing at Eurochemic: 1968, 69
Burn-up: 1200-1500

EDF3
Site and country: Chinon, France
Owner: EdF
Builder: EdF, CEA
Date of commissioning: 1967
Power (MW): 480
Reactor category: PR
Family: GGR
Type: NUGG
Fuel manufacturer in 1969: CERCA-SICN-Tréfimétaux
Dates of reprocessing at Eurochemic: 1968, 69
Burn-up: 1200-1500

EL1 (ZOE)
Site and country: Fontenay-aux-Roses, France
Owner: CEA
Date of commissioning: 1948
Power (MW): 0.15
Reactor category: RR then PRR (1953)
Family: HWR
Type: HWCHWR
Fuel type: Natural U metal
Cladding type: Al
Dates of reprocessing at Eurochemic: 1966

EL2
Site and country: Saclay, France
Owner: CEA
Builder: CEA, CENS
Date of commissioning: 1952
Power (MW): 2.5
Reactor category: PRR
Family: HWR
Type: GCHWR
Purpose: Research, materials testing
Fuel manufacturer in 1969: SICN
Fuel type: Natural U metal
Cladding type: Mg-Al
Dates of reprocessing at Eurochemic: 1966

EL3
Site and country: Saclay, France
Owner: CEA
Builder: CENS, Loire Penhoët
Date of commissioning: 1957
Power (MW): 17.5
Reactor category: PRR
Family: HWR
Type: HWCHWR
Purpose: Materials research, radioisotope production
Fuel manufacturer in 1969: CERCA-SICAF
Fuel type: U (1.6 to 4.5% enr), "cristal de neige"
Cladding type: Al
Dates of reprocessing at Eurochemic: 1966, 67;
CdN 68, 70, 74
Burn-up: 1800-15 000

FR2
Site and country: Karlsruhe-Leopoldshafen, Germany
Owner: Gesellschaft für Kernforschung mbH (GfK)
Date of commissioning: 1966
Power (MW): 44
Reactor category: PRR
Family: HWR
Type: HWCHWR
Purpose: Research, fuel elements, isotope production
Dates of reprocessing at Eurochemic: 1967
Burn-up: 900

FRG1
Site and country: Geesthacht (Hamburg), Germany
Owner: GKSS
Builder: Babcock-Wilcox (USA), Deutsche Babcock und Wilcox
Date of commissioning: 1958
Power (MW): 5
Reactor category: RR
Family: Swimming pool
Remarks: Research studies for the future ship propulsion reactor
Fuel manufacturer in 1969: Nukem
Dates of reprocessing at Eurochemic: 1967, 71, 72, 74

FRJ1 (Merlin)
Site and country: Jülich, Germany
Owner: Kernforschungsanlage Jülich (KFJ)
Builder: AEI-John Thompson Nuclear Energy Corporation (UK), Stadtwerke Düsseldorf
Date of commissioning: 1962
Power (MW): 5
Reactor category: RR
Family: Swimming pool
Type: Merlin
Dates of reprocessing at Eurochemic: 1973

FRJ2 (Dido)
Site and country: Jülich, Germany
Owner: Kernforschungsanlage Jülich (KFJ)
Builder: Head Wrighton Processes Ltd (UK), AEG, Rheinstahl
Date of commissioning: 1962
Power (MW): 10
Reactor category: RR
Family: MTR
Type: Dido D_2O_2- D_2O_2
Purpose: Materials testing
Fuel type: Enriched U
Cladding type:
Dates of reprocessing at Eurochemic: 1967, 71, 72, 74

FRM
Site and country: München Garching, Germany
Owner: Technische Hochschule
Builder: AMF Atomics (USA), MAN
Date of commissioning: 1957
Power (MW): 4
Reactor category: RR
Family: Swimming pool
Purpose: Research and isotope production
Fuel type: U-Al
Cladding type: Al
Dates of reprocessing at Eurochemic: 1967, 71, 72, 74

Halden
Site and country: Halden, Norway
Owner: OCDE-Norway (IFA in 1960)
Date of commissioning: 1959
Power (MW): 20
Reactor category: PRR
Family: HWR
Type: BHWR (Boiling Heavy Water)
Remarks: Boiling D_2O
Purpose: Research into boiling heavy water reactors
Fuel type: Enriched U
Cladding type: Zr
Dates of reprocessing at Eurochemic: 1968
Burn-up: 380

HFR
Site and country: Petten, Netherlands
Owner: EURATOM Joint Research Centre (JCR)
Builder: Allis Chalmers (USA)
Date of commissioning: 1961
Power (MW): 30
Reactor category: RR
Family: ORR tank
Type: Modified ORR
Purpose: Research, materials testing
Fuel type: U (93 % enr) - Al
Cladding type: Al
Dates of reprocessing at Eurochemic:
1967, 68, 70, 71, 72, 74

HOR
Site and country: Delft, Netherlands
Owner: Technische Hogeschool Delft
Builder: AMF Atomics (USA)
Date of commissioning: 1963
Power (MW): 0.5
Reactor category: RR
Family: Swimming pool
Dates of reprocessing at Eurochemic:
1967, 68, 70, 71, 72, 73, 74

KRB
Site and country: Gundremmingen (Kreis Günzburg, Bayern), Germany
Owner: Kernkraftwerk RWE/Bayernwerk GmbH
Builder: AEG, General Electric (USA), Hochtief
Date of commissioning: 1967
Power (MW): 237
Reactor category: PR
Family: BWR
Fuel type: U (2.22% enr)
Cladding type: Zircaloy-2
Dates of reprocessing at Eurochemic: 1973
Burn-up: 14 600

LUCENS
Site and country: Lucens (canton of Vaud), Switzerland
Owner: SNA (National Company for the Promotion of Industrial Nuclear Engineering)
Builder: Thermatom, AGL
Date of commissioning: 1966
Power (MW): 5
Reactor category: PPR
Family: HWR
Type: GCHWR
Remarks: HWR
Purpose: Planned in 1960, began production in 1968, closed in 1969 following an accident (rupture of a power tube)
Fuel manufacturer in 1969: CERCA-SICN-Tréfimétaux
Fuel type: U (0.98% enr) (MAGNOX)
Cladding type: Mg-Zr
Dates of reprocessing at Eurochemic: 1971, 72
Burn-up: 14

NPD
Site and country: Rolphton, Canada
Owner: Ontario Hydro, AECL, Canadian General Electric
Builder: AECL, Atomic Energy of Canada Limited
Date of commissioning: 1962
Power (MW): 23
Reactor category: PPR
Family: HWR
Type: CANDU
Purpose: Nuclear Power Demonstration Reactor
Fuel type: Natural U, oxide pellets
Cladding type: Zircaloy-2
Dates of reprocessing at Eurochemic: 1971,72
Burn-up: 5000
Remarks: Contract via the CEA

OSIRIS
Site and country: Saclay, France
Owner: CEA
Date of commissioning: 1966
Power (MW): 50
Reactor category: RR
Family: Swimming pool
Type: High flux reactor
Dates of reprocessing at Eurochemic: 1967, 68, 70-73

Otto Hahn
Site and country: Geesthacht (Hamburg), Germany
Owner: GKSS
Date of commissioning: 1968
Power (MW): 7.45
Reactor category: Ship reactor
Family: PWR
Fuel manufacturer in 1969: Nukem, CERCA
Fuel type: Enriched U
Cladding type: Stainless steel
Dates of reprocessing at Eurochemic: 1974
Burn-up: 7300

PEGASE
Site and country: Cadarache, France
Owner: CEA
Date of commissioning: 1963
Power (MW): 30
Reactor category: RR
Family: Swimming pool and tank reactor
Remarks: Tank, H_2O, CO_2
Purpose: Irradiation of power reactor fuel elements
Fuel manufacturer in 1969: CERCA
Dates of reprocessing at Eurochemic:
1967, 68, 70, 71, 72, 74

R2
Site and country: Studsvik, Sweden
Owner: AB Atomenergi
Date of commissioning: 1960
Power (MW): 30
Reactor category: RR
Family: Tank
Remarks: Tank, H_2O
Purpose: Materials testing, research
Fuel type: U-Al
Cladding type: Al
Dates of reprocessing at Eurochemic: 1967, 68, 71, 72

R3 Adam
Site and country: Ågesta, Sweden
Owner: AB Atomenergi, State Power Board
Builder: ASEA
Date of commissioning: 1964
Power (MW): 12 + district heating
Reactor category: PPR
Family: HWR
Type: Pressurised HWR
Remarks: HWR enriched UO_2 in 1960, natural UO plus D_2O in 1966
Dates of reprocessing at Eurochemic: 1969
Burn-up: 1500-5200

RAPSODIE
Site and country: Cadarache, France
Owner: CEA
Builder: CEA, GAAA
Date of commissioning: 1966
Power (MW): 20
Reactor category: RR
Family: FBR
Purpose: Research into breeder reactors, fast neutron irradiation
Fuel manufacturer in 1969: SICN
Fuel type: U blanket
Cladding type: Stainless steel
Dates of reprocessing at Eurochemic: 1974
Burn-up: 1300

SAPHIR
Site and country: Würenlingen, Switzerland
Owner: Reaktor AG in 1960, then the Federal Institute for Reactor Research (EIR)
Builder: ORNL (USA) 1954
Date of commissioning: 1957
Power (MW): 1
Reactor category: RR
Family: Swimming pool
Purpose: Materials testing, research
Dates of reprocessing at Eurochemic: 1972

SENA (Chooz A1)
Site and country: Chooz, France
Owner: EdF/Centre et Sud (Belgium)
Builder: Westinghouse, Framatome, ACEC, MMN, Cockerill
Date of commissioning: 1967
Power (MW): 266
Reactor category: PR
Family: PWR
Fuel manufacturer in 1969: MMN-CERCA-Westinghouse
Fuel type: U (2.96 to 3.36% enr)
Cladding type: AISI 304 steel
Dates of reprocessing at Eurochemic: 1971, 72, 73
Burn-up: 12 900-19 800

SEP
Site and country: Dodewaard , Netherlands
Owner: Samenwerkende Electriciteits-Produktie-Bedrijven
Date of commissioning: 1968
Power (MW): 54
Reactor category: PR
Family: BWR
Fuel manufacturer in 1969: N.V. Philips Gloeilampenfabrieken (experimental installation at Petten)
Fuel type: U (2.5% enr)
Cladding type: Zircaloy
Dates of reprocessing at Eurochemic: 1971-74
Burn-up: 13 600-17 300

SILOE
Site and country: Grenoble, France
Owner: CEA
Builder: CEA, Indatom
Date of commissioning: 1963
Power (MW): 15
Reactor category: RR
Family: Swimming pool
Fuel manufacturer in 1969: CERCA
Fuel type: U (90% enr)-Al
Cladding type: Al
Dates of reprocessing at Eurochemic: 1968

Trino (E. Fermi)
Site and country: Trino Vercellese, Italy
Owner: SELNI-EDISON-VOLTA, then ENEL (1964)
Builder: Westinghouse

Date of commissioning: 1965
Power (MW): 257
Reactor category: PR
Family: PWR
Fuel manufacturer in 1969: COREN
Fuel type: UO_2 (2.7 to 4% enr)
Cladding type: Stainless steel
Dates of reprocessing at Eurochemic: 1969, 70, 72, 74
Burn-up: 8500-21 000

TRITON I
Site and country: Fontenay-aux-Roses, France
Owner: CEA
Builder: CEA, Indatom
Date of commissioning: 1959
Power (MW): 0.001
Reactor category: RR
Family: Swimming pool
Purpose: Shielding studies
Fuel manufacturer in 1969: CERCA
Fuel type: U (20% enr)
Dates of reprocessing at Eurochemic:
1967, 68, 70, 71, 72, 74

VAK (Versuchs Atom Kraftwerk)
Site and country: Kahl, Germany
Owner: Versuchsatomkraftwerk Kahl GmbH in 1966
(RWE, AEG, Bayernwerk in 1960)
Builder: AEG, General Electric (USA), Hochtief
Date of commissioning: 1962
Power (MW): 15
Reactor category: PPR
Family: BWR
Fuel type: UO_2 (2.3 to 2.6% enr)
Cladding type: Zircaloy
Burn-up: 3800-17 300

Chronology of Part III
November 1965 to January 1975

1965

winter 1965-66: **tests on extraction columns**

8 December 1965: creation of the non-profitmaking association Radioprotection (ASBL Radioprotection) which on 24 January 1966 became Contrôle Radio-protection, or CORAPRO

20 December 1965: *the CNEN signed a contract with Bomprini Parodi Delfino (BPD) for the complete construction of EUREX at Saluggia*

22 December 1965: **Degussa transferred its share to Nukem**

December 1965: *SGN signed a 10 MF contract for Tokai-Mura. The preliminary project had been prepared by Nuclear Chemical Plant Ltd. (United Kingdom), SGN (France) and Weinrich and Dart (United States)*

1966

1966: start of discussions on the Doel power plant (bid by Traction électricité)

1966: *closure of the Hanford REDOX plant*

6 January 1966: *the first load of spent fuel from Latina was shipped to Windscale (262.5 tonnes of U)*

11 March 1966: **Glenn T. Seaborg visited the plant. Start of discussions between EURATOM and the USAEC aimed at resolving the problem of security control at Eurochemic**

24 March 1966: **creation of a special Working Group on finance for the period 1968-1971, which recommended borrowing 1.335 million dollars to cover the additional construction expenditure**

1966: *commissioning of NFS at West Valley*

June 1966: *active tests on UP2 at La Hague*

7 July 1966: **official opening of the plant and start of the first reprocessing campaign**

July 1966: **completion of the high level liquid waste treatment installations**

8 au 14 September 1966: *international nuclear industry exhibition in Basle (NUCLEX 1966).* **Eurochemic took part**

29 November 1966: first electricity generation at BR3-Vulcain

1967

1967: commissioning of the SENA Franco-Belgian power plant at Chooz

1967: *start of construction of WAK (Wiederaufarbeitungsanlage Karlsruhe). The plant was to come into service in 1971*

1967: *Japan created the governmental organisation PNC (Power Reactor and Nuclear Fuel Development Corporation)*

May 1967: **the Special Group approved the Transitional Programme for 1968-1969, submitted to the Board of Directors on 2 March**

August 1967: *the IAEA carried out its first inspection of a reprocessing plant at West Valley (NFS)*

1967: *an initial programme of waste dumping in the Atlantic carried out under the auspices of the OECD. Further programmes were to be carried out until 1982*

1968

26 March 1968: *EUREX handed over to CNEN by its builders*

1 July 1968: *62 countries signed the nuclear Non-Proliferation Treaty in Washington. The NPT came into force on 5 March 1970 after ratification by 43 countries*

2 August 1968: *the Italian Economic Planning Committee (CIPE) decided that Italian fuels should be reprocessed in a single national plant. Launch of the EUREX 2 project*

1968-1972: first Belgian nuclear five-year plan

1968: *commissioning at Fontenay-aux-Roses of a reprocessing unit using a dry process: ATTILA*

1969

1969: **after leaving Eurochemic, E. Shank became director of the Nuclear Project Engineering Division at Allied Chemical**

1969: the Belgian electricity generating companies set up SYNATOM, an organisation charged with co-ordinating certain activities of common interest

May 1969: *signature of a contract between CNEN and SNAM Projetti (ENI) to built EUREX 2*

March-April 1969: **the Board of Directors and the Special Group approved the "Future Programme" or Five-year Programme for the period 1970-1974. On 22 July the OECD Council agreed to the financial arrangements**

1969: *commissioning at La Hague of an AT1 unit for fast reactor fuels (Rapsodie). It was to operate until July 1979*

21 January 1969: *accident at the Swiss Lucens power plant (canton of Vaud) which finally put an end to the national reactor system*

July 1969: *the first charge of Japanese spent fuel arrived at the Windscale plant for reprocessing*

April to July 1969: **an incident in a plutonium evaporator delayed the campaign by 2 months**

22 July 1969: **creation of "founders' shares" by modification to the company's statutes**

September 1969: **creation of the Société de fluoration de l'uranium (SFU) intended to convert liquid uranyl nitrate into solid uranium tetrafluoride on the site**

20 October 1969: *first delivery of fuel from the Latina power plant to EUREX*

November 1969: *France made public its decision to abandon the natural uranium graphite-gas type of reactor in favour of the PWR*

31 December 1969: **signature of the Mühleberg contract**

1970

February 1970: *publication of the FORATOM Report*

4 March 1970: *Treaty of Almelo (Netherlands) between the Federal Republic of Germany, the United Kingdom and the Netherlands on isotopic separation by centrifuge. In July 1971 the CENTEC was established at Bensberg (Federal Republic of Germany), a project preparing the way for the URENCO plant at Capenhurst (United Kingdom) which became operational in September 1977*

5 March 1970: *the NPT came into force for the USSR, the United States, the United Kingdom and 40 Third World countries*

March 1970: *opening of negotiations between the UKAEA, the CEA and the German undertakings to co-ordinate their commercial activities in reprocessing. This led in October 1971 to the creation of UNIREP*

1970: *commissioning of the oxide head-end at Windscale*

1970: *creation of Allied Gulf Nuclear Services (AGNS) owned half by Allied Chemical and half by the nuclear division of Gulf Oil (which later sold its shares to General Atomics). AGNS was responsible for the Barnwell plant project in South Carolina. The project was interrupted in 1977*

1970: *SGN was sold by Saint-Gobain to Pont-à-Mousson*

17 May 1970: **installations shut down after the spring campaign. Reprocessing only started again on 14 February 1971**

summer 1970: **commissioning of the SFU unit**

September-October 1970: **sub-contracting agreement between Eurochemic and the JEN for the reprocessing of Saphir fuel in the "Planta Caliente M1" at Moncloa, the only installation capable of reprocessing HEU fuels with 20 to 100% enrichment in Europe, opened in 1967**

October 1970: *active operations commenced at the Enriched Uranium Extraction Plant (EUREX)*

14 October 1970: **creation of the fourth and final joint undertaking of the ENEA, in conjunction with the IAEA and the FAO, the "International Food Irradiation Project" at Karlsruhe (the project ran until December 1981)**

1971

1971: *reflecting the Treaty of Almelo, France announced that she would build a new gaseous diffusion enrichment plant and was looking for partners. This led in 1972 to the establishment of EURODIF and in 1979 to the commissioning of the Tricastin plant*

1971: closure of the Anglo-Belgian Vulcain programme following the British withdrawal

1971: *in Japan, start of construction of Tokai Mura by SGN and the Japan Gasoline Corporation (JGC) under the supervision of PNC*

8 January 1971: *reprocessing commenced at EUREX*

1971: *in the United States, the Atlantic Richfield Corporation (ARCO) initiates a project for a PUREX plant to be built near Leeds, South Carolina, to be known as the Atlantic Richfield Reprocessing Center (ARRC). These plans did not come to fruition*

March 1971: **the SFU planned to build a tetrafluoride-hexafluoride conversion installation on the site**

spring 1971: **opening of a new storage building for intermediate level liquid wastes**

26 May 1971: **the Board of Directors held captive at Mol for seven hours by the staff**

23 August 1971 : *creation of Kernbrennstoff-Wiederaufarbeitungs-Gesellschaft (KEWA) to prepare the way for UNIREP. The shareholders were Bayer 25%, Hoechst 25%, Gelsenberg 25% and Nukem 25%*

7 September 1971: *commissioning of WAK with fuel from FR2 delivered in 1970. The plant was operational until 1990*

September 1971: *fourth Geneva Conference on the Peaceful Uses of Atomic Energy*

13 October 1971: *the CEA, British Nuclear Fuels Limited (BNFL) and KEWA set up a company under German law, UNIREP, with its head office in Frankfurt (1/3 shares) to market reprocessing services*

19 November 1971: **the Board of Directors decided upon rapid closure of the reprocessing plant**

1972

1972: *the CEA decided to adapt UP2 at La Hague for PWR fuel reprocessing by means of a high activity oxide (HAO) head-end unit. As and when PWR and BWR fuels arrived, fuel from natural uranium graphite-gas reactors was to be reprocessed by UP1 at Marcoule*

January 1972: *first delivery of spent fuel to the Morris MFRP, which was to be commissioned in mid-1972*

1972 : *creation in Switzerland of CEDRA (National Co-operative Venture for the Storage of Radioactive Wastes)*

1972: **leak of slightly radioactive liquid waste following a problem in the pipeline transferring liquid waste to the CEN plant**

25 February 1972: *signature of an agreement for the construction of EURODIF. An Economic Interest Grouping was set up in June 1972. EURODIF became a limited company on 9 October 1973. The Tricastin site was selected in 1973*

March 1972: *the NFS plant at West Valley closed for modernisation and extension. In June the USAEC indicated that because of these modifications, a new licence would be required for the plant to restart. No new contract could be signed. In October, NFS decided not to reopen*

16 March 1972: **review of the Five-year Programme by the Board of Directors, providing for operations to terminate on 1 July 1974, operations on waste solidification to begin, the management to be reorganised and staff numbers to be reduced. This decision was approved by the Special Group in November 1972**

20 April 1972: **ENEA became NEA following the accession of Japan to membership**

November 1972: **negotiations opened with the Belgian government for possible continuation of reprocessing activities on a national basis**

November 1972: **start of construction of a bituminisation installation (civil engineering work completed at the end of 1973)**

December 1972: *BNFL and SGN announced an agreement to combine their reprocessing technology*

1973

1973: **Australia acceded to membership of NEA**

1973: *first deliveries of fuel from light water reactors to La Hague*

1973: in the negotiations about the EURODIF site, Belgium proposed a site near the Tihange power plant

26 September 1973: *accident in the Windscale reprocessing plant, rendering the oxide head-end of the installation unserviceable. It was to be out of service for five years. In 1979 a fire led to the definitive closure of the building*

1974

1974: *creation of PWK by the 12 main German electricity companies involved in nuclear power*

1974: *the pilot plant (SAP) at Marcoule is ready to process fuel from fast reactors in the Plutonium Oxide Process (TOP); these reactors included Rapsodie, KNK II and, in part, Phénix, for which the Fast Oxide Process (TOR) unit was under construction*

1974: *the 1973-1974 oil crisis gave new impetus to reprocessing development schemes*

Germany announced a waste management programme, the Projekt Wiederaufarbeitung und Abfallbehandlung. In February 1974, UNIREP announced the launch of a project for a large reprocessing plant to be built under the responsibility of KEWA. The preliminary design work was to be handled by Uhde and Lurgi

May 1974: *explosion in the Rajasthan desert, India, of a nuclear device produced using plutonium supplied by fuel from a CANDU reactor reprocessed in the Trombay plant*

July 1974: *abandonment of the Morris MFRP plant project after more than two years of fruitless testing. The Aquafluor process did not work. An investment of 64 million dollars had been lost. A further 60 to 130 million dollars had to be spent for adapting the plant to a different process*

1 July 1974: **reorganisation of Eurochemic with a view to termination of reprocessing**

27 July 1974: **the company's life extended by five years to allow it to deal with the problems of wastes prior to liquidation**

19 September 1974: creation of a "Belgoprocess Study Group" to examine the possibility of a takeover by the Belgian government or by a new international group

30 September 1974: establishment of a "Belgoprocess Research Federation"

19 October 1974: *the Tokai Mura plant, under construction since 1971, handed over to its owner. Commencement of operations planned for March 1976*

1975

January 1975: *the USAEC was replaced by the Energy Research and Development Administration (ERDA) which in 1977 became the Department of Energy (DOE), and by the Nuclear Regulatory Commission (NRC)*

10 January 1975: **extraction cycles at the Eurochemic plant taken out of service**

24 January 1975: **end of plutonium purification and oxide conversion operations and hence of reprocessing at Eurochemic**

PART IV

The second life of the company:

From international reprocessing to the management of Belgian radioactive wastes

February 1975 - November 1990

Chapter 1

Shutdown of the plant and preparations for the transfer to Belgium under the Convention of 24 July 1978

January 1975 - January 1980[1]

Once the decision to shut down the plant had been taken, preparations were made for the post-reprocessing period. A programme of work was drawn up, and co-operation continued despite the defection of one country; negotiations also began for the handover of the site.

However the progress of the discussions was delayed by domestic problems in Belgium and by the international criticism of reprocessing. The question of reprocessing had in fact been directly involved in the moves against the proliferation of nuclear weapons in the second half of the 1970s, led by the United States, which promoted a fuel cycle approach that did not include reprocessing. Hence the threat of an international ban on reprocessing, together with the widely varying rate at which nuclear power was being developed in member countries, thoroughly complicated the process of reaching agreement on the future of the installation, a process which in fact was not concluded until 24 July 1978.

During this time the restructured company carried out a safety programme on the site, designed to leave open the possibility of a resumption of industrial activities, and which extended technical operations into fields as yet unexplored on a European co-operative basis.

Preparations for the post-reprocessing period: initial principles, extension of the life of the company and opening of negotiations with Belgium 1973-1975

The programme of work subsequent to 1 July 1974

On 8 November 1972 the Special Group decided that operations should come to an end on 1 July 1974, whereupon discussions began with the Belgium authorities. A joint Eurochemic-Belgian working group was set up with the task of reaching the agreement provided for in Article 32 of the Statutes[2] so as to bring the joint undertaking to an end by fully liquidating the company.

The first issue was to determine the safety conditions that would govern the operations of storage and decommissioning. In June 1973[3] the Managing Director, Emile Detilleux, proposed an overall programme of work which was discussed at the meeting of the Board of Directors held on 26 June 1973, and approved in outline by the Special Group at its meeting on 27 November 1973.

The Board of Directors identified a number of principles as a basis for the work to be carried out following the completion of reprocessing.

The overall aim was to clean up the site, i.e., to decommission the plant and condition the wastes so as to leave the site in a safe condition. However considerable uncertainty remained as to how these operations

---

[1] The chronological arrangement of the chapter follows a technical logic: January 1975 saw the end of the purification of the plutonium resulting from the final reprocessing campaign. In January 1980, decontamination of the plant and various dismantling operations were completed. The chapter could also have been subdivided according to political events, although this would have been less practical: July 1974 was the end of the fifteen-year period foreseen by the Convention for the initial life of the company, and October 1979 saw the first stage of the transfer of the site to Belgium by virtue of the Convention of July 1978.

[2] Article 32: "Upon the liquidation of the company an agreement shall be concluded with the government of the headquarters state, and possibly with the governments of countries in which installations of the company are situated, as regards the possible taking over of all or part of the installations as well as the storage and control of radioactive wastes."

[3] CA(73)7 and CA/M(73)2. NE/EUR/M(73)1.

would be carried out in fact, this being the first time an undertaking of this kind had been applied to a reprocessing installation of this size[4].

It was envisaged at the time that decommissioning and the conditioning of the radioactive wastes would begin in 1974 and be essentially complete by 1981. As far as the wastes were concerned, the lack of any final storage site meant that they had to be conditioned in a transportable form and that temporary storage facilities designed to last at least 50 years had to be constructed. The work would be done under the financial responsibility of Eurochemic, preferably carried out by some of its employees. Although consideration had originally been given to sub-contracting these operations, the Technical Committee had stressed the advantages of using personnel familiar with the site. Work therefore began on this basis as the installations were shut down during the final LEU campaign of 1974-1975.

The extension of the life of the company and the departure of the Netherlands

It soon became clear that liquidation of the company immediately after 27 July 1974 – on completion of the 15-year period originally envisaged by the Convention – would be difficult in view of the amount of work that still had to be done. Therefore on 10 July 1973, the Special Group instructed the Director-General of NEA to undertake discussions with the relevant national authorities with a view to extending the life of the company beyond 1974. These negotiations resulted – after some difficulties[5] – in the Special Group's decision on 10 July 1974 to extend the life of the company for a further five years, until 27 July 1979. However this resolution was not unanimous: Spain and Turkey abstained, although it was accepted that these abstentions would not stand in the way of the decision[6].

The Netherlands had notified the Special Group as early as 4 June 1974 that it wished to terminate its participation, and on 10 July confirmed that it would leave the company before 1 January 1975. The Dutch departure raised the problem of what should become of the country's shares. Since no new shareholder was in the offing, it was clear that they would have to be returned to the company although this would have the effect of reducing the company's capital. Negotiations began on the amount of the "ticket de sortie" [leaving payment], the principle of which had been pointed out by Belgium. The approach to the evaluation was that there should be a contribution to the costs of rendering the site safe. The negotiations were conducted in La Haye by NEA and the ministries concerned. The Dutch authorities insisted that in return for their payment of the "ticket de sortie" their country should continue to receive information, notably on safety problems, and that the Dutch employees of the company would not suffer by the country's withdrawal. They received assurances on these points. In September 1975 the Board of Directors[7] agreed to the sum of 185 million BF as the negotiated cost of the "ticket de sortie". However arguments continued about how this amount should be paid. By the time of the twenty-fourth meeting of the Special Group held on 22 July 1975, during which it was noted that the Netherlands had ceased to be a shareholder on 27 June, the problem had still not been finally resolved. In 1977 the price of the "ticket de sortie" had been set at a non-revisable amount of 180 million BF and the methods of payment up to 1981 were defined.

The Dutch departure was a precedent that caused a great deal of concern to Belgium, which feared a breakdown of that solidarity between member countries it saw as essential for satisfactory negotiation of the agreement which had to be found to terminate the company's activities. In fact the departure of the Netherlands led Belgium to make concessions in order to ensure that agreement was reached. Finally only Turkey, by then a very minor shareholder, left the company in 1980[8]. The remaining countries paid all their dues and remained in the company until it was dissolved.

---

4 Although NFS ceased its activities two years before Eurochemic, its decommissioning encountered substantial financial obstacles. It was not until 1982 that a first decommissioning plan was established by the public authorities, i.e., the state of New York, then owner of the site, and the DOE. In November 1991 the project did not seem to have made a great deal of progress because it was the target of severe criticism in an American report, cf. RGN 1992, 1, 61. In France, decommissioning of AT1 began following its shutdown in July 1979.

5 The meeting of 14 June was unable to reach a decision.

6 On this occasion the Secretariat developed a legal argument concerning "positive abstention".

7 CA/M(75)4, p. 8.

8 Transfer of shares approved by the Board of Directors on 2 July 1980. Turkish participation effectively ended on 26 April 1981. The "ticket de sortie" on this occasion cost 7 996 430 BF. Source: "Ticket de sortie" documentation.

Initial negotiations of the Convention with the Belgian government 1973-1975

The initial position of Belgium

In May 1973 the Belgian Ministerial Committee for Economic and Social Co-ordination (CMCES) held its first meeting to "determine the position to be adopted regarding the desire to liquidate the Eurochemic company". A working group involving the various ministerial departments concerned[9] was established to deal with the question on the Belgian side, in parallel with the Eurochemic-Belgian working group. Marcel Frérotte, who had been Director for Energy since December 1974 and who *ex officio* held the first Belgian seat on the company's Board of Directors, was also the Chairman of the Belgoprocess Federation. He thus played a key role throughout the negotiations[10].

On 27 June 1974[11] the Minister for Economic Affairs, André Oleffe, set out the overall position of Belgium with regard to the programme of work to be carried out, in a letter to the chairman of the Board of Directors. The letter approved the points previously set out in the negotiations[12], continued by the Board of Directors and the Special Group[13] but added two essential conditions for the future of the negotiations, one concerning the long term, the other the short term.

In fact the minister believed that in the long term Eurochemic's responsibility as regards the surveillance and storage of wastes should encompass final storage, hence "in fact extend over many decades, well beyond 1981", which at the time was the date planned for the works to be completed. "The Belgian government considers however that Member countries could fulfil this obligation by depositing a lump sum, to be negotiated, to finance the work and expenditure of the programme planned at present".

In the short term the minister requested that "for a period of three years, as of July 1974, the installations [should be] maintained in a condition enabling start-up of the plant again", a period that would enable the creation of a new company. The "placing on standby", to use the expression that subsequently became current, was thus requested, which had an impact on the way in which the technical operations were carried out in the plant after the end of reprocessing.

Exploration of the possibility of restarting reprocessing on the site in 1973-1975

Creation of the Belgoprocess Study Group

The request for the plant to be placed on "standby" reflected the search by the Belgian government for ways in which operation of the plant could be continued. It was based upon a preliminary study of the possible extension of the plant carried out at the request of Belgium in 1973 by the ERSA group which in addition to Pechiney involved AGIP (Italy), STEAG (Germany), Comprimo and Interfuel (Netherlands), ASEA ATOM (Sweden) and Belgonucléaire[14].

On 19 September 1974 therefore, the CMCES authorised the establishment of a Belgoprocess Study Group, made up of equal numbers of government representatives and representatives of private electricity producers, the latter being grouped together in SYNATOM[15]. The "Syndicat" was set up on 30 September 1974 and actively assisted the Belgian delegation in the negotiation of the Convention. Thus at the meeting of the Board of Directors held on 12 November 1974[16], the Belgian delegate was able to state that Belgium was "very seriously considering" the creation of "a new company which would take over the existing plant with a view to extending its capacity and operating it in an industrial and commercial structure"[17]. The Belgian share in the

---

9 Known as the "interdepartmental working group".

10 Marcel Frérotte was to become the first Chairman of ONDRAF/NIRAS, a post he occupied until the end of his term in 1991. He retired as Director-General for Energy in 1985. His view of the events is a valuable historical overview of the period, FREROTTE M (1987). *Intercommunale* is a quarterly review published by Intermixt, an association of administrators representing the communes and provinces in the joint intercommunal organisations concerned with gas, electricity and communications. The article itself and the final manuscript were provided by its author.

11 Letter reproduced in NE/EUR(74)3.

12 CA(73)7, p. 5 to 7 of 12 June 1973.

13 See above.

14 According to Marcel Frérotte, p. 6, Earl Shank, then a Director of the Belgian subsidiary of Comprimo, believes that a number of European industrialists might have been interested in a restart of activities.

15 90% of Belgium's electricity was generated by private producers. Since 1969 SYNATOM served to group their joint interests in managing the fuel cycle for the Belgian power plants. For example in February 1972 SYNATOM was, together with the CEN and BN, a member of the Belgian Federation for Isotopic Separation which signed the protocol with a view to the construction of EURODIF. In 1984 the Belgian government took a 50% holding in SYNATOM.

16 CA/M(74)2, p. 9.

17 Ibid., p. 10.

new company would be at least 51%, divided equally between the government and SYNATOM. The reprocessing capacity would be raised to 300 t/year. The main objective of the plant would be to reprocess fuel from the five Belgian power reactors, Doel 1, 2 and 3 and Tihange 1 and 2. Studies carried out at the time showed that the small scale of the plant also made it attractive for reprocessing fuels which could not or were unlikely to be processed elsewhere[18], for exotic fuels, and small tonnages[19].

This desire to see a continued domestic reprocessing capacity was entirely in line with the position Belgium had taken earlier. However the wish to turn it into an undertaking in which Belgium had a majority holding was related to the immediate international context, which appeared to be stimulating the growth of the nuclear power industry.

The Belgoprocess Study Group was the Belgian manifestation of the nuclear euphoria of 1974-1975 and the then favourable economic context to European reprocessing

In most European countries having a nuclear power technology, the rise in oil prices following the first oil shock of October 1973-January 1974 led to a decision to accelerate their nuclear power programmes[20]. Belgium was no exception, and in 1974 the electricity companies published a substantial programme for extending their power plants. In certain European countries, the foreseeable increase in capacity renewed hopes of "closing the loop" in the national fuel cycle by building reprocessing plants.

In February 1974, UNIREP had announced the start of preliminary design work on the German reprocessing plant, which was ordered from Uhde and Lurgi under the responsibility of KEWA[21]. A major project combining reprocessing with waste management units was launched within the year and introduced the idea of "Entsorgung"[22]. In July 1975 this resulted in the creation of the PWK[23] company, which brought in most of the German electricity companies to finance the construction by KEWA of a large plant since a capacity of 1500 t/year was envisaged.

Spain and Sweden also dusted off their projects in 1975 and officially announced plans in mid-1976[24] for large plants of 1000 and 800 tonne capacity respectively.

Belgium, which in Eurochemic had a plant and qualified personnel[25] to run it, saw this as an opportunity not to be missed.

From the purely economic point of view there was also a sort of "reprocessing gap" that needed filling rapidly. In fact between September 1973 and May 1975 the Eurochemic plant was the only one in the world capable of reprocessing spent fuel from light water reactors.

The United Kingdom no longer had any workable capacity as a result of the accident of 26 September 1973. France was in the same position, since the work to graft the high activity oxide (HAO) unit on UP2 was still going on at La Hague[26]. The WAK plant was in an extended shutdown and was not returned to service until May 1975, but its reprocessing capacity was less than that of Eurochemic.

In the United States, civilian reprocessing was in a catastrophic state: it had been decided in October 1972 to close the NFS plant in West Valley[27]. In July 1974, General Electric threw in the sponge after spending more

---

[18] Review in SYBELPRO (1983), exhibit 3 p. 3.

[19] This is a return to the "low profile" that certain partners envisaged as the solution during the crisis of 1970-1972, see Part III Chapter 3.

[20] On 5 March 1974 for example, the French Prime Minister, Pierre Messmer, launched the French nuclear power programme. For the origins of the "Messmer plan", see BOITEUX M. (1993), pp. 149-150.

[21] AtW, 1974, 60. However this project soon stalled.

[22] AtW 1974, 340 to 346: "Projekt Wiederaufarbeitung und Abfallbehandlung". The term "Entsorgung" was adopted as an index keyword in AtW for the first time in 1975.

[23] Projektgesellschaft Wiederaufarbeitung von Kernbrennstoffen.

[24] For Sweden see RGN, 1977, 1, 61. For Spain see RGN 1976, 4, 361.

[25] By then, Belgians made up a very large majority of the staff: 122 out of 170 in 1975.

[26] HAO entered service in May 1976.

[27] There are two versions of the West Valley closure. The first puts it in October 1972 when the licence to recommence reprocessing following enlargement of the plant was made conditional upon further work concerning in particular the solidification of plutonium and measures to combat earthquakes and tornados. The cost of this additional work increased the original investment by a factor of 10 and made the viability of the enterprise even more illusory. According to the second version, West Valley was one of the victims of the anti-reprocessing policy of the Ford and Carter governments. This version made it possible to attribute to political action what was in fact the result of an economic and primarily technical failure, as demonstrated by the radiological disaster described in Part III Chapter 4. The requirements of economic viability were also a cause of the unduly high doses.

than two years on fruitless testing of the new Aquafluor process in the MFRP plant at Morris. The AGNS plant at Barnwell (1500 tonnes) was then the only one under construction and the initial cold tests were planned for 1976[28].

Finally in Japan the Tokai Mura plant entered the cold testing phase in November 1975.

Changes to Belgium's initial position

This proposal to create a new Belgian company brought about a change in the country's position regarding the financial obligations of Eurochemic. While those relating to the conditioning and storage of radioactive waste were maintained in their entirety, the decommissioning costs were reduced to one-third, since the new company could consider paying the rest. In 1975 therefore the Eurochemic-Belgian working group explored this new avenue.

The future of Eurochemic in a context of Belgian political uncertainty and international upheaval. The long negotiation of the Convention 1975-1978

However changes in the Belgian domestic situation prevented any decisions from being taken until the autumn of 1976. In addition, at the end of 1975, the international reprocessing situation began to evolve and had completely reversed itself by 1976 and 1977. It is understandable that in these circumstances it was difficult to make progress with the negotiations.

A year awaiting the Belgian parliamentary debate on energy: the first Nuclear Energy Evaluation Committee and the "Report of the Wise Men" of March 1976

The promise of a parliamentary debate

By 1973 the development of ecological sensitivities in Belgium[29] led the government to promise a parliamentary debate on energy policy. The acceleration of the nuclear power programme following the first oil shock led in 1974 to the publication of the "Manifesto of the 400", a critical document signed by 400 scientists and university lecturers.

Establishment of a committee

In preparation for this debate[30], the Minister of Economic Affairs, André Oleffe, established a Nuclear Energy Evaluation Committee, under the co-chairmanship of two university professors from Brussels and Ghent, André Jaumotte and Julien Hoste. The Committee was given six months to "report on the economic aspects of utilising nuclear power, the alternative approaches, health and safety, ecology and finally the fuel cycle. The report was to cover all the relevant problems, including questions, criticisms and concerns both in Belgium and abroad". Besides conclusions, the report was to make recommendations[31].

In addition to its two chairmen, the committee had nine assessors, seventeen rapporteurs and sixty-eight experts. These were essentially university staff, civil servants or employees of nuclear organisations. The public at large was associated with the report only through the right to put questions exercised by "three major trades union organisations" and by "Interenvironnement". The committee was subdivided into eight specialist groups. The eighth group, responsible for the fuel cycle, was made up of 14 Belgians. Of these 12 came from CEN/SCK, and two from Eurochemic: Emile Detilleux and André Osipenco[32]. The time allotted proved too short and the report was not submitted until March 1976.

---

28 The hot tests never took place. The US Nuclear Regulatory Commission (NRC) required the problem of solidifying plutonium and wastes to be resolved as a condition for initiating the procedure for granting the operating licence. The cost of meeting these requirements was put at one billion dollars, in addition to the 360 million already invested. The shareholders then abandoned the project, particularly as the operating licence had not been obtained, largely owing to the general political background described later.

29 We have used the 15 January 1992 version of the text prepared by Baron André Jaumotte for the book prepared under the direction of J. Vanderlinden, *The Nuclear Energy Evaluation Committee Known as the Committee of Wise Men,* as well as HOSTE A., JAUMOTTE A., (1976). The report of Group VIII is 150 pages long, the volume being paginated by Group.

30 The extremely long lapse of time between the announcement of the debate and its actually being held leads one to wonder whether the aim was not to delay it...

31 Draft of the article by A. Jaumotte, p. 1, for VANDERLINDEN, J. ed. (1994).

32 The former was Managing Director, the latter Head of the Health and Safety Division, seconded to Eurochemic by CEN/SCK.

Conclusions in favour of Belgian reprocessing

The conclusions concerning reprocessing did not come as a surprise, owing to the very make-up of the committee: Belgium should reprocess or arrange to have reprocessed all its fuels and develop technologies for processing and storing wastes. The summary report of the groups[33] refers to the scarcity of reprocessing capacity in Europe and the impact on the "small countries" of European reprocessing being concentrated in France, the United Kingdom and Germany: "owing to the shortage of capacity and the lack of enthusiasm of these large countries for keeping other countries' wastes for storage for no return, a few smaller European countries find themselves in a difficult situation with regard to reprocessing, particularly in the short and medium term"[34].

The report evaluates "the requirements to meet the demand from reactors currently ordered (6000 MWe) as 180 t/year in 1985 and 300 tonnes in 1990". It stresses the value of the Eurochemic plant and of the plans for extension "by a Belgian group to a capacity that would meet domestic needs. This could be a very important step in preserving the Belgian nuclear programme".

It was considered that the 300 tonne plant[35] then envisaged would cover the needs of 10 000 MWe of nuclear power plant and could reprocess at a price of 5000-6000 BF/kg, a price "of the same order of that expected in other countries".

The most important point is the change from a proposal for renewed international co-operation to the formulation of a plan for a domestic semi-public company.

The general report, which repeated the conclusions of the specialist reports, provided some further clarification about reprocessing[36]: "steps must be taken to ensure that the spent fuel can be reprocessed reasonably quickly for the following reasons:

- reprocessing is an essential step in the conditioning of radioactive wastes;
- the possibility of using plutonium in reactors means that adequate reprocessing capacity must be developed in due course [national needs were to be of the order of 300 t/year in 1988];
- it appears very difficult at present to conclude long-term reprocessing contracts with overseas firms; in the absence of such contracts, Belgium will need its own reprocessing system with a capacity matching the installed generating capacity;
- further improvements can be made to fuel reprocessing, particularly as regards the emission of volatile radioactive fission products, the conditioning and recovery of plutonium, the conditioning of highly active wastes, the extraction of tritium and noble gases, residual gases, the reprocessing of new types of fuel, and so on. It is essential to continue research in this field with a view to substantial expansion of this sector".

A parliamentary debate quietly forgotten?

The impact of the report was limited, as André Jaumotte writes in his article: "Unfortunately the document submitted to the minister in March 1976 was not followed by the parliamentary debate on energy policy for which it was supposed to prepare the way". It was nevertheless given to the members of parliament and provided a basis for government action.

The most significant result as regards the negotiations about Eurochemic was to delay their progress. However, during the year in which the report was being prepared, the international reprocessing situation had profoundly altered.

Reprocessing and the international debate: American opposition

The impact of the Indian bomb and the first oil shock

India's accession to the club of nuclear powers coincided with the expansion of European technological exports, particularly in the nuclear field, against the background of the first oil shock. This situation revived American fears, started the "battle of plutonium"[37] to use the express of Bertrand Goldschmidt, and created "a gulf in the nuclear policy of the non-communist world".

---

33 Report of group VIII, p. 5 to 7.

34 Ibid, p. 6.

35 This estimate, clearly too optimistic, was intended to "give impetus" to the issue. In fact the tonnage predicted in December 1976 to the Board of Directors of Eurochemic was only 60 t/year (see below).

36 Quoted by FREROTTE M. (1987), p. 9.

37 GOLDSCHMIDT B. (1980) devotes a chapter to the subject, pp. 414-444.

American concern about the dangers of proliferation

India became a nuclear power following the underground explosion that took place in the Rajasthan desert in May 1974. This development was clearly linked to imports of western technology[38]. The United States became more concerned than ever about the dangers of proliferation resulting from the export of fissile materials, reprocessing plants and laboratories.

They had tried, since the middle of the 1960s, together with the other two leading nuclear powers, to slow down and control this threat by negotiating the Non-Proliferation Treaty[39]. Subsequent to March 1970 they redoubled their efforts to increase the number of countries signing the Treaty[40], but the Indian explosion caused them also to use different channels, notably to hold back the enthusiastic exporting countries in continental Europe.

France, Germany and Italy, as exporters of nuclear cycle technology, heightened American concern

During the summer of 1975 in fact, Germany and France signed two contracts directly involving reprocessing that were sensitive from the point of view of proliferation risks, while Italy reached a more general agreement concerning the fuel cycle.

In a nuclear association agreement concluded on 27 June 1975 in Bonn between the Federal Republic of Germany and Brazil, Nuclebras and a German consortium involving KEWA and F. Uhde GmbH in fact signed a contract to build a pilot reprocessing installation with a capacity of 10 kilogrammes of uranium a day using the know-how and experience from the German WAK pilot plant. The announcement of this contract led to a veritable outcry in the American press[41].

At the end of the summer, France negotiated an agreement with South Korea providing in particular for the supply of a pilot reprocessing plant[42]. The pressure put on Korea by the United States was such that Korea unilaterally pulled out of the contract before the end of the year[43]. However France found a new partner and signed an agreement with Pakistan in March 1976[44].

In January 1976, Italy in turn announced that it had concluded an agreement with Iraq, covering the fuel cycle in particular.

On the European side these contracts were primarily motivated by the need to obtain foreign currency to meet the rise in the cost of oil. As such they formed part of the long list of exports of turnkey plants or engineering services of the period. For this reason the American intervention in the Korean affair was seen by France as just one episode in the commercial, monetary and financial war fought around the world after the oil shock[45]. The motivations of the client countries however were probably more in tune with the American concerns, as shown by the yielding of the Franco-Pakistan negotiations described below.

However bilateral pressures were only one aspect of an American policy which was to become more stringent and encompassing both at home and abroad, in bilateral and multilateral relations.

---

38 The first Chinese bomb of 1964, which used U-235 unlike the other early weapons, was probably based on Soviet technology exported during the 1950s, with the construction of an enrichment plant in Sin Kiang. See GOLDSCHMIDT B. (1980), p. 172-173.

The Indian plutonium plant is described in SETHNA H. N., SRINIVASAN N. (1965). The technology of the Trombay reprocessing plant is conventional, based on the PUREX process. Its development was entirely Indian but based upon published international sources. The plutonium for the bomb had been extracted from a Canadian fuel. Homi Sethna, who directed the construction of Trombay, became the Chairman of the Indian Atomic Energy Commission in 1972 and led India's plutonium programme. In 1974 a new reprocessing plant was under construction at Tarapur. India announced in 1978 that it had been put into service the previous year. For the Indian explosion see GOLDSCHMIDT B. (1980), p. 220-222.

39 The fissile materials produced by reprocessing together with the relevant plant items are covered by Article III, § 2 of the NPT: "Each State Party undertakes not to provide: (a) source or special fissionable material, or (b) equipment or material specially designed or prepared for the processing, use or production of special fissionable material, to any non-nuclear weapon state, unless the source or special fissionable material shall be subject to the safeguards required by this article" (§ 1 of the Article).

40 By the end of the 1970s, 111 countries had signed the Treaty.

41 Analysed in LOWRANCE W. L. (1977), pp. 201-221.

42 RGN (1975) 4, p. 301. According to R. Rometsch, a contract was signed at the same time in the greatest secrecy between SGN and Israel for a plant at Dimona.

43 In 1982, Korea returned to the fray when it announced its intention to build a 200 t/year unit. RGN (1982) 3, p. 297.

44 There was in fact a Franco-Pakistan contract to which was added a special trilateral agreement including the IAEA covering inspection of the specific use of the installation, since neither France nor Pakistan were signatories of the NPT.

45 See for example RGN (1977), 2 p. 129 *et seq.*, for the French reaction to the Carter Declaration (see below).

From the "Ford moratorium" to the "Carter abandonment". The end of American civilian reprocessing and the Nuclear Non-Proliferation Act (NNPA), 1976-1978

The American position hardened still further during the summer following a CIA report which threw suspicion on Taiwan when IAEA inspectors had noted the disappearance of ten fuel pins from the country's research reactor.

At the end of October 1976, President Gerald Ford drew the political conclusion as far as domestic activities were concerned when he announced that the United States were abandoning reprocessing and the recycling of plutonium "unless there is a valid reason showing that the world community can in fact overcome the risks of proliferation". This decision ended ongoing American projects. Tests were halted at the Barnwell plant and the announcement made by Exxon Nuclear at the beginning of the year concerning the construction of a major plant – 1500 tonne capacity subsequently to be expanded to 2100 tonnes – at Watts Bar near Oak Ridge, remained a dead letter. As Gerald Ford had stated, "eliminating the risk of proliferation must take priority over economic interests"[46].

The election of President Carter in no way softened this position, indeed it reinforced it.

On 7 April 1977 the new President announced, in the preamble to his energy programme[47], that "the United States is very concerned about the possible consequences to all countries of greater dissemination of nuclear weapons and explosives. We believe that these risks would be much worse if "sensitive" technologies giving direct access to plutonium, high enriched uranium or other materials which could be used to make weapons were to spread". Reprocessing was thus a direct target. This situation led to six decisions being taken. The first was the following: "First of all we shall postpone indefinitely the reprocessing and recycling for commercial purposes of plutonium produced in the United States. We have reached the conclusion, on the basis of our own experience, that *a viable and economic programme of nuclear power can be maintained without either reprocessing or recycling*[48]. The reprocessing plant at Barnwell (South Carolina) will not receive federal support or funding"[49].

The sixth decision concerned the overseas impact of this new policy: "We shall continue to embargo the export of equipment and technologies that could be used for enriching uranium and for chemical reprocessing...[in order to allow countries] to achieve their energy objectives while reducing the risk of the spread of nuclear weapons [...] we shall look into the possibility of establishing an international nuclear fuel cycle evaluation programme with the aim of developing other cycles".

This statement however was somewhat late in the day, because it was a long time since the United States had the monopoly of reprocessing technology. Its aim of course was to freeze the situation.

The natural sequel to this speech was the "Nuclear Non-Proliferation Act" which came into force on 10 March 1978: in the reprocessing field the Act had the result of stopping the transfer of American fuels for reprocessing for a number of years[50].

However the domestic provisions, even though they had an impact on the countries who had signed bilateral agreements, were to be extended by the efforts made at international level. These were deployed around the world.

International doubts about reprocessing

Bilateral pressures

In fact the United States' abandonment of reprocessing was accompanied by an intensification of the pressures it exerted directly on other countries. These were relatively effective and led to a *de facto* moratorium, because on 16 December 1976 the French government announced its decision that it would "until further notice no longer authorise [...] the sale of industrial installations for reprocessing spent fuel to other countries"[51]. However this decision had no retroactive effect and therefore did not concern Pakistan.

---

[46] RGN (1976) 6.

[47] Extract from the statement reproduced in RGN (1977) 2, p. 129.

[48] This statement was effectively the birth of the "Direct Cycle": see below with regard to INFCE.

[49] The statement was the death warrant for the plant. All attempts to save it failed.

[50] A long and excellent on-the-spot analysis of the history and content of the NNPA is given in WIESE W. (1978). It underlines the fact that the provisions of the Act stimulated European enrichment, particularly EURODIF, but was a particular threat to reprocessing, notably the German projects.

[51] Quoted in RGN (1976) 6, p. 356.

The Franco-Pakistan contract was nevertheless not carried through. Discussions dragged on at length and in November 1978[52], negotiations foundered because of the French demand that modifications should be made to render the installation less sensitive, thus reflecting an awareness of the dangers of proliferation denounced by the United States. Pakistan had already paid 88 million French francs out of a total of 168 million, and wished to adhere to the terms of the agreement concluded. In the end, France unilaterally broke the contract, resulting in a financial dispute between the two countries which persisted for many years[53].

The result in Germany was the same but events were in the reverse order. Initially, in March 1977, as a result of joint pressure from the United States and the USSR[54], Germany abandoned its intention to supply the reprocessing plant to Brazil. Later, on 17 June 1977, it abandoned all contractual schemes related to reprocessing.

However American opposition to reprocessing was not limited to transfers of technology and equipment. Thus Japan concluded reprocessing contracts not only with COGEMA, which since 1976 had taken over the fuel cycle activities of the French Atomic Energy Commission (CEA), but also with BNFL on 30 September 1977. During the same month, the United States gave permission for American fuels to be reprocessed in the Tokai Mura pilot plant for a two-year period, on condition, it is true, that installations should be built to treat the plutonium produced in order to reduce the risks of diversion[55]. It was in these conditions that the first Tokai campaign began in November 1977. Finally in 1978 Georges Besse, then chairman of COGEMA, was able to get 32 electricity companies to sign a contract, known in the press as the "UP3 Service Agreement", holding out hopes that a large reprocessing plant would be built next to UP2 at La Hague for reprocessing foreign fuels. The plant was French, but the future clients who contributed to its construction were foreign[56].

The intense American efforts concerning technology exports also made themselves felt through the Club of London.

Multilateral pressure within the international bodies and organisations: the Club of London, IAEA, NEA, IEA

The problems of exporting nuclear technologies were raised during negotiations which had begun in the Club of London[57] in 1974 to result in September 1977 in the adoption of a "code of good practice"[58]. This code was approved by the 15 members of the Club[59] and made public by the American State Department on 11 January 1978[60].

The American abandonment of reprocessing was also accompanied by efforts aimed at organising co-operative and international control structures in the field of the fuel cycle. The chosen forum was the IAEA, but in October 1976 the United States had also become a member of the NEA.

These efforts resulted between 1975 and 1977 in the scheme for regional fuel cycle centres and then, in 1977 to 1980, by the organisation, at the initiative of the United States, of the International Nuclear Fuel Cycle Evaluation, which involved forty countries and four international organisations: the IAEA, which published the results, the European Communities represented by the Commission, and the two specialist agencies of the OECD, the International Energy Agency (IEA), created within the OECD following the first oil shock, and of course the NEA.

---

52 RGN (1978) 5, p. 434. According to oral reports, certain deliveries had already been made, notably a shear unit and drawings. Shortly afterwards, a Swiss firm received an order for the ventilation installation.

53 This was resolved – in fact by changing the problem – during a visit President François Mitterand made to Islamabad on 21 February 1990, through a scheme to sell a nuclear power plant [RGN (1990) 1, pp 46-47], which itself was to lead to a further dispute.

54 Denounced in the German press as "Querschüsse", literally "cross-fire", or underhand actions.

55 RGN (1980), 4, pp. 363-364. In fact Tokai Mura was to reprocess fuels of American origin for which the contracts included a right of American veto on reprocessing. The United States attempted to exercise this right in order to prevent Japanese reprocessing. The IAEA acted as honest broker and proposed a special inspection regime.

56 See Part V Chapter 2.

57 The origin of the Club of London stems from Article 3.2 of the NPT regarding exports. The control of exports by an international organisation came up against the principle of national sovereignty. A committee of the IAEA recommended a simple association.

58 RGN (1977) 6, p. 528.

59 United States, USSR, France, United Kingdom, Japan, Federal Republic of Germany, German Democratic Republic, Canada, Belgium, Netherlands, Italy, Poland, Czechoslovakia, Sweden and Switzerland.

60 The London Agreement encompassed five main provisions which, while not banning exports, fenced them with strict control measures to ensure that they were not used for military purposes. However application was difficult because control concerned specifically nuclear installations, not spare parts or non-nuclear systems.

An attempt to combat the danger of proliferation through internationalisation. Eurochemic at the centre of the project for "Regional Nuclear Fuel Cycle Centres" (RNFCC[61]) of the IAEA

At the end of 1975, as a reaction to the German and French contracts and in view of the difficulties encountered by domestic commercial reprocessing, the United States had launched the idea of setting up "multinational reprocessing centres" which would have enabled reprocessing to be both controlled and financed[62]. In 1975 the IAEA had set up a "Study Project on Regional Nuclear Fuel Cycle Centres"[63], with financial support from the United Nations Environment Programme, the IBRD and the United States government.

Such plans for internationalising reprocessing could only be favourable to the survival of the Mol undertaking and the resumption of international reprocessing, for example under the auspices of the IAEA or EURATOM.

The project was largely structured by Eurochemic or former Eurochemic staff, whose experience was put to considerable use and who held a substantial number of key posts in the different committees[64]. Rudolf Rometsch, a Deputy Director-General of the IAEA, was responsible, with a team from the Vienna Agency, for the final revision of the summary report. The "Eurochemic group" headed five of the twelve committees: E. M. Lopez-Menchero, who by that time had returned to Spain, was the main consultant on the problem of controlling nuclear materials, Pierre Huet and Pierre Strohl took care of the institutional and legal aspects, Emile Detilleux reprocessing, Franz Marcus, who had returned to Denmark, waste management, and H. Keese, who had returned to Germany, the transport of radioactive materials.

The aim of the RNFCC project was: "to examine the economic safety, safeguards and security aspects of *multinational as opposed to a wholly national nuclear fuel cycle approach*, and to develop a methodology whereby member states might make decisions together as appropriate in order to meet their fuel cycle requirements"[65].

The project was published in 1977. It stressed the many advantages of the international management of the sensitive parts of the fuel cycle, similar in many respects to the approach to control by internationalisation already existing in the Lilienthal-Acheson proposal[66] which had remained on the table after its rejection throughout the 1950s until it became, it will be remembered, one of the causes of the birth of the Eurochemic project.

However the United States position with regard to reprocessing had changed between the start of the project in 1975 and its completion in 1977. While the USA initially pushed the RNFCC project, they later attempted to prevent its publication, preferring simply to abandon reprocessing – a view they attempted to promote during the INFCE[67], to the fallback solution involving "international reprocessing". However publication of the report was nevertheless achieved through the joint efforts of the Director-General of the IAEA, Sigwald Eklund, and Rudolf Rometsch. It naturally had no sequel, but this episode, as far as former Eurochemic staff were concerned, was a sort of "honourable struggle" aimed at saving international reprocessing.

Emergence of the "once-through cycle"[68]. The International Nuclear Fuel Cycle Evaluation (INFCE[69])

Discussion of the RNFCC was taken over by the INFCE which lasted from October 1977 to February 1980[70]. The final communiqué of the preparatory conference which was held in Washington from 19 to 21 October 1971 described its organisation and limitations[71]:

---

61 Regional Nuclear Fuel Cycle Centres.

62 This idea had been launched by Henry Kissinger in the early 1970s, as a complement to the NPT.

63 Study Project on Regional Nuclear Fuel Cycle Centres (RFCC). IAEA (1977).

64 IAEA (1977), Vol. 1, pp. 105-123.

65 IAEA (1977), Vol. 2, Foreword, not paginated.

66 See Part I Chapter 1.

67 See below.

68 To apply the term "cycle" to a series of operations ending simply in the storage of the irradiated fuel is something of a misnomer. After all it is the existence of the reprocessing which closes the loop, and without reprocessing there is no cycle but merely a linear series of operations ranging from the mining of the ore through to the storage of the spent fuel.

69 International Nuclear Fuel Cycle Evaluation.

70 IAEA (1980).

71 IAEA (1980), p. 311. The same month, Argentina announced its intention to commence construction of an experimental reprocessing plant at the Ezeiza Centre (Buenos Aires). The country had operated a small laboratory unit from 1960 to 1970.

"The participants [...] are aware of the urgency of world energy requirements and of the need, if these are to be met, to expand the peaceful applications of nuclear energy. They are also convinced that effective steps can and should be taken, both nationally and through international agreements, in order to minimise the risk of proliferating nuclear weapons without compromising energy resources and the development of nuclear energy for peaceful purposes. [...] The participants have agreed that the evaluation should be a technical and analytical study and not a negotiation. Its results will be communicated to governments so that they may take them into account in formulating their energy policies and in the international discussions relating to co-operation on nuclear energy and on relevant controls and guarantees. The results of the evaluation will not be binding upon participants".

The conference was organised into eight international working groups and involved 519 experts from 46 countries and five international organisations. Group Four was concerned with reprocessing, plutonium management and recycling. The first plenary conference was in Vienna in November 1978; the final conference was in Vienna's Hofburg, 25 to 27 February 1980[72].

One of INFCE's most important aspects was that it accepted the existence, amongst the three types of cycle examined, of a once-through cycle not involving reprocessing, the other two being the recycling of plutonium from reprocessing in thermal reactors and recycling in fast breeder reactors (see the cover of the summary report shown in Figure 113, which sets them out in this order, showing the links between them).

Reprocessing[73] is presented as one of the two ways of resolving what is now regarded – following a reversal of values compared with the criteria that had been in force since 1945 – as the "plutonium problem".

In fact until the 1970s, plutonium tended to be regarded as an asset for military and energy purposes. It remains so for countries pursuing breeder projects.

However the INFCE marked a turning point and led to a different approach. A feature of the current context is the prospect of dismantling nuclear weapons and hence the problem of dealing with substantial stocks of plutonium[74]. Another characteristic is that many countries have abandoned the fast breeder reactor. These factors place element number 94 in the category of wastes. However it is a special type of waste because it can be a source of energy. In this context the fabrication of mixed oxide (MOX) fuel[75] can appear as an attempt to utilise a secondary energy source.

At the end of the 1970s, INFCE put the problem in the following way: "plutonium production is inevitably linked to the operation of nuclear power plants. Hence the problem is not that of avoiding such production but rather what to do with this plutonium once it has been produced. There are then two main options: to leave the plutonium as it is in spent fuel elements which are then stored; alternatively to reprocess the irradiated fuel elements and to store the extracted plutonium, or to recycle it in thermal or fast breeder reactors".

Reviewing the stages of the fuel cycle that are most exposed to the threat of fissile material diversion, INFCE regards reprocessing as being the portion most vulnerable to clandestine diversion by governments. This applies to transport, where groups might attempt theft, and plutonium storage as concerns diversion not concealed by governments. INFCE therefore stresses that "it is important in the future, if reprocessing becomes common, to take the most effective steps from the technical, institutional and guarantee points of view, in order to enhance the protection of these materials against diversion"[76].

However as far as wastes are concerned, INFCE stressed the low degree of risk that reprocessing wastes (depleted uranium, vitrified fission products) might be diverted, and suggested that "only spent fuel from once-through fuel cycles should be regarded as liable to diversion, particularly as underground storage facilities are increasingly a preferred target owing to the large amount of fissile materials stored and their declining radioactivity"[77].

In its comparison between the once-through cycle and the cycle involving reprocessing, INFCE simply displaced the point of vulnerability.

---

72 Thus the Eurochemic-Belgian Convention was signed between the preparatory conference and the first plenary meeting of INFCE.

73 INFCE (1980), p. 41 et seq.

74 See Part V Chapter 2.

75 This type of fuel can be used in light water reactor power plants. A present-day MOX composition consists of 92 to 93% of uranium depleted by 0.2 or 0.3%, obtained from enrichment and not reprocessing plants. It will be recalled that the U-235 content in natural uranium is 0.7%. It is combined with 7 or 8% plutonium.

76 Ibid, p. 42.

77 Ibid, p. 43.

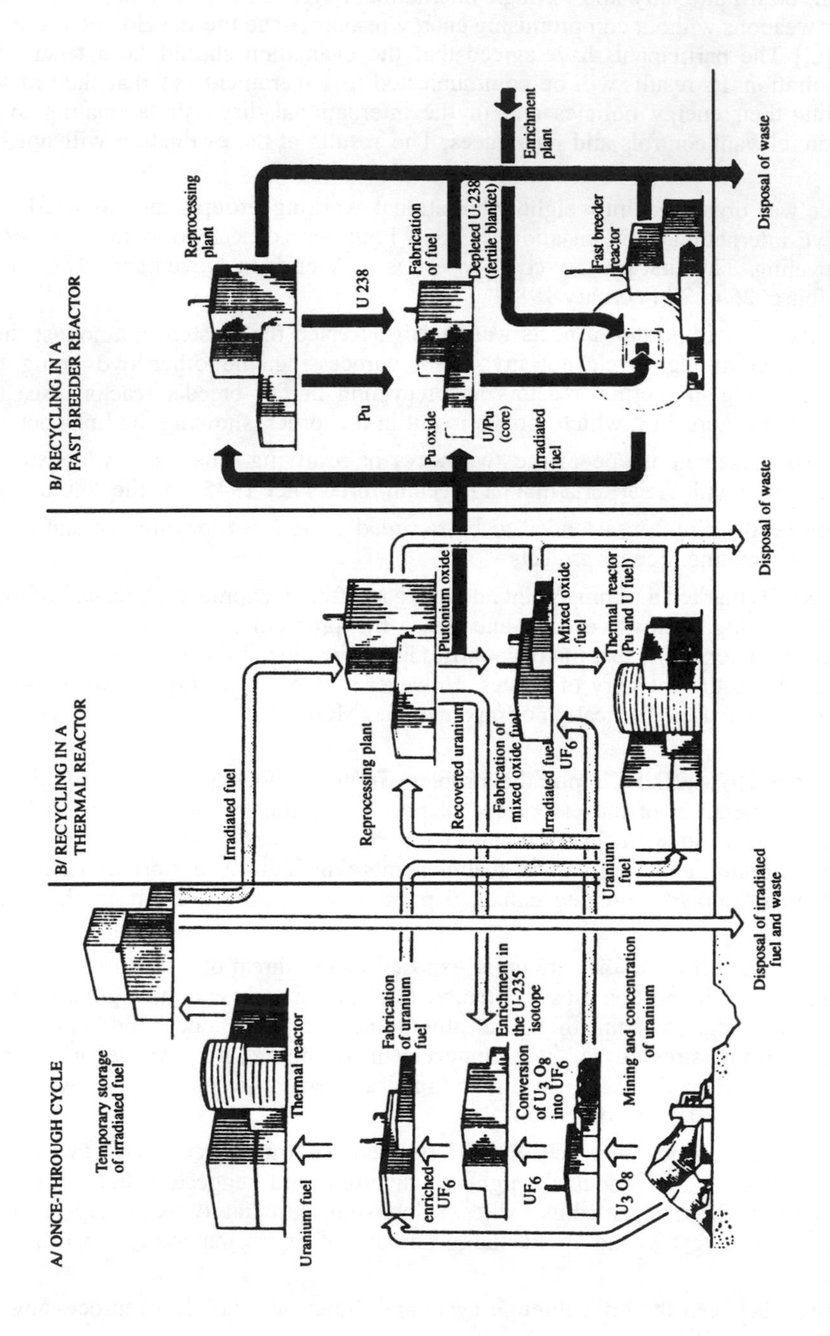

Figure 113. The three fuel cycles envisaged by INFCE (Source: IAEA (1980), cover page from the summary of the study).

INFCE also reviewed technical measures whereby the risks could be reduced. It highlighted the immediate advantages of placing several fuel cycle installations on a single site, and those of mixed conversion whereby mixed U-Pu oxides are produced. It recommended that ultimately uranium and plutonium should not be separated in the reprocessing plants, and that materials containing uranium 235 should be degraded by adding uranium 238 or other neutron poisons. It also proposed, underlining the ecological, radiological and economic hazards, that fissile materials should be protected by the creation of special radiation barriers, such as those resulting from partial reprocessing whereby certain fission products are retained with the plutonium.

After INFCE therefore, the reprocessing cycle and the once-through cycle appeared to have shared the honours.

In fact, thanks to INFCE, the new cycle had gained international legitimacy. However the United States had not given up its fight against proliferation. In any event, as from the end of the 1970s, although reprocessing had not ceased to exist, at least another approach appeared possible. From then on the nuclear power countries fell into two camps, those who reprocessed and those that did not, each camp proclaiming the advantages of its own approach.

To simplify the arguments, the supporters of the once-through cycle pointed to economic arguments (it costs less in the short term), political arguments (reprocessing contributes to proliferation) and technical arguments (it is a dangerous activity). This was the attitude of the United States and was taken up by the opponents of nuclear power in those countries applying reprocessing.

The supporters of reprocessing, for their part, invoke the economic value of recycling raw materials together with the "conservation of resources" aspect since spent fuel still contains over 90% of usable uranium. They also point to the importance – from the ecological and safety standpoints – of separating the different elements so that they can be treated and conditioned in specific ways. In responding to criticism they point to the short-term nature of their opponents' economic views, the lack of experience as to the costs and technical feasibility of the long-term storage of spent fuels left unprocessed. At present France, the United Kingdom, Japan and India are supporters of reprocessing. The clients of the Windscale-Sellafield and La Hague plants are gravitating to the first two.

In view of these profound changes in the international context that began in the mid 1970s, the hesitation of Eurochemic members to continue the reprocessing experiment is easily understood.

Besides these international problems there were also the domestic debates affecting several member countries which caused some of them to decide against developing nuclear power. These internal developments substantially weakened the basis for international co-operation on any reprocessing project.

National doubts amongst Eurochemic members in the second half of the 1970s

The summary table (Table 85) is an attempt to utilise the principal national events to illustrate the positions of the different Eurochemic member countries in 1974 on the prospect of resuming reprocessing[78].

Negotiation of the Eurochemic-Belgian Convention

In these circumstances the negotiations were difficult and progress was slow. On the Belgian side, there was movement only after the Committee of Wise Men had submitted its report.

There was no meeting of the Special Group between 22 July 1975 and 14 April 1978. The meeting planned for 3 June 1976 was first postponed by a month, then cancelled. In fact it took the whole of 1977 to reach agreement. The initiative was no longer with the Special Group, but the Belgian authorities.

The Belgian move of December 1976

On 13 September 1976 Fernand Herman, who had taken over as Minister for Economic Affairs from André Oleffe who died in August, working in conjunction with an *ad hoc* interministerial group, authorised the government's representative on the Board of Directors to negotiate a preliminary agreement before the end of November for submission to the government in early December. The preliminary agreement was to propose two main scenarios. The first envisaged Belgium's taking back the site, either to maintain a reprocessing business on it, which would require transfer of the operating licence, or in order that Belgium could embark upon decommissioning.

According to the second scenario, it was up to Eurochemic to undertake decommissioning.

---

78 The data have been taken from the journals AtW and RGN.

Table 85. **Positions of the different member countries as regards nuclear energy in the second half of the 1970s**

Eurochemic member country in 1974	Principal domestic developments having an impact on Eurochemic co-operation	Position in 1980
Belgium	See developments detailed below	Highly favourable to the resumption of reprocessing on the site
Germany	Autumn 1976: Selection of sites suitable for a major reprocessing plant March 1977: Choice of Gorleben. Increasing protests about Gorleben and the storage site in the Asse mines in 1978. Gorleben reprocessing site abandoned following the demonstration in Hanover on 31 March 1979. Search for new sites in Hesse and then in Bavaria in 1981	Support for projects insofar as they are in line with the current programme for a major reprocessing plant. Assumed its responsibilities. Uses the site for co-operative experiments relating to parts of its own nuclear development programme
France	1976: Commissioning of the HAO head-end of UP2 for reprocessing light water reactor fuels 1978: UP3 Service Agreement. Start of the programme to build the large reprocessing plant for foreign fuels	Support for projects insofar as they are in line with current projects for La Hague. Assumes its responsibilities
Switzerland	April 1978: Law limiting plans for power stations, organising waste management. Reprocessing contracts with COGEMA. Increasing anti-nuclear protests (1979, Kaiseraugst project)	Support for a possible resumption. Still attached to the possibility of diversifying the reprocessing site. Reprocessing on Swiss territory never considered
Austria	5 November 1978: Referendum hostile to the commissioning of the Zwentendorf nuclear power plant. Nuclear power development plans abandoned	Little involved. Assumes its responsibilities
Sweden	1976: Plans for national plant. Public debate on the future of nuclear power. Prior to the referendum of 23 March 1980: agreement for 12 new power plants but plans for leaving the nuclear sector in 2010. Experiments with direct storage	Little involved. Assumes its responsibilities
Denmark	No nuclear power	Little involved. Assumes its responsibilities
Norway	No nuclear power	Little involved. Assumes its responsibilities
Portugal	No nuclear power	Little involved. Assumes its responsibilities
Spain	Hostile to the programme following the decision to close the plant. Plans for a national plant abandoned when the nuclear programme was cut in 1978 End 1979: Signature of reprocessing contracts with BNFL and COGEMA	Limited support. Payment arrears
Italy	National development plans based upon EUREX and ITREC Holds to positions of the 1960s	No support. Assumes its responsibilities within the limits defined in the early 1960s
Turkey	No nuclear power Asked to leave in 1980	No longer feels involved
Netherlands	Withdrawal at the time of the first extension of the company in 1974	No longer involved

The Belgian delegate introduced these new elements at the meeting of the Board of Directors held on 15 September 1976, and a draft exploring the two scenarios was drawn up on this basis. This was submitted to the Board of Directors on 17 December 1976. Immediately beforehand, the Belgian CMCES had decided to opt for the take-back proposal, which left open the possibility of resuming reprocessing. The Belgian idea at the time was to restart a plant of modest size (60 t/year) for reprocessing Belgian fuels.

The CMCES took two other decisions of a similar nature. The first authorised the Minister for Economic Affairs to take all necessary steps to immediately establish the Belgoprocess company with the task of "devising, negotiating and co-ordinating the actions related in particular to the takeover of the Eurochemic installations". The second confirmed that all the operating and other costs of Belgoprocess would be levied on the price charged for nuclear electricity generated in Belgium, thus transferring the costs to the consumer.

Preparations of the Convention in 1977

Accordingly a draft Convention was drawn up by a group of negotiators, comprising the representatives appointed by the Board of Directors – the Chairman of the Board of Directors, Lars A. Nøjd, accompanied by the Director of the company, Emile Detilleux, Pierre Huet as adviser to Eurochemic and Pierre Strohl for the NEA – together with the representatives of the governmental authorities and of the Belgoprocess Study Group. An initial progress report on the task was submitted to the Board of Directors on 22 June 1977[79] and a first draft of the Convention was reviewed on 12 October[80]. A revised draft was submitted to the Board of Directors on 21 December 1977 and the final text on 27 January 1978[81].

Signature of the Convention and its entry into force

The financial arrangements which were to accompany the agreement were the subject of difficult discussions. A preparatory meeting of the Special Group was held on 14 April 1978 in an attempt to resolve the problem. The meeting planned for 2 June was finally postponed to 28 June 1978 since it was not found possible to reach an agreement. This meeting approved the Convention, recommended that governments enter into the financial undertakings necessary for proper execution of the Convention, and decided to extend the life of the company for a further three years until 27 July 1982. The CMCES authorised the ministers involved to sign the Convention on 13 July 1978. The agreement was initialled on 24 July 1978 and became effective only on 30 October when Belgium decided that the financial guarantees were satisfactory.

In fact the coming into force of the Convention was conditional upon the governments entering into financial commitments the total of which had to be considered sufficient by Belgium. It was therefore a rather special bilateral Convention between Eurochemic and Belgium because it was subject to financial commitments by the member states. This was the first time in the history of Eurochemic that government undertakings had been manifest. Hitherto, the company alone had been responsible for its wastes. The recognition of governmental responsibilities for wastes produced by an international company reflects the profound originality of nuclear undertakings compared with enterprises in other sectors, which from the legal standpoint goes beyond the traditional distinction between contractual obligations and legal obligations, extending, beyond "international positive law", into the principles of "international ethics"[82].

The Convention of 24 July 1978, the exchange of letters and the related expenditure

Structure of the Convention

The Convention between the Government of Belgium and Eurochemic on Takeover of the Installations and Execution of the Legal Obligations of the Company was signed on 24 July 1978 in Brussels between the Belgian government, represented by the Minister of Foreign Affairs, H. Simonet, and the Minister for Economic Affairs, Willy Claes, on the one hand, and the Eurochemic company represented by the Chairman of the Board of Directors, Lars A. Nøjd, and the Manager, Emile Detilleux, on the other hand. The text is written in French and Dutch. It comprises 26 Articles grouped into six parts, respectively covering the takeover of the installations (Articles 1 to 9), radioactive waste (Articles 10 to 15), transfer of the operating licence (Article 16), the financial contribution to dismantling expenditure (Article 17), and relations between Eurochemic and the new company

79 CA(77)7.

80 CA(77)9.

81 CA(78)1. The version of 27 January does not include the annexes but a seven-point appendix. It includes an article numbered "11 bis" which was to become Article 12 in the official version of the Convention, which moved one article to the end of the text.

82 Pierre Strohl saw this problem of Eurochemic wastes as demonstrating a new theory of international obligation, Strohl P. (1991).

(Articles 18 to 23). The final clauses (Articles 24 to 26) cover the settlement of disputes and the conditions governing the entry into force of the Convention. The Convention is supplemented by six annexes, the principal ones being:

- a list of buildings and other immovables constructed or to be constructed, for each transfer zone, including a map showing the boundaries of zones and buildings;
- a list of moveable property, with detailed tables showing the work to be carried out by Eurochemic in the zone to be transferred first;
- a list of "works, conditioning and storage of radioactive wastes: work to be carried out by the new company, at the request of Eurochemic, against payment";
- finally a list of services to be supplied and paid for, as appropriate, by Eurochemic or by the new company.

Piecewise transfer to allow rapid resumption of reprocessing

Under the Convention[83], Eurochemic transfers the site (land, installations, buildings, equipment, materials, stocks) to the Belgian government, acting on behalf of the new company to be created. The transfer of ownership, the operating licence and liabilities was to be done in two stages, each covering two of the four zones defined on the site (Figure 114).

The conditioning of high activity liquid wastes from reprocessing

For the high activity wastes, a dual process was planned, adapted to each type of high activity liquid waste.

As concerns the wastes from the reprocessing of highly enriched uranium fuels, their treatment, conditioning and the construction of the corresponding facilities could be entrusted to "another company"[84], on behalf of and under the responsibility of Eurochemic, under conditions to be covered in a separate contract, otherwise by the new company on behalf of Eurochemic, or vice versa.

Wastes from the reprocessing of natural or slightly enriched uranium fuels were to be treated, conditioned and stored in new facilities, these tasks being carried out by the new company on behalf of Eurochemic until all the Eurochemic wastes had been dealt with. This industrial facility would later have the job of conditioning waste from the new reprocessing plant.

These arrangements were at the origin of the subsequent parallel development of two separate vitrification projects. Eurochemic gave the new company the same advantages regarding access to information and patent licensing as it did to its shareholders. The Convention also made provision for the transfer of operating licences and for the preservation of the archives.

The new company running the site on behalf of Belgium would take on staff from the Eurochemic company, according to its needs and as the property was transferred, and undertook to offer equivalent jobs. However a degree of flexibility in the distribution of tasks was allowed.

The "Carter clause" and the exchange of letters

In the January 1978 version[85], an appendix to the annex sets out seven provisions to be covered by an exchange of letters between the Belgian government and Eurochemic, the first of which, which was particularly important, became known as the "Carter clause". Its purpose was to provide for the eventuality that it would not be possible to recommission the reprocessing plant, for reasons related to "a fundamental change in the circumstances existing at the time the present Convention was signed, arising from public provisions in Belgian law or from provisions of international law binding the government or other similar provisions such as to render impossible the recommissioning of the plant for reprocessing irradiated fuels". This was clearly an allusion to the overall world reprocessing situation since the American declaration of 1977 and, more precisely, to the fear of reprocessing being the subject of an international ban. The effect of such an eventuality would not be to terminate the Convention but to modify certain of its financial provisions, in particular the amount of Eurochemic's contribution to decommissioning expenditure[86].

---

83 The following analysis is based upon the Convention, comments made by Marcel Frérotte and on NE/EUR(78)1, Annex II. The buildings shown on the following diagram are identified in Figure 152 at the end of Part IV Chapter 2.

84 Gelsenberg AG had already been selected.

85 CA(78)1.

86 Under Article 17 of the Convention this amount was set at 490 million BF (1977), indexed and paid in six instalments up to the end of 1982. A ban on reprocessing would have increased this amount to 1087 million BF

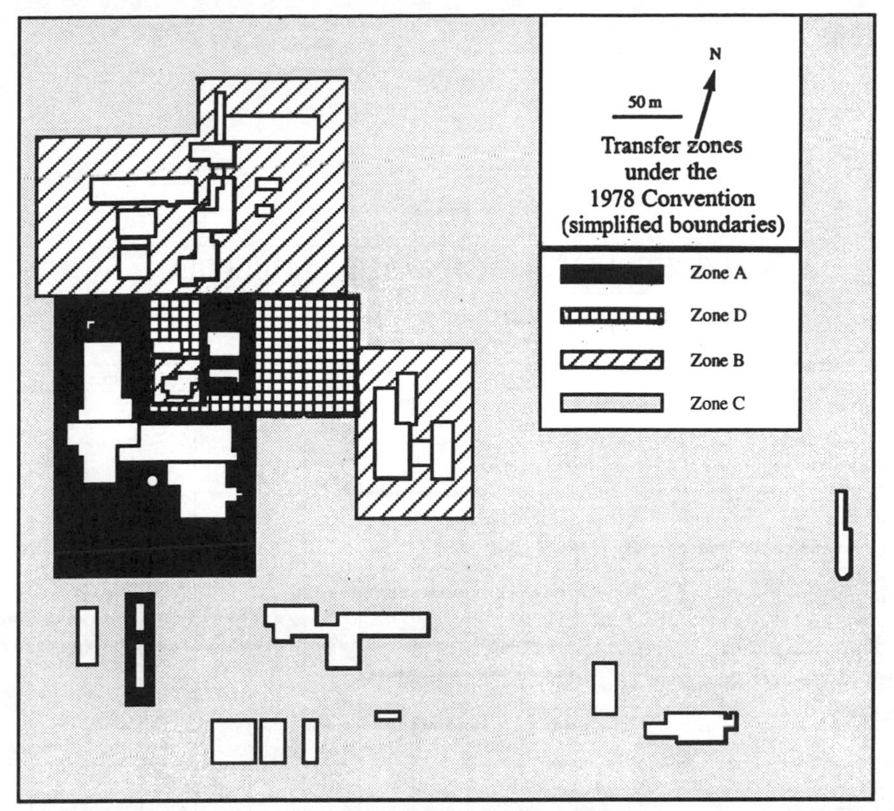

Figure 114. The reprocessing installations (zone A) and a small part of the nearby land (zone D, intended for the installations for treating and storage high activity wastes) were to be transferred on 1 October 1979 (1 April as regards the irradiated fuel storage building). The remainder of the site, comprising in particular the facilities for the treatment of low and intermediate activity waste, the research laboratory (zone B) and the general services buildings and related facilities (zone D) were to be transferred to Belgium on 31 December 1981.

Prior to transfer, Eurochemic was to carry out a number of works and tasks involving cleaning and decontamination (in zones A and B) together with all the operations of conditioning and storage of intermediate activity wastes (the low activity wastes being treated by the CEN). The transfer of zone A was subject to a delay penalty if it affected the restarting of the reprocessing installation.

Advantages of the agreement

The costs to the company arising from the application of the Convention were assessed at 3.656 billion BF at 1977 values[87]. A separate estimate in the same document put the costs to Eurochemic if the Convention were not to be applied at 8.744 billion BF. The difference was that if agreement was not reached, Eurochemic would be required to completely decommission the installation. Under the other hypothesis, the installations would simply be placed in a condition allowing possible resumption of reprocessing. Thus it appeared more advantageous to conclude an agreement, particularly as in those circumstances, Eurochemic could commence liquidation, as planned, after 27 July 1982, and thus discharge its legal obligations at the end of 1988 (Figure 115).

If agreement had not been reached, the company would have had to manage the site, at least up to and including the completion of dismantling and the removal of radioactive wastes, thus considerably extending its life, which its participants no longer desired.

By the beginning of the 1980s, Belgium could hope to inherit a site and facilities allowing her to resume reprocessing. For this purpose she could count on the experience built up in the international undertaking, and was assured of financial contributions for managing the "inherited reprocessing wastes".

After reprocessing had ended and while the Convention was being negotiated, work had been done on the plant and facilities to permit either a resumption of reprocessing or a rapid disengagement by the company.

(indexed). These figures should be compared with those, much higher, put forth by ONDRAF for its proposed dismantling plan in 1987. See the following chapter.

[87] NE/EUR(78)1, Annex III.

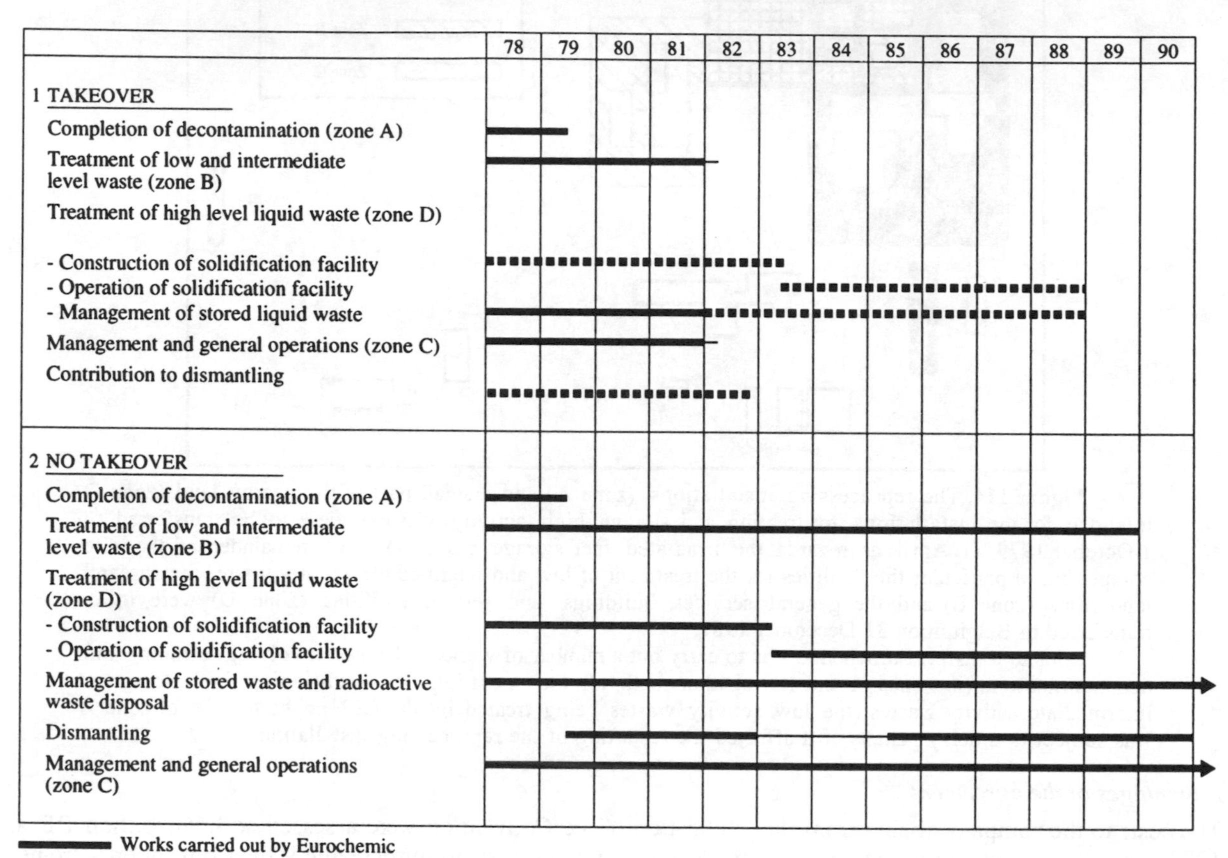

Figure 115. Two work schedule scenarios envisaged in 1978 by the Special Group. At the top, takeover by Belgium would mean no further Eurochemic presence on the site as from 1982 and the end of its financial commitments in 1989. At the bottom, the absence of takeover would prolong Eurochemic presence on the site and funding beyond 1990 (Source: NE/EUR(78)1, p.29).

Shutdown of the plant. Rinsing, decontamination, dismantling and waste production from January 1975 to January 1980[88]

The new company organisation

A new organisation came into force on 1 July 1974, and remained virtually stable for 10 years (Figure 116)[89].

The post of Technical Director disappeared now that reprocessing had come to an end. André Redon left for Marcoule to manage the pilot reprocessing units for fast breeder reactor fuel. The Manager, still Emile Detilleux, was now assisted by a Deputy, Hubert Eschrich (Germany)[90]. Apart from two small Documentation and Programming Sections, to which a Technical Secretariat was added in 1977 and a Legal Office in 1978, the company had five functional departments, formerly known as divisions – with the following staff numbers as at 31 December, in decreasing order:

Table 86. **Eurochemic staff numbers by function, 1975 to 1979**

	1975	1976	1977	1978	1979
Plant operations department (POD)	59.5	59.5	63	65	83
General services department	44.5	44.5	48	47	56
Industrial development department	25.5	25.5	26	25	28
Health and Safety department	21.5	21.5	26	25	31.5
Administrative department	13	18	18	20	18
Management and miscellaneous	6	6	9	10	9.5

Table 87. **Eurochemic staff numbers by function, 1975 to 1979 (percentages)**

	1975	1976	1977	1978	1979
Plant operations department (POD)	35.0	34.0	33.2	33.9	36.7
General services department	26.2	25.4	25.3	24.5	24.8
Industrial development department	15.0	14.6	13.7	13.0	12.4
Health and Safety department	12.6	12.3	13.7	13.0	13.9
Administrative department	7.6	10.3	9.5	10.4	8.0
Management and miscellaneous	3.5	3.4	4.7	5.2	4.2

In 1977 there was so much work to be done that staff numbers went up for the first time since 1966, mainly in the plant operations and general services departments, with increasing use being made of outside staff. For example in 1978 the company had 193 employees, of whom 160 were working in controlled areas. Also in that year, 52 outside staff were working in controlled areas, nearly one-third of the workforce, an unprecedented proportion.

Staff numbers thus evolved as shown in Table 88.

Table 88. **Variation in Eurochemic staff numbers, 1975 to 1979**

Year	1975	1976	1977	1978	1979
Staff numbers	170	175	190	193	208
Change	0	+5	+15	+3	+15
% change over previous year	0%	+3%	+8.6%	+1.6%	+7.7%

88 Sources: RAE 1975-1980, DETILLEUX E. (1978).

89 Liquidation of the company led to further simplification of the organisation chart and the merger of the General Services Division with the Operations Division.

90 A refugee from East Germany at the end of the 1950s, Eschrich was a scientist, initially self-taught. He worked at Kjeller and obtained his doctorate at the University of Gothenberg based upon the work he had done at Eurochemic.

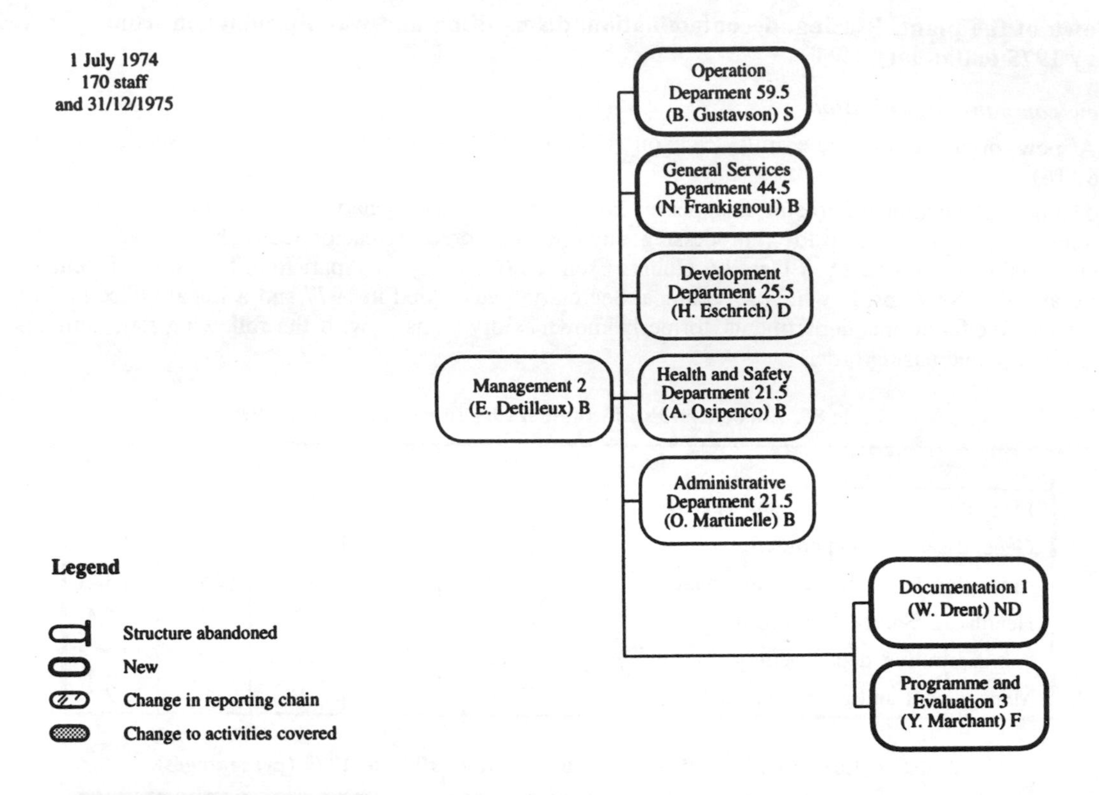

Figure 116. Organisation chart of the company from 1 July 1974 to the end of 1975 (Source: RAE 1974 and 1975).

Preparations for the technical programme

General principles

Preparations for the technical safety programme had been preceded by the definition of principles already described above. The programme involved dismantling the plant followed by temporary storage of the wastes for a period of at least 50 years. The intention at the time was that the long-term management of the site would be entrusted to the specialist departments of the CEN/SCK[91].

The closure programme provided for a level corresponding to level two on the IAEA decommissioning scale to be reached, although with arrangements allowing level three to be attained subsequently[92].

Table 89. **The three IAEA levels of decommissioning**[93]

Decommissioning level	Nature
1	Barriers
2	Partial release
3	Access to the installation or land without restriction

Thus level 2 corresponds to a partially decommissioned site or "restricted site release"

The choice of this intermediate level of decommissioning reflected the uncertainties about the resumption of reprocessing. It also meant that all the wastes resulting from operations and decontamination, together with all the decommissioned facilities, were conditioned with a view either to sea dumping (low activity waste) or storage. It also required the decommissioned buildings to be cleaned and decontaminated to an extent allowing

[91] In fact the legislation which subsequently – in 1980 – created the Belgian National Agency for Radioactive Waste Management, ONDRAF/NIRAS, provided that ultimately these services would be transferred from the CEN to the Agency. Thus finally the management of the radioactive wastes was handed over to Belgoprocess, a wholly-owned subsidiary of ONDRAF, which thus became Eurochemic's successor on the site.

[92] Completely decommissioned site or "unrestricted site release".

[93] Source: Blanc D (1991), p. 105.

the ventilation system to be shut down, and not subjected to continuous surveillance and maintenance; finally that they were kept sealed and inaccessible without permission.

Phases of the programme

The technical programme essentially involved three tasks spread out over time: in order, cleaning and decontamination, dismantling of equipment, and conditioning and storage of the wastes. Cleaning and decontamination were themselves to be carried out in three stages:

Thorough rinsing of the process systems, using the actual reprocessing reagents, in order to recover as much of the residual fissile materials as possible. The decontamination of the process circuits and the outer cell surfaces would allow personnel to enter the active areas and cells and identify any areas still highly contaminated requiring the installation of local shielding.

Finally this would be completed by partial dismantling and the installation of specific decontamination loops in order to eliminate any residual "hot spots".

In fact these activities were fairly similar to the preparations for direct maintenance carried out several times during the reprocessing period, particularly during the May 1970-February 1971 shutdown.

Practical organisation and full-scale experiments

The practical organisation of the work was the same. The interventions were carefully prepared and organised by the HWP and PREP, facilitated by the fact that the personnel involved had taken part in plant operations and so had a good "practical knowledge of the installations and the technical operating records (drawings, modifications, operations reports, incident reports)"[94].

The amount of work to be done was estimated with reference to three particular cells for which a detailed study was carried out. This showed that it would be necessary to remove a 5 centimetre thick layer of concrete from the entire back surface of the cell and the inside walls to a height of 2 metres in order to attain the ambient radiation level of 0.2 mSv/hour originally planned.

It also became clear that it would be necessary to use a team of 40 people divided into five groups, assisted by ten people responsible for radiation protection, logistics and co-ordination, i.e. 50 people from a total workforce which totalled 170 at the time, including 59 in the Operations Department and 21 in Health and Safety. Each group had a team manager, four specialists responsible for cutting and dismantling and three workers whose job was to remove waste, clean the site and bring in supplies.

It was then estimated that the job would take 62 months, or 240 000 man-hours of actual work, operations in the cells accounting for 60% of the total.

It was estimated that the amounts of liquid and radioactive wastes resulting from the dismantling process would be between 300 and 500 m^3 of concentrated decontamination solutions in intermediate activity waste. Similarly the quantity of solid waste arising from the work and dismantling was predicted at 4000 m^3 prior to conditioning.

Programme sequence from January 1975 to January 1980

The programme began in January 1975 with rinsing and decontamination operations.

Rinsing was carried out at the beginning of 1975. The first phase of decontamination took up the remainder of 1975 and continued in 1976 and 1977. It was completed in 1978, followed by final decontamination and dismantling. The rate of progress was such that the work could not be completed by October 1979, the date set out in the Conventions for transfer of the first tranche. Operations came to an end at the end of January 1980.

Belgium's decision to resume reprocessing, which was officially announced to the Board of Directors on 17 December 1976, resulted in modifications to the programme during the initial decontamination phase. The extent of dismantling was reduced and limited to those systems and parts of systems whose condition made them unsuitable for re-use. As a result the ambient levels to be attained in the cells where equipment was to remain was allowed to exceed the originally set target of 0.2 mSv/h. In practice it was 1 mSv/h.

---

[94] Detilleux E. (1978), p. 185.

Rinsing operations and the recovery of residual fissile material. January - June 1975. The MUF rediscovered

These operations resembled the rinsing carried out between HEU and LEU campaigns during reprocessing, although it was more thorough.

The reagents used were essentially dilute and concentrated nitric acid, uranous nitrate in a nitric solution as a plutonium reducing agent, the solvent used for extraction, TBP, its hydrocarbon diluent, Shellsol T, caustic soda and sodium carbonate.

These liquids were circulated, stirred and heated by the same systems that were used during the reprocessing process, such as ejectors, air lifts, syphons, pumps, stirrers, etc.

The progress of rinsing was monitored by an intensive programme of sampling and analysis. After four months, the concentrations in the rinsing solutions had stabilized at under 5 mg/l of uranium and less than 1 mg/l for plutonium. Rinsing was halted at this point. In fact to recover the remaining fissile materials would have required the use of more corrosive solutions, but these might have threatened the integrity of the systems if reprocessing was resumed.

Stabilisation of the plutonium content in the rinsing solutions also meant that a considerable mass of the element, probably equivalent to the amounts unaccounted for, would remain inside the installation until reprocessing was resumed or decommissioning completed. It also meant that decommissioning would have to be done at least in part in the special conditions required by the presence of alpha emitters.

However the rinsing process did reduce the MUF. In fact 38 kilogrammes of uranium and 1275 grammes of plutonium were "removed" from the plant, reducing the quantities of fissile materials not accounted for at the end of reprocessing from 2.37% to 1.88% for plutonium and from 0.38% to 0.22% for uranium[95]. Over 60% of the plutonium recovered arose from rinsing the purification cycles and 37% from the oxide conversion unit alone.

The first phase of decontamination

From the outset the strategy followed was to use the "mildest" possible reagents. In fact there were two disadvantages in using too highly corrosive agents such as fluorides, sulphuric acid or hydrochloric acid: it could lead to leaks in the installations during cleaning and to contamination which would complicate the operations in progress. Also their presence in the decontamination solutions could make it more difficult to condition the wastes.

The announcement of Belgium's decision to take back the installations did of course confirm to the technical teams that this choice was the right one.

The agents mostly used were nitric acid and caustic solutions – together with oxidising or reducing agents and complexing agents as appropriate[96]. As in the rinsing phase, the greatest possible use was made of plant systems for circulating, stirring and heating the solutions. The decontamination campaigns were arranged in 120-hour weekly cycles, with three shifts working five days a week.

The results of this initial decontamination phase expressed in terms of contact and ambient dose rates differed widely from cell to cell. Table 90 shows these results in decreasing order of ambient maxima in mid-1978[97].

The reason for the high dose rate in the dissolver cells, now fitted with remote manipulators, is that the dissolver tank had not yet been decontaminated "owing to various operations carried out in the loading cell connected to it"[98]. In fact the loading cell was used for reconditioning (in sealed canisters and baskets) the high activity solid wastes which had been stored there as they were produced in the Solid Waste Pond, and for cutting up highly contaminated heavy items such as the insoluble baskets from the second dissolver and its neutron shields. The ambient dose rate in the fission product concentration cell was due essentially to the high dose found on the bottom of the evaporator (1200 mSv/h), which made it necessary to install local shielding.

All in all the rinsing and initial decontamination operations resulted in some 140 000 Ci of activity being removed from the plant.

---

95 As far as uranium was concerned, the conditions in which the operations were carried out made it impossible to distinguish between HEU and LEU.

96 Detilleux E. (1978), p. 181 lists the decontamination agents used and the quantities: 12 m^3 nitric acid: 100 m^3; 8.5 m^3 caustic soda: 122 m^3; 30% hydrogen peroxide: 1 m^3; SST: 0.180 m^3; hydrazine: 2350 kilogrammes of nitrate; oxalic acid: 1250 kilogrammes; potassium permanganate: 750 kilogrammes; sodium carbonate: 1000 kilogrammes; sodium citrate: 800 kilogrammes; sodium tartrate: 1800 kilogrammes.

97 Ibid p. 181.

98 Ibid p. 182.

Table 90. **Contact and ambient dose rates in some of the plant cells during the decontamination phase**

Cells	Maximum in contact with equipment	Dose rate (mSv/h)	
		Maximum ambient	Minimum ambient
Dissolving	300	30.00	20.00
FP concentration	1200	30.00	10.00
Solvent treatment	170	4.00	0.35
Co-decontamination cycle	25	3.00	0.50
Separation cycle and second U cycle	6	3.00	0.10
Dissolution gas treatment	30	1.00	0.20
Second Pu cycle	8	0.75	0.02

The final decontamination phase using mild methods

The final stage of decontamination was in progress at the time the Convention was signed, and work continued until the end of January 1980[99], i.e., a few months later than the transfer date provided for in the Convention.

The initial requirements had been relaxed owing to the then likely prospect of the plant being recommissioned, which required the risks of corrosion to be minimised. The methods used were made even less harsh and the use of aggressive reagents was banned. Dismantling was limited to the strict minimum necessary to give access to contaminated areas in order to avoid subsequent rebuilding.

Therefore the final stage of decontamination of tanks and pipes was essentially done with water jets at high pressure (up to 400 bars [568 psi] and 20 m^3/h). This was done by an outside contractor, Smetjet, a specialist in the use of high pressure water jets to clean industrial equipment and which had mobile systems mounted on trucks for the purpose. The decontamination factors varied widely, from 20 to 100, depending on the degree of accessibility to surfaces. However the large amounts of effluent were very easy to concentrate and the concentration factors substantial owing to their very low salt content.

The outer surfaces of certain systems, cell walls and areas contaminated in incidents or partial dismantling were cleaned by brushing with water, with high pressure jets or by removing the contaminated surface layer mechanically, depending on the degree of contamination.

Dismantling operations

The dismantling programme, having been reduced to the minimum, involved only those installations where there was a risk of contamination or irradiation, or which could not be, or were no longer, considered for re-use/

In the plant it was necessary to dismantle the loading tubes of the first and third dissolvers (Figures 117-119)[100], a multipurpose evaporator which was heavily corroded and a phase separator which could not be decontaminated *in situ*. Also some 8000 litres of boron glass Raschig rings which had been used as neutron poisons in 16 tanks had to be removed, to allow the tanks to be decontaminated[101]. In addition a number of pipes and ejectors blocked by solid deposits were cleared or cut out.

In the analytical laboratory, the various internals were removed from gloveboxes and other enclosures. Following initial local decontamination, the gloveboxes were taken to a special area where they were further cleaned until they were suitable for re-use.

This was not the case of the plutonium oxide production unit which had caused many problems in operation and which was given special treatment owing to its high plutonium contamination. The amount of plutonium remaining in the enclosures and systems owing to operational problems was calculated to be 3.1 kilogrammes. The entire system was cleaned, dismantled, cut up and placed in plastic bags which were then hermetically sealed by welding. The bags were then placed in 190 28-litre metal canisters[102]. These operations involved 12 men for three months (1000 man-hours), their total absorbed dose being 60 mSv.

---

99 CL(86)1 examines the financial consequences of the operations.

100 The first dissolver was no longer suitable for the fuel characteristics as regards burn-up, and MTR fuel could no longer be processed in the third. See Figure 113. Figures 118 and 119 show other aspects of the work in the plant at this time.

101 HILD W., DETILLEUX E., GEENS L. (1980), p. 203.

102 DETILLEUX E. (1978) p. 182: The unit "had six alpha enclosures protected by gamma shields. These enclosures, which were interconnected, contained the systems for the precipitation and calcination of plutonium oxalate, together

Figure 117. Using a saw to dismantle the main tube in the third dissolver in 1978 (Source: Eurochemic colour slide, undated).

Conventional tools[103] such as diamond-studded saws, metal saws and twist drills were used for dismantling, together with normal working and protection systems – vinyl suits, ordinary or breathing air masks, moveable local shielding.

Generally speaking, the operations required workers to be in the cells since there were no remotely controlled machines or robots for technical and cost reasons.

This predominantly manual work is the reason for the doses received, which were carefully controlled by the planning of the operations and monitored by the radiation protection team.

Radiological review of the decontamination and dismantling work at Eurochemic[104]

Radiation protection principles: the rinsing and decontamination phase

The procedures applied during these phases were not fundamentally different from the maintenance operations carried out between the reprocessing campaigns. Each operation was planned by an HWP or PREP and was carefully monitored. As a general rule, any unexpected circumstances resulted in the work being

---

with those used for conditioning and sampling the plutonium oxide. After being cleaned and decontaminated, the systems were dismantled and cut into pieces smaller than 30 centimetres in size. The cutting operations were carried out inside the enclosures using normal glovebox techniques (...). After removal of internals and a final cleaning, three coats of plastic paint were sprayed on the inside walls of the enclosures; the outer walls were then given two coats of paint. To prevent any contamination during separation of the enclosures, the box being removed was slid into a plastic bag, the end of which was sealed on the other box. After separation of the two boxes, the bag was closed by a double weld. Each box was then wrapped in a second plastic bag before being placed in a wooden box for shipment to the dismantling shop" [from the CEN/SCK].

[103] A short review of the techniques used – both routine and innovative – for decontamination and dismantling in the decommissioning of nuclear installations at the end of the 1970s is given in OECD/NEA (1980 a and b). Automated techniques are at the experimental stage, for example the VIRGULE prototype in France described in CREGUT A. (1978), p. 171.

[104] OSIPENCO A., DETILLEUX E., FERRARI P. (1980).

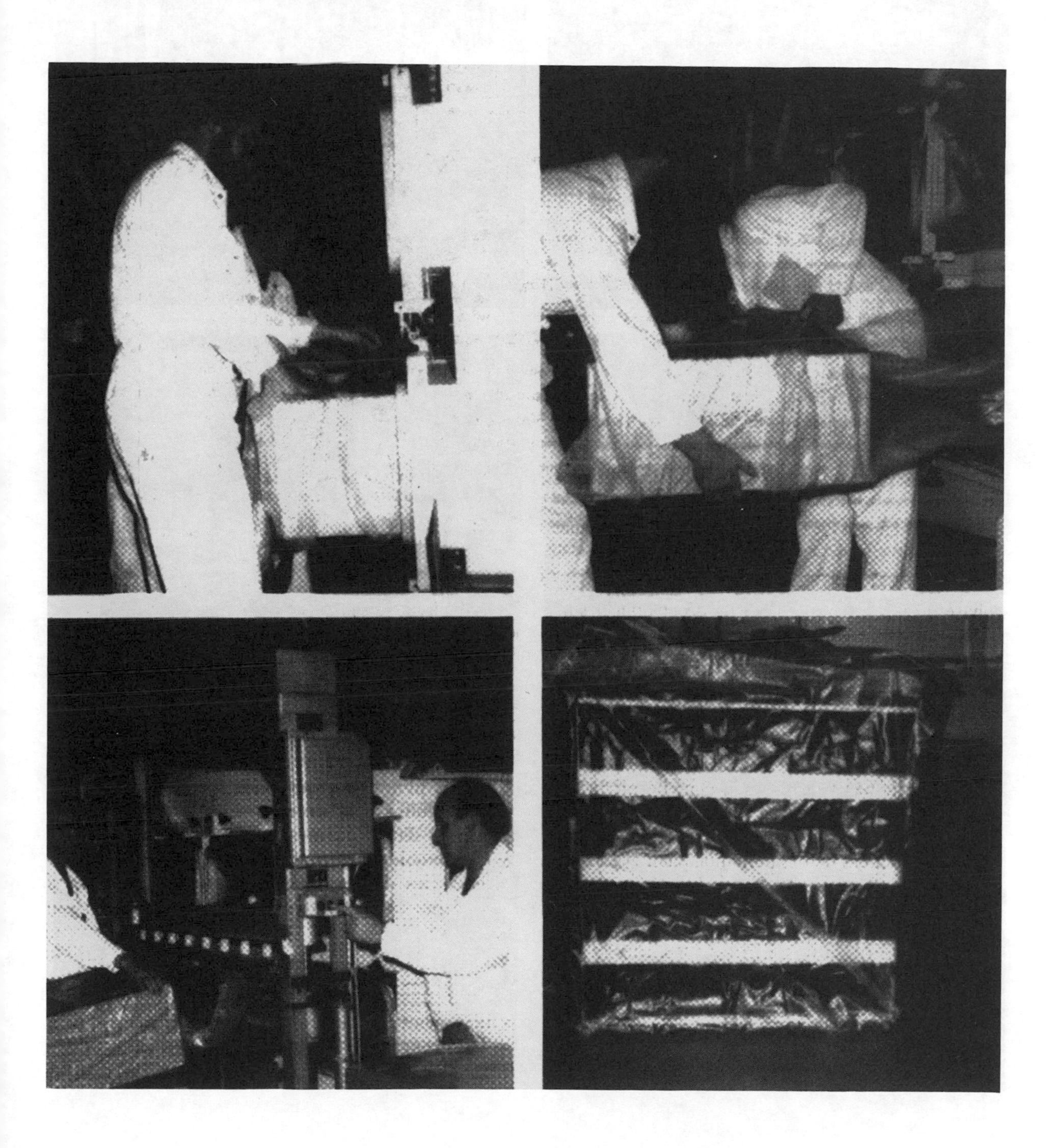

Figure 118. Replacing an absolute filter in the plant ventilation system room. Two operators slip the used filter into a plastic bag which is then hermetically sealed. The filter conditioned in this way is removed and conditioned as waste (Source: RAE 1981, Figure between pages 8 and 9).

Figure 119. Some areas of the plant after decontamination, repairs and painting with a view to the resumption of reprocessing. The left-hand column shows the condition of the areas after decontamination; the right hand column shows the condition after refurbishment and painting. At the top, in the former dry plutonium separation zone, n° 29. In the centre, Cell 13 which contained the uranium process tail-end pumps. At the bottom, plutonium preparation zone 22 (Source: RAE 1981, photographs between pages 8 and 9).

temporarily halted for an evaluation, with the work plan being modified if appropriate. During a five-year period 5238 PREPs and 215 HWPs were issued[105].

The work teams were equipped with hooded suits made of PVC or "tyvec" for operations in dry conditions, rubber gloves and boots and a mask fitted with a dust filter or, if the air contamination exceeded 10^{-7} Ci/m^3, a mask fitted with an external breathing air supply. However in certain congested cells or for tasks where access was difficult, a mask without an air line had to be used.

For external dosimetry the methods used were the same as during the period of reprocessing, with internal dosimetry based on analysis of urine samples. If unduly high results were found, the CEN/SCK services made further analyses of faecal matter or carried out lung and whole body scans. During the decontamination period, "no significant internal contamination was found for the workers involved" [in the operations of rinsing and decontamination][106].

Doses received

Table 91. **Annual collective doses, 1975 to 1980**[107]

Year	Collective annual doses (mSv)
1975	1566
1976	1462
1977	1746
1978	2250
1979	1963
1980	1264

Table 92. **Mean doses, 1975 to 1980**

Year	Mean annual dose (mSv)
1975	9.3
1976	8.4
1977	9.7
1978	10.6
1979	9.3
1980	6.0

There was a rising trend in collective dose between 1976 and 1978, reflecting the intensification of the decontamination programme. The relative stability of the mean doses is due to the increased workforce and the use of contract staff[108].

Table 93. **Eurochemic and contractors' staff and mean doses, 1975 to 1978**

Year	Eurochemic staff: Number	Eurochemic staff: Mean dose (mSv/year)	Contract staff: Number	Contract staff: Mean dose (mSv/year)	Total : Number	Total : Mean dose (mSv/year)
1975	152	9.6	17	6.4	169	9.3
1976	153	8.2	22	9.4	175	8.4
1977	160	9.6	20	10.4	180	9.7
1978	160	8.3	52	17.9	212	10.6

105 Ibid p. 258 for quarterly breakdowns.

106 Ibid p. 259.

107 RAE 1975 to 1980. It should be remembered that the legal limit was 50 mSv and that the current proposals of the ICRP are for 20 mSv.

108 There are no data for 1979 and 1980 other than the collective doses in the RAE. See the Annex regarding radiation protection review.

From 1976 to 1978, contract staff were used in increasing numbers to meet the requirements and urgency of the programme, and received average doses more than twice as high as Eurochemic personnel, contrary to the situation that obtained during the reprocessing period, when the doses received by contract staff were comparable with those received by all staff. The reason for this greater exposure is probably the fact that the contract staff were not as familiar with the site, which exposed them to more accidental contamination.

However comparison with the period of reprocessing as shown in Table 94 reveals a net drop in both mean and collective dose.

Table 94. **Mean doses, 1969 to 1979**

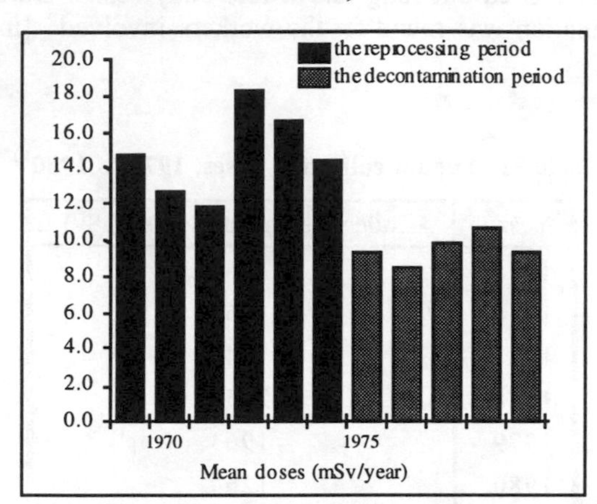

Table 95. **Annual collective doses in major periods**

Period	Mean annual collective dose (mSv)	Average of mean annual dose (mSv)
1969-1974 (reprocessing)	3495.0	14.7
1975-1979 (decontamination)	1797.4	9.4

The difference between the reduction of nearly half in collective dose and only a little over a third in mean dose can be ascribed to the drop in staff numbers, but there is no doubt about the overall reduction.

The underlying reason is a combination of several factors. The different nature of the work, involving many operations carried out directly in the cells, would have tended to increase the dose. The fact that this was not so is probably because of the careful way the tasks were prepared, the skill of the teams involved and also the fact that the plant was shut down and that there were far fewer incidents.

In fact in five years only nine incidents – using the same criteria as those in the reprocessing period – were recorded compared with 99 during the previous 10 years.

Table 96. **Annual breakdown of incidents, 1975 to 1979**

1975	1976	1977	1978	1979
2	0	2	2	3

These involved a fire in the low activity liquid waste treatment unit, two incidents of ground contamination, one of air contamination, two of contamination affecting staff, a loss of electrical supplies and two unclassifiable incidents[109].

The reasons for the doses received

The doses received were primarily linked to particular problems, the review of which led to recommendations which are set out in the conclusion of the article by Osipenco A., Detilleux E., Ferrari P. (1980).

[109] VIEFERS W. (1986), lists the incidents in an annex without page numbers. The two unclassifiable incidents were the accidental spillage of caustic solution and the disconnection of the nitric acid feed pipe to a glovebox.

The need to send workers inside cells to measure changes in radiation levels as decontamination proceeded was the cause of a collective dose of 230 mSv absorbed by 11 workers over an 18-month period. This was the price paid for there being no systems for making these checks without people having to enter the cells.

The layout and small size of the entry airlock made access to the cells difficult, causing a number of accidental contamination incidents, notably due to objects being dropped or liquids spilled.

The large number of systems accommodated in certain cells in order to limit their volume made movement difficult and was the cause of high background radioactivity. In certain cells through which corrosive solutions were piped, unexpected damage was found to certain systems which had been considered of secondary importance and given insufficient or inadequate protection. Also the geometry of some pipework and items of equipment was a further cause of irradiation owing to the formation of "hot spots" due to the accumulation at sharp bends and low points, and in tank overflows of highly active solid residues formed from precipitates and degradation products, known in the reprocessors' jargon as "crud"[110].

In general, the problems encountered made the work longer and more complicated but never brought it to a halt.

These difficulties were due essentially to a combination of two kinds of factor: first that the system design made inadequate allowance for the requirements of dismantling; secondly the use of lower quality materials for equipment items that were not directly involved in the process, such as water, steam and compressed air pipes, system supports, and so on. This had effects not foreseen when the plant was designed. A particularly significant example was that of the identification labels which had been fitted to pipework during construction. In some cells the iron wire fastening the labels in place had not withstood the corrosive atmosphere, so the pipes concerned could not be immediately identified[111].

The decontamination and dismantling phase – like the reprocessing period – produced wastes, although these were sometimes of a special nature.

Novel and conventional wastes adding to the "nuclear liability" of the company

The wastes produced

Considerable increase in output of solid wastes

By comparison with the period of reprocessing, the amount of conventional low activity solid wastes went up substantially during the decontamination operations. Plastic sheeting and bags, absorbent paper, cloths, protective clothing, gloves and shoe covers were collected from the worksites in plastic bags, compacted by Eurochemic and conditioned in standard drums, taken over by CEN/SCK with a view to sea dumping (Figure 120).

However there were also large quantities of a new type of waste: incompressible and incombustible solids produced by the decontamination of surfaces and by dismantling. These included concrete rubble, pipework sections, tank parts, valves and lead bricks. The thorough decontamination of concrete surfaces in particular produced substantial amounts of non-compactible solid wastes.

The activity levels of these wastes was very variable and they had to be sorted. Low activity beta and gamma wastes were usually encased in concrete before being shipped to the CEN/SCK with a view to sea disposal.

It is difficult to estimate the total volume of solid wastes produced because the inventories are very detailed and the units employed often specific – for example the number of filters and prefilters – and vary from year to year, thus preventing any summation. Production in the year 1977 is given below as an example: [112]

Table 97. **Production of solid wastes in 1977**

Type of waste	Conditioning	Number of items
Beta-gamma combustible	220-litre drums	614
Beta-gamma non-combustible	220-litre drums	393
Beta-gamma non-combustible	30-litre canisters	20
Plutonium-contaminated equipment	30-litre canisters	305
Plutonium-contaminated equipment	220-litre drums	78
Absolute filters	cardboard and plastic	199
Prefilters	cardboard and plastic	200

110 HILD W., DETILLEUX E., GEENS L. (1980), p. 200.

111 For a more comprehensive analysis, see HILD W., DETILLEUX E., GEENS L., (1980) p. 203.

112 RAE 1977, p. 27.

Figure 120. Stacking drums of low activity wastes at CEN with a view to disposal in the Atlantic (Source: Eurochemic colour slide).

Liquid waste

The liquid wastes arising from the decontamination and dismantling operations were removed, stored and conditioned just like the liquid wastes generated during reprocessing. After concentration they formed about 200 m^3 of intermediate activity liquid waste which as a general rule was added either to the stocks of liquid waste from aluminium decladding or to the hot waste concentrates (HWC)[113]. Some concentrates produced after June 1978 were not stored but sent directly to the bituminisation installation then in service[114].

The table below gives the statistics for the production of intermediate activity liquid waste for the site as a whole. This is the reason why the total amounts of HW concentrated into HWC are higher than those shown above for the decontamination operations alone.

Table 98. **HW and HWC concentrate production in m^3 from 1975 to 1980**[115]

	1975	1976	1977	1978	1979	1980
HW	374.0?	1730.0	1230.0	1813.0	2928.0	1013.0
HWC	70.0?	48.0	27.5	18.9	52.7	27.6
HWC/HW concentration factor	5.34	36.04	44.72	95.92	55.55	36.70

The increase in the concentration factor up to 1978, when the bituminisation unit came into service, reflects the falling effectiveness of decontamination, linked to the declining amount of contamination that could be removed without damaging the plant.

113 ETR 324, p. 4.

114 See Part IV Chapter 2.

115 For 1975, for decontamination of the plant only. RAE 1975, 22.

Disposal of the wastes produced from 1975 to 1979[116]

Low activity solid waste

The low activity wastes were conditioned by Eurochemic and then taken over, at this time as during the reprocessing period, by the CEN/SCK which was responsible for the dumping in the Atlantic Ocean of all wastes of this type produced in Belgium.

Sea dumping campaigns took place every year during the period[117]. Belgium took part in all of them except that of 1977.

During the first voyage in 1976, an incident occurred which threw doubt on the quality of the conditioning of Eurochemic containers. According to the annual report of the NEA[118], "one [of the] container[s] [conditioned by Eurochemic] floated on the surface, while another burst on striking the water. The first sank after holes had been made in it, and all the materials released from the burst container were recovered under the responsibility of the operations supervisor [appointed by the NEA], repackaged and taken for examination to the Harwell research establishment, where it was found that the activity level of the wastes was very low".

This incident led to a review of the preparation and surveillance procedures. An enquiry at Eurochemic established the cause of the incident as "human error".

Some of these low activity solid wastes could not be taken over by the CEN/SCK because they did not satisfy the sea dumping conditions. They were therefore stored on the Eurochemic site, some encased in concrete, the remainder as they were with a view to later conditioning with wastes arising from the dismantling or restart of the installations. Similarly the alpha-contaminated solid wastes and the high activity wastes were stored temporarily on the site.

Low activity liquid wastes

As before, the low activity liquid wastes were sent to the CEN facilities for processing before disposal (Table 99). The quantities of warm and cold liquid wastes were significantly lower than during the reprocessing period, but the volumes of inactive condensate were fairly comparable, particularly owing to the considerable use of high pressure water jets.

Table 99. **Production of low activity liquid wastes, 1976 to 1980, expressed in m^3/year**[119]

	1976	1977	1978	1979	1980
Warm (WW)	3246	3610	3610	8550	3096
Cold (CW)	4224	6353	7653	6915	5187
Inactive condensate	21 648	18 750	17 046	20 284	15 037

The intermediate activity liquid wastes arising from the concentration of the rinsing and decontamination solutions were stored with the reprocessing liquid wastes. Conditioning of these wastes in bitumen began in June 1978 in the new EUROBITUM facility, the drums produced in this operation being stored in the adjacent EUROSTORAGE installation. The commissioning of EUROBITUM in 1978 marked the beginning of the conditioning of the reprocessing and decontamination wastes which previously had simply been kept on the site. Work in the plant was suspended pending a decision and the plant was placed on "standby", all discharges being prevented by keeping the buildings at a reduced pressure. The plant workforce was retrained in waste management operations.

Thus the change to waste management was completed in the early 1980s, when the site became the property of Belgium.

---

[116] The management of the wastes stored on the site is dealt with systematically in Part IV Chapter 2. The following section therefore refers only to the disposal of low and intermediate activity wastes during the shutdown period.

[117] In 1976 the OECD proposed that a "multilateral consultation and supervision mechanism for the sea dumping of radioactive wastes" should be established (NEAR 76 p.40). This mechanism was set up on 22 July 1977 (NEAR 77 p.54) for immediate application.

In 1978 a new dumping site was chosen, close to the previous one. This was a rectangle bounded by longitudes 16°W and 17°30'W and by two lines of latitude located ten nautical miles to the north and south respectively of 46°N. The provisions of the MCSM were applied during this operation on a voluntary basis. As from that year, two separate operations were carried out, the first involving Belgium, the Netherlands and Switzerland, the other the United Kingdom only. The suitability of the site was reassessed by an expert group in 1979-1980 whose evaluation was published in April 1980. This authorised continuation of the operations for the next five years.

[118] NEAR, 1976, 41-42. The figures for the sea dumping operations are assembled in Part IV Chapter 2.

[119] No data given in the RAE for 1975.

Chapter 2

From Eurochemic to Belgoprocess

Waste management and the beginning of decommissioning

January 1980 - November 1990

Belgium had difficulties in setting up an organisation to take over the site, as a result of which the international company, although liquidated in 1982, continued its industrial activities until the end of 1984. An agreement concluded in April 1986 allowed Eurochemic to determine its final liabilities and in November 1990 brought to an end the international co-operation which had begun 35 years earlier. It was now acknowledged that reprocessing would never resume at Mol.

Work on the waste conditioning programme, which had become the priority issue following the end of reprocessing, gained momentum once the operations to place the reprocessing plant on standby had been completed. The 1980s was a period of technical co-operation involving the sorting of wastes, notably those contaminated by plutonium, the bituminisation of the intermediate activity liquid wastes and the conditioning in concrete of the solid wastes which had accumulated on the site.

However the conditioning of the two types of high activity liquid wastes, in which most of the radioactivity accumulated on the site was concentrated, raised new technical problems that had to be resolved. This involved intense international R & D activity resulting in the development of an original method of vitrification which was successfully applied in a pilot facility.

Dismantling of the plant, following the official announcement of the abandonment of reprocessing on the site, and which was to terminate the life of the undertaking, was in progress when the international co-operation initiated 35 years earlier came to an end.

Belgian takeover of the site and the liquidation of the international company

The application of the Convention was complicated first by the fact that Belgium found it impossible to set up a company to take over the business and, secondly, by some delay in the implementation of the technical programme. The Belgian government was obliged to ask Eurochemic to carry out a number of operations on its behalf, not only during the transfer period, for which provision had been made, but also thereafter.

The Belgian logjam and the Agreement of 17 December 1981

Awaiting the Belgian parliamentary debate on energy

The signature of the Convention in July 1978 was followed by the voting through of the budget law on 5 August, Article 89 of which authorised the government to entrust the fuel cycle activities to a joint-stock company which would take over the site from Eurochemic. However parliament placed two conditions on the process: first that the government should have a majority holding and, secondly, that the company should be established before 31 December 1978.

However the government had also committed itself not to set up the company until after the planned parliamentary debate on energy policy, which did not take place either in 1978 or in 1979.

The stalemate continued in 1980 despite the fact that the principles set out in the 1978 Law were repeated in Article 179 of the Law of 8 August 1980[1]. However the condition which this time was explicitly expressed by parliament was that no reprocessing of irradiated fuel should take place in Belgium until both chambers had reached a decision on energy policy. In its second paragraph the same article of the 1980 Law also raised the principle of creating a national agency for radioactive waste management[2].

---

1 *Moniteur belge* of 15 August 1980, p. 9520. Law on budgetary proposals, Article 179 § 4.

2 The Royal Order of 30 March 1981 created, in application of the Law of 8 August 1980, the ONDRAF/NIRAS. *Moniteur belge* of 5 May 1981, pp. 2654-2661.

Although discussions had been started by the Minister for Economic Affairs, Willy Claes, in a parliamentary committee, the year 1981 continued to go by without a decision being possible. However on 24 April 1981 an agreement was signed between the government and the electricity producers, providing for the costs of managing irradiated fuel to be incorporated in the price of electricity generated in nuclear power plant. This measure was based upon the decision taken by the CMCES on 16 December 1976, broadened to cover the whole issue of spent fuel management. It was to have the result of modifying the position of the electricity producers on plans for national reprocessing, moving it farther from that of the government.

Negotiation of the 1981 Agreement

Accordingly when Belgium became owner of the site on 31 December 1981, it found it impossible not only to resume reprocessing but even simply to manage the proportion of wastes for which it was responsible by virtue of the 1978 Convention. In fact there was neither joint stock company nor public organisation capable technically and legally of coping with these functions. The original plan was for Eurochemic to terminate all technical activities on the site once this had been transferred to Belgium, but in fact the company was to be liquidated on 27 July 1982 after a further and final three-year extension to its life.

This foreseeable situation had led the Belgian delegation on the Board of Directors to request negotiation of an agreement aimed at extending the presence of Eurochemic on the – now Belgian – site and requesting ONDRAF, which had held its constituent assembly on 5 October 1981, temporarily to discharge a number of tasks incumbent on the Belgian government under the 1978 Convention.

The 1981 Agreement

The *"Agreement Relating to the Implementation of the Convention Between the Belgian Government and the Eurochemic Company"* was signed on 17 December 1981. Its preamble noted that "since the new company has not been set up, ownership has been transferred to the government..., [that] Eurochemic has not completed all the work it was to carry out itself[3], [...] and that since the new company has not been formed Eurochemic is unable to request the said company [as provided for in the 1978 Convention] to take over part of its tasks; [that] it is in the mutual interest of the government and of Eurochemic that the implementation of the Convention should not be interrupted, in the best possible technical and economic conditions".

The agreement therefore provided for the industrial activities of Eurochemic to continue on the site up to 31 December 1983, partly on behalf of the Belgian government[4]. In return the government made an annual payment to Eurochemic fixed at 72 million BF for 1982 on a cost price basis[5].

As far as the staff were concerned, the Belgian government undertook under Article 7 to recruit "at least 160 members of Eurochemic staff" out of the 226 employed at that time. These staff were to be recruited by 1 January 1984 at the latest, with open-ended contracts and retention of seniority acquired since they joined the international company.

If the Belgian company was not set up before 31 December 1983, Eurochemic would pay a lump sum to the government as a contribution to the compensation to be paid to the staff if dismissals should be necessary[6].

The energy debate, the liquidation of the Eurochemic company and the request for a second agreement (1982 - early 1983)

The debate – finally

At the beginning of 1982 there began the process which finally led to the debate on energy, promised in 1974, being held. The Secretary of State for Energy, Etienne Knoops, asked the chairmen André Jaumotte and Julien Hoste for an update of the "Report of the Wise Men". This was submitted in March 1982. It confirmed and added to the conclusions and recommendations of 1976. With regard to reprocessing[7], it drew attention to

---

3 This applied in particular to the bituminisation of intermediate activity liquid wastes.

4 Article 3a.

5 Article 5a.

6 The social problems thrown up by this situation which, at the very least, was complicated and painful for the staff, were followed up and negotiated, throughout the entire period from 1978 until the final takeover, by a group made up of representatives of the Eurochemic trades unions and management, and the Belgian Delegate to the Board of Directors (subsequently the Board of Liquidators).

7 Final report, updated sections, Chapter VIII, p.4.

the reversal in the official position of the United States[8] and referred to the safety assurances provided by the international system of guarantees to back up the return of the reprocessing plant to service at its nominal capacity of 300 kg/day. It stated that "although this capacity would be insufficient for our long-term needs, its importance goes well beyond its nominal value because it allows Belgium to demonstrate that it is keeping control of most fuel cycle techniques. This is an essential condition for the effective and independent determination of our energy policy"[9]. The debate then began in the chamber in June 1982 and ended in the Senate at the beginning of March 1983.

Liquidation of the company as from 28 July 1982[10]

At the end of the three-year extension from 27 July 1979, decided upon in June 1978, Eurochemic went into liquidation. A Board of Liquidators took over from the Board of Directors. Its make-up was the same as its forerunner. Each member country was represented by a "liquidator". Thus there was no longer a second seat for Germany, France and Belgium[11]. The Board of Liquidators decided to keep the Restricted Group and the Technical Committee in operation. Naturally its main objective was to liquidate the company's assets as quickly as possible. However the difficulties facing the creation of a takeover company had the paradoxical result that Eurochemic, although in liquidation, continued for nearly two and a half years discharging a technical programme defined by the 1978 Convention. In this way it carried out, with all its staff, until 31 December 1984, work on behalf of the Belgian state which had actually owned the site since 31 December 1981[12].

The request for a second agreement

By the end of 1982 it became clear that the "new company" could not be set up by the deadline provided for in the Agreement, i.e. 31 December 1983. The Secretary of State for Energy, Etienne Knoops, therefore notified the chairman of the Board of Liquidators in a letter dated 3 December 1982 that his country had found it impossible to meet the obligations of the agreement, notably as regards jobs. He was therefore considering seeking a further extension of the presence of Eurochemic on the site.

The principle of negotiating a second agreement was accepted, with difficulty, by the Board of Liquidators on 13 April 1983. However the Board made its acceptance conditional on Belgium's accepting the principle of future discussions on an agreement relative to a lump sum payment that would release Eurochemic once and for all from its financial obligations arising from the 1978 Convention. In fact the member countries wanted to put an end to their co-operation by a final settlement of accounts. It must be remembered that Belgium had proposed such an agreement as early as 1974 in the letter of André Oleffe, then Minister for Economic Affairs[13].

The *"Second Agreement Relating to the Implementation of the Convention Between the Belgian Government and the Eurochemic Company"* was therefore negotiated as from April 1983.

However between the request for an extension and the signature of the second agreement in November, it appeared that the Belgian situation was finally resolving itself.

8 On 8 October 1981, President Ronald Reagan lifted the ban on reprocessing in the country [AtW, (1981) 12, p. 652]. However this decision, which would have allowed the Barnwell plant to startup, hung fire. In fact a few days later the President of Allied Corporation, owner of the plant together with General Atomic Co., notified the Secretary of State for Energy that he was abandoning all idea of resuming reprocessing "owing to continuing regulatory uncertainty" and because "reprocessing is commercially non-viable". Since 1968 the owners had spent 400 million dollars on the project.

9 Final report, updated sections, Chapter VIII, p. 4.

10 For the legal aspects of liquidation, see in particular STROHL P. (1991) and the note by Otto von Busekist in CL(90)6/AG(90)3, 3: *Brief Report on the Main Activities of the Company During the Period of Liquidation.*

11 See list of liquidators in the Annex.

12 Pierre Strohl, in his article on the liquidation of the company [STROHL P (1991), p. 736-737], describes the unique legal situation it produced. It is true that some aspects of waste management came under the normal liquidation of the company, as part of the process of "settling liabilities". But for a company in liquidation to undertake new activities contravened the normal legislation governing liquidation and raised the problem of the competence of the Board of Liquidators in taking decisions. Here again in fact, the practical requirements arising from the special features of a nuclear undertaking and from the shareholders' sense of responsibility won the day over the letter of the law. In fact it was impossible to close down the business before the problem of "nuclear liabilities" was settled.

13 See Part IV Chapter 1.

Resolution of the situation in Belgium and the SYBELPRO project, March 1983 - May 1984

The decision to resume reprocessing, the SYNATOM joint-stock company and plans for a new international plant

The debate on energy policy ended in the Senate at the beginning of March 1983. Both chambers had come out in favour[14] of a resumption of reprocessing in the country. This decision enabled the provisions of the Law of August 1980 to come into force.

Accordingly on 8 March 1983 a Royal Order authorised the National Investment Company (SNI) to set up a subsidiary specialising in managing fuel cycle activities. This subsidiary then became part of SYNATOM, thus creating on 11 April 1983 a joint-stock company owned half by the Belgian government and half by the electricity producers, the "Société belge des combustibles nucléaires, SYNATOM".

However within the new company, the attitude of the electricity producers to the question of national reprocessing had evolved during 1982, probably owing to the content of the Convention they had signed with the government on 24 April 1981[15]. In fact the Belgian electricity producers now made their financial contribution to resumed activities at the Eurochemic facility subject to two conditions. The first, a *sine qua non*, was that there should be a foreign shareholding, French in particular, to guarantee the quality of the transfer of technology. The second, related to the first, was that an international research group should prepare a review based upon a preliminary safety analysis approved by the national authorities[16].

These new requirements complicated relations between the government authorities and the electricity producers, the former tending to believe that these were more or less dilatory manoeuvres on the part of the latter to escape what they considered to be an obligation on their part. On the other hand the electricity producers considered that they alone should not have to bear – or have the electricity consumers bear – the costs of a strategy based partly on political considerations, and that in the future they could very well go on having their reprocessing done elsewhere[17].

The private partners in SYNATOM therefore demanded that a Study Group known as SYBELPRO[18] should conduct a study of the technical and economic feasibility of restarting a reprocessing plant by re-using and extending the Eurochemic facilities.

The CMCES, taking its cue from the parliamentary decision, therefore gave authorisation in principle on 5 May 1983 for the restart of Eurochemic[19] and also gave permission for SYNATOM to conduct the feasibility study. SYNATOM was subsequently to take over the site under procedures set out in a later Convention. The CMCES also followed SYNATOM in the scheme for a plant of greater capacity. Germany and France responded favourably.

SYNATOM had been in touch with the main European reprocessing undertakings. The contract setting up SYBELPRO was signed between SYNATOM, DWK and COGEMA on 27 May 1983[20]. The parties contributed 60%, 20% and 20% respectively to the Study Group. BNFL was to join the project at the beginning of 1984, taking a 5% holding bought from SYNATOM.

The second agreement: a final conventional delay

It was therefore clear that the creation of a takeover company was unthinkable before mid-1984, after the deadline of 31 December 1983.

---

14 At the same time in France, the Castaing Commission whose work had been closely followed in Belgium submitted its report, see RGN (1983) 2, p. 162.

15 See above. It appears that another factor weighed in the balance: the choice made necessary, by the sheer extent of the necessary investment, between supporting the Belgonucléaire MELOX plant, upstream of nuclear power production, and that of the reprocessing project, downstream.

16 This was not a safety analysis such as the law required for a project actually being carried out (see Part III Chapter 1), but a preliminary review. In fact it was difficult to obtain even a simple opinion from the authorities which did not imply some sort of approval. The Belgian authorities therefore gave a "provisional opinion" to be confirmed or abrogated in due course, on the basis of the legal safety dossier.

17 As signatories of contracts that gave rise to UP3 in 1978, they knew that they could have their spent fuel reprocessed at La Hague until the year 2000.

18 SYBELPRO should not be confused with the Belgoprocess syndicat d'études set up on 19 September 1974, which was solely Belgian.

19 Essentially the restart was supposed to "obtain a return on the investment, and increase the capacity of the installations in order to work at a competitive price, while abiding by the discharge standards".

20 SYBELPRO (1983). The representatives were R. De Coort and R. Cayron (SYNATOM), G. H. Scheuten and G. Otto (DWK) and François de Wissocq and Jacques Couture (COGEMA).

The second agreement which had been negotiated since April was signed on 23 November 1983. It reflected the new situation and pushed back the deadlines set out in the first agreement. However it was then made clear that this was the last chance.

The extension of Eurochemic's transitional presence on the site was "up to 31 December 1984 *at the latest*"[21]. Article 1 is very firm: "by 30 April 1984 at the latest, the government will notify Eurochemic of the decision taken about the future of the plant and of the way in which the obligations incumbent on the new company by virtue of the Convention will be implemented. In any event, if the government has not notified Eurochemic of the decision to restore the plant to service by 30 June 1984, it shall be regarded as having abandoned the application of the relevant provisions of the Convention and of the present agreement".

The principle of negotiating a "Lump Sum Agreement" was laid down in Article 3: "the government and Eurochemic will as soon as possible undertake negotiations for concluding an agreement with a view to settling, in a lump sum, the amounts that remain due by Eurochemic after 31 December 1984 by virtue of the Convention. This agreement will determine the amount and schedule of payments to be made by Eurochemic, on the basis of an evaluation made with a common accord".

The SYBELPRO project. Resumption of reprocessing in an extended plant

The general specification[22]

The new plant was to be capable of reprocessing fuels from standard and non-standard light water reactors, as well as mixed oxide fuels. The plant might possibly reprocess other fuels and accommodate the conditioning of equipment components.

The cost of the feasibility study, estimated at 250 million BF, was shared *pro rata* between the members of the Study Group. In fact SYNATOM believed that foreign technology should be brought in to some extent and considered from the outset that the plant, which it would fund to a maximum of 55%, should reprocess at least 45% of foreign irradiated fuels.

The SYBELPRO study

The study was completed in April 1984. It covered the extension of the plant to a nominal capacity of 120 t/year, approximately twice its previous size, at a cost of 27 billion BF[23]. According to the study, the plant would be competitive with other similar installations abroad. The new plant would differ from the old one through its higher capacity, but also in the following aspects (Table 100), which drew upon the experience of problems in the operation of Eurochemic.

If the project were to be carried through, the existing buildings and facilities would have to be adapted, and some new facilities constructed (Figure 121).

Table 100. **Comparaison between the Eurochemic plant and that of the SYBELPRO project**

EUROCHEMIC (period of reprocessing)	Belgoprocess plant planned by SYBELPRO	Comments
chemical dissolution	mechanical decladding	less decladding waste and effluent
storage of solid wastes in the pond	decladding wastes embedded in concrete	on-line conditioning
LEU and HEU circuits, variety of fuel types	only LEU - PWR fuels 1st co-decontamination/separation cycle 2nd and 3rd U extraction cycles (capacity 600 kg U/day) 2nd and 3rd Pu extraction cycles (mixer-settlers for the 3rd)	adapted to predominant types of fuel, better performance from fission product separation and extraction
fission product concentration	concentration of fission products and removal of nitrates	less volume of vitrified wastes[24]
acid and solvent recovery	circuits allowing tritiated acid solutions to be recycled in the process; used solvent to be conditioned by EUROWATT	special treatment of tritium; on-line conditioning of liquid wastes
storage of ILLW and HLLW	intermediate storage of liquid waste, conditioning (bituminisation by EUROBITUM or vitrification by AVB), intermediate storage of conditioned liquid wastes	delayed conditioning compared with simple storage "as is"
final products: uranium nitrate and plutonium oxide	uranium nitrate and mixed U-Pu oxide	takes into account the require-ments of non-proliferation

[21] "At the latest" was added by hand to the original copy of the agreement.

[22] SYBELPRO (1983), "Exhibit 3".

[23] The specification was for 18.6 billion. SYBELPRO (1983), Exhibit 4. To clarify the situation, the updated cost of designing and building the original plant could be regarded as being of the order of 10.4 billion BF (1984). For the overall balance sheet, see Part V Chapter 2.

[24] The presence of nitric acid reduces the specific fission product conditioning capacity per gramme of glass.

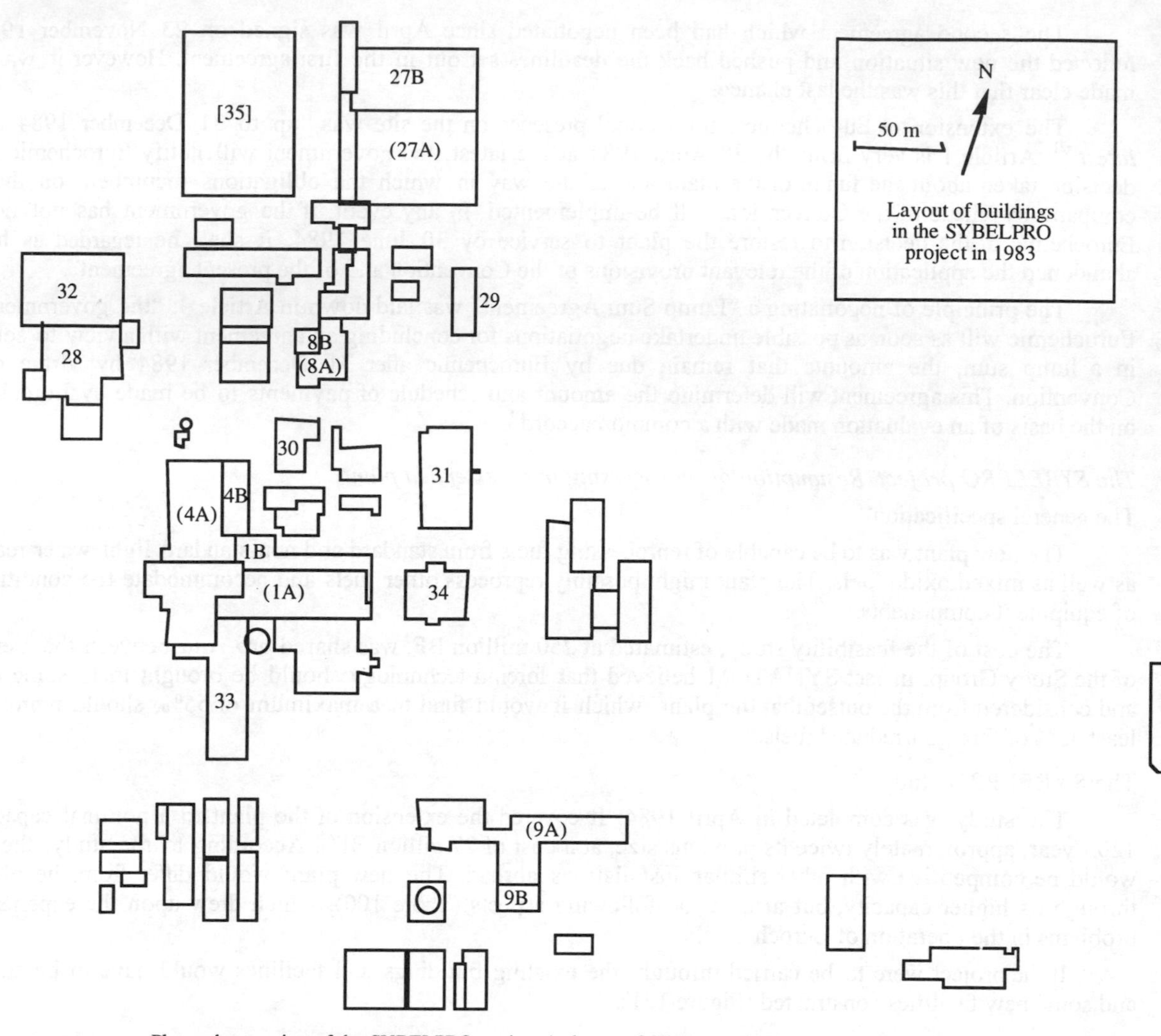

Planned extension of the SYBELPRO project (scheme of 28 August 1983)

1B Extension to the extraction building
4B Extension to the ventilation building
8B Extension to the LLLW treatment building
9B (Possible) extension to the general services building
27B Extension to the intermediate activity waste storage building
28 HLLW vitrification building (Belgian AVB vitrification unit)
29 Storage building for vitrified Eurochemic waste (completed)
30 High activity liquid waste storage building
31 Building for high activity liquid waste conditioning (PAMELA), under construction
32 Storage building for wastes vitrified by AVB
33 Head-end building
34 Plutonium conversion building
[35 Storage building for conditioned low activity wastes; plan of December 1983]
For the other buildings, see the general drawing at the end of the chapter.

Figure 121. Layout of buildings in the SYBELPRO project in 1983. The scheme given here is taken from a document attached to the study, dated 29 August 1983, which very largely resembles an annex to the contract dated December 1982. The latter provided for Building 30 to be joined to Building 32, the vitrified Pamela waste to be stored to the north of the AVB (Buildings 28 and 32 no longer being joined), and finally for a Building 35 for further storage of low activity conditioned wastes to the north of Building 23.

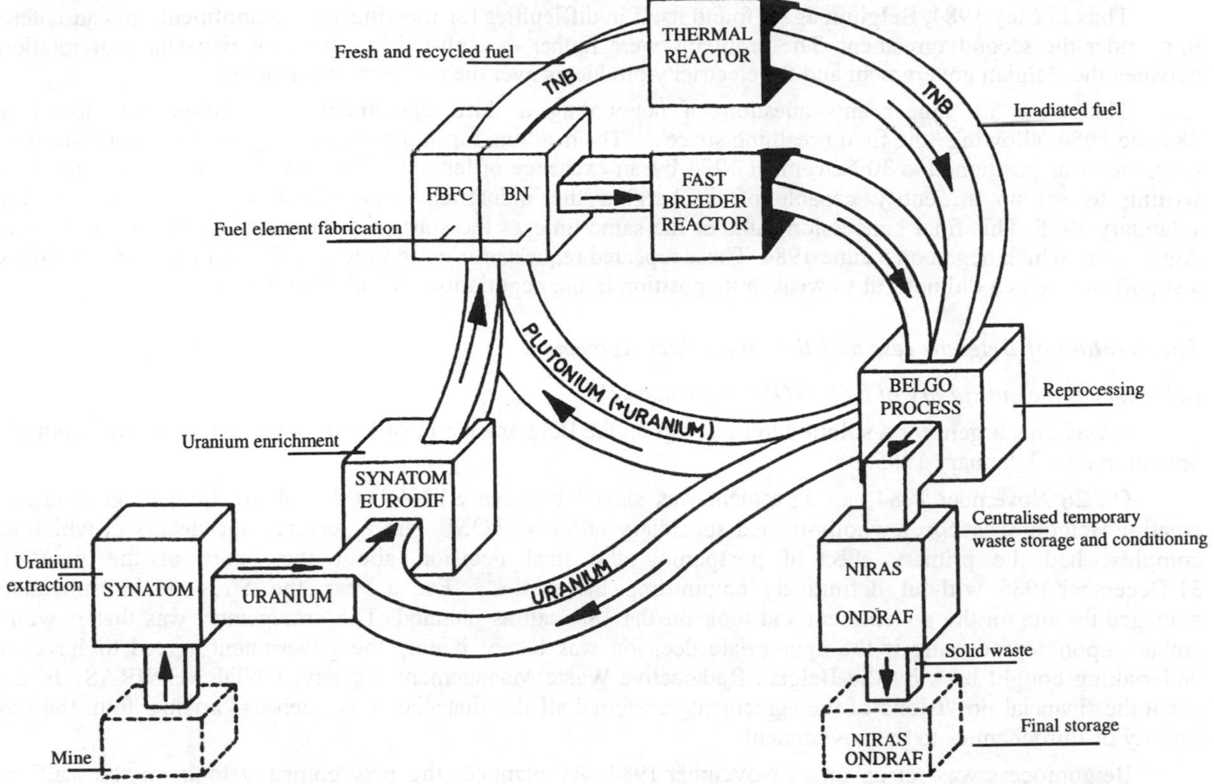

Figure 122. Organisation of the fuel cycle in Belgium as it was still being envisaged at the end of 1985. The takeover of reprocessing by Belgoprocess would complete the loop of a cycle incorporating a fast breeder reactor, using imported uranium enriched by EURODIF, in which Belgium was participating (Source: NEW, December 1985, p. 22, which in fact reproduces a Belgonucléaire document).

For example, for the reprocessing plant proper two new buildings were planned, one to house the head-end and its mechanical decladding facility, and another to accommodate the plutonium-mixed oxide conversion unit.

The defection of France and Germany; failure of the SYBELPRO project and disputes between the government and the Belgian electricity producers about the financial charges. The new impasse and the exchange of letters of 26 June 1984.

A few weeks after the submission of the study, COGEMA made it known that it was not in a position to guarantee that it would utilise 20% of the plant's capacity. DWK announced that it was obliged to await the decision about the German plant. At the beginning of the year DWK had asked two groups, one of which was directed by KWU and involved SGN, to submit a bid for the design, construction and commissioning, on a turnkey basis, of a reprocessing plant[25] with a capacity of 350 t/year, to be built either at Gorleben or Wackersdorf[26]. SYBELPRO was therefore dissolved in May 1984, although contacts continued between Belgium and Germany (Figure 122)[27].

[25] The Belgian expression is "key in the door".

[26] RGN (1984) 2, p. 188; 3, p. 279. The size of the German plant envisaged for Gorleben, originally set at 1400 t/year in 1977, had been lowered following a compromise agreed with the Federal government in October 1979 to attempt to put an end to local opposition, see RGN (1980) 1, p. 108 and AtW (1980) 1, p.61. In the end, DWK abandoned the Gorleben site, which went on to be used only for waste storage, and chose Wackersdorf, in Bavaria, in January 1985. When local opposition became vociferous there too, the SYBELPRO solution appeared as a sort of consolation approach or a possible fall back position, hence the German desire not to "burn its boats".

[27] These links were strengthened by the organisation of the PAMELA vitrification project. The federation of Swiss electricity producers, clients of the CEDRA, at that time under the chairmanship of Rudolf Rometsch, himself in favour of a resumption of reprocessing in a facility he knew well, then made it known that it was prepared to join a project on condition that Germany would participate to form a three-way undertaking. Hence in 1985 Belgonucléaire was again able to look forward to having a complete fuel cycle in the country in the near future, see Figure 122.

Thus in May 1984, Belgium again found itself in difficulties for meeting the commitments it had entered into under the second agreement. The problems were further exacerbated by the clear worsening in relations between the Belgian government and the electricity producers over the problem of financing.

There was no longer any question of negotiating a third agreement. A solution was found on 28 June 1984, allowing "the final breathing space"[28]. The deadline for notifying the decision to restore the plant to service was postponed to 30 November 1984 by an exchange of letters[29]. The government also undertook in writing to set up an entity capable of discharging the industrial responsibilities on the site as from 1 January 1985. This final concession came at the same time as the start of discussions about the Lump Sum Agreement, which began on 5 June 1984. These repeated requests for more time put Belgium in the position of a supplicant, which did not fail to weaken its position in the negotiations which then began.

The creation of Belgoprocess and the Lump Sum Agreement

Belgoprocess, a subsidiary of SYNATOM, November 1984

It was thus urgent for a solution to be found on the Belgian side in order that a "host structure" could be operational by 1 January 1985.

On 26 November 1984, an agreement was signed between SYNATOM and the Belgian government, creating Belgoprocess SA, a wholly-owned subsidiary of SYNATOM. The agreement, the details of which are complex, had the primary effect of postponing the final decision about the future of the plant to 31 December 1985 without definitively committing the parties[30]. For a year, the SYNATOM subsidiary managed the site for the government and took on the 180 staff as planned. The arrangement was that it would embark upon reprocessing if the appropriate decision was taken; if not, the government agreed to have the undertaking bought back by the Belgian Radioactive Waste Management Agency, ONDRAF/NIRAS. In any event the financial provisions of the agreement transferred all the financial consequences "arising from the past activity of Eurochemic" to the government.

Belgoprocess was set up on 29 November 1984. As planned, the new company took on the staff and began direct management of the site as from 1 January 1985. The changeover from Eurochemic to Belgoprocess marked the end of the unusual linguistic situation on the site. The official working language was now Dutch[31].

Negotiations on labour relations from October to December 1984 and the transfer of most of the Eurochemic staff to Belgoprocess[32]

The difficulties encountered in establishing the Belgian takeover organisation naturally caused great concern amongst the staff. This resulted in a one-day strike on Monday 24 September 1984 on the grounds that the promise to take on 160 employees appeared insufficient. Negotiations began in October 1984 and concluded on 16 December 1984 with an agreement between the trades unions and the new company.

Belgoprocess agreed to take on 170 of the 226 staff[33] and offered them new contracts of employment which contained less social advantages than those the staff had enjoyed at Eurochemic. The latter agreed to compensate the staff for the loss of certain rights, notably access to the Eurochemic Provident Fund[34].

On completion of the negotiations, 159 employees accepted the new conditions and formed the nucleus of the Belgoprocess staff[35]. Of the 51 non-Belgian staff, 14 refused the new conditions and were therefore dismissed by Eurochemic[36]. Thirty-one staff aged over 55 accepted early retirement. Emile Detilleux resigned after being appointed Managing Director of ONDRAF.

---

28 Frérotte, p. 21.

29 Letter from Mr. Etienne Knoops dated 26 June 1984, in CL/M(84)3 Annex.

30 The period was further extended to 31 March 1986.

31 Dutch had been imposed as the working language in contacts between the trades unions and the management since the 1970s.

32 Sources: RAE 1984, CL archives, communication from Oscar Martinelle.

33 The Eurochemic workforce had remained stable from 1979 to 1982, but rose that year as a result of the increased workload related to the development of the waste conditioning activities. See below.

34 The employee had contributed 7% of salary to the Provident Fund, to which the company added a further 14%. Thus each employee received a capital sum amounting to 21% of total salary on leaving the firm.

35 These 159 joined an existing group of 19 people in 1985. Belgoprocess had 178 employees in 1985, 175 in 1986, 169 in 1987 and 182 in 1988.

36 Of the company's 226 employees, 175 were Belgian, 18 Dutch, 13 Spanish, 7 German, 5 French, 4 Italian, 3 British and 1 Swedish.

During 1985 therefore, Eurochemic staff numbers were reduced to five, of whom two took early retirement at the end of the year[37]. This left a team of three to handle the company's remaining problems on the site: the Secretary of the Board of Liquidators, Otto von Busekist, the Financial Director, Oscar Martinelle, and one secretary. This small team continued, until the end of the liquidation period, to handle the tasks for which Eurochemic remained responsible on the site: production of the final technical reports[38], management of the Provident Fund and of the hostel facilites, which were sold on 22 June 1987 for 40 million BF, and a small number of employment disputes which were mostly settled amicably. The Mol office also had to start putting the archives in order, the production of a history of the company having been approved in principle in 1985 following a seminar on the Eurochemic experience which had been held in June 1983 at BCMN in Geel. The office also provided a liaison function for the negotiations of the Lump Sum Agreement provided for under the second agreement, and then for its application, which was to bring about the end of the liquidation period.

The "Lump Sum Agreement" of 10 April 1986 and the end of the liquidation period of the Eurochemic company. June 1985 - November 1990

Issues and principles

Negotiations had begun on 5 June 1984 about the agreement which had been defined in principle during preparation of the second agreement.

The issues being negotiated were substantial and are set out in the final section of the preamble to the Lump Sum Agreement:

"CONSIDERING the importance the parties place on settling on a lump sum basis the costs for which Eurochemic remains responsible by virtue of the Convention and the agreements intended to bring about the liquidation of Eurochemic as soon as possible;

HAVE AGREED as follows".

The participants in Eurochemic now firmly desired to put an end to their co-operation which was now in a form very remote from the original project, and it was well known publicly that many of the partners believed that it had simply gone on too long[39]. However they wanted to do so at the least cost, while not evading their responsibilities.

Belgium wanted to cover as much as possible of the costs it faced in settling the "Eurochemic liabilities". However it did inherit a nuclear site, at a time when it was proving virtually impossible to create new installations in Europe owing the changes that had taken place in the decision-making processes concerned with nuclear establishments.

The financial aspects were somewhat delicate. How is it possible for example to estimate costs extending over several tens or even hundreds of years, the details of which – such as the cost of decommissioning – are not fully known? The parties surrounded themselves by advisers: SYNATOM and then Belgoprocess on the one hand, the Technical Committee on the other.

From the outset however the financial negotiations were confined within the limits of what was politically acceptable, as Marcel Frérotte describes retrospectively in his history[40]. It would be difficult to obtain contributions to Eurochemic beyond 1990, and it was necessary to remain within the envelope of the usual annual contributions, which prior to liquidation were around 400-500 million BF a year.

The fact that the company had progressively made provisions[41] with the Lump Sum Agreement in mind, which at the time amounted to 1.3 billion BF, gave Eurochemic a certain margin for manoeuvre and considerably facilitated conclusion of the agreement.

The negotiating process

Belgium's initial proposal was for 4.8 billion BF (1984) on the basis that reprocessing would be resumed. This figure was obtained by taking into account the contribution of Eurochemic to the construction and operation of an industrial facility for the vitrification of high activity wastes produced by the reprocessing of

---

37 The former Deputy Director, Hubert Eschrich, and the former Head of Documentation, Willem Drent.

38 During the liquidation period, 16 reports were published.

39 In 1983 Otto von Busekist wrote and issued to the members of the Board of Liquidators meeting at Mol a spoof copy of the *Minutes of the One-hundred Fiftieth Meeting of the Board of Directors held in Mol on 1 April 2007* which poked fun at the lassitude of the partners; this document was subsequently circulated within NEA.

40 FREROTTE M. (1987), p.24.

41 Communication from Oscar Martinelle.

natural or slightly enriched uranium fuels[42], subsequently intended to process the liquid waste from the future plant, derived from the Atelier de vitrification de Marcoule (AVM), hence its name – Atelier de vitrification belge (AVB) – and the construction of which naturally depended on the decision to reprocess. Its total cost was 5.2 billion BF, of which 60%, or 2.972 billion BF, would be paid by Eurochemic. In addition there were the running costs involved in conditioning 800 m^3 of high activity liquid wastes arising from the reprocessing of the highly enriched uranium fuels which was to be carried out in a pilot German installation known as PAMELA[43], amounting to 1.3 billion BF. Finally it was necessary to add the costs of storing and treating the other wastes remaining on the site. Eurochemic initially proposed 4.4 billion BF on the assumption that reprocessing would resume.

However the company soon demanded that consideration be given to the increasingly likely hypothesis that reprocessing would not be resumed. Under this hypothesis, AVB would not be required, PAMELA would be adapted and the total cost would be reduced to 2.7 billion BF.

Belgium's position in the negotiations was made increasingly less tenable by the unfavourable outlook for the resumption of reprocessing[44]. The argument for the AVB had to be abandoned once the resumption of reprocessing became highly unlikely. The Belgian negotiators then found themselves in a difficult situation if they were to avoid a sharp reduction in the amount they were considering requesting. This would have been extremely difficult to carry off politically.

In these circumstances, Belgium could have invoked decommissioning costs. According to Emile Detilleux however, this was not possible because although the Belgian electricity producers had already abandoned the project, the government, notably the Secretary of State for Energy, Firmin Aerts, still hoped to convince Germany to join in a later restart at the plant. In that context it was impossible to use the decommissioning cost as an argument for increasing the amount that could be sought from Eurochemic's Board of Liquidators, which would have meant acknowledging on the one hand that reprocessing was over, while trying on the other hand to persuade Germany to join Belgium (and Switzerland) to resume it. Germany, for its part, kept a cautious silence. It is true that in February 1985 they had announced that Wackersdorf would be the site for their 350 t/year plant, and the Bavarian Ministry for the Environment had given partial authorisation for work to commence at the beginning of October. However local opposition was as strong as ever. On 12 October 27 000 people protested in Munich against the plant being located at Wackersdorf, and the commencement of work on 11 December was interrupted by a demonstration on the site resulting in its being occupied by militant ecologists, who were ousted by the police on 7 January 1986[45]. At the same time, legal moves were begun to prevent work from starting. At this time therefore, DWK was uncertain that its project would go through.

Hence both parties found themselves being pushed towards a larger sum, even though Eurochemic had made provision to cover it within a short period without increasing the annual payments by the member countries.

In these circumstances a compromise was found by increasing the German estimate of the cost of adapting PAMELA to vitrifying the wastes intended for AVB. Eurochemic made an offer of 3.4 billion BF, with an indexing clause.

In a final phase, the lump sum was raised to 3.690 billion BF (indexed), the additional amount coming essentially from Germany's renouncing 220 million BF of royalties to which she was entitled from Eurochemic for the use of PAMELA, the saving benefiting Belgium.

The Lump Sum Agreement of 10 April 1986 and the end of the Eurochemic liquidation period

Finally, the Board of Liquidators adopted the draft agreement on 12 December 1985. It was approved by the Special Group on 9 April 1986 and was signed in Brussels on 10 April 1986 between the Secretary of State for Energy, Firmin Aerts, and the chairman of the Board of Liquidators, Lars A. Nøjd.

The "Lump Sum Agreement" consists of 5 Articles. The first sets the amount, the dates of payment and the indexing formula (Table 101), which is a combination of 40% of variations in the retail prices index and 60% of those in the wages and salaries index. The second Article states that this payment releases Eurochemic from all the obligations arising from the Convention and the other Agreements. Article 3 maintains Eurochemic's right to information on the activities carried out by Belgoprocess on its behalf[46], gives it right of

[42] For which provision was made in the 1978 Convention.

[43] For AVB and PAMELA, see below.

[44] The electricity producers had already given up, but the Belgian government was continuing to negotiate for the co-processing in Belgium of fuel from the Kalkar fast breeder reactor.

[45] Sources: RGN (1986) 1, pp. 67-69 and a search through AtW.

[46] This was already provided for by the 1978 Convention.

access to the technical records and the right to use premises to enable it to carry out tasks relating to its own liquidation. The other Articles cover the settlement of disputes and the entry into force of the Convention.

Table 101. **The schedule of payments under the Lump Sum Agreement**

Date	Amount (million BF)
1985	Advance of 220
within 3 months of signature	1000
31 December 1986 at the latest	470
31 December 1987 at the latest	500
31 December 1988 at the latest	500
31 December 1989 at the latest	500
31 December 1990 at the latest	500

An exchange of letters dealing mainly with payment guarantees completed the agreement, which came into force on 1 October 1986. The first payment was made in October 1986 and the others followed as arranged[47].

International co-operation in the Eurochemic company came to an end following the final meeting of the Board of Liquidators on 28 November 1990.

By that date, the idea of resuming reprocessing in Belgium had officially and definitively been abandoned for nearly four years.

Belgium's final abandonment of reprocessing: Belgoprocess, an ONDRAF subsidiary since 1986

The last hope of seeing further co-operation on reprocessing at Mol ended in fact in 1985. The electricity producers in SYNATOM had ceased to believe in it by the beginning of that year[48].

DWK announced officially after the meeting of its Board of Directors on 5 November 1986 that it was giving up its interest in the resumption of reprocessing in Belgium "for reasons connected with the situation in the Federal Republic of Germany"[49].

As a result, the CMCES decided on 22 December 1986 not to resume reprocessing in Belgium[50]. Four days earlier, in application of the Convention binding it to SYNATOM, the government had charged ONDRAF with the decontamination and decommissioning of the plant and had signed a Convention with it for the takeover of Belgoprocess.

On 23 December 1986 the new Board of Directors of Belgoprocess, now a subsidiary of the Belgian radioactive waste management organisation, ONDRAF/NIRAS, duly met.

---

47 For Spain however, the financial story of Eurochemic continued until 30 September 1992. As we have seen, Spain had been angered by the way in which the end of reprocessing had been decided upon, and showed bad grace in paying the amounts it owed. However in order to bring co-operation to an end, the accounts had to be cleared. Negotiations therefore took place between the Chairman of Eurochemic's Board of Liquidators, assisted by the Board's secretary and the Director for International Affairs of the Centro de investigaciones energeticas medioambientales y tecnologicas (CIEMAT), the organisation which had succeeded the Junta de energia nuclear. The agreement reached in Madrid on 20 January 1988 provided for the payment of the 83.7 million BF of arrears built up between 1974 and 1986, together with the 46.3 million BF representing Spain's share under the Lump Sum Agreement, in five annual payments, the last of which was to be by 30 September 1992 at the latest. To enable the liquidation period to be brought to a conclusion, the debt corresponding to the final two years was sold to the bank handling the company's accounts. CL/M(90) 1, p. 3.

48 The directors of Belgoprocess submitted their resignations to the Chairman by 28 May 1985, so that the government might proceed quickly to a takeover.

49 The sources are too sparse for one to see whether there was any particular reason why DWK chose this time to take its decision. It is likely that progress with the work at Wackersdorf had led the DWK directors to believe that the construction of the plant was irreversible. In fact a complex legal procedure had begun which was to delay the project and above all to make its completion hypothetical, AtW (1988) 2, p. 75 and AtW (1988) 8-9, p. 410. Ultimately it was the signature on 6 June 1989 of a reprocessing contract between COGEMA and VEBA, involving 11 electricity companies out of the 12 power plant operators in Germany, providing for the reprocessing of Germany fuel in UP3 until the year 2008, which gave the *coup de grâce* to the plans for a national reprocessing plant, RGN (1990) 1, p. 56.

50 The abandonment of reprocessing had an impact on the numbering of the buildings on the site. Today this shows gaps between 28 and 35 corresponding to the buildings planned for SYBELPRO which were never built.

Belgoprocess was to continue work on the site (renamed BP1 in 1990) on behalf of the Belgian government[51].

In fact the Mol-Dessel facility had been converted into a centre for the conditioning and temporary storage of radioactive wastes from the whole country, pending a decision regarding final disposal (by 1980 the image of Eurochemic was that of an installation responsible for nuclear wastes: see Figures 123a and 123b)[52]. The wastes which had been produced during the operation of the plant were given specific treatment between 1978, when the bituminisation facility was commissioned, and September 1991, when the vitrification of fission products was completed. In some cases proven techniques were used, in others they were novel. Some were carried out internally by Eurochemic, others were the opportunity for further international co-operation on the site, although in very different forms from those which had existed in the international company.

Managing Eurochemic's wastes[53]

Background to waste management on the Eurochemic site. Introduction to and radiological review of the years of waste management

The many people and organisations involved, and the limited role in time of the Eurochemic company necessitate a product-based approach

The Eurochemic wastes were conditioned in such a way as to permit either their disposal off site or their intermediate storage under a programme outlined in the early 1970s. This programme was completed a few months after the company had ceased to exist, in fact on 5 September 1991, while Eurochemic had had no operational staff since the end of December 1984.

The implementation of this programme involved various actors – naturally Eurochemic and Belgoprocess – but also enterprises from the two main shareholders in the company: France and, especially, Germany.

The Eurochemic structures which had been established in July 1974 continued without major change, except for the absorption of the General Services Department by the Operations Department in 1982.

Table 102. **Changes in Eurochemic staff numbers, 1980 to 1984**

Year	1980	1981	1982	1983	1984
Staff numbers	205	200	226	226	159
Change	-3	-5	26	0	-67
Percentage change over previous year	-1.4%	-2.4%	+13%	0%	-29.6%

51 The Transnuklear affair, in which the managers of the CEN Waste Management Division were involved, accelerated the restructuring of the waste sector which had been in hand since the creation of ONDRAF in 1981. The conclusions of the parliamentary enquiry into this affair, which lasted from 17 March to 14 July 1988, put an end to the discussions between ONDRAF and the CEN/SCK aimed at establishing joint management of the activities of the Waste Division of the Mol Centre. The CEN/SCK division and the site it managed was transferred to ONDRAF as from 1 March 1990 with no swap arrangement for the Centre. On this occasion the Eurochemic site was renamed Belgoprocess 1 (BP1) and the CEN site BP2. The numbers of Belgoprocess employees rose owing to its new tasks and the incorporation of the CEN Waste Department. They increased from 182 at the end of 1988 to 274 at the end of 1989 and 297 at the end of 1990.

52 Although Belgium still does not have a storage facility for low activity wastes comparable with the Centre de l'aube at Soulaines in France, as from 1974 it did carry out pioneering research on deep burial in an underground laboratory in Boom clay some 200 metres beneath the CEN/SCK. However no final decision has yet been taken.

53 This part and the next are based upon special sources. As from January 1985 and during part of 1986, the documents of the Board of Liquidators include a series of reports in English, prepared by Belgoprocess, on the work done on behalf of Eurochemic during the periods between the meetings of the Board of Liquidators: *Progress Report on the Work Carried Out by Belgoprocess on the Account of Eurochemic.* They also include progress reports in French produced by DWK on the vitrification of LEWC wastes from Eurochemic. However for the subsequent period it is necessary to use sources outside Eurochemic: reports by ONDRAF and Belgoprocess, the annual Pamela reports and various internal documents written in German and the secondary sources listed in the bibliography. The study by Belgoprocess, which today is responsible for the centralisation and temporary storage of all Belgian radioactive wastes, is based only upon parts of the ONDRAF and Belgoprocess annual reports that deal with what they call the "nuclear liability of Eurochemic". The other activities are mentioned only to the extent that they concern the Eurochemic site.

Figure 123a. Eurochemic seen by the opponents of nuclear power. Cartoon taken from a pamphlet distributed on the occasion of the international anti-atomic demonstration (Internationale antiatoom-manifestatie) which took place on 25 October 1980 at Mol. The word Mol means a mole in Dutch (Source: Eurochemic archives).

Figure 123b. "The activities of ONDRAF and its subsidiary Belgoprocess are closely linked. Together we are working to manage the radioactive wastes and clean up the nuclear liability of Mol-Dessel". First sentence of the first annual joint report of ONDRAF and Belgoprocess, accompanied by the drawing shown here (Source: Annual ONDRAF-Belgoprocess report 1990, p. 1).

The 1982 increase was related to the development of plans for new treatment and storage facilities, which led in the same year to the setting up of a Drawing Office employing five people. Nearly 30% of the workforce left the company in December 1984, following the negotiations for personnel to be taken over by Belgoprocess.

Staff distribution by function remained fairly stable until the last few days of 1984, except for the Drawing Office and the 1982 structural changes involving the merger of two departments which had in any event always worked together (Tables 103 and 104)[54].

A variety of problems had to be resolved, since each type of waste required particular treatment and hence special installations. The small amounts of certain wastes required processing on a laboratory scale; others required the construction of semi-industrial or industrial facilities.

Since the decision to close the plant, research and development had been focused on problems related to the conditioning of radioactive wastes[55]. The R & D work was carried out both internally and on external contract for the small Eurochemic R & D team, which was about 30 strong. Thus the treatment of contaminated solvents[56] was a research topic dealt with internally by the EUROWATT installation. Co-operation was particularly intense with Germany in developing the ALONA project for treating plutonium wastes, and especially in developing the PAMELA project for the vitrification of high activity liquid wastes arising from the reprocessing of natural and slightly enriched uranium fuels. The plans for the Belgian vitrification unit (AVB), which was in fact a transfer of French technology, did however necessitate a few adaptations in order to be able to treat high activity liquid waste arising from the reprocessing of highly enriched uranium fuels, such as for example the prior removal of mercury using a process also developed by Eurochemic.

The history of these treatment processes is given here for each type of waste, ranging from low activity wastes – for which the treatments did not differ fundamentally from those used during the reprocessing period – to high activity liquid wastes, for which novel approaches were used. The extent of the high activity liquid waste management programme and the complexity of the organisation required to carry it out justifies the fact that a special part of this book is devoted to it.

The complicated institutional arrangements characterising these processes makes summaries difficult, notably as regards the financial aspect and that of worker doses as from 1984. An estimate of costs is given in the general conclusions. The dose statistics for the period 1980-1984 are given below. They mainly show the evolution resulting from the changes in the predominant activities on the site.

Table 103. **Functional structure of Eurochemic, 1980 to 1983 (by number)**

	1980 and 1981	1982 and 1983
Operations	83	127
General services	56	0
Industrial development	28	31
Health & Safety	31.5	31.5
Administration	18	21.5
Other	9.5	15
Total[57]	226	226

Table 104. **Functional structure of Eurochemic, 1980 to 1983 (as percentages)**

	1980 and 1981	1982 and 1983
Operations	36.7	56.2
General services	24.8	0.0
Industrial development	12.4	13.7
Health & Safety	13.9	13.9
Administration	8.0	9.5
Other[58]	4.2	6.6

54 The 1984 structure is not given in RAE 1984, but it remained the same as that of 1983.

55 A summary is given in ETR 318 by J. Van Geel, *Development Work on Waste Conditioning,* pp. 121-128.

56 The literature is frequently ambiguous about the meaning of the term "solvent". It can refer to the TBP-kerosene mixture, kerosene itself (as opposed to the TBP which is then the complexing agent) or TBP (then as opposed to kerosene which is a diluent).

57 Including the Drawing Office as from 1982.

58 Idem.

Waste management on the Eurochemic site led to a reduction in radiation doses received

The overall change throughout the period when doses were measured at Eurochemic, i.e., 1969 to 1984, shows the downward trend in collective and mean doses (Tables 105 and 106). This trend can be illustrated by calculating the mean doses per period (Table 107).

It is highly likely that the downward trend in collective and individual doses continued on the site after the departure of Eurochemic, but only fragmentary data could be found on this topic, since some of the operations on Eurochemic wastes took place in the broader frame of the management of all Belgian wastes[59].

For the PAMELA installation, in which most of the radioactivity on the site was treated, the collective dose received during the six months of LEWC treatment was 40 mSv, and 300 mSv from 1 October 1986 to 15 February 1991. The maximum cumulative individual dose during the period from 1 October 1985 to 15 February 1991 was 20 mSv[60], or a maximum annual mean of 3.8 mSv, the average dose to all the personnel necessarily being lower. There is no doubt that the main reasons for the lower doses were the experience gained and the improvement in techniques, notably as regards handling operations, remote maintenance and organisation[61].

Table 105. **Overall trend in doses received by the staff, 1969 to 1984**

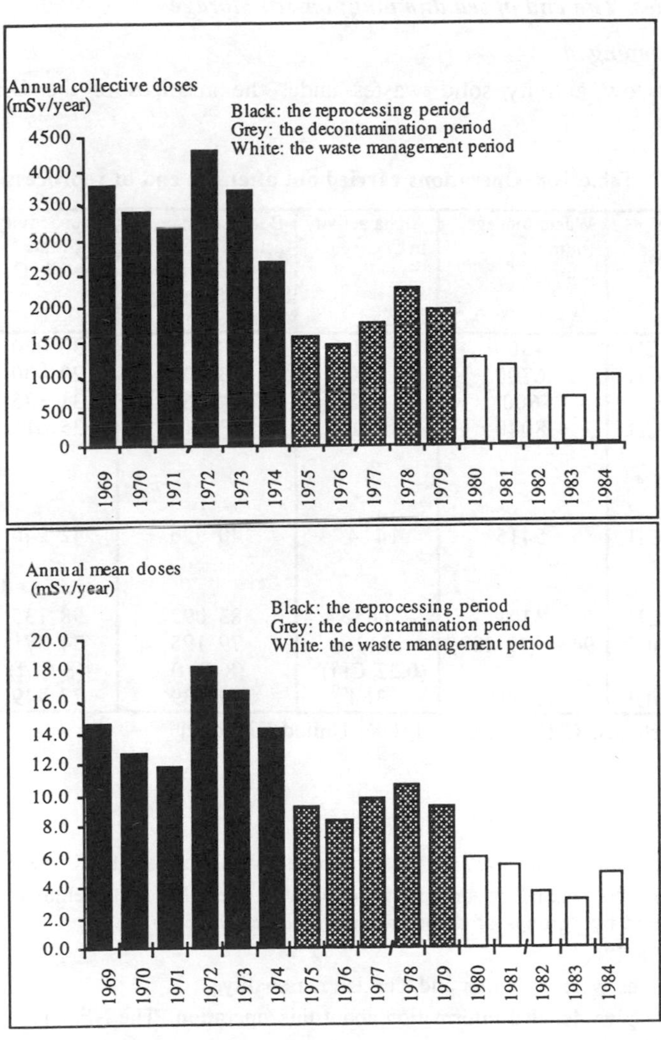

59 The first joint annual report of ONDRAF and Belgoprocess, published in 1991, reports a maximum dose of 14.7 mSv and a mean dose of 1.5 mSv for 1990. Source: *1990 ONDRAF/Belgoprocess Annual Report*, p. 12.

60 KUHN, WIESE in RECOD 91, p. 274 and 275.

61 Part V Chapter 1 describes and analyses the lower doses in the most modern reprocessing plants.

Table 106. **Summary of doses during the waste management period, 1980 to 1984**

	Annual collective dose (mSv)	Annual mean dose (mSv)
1980	1264.0	6.0
1981	1141.1	5.4
1982	798.0	3.7
1983	677.0	3.2
1984	1005.3	5.0

Table 107. **Mean doses during major periods**

	Mean annual collective dose	Mean of annual mean doses
1969-1974	3495.00	14.7
1975-1979	1797.40	9.4
1980-1984	977.08	4.6

Low activity solid wastes. The end of sea dumping; on-site storage

Continuation of sea dumping

Sea dumping of low activity solid wastes under the auspices of the OECD continued until 1982 (Table 108).

Table 108. **Operations carried out after the end of reprocessing**

Year (number of voyages)	Participating countries	Waste tonnage (number of drums)	Alpha activity in Ci	Beta-gamma activity in Ci (tritiated waste not included as from 1972)	Beta activity of tritiated waste in Ci	Belgian data or remarks
1975	B,NL,UK,CH	4500	770	28 000	30 000	
1976	B,NL,UK,CH	6700	880	33 000	20 000	
1977 (2)	NL,CH,UK	5600[62]	958	44 565	31 885	
1978 (3)	B,NL,UK,CH	8040	1101	43 015	35 614	Belgium took part in the first and was the only country involved in the second
1979 (2)[63]	B,NL,UK,CH	5415	1414[64]	40 926	42 240	Waste from B, NL and CH during the first voyage
1980 (2)[65]	B,NL,UK,CH	8391	1853	83 092	98 135	
1981 (2)[66]	(B,NL,CH)UK	9434 (12 407)	2117 (0.22 Ci/t)[67]	79 195 (8 Ci/t)	74 371 (8 Ci/t)	
1982 (2)[68]	(B,NL,CH)UK	11 693	1364[69]	49 539	77 449	

B: Belgium; NL: Netherlands; CH: Switzerland; UK: United Kingdom

62 Source of data in this column from 1977 to 1980: retrospective statistics NEAR 1982, p. 36.

63 During this operation, the second governed by the MMCS, the NEA representatives were of German and Dutch nationality. It was disturbed by "groups of demonstrators". NEAR 1979, p. 41.

64 Including 0.3 Ci for Ra-226.

65 The NEA representatives were of Italian and Turkish nationality.

66 The NEAR chooses to give detailed information about this operation. The NEA representatives were of American, French and Swedish nationality.

67 Including 60 Ci of Ra-226 (0.006 Ci/t), NEAR 1981. The table on page 36 of NEAR 1982 contradicts this by indicating "plus 60 Ci".

68 This was the final campaign. The report gives a retrospective table of the dumping operations, which we have used as appropriate to give the missing data for tonnages for 1977 to 1980. The report notes the growing opposition. The sea dumping operations gave rise "to demonstrations and interference by groups of objectors, thus generating some risk that the operation would be affected". NEAR 1982 p. 37.

69 Plus 64 Ci of Ra-226.

By the end of the 1970s however, the opposition[70] to sea dumping of waste became more virulent. The 1982 campaign was therefore the last one to be organised under the auspices of the OECD[71].

Between 1967 and 1982 a total of 1 million curies in 95 000 tonnes of wastes had been dumped in the sea. These included the suitable low and intermediate activity solid wastes from Eurochemic.

Local storage after 1982

The *de facto* cessation of sea dumping as from 1982 resulted in the accumulation of drums of wastes on the CEN/SCK site where they had been transferred pending disposal.

There was a pressing need to build an intermediate surface storage facility for the low activity wastes that were not alpha-contaminated, and also to minimise their volume, pending the choice of a final storage site. In 1984 ONDRAF began a research programme on this subject. In the meantime, two warehouses were built in the north-east part of the site (Buildings 50 and 51). The first of these was put into service in 1988. Since the CEN treatment and conditioning facilities which ONDRAF inherited in March 1989 were obsolete, it was decided in 1988 to build a central facility for low activity wastes, the Centrale Infrastruktuur voor Laagaktief Vast Afval (CILVA). This facility, which cost nearly 4.5 billion BF, includes installations for supercompacting, low temperature incineration, ash processing and concreting. CILVA was officially opened on 6 May 1994 and was planned to enter active service in the autumn of 1994 (Figure 124)[72].

Treatment of alpha-contaminated solid wastes by acid digestion. EUROWETCOMB, ALONA; initial Eurochemic-German co-operation[73]

The plutonium-contaminated solid wastes were the subject of intense R & D effort, which resulted in the development of a selective treatment method unique in Europe.

The problem of treating plutonium-contaminated solid wastes

Once reprocessing ended, it was clear that the plutonium contaminated wastes would have to be treated quickly. By 1980, 6000 28-litre canisters had been produced, the great majority containing less than 10 grammes of plutonium. Those which contained less that 2 grammes had been dumped at sea[74], and 300 containing between 2 and 10 grammes were compacted and coated in bitumen. There remained 300 canisters temporarily stored in Buildings 25 and 5/22. It was decided first of all to separate the combustible wastes, which were in the majority, from the non-combustible and mixed wastes. It was considered that 164 28-litre canisters could be processed[75]. The idea of incinerating the combustible wastes on the site was rejected owing to problems of gas filtration, tar formation and the expected difficulties in recovering the plutonium they contained. Consideration was given to a chemical process developed by Hanford Engineering Development Laboratory (HEDL) in the United States on the scale of a non-active pilot facility with a capacity of 5 kg/day. This involved breaking down the organic material present using a hot mixture of sulphuric and nitric acids, hence its name of the acid digestion or wet combustion process[76]. It was planned, after washing the residue with nitric acid, to purify the solution using a chromatography process. Tests on this method carried out at Mol in 1974 led in 1975 to the initiation of a project for a facility known as EUROWETCOMB, capable of treating 15 to 20 kilogrammes of waste a day.

At the same time work went on, on a co-operative basis between the CEA, the IAEA – which gave the contract – and EURATOM – which carried out the non-destructive measurements – to compare methods of measuring the quantity of plutonium contained in the wastes. The IAEA wanted to confirm the reliability of the non-destructive methods they used in the tests related to the application of the system of guarantees. For this

---

70 The main opponents were Greenpeace, Spanish fishermen and the Irish government.

71 In 1984, a consultative meeting decided upon a voluntary moratorium on the sea dumping of radioactive wastes. Belgium was the last to accept the moratorium on the grounds that, lacking any arrangements for the final surface storage of the wastes hitherto dumped at sea, "a door must be left open". The Belgian acceptance of the position was therefore accompanied by the initiation of research into the terrestrial storage of this category of wastes. Under the London Convention of 12 November 1993, the sea dumping operations were to be banned for 25 years.

72 For the intermediate activity solid wastes, a special conditioning facility was put into service around 1980.

73 Sources: RAE, SWENNEN R., CUYVERS H., VAN GEEL J., WIECZOREK H. (1984), WIECZOREK H., OSER B. (1986).

74 RAE 70-74, p. 199.

75 The canisters containing non-combustible wastes are stored in Building 5/22 awaiting subsequent treatment.

76 RAE 1975, p. 56.

Figure 124. Operators using remote manipulators in the conditioning facility for intermediate activity solid wastes in Building 23 (Source: Eurochemic slide, no date).

purpose they had to be compared with the chemical analyses carried out after incineration. The CEA conducted this operation at Marcoule on 28 reference drums[77]. The results suggested that the plutonium wastes contained about 8 kilogrammes of element 94[78].

Ad hoc *international co-operation*

1976 saw the determination of the characteristics of a pilot installation to be built on a co-operative basis between Nukem (as architect engineer), the Institut für nukleare Entsorgungstechnik (INE) of the Karlsruhe Nuclear Centre (for R & D into acid digestion and acid recovery), and Eurochemic (responsible for pretreatment), for the recovery of the plutonium and the management of the installation. The European Community supported the project as part of its own research programme.

The initial cost estimates considerably exceeded the initial plans and unexpected technical problems were encountered. It was decided to send a joint mission to the United States to collect more information from the Hanford Engineering Development Laboratory (HEDL) and to examine other methods developed in laboratories in the United States. This mission took place at the beginning of 1977 and confirmed that acid digestion was the best way of resolving the problem.

The installation comprised three units: the pretreatment system, the ALONA digester and the plutonium recovery unit.

The pretreatment system

Construction of the pretreatment installation began in 1978 in the research laboratory. It consisted of a series of four sealed alpha gloveboxes intended for sorting and crushing the wastes so that they could be loaded into the digester. The installation was completed at the end of May 1980 and entered active service in June. Initial operations revealed a series of faults and the facility had to be shut down in November because the two operators had received the maximum dose allowed over any 13-week period. Work was started therefore to

---

77 RAE 1975, p. 67-68.

78 The initial uncertainty was fairly wide at ± 2 kilogrammes. The subsequent chemical analyses reduced this spread.

install additional shielding[79]. Operations restarted at the beginning of 1981 and were completed during the year, by which time 803 kg of combustible wastes had been separated from 167 kg of non-combustible wastes. The project then moved on to the next phase: construction of the ALONA acid digestion unit.

ALONA (Figure 125)[80] and the plutonium recovery unit

Construction of the ALONA acid digestion unit began in May 1981 for the Karlsruhe Waste Management Institute. The unit comprised seven gloveboxes; cold tests began in November. In September the Eurochemic team began to install the plutonium recovery unit, also with seven gloveboxes, in Laboratory 105 located in the hot wing of the research laboratory. Work was held up in 1981 and 1982, first because the tests had shown the need for some adaptations such as improvements to radiation shielding and, secondly, owing to priority being given to other tasks, notably the consequences of the EUROBITUM fire (see below). The cold tests nevertheless began in 1982.

ALONA went active on 28 February 1983. Operations revealed a number of problems although the recovery unit functioned well. Modifications were made at the end of the year and ALONA was restarted at 60% capacity in 1984. At the end of the year, when the facility was handed over from Eurochemic to Belgoprocess, 63% of the wastes had been treated, containing 74% of the plutonium, 93% of which had been recovered (Table 109).

Table 109. **Material balance for all the plutonium recovery operations at ALONA up to the end of 1984**[81]

	Measurement in kg	uncertainty in kg	% uncertainty/measurement
Input	4.78	± 0.72	15.1
Process inventory	0.36	± 0.04	11.1
Losses in effluent and waste	0.22	± 0.04	18.2
Output	3.5	± 0.07	2.3
MUF	0.7	± 0.35	50.0
Ratio MUF/input	14.6 %		

The ALONA experience confirmed the difficulty of establishing any absolutely reliable accounting system for fissile materials owing to the cumulative nature of the uncertainties, and can only enhance the necessity of local inspections in order to limit the risks of proliferation.

Operations continued until July 1985. All in all, 6.3 kilogrammes of plutonium were extracted from 800 kilogrammes of wastes. The throughput figures shown on Figure 126 show that the startup problems were resolved within 18 months and that the faster rate of working as from the autumn of 1984 led to more efficient waste treatment during the final year of operation. The installation has now been decommissioned.

Acid digestion of plutonium-contaminated solid wastes: technical review

The acid digestion experiment carried out at Eurochemic[82] was the most thorough of the two experiments conducted in the world. It demonstrated the technological feasibility of the process but also drew attention to the technical problems that any unit intended to operate for long periods would encounter. The main difficulties were the blocking of the feed system by plutonium particles, and problems of corrosion and strength of the materials used: glass and special steels[83].

The summary figures show the extent of the low activity, non-alpha process wastes. On the average, 2.3 kilogrammes of process wastes were produced for every 1 kilogramme of wastes treated by acid digestion. However the experiment managers believed that modifications to the process tested in the laboratory could reduce this figure to 1.3 kg/kg.

The problem of recovering plutonium from solid reprocessing wastes is today still the subject of laboratory research.

With EUROWATT, Eurochemic also did pioneering work on the treatment of a very special category of waste arising from reprocessing – contaminated solvent – of which about 21 m^3 had been stored since the end of reprocessing.

---

79 RAE 1980, 37.

80 Aktive Labor-Anlage für die Optimierung der Nassverbrennung: Active Laboratory Facility for the Optimisation of Wet Combustion.

81 RAE 1984, 9. Data in kilogrammes, calculations made from the raw data in the source.

82 OECD/NEA (1993), p. 165; WIECZOREK H., OSER B. (1986), p. 82

83 WIECZOREK H., OSER B. (1986) recommend the use of Tantalum and Teflon in the future.

Figure 125. General view of the ALONA installation (Source: Eurochemic slide).

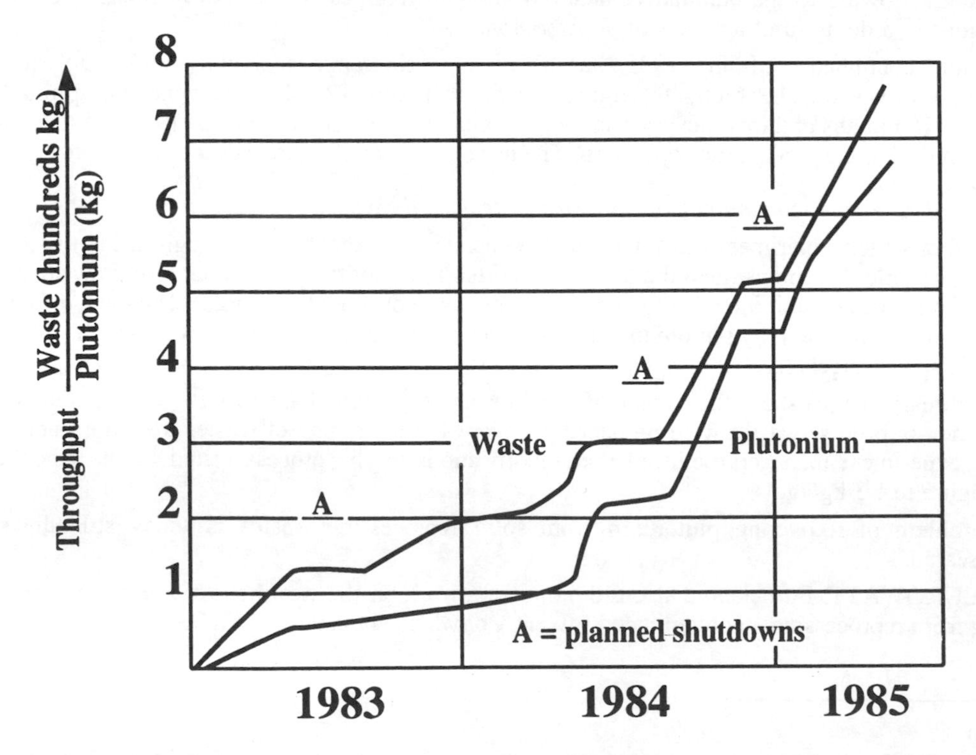

Figure 126. Rate of waste treatment and plutonium production in the ALONA installation from 1983 to 1985. The graphs show the throughput of waste (hundreds of kilogrammes) and of plutonium (kilogrammes). The letter A indicates programmed shutdowns (Source: WIECZOREK H., OSER B. (1986), p. 82).

EUROWATT: treatment of contaminated solvent and co-operation between Eurochemic and France[84]

The treatment of contaminated solvents was part of the R & D work during the reprocessing period, and included tests of separation techniques involving chromatography and phase separation[85]. For example with regard to the chromatography method, tests had been done with columns containing alumina, manganese dioxide, cellulose powder, carbon, silica gel, kieselguhr[86], asbestos and ion exchange resins. Phosphoric acid was used for the organic phase separations. Pilot experiments took place during the first half of the 1970s[87], resulting in 1974 in the granting of a Belgian patent for a process for treating Eurochemic organic wastes, known as EUROWATT[88]. Patents were also taken out in the United Kingdom in 1974 and the United States in 1978.

The aim of the process was to avoid incinerating the TBP-kerosene solvent-diluent mixture, owing to the risks of corrosion, and to recover the diluent by purification for recycling in reprocessing plants. It was demonstrated in the laboratory by tests carried out until 1978. It involved a phase of separating the TBP and fission products from the kerosene, in mixer-settlers, followed by thermal decomposition of the TBP in the presence of phosphoric acid, the fission products entering the acid phase before being conditioned as high activity wastes.

Construction of the pyrolysis unit for the thermal decomposition of the mixture at 200°C encountered material problems. The unit had to withstand the phosphoric acid which was used as a catalyst. Finally a special alloy, Hastelloy B, was chosen for the pilot installation the construction of which was decided upon in 1976 for treating the 20.8 m^3 of used solvent then being stored.

In 1977 an agreement was signed with SGN for the construction of a demonstration unit in Room 103 of the research laboratory building. Construction started at the end of 1978 and cold tests began in 1979. Treatment began on 3 November 1980 and was completed on 13 March 1981. The 20.8 m^3 of used solvents produced less than 1 m^3 of concentrated phosphoric acid and 3.5 m^3 of organic solutions which were conditioned by being incorporated into bitumen; the separated kerosene was regarded as cold effluent and transferred to the CEN for incineration.

The experience acquired revealed the difficulties in controlling the mixer-settlers and the pumps in the installation, but did suggest that larger facilities were feasible.

With this in view COGEMA got in touch with Eurochemic in 1982 about applying the process to the organic wastes from UP2 at La Hague. Laboratory tests conducted on two litres of French solvent showed that although the installation could treat these solvents, adaptations would be needed, notably as regards radiation shielding owing to the high burn-up values of the materials treated at La Hague[89]. These adaptations were made in 1983 and 4 m^3 of solvent from La Hague were treated[90]. However the quality of the diluent obtained was such that it could not be re-used in UP2, which resulted in further improvements to the process being planned.

Discussions then began about the possibility of COGEMA having 100 m^3 of used solvent – 12 months' production – processed in a one-year campaign. This would have provided the industrial experience needed to facilitate adoption of the process by COGEMA. The negotiations continued into 1984, at the beginning of which year the installation was used to treat 60 litres from Kjeller and 200 litres from the Eurochemic plant which were highly degraded and had consequently not been treated during the initial phase. The necessary improvements were then carried out.

Pending the outcome of the discussions with COGEMA, the installation was decontaminated and placed on standby. However the contract was never signed, and today a different process is used in France[91].

Bituminisation of intermediate activity liquid wastes and the storage of drums of bituminised waste

By the end of the 1960s, the bituminisation of intermediate activity wastes was a routine practice at Marcoule. It was a rapid process and appeared a safe way of stabilising the plant's effluent. Plans to build a

84 See ETR 287 for the development of Eurowatt and ETR 312 for the pilot installation.

85 These are reported in 1969 in the RAE 70-74.

86 A type of silica-based pumice stone.

87 Technical details in RAE 70-74, pp. 185-187.

88 Eurochemic Organic Waste Treatment.

89 During the 1980s, fuels with burn-up values of 35 - 40 000 MWd/t were reprocessed.

90 Details of the operations are given in RAE 1983, 17-18.

91 At UP2 and UP3, organic wastes are treated by distillation at low pressure with a film evaporator, known as "flash distillation".

bituminisation unit and an adjacent intermediate storage facility at Mol had been decided as early as 1971[92]. Its construction by Comprimo and Belgonucléaire was problem-free, but its operation did reveal certain difficulties.

The principles of EUROBITUM and EUROSTORAGE (Figures 127 to 129)

The bituminisation process used at Eurochemic starts with a chemical pretreatment to convert the liquid into slurry, which is then incorporated into bitumen.

The pretreatment is a neutralisation-precipitation operation – usually in the proportion of 60% water and 40% salts in the case of the decladding effluent for example – which renders the fission products and actinides insoluble and entrains them in the precipitate.

The slurry is incorporated into the bitumen by means of four mixing screws in a horizontal evaporator below 200°C, which removes water from the slurry. At the outlet, the salt-bitumen mixture with proportions 45% salts, 54.5% MEXPHALT 85/40 bitumen, with less than 0.5% water, and activity of the order of 1 Ci/l, is poured into 220-litre drums in a special cell at a rate of 160 l/h. The installation had a planned annual capacity of 3000 to 3500 drums.

The filled drums are loosely closed by a lid fixed at three points, allowing any gas that might be produced subsequently to escape. They are then transferred automatically to the EUROSTORAGE installation.

EUROSTORAGE is a modular structure made up of casemates. Two of these were built in 1978 when EUROBITUM was started up. Each casemate is 64 x 12 x 8 metres and can accommodate 5000 drums, stacked in four layers. The drums are positioned by a remotely controlled overhead crane.

Startup of EUROBITUM and EUROSTORAGE (Figures 130 to 132)[93]

The bituminisation and storage of drums in Buildings 26 and 27 began in June 1978. The facility was automated and controlled from a central control room. It operated without problems for more than three and a half years.

By the end of 1981, 1440 m^3 of intermediate activity liquid waste had been conditioned in 7908 drums. In 1981 the installation was operating at 88% capacity, 10% of the remainder being taken up by extruder rinsing and 2% on maintenance and repairs.

The fire of 15 December 1981[94]

On 15 December 1981, fire broke out in three drums as they were being filled on the carousel (Figures 133-135).

Although the effects of this incident were confined to the filling cell, the fire demonstrated the importance of the human factor in the safety of nuclear installations, and revealed one of the delicate features of incorporating wastes into bitumen.

A brief description of the incident is given in a paper attached to the minutes of the eighty-fourth meeting of the Board of Directors held on 15 January 1982:

"At about 2:30 p.m. on Tuesday 15 December, the operators of the facility in which the intermediate activity liquid wastes are incorporated into bitumen noted an exothermic reaction in one of the 220-litre drums being filled...Corrective action was immediately taken by the operators and by the establishment's fire brigade in accordance with the relevant standing rules and instructions. Subsequently, assistance was given by the Mol fire brigade. The fire remained contained inside the drum and was extinguished after a few minutes by the use of water. Over the next few hours the same phenomenon of combustion occurred in three other drums and was dealt with in the same way ... The cause of the exothermic reaction that led to this combustion in the drums will be identified. The installation will not be restarted until the causes have been determined and the systems thoroughly reviewed".

---

92 Today, there are differing views about bituminisation. There are plans to replace it at UP3 by vitrification, which requires the intermediate activity wastes to be concentrated. Bituminisation seems suitable for temporary storage, but it appears likely that further conditioning in sealed containers will be necessary for final storage. In any event, these are modern issues that did not exist when Eurochemic opted for bituminisation.

93 For EUROBITUM, see ETR 324, published in January 1990.

94 The incident was covered by three reports: a five-page report dated 17 December 1981 prepared at the request of the Turnhout court, forwarded on 18 December to the members of the Board of Directors; a one-page summary attached to the Board dossier of 15 January 1982; and report ETR 314 published in December 1990.

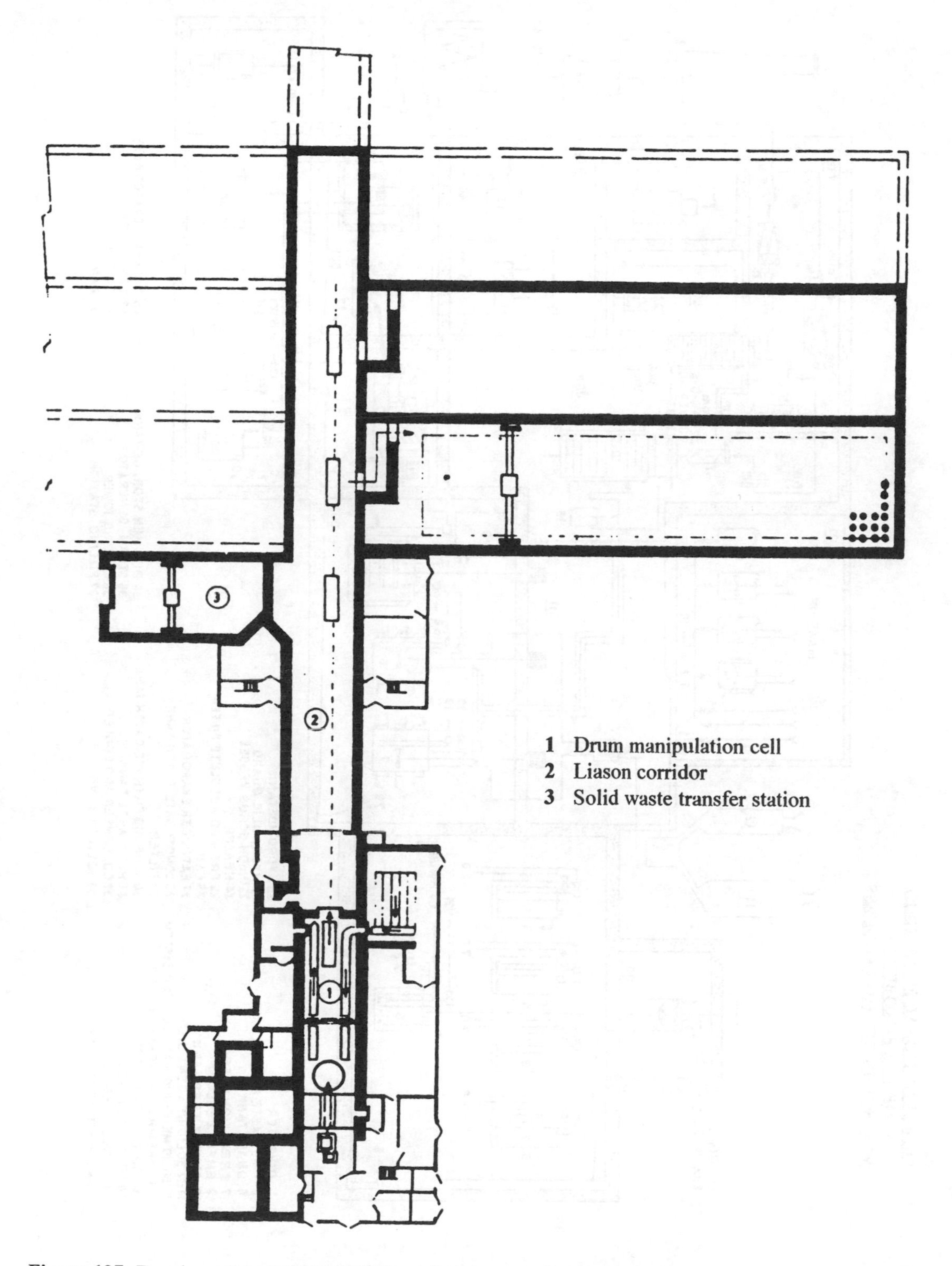

Figure 127. Drawing of the EUROBITUM conditioning installation (at the bottom) and the EUROSTORAGE modular storage facility (at the top) for intermediate activity wastes. The connecting corridor runs between the two (Source: RAE 1975, p. 39).

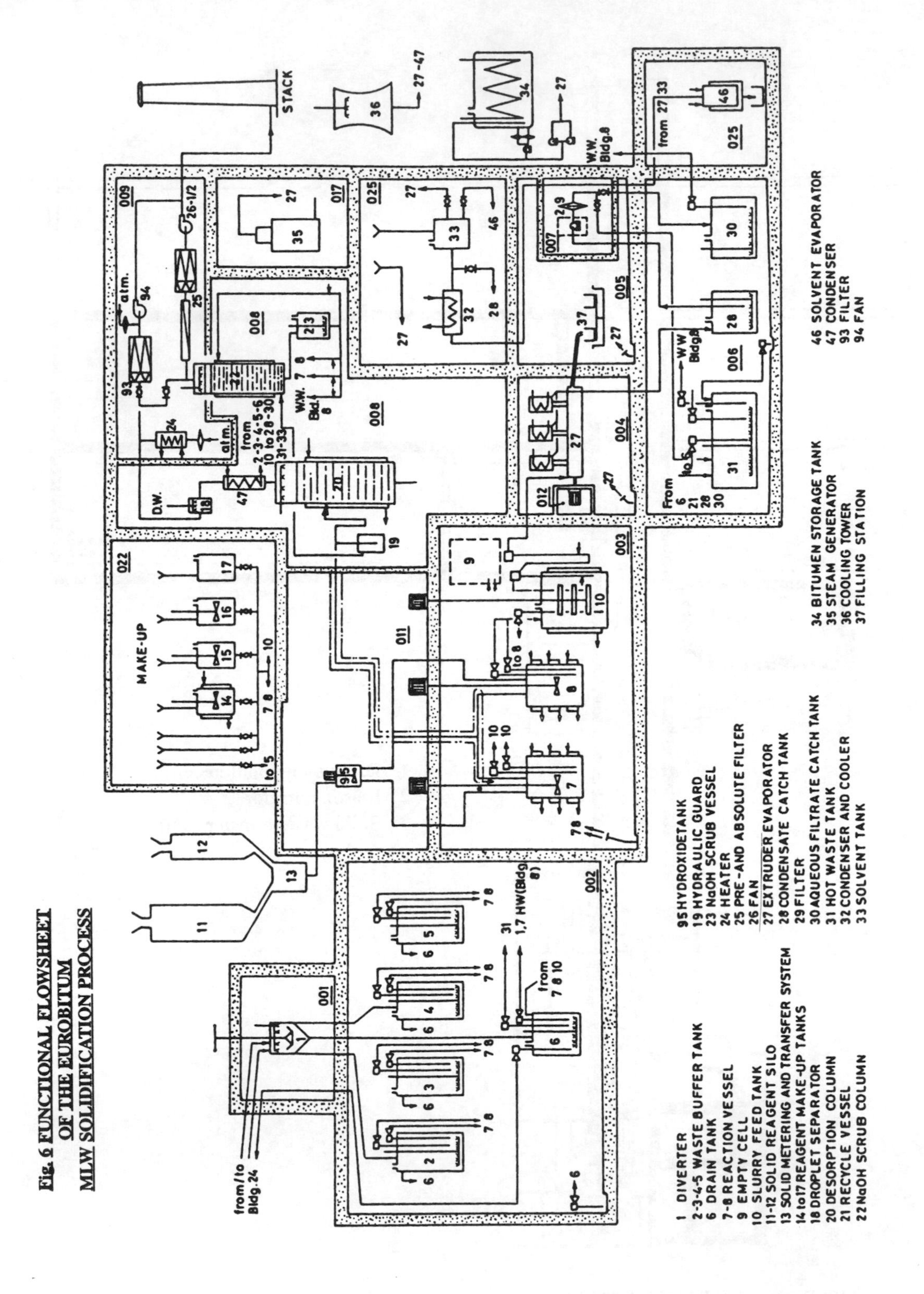

Figure 128. Flow chart of the EUROBITUM process (Source: ETR 324, Figure 6).

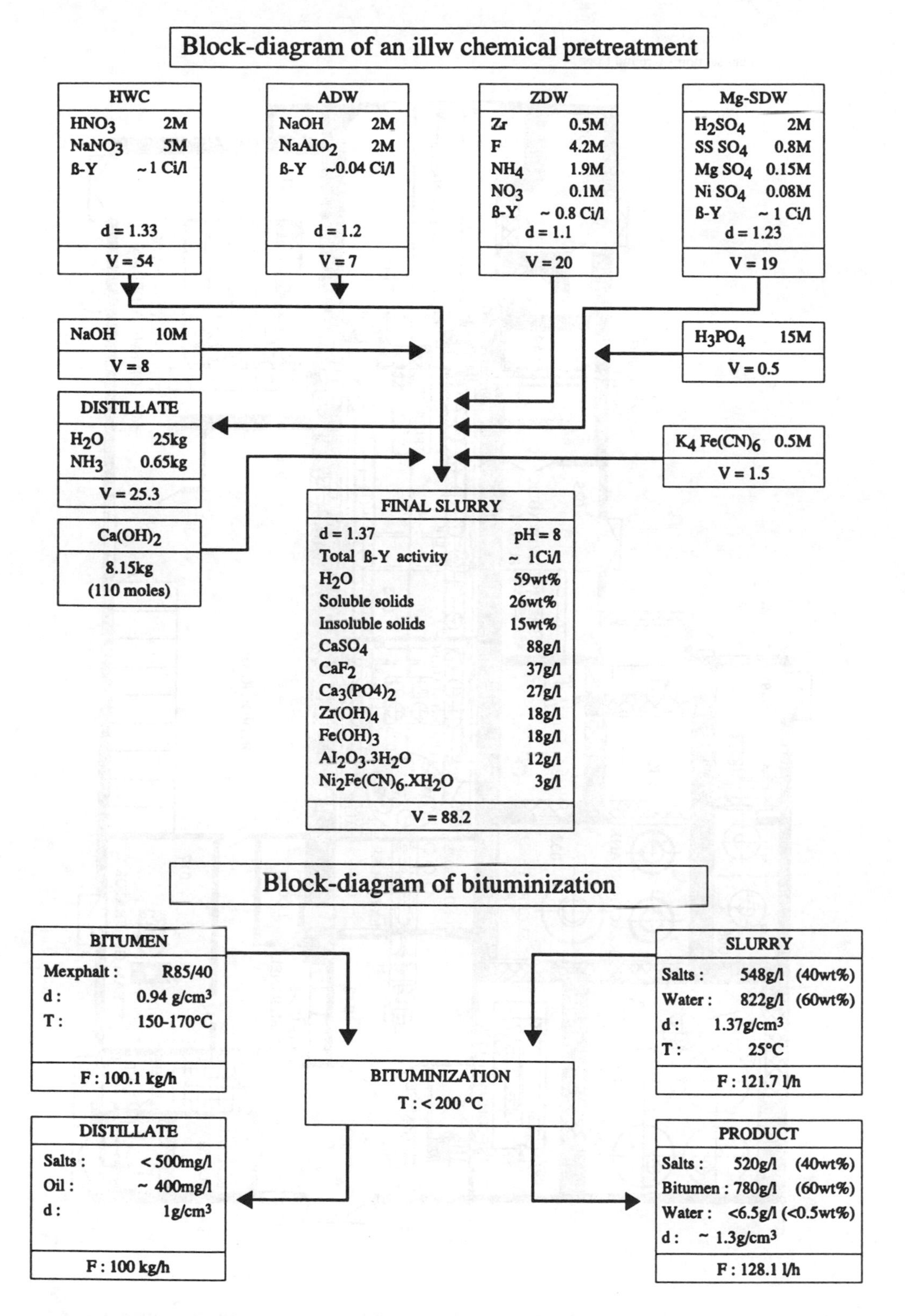

Figure 129. Block diagram of a pretreatment operation on intermediate activity liquid waste and the bituminisation of the resulting slurry (Source: ETR 324, Figures 3 and 4).

Plan section - Ground floor

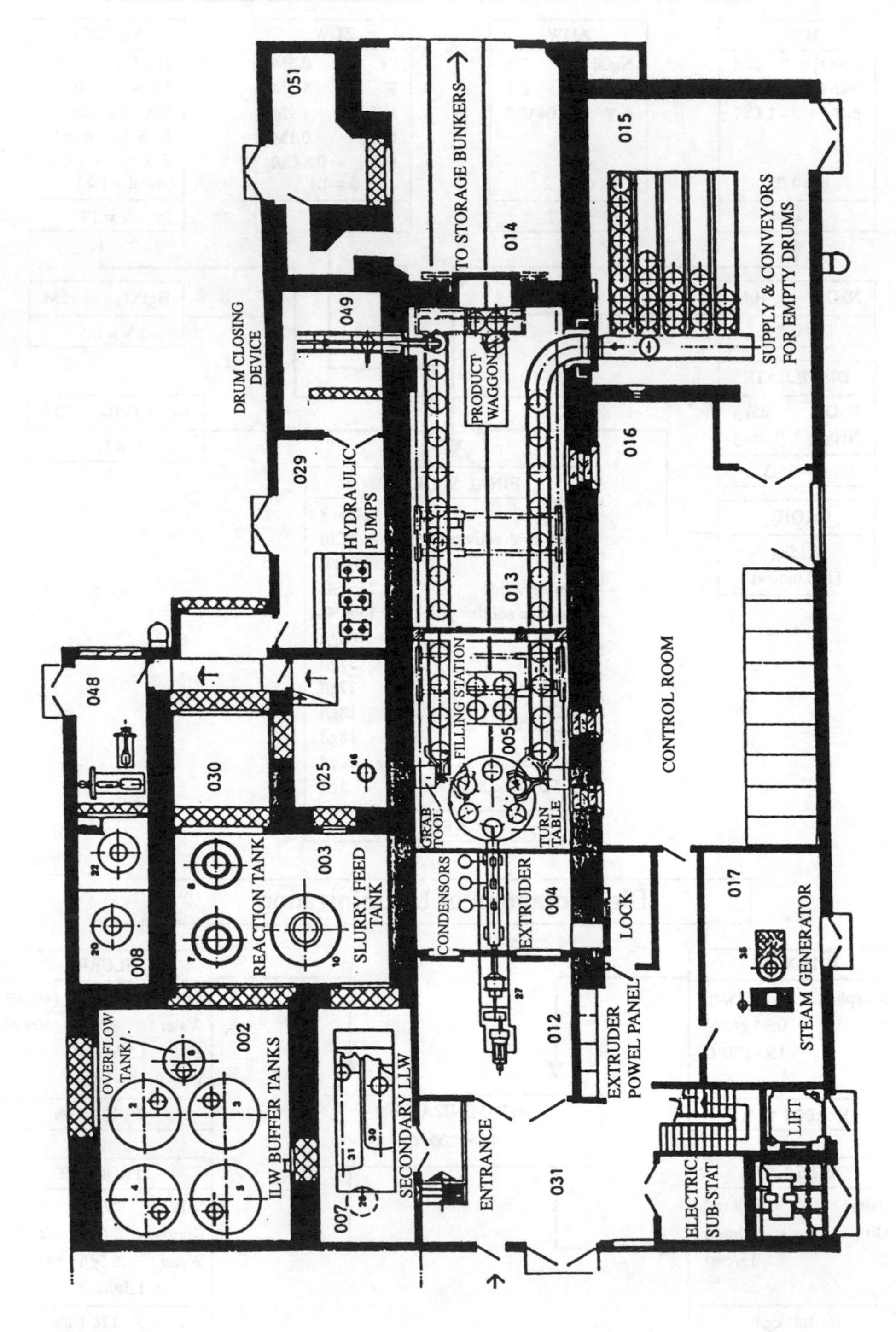

Figure 130. Plan drawing of the ground floor of the bituminisation installation (Source: ETR 324, Figure 12).

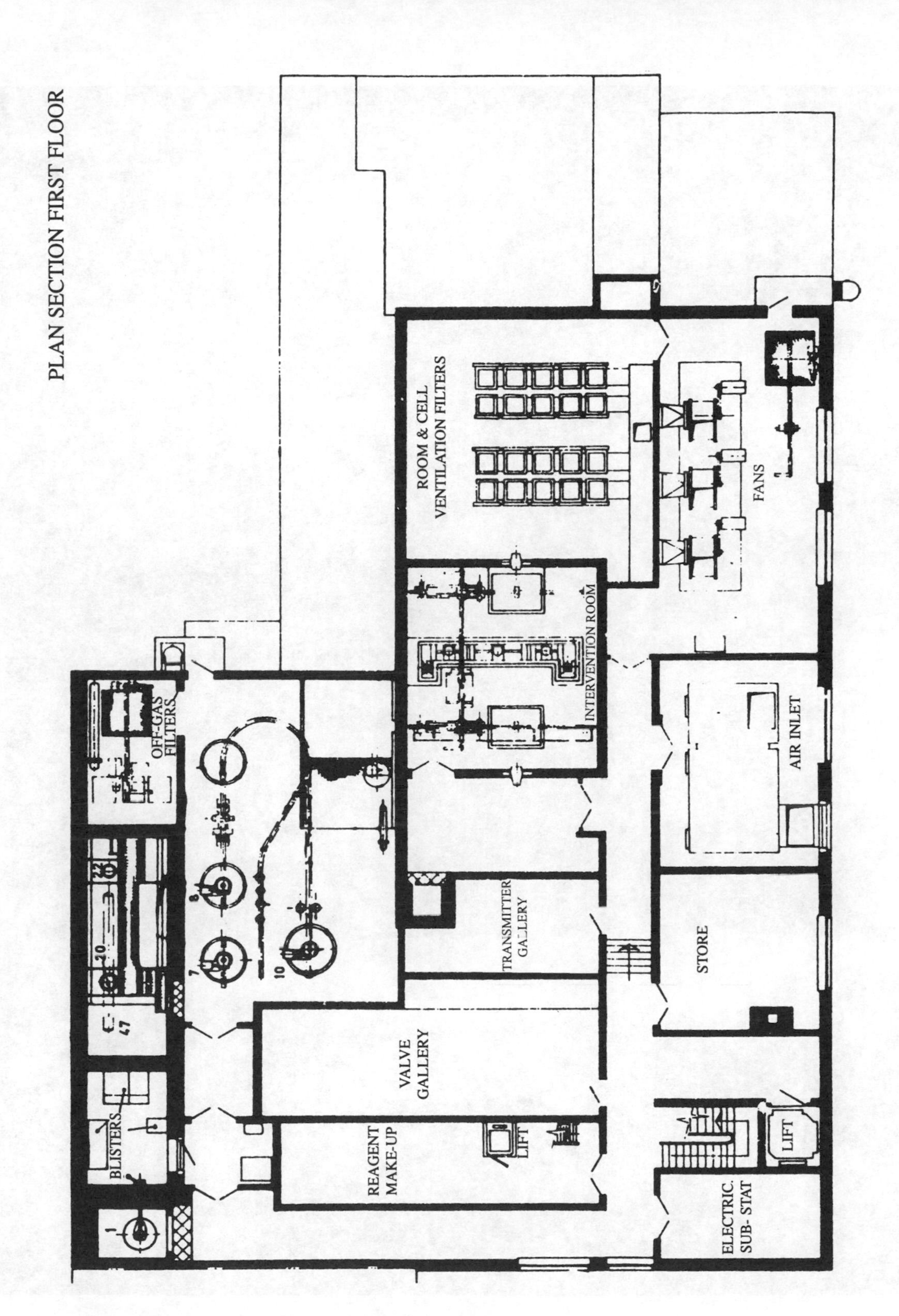

Figure 131. Plan drawing of the first floor of the bituminisation facility, showing the ventilation and filtration systems for gaseous effluent (Source: ETR 324, Figure 13).

Figure 132. Filling the first module of EUROSTORAGE, with its overhead crane and drum grab (Source: Eurochemic colour slide, no date).

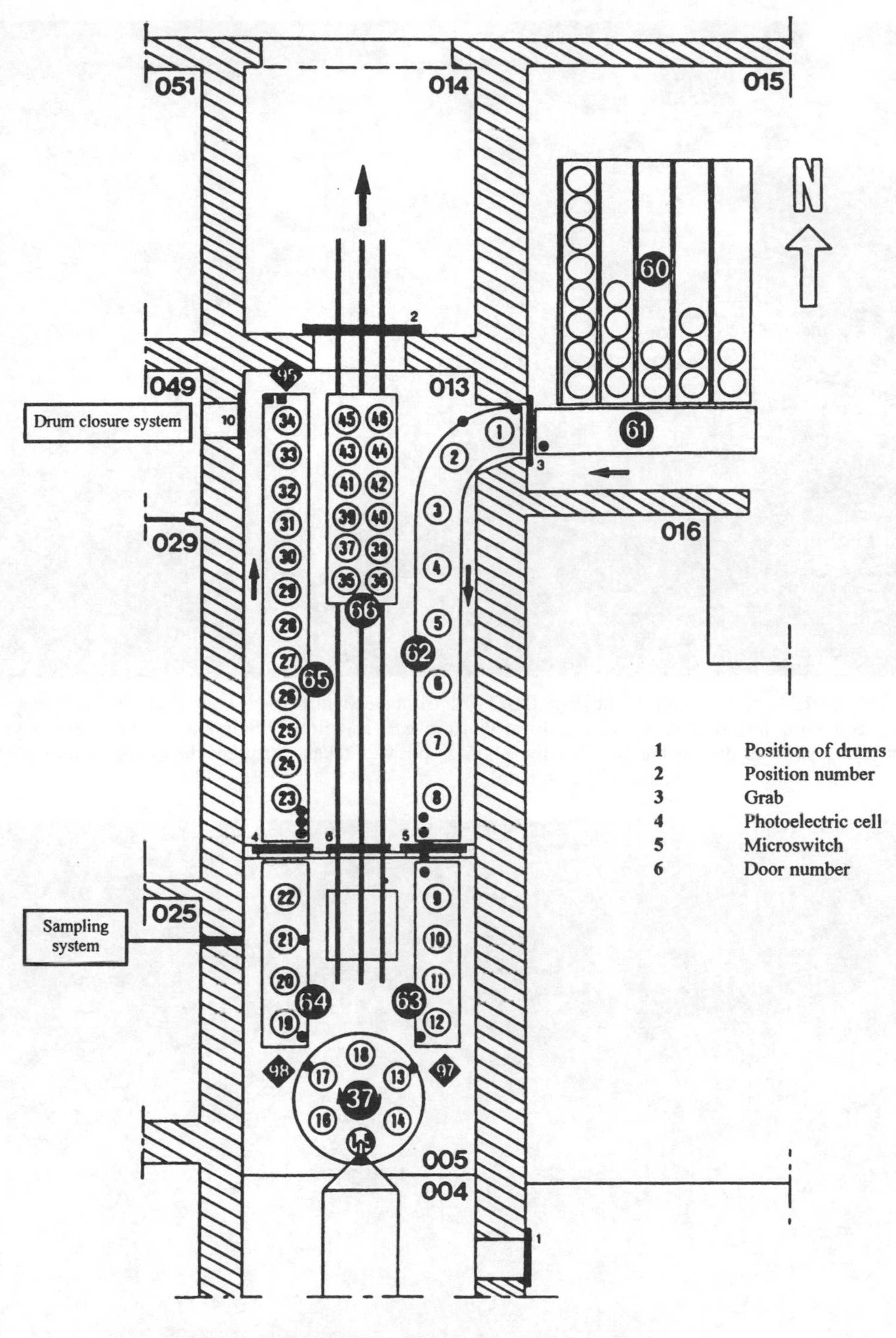

Figure 133. Diagram of the EUROBITUM filling carousel and the drum transfer system. The empty drums stored in 60 are automatically fed to the Carousel 37 and then loaded onto Transfer Trolley 66 which takes them to EUROSTORAGE (Source: ETR 324, Figure 8).

Figure 134. Photograph of the EUROBITUM filling cell in normal operation. In the foreground, the carousel with four drums, one of which, at the left rear, is being filled. The predominantly yellow colour of the slide is due to the fact that the photograph was taken through the thick window of the control room (Source: Eurochemic slide, no date).

Figure 135. Photograph of the EUROBITUM filling cell after the fire of 15 December 1981 had been extinguished. The self-ignition of three drums on the carousel caused them to overflow and another drum to tip over (Source: RAE 1982, p. 21).

The enquiry into the incident was not published as an ETR until 1990, the reason given being the backlog in the programme of publishing technical reports. It revealed that not all the test procedures had been followed. Apparently the operations staff, wanting to save time in order to complete a series of conditioning operations before the Christmas holidays, had not strictly applied the verification procedure allowing a particular mixture of bitumen and salt – which should have necessitated a more careful check on temperature and flammability – to be fed into the bituminisation unit, causing the self-ignition of the drums. The cost of the incident was evaluated at 30 million BF.

The completion of conditioning

However the fire of 15 December had caused only very superficial damage to the installation. The restart of conditioning was authorised by CORAPRO on 23 April 1982 and continued without difficulty under the responsibility of Eurochemic. By the time the company handed over to Belgoprocess on 1 January 1985, 2270 m^3 of intermediate activity liquid waste, essentially decladding effluent, had been conditioned in 11 965 drums, totalling 843 000 Ci, of which 6800 Ci was alpha activity. The installation was subsequently used for bituminising the intermediate activity liquid waste produced by the vitrification of high activity liquid wastes (see below). It is still in service today.

Review of the bituminisation of intermediate activity liquid wastes

Annual throughput

Table 110. **Production of drums of bituminised waste produced, 1977 to 1984**

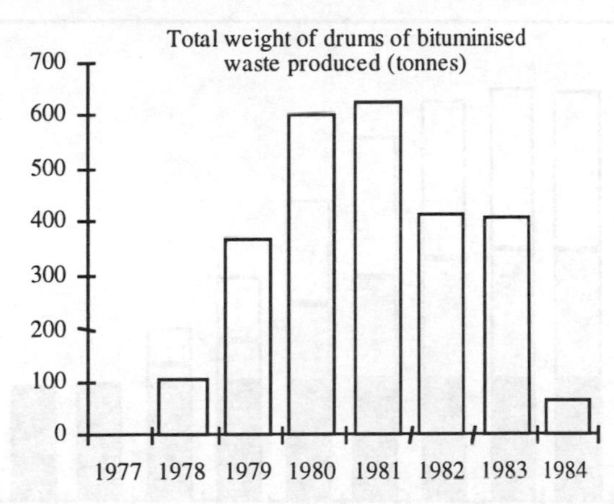

Table 111. **Trends in stocks of intermediate activity liquid wastes, 1977 to 1984, and production of drums of bituminised waste produced**

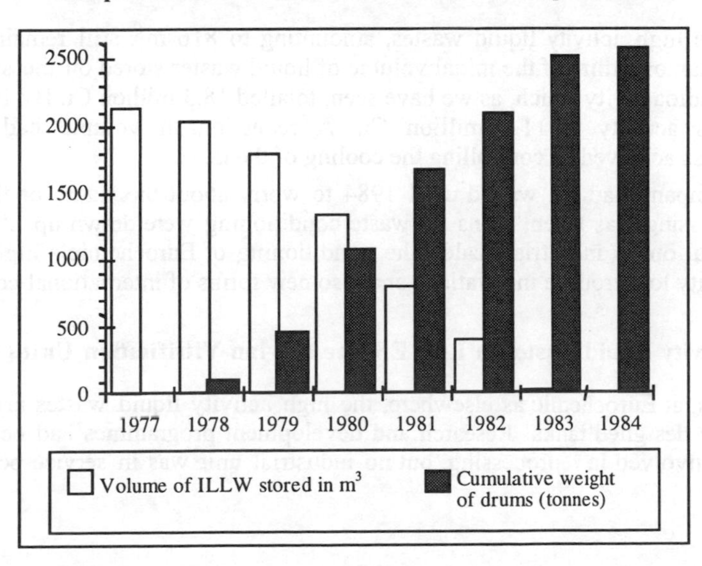

The waste situation on 31 December 1984 on completion of Eurochemic's management of the site

The overall position in 1984, upon completion of the bituminisation of the intermediate activity liquid wastes, and when Eurochemic ceased to be operator of the site, is shown in Tables 112a and 112b.

Table 112a. **Changes in stocks of liquid radioactive wastes, 1974 to 1984, in m³ (rounded values)**

Year	LEU high activity liquid wastes LEWC	HEU high activity liquid wastes HEWC	Liquid decladding wastes JDW	Hot liquid waste concentrates HWC	Total volume
1974	65.0	800	950	1050	2865
1975					
1976	65.0	806	939	1165	2975
1977	65.0	805	948	1176	2994
1978	61.8	795	892	1142	2890
1979	61.8	795	768	1009	2634
1980	57.3	782	572	757	2169
1981	55.5	781	316	471	1624
1982	51.4	774	153	242	1221
1983	51.5	774	0	13	838
1984	48.4	768	0	0	816

This can be shown graphically as follows:

Table 112b. **Changes in stocks of liquid radioactive wastes, 1974 to 1984, in m³**

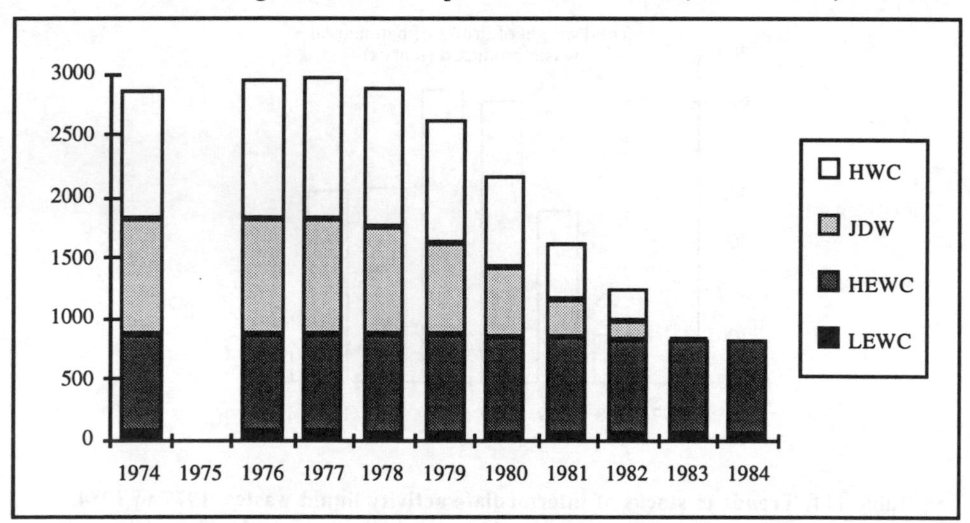

In 1984, only the high activity liquid wastes, amounting to 816 m³, still remained to be conditioned. They represented less than one-third of the initial volume of liquid wastes stored on the site, but they contained the greater part of the radioactivity which, as we have seen, totalled 18.3 million Ci. By 1984 natural radioactive decay had reduced this activity to 11.9 million Ci. A reduction in volume had been obtained by a self-concentration process achieved by controlling the cooling of the tanks.

Of course the company had not waited until 1984 to worry about the future of these wastes. Once the decision to halt reprocessing was taken, plans for waste conditioning were drawn up. At that time in fact no approach was operational on an industrial scale. The conditioning of Eurochemic's high activity reprocessing wastes was an opportunity to introduce innovations and also new forms of international co-operation on the site.

Treatment of high activity liquid wastes in LOTES, the Belgian Vitrification Unit (AVB) and PAMELA

In the early 1970s, at Eurochemic as elsewhere, the high activity liquid wastes arising from reprocessing were stored in specially designed tanks. Research and development programmes had been initiated during the sixties in the countries involved in reprocessing, but no industrial unit was in service before 1978. The Atelier de vitrification de Marcoule (AVM) was the first such installation to be put into service anywhere in the world. PAMELA, the Mol installation, was the second[95].

The situation in 1972-1974 and the selection of projects

In its treatment of high activity liquid wastes, Eurochemic was in fact a world pioneer in the search for processes to stabilize high activity liquid wastes. The most up-to-date techniques were brought in to resolve the conditioning problem. In the early 1970s, all research was focused on vitrification, with work going on in France and Germany as well as in the United States. The United Kingdom is outside the scope of this study but she was no exception to the rule, with the FINGAL and HARVEST projects, and had provided inspiration for Eurochemic's research from the 1960s onwards[96].

R & D in France, Germany and the United States

France's GULLIVER and the Marcoule vitrification unit (AVM)

In France, the problem of solidifying high activity liquid wastes had been researched since the end of the 1950s. Laboratory studies of vitrification had begun in France at Fontenay in 1957 and a pilot vitrification unit, named GULLIVER, operated there from 1965 to 1967[97]. As from 1968, research was concentrated at Marcoule, where most of the fission product solutions were located at that time. Research into the different kinds of glass resulted in the choice of borosilicate glasses. At the same time research had been carried out on two vitrification processes whereby fission products were incorporated into a mass of molten glass which was then cast into blocks. This led to the construction of the PIVER pilot unit and the development of the Atelier de vitrification de Marcoule (AVM) project.

PIVER investigated a discontinuous process in which calcination and vitrification took place in the same vessel which was heated in an induction furnace. The project began in 1965 and resulted in the pilot unit being commissioned in 1969. It ran until 1973 and produced 12 tonnes of glass containing 5 million Ci[98]. However the process which was applied on an industrial scale in France was not this discontinuous, integrated system, but a continuous process involving separate installations.

In the continuous process that resulted in AVM the two operations of calcination and incorporation into glass were separate. The former took place in a rotary calciner, the second in a metal furnace. This process was first tried out in a half-scale pilot version. The design of AVM was entrusted to SGN which began work in 1973.

The AVM (see diagram in Figure 136) was a demonstration facility which operated as from 1978. It was finally selected for use at the French reprocessing plants for two main reasons: first the calcinate had a highly uniform grain size, producing extremely homogeneous glass blocks, and secondly because separating the two operations made the installation easier to control, the process parameters being simpler. The main problems that appeared during testing concerned maintenance, the filtration of dust from the calciner and the formation of a crust on its walls. These were resolved by developing remote maintenance tools, trapping the dust and dissolving it in a bath of nitric acid before recycling, and by introducing an additive to prevent crust formation.

Scattered research in Germany from 1965 to 1973

The initial work had been done in Germany as from 1965 in Karlsruhe with the experimental development of a vitrification process known as VERA[99]. This was originally in three separate stages: the removal of nitrates from high activity waste concentrates (HAWC) using formic acid; the calcination of the resulting solution in a steam-heated calciner[100]; finally vitrification, by incorporating the calcinate into a molten mass of borosilicate glass. The Karlsruhe project benefited, during its development, from fundamental research done in Berlin.

In 1967 the Hahn Meitner Institut in Berlin (HMI) began research into the use of borosilicate glasses and ceramics for vitrifying high activity liquid wastes.

---

[95] To sum up, the main problem in the conditioning of high activity liquid wastes is how to stabilise the mixtures which can include up to 40 different chemical elements. Vitrification appears the most straightforward way of doing this, but the operation is complicated by the high temperatures needed and by the radioactivity. See KRAUSE H. (1982) on the alternatives to glass.

[96] The United Kingdom was to adopt the French AVM process in the early 1980s.

[97] See CREGUT A., LURIE R. (1985), p. 134 concerning its decommissioning as from 1981.

[98] It remained in service until 1979 and was used to condition HLW from the reprocessing of fast reactor fuels.

[99] VERA: Versuchsanlage für die Verfestigung von hochaktiven Abfallösungen; Experimental Installation for the Solidification of High Activity Liquid Wastes.

[100] "Dampfbeheizter Sprühkalzinator".

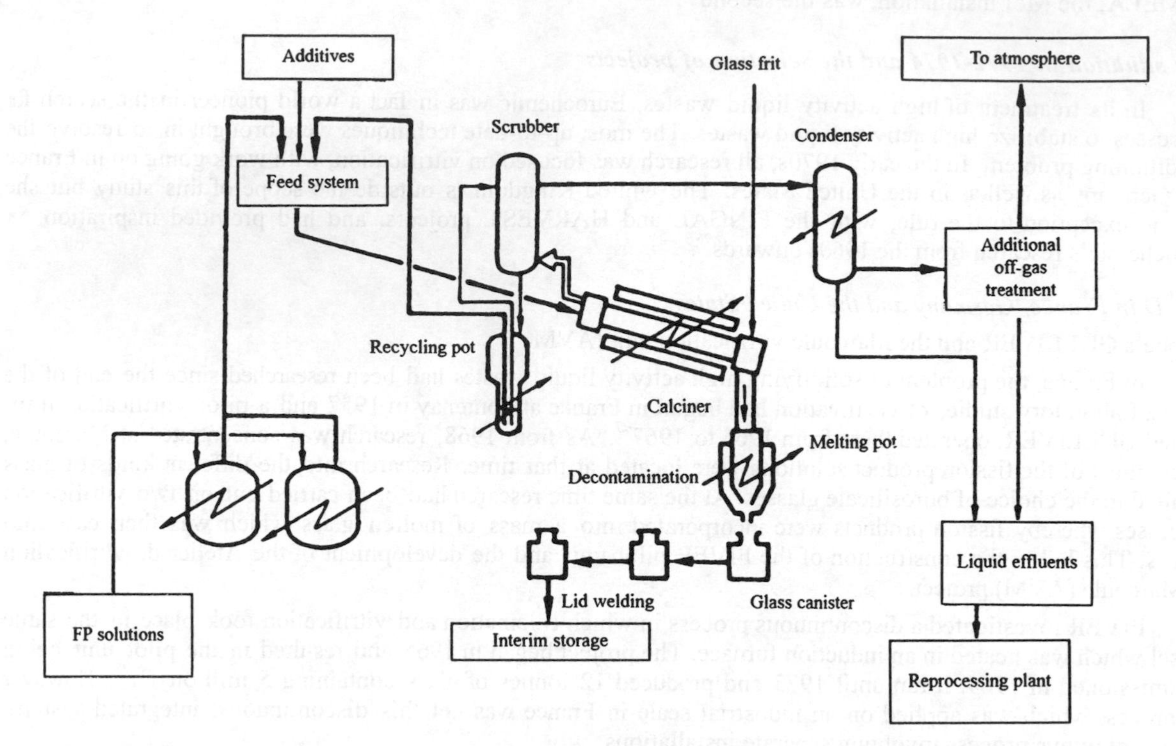

Figure 136. Block diagram of AVM vitrification, showing its two separate stages: the rotary calciner for the fission product solutions and the furnace for mixing and melting the calcinate and the glass frit (Source: ALEXANDRE D., CHOTIN M., LE BLAYE G. (1987), p. 290).

In 1968 another programme was started at the Juliers centre (Jülich) involving two projects, FIPS and PHOTHO[101]. The latter project was a co-operative venture with the firm Gelsenberg AG, which in 1973 began work together with Eurochemic in an initial PAMELA[102] product in order to develop it by adapting it to the PUREX wastes. As its name indicates, the conditioning process then involved phosphate glasses, which it was proposed to incorporate into metal matrices. Generally speaking, research at that time was less advanced in Germany than in France.

Research at the Idaho centre

Research into vitrification was also being carried out at Hanford, where pilot units were in service. However for our purposes the so-called Idaho process was much more important: at that time this was limited merely to the calcination of the high activity wastes produced by the ICRP (formerly ICPP) plant. Since this plant reprocessed only highly enriched uranium fuels, its high activity liquid wastes had a high aluminium content. The process, also known as WCF[103], involved calcination of the waste in a fluidised bed furnace and produced granules of alumina (Al_2O_3) in which the radioactive substances were coated.

Idaho's objective was the final storage of these calcinates in large underground silos. A large installation consisting of a calciner and a storage building was in fact built and operated. The Idaho process was to provide inspiration for the Eurochemic researchers, who were also working on the solidification of high activity liquid wastes.

---

[101] FIPS: Fission Product Solidification Process. PHOTHO: Phosphatglasverfestigung von Thorex Waste.

[102] PAMELA was then developed as "Phosphatglasverfestigung mit anschliessender Metalleinbettung zur sicheren Endlagerung von hochaktiven Spaltprodukten", or "Solidification of high activity fission products in phosphate glasses, followed by their incorporation in a metal matrix for safe final storage".

[103] WCF: Waste Calcination by Fluidised bed.

Eurochemic: the research programme into the solidification of high activity liquid wastes. From LOTES to pilot tests for separating mercury from HEU campaign wastes

Research at Eurochemic

The LOTES[104] process (Figure 137) was derived from research made necessary by the special nature of the high activity liquid wastes produced from highly enriched fuels (HEWC), which it had originally been proposed to mix with those from natural and slightly enriched uranium (LEWC) for solidification in aluminium phosphates[105]. It had certain characteristics similar to those applied at Idaho. It was a low temperature solidification project, with two particular features:

- the low temperature conversion of the nitrates in the high activity liquid wastes into phosphates, and their incorporation into a matrix of aluminium phosphate in the form of spherical particles 3 to 15 milimetres in diameter. These operations were to take place in a fluidised bed reactor at a temperature of 350°C;
- incorporation of the particles into metal matrices. Matrices made of lead-antimony-zinc alloy, aluminium, and graphite were tried out in the laboratory.

Consideration was also given to vitrifying the product of the LOTES process by forming a phosphate glass, which led to contacts with the German Gelsenberg project, the prelude to valuable co-operation which was to continue fruitful subsequently.

The 1972 choice

All these different national programmes, well known to the members of the Technical Committee, served as a basis for the choices which had to be made for the solidification of Eurochemic's high activity liquid wastes.

The Eurochemic research programme into the solidification of high activity liquid wastes, defined in June 1972, therefore explored four solidification processes[106]:

- the WCF fluidised bed calcination process developed at Idaho;
- the PIVER process developed in Marcoule;
- the VERA process developed in Karlsruhe;
- the LOTES process developed at Eurochemic with the co-operation of Gelsenberg.

The first stage of the research, between 1972 and 1974, involved experimental work to determine the applicability of these processes specifically to Eurochemic wastes.

For testing the WCF process, a 16 centimetre-diameter reactor was built in which the calcination of simulated solutions was tested. It soon became clear that the process had two disadvantages: first of all, a substantial fraction of the calcination products was carried over into the off-gas filtration system from which the extremely fine powder was difficult to remove; secondly the granular products obtained were extremely easy to leach out[107]. However it was possible to make certain improvements. With regard to the second drawback, it was possible to link calcination with the LOTES process, since LOTES was used to convert metal oxides into phosphates.

For the vitrification of phosphates, Eurochemic developed a technique of producing phosphate glass beads in which a thin trickle of molten glass into which the radioactive wastes had first been incorporated was poured on to a moving belt or a rotating metal disc. These glass beads were produced and incorporated into low melting point metal alloys in co-operation with the German firm Gelsenberg AG, research being carried out both at Juliers and Mol. In December 1974, Gelsenberg conducted a first feasibility demonstration by producing 400 grammes of glass beads incorporating a 600 Ci/l solution from Eurochemic, with a specific activity of 100 Ci/kg.

---

[104] Low Temperature Solidification.

[105] See for example the R & D programme dated 18 February 1974, an unreferenced document attached to the Board of Directors dossier for the same year.

[106] CA/M(72)2, p. 3-7 of 8 August 1972

[107] Leaching is a phenomenon of partial and selective dissolution of certain components.

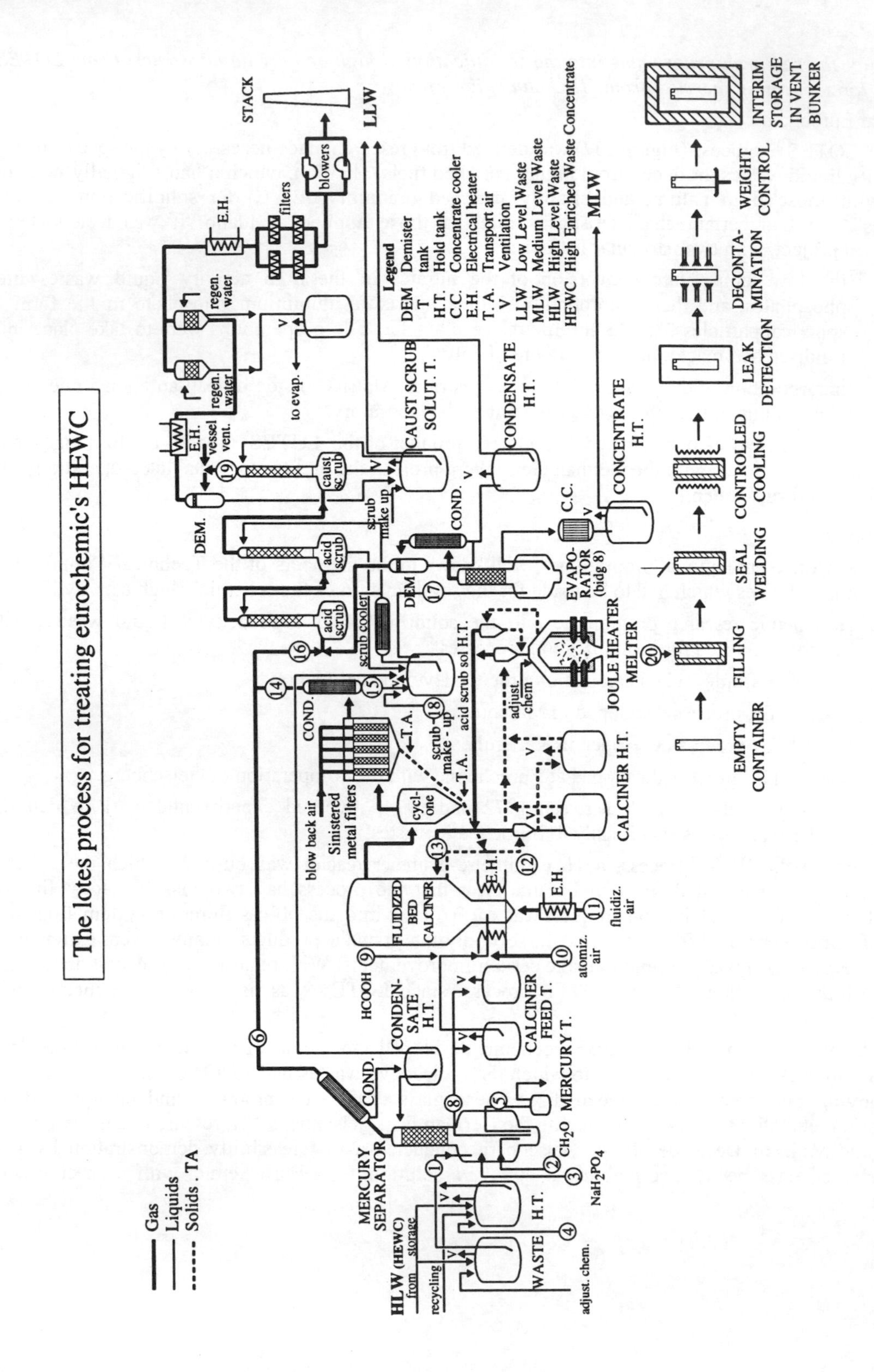

Figure 137. Block diagram of the LOTES process, with its fluidised bed calciner and its Joule effect fusion furnace (Source: RAE 1977, p. 44).

From the options of 1974 to the choice of 1977. High activity liquid wastes and LOTES, VITROMET, PAMELA and AVM

The 1974 report on waste management

By the end of 1974, research and development work had made good progress but had not yet resulted in a choice being made. The 1974 report summarised the results of the research into existing processes, and proposed a demonstration programme on Eurochemic's high activity liquid wastes considering two possibilities which, although possibly complementary, both resulted in incorporation in a metal matrix. The HEWC, containing substantial amounts of aluminium, would be calcinated and/or passed through the LOTES process, while the LEWC would be vitrified.

Research continues

Research on the fluidised bed-LOTES process continued in 1975.

With regard to vitrification, the first sample of VITROMET was produced in the laboratory in September 1975; it contained 1000 cm^3 of high activity liquid wastes incorporated in a 28% proportion in phosphate glass beads[108], themselves integrated in a lead alloy[109]. The entire mixture was cast into a canister[110] and weighed 13.4 kilogrammes[111]. In 1975, these glass beads were to be cast experimentally into a metal matrix.

In 1976, consideration was being given in the German programme to building a ceramic furnace with direct Joule effect heating in which the wastes were not only calcinated but also incorporated into the mass of glass, whereas previously an approach closer to the French method had been considered.

The first simplification of 1976

In 1976 a preliminary design study for solidification had been entrusted to Nukem; this study reviewed the process involving calcination of the HEWC followed by vitrification of the calcinate in blocks of borosilicate glass and then incorporation into phosphate glass beads which in turn were cast in a metal matrix.

At its meeting on 17 December 1976, the Board of Directors decided to apply the LOTES process to the high activity wastes, but also to allow Gelsenberg to build a demonstration unit for the PAMELA process[112], in which part of the high activity liquid wastes arising from slightly enriched fuels would be conditioned.

The year 1977 was therefore given over to improving the process, especially its flow chart, with the aim of issuing calls for tender for the engineering design services and selecting an industrial architect. Contacts were made with four European architect engineers, but none of them was considered suitable.

This undertaking was underpinned by a project for an international programme in high activity waste management under the auspices of the OECD Nuclear Energy Agency. The "International LOTES Programme" was the subject of a memorandum prepared by Yves Sousselier, Emile Detilleux and Pierre Strohl, and was intended to involve the new members of the NEA, particularly Japan and Australia, in co-operation on the Mol site[113]. However it did not have the expected success.

The second simplification of 1977. The abandonment of LOTES and the emergence of AVM

The LOTES project was abandoned at the end of 1977 in favour of AVM, developed by the CEA. The company's 1977 report sets out the reason as "the anticipated transfer of Eurochemic's installations to the Belgian authorities, who prefer the application of the industrially more advanced AVM process"[114]. Belgium had in fact indicated its preference for the French system, with the prospect of a rapid resumption of reprocessing.

The abandonment of LOTES naturally led to fluidised bed calcination of the Idaho type also being dropped.

All that remained of the LOTES development programme was one item – essential both to LOTES and to AVM – the separation of the two tonnes of mercury contained in the HEWC by reduction with formaldehyde. Research into the mercury removal was completed in 1979.

---

108 P_2O_5 - 60%; Na_2O - 8%; Al_2O_3 - 3%; Fe_2O_3 + CaO - 1%.

109 Pb(84%), Sb(12%), Sn(4%).

110 Translation adopted for the German "Kokille".

111 RAE 1975, 57 to 60.

112 CA/M(77)1, p. 8 for the decision. The scenario is described in CA(76)13, Part 2.

113 The project was launched at the beginning of March 1976: CA(76)5. It proposed a budget of 525 million BF over a seven-year period. For an interpretation of this programme, see Part V Chapter 2.

114 RAE, 1997, 48.

The next step was the construction in the Eurochemic research laboratory, in conjunction with SGN and in relation to the AVB project, of a 1/20th-scale pilot unit which came into service in 1982[115].

The replacement of LOTES by AVB and the vitrification of LEWC by PAMELA 1978-1986

Thus at the end of 1977 it appeared that the vitrification of the high activity liquid wastes was to be based on two projects emanating from the company's two main shareholders, France and Germany.

This was echoed in the 1978 Convention. France was to build the vitrification unit for the HEWC and, subsequently, the high activity wastes from the future Belgian reprocessing plant. Co-operation between Germany and Eurochemic continued in the PAMELA project, intended to vitrify the LEWC.

During the previous year in fact, the German Ministry for Research and Technology (BMFT) had incorporated all the German vitrification schemes into the PAMELA project, while management of the project passed from Gelsenberg to DWK in conjunction with the BMFT, against the more general background of building a reprocessing plant in Germany. The acronym PAMELA then came to mean Pilotanlage Mol zur Erzeugung lagerfähiger Abfälle[116]. Phosphate glasses were abandoned in favour of borosilicate glasses. The LEWC wastes were also incorporated in these in a directly heated ceramic furnace from which the finished product could be cast either in the form of blocks or as beads to be incorporated in a metal matrix. This option satisfied political as well as technical concerns. First of all it took up research carried out with Eurochemic and extended it even when LOTES was abandoned and, secondly, appeared as the most suitable for the future vitrification of high activity wastes generated in reprocessing fuel from fast breeder reactors.

The plan to vitrify the HEWC using the AVB and the storage installation for the vitrified product

The French model

The AVB was an exported version of the French AVM unit[117], which had been in production since 27 June 1978 and which was running very satisfactorily[118].

However the choice of French technology for the AVB, which had been written into the 1978 Convention, did not result in the signature of a contract until 23 March 1981. In fact construction of the AVB was conditional on the resumption of reprocessing by Belgium. The contract was signed between SGN, Eurochemic and the Belgian Ministry for Economic Affairs. The preliminary project design work officially began on 15 June 1981.

Franco-Belgian co-operation and Eurochemic's place in the bilateral arrangement

The principal objective of AVB was not to treat Eurochemic wastes, even if that was its first aim. It was a question of building the vitrification unit capable of treating liquid wastes from the future Belgian plant.

In this context, Eurochemic had three functions: it funded 60% of the design and construction work[119]; it provided SGN with details of the process for mercury separation which was absolutely essential prior to calcination using the French process and which, as we have seen, stemmed from research carried out at the time of LOTES; finally Eurochemic was to provide SGN, under a contract signed with the CEA, the special process for vitrifying the HEWC, which differed from the wastes dealt with at Marcoule particularly in their high aluminium content[120]. The two studies were postponed until 1983. This delay was aimed at facilitating, by transferring information on the specific characteristics of its own wastes, the adaptation of the French process to those wastes.

---

115 Review in ETR 311.

116 "Mol Pilot Installation for Producing Wastes Suitable for Storage".

117 Sources: RGN, 1978, 1, 54; DAMETTE (1985).

118 During the first thirty months of operation, the unit vitrified 380 m^3 of liquid waste and produced 170 tonnes of glass. However, the electrodes in the metal furnace suffered premature wear and had to be replaced on six occasions during this period, which was of some comfort to Germany in its development of ceramic furnace technology. At that time identical projects were in hand in the United States, for an installation with an annual capacity of 250 m^3 intended for treating the liquid wastes from Savannah River, and also in Japan.

119 In fact the arrangement was more complicated: Eurochemic paid for 100% of those parts of the AVB intended exclusively for conditioning its wastes, 60% of those parts common to the vitrification of Eurochemic wastes and the future Belgian wastes, and nothing for the installations intended exclusively for Belgian wastes. The final average figure was 60% of the total cost.

120 18.09% of Al_2O_3.

The storage installation

The signature of the contract also initiated the procedure for building the storage facility for the vitrified wastes, which was to be in the immediate vicinity of the AVB. A group made up of representatives of Eurochemic, the Ministry for Economic Affairs and the Belgoprocess Study Group monitored the progress of the SGN engineering design work. The preliminary safety study and the overall construction bid were submitted on 29 April 1983. For this occasion SGN was associated with Tractionel Engineering International (TEI), a subsidiary of the Belgian firm, Tractionel. The conclusions of the technical study for adapting the AVB to Eurochemic's HEWC provided for these to be solidified in 369 tonnes of glass over a five and a half-year period[121]. The preliminary safety report was approved by the Belgian authorities in 1984; the intended site was cleared for the AVB and for the storage facility, but the decision on the resumption of reprocessing was awaited before work began.

Epilogue

As we know, this decision never came and the AVB was finally abandoned in 1986. The cleared site remained as it was until work began on the CILVA installation for the conditioning of the low activity wastes.

Hence the transfer of French technology did not take place, unlike the German project.

PAMELA, a second co-operative experiment between Eurochemic and Germany[122]*. Vitrification of the LEWC*

PAMELA was to be Germany's entry into the technology of vitrifying radioactive substances, which was essential for its programme of integrated waste management which also included the construction of a national reprocessing plant.

PAMELA; an element in the integrated German programme of fuel management

The German developments which, as we have seen, were somewhat dispersed after 1965, had been combined and supported by the BMFT in 1977 under the "Technologieprogramm zur HAW Verfestigung"[123], which itself formed part of the plans to provide the Federal Republic of Germany with a comprehensive reprocessing and waste management system. The work was managed jointly by the BMFT and DWK[124].

DWK, a joint subsidiary of twelve German electricity companies, was in the process of concentrating all the reprocessing activities into its own hands. In July 1978 it was to acquire the Gorleben site. The same year it had taken over the vitrification activities of Gelsenberg and thus continued the agreement that existed between that firm and Eurochemic.

Co-operation between Germany's industry and public research system and Eurochemic

The collaborative procedures for PAMELA were laid down in the 1978 co-operation agreement between Eurochemic, the BMFT and DWK. Eurochemic provided a site and worked with the German R & D centres with the structure shown in Figure 138.

At Steering Committee level, Eurochemic was represented by its Deputy Director, Hubert Eschrich, who was of German nationality. The construction and operation of the demonstration facility was entirely funded by DWK and BMFT in the proportions 20%-80%; they co-ordinated the efforts of a number of German and Belgian research institutes and companies as well as the Eurochemic research department. An engineering design subsidiary of DWK, HAW-Technikum, also known as DWK-Mol, was established at Mol. The technological programme itself involved the pilot plant and the Karlsruhe Nuclear Centre (WAK and the KfK), the Berlin Hahn-Meitner Institute (HMI), DWK-Mol and Eurochemic.

---

[121] RAE 1983, 73.

[122] Sources other than the RAE: DWK (1985); HÖHLEIN G., TITTMANN E., WIESE H. (1985).

[123] "Technical Programme for the Solidification of High Activity Wastes".

[124] In 1971, when UNIREP was created (see Part III Chapter 3), Bayer, Hoechst, Gelsenberg and Nukem had combined their reprocessing interests in KEWA. During the summer of 1975 KEWA, which at that time was essentially an engineering design company, got together with the electricity producers, grouped for the occasion in PWK, for the reprocessing plant project (the letter P in the acronym means Projektgesellschaft). During the first quarter of 1977, PWK modified its objectives, because the installation of an integrated waste management centre (Integriertes Entsorgungszentrum) had taken shape at Gorleben, and became DWK (the D for Deutsche Gesellschaft). The chemical firms then wished to dissociate themselves from reprocessing and sold their shares in KEWA to DWK, which also bought WAK in 1977. Thus a focus for the back-end of the fuel cycle formed around the German electricity producers. KEWA became the engineering design centre for this interest. Sources: various AtW, particularly SCHEUTEN G. H. (1985), pp. 362-363.

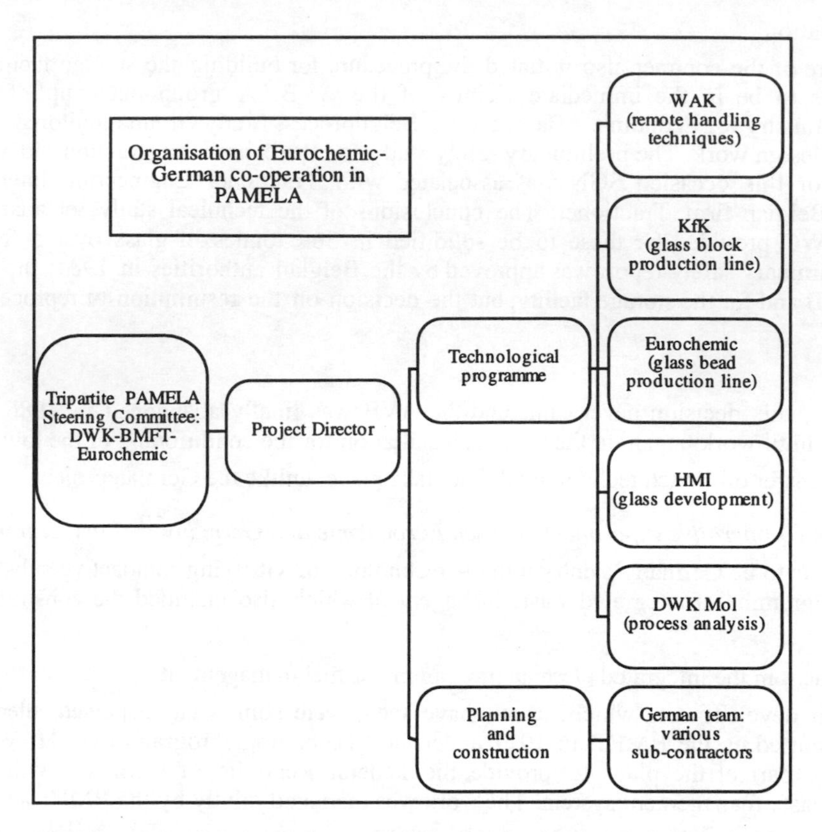

Figure 138. Eurochemic structure linked to the German R & D centres

The objectives of PAMELA

The objective was to build a demonstration facility with the task of vitrifying the 64 m^3 of Eurochemic's LEWC. Thus Pamela served as an industrial pilot for the vitrification unit which it was then hoped to build in the immediate vicinity of the major German reprocessing plant planned for Gorleben. The project also included demonstrations of the maintenance and decommissioning of its main systems, both operated remotely, which provided a guarantee in terms of radiation protection.

Parallel R & D progress on the different sites

- An initial ceramic furnace was tried out in Karlsruhe[125] as from 1977 and vitrified simulated solutions over a two-year period. It proved the reliability of the ceramic material used as a lining[126]. It was followed by a second unit which was operated for two years using solutions identical with those produced by the Eurochemic installation, then by a third in 1982 which was enclosed in a simulated hot cell and fully remotely controlled. The lifetime of such a furnace was estimated to be two or three years.
- In Berlin the HMI carried out experiments on different compositions of borosilicate glass, within the following ranges:

Table 113. **The types of borosilicate glass tried out in Berlin**

Component	Percentage by weight (range)
SiO_2	30 to 42
TiO_2	3.2 to 8
Al_2O_3	0 to 8
B_2O_3	8 to 15
CaO	0 to 12
Na_2O	16 to 22
High activity liquid wastes (in the form of oxides)	15 to 30

125 KRAUSE H. (1982), pp. 620-621.

126 30% Cr_2O_3, 30% ZrO_2, 30% Al_2O_3, 10% SiO.

– Eurochemic and DWK developed close co-operation at Mol. A joint team worked on aspects of the project in the Eurochemic laboratories throughout this period.

Work on testing different glasses continued at Eurochemic as from 1979, no longer on phosphate glasses but on borosilicate materials. The Eurochemic team was responsible for tests on the VITROMET glass bead production line and their incorporation in lead matrices. A pilot facility for producing glass beads was built in 1980. For example during 13 months in 1981-1982, the team tried out a glass bead production installation on a reduced scale (Figure 139). At the same time the DWK team utilised the pilot hall of the research laboratory for building a full-scale model of the feeding device, furnace and off-gas systems.

In 1982 the two teams began to work separately, the Eurochemic group working on a pilot system for producing glass beads and conditioning the VITROMET in full-scale containers. A number of production campaigns took place between 1982 and the first quarter of 1983. As experience was accumulated, a number of modifications were made, aimed mostly at simplifying the drive mechanisms and the construction of the rotating disc. Changes were also made to the geometry of the glass bead production pot, the material of which was defined as Inconel 690. The tests for which Eurochemic was responsible came to an end on 31 December 1983, when the PAMELA installation was undergoing cold testing.

The construction of PAMELA

Technical principles (Figure 140)

The PAMELA process was technically different from AVM in a number of ways. The process used only one furnace lined with ceramic bricks and heated by the Joule effect through metal electrodes directly in contact with the molten glass. The solutions to be vitrified were introduced directly into the furnace where the processes of evaporating the water, calcining the salts and digesting the calcinate in the mass of glass took place all together in contact with the molten glass.

The furnace could produce either glass blocks or, through an overflow system, glass beads intended for producing VITROMET. Although the amount of glass per container is higher in the case of blocks, the embedding of the beads in the metal matrix, by allowing better dispersion of heat, allows a higher concentration of fission products per unit volume of glass. This version was intended to try out a type of conditioning which was being considered for use for the vitrification of very high activity wastes arising from the reprocessing of fast breeder reactor fuels, notably from Kalkar.

In view of the small amount of LEWC available, it was decided to utilise small containers – 59 litres in volume and 30 centimetres in diameter – so there would be more of them, thus increasing the number of operations downstream of the furnace for a better test of the reliability of the processes and systems used, as well as handling, checking the tightness of the welded lids, decontamination, quality control, and so on.

The glass beads were incorporated into a lead matrix.

The containers were shifted automatically out of the filling cell; following an initial cooling period and checks on the seal, weld quality and any external contamination, they were transferred to the intermediate storage facility.

Eurochemic-PAMELA co-operation on the construction of the demonstration pilot unit

During the construction period, Eurochemic worked together with DWK on the connections to be established between PAMELA and some of the facilities existing on the site.

Although the construction of PAMELA was entirely funded by DWK and the BMFT, the job was considerably facilitated by the presence of the old Eurochemic reprocessing installation. PAMELA was able to benefit in particular from the transfer of the operating licence granted to Eurochemic by the Royal Order of 9 April 1981, once the commune of Dessel had given its permission for construction on 15 December 1980. Obtaining such a licence would have taken much longer – and the outcome much more risky – if the demonstration unit had had to be built in Germany.

Work began on 4 August 1981 and the civil engineering stage completed on 28 February 1983. PAMELA is a self-contained building located to the east of the reprocessing plant, 58 metres long, 30 metres wide and 25 metres high, with a volume of about 32 450 m^3 (Figures 141 to 145). All equipment was installed by 31 July 1984. Cold testing lasted until 30 September 1985. The conditioning of Eurochemic's high activity liquid wastes then began as Belgoprocess took over the management of the site.

The total cost of building PAMELA was 108 million DM. Belgian contractors accounted for a total of 30 million DM, a little under 30% of the work involved. Clearly therefore PAMELA was a German demonstration unit on Belgian territory.

Figure 139. The VITROMET process: the production of glass beads and their conditioning in metal. At the top, the rotating disc for cooling the beads. At the bottom, section of a VITROMET container showing how the beads are distributed in the lead matrix. Photographs taken during functional tests at Eurochemic (Source: RAE 1982, p. 58).

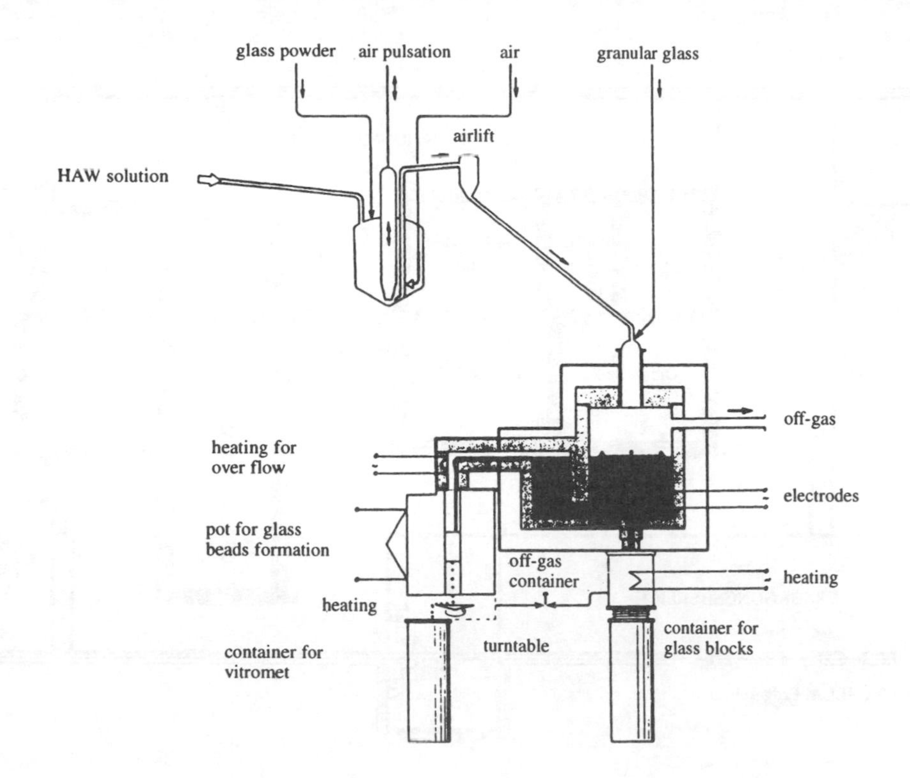

Figure 140. Flow chart for the vitrification process in PAMELA. The solutions of fission products mixed with glass powder are fed into the ceramic electrode furnace where calcination and glass melting take place. The mixture is poured from the bottom of the furnace to fill the block containers, with an overflow system being used for producing glass beads (Source: RAE 1982, p. 48).

Figure 141. Overall view of the PAMELA building on completion of the civil engineering works in 1984. In the background, the container storage building is beginning to rise (Source: DWK slide).

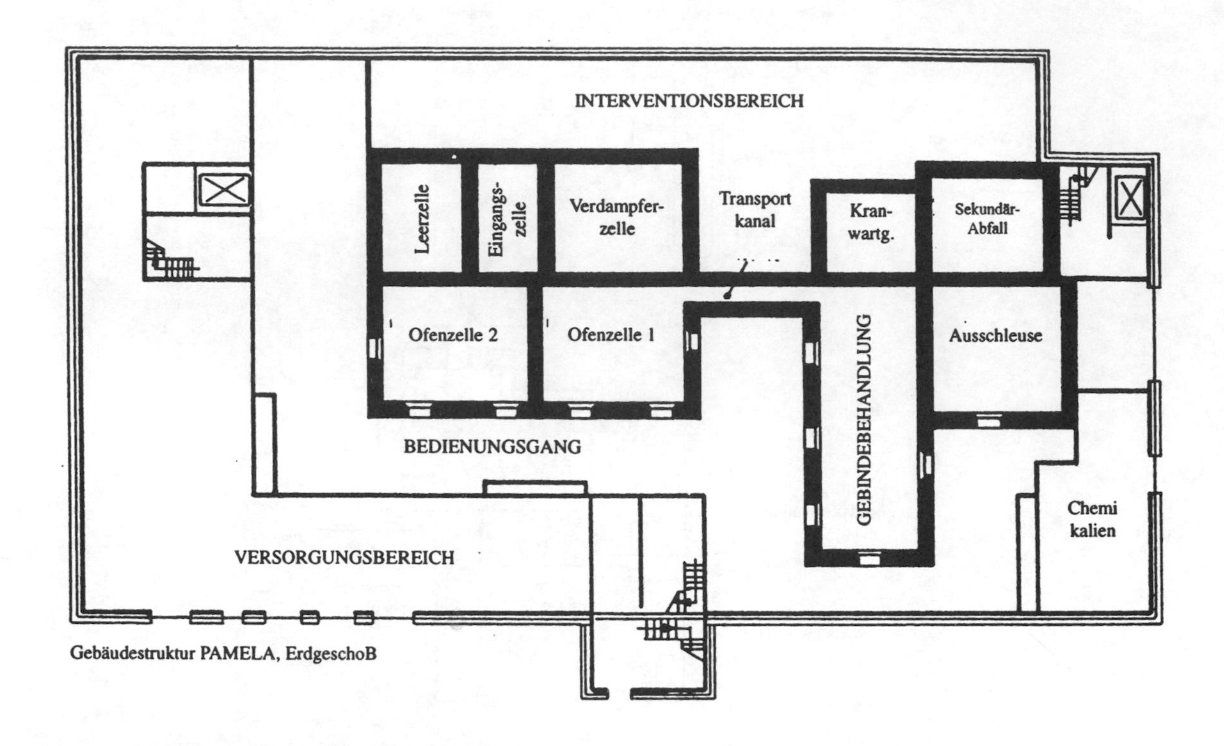

Figure 142. Plan drawing of the ground floor of the PAMELA installation. The hot cells are surrounded by operations and service areas. Two cells are provided, one for the furnace in operation and the other for the furnace on standby. The container handling cell (Gebindebehandlung) is linked to the cell containing the operating furnace by a transfer corridor (Source: DWK (no date), p. 23).

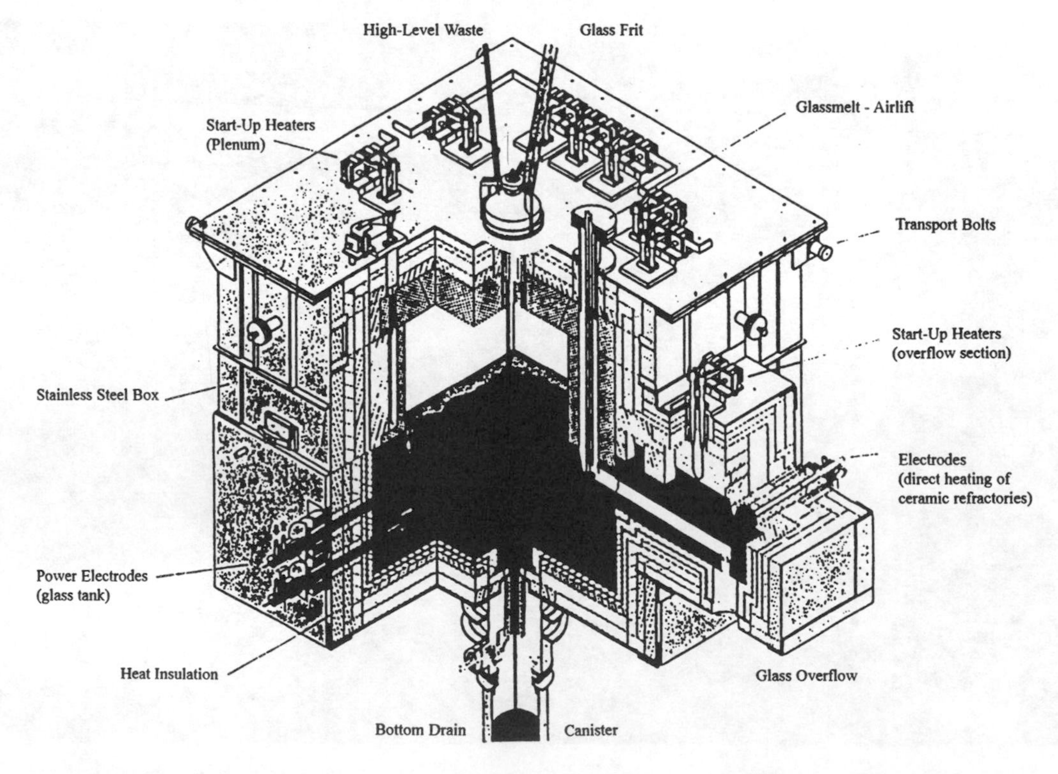

Figure 143. Main components of the PAMELA furnace (Source: WIESE H., DEMONIE M. (1990), Figure 3).

Figure 144. A DWK worker carrying out installation work on the top of the furnace installed in its cell before the commencement of testing (Source: DWK slide).

Figure 145. View of the inside of the container handling cell, taken from the transfer corridor before the commencement of operations. In the centre, the storage and temporary cooling systems; in the background, in front of the window and under the manipulator arms, two containers awaiting transfer. To the left, two decontamination stations which are loaded from above. The cell is served by two overhead cranes running on two separate pairs of rails (Source: DWK (1985), p. 23).

Building 29: the storage of wastes vitrified in PAMELA[127]

Although the construction of PAMELA was funded by Germany, this was not the case of the temporary storage unit for the vitrified wastes arising from the treatment of Eurochemic effluent.

Belgonucléaire had provided Eurochemic with an initial design for the installations for storing Eurochemic wastes from AVB and PAMELA in 1981. However compared with the initial project, the uncertainties about the resumption of reprocessing by Belgium and hence the construction of the AVB had resulted in two substantial modifications to the project; first a change of location, since it was close to PAMELA (only 52 metres to the north) and secondly a reduction in size, since it was now only part of the installation, intended only for the containers vitrified by PAMELA. However the facility had to be capable of extension to accommodate containers of vitrified HEWC and so a modular design was needed.

The civil engineering works were practically completed by the end of 1984 and the installation was able to receive the first containers of glass produced in PAMELA by 1 October 1985. Belgoprocess was now the operator.

The design is very close to the one adopted for the glass storage facility adjacent to the AVM, but at Mol it is built above ground owing to the high level of the water table and is completely separate from the vitrification unit – the lifetime of the latter being much more limited than that of the storage facility. It consists of a reception hall for the containers coming from PAMELA, and a remotely-controlled machine which not only unloads the transport flask but also loads the containers into ventilated pits enclosed in a storage cell (Figures 146 and 147). The containers are stacked six high in each pit. The forced ventilation cooling system is adjusted to ensure that the temperature in the cell does not exceed 50°C. The plan is for the containers to be stored in this way for at least 50 years, by which time the activity will have decayed to such an extent that disposal in a deep geological repository will be feasible.

Operation of PAMELA for Eurochemic. Vitrification of the LEWC[128]

The vitrification unit was operated by a team of 50 of whom 18, working in six shifts, were assigned directly to the glass production lines which operated continuously during the campaigns.

The first transfer of solution took place on 23 August 1985; the first vitrification campaign began on 1 October and was completed on 4 December.

At that time the activity of the 48.9 m^3 of LEWC arising from the 181 tonnes of uranium reprocessed by Eurochemic was about 7.9 million Ci, out of a total of 11.9 million Ci for all the high activity liquid wastes. Thus the vitrification of the LEWC stabilized the greater part of the radioactivity still present on the site. The process was successfully accomplished between October 1985 and May 1986.

During the first campaign, 22.5 m^3 of waste were vitrified. The 30.6 tonnes of glass blocks produced were conditioned in 199 containers stored in Building 29[129]. Two other campaigns followed, 13 January-14 March 1986 and 3 April-1 May 1986, the second one producing VITROMET. These campaigns completed the vitrification of the LEWC.

The cost to Eurochemic was 15.6 million DM, or 319 DM/litre.

Vitrification of the 48.9 m^3 of Eurochemic's LEWC produced a total of 540 containers, of which 440 contained glass blocks and 100 VITROMET.

The volume of glass produced was 30 m^3 and weighed 77.7 tonnes. Figure 148 shows the material data for the vitrification of one litre of LEWC. The vitrification process reduced the volume of waste, since one litre of LEWC produced 0.68 litre of glass, weighing 1.77 kilogrammes. This enclosed 195 grammes of radioactive waste, with an activity of 176 Ci. However a small quantity of secondary waste had been produced, at the rate of 0.44 litre per litre of effluent. This was sent to the appropriate treatment units on the site.

Vitrification of the HEWC, which contained about 4 million Ci in a volume of 800 m^3, was done in a somewhat different manner.

127 For a drawing and initial storage review see DE CONINCK A., DEMONIE M., CLAES J. (1986).

128 For a review of the LEWC campaign, see TITTMANN E., SCHEFFLER K. (1987).

129 CL/M(85)3; CL(86)5.

Figure 146. An operator remotely controls the loading machine to position it above the seal plugs in the storage pits of Building 29 (Source: Belgoprocess archives).

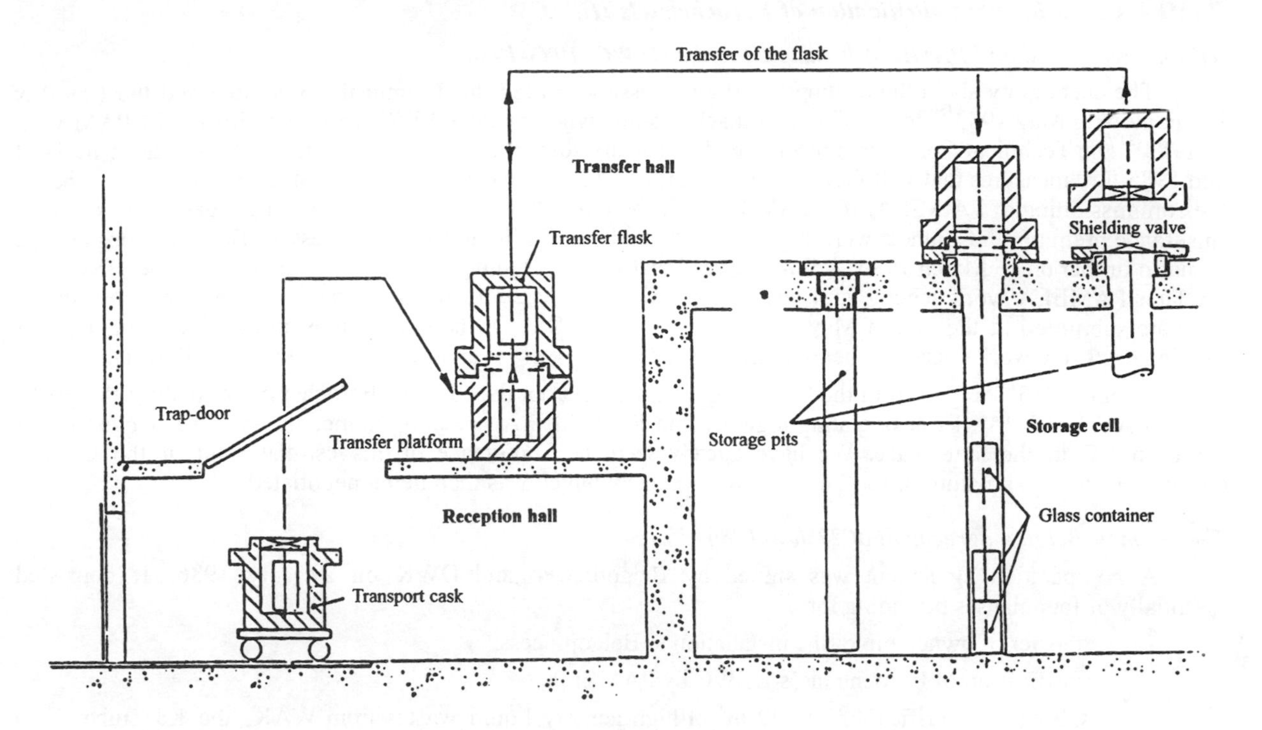

Figure 147. Diagram showing the transfer of vitrified waste containers from PAMELA to the temporary storage building (Building 29). The transport flask from PAMELA is lifted in the reception hall. The container is taken from the flask to the loading-unloading machine which is positioned by remote control above one of the storage pits; the container is then lowered into position to undergo at least 50 years cooling (Source: RAE 1984, p. 47).

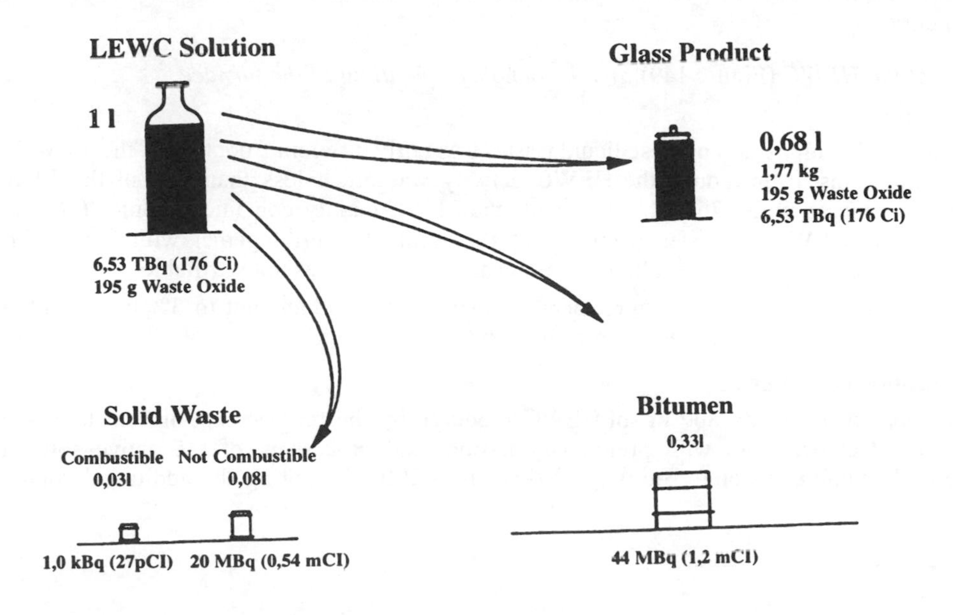

Figure 148. Data for the vitrification of one litre of high activity liquid waste concentrates (LEWC) arising from the reprocessing of natural or slightly enriched uranium fuel (Source: KAUFMANN F., WIESE H. (1987), p. 287).

PAMELA goes Belgian; vitrification of Eurochemic's HEWC

AVB is abandoned and PAMELA is adapted to the HEWC, 1983-1986

The uncertainty about the resumption of reprocessing had led the Technical Committee and the Board of Liquidators in May 1983[130] to consider a fallback position whereby the HEWC would be vitrified in PAMELA. In 1984[131] the Technical Committee recommended that the sum of 50 million BF should be set aside in 1984 and 1985 for "measures that will facilitate the modifications required for the treatment of the HEWC"[132], before the commissioning of PAMELA. If the AVB were to be built, "this expenditure could be regarded as a sort of 'insurance premium'". Contacts were made between BMFT, DWK and Belgoprocess in Bonn to consider the scenario further on 15 March 1985[133]. DWK agreed that priority should be given to reprocessing the HEWC in the later PAMELA work scheme and began a study on the costs of the necessary modifications. An initial estimate submitted at the end of May was for 74.8 million DM for the conversions and 22.6 million DM for running costs, the work necessary being estimated at 55 man-years, over a period of five and a half years.

In June 1985[134] a note from the German government set out the costs to Eurochemic of treating the HLW both for AVB and PAMELA operating together and for PAMELA operating alone, showing a savings of over a billion BF in the latter case. We have already seen the importance of this estimate and of the German concessions to the signature of the Lump Sum Agreement which was then being negotiated.

The German-Belgian agreement of 27 July 1986

A co-operation agreement was signed by Belgoprocess and DWK on 27 July 1986. It consisted essentially of four clauses providing for:

- transfer of ownership of the installation to Belgoprocess;
- vitrification of Eurochemic's HEWC by PAMELA;
- subsequent vitrification of 100 m^3 of high activity liquid wastes from WAK, the Karlsruhe pilot reprocessing facility;
- procedures for transferring vitrification experience to the planned installation at Wackersdorf. In particular, DWK obtained a guarantee that its employees would retain posts of responsibility in the installation that was now Belgian.

This agreement came into force on 16 August 1986 and brought together the Belgian and German teams working in the unit under the authority of Belgoprocess. More broadly, it put an end to a situation that was looking increasingly singular, with a German installation operating on Belgian territory. It provided Belgoprocess with a comprehensive system for treating radioactive wastes. As far as Eurochemic was concerned, this move held out hope that the 800 m^3 of liquid waste still present on the site could be conditioned within a reasonable period.

Vitrification of the HEWC (Figure 149) *and the remote dismantling of the furnace*

Adaptation works

The chemical composition of these liquid wastes was fairly different from that of the LEWC, which raised some special problems, even though the HEWC activity was much less than that of the LEWC, averaging 4.43 Ci a litre compared with 176 Ci/l. Also, while the LEWC mainly contained sodium (46 g/l together with 15 g/l of iron), the HEWC consisted primarily of aluminium (45 g/l together with 2 g/l of mercury). The mercury had to be removed from the solutions beforehand because it was not vitrifiable.

The adaptation work, which the agreement had estimated would amount to 3% of the initial construction costs, was carried out between May and September 1986.

HEWC vitrification campaigns

The campaigns to vitrify 800 m^3 of HEWC produced by the reprocessing of 30.6 tonnes of alloy MTR fuel began on 1 October 1986 with preliminary testing, and processing of the active solutions started in January 1987 when an extension to Building 29 (known as 29B) for storing the additional containers had been completed.

---

[130] CL/(83)6, § 7.

[131] CL/(84)3, p. 3.

[132] It was essentially a matter of separating out the mercury and changing from 50-litre to 150-litre containers.

[133] CL/M(85)1, p. 5.

[134] CL(85)10.

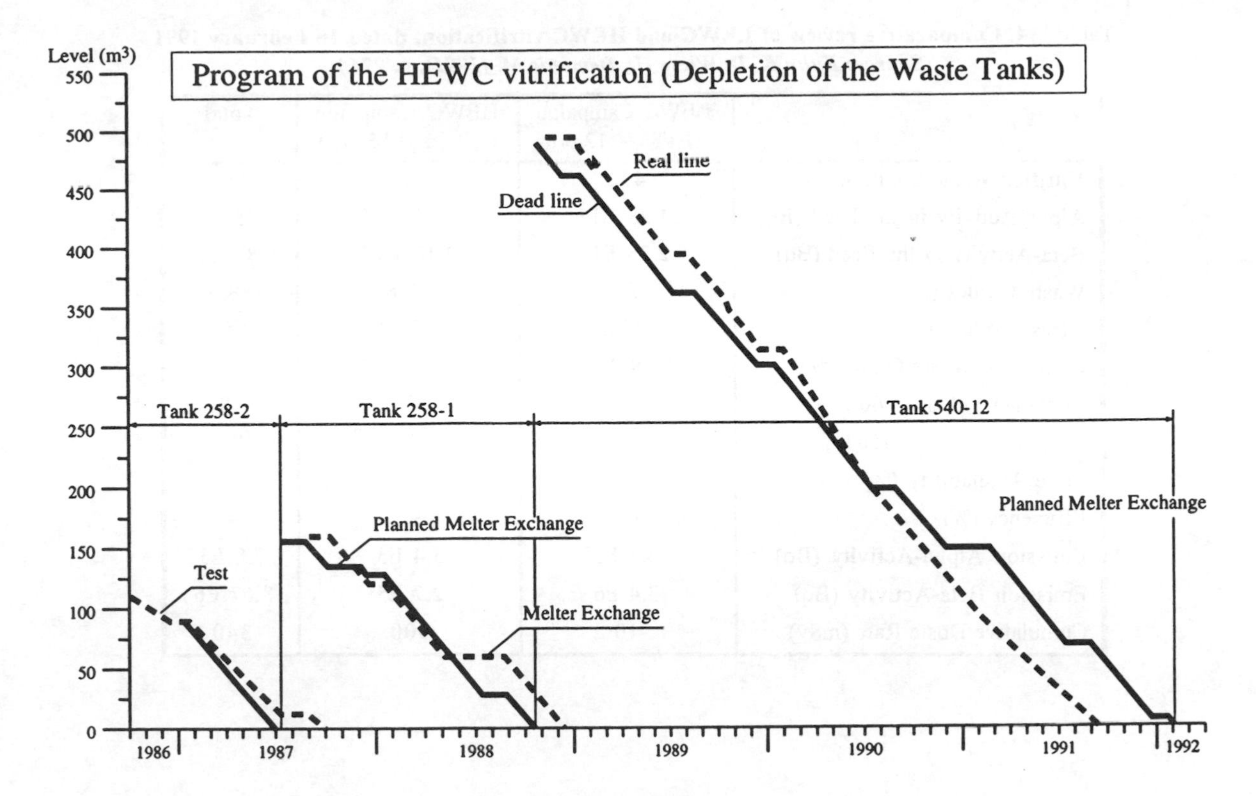

Figure 149. Progress in PAMELA's vitrification of high activity liquid waste concentrates arising from the reprocessing of highly enriched uranium fuel (HEWC) measured by the decrease in the liquid levels in the three HEWC storage tanks. Target figures are shown in solid lines, actual results are shown dotted (Source: Internal PAMELA document, supplemented by KUHN K. D., WIESE H., DEMONIE M. (1991), p. 274 up to the end of January 1991, and extended to the completion date of vitrification: 5 September 1991).

Dismantling of the first furnace and its replacement

Besides vitrification, the PAMELA demonstration programme included the experimental remote dismantling of the furnace and the installation of a new one. The purpose of this programme was to demonstrate the feasibility of dismantling and the benefit in terms of dose of the remote maintenance and standardisation options which were to be applied throughout the Wackersdorf plant[135]. It had been planned to replace the first furnace by the end of 1987, but it operated so well that its life was extended until 6 May 1988. By that time the furnace had vitrified – in addition to the LEWC – 218 m^3 of HEWC solutions, incorporated into 129.6 tonnes of glass.

However there were some problems: the electrodes in contact with the molten glass progressively corroded, and noble metals – particularly platinum – were deposited on the flat bottom of the furnace. The planned shutdowns had also played a part: the use of special glass frit for each restart had enabled the furnace to keep its integrity but had limited its lifetime to two years and ten months.

The remote replacement of the first furnace and the installation of the second one took nine weeks.

Completion of vitrification and overall review of HEWC vitrification

The second furnace, identical with the first, was commissioned in August 1988 and was used to vitrify the remaining HEWC ahead of schedule. A further planned replacement appeared unnecessary. The third furnace supplied therefore remained unused and vitrification was completed on 5 September 1991. Table 114 compares the characteristics of the LEWC and HEWC campaigns. The material balance is shown on Figure 150. The fact that a substantial amount of aluminium was present limited the incorporation of fission products despite their lower specific activity.

---

[135] See Part V Chapter 1.

Table 114. **Comparative review of LEWC and HEWC vitrification, dated 15 February 1991**
(Source: Kuhn K.-D, Wiese H., Demonie M. (1991), p. 274.)

	LEWC Campaign 1/10/86 - 12/6/86	HEWC Campaign 1/10/86 - 15/2/91	Total
Vitrified Waste Solution (m^3)	47.2	752.4	799.6
Alpha-Activity in the Feed (Bq)	1.28 E15	2.75 E14	1.55 E15
Beta-Activity in the Feed (Bq)	2.78 E17	4.10 E17	6.88 E17
Waste Oxides (t)	7.7	80.8	88.5
Glass Product (t)	77.8	377.2	455.0
Content of Waste Oxide (%)	9.9	21.98	-
Container Filled 60 l	542	939	1481
150 l	-	605	605
Time Availability (%)	88	92	-
Efficiency (%)	69	97	-
Emission Alpha-Activity (Bq)	4.4 E3	3.4 E3	7.8 E3
Emission Beta-Activity (Bq)	3.4 E6	2.7 E5	3.67 E6
Cumulative Doste Rate (mSv)	40	300	340

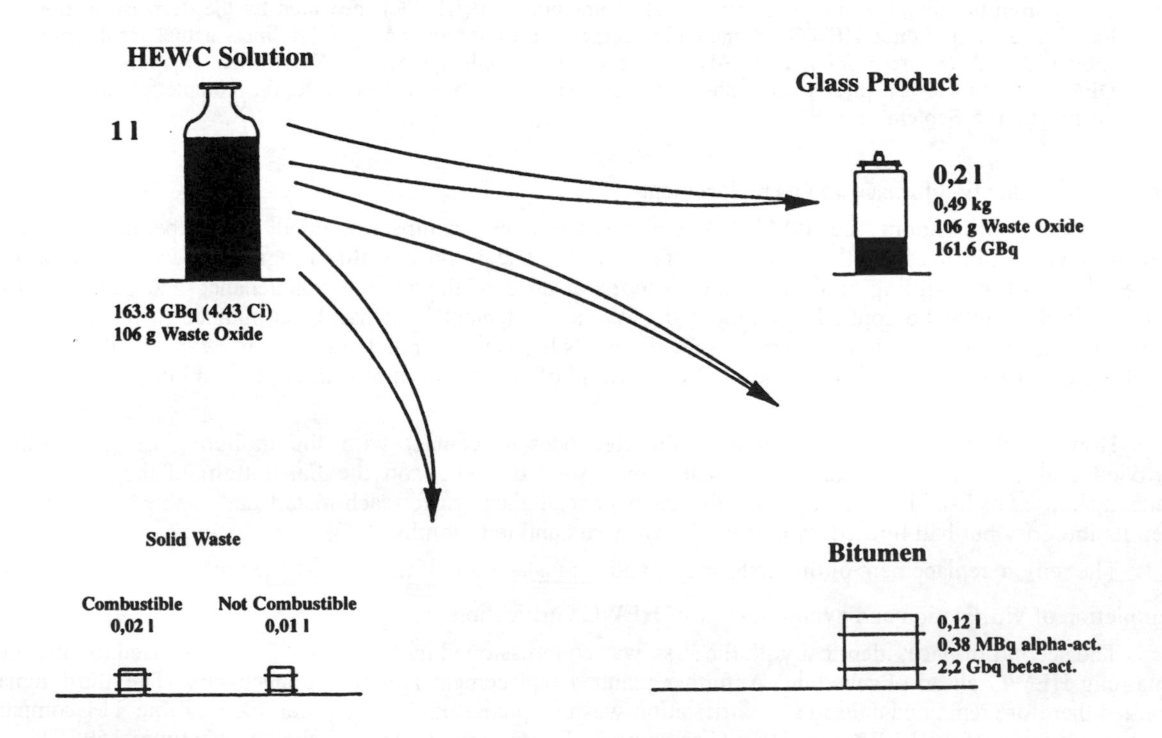

Figure 150. Data for the vitrification of one litre of high activity liquid waste concentrate (HEWC) produced by the reprocessing of highly enriched uranium fuel (Source: WIESE H., DEMONIE M. (1990), Figure 2).

PAMELA after Eurochemic

As we have seen, the principle of vitrifying high activity liquid wastes from WAK in PAMELA had been adopted in the 1986 agreement, so design work for the reception and storage of these wastes began accordingly in 1988. However the entire process was held up by the consequences of the Transnuklear affair[136]. Even today, vitrification still encounters problems of transport. PAMELA is therefore on standby and some of its cells are being used for decommissioning or remote conditioning work (see the HAVA project below).

Overall review of the Eurochemic waste conditioning programme

The programme of stabilising the wastes generated by reprocessing and decontaminating the installations was therefore completed on 5 September 1991. The solid and liquid wastes not dumped at sea were encased in concrete or bitumen, or vitrified. Table 115 is an attempt to summarise all the operations and programmes that were necessary to achieve this result.

Table 115. **Overall view of Eurochemic's waste management operations and programmes**

Waste category: nature	Name of installation or treatment Institutional form Eurochemic contact	Name and type of treatment	Treatment started	Treatment ended
Low activity: liquid waste	STE (Building 8) and STE CEN/SCK. Bilateral co-operation. Eurochemic - CEN/SCK	liquid waste treatment units	treatment as and when produced	n.a.
Low activity (non-alpha): combustible and non-combustible solids	Building 23 and storage at CEN/SCK until 1982 Building 23 and storage in Building 50	incineration or compacting, conditioning for sea dumping; conditioning and temporary storage on site	1969 1982	final sea dumping campaign in 1982
Low and medium alpha activity: solids, from Pu production and purification facilities	ALONA Co-operation between Eurochemic/Belgium (Belgoprocess)/Germany (KfK, Universität Darmstadt, Nukem)/ EURATOM Eschrich, Van Geel, Swennen	selective sorting, pretreatment and acid digestion	June 1980 (pretreatment), 28 February 1983	July 1985
Intermediate activity: contaminated solvent	EUROWATT Eurochemic/ SGN Van Geel, Humblet	thermal decomposition	3 November 1980	13 March 1981
Intermediate activity: solid	Building 23 Eurochemic-Belgoprocess	cell for embedding in concrete (Building 23). Exceptionally bituminisation	1980	still in service
Intermediate activity: Concentrates of: decladding solutions (JDWC), concentrates of hot liquid wastes (HWC), concentrates of rinsing and decontamination solutions	EUROBITUM-EUROSTORAGE Eurochemic-Belgonucléaire-Comprimo Hild	Chemical pretreatment and bituminisation of slurry, temporary storage.	June 1978	still in service (end of 1984 for Eurochemic waste)
High activity: LEWC	PAMELA Eurochemic/ Belgium (Belgoprocess)/ Allemagne (BMFT, DWK, KfK, HMI et al) Eschrich (VITROMET: Van Geel)	continuous vitrification in ceramic furnace, production of blocks or Vitromet, temporary ventilated storage. PAMELA and Building 29	1 October 1985	1 May 1986
High activity: HEWC	Same, except Eurochemic	See above	Third quarter 1986	5 September 1991
High activity: Solid reprocessing wastes in pond	Belgoprocess	Still stored in pond. HAVA project (Hoog aktieve vaste afval), cutting up and embedding in concrete in a PAMELA cell.	1995	-
Various activity levels: Solid wastes from buildings and reprocessing equipment	Eurochemic Belgoprocess	Decontamination of surfaces /dismantling of equipment (see Part IV Ch. 1); Pilot dismantling experiment (see below); Dismantling (see below)	1976 1987 1990	1980 1990 2005 ?

[136] For details of the Transnuklear affair and the resulting crisis in Germany and Belgium, see the Chronology of Part IV.

The last two lines of the table correspond to two special types of waste.

The first consists of high activity solid wastes stored in the pond specially built for the purpose in 1968. They are the subject of a project for embedding in concrete in a PAMELA cell, on standby pending delivery of high activity liquid wastes from WAK. This is the so-called HAVA project, which could be carried out at short notice.

The second concerns all the reprocessing buildings which had remained on "standby" since the completion of decontamination and dismantling in January 1980[137]. The problem, which had not been dealt with during negotiation of the Lump Sum Agreement for the reasons that were described, came to the fore immediately after the official announcement that reprocessing was abandoned on 22 December 1986.

Initial dismantling of the reprocessing facilities

The 1987 overall dismantling scheme

The final decision to abandon reprocessing led to the implementation of a decommissioning programme entrusted to ONDRAF. A report on the technical and financial aspects was submitted in July 1987[138]. This estimated the total cost of the operation at 11 billion BF, including 5.725 billion BF[139] for the installations not used at the time and which represented the first stage of the work to be done. This was the reprocessing plant (Building 1), its analytical laboratory (3), the storage facility for the final products of reprocessing (6A and 6B), and the storage tanks for high and intermediate activity liquid wastes (5, 22, 21 and 24).

The cost of managing all the wastes was estimated at 10 billion BF, spread over a period of about fifty years, since disposal into a geological formation was not being envisaged before 2030.

Manpower requirements were estimated at 835 man-years, costing 54% of the total. The dose limit envisaged for the study was a maximum of 30 mSv/year. The target levels of radioactivity to be reached on surfaces were set at 108 x 10^{-12} Ci/cm^2 beta-gamma and 10.8 x 10^{-12} Ci/cm^2 alpha (4 and 0.4 Bq).

Using these parameters, a detailed estimate was made of the quantities of radioactive waste that would be generated by the first stage of decommissioning. The amounts of primary radioactive wastes were of the order of 4600 tonnes, distributed as shown in the table below. These are essentially solids, of which a little over 260 tonnes would have to be treated as high activity wastes, particularly owing to their plutonium content.

Table 116. **Production of primary radioactive wastes foreseen in the 1987 decommissioning plan**

Waste category	Nature	Quantity
Primary wastes		
to be disposed of in deep geological formations (content of alpha emitters above 10 Ci/tonne of conditioned wastes)	metal waste	230.6 t
idem	waste in concrete	30 t
idem	special wastes	12 m^3
to be buried at shallow depth (content of alpha emitters below 10 Ci/tonne of conditioned waste)	metal waste	1095.3 t
idem	waste in concrete	3235 t
idem	special wastes	180 m^3
Secondary wastes (activity low or zero)		
idem	low activity liquid wastes	14 171 m^3
idem	combustible wastes	903 m^3
idem	compactible wastes	10 412 filtres
idem	non-active liquid wastes	40 083 m^3

137 The "standby" arrangement required the ventilation system to be kept running.

138 FREROTTE M. (1987), pp. 70-71.

139 OECD (1991), p. 83-85 and 89-93. This is equivalent to $172.9 million (January 1990 values). The estimate covered all costs, from preparations for decommissioning through to cleaning up and rehabilitating the site, but did not include the "costs related to the final part of the fuel cycle" (final storage), nor the costs of demolishing the main building using conventional methods. The estimate included a 30% margin for "contingencies".

For the relationship between decommissioning costs and the initial investment and running costs, see below.

The pilot experiment for the dismantling of Building 6A and 6B

Although the July 1987 study proposed a comprehensive decommissioning scheme, it did draw attention to the many cost uncertainties affecting the estimates and recommended that a pilot experiment up to level three should be conducted on Building 6A and 6B, which were then no longer in use[140].

The estimated costs deduced from the general study amounted to 94.74 million BF, including 21.86 million BF for contingencies. These costs were subdivided under various headings. The aim of the pilot experiment was first to demonstrate the technical feasibility of decommissioning and, secondly, to compare actual costs with expected costs.

The pilot experiment on decommissioning the two buildings began in 1987[141].

The initial situation

The two buildings had been used for storing uranyl nitrate, plutonium dioxide and used solvents. The total volume of the buildings was 3200 m^3, of which 946 were concrete with a total area of 4860 m^2. The process equipment weighed 23 485 tonnes and the other systems 15 825 tonnes.

The mean residual level of contamination was below 5 Bq/cm^2 (135.10^{-12} Ci/cm^2), but there were patches of contamination at 100 Bq (2.7×10^{-2} Ci) of alpha and beta emitter. Contact doses varied from 300 to 500 mSv/h.

Dismantling of the equipment and destruction of the buildings

By comparison with the 1987 study, the decontamination limit had been reduced by a factor of ten, i.e., to 0.4 Bq/cm^2 beta and 0.04 Bq/cm^2 alpha (10.8 and 1.8×10^{-12} Ci).

The buildings were isolated from the outside by mobile units and airlocks. Pipework was removed using pipe cutters, saws, grinders and hydraulic shears to prevent contamination from spreading. Small storage tanks were cut up using a plasma torch, while other, bigger tanks were removed and decontaminated chemically with nitric acid, sodium hydroxide solutions and hydrofluoric acid paste.

Once emptied, the concrete structure was decontaminated in its turn by surface chipping (Figure 151).

Since there was no precedent for simply dumping the rubble of such buildings, the remaining structures were comprehensively inspected twice by the Health and Safety Department and subjected to a random verification by CORAPRO, with deep samples being taken and alpha and beta destructive analyses carried out.

In mid-1989 the level of radioactivity of the buildings, verified by CORAPRO, was such that consideration could be given to destroying them by conventional means with the resulting rubble being placed on normal industrial dumps. The structure was demolished in 1990 using traditional methods and the debris dumped. All traces of the buildings were then removed from the site.

Review

The result of the pilot dismantling experiment in terms of dose was very low, even though manual methods predominated[142]. The cumulative dose was 7 mSv for 22 482 man-hours, or 13.8 man-years. Manpower accounted for 50% of total costs.

The costs of the pilot experiment confirmed the estimates, bearing in mind the 30% margin included for contingencies and costs related to the pioneering nature of the task.

The main phases of the operation were recorded using a video camera; the film was then analysed and commented upon by the teams responsible for the programme in the plant itself.

Initial dismantling of the reprocessing plant 1989-2005?

The demolition of the two final product buildings was also the first stage of decommissioning the installation.

In parallel with the pilot experiment, some decommissioning work had been carried out in 1988 in the research laboratory, for example the dismantling and decontamination of gloveboxes and ventilation pipework. These tasks continued in 1989 in other laboratory rooms. Preparatory work was also done in the plant's analytical laboratory so that solid wastes containing plutonium could be stored there.

---

140 ONDRAF report, 1987.

141 ONDRAF reports 1987 to 1989, and TEUNKENS L. (1993).

142 Only a remotely-controlled chipping hammer is referred to in the sources.

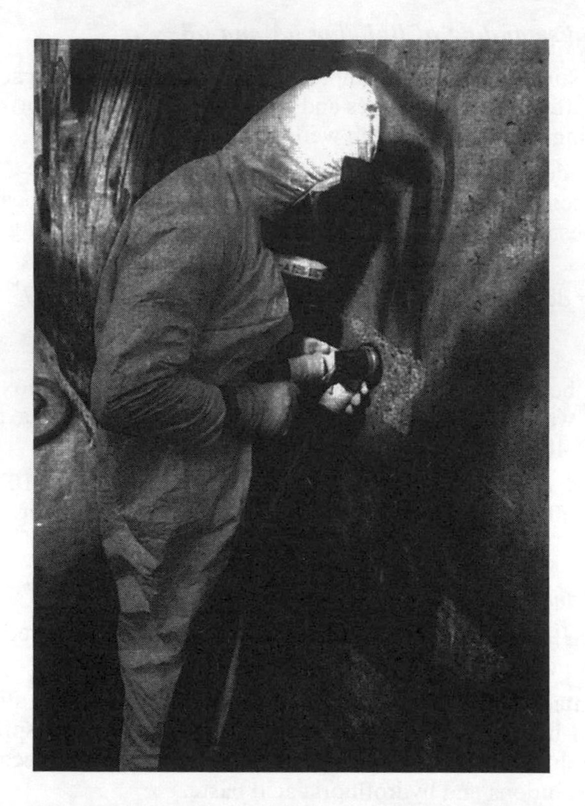

Figure 151. Manual cleaning of vertical contaminated concrete surfaces during the dismantling operations of Buildings 6A and 6B (Source: Belgoprocess report 1989, p. 4).

Figure 152. Remotely-controlled chipping hammer being used to remove the contaminated layer of concrete from the floor in Buildings 6A and 6B (Source: ONDRAF report 1989, p. 8).

Figure 153. Tanks for non-radioactive substances still in position and decontaminated tanks (on the ground, showing surface scratch marks) awaiting removal and recovery by conventional industrial methods, in one of the plant's service corridors (Source: Photograph taken by the author in August 1992).

Work is still going on today in the reprocessing building (Figure 153). It progressed slowly and unevenly, as funds were released by the Belgian government. This was in fact unproductive expenditure which was in no way urgent, unlike other decommissioning tasks for which ONDRAF was responsible, related in particular to the takeover of the CEN waste department in 1989[143].

The plant building appeared robust and decontamination was very thorough with the residual activity being contained by keeping the plant at slightly reduced pressure. The cost of decommissioning is also very high. All these factors suggest that the timescale of 15 years, proposed in the 1991 survey by OECD/NEA[144] for completing the decommissioning of the first tranche, will not be achieved and that the Eurochemic plant will continue to be a feature of the Campine countryside for a long time to come (Figures 154 to 156).

The contribution of waste management to the overall financial results of Eurochemic

In concluding this history of Eurochemic, it is worth attempting to produce an overall financial balance sheet of this experiment in European co-operation in order first to evaluate the different national contributions to the undertaking and, secondly, to assess the relative cost of the major phases of construction, operation and waste management.

National contributions to the co-operative venture can be measured only up to 1984, the last year to be covered by an annual report. The 1984 report, in one of its annexes[145], gives a breakdown of the founder shares which were issued to contributors after 1969 at a price of 5000 EMA/UA per share, the contributions being paid directly into the OECD budget. If we add the capital contributions prior to 1964, the founder shares allocated to

---

[143] This takeover, accelerated by the Transnuklear scandal, revealed shortcomings in the Centre's waste management and the obsolescence of a number of items of equipment, necessitating as a result a programme for the rapid clean-up of the "nuclear liability of the CEN".

[144] OECD (1991), p. 84.

[145] RAE 1984, p. 60.

Figure 154. Aerial view of the Belgoprocess site in 1987 (Source: 1987 ONDRAF report, cover page).

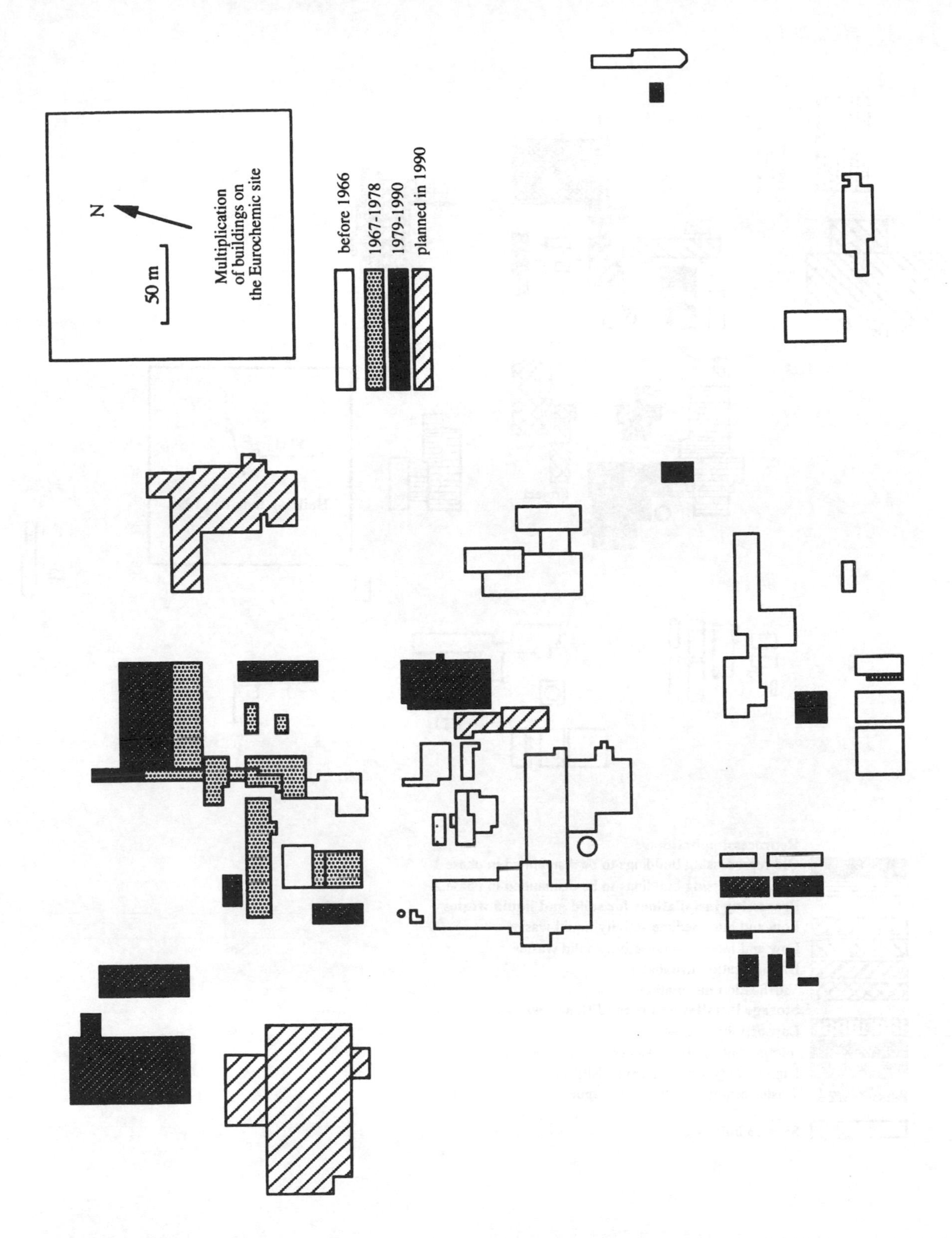

Figure 155. Showing how buildings multiplied on the Eurochemic site. The situation in 1965, 1978 and 1990 and the construction programme planned in 1990.

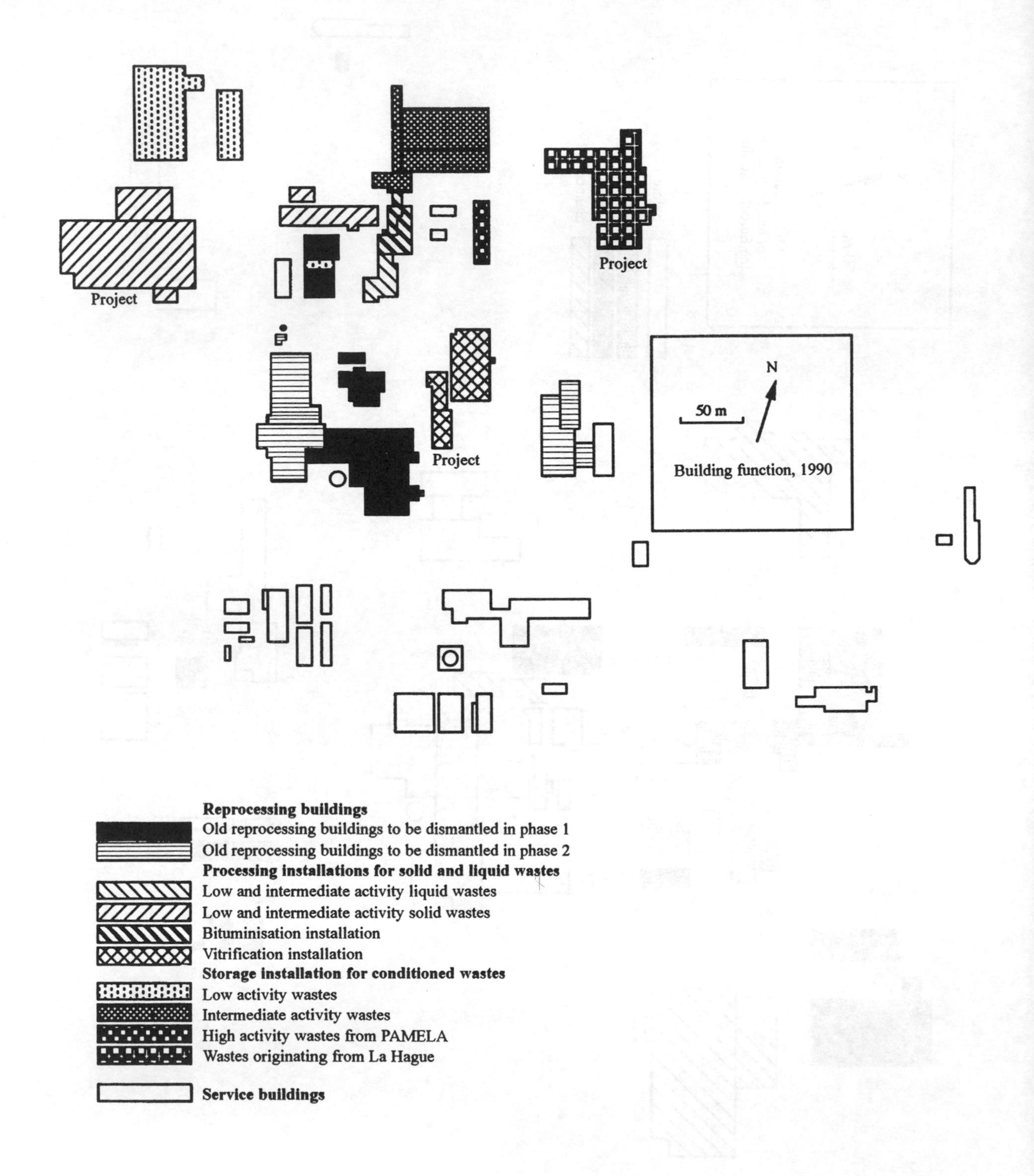

Figure 156. Actual and intended functions of buildings existing or planned on the site of BP1 in 1990.

the Netherlands until 1975 and Turkey until 1980, together with their "tickets de sortie", it is possible to evaluate total contributions in EMA/UA and the breakdown by country. The total amount spent on the project is obtained by adding these figures to the amount of the Lump Sum Agreement covering expenditure for the years 1985-1990. These data are shown in Table 117.

Cost of the Eurochemic project is denominated in million EMA/UA (today's value) and national breakdown from the outset until 1984.

Table 117. **Contributions by member countries to financing Eurochemic, 1957 to 1984**

Country	Capital million EMA/UA	% capital	Founder shares million EMA/UA	% Founder shares	Ticket de sortie million EMA/UA	Total million EMA/UA	% contribution
Germany	6.425	18.0	48.354	33.5		54.779	29.8
France	6.425	18.0	39.460	27.3		45.885	25.0
Belgium	5.100	14.3	7.961	5.5		13.061	7.1
Spain	2.425	6.8	9.695	6.7		12.120	6.6
Sweden	2.700	7.6	8.470	5.9		11.170	6.1
Italy	2.200	6.2	8.238	5.7		10.438	5.7
Switzerland	2.350	6.6	7.196	5.0		9.546	5.2
Netherlands	2.475	6.9	2.090	1.4	3.600	8.165	4.4
Denmark	1.625	4.5	4.385	3.0		6.010	3.3
Austria	1.500	4.2	4.006	2.8		5.506	3.0
Norway	1.425	4.0	3.323	2.3		4.748	2.6
Portugal	0.300	0.8	1.187	0.8		1.487	0.8
Turkey	0.800	2.2	0.020	0.0	0.160	0.980	0.5
Total in December 1984	*35.750*		*144.385*		*3.760*	*183.895*	
Lump Sum Agreement 1985-1990:						73.800	
General total 1959-1990:						*257.695*	

Although comparisons are difficult owing to exchange rate fluctuations, the total sums involved are clearly modest compared with the major aviation or space projects, such as the Concorde, the real cost of which in 1973 was 1.065 billion dollars, or ELDO, estimated at 720 million dollars in 1968[146].

The breakdown of contributions shows a skewed tendency, with Germany, France and Belgium accounting for 61.9% of the total effort and 66.3% of funding in the form of contributions, Germany providing nearly 30% and France one quarter.

Any evaluation of the relative cost of the different phases of the project is more difficult since it is necessary to correct for inflation over 30 years. The estimate given here is approximate. For the period of construction, regarded as continuing up to 1966 owing to the substantial installation work still going on that year, the figures have been taken from the table in Part II Chapter 2 and converted into contemporary Belgian francs. The later figures have been taken from the balance sheets[147]. These figures in contemporary BF have been updated into 1990 BF using the five-year index of construction prices prepared by the Belgian National Statistics Institute. For the conversion into 1990 values, it was assumed that the change each year was equal to one-fifth of the five-year variation. The estimate is based only on actual Eurochemic receipts. It does not include the evaluation of decommissioning done by ONDRAF in 1987 (11 billion contemporary BF), nor the estimated cost of managing the decommissioning wastes, set at 10 billion BF.

The statistical work described above produced the evaluation shown in Table 118, rounded to the nearest 10 million BF and expressed in billion 1990 BF. Despite its approximate nature, it nonetheless does give an idea of the sums involved.

In view of the very short period of operation, the striking feature of this table is that waste management cost more than planning and executing the project. Even though this is a very rough estimate, it does draw attention to the care that must be taken in approaching industrial projects in a way that encompasses their entire life cycle. The early completion of this by Eurochemic permits a retrospective approach to a problem which for most nuclear establishments is still in the future[148].

---

146 KOENIG C., THIETARD R. A. (1988), p. 55 and 62.

147 See the statistical summary of the balance sheet in an Annex.

148 Decommissioning scenarios have been drawn up by COGEMA. The methodology used is described in BARBE A., PECH R. (1991), but without any financial data. See in particular page 957. The methods for evaluating future costs that are generally used in industry are radically different from the historical approach which is necessarily retrospective.

Table 118. **Estimated cost of the different phases of operation of the plant, 1958 to 1990**

Period	Cost in billion BF 1990	%
Project planning and construction 1958-1966	13.06	35.8
Operation of the reprocessing plant 1967-1974	6.62	18.1
Waste management 1975-1990	16.80	46.1
Total	36.48	100.0

However it is important not to apply these conclusions too hastily to other enterprises, particularly the most recent. At Eurochemic, reprocessing and waste management were two operations separated in time, and the direct maintenance approach made the undertaking largely dependent on human labour, which costs dearly in Belgium as in other developed countries. However reprocessing, like other industries, has made considerable progress which is not without having an effect on the costs of decommissioning, and which the final part of the book will attempt to measure by placing Eurochemic in a broader historical context, from the technical and political standpoints.

Chronology of Part IV

February 1975 - September 1991

1975

early 1975: *creation of the Franco-Belge de fabrication de combustibles (FBFC) company, a subsidiary of MMN and EUROFUEL (combining PUK, Framatome, Westinghouse and Creusot-Loire), which took over the Dessel fuel fabrication installations next to Eurochemic*

April 1975: Canada joins the OECD/NEA

April 1975: creation of a Belgian Committee for Nuclear Energy Assessment (known as the Committee of Wise Men)

May 1975: *WAK is restarted after an extended shutdown*

summer 1975: *under the FRG-Brazil nuclear co-operation agreement of 27 June, signature between Nuclebras and a German consortium involving KEWA and F. Uhde GmbH, of a contract for the construction of a pilot reprocessing plant*

summer 1975: *agreement between France and South Korea providing in particular for the supply of a pilot reprocessing plant. American pressure led South Korea to abandon the contract at the end of 1975*

30 June 1975: the Netherlands withdraws from Eurochemic

July 1975: *creation of the Projektgesellschaft Wiederaufarbeitung von Kernbrennstoffen (PWK) intended to finance the construction of a major German reprocessing plant rated at 1500 tonnes a year by Kernbrennstoff-wiederaufarbeitungs-gesellschaft MbH (KEWA)*

30 August 1975: *entry into force of the London Convention, opened for signature since 30 August 1972, on "The Prevention of Marine Pollution by the Dumping of Wastes and other Materials" notably restricting sea dumping to slightly radioactive wastes, requiring authorisation*

1975: approval of the "basic technical programme" on the management of intermediate and high activity wastes and attempt to launch an international R & D programme on wastes at Eurochemic

November 1975: *cold tests at Tokai Mura*

end 1975: start of cold tests in bituminisation installations

1976

1976: Finland accedes to OECD/NEA

1976: *transfer of security control from OECD/NEA to IAEA*

1976: *creation of TRANSNUBEL, a Belgian nuclear transport company (60% Belgonucléaire, 20% Transnucléaire, 20% Transnuklear GmbH)*

1976: *testing of the Barnwell plant in South Carolina, planned capacity 1500 t/year, belonging to Allied General Nuclear Fuel Services (AGNS)*

early 1976: *announcement of an Exxon Nuclear (wholly-owned subsidiary of Exxon) project for a 1500-tonne reprocessing plant, subsequently raised to 2100 t/year at Watts Bar (Tennessee) near Oak Ridge on an ERDA site*

19 January 1976: *creation of the Compagnie générale des matières nucléaires (COGEMA; fuel cycle subsidiary of the CEA). Its statutes were approved by the decree of 4 March 1976*

January 1976: *signature of an agreement between Italy and Iraq, covering the fuel cycle in particular*

January 1976: *IAEA inspectors note the disappearance of ten fuel rods in the Taiwan research reactor*

1976: *launch of plans for a new British reprocessing plant, the Thermal Oxide Reprocessing Plant (THORP), intended to replace the oxide head-end at Windscale. The site is renamed Sellafield*

February 1976: *the European Community approves the investment coordination agreement signed between the members of UNIREP, under which the German reprocessing plant would be commissioned once La Hague and Sellafield were no longer able to deal with all reprocessing contracts*

March 1976: **termination of the Dragon joint undertaking**

March 1976: *signature of an agreement between France, Pakistan and the IAEA – charged with monitoring its civilian use – for France to supply a reprocessing installation to Pakistan*

March 1976: in its report, the Belgian Committee of Wise Men sets out the view that Belgium should reprocess all its fuel or have it reprocessed and develop technologies for treating and storing wastes. It provides for a review in five years' time

16 May 1976: *experimental start-up of the high activity oxide (HAO) unit added to UP2 at La Hague. Reprocessing during the year of 16 tonnes of fuel from Mühleberg*

mid-1976: *the "Swedish Committee for the Investigation of Radioactive Wastes" set up in 1973 considers that Sweden should develop plans for a national reprocessing plant rated at 800 t/year and recycle its plutonium as fuel in its power plants*

mid-1976: *Spain considers a 1000 t/year reprocessing plant for the end of the 1980s*

August 1976: *start of negotiations between UNIREP and Japan for the reprocessing of 4000 tonnes of fuel by COGEMA and BNFL*

end October 1976: *President Gerald Ford announces that the United States is abandoning reprocessing and plutonium recycling*

autumn 1976: *three sites in Lower Saxony are considered for the German reprocessing plant*

1 October 1976: **the United States becomes a member of the NEA**

November 1976: *after the submission of a preliminary project, KEWA requests a consortium consisting of Uhde, Lurgi, Nukem and SGN to prepare a detailed plan for the major German reprocessing plant*

16 December 1976: *the French government announces its decision "to no longer permit the sale of industrial facilities for reprocessing spent nuclear fuels to other countries until further notice"*

16 December 1976: **the Belgian government agrees to take over the Eurochemic installations to resume reprocessing**

1977

early 1977: *in its recommendations the ICRP adopts the ALARA (As Low As Reasonably Achievable) principle*

1977: *commissioning of the Tarapur reprocessing plant in India*

1977: *OECD/NEA launches the Study Programme on the Burial of High Activity Wastes in "geological formations in the deep ocean" (the SEABED project). A feasibility report was published in 1988*

February-March 1977: *PWK becomes the Deutsche Gesellschaft für Wiederaufarbeitung von Kernbrennstoffen mbH (DWK), the head office moving to Hanover as from 1 July 1977. DWK is responsible for the plans for the plant to built at Gorleben in Lower Saxony (1400 t/year). Demonstration by objectors on 12 and 13 March*

1 March 1977: *start of the licensing procedure for the THORP plant*

March 1977: *postponement of the sale by the FRG of a 10 kg/day reprocessing plant to Brazil following pressure from the United States and the USSR*

7 April 1977: *in presenting his energy programme, President Jimmy Carter announces that the United States is abandoning reprocessing and proposes a Non-Proliferation Bill which was adopted one year later*

1977: SYNATOM is charged with merging Belgian fuel cycle activities

1977: abandonment of the LOTES process and adoption, along with PAMELA, of the French process from the Marcoule vitrification unit

14 June 1977: *start of "Parker inquiry" on Windscale-Sellafield*

17 June 1977: *Germany announces its decision to grant no further licences for the export of spent fuel reprocessing installations and technologies until further notice*

June 1977: *opening of the public inquiry into the construction at Windscale of THORP (enquiry published in March 1978. The House of Commons voted in favour of extending Windscale the same month)*

June 1977: *commencement of the International Fuel Cycle Evaluation (INFCE) which was to last until February 1980*

22 July 1977: *the OECD Council votes in favour of the Multilateral Mechanism for Consultation and Surveillance for the Dumping of Radioactive Wastes at Sea (MMCS)*

September 1977: *agreement on a "code of good practice" signed by the members of the "Club of London"*

September 1977: *official opening of URENCO at Capenhurst*

22 September 1977: *startup of the Tokaï Mura plant following the conditional agreement of the United States for two years. During the first half of 1978 the plant was to process 19 tonnes of fuel and extract 64 kilogrammes of plutonium*

1 September 1977: *DWK bought shares from four shareholders of KEWA and took 20% of their holding in GWK, the operator of the WAK plant in Karlsruhe*

September 1977: *Saint-Gobain Pont-à-Mousson sells 60% of its wholly-owned subsidiary Saint-Gobain techniques nouvelles (SGN) to COGEMA. It retained 40% until 1979*

30 September 1977: *signature of a reprocessing contract between COGEMA and ten Japanese electricity producers covering 1600 tonnes of spent fuel. On 24 May 1978 a contract covering the same quantity was to be signed with Windscale*

October 1977: *the EEC Council of Ministers decides to build the Joint European Torus (JET) at Culham in the United Kingdom*

October 1977: *INFCE opens in Washington. The ERDA is absorbed by the Department of Energy (DOE)*

1977: postponement of the international R & D programme on the conditioning of high activity wastes, planned in parallel with the 1975 basic technical programme

autumn 1977: commissioning of the EUROBITUM bituminisation installation which was to begin active operations in 1978

1978

10 March 1978: *the United States adopts the Nuclear Non-Proliferation Act (NNPA)*

1978: SYNATOM signs a reprocessing contract covering 530 tonnes of fuel with COGEMA

1978: *"UP3-Service Agreement": COGEMA and 32 electricity companies sign an agreement providing for the construction of a plant at La Hague to focus solely on the reprocessing of foreign fuels during the first ten years of its life. Launch of the UP3 project*

1978: *sharp reduction in the Spanish nuclear power programme*

March 1978: *abandonment of plans to restart the PUREX plant at Hanford. Further plans to convert the Barnwell plant for storing spent fuel away from the power station site. In September the question of doing the same for Morris and West Valley was discussed*

early April 1978: *signature of a contract between COGEMA and DWK for reprocessing 1705 tonnes of German fuel in addition to the 1600 tonnes of Japanese fuel and the 620 tonnes of Swedish fuel already negotiated under the UP3 project*

April 1978: *the Swiss National Council reviews its atomic law, restricting plans for power stations, making nuclear power plant operators responsible for waste management and making provision for the establishment of a common fund for the decommissioning of nuclear installations to be financed by the electricity producers. In May, signature of a reprocessing contract with COGEMA covering 470 tonnes to be reprocessed from 1980 to 1990*

27 June 1978: *commissioning of the Atelier de vitrification de Marcoule (AVM; Marcoule vitrification unit) with a capacity of 50 m^3/year for vitrifying wastes arising from the reprocessing of fuel from natural uranium graphite-gas reactors*

June 1978: **life of the company extended to July 1982**

24 July 1978: **signature of the Convention transferring the site to Belgium. This came into force on 30 October 1978**

5 August 1978: Article 78 of the budget bill makes the establishment by the government of a joint stock company for the fuel cycle conditional upon a parliamentary debate on the country's energy policy

July 1978: *DWK acquires 400 hectares of land at Gorleben in Lower Saxony for the purposes of reprocessing and nuclear waste storage*

24 August 1978: *a leak in the Japanese reprocessing plant at Tokaï Mura results in the installations being shut down for a year after reprocessing 19 tonnes of fuel*

17 October 1978: *criticality accident at ICRP. No victims owing to effective shielding*

October 1978: *Argentina declares its intention to start building an experimental reprocessing plant at the d'Ezeiza centre near Buenos Aires, where a small laboratory unit operated from 1960 to 1970*

5 November 1978: *the Austrian electorate votes in a referendum against the commissioning of the Zwentendorf nuclear power station, a BWR built by KWU and completed in the autumn of 1977*

1979

1979: **completion of the two Eurostorage silos**

1979: **signature of an agreement for the vitrification of the high activity wastes (HEWC) from PUREX using the PAMELA process built by DKW with the assistance of the German government**

1979: *first antinuclear "initiative" in Switzerland, after 10 years of a growing antinuclear movement following the plans for a power plant at Kaiseraugst*

1979: *Saint-Gobain sells its 40% holding in SGN, 34% to Technip and 6% to COGEMA*

March 1979: *discovery of leaks of radioactive liquid in a tank at the Windscale plant used for the military programme since 1958. The investigation reveals that contamination had continued over several years and spread about 100 000 curies on the ground. Some 700 m^3 of soil had to be removed and the subsoil isolated. A 1980 report by the Health and Safety Executive criticised BNFL for not having applied the requisite safety standards*

28 March 1979: *accident at the Three Mile Island 2 power plant (TMI) at Harrisburg (Pennsylvania).*

31 March 1979: *50 000 people demonstrate against Gorleben in Hanover*

1 April 1979: *opening of the EURODIF plant at Tricastin*

2 April 1979: **in application of the 1978 Convention, first transfer of property to the Belgian government since no Belgian company has been established**

May 1979: *opening of the public enquiry into the extension of La Hague, comprising two reprocessing plants, UP2-800 (doubling the capacity of the UP2 plant) and UP3-A (an entirely new plant), a vitrification plant and three cooling towers for which construction was planned for 1984-1985*

21 May 1979: *agreement between the CEA and SGN-GWK on the provision of know-how and engineering data for the construction of the vitrification unit at the Karlsruhe centre which adopts the technology used at Marcoule. The waste from WAK would be conditioned in the HOVA unit with a capacity of 25 tonnes a year*

July 1979: *AT1 at La Hague shut down and beginning of decommissioning*

1 September 1979: *DWK becomes sole owner of GWK the operator of WAK*

October 1979: *the federal German government and the local authorities involved at Gorleben reach a compromise. The decision to build a reprocessing plant is postponed until at least 1985, the capacity of the plant reduced to 300-350 tonnes/year and the site would not automatically encompass storage. Subsequently the proposal was to be reversed, for it to become a storage site and the idea of reprocessing abandoned. Another site was considered at Hesse in the Rhineland or in Bavaria. In the end it was at Wackersdorf*

November 1979: *restart of Tokai Mura after a 15-month shutdown*

7 November 1979: *France establishes the Agence national pour la gestion des déchets radioactifs (ANDRA; National Agency for Radioactive Waste Management) to succeed the Waste Management Office set up in May 1978*

13 December 1979: *Japan announces the establishment (for 1 March 1980) of the Japan Nuclear Fuel Service Company (JNFS), a private company involving 10 electricity companies and 90 others charged with building a commercial 1200 tonnes a year reprocessing plant, to be commissioned at the end of the 1990s (an outline of the plans for the future plant at Rokka Shô Mura, but at the time an island in Kyushu was considered). Head Office in Tokyo, registered capital 10 billion yen*

end December 1979: *publication of the Plan particulier d'intervention (PPI; Specific Emergency Plan) for La Hague. This is the second French nuclear PPI following that at the Fessenheim nuclear power station*

1980

1980: **signature of a "memorandum of acceptance" confirming completion of the clean-up and decontamination work in the zones transferred**

1980: **Turkey withdraws from the company**

1980: *the 96th US Congress authorises the solidification of high activity liquid wastes at West Valley, the site of which then belonged to the state of New York. Discussions are opened between COGEMA, the DOE and Westinghouse for the vitrification of 2400 m^3 of fission product solution produced by the West Valley plant, using the AVM process. In 1983 the AVM project is in competition with a domestic process developed by the Batelle Pacific Northwest Laboratories (Battelle PNL) using a ceramic crucible like PAMELA. This is the one that was chosen, but was not completed*

25-27 February 1980: *final session of INFCE*

February 1980: *decommissioning of the German nuclear merchant ship Otto Hahn*

23 March 1980: *Sweden holds a referendum which decides that 12 power plants should be built but that the operation of nuclear plants should cease in the year 2010*

15 April 1980: *a loss of electricity supplies causes a fire and slight contamination of installations at La Hague*

25 May 1980: *publication of the declaration of public utility of UP3*

May-June 1980: **DWK awards a contract to Nukem for the design and planning of PAMELA at Mol**

17 May 1980: *slight radioactive contamination at WAK linked to a leak from a dissolver during the heat-up period. Repairs necessitate shutdown for decontamination. While the operations are in progress, a leak of radioactive steam condensate contaminates part of the roof and some of the site*

1980: *fire at La Hague in a silo storing graphite shells and end fittings from magnesium-clad fuel*

summer 1980: *extension of a ten-year contract for BNFL to reprocess Magnox fuel from the Italian Latina power plant*

April 1980: *the American government authorises Tokai Mura to continue its activities for a further year*

8 August 1980: under Article 79 of the finance bill, the Belgian parliament decides that reprocessing cannot be restarted in Belgium until Parliament has reached a decision about it in the debate on energy policy (same terms as the 1978 law)

8 August 1980: after long negotiations, agreement in principle is reached for the government to take a 50% holding in the joint-stock company formed from SYNATOM and the Société belge des combustibles nucléaires

On the same day legislation is passed providing for the creation of a radioactive waste management organisation. This was ONDRAF/NIRAS, established under the Royal Order of 30 January 1981

25 October 1980: an international anti-nuclear demonstration at Mol protests against the resumption of reprocessing and of waste management

October 1980: *ASEA-Atom concludes a contract with the Swedish power plant operator for a project for intermediate underground (at a depth of 25 metres) storage known as Away From Reactor (AFR) prior to reprocessing, or final storage without reprocessing. The site proposed is on the Simpevarp peninsula, near the Oskarsham power plant*

November1980: *BNFL proposes to adopt the French AVM process for the vitrification of high activity liquid wastes at Windscale. Thus AVM is preferred to the British HARVEST process*

1981

1981: *a dissolver leak shuts down WAK which does not restart until 1983*

1981: a Convention signed between the Belgian government and the electricity producers under which all the costs of fuel management would be incorporated in the price of nuclear electricity

toward 20 January 1981: *just before Ronald Reagan takes over as President, the American government authorises Switzerland to arrange for fuel of American origin to be reprocessed. This decision is regarded as a softening of American policy*

22 April 1981: **launch of the international STRIPA research project on the storage of radioactive wastes in geological formations, specifically in an abandoned Swedish iron mine. Co-operation between Sweden, Finland, Switzerland and the United States under the auspices of OECD/NEA with planned expenditure of 10 million dollars up to 1984**

May 1981: *just before the second round of the French presidential elections, the Minister for Industry, André Giraud, authorises the construction of UP2-800 and UP at La Hague as well as STE-3, a radioactive waste treatment unit. The relevant decree is published in the French Official Gazette on 16 May*

May 1981: **agreements between the French CEA and Ministry for Foreign Affairs with SGN and the Belgoprocess Syndicat d'études concerning the Belgian vitrification unit (AVB) using the AVM process**

3 August 1981:	*the shares in BNFL held by the UKAEA are transferred to the Secretary of State for Energy*
16 July 1981:	*Ronald Reagan announces the lifting of the ban on commercial reprocessing, a review of the 1978 NNPA and the implementation of a high activity waste management policy*
July 1981:	*discussions between DWK, Japan and AGNS for possible participation in the Barnwell plant, the commissioning of which is again being considered in the light of President Reagan's softer nuclear policy*
July 1981:	*engineering design contract between BNFL and SGN for the construction of a vitrification model based upon AVM at Windscale*
7 August 1981:	creation of the Safety of Nuclear Installations Department (SSTIN) within the Belgian Ministry of Employment and Labour
14 August 1981:	creation of the Protection Against Ionising Radiation Department (SPRI) within the Belgian Ministry of Public Health and the Environment
18 August 1981:	*the proposed site for the German reprocessing plant in Hesse, the plant capacity of which is now 350 t/year, is declared as unsuitable "for geological reasons"*
September 1981:	**laying of the foundation stone of the PAMELA demonstration unit**
October 1981:	*France sets up the 12-member Castaing Committee with the task of supervising the use of reprocessing technologies in France*
5 October 1981:	initial session of ONDRAF
8 October 1981:	*President Reagan lifts the ban on reprocessing but, a few days later, Allied Corporation (linked with General Atomics Co.) announces, in a letter from its President to the Secretary for Energy, that it is abandoning the Barnwell plant "owing to the continued regulatory uncertainties" and since "reprocessing is not commercially viable". The cost of the project since 1968 is estimated at 400 million dollars*
30 October 1981:	*signature of an American-Japanese agreement authorising reprocessing at Tokai Mura until the end of 1984*
15 December 1981:	**three drums catch fire in the conditioning cell of the EUROBITUM installation**
17 December 1981:	**signature of the first Eurochemic-Belgian agreement extending Eurochemic's management of the site until 31 December 1983. Transfer of property to Belgium completed on 31 December**

1982

1 February 1982:	*publication of a map showing the sites being considered for the German reprocessing plant in Bavaria*
1982:	debate in the Belgian Chamber of Representatives approving resumption of activities in the old installation. In March 1983 the Senate also votes in favour
March 1982:	submission of the updated report by the Committee of Wise Men, ordered in January
1982:	*the Italian CNEN becomes the ENEA*
1982:	*announcement of the start of decommissioning of the NFS plant in West Valley. The MFRP at Morris is used as a storage pond for spent fuel*
1982:	*the Indian Department of Atomic Energy (IDAE) builds a new fuel reprocessing installation at Trombay. Entry into service in 1985*
27 July 1982:	**start of liquidation of the Eurochemic company**
August 1982:	**Pierre Strohl becomes Deputy Director General of the NEA**

1983

1983: *19 signatory countries of the 1972 London Convention declare themselves against the dumping of any nuclear wastes in the sea*

1983: *WAK resumes reprocessing after a pause of two years*

18 February 1983: *Tokai Mura shut down following the detection of leaks at the dissolver and acid evaporator. Resumption does not seem possible until September or October 1983*

March-April 1983: *signature of a reprocessing contract covering 220 tonnes of fuel between the Swiss electricity producers and BNFL which covers, together with that signed with COGEMA, all Swiss production until the end of the 1990s*

11 April 1983: creation of the semi-public company "Société belge des combustibles nucléaires SYNATOM" [SYNATOM-Belgian Nuclear Fuels Company]

5 May 1983: the Belgian CEMS agrees in principle to the restart of Eurochemic

May 1983: formation of the Syndicat d'études SYBELPRO

June 1983: operation of the bituminisation installation comes to an end

23 November 1983: signature of the second Eurochemic-Belgian agreement extending the transitional period until 31 December 1984

1983: SGN submits the preliminary detailed plans for AVB

1983: PAMELA civil engineering works completed

1983: Eurochemic starts construction of a storage site for the vitrified products expected from AVB and PAMELA

end 1983: *discovery of leaks of radioactive effluent in the Irish Sea and along 200 metres of coast line, the result of negligence at Sellafield. Legal proceedings are considered in January 1984*

1984

1984: *international moratorium on the sea dumping of radioactive wastes, suspending these operations* sine die

1984: BNFL joins SYBELPRO

April 1984: the feasibility report submitted by SYBELPRO concludes that a resumption of activities is possible from the technical, economic and safety standpoints. SYNATOM decides to take up 55% of the planned capacity of the plant (120 t/year)

March 1984: *further cuts in the Spanish nuclear power programme*

April 1984: *official opening of JET at Culham*

April 1984: *DWK approaches two consortia, one of which is managed by KWU, for a bid to plan and construct a "turnkey" reprocessing plant rated at 350 t/year to be submitted before 30 September 1984, the plant to be built at Gorleben or Wackersdorf*

10 May 1984: COGEMA decides not to participate in SYBELPRO, which is dissolved and SYNATOM searches for partners until 1986

June 1984: *General Electric starts decommissioning the first American power reactor, Shippingport. Work was to take four and a half years with wastes being sent to DOE storage sites such as that of Richland*

5 June 1984: opening of negotiations on the "Lump Sum Agreement" covering Eurochemic liabilities

July 1984: *publication of Sir Douglas Black's report on leukaemia in the Sellafield area, requested in November 1983*

23 September 1984: *the Swiss reject two anti-nuclear initiatives*

5 October 1984: *the Japanese ship Seishin Maru leaves Cherbourg with 251 kilogrammes of plutonium extracted at La Hague to be taken to Japan. It passes through the Panama Canal and arrives in Tokyo on 15 November*

15 November 1984: *the French Higher Council for Nuclear Safety reviews the third Castaing report on reprocessing which recommends the continuation of PUREX reprocessing and research into advanced reprocessing*

26 November 1984: the Belgian government and SYNATOM agree to create Belgoprocess S.A.

29 November 1984: **creation of Belgoprocess S.A.**

1985

February 1985: *DWK announces the choice of Wackersdorf (Bavaria). In early October the Bavarian Ministry for the Environment gives partial authorisation for work to start on the plant the completion of which is then planned for 1993, with a capacity of 350 tonnes built by DWK. Strong local opposition*

5 June 1985: *commencement of legal proceedings against BNFL following the 1983 contamination. At the end of August BNFL is found guilty and ordered to pay a fine of £10 000 and costs of £60 000*

summer 1985: *Saclay experiments on the project to re-enrich uranium obtained from reprocessing using the SILVA laser process*

1 January 1985: **Belgoprocess, a wholly-owned subsidiary of SYNATOM, takes control of the industrial site (known as "Dessel" and then BP1) and the majority of Eurochemic staff**

1 October 1985: **PAMELA, under test since January, commences vitrification**

12 October 1985: *27 000 people demonstrate in Munich against Wackersdorf*

end 1985: *the Swedish government announces the stages in which the country will conditionally abandon nuclear power*

11 December 1985: *start of development work on the Wackersdorf site, interrupted by a demonstration involving 25 000 people and by an occupation of the site by 200 to 400 militant ecologists. The latter are removed by the police on 7 January 1986*

1985: **abandonment of the AVB project and negotiations between DWK and Belgoprocess for the use of PAMELA for vitrifying the fission products originally intended to be vitrified in AVB**

6 November 1985: **end of negotiations on the Lump Sum Agreement**

1986

10 April 1986: **signature in Brussels of the Agreement between the Belgian government and Eurochemic on the lump sum settlement of Eurochemic's financial liabilities. The Agreement came into force on 1 October 1986, providing for the payment of 3.69 billion BF in five annual payments up to 31 December 1990**

25-26 April 1986: *major accident at the Chernobyl 4 power plant*

1986: **agreement between Belgoprocess and DWK to use the PAMELA process for vitrifying 800 m^3 of high activity wastes originally intended for AVB. Operations commence in October 1986**

5 November 1986: **following a meeting of the DWK Board of Directors, the company officially announces that it has decided not to participate in restarting Eurochemic**

22 December 1986: the CEMS decides officially not to resume reprocessing on the Dessel site. A few days beforehand, on 18 December 1986, the Belgian government had charged ONDRAF with decontaminating and decommissioning the plant. Under an agreement between the Belgian government and ONDRAF, Belgoprocess becomes a wholly-owned subsidiary of ONDRAF

1987

1987:	*UP2 ends its alternate campaigns on FR, GCGR and PWR fuels and now reprocesses only PWR fuels. From 1966 to 1987 it had reprocessed 5000 tonnes of GCGR fuels and 3000 tonnes of PWR fuels up to 1991*
1987:	*SGN concludes an agreement for technology transfer with COGEMA and the CEA and the Japan Nuclear Fuel Service (JNFS) for building the planned reprocessing plant at Rokka Shô Mura (800 t/year)*
1987:	*the Junta de Energia Nuclear (JEN) becomes the Centro de Investigaciones Energeticas Medioambientales y Tecnologicas (CIEMAT)*
March 1987:	*start of the "Transnuklear (TN) scandal". A new Commercial Director at Transnuklear (2/3 Nukem, 1/3 Transnucléaire – a French company) discovers a slush fund and forged invoices in the accounts. He commences legal action against "persons unknown" on 8 April. It appears that the firm was paying bribes to ensure that it transported German low activity wastes from nuclear power plants to the treatment centres in Karlsruhe, Studsvik and Mol. However in November 1987 the scandal moves from Transnuklear to the CEN in Mol, which was a TN subcontractor for conditioning German wastes which were then returned to Germany. It appeared that the drums brought in did not contain the original materials from the power plants. Labels had been forged and the contents changed*
July 1987:	**a first estimate of the total cost of decommissioning the reprocessing installations is 11 billion BF**
November 1987:	*following a series of referenda on 8 November (one of which concerned nuclear questions) Italy decides a five-year moratorium on any further investment in the nuclear field*
22 December 1987:	*temporary suspension of all transport licences issued to Transnuklear. On 8 January the chairman of the TN Board of Directors, a director of Nukem, resigns*

1988

1988:	the Belgian nuclear programme is frozen. Cancellation of the fifth unit at Doel, withdrawal from the Kalkar fast breeder programme, re-organisation of the Mol Centre and creation of the Office public de contrôle et de sécurité des installations nucléaires (OPCSIN; Public Agency for the Control and Safety of Nuclear Installations)
14 January 1988:	*the Transnuklear (TN) scandal spreads to Nukem. The enquiry reveals that Nukem had known since at least October 1987 of the changes made in drum contents at Mol. The federal government suspends all licences permitting Nukem to work with fissile or radioactive materials. The entire management team is replaced. DEGUSSA which has a 35% holding in Nukem is given the task of correcting the situation. The scandal reaches the media when the press alleges that TN had transported fissile material from Mol through Germany to Libya or Pakistan; this was subsequently categorically denied but given credence at the time as a result of the speed of the action taken by the German government*
19 January 1988:	*commissioning of the HAGUENET computer network at La Hague*
20 January 1988:	**agreement with CIEMAT on Spanish arrears from 1974 to 1986 and on a timetable of payments up to December 1992**
January-February 1988:	*the Bavarian administrative court cancels the construction permit for the central building at Wackersdorf. A new modified permit is lodged but these legal manoeuvres do not affect the site*
early March 1988:	*abandonment of the project to build the Swiss nuclear power plant at Kaiseraugst after 20 years of uncertainty, for political and economic reasons*

17 March 1988: creation of a Belgian Parliamentary Committee of Enquiry into the Transnuklear Affair. Its report is submitted on 14 July 1988. Following this report the CEN is relieved of its responsibilities for waste management which pass to ONDRAF-Belgoprocess. The transfer took effect on 1 March 1990

July-August 1988: Nukem abandons its fuel fabrication activities following the TN scandal. These are transferred to CERCA (50/50 Péchiney Framatome) and to a new company involving Siemens and ABB (Asea Brown Boveri, a Swedish-Swiss company). The TN subsidiary is liquidated during the spring. Nukem then concentrates on engineering consultancy, and nuclear services other than transport and environmental protection methods

10 August 1988: the Italian government adopts a new energy plan which abandons nuclear power and gives preference to coal

September-October 1988: Japanese fuel arrives at THORP

November-December 1988: UP3 commences the testing phase

1988: Belgoprocess joins the NEA "International Co-operative Programme for the Exchange of Scientific and Technical Information on Decommissioning Projects for Nuclear Installations" which then included 14 projects

1989

30 March 1989: the safety case for the Rokka Shô Mura reprocessing plant is submitted

7 June 1989: the Franco-German agreement (COGEMA-VEBA) on the reprocessing of German fuels at UP3 signs the death warrant of the Wackersdorf project and of reprocessing in Germany. Eleven of the twelve electricity companies entrust the reprocessing of their fuel to COGEMA, covering 15 of the country's 21 reactors, for fuel discharged up to 2005 and reprocessing up to 2008 in UP3. The other six German reactors, accounting for 31% of installed nuclear generating capacity, will be reprocessed by BNFL. In the wake of the decisions it is decided at the beginning of 1990 to cease all reprocessing in Germany, which leads to the closure of WAK and two small units in Karlsruhe in 1991-1992

September-October 1989: completion of the Sellafield vitrification unit using the AVM model

September-October 1989: the R7 vitrification unit at La Hague, based upon AVM, enters service

mid-November 1989: the electricity industry in the United Kingdom is privatised, excluding the nuclear plants which remain under government control in Nuclear Electric. The government decides on a five-year moratorium for the construction of any new plants

17 November 1989: UP3 goes active

1989: COGEMA restructures its engineering activities under six headings combined in the "EURISYS network" (Consultancy, Engineering Design, Computing, Mechanical Engineering and Maintenance, Documentation, Testing and Technical Assistance). This system is headed by SGN

1990

1990: conditioning of the final batch of intermediate activity solid wastes from Eurochemic

early 1990: France offers UP3 and its ponds to control by EURATOM and the IAEA

February 1990: the Gardner report confirms that there is a correlation between the statistical occurrence of leukaemia cases around the Sellafield plant and parental exposure to low doses of radiation

May-June 1990: OECD/NEA and the IAEA introduce the International Scale of Nuclear Events comprising 7 levels, to be tested up to the end of 1991

23 September 1990: a vote following a Swiss initiative approves a 10-year moratorium on any further construction of new nuclear power plants

26 September 1990: *UP3 enters full service (only the T7 vitrification unit is not yet operating)*

1990: **complete demolition of Buildings 6A and 6B by Belgoprocess as part of the "pilot decommissioning project"**

1990: signature of an agreement between the Belgian government, ONDRAF and the electricity producers on the funding and management up to the year 2000 of clean-up and decommissioning operations on BP1 and BP2

end 1990: *THORP has firm contracts worth 6 billion pounds covering 6700 tonnes up to the year 2002. The predicted cost of the plant is estimated at 2.6 billion pounds (the plant came into service in March 1994)*

28 November 1990: **final session of the Board of Liquidators of Eurochemic. The company is dissolved**

Following the liquididation of the company

31 December 1990: *end of reprocessing at WAK. The cost of maintenance, pending decommissioning is estimated at 128 million DM (1990 values)*

1991

early 1991: *the ICRP publishes new radiation protection standards known as "Publication 60"*

June 1991: *25th anniversary of La Hague. The plant has reprocessed 8500 tonnes of spent fuel, of which 5000 tonnes came from natural uranium graphite-gas reactors and 3500 tonnes from light water reactors*

5 September 1991: **completion of vitrification in the PAMELA installation of the final high activity liquid wastes produced by Eurochemic**

PART V

Final considerations

Chapter 1

Eurochemic and the history of reprocessing
The technical perspective

In the half-century since the first reprocessing installation was commissioned, during which three generations of plant have emerged[1], Eurochemic occupies a central position. It has features in common with its direct or indirect descendants which it passed on to them, in line with its primary calling, in particular ways that merit examination.

The change to the third generation stemmed largely from a learning process in which Eurochemic played a significant role. The principal problems that had appeared in the operation of the second generation plants were resolved by technological advances arising from the "third industrial revolution", concerned notably with materials and information, in their cybernetic and computer combinations. The fields particularly affected by this industrial revolution were systems that were most sensitive to corrosion, the methods of maintenance and the system of operating plants.

However Eurochemic was not just a reprocessing plant: it was also a place of training in international co-operation for a whole generation of scientists, engineers and technicians who thereby acquired skills in the nuclear field or were converted to it. Accordingly, when the international approach to reprocessing which Eurochemic represented finally came to an end, the abilities and experience accumulated in a large number of sectors were hived off. By monitoring a few careers it is possible to plot the networks of influence of the "Eurochemic family" and thus get a better grasp of what was and still is the culture of the joint undertaking's managers.

Eurochemic's place in the history of reprocessing. An attempt at a technical genealogy

Eurochemic appears to occupy a strategic position at the centre of the intermediate generation, between the large military plants of the 1950s and the major civilian plants of the 1990s. Eurochemic inherited both American and French experience, benefiting from a twofold transfer of technology: this enabled the company to extend reprocessing techniques[2] and to fulfil its task of training engineers and technicians. The experience gained and fed back played a part – first directly and then in a more diluted fashion as time went by – in the technological development of reprocessing from the 1960s onwards.

The first generation: plants specialising in plutonium production. 1943-1954

The first generation of plants were built during the war with essentially military objectives. They reprocessed large quantities of nuclear fuel[3] supplied from reactors built nearby. Their primary activity was to produce plutonium in the largest possible quantities. The materials and systems used were very similar to those routinely used in the chemical industry by which, through Du Pont de Nemours, they had in fact been designed

---

1 At the 1983 seminar on Eurochemic experience, Michel Lung suggested that there had been five generations of plant around the world. ETR 318, p. 147. The "canyon" plants (Hanford and SRP) are regarded as "one-off systems"; the first generation illustrated by Windscale 1, UP1, SAP and Idaho; the second includes Eurochemic, NFS, Windscale 2 and UP2; the third Tokai Mura, WAK, Barnwell, HAO, Eurex and AT1; the fourth comprises UP3, Wackersdorf, Tokai and THORP. On the basis of this article and discussions with its author I have been led to suggest a different breakdown based on the idea of major advances in technology. See also the use of the "generation" concept by
M. Delange with regard to French reprocessing, DELANGE M. (1985). RGN, 1985, 6, pp. 559-568. The latter distinguishes four generations but in France alone, with one generation per plant: UP1, UP2, HAO (and AVM), UP3 and finally UP2-800.

2 GOLDSCHMIDT B. (1980), p. 397 sees Eurochemic as "a formidable purveyor of technology".

3 Fuel of the natural uranium type, with low burn-up. The specific activity per unit volume of solution was relatively low: 10 to 20 Ci/litre.

and built. These were the so-called "canyon" plants, mostly underground, fully remotely controlled and consisting of large cells capable of accommodating all kinds of equipment, served by an overhead crane. Maintenance was done remotely because the levels of radioactivity in the cells were very high, even though the fuels being reprocessed had a relatively low burn-up of hundreds or a few thousand MWd/tU. Information now emerging about the Soviet plants suggests that they use the same operational approach.

A feature of this generation of reprocessing plants was that they employed a great variety of processes, which sometimes succeeded one another in the same plant, as at Cheliabinsk. At Hanford and Savannah River this sequence culminated in the choice of the PUREX process around 1953-1954.

Waste management was in an embryonic state: most liquid high level wastes were simply stored as they were[4] or partially disposed of in the ground (as in the United States), into surface or ground waters (as in the Soviet Union), or at sea (as done by the British). This was the apogee of the environmental dilution approach.

The second generation: diversified PUREX plants. 1955-1977

The second generation involved a greater variety of plant types even though the PUREX extraction process was almost universally adopted. The plants differed in the other stages of treatment: the head-end of the process, and plutonium and uranium purification. These plants were either military, civilian or a combination of the two, covering a wide range of capacities, usually with fairly low annual throughput, which can be explained by the higher burn-up of the fuels being reprocessed. They were intended to separate both uranium and plutonium. They used more specific materials and systems, including special steels and alloys.

More attention was paid to the wastes and their treatment[5]. The principles applied were those of storage or delayed treatment, with a trend towards dividing wastes into categories and attempting to reduce its volume. Architecturally, more cells were used since maintenance was done directly and it had to be possible to decontaminate the working areas before people entered. Hence the cells were more specialised. The detail of operations varied according to the type of fuel being processed, the variety of which reflected the increasing numbers of reactor and fuel types. Fuels made of natural uranium and uranium enriched in proportions extending up to 100% were reprocessed, and preparations were made for reprocessing fuel from fast breeder reactors. This second generation covered the period 1955-1977 approximately.

Eurochemic was not simply an excellent example of this generation: its design for manifold applications not only made it the epitome of the second generation plants but also set it clearly apart from the others.

The third generation: present-day plants

The third generation is subsequent to 1977 (Figure 157)[6]. These civilian plants have high capacity, although less than those of the first period, but they are capable of reprocessing fuels of very high burn-up, of the order of 40 000 MWd/tU. These plants specialise in reprocessing only one type of fuel[7], the oxide fuels which have become widespread since the light water reactor, especially the pressurised water type, superseded other designs in the 1960s[8].

---

[4] The long-term storage of corrosive solutions resulted in many leaks. If the cooling systems fail, the solutions can evaporate, leaving unstable and possibly explosive residues. These hazards gave impetus to research into the stabilisation of wastes. At the most modern plants, the high activity solutions are vitrified as and when they are produced. Today there exists some fifteen years of experience in the temporary storage of vitrified wastes, but final disposal into geological formations, which is planned and technically investigated, is encountering complex political problems.

[5] Also the question of space is seen differently in Europe: there is less of it and the population density is higher. These overall geographical and cultural factors contributed to the development of the "concentrate and confine" approach as a necessary alternative to that of "dilute and disperse" which characterised only the early days of nuclear power generation.

[6] UP3 is in service and THORP has just started up; Rokka Shô Mura is under construction, as is UP2-800. Wackersdorf was abandoned in 1989 after more than ten years of design work. Figure 157 shows how the French CEA saw the objectives of R & D on these "plants of the future" in 1979.

[7] However fuel design and composition still varies according to the reactor, its builder, the date of commissioning, and so on.

[8] Projects concerned with fuels irradiated in fast breeder reactor cores or their fertile "blankets" are continuing on a pilot scale (and are coming to an end as the system is gradually abandoned). Some current projects also envisage the reprocessing of mixed-oxide fuels (MOX).

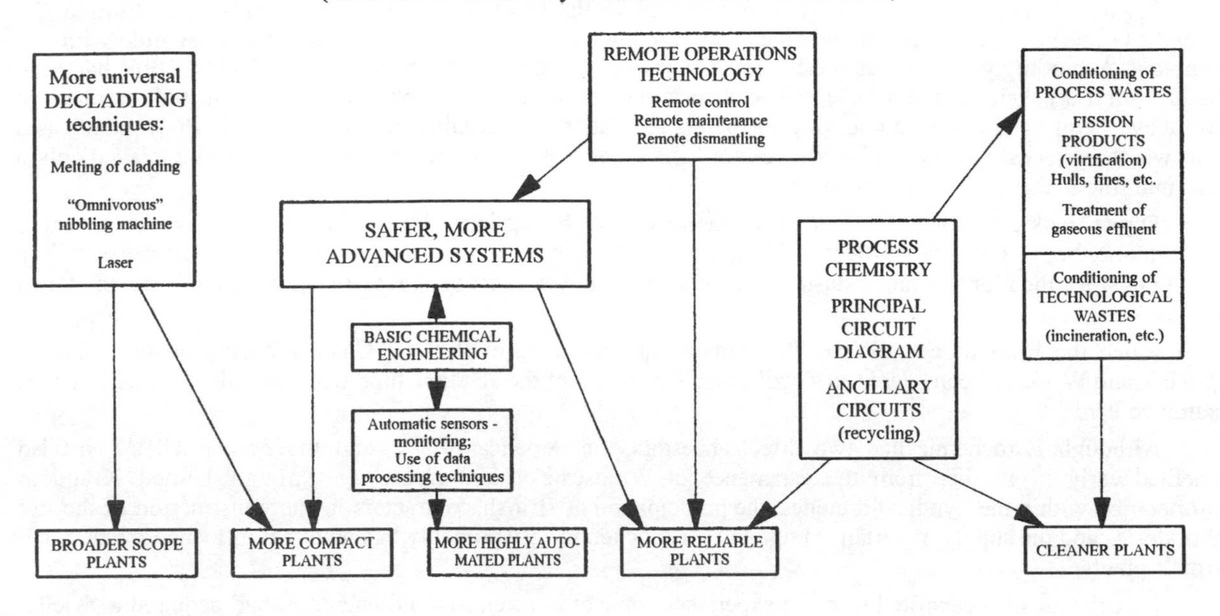

Figure 157. Major objectives of reprocessing R & D at the CEA in 1979 (Source: PIATIER H. (1979), p. 162).

There are obvious technical advances compared with the second generation plants, involving both the materials used and the remote maintenance approach[9]. The computer revolution has led to centralisation of the system for collecting and processing data in process control and instrumentation. Human intervention in the cells is becoming rare, resulting in a substantial fall in dose received. As far as wastes are concerned, the objective is to condition them as quickly as possible into the smallest possible volume. In the process, this has brought about much more internal recycling, the use of reagents that can be completely evaporated[10], the practice of in-line conditioning, and the greatest possible degree of concentration into high activity liquid wastes intended for vitrification[11], since the inactive condensate can be discharged harmlessly into the environment. Thus there is a trend towards simplifying the types of liquid waste. With regard to gaseous waste, highly sophisticated filtration and purification procedures have been developed, particularly for tritium and iodine.

Naturally there were internal reasons for these developments, such as the incorporation of technical advances, feedback of experience, trades union pressure, and so on. However the great emphasis placed in the third generation projects on reducing doses and the amounts of wastes produced is also linked to external – notably environmental – pressures, both from governments and others. Wackersdorf is a particularly eloquent example in this context.

Eurochemic and the second generation of reprocessing plants. Family histories

Historically and technically, Eurochemic was central to the second generation. The plant was the direct heir of Marcoule and the ICPP and, to a certain extent, THOREX, as described in Part III Chapter 1.

In what follows we shall attempt first to define the links between Eurochemic and the other plants of the second generation. We shall then show how the Mol plant passed on its experience to two of its descendants,

---

[9] Hence a return to some characteristics of the first generation, but with much improved technical design and construction.

[10] "Salt-free reagents", such as those containing ammonia.

[11] As a result, UP3 is considering abandoning bituminisation in 1995.

one in France, the other in Germany (Eurochemic's influence on the Italian pilot plant EUREX is only slightly referred to here; see the perspective sketch in Figure 158)[12].

The position of Eurochemic in the family tree of reprocessing plants

Figure 159 is an attempt to position Eurochemic in the family tree of reprocessing plants in Europe, the United States and Japan. Not all establishments are shown, but only those which had some link with the company. Accordingly it excludes the Soviet, Chinese and Indian plants in respect of which no link has been found with Eurochemic, and whose characteristics are not well known, except for the first Indian plant at Trombay[13]. Only a limited number of pilot plants and laboratory installations are included. Pilot plants and units which reprocessed only fast breeder reactor fuels are absent, this aspect of reprocessing having played only a marginal role in the history of Eurochemic.

Direct links are shown as bold lines, indirect links by ordinary lines. Eurochemic is given a central position only because it is the subject of the figure, and there is no suggestion that Eurochemic was necessarily a staging post for the reprocessing industry in the countries shown. Projects that were not completed are shown in brackets[14].

Since the Belgoprocess SYBELPRO plant was described in Part IV Chapter 2 and those at Tokai, EUREX and WAK are considered in detail later, the nature of the relationships with the other plants must be examined here.

Although Eurochemic had two direct ancestors, the American ICPP and the French UP1[15], it also benefited early in its life from the experience of Windscale 1 and Dounreay, through United Kingdom co-operation within the Syndicat d'études, the participation of British contractors in the construction of the hot laboratories and the supply of certain equipment[16]. Eurochemic very probably had no direct influence on the later British plants.

Eurochemic also benefited from the experience some of its scientists and engineers had acquired at Kjeller and indeed made use of this small international installation for testing some aspects of the process. Kjeller provided some of its senior staff, such as Teun Barendregt. In fact the Norwegian pilot plant was shut down at the time the Mol plant came into service, with the learning process continuing there for the Scandinavian countries and the Netherlands.

As regards its progeny, and although there was exemplary co-operation between the United States and Eurochemic from the very beginning, the plant's influence on the reprocessing industry in the United States is no more than tenuous. For this there are several reasons.

According to Earl Shank[17], the information collected by the Americans during the construction phase and the first few years of operation was communicated only to the research centres of the USAEC and did not reach the private companies concerned with reprocessing[18]. As a result, NFS had access to data only through its own direct links with Eurochemic[19].

---

[12] For many reasons we shall merely touch on Eurochemic's influence on the Italian EUREX pilot plant. First of all Eurochemic, unlike the German pilot plant, played only a limited role together with bilateral co-operation with the United States, in Italy's acquisition of reprocessing techniques. Secondly, the building of EUREX did not lead on to a more extensive project. EUREX 2 was stillborn. Also the documentation found available was very sparse, despite the quick summary made by S. Cao in ETR 318, pp. 159-160, and searches in Italian sources (see the bibliography at the end of the book). Thirdly, scrutiny of these different elements did not seem to lead to conclusions any different from those obtained by studying the German pilot plant.

[13] Three maps showing the world's reprocessing installations are given in an Annex.

[14] Barnwell was built but never used.

[15] See Part III Chapter 1.

[16] In particular, a vacuum evaporator installed in the effluent treatment plant, which always gave complete satisfaction. According to Michel Lung, the robust design of the British evaporators of this type led the designers of Rokka Shô to order their own from the British.

[17] Communication of 19 April 1994.

[18] There are two reasons for this. To begin with, the very loose links between the USAEC and private industry. Secondly, the fact that the USAEC respected the bilateral nature – in the institutional rather than the national sense – of its link with Eurochemic.

[19] Visits and exchanges of information took place in both directions, but the effects were limited by the fact that the two construction projects were proceeding in parallel, with each party observing the other. Eurochemic visitors to the West Valley plant retain a mixed impression, dominated by the memory of leakages and contamination incidents ("there were leaks everywhere").

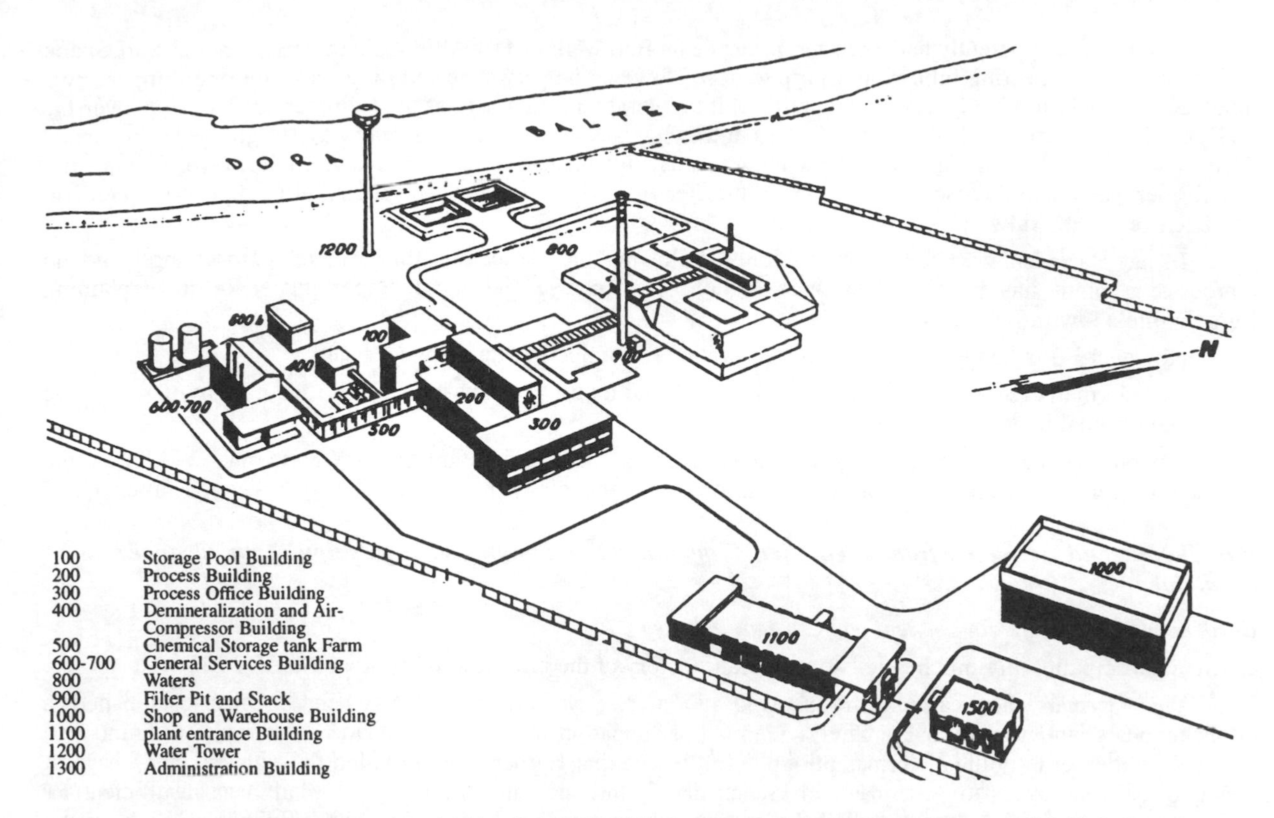

Figure 158. Perspective sketch of the EUREX site. As regards building layout, EUREX is based directly upon Eurochemic (Source: CALLERI G., CAO S., et al (1971), p. 378).

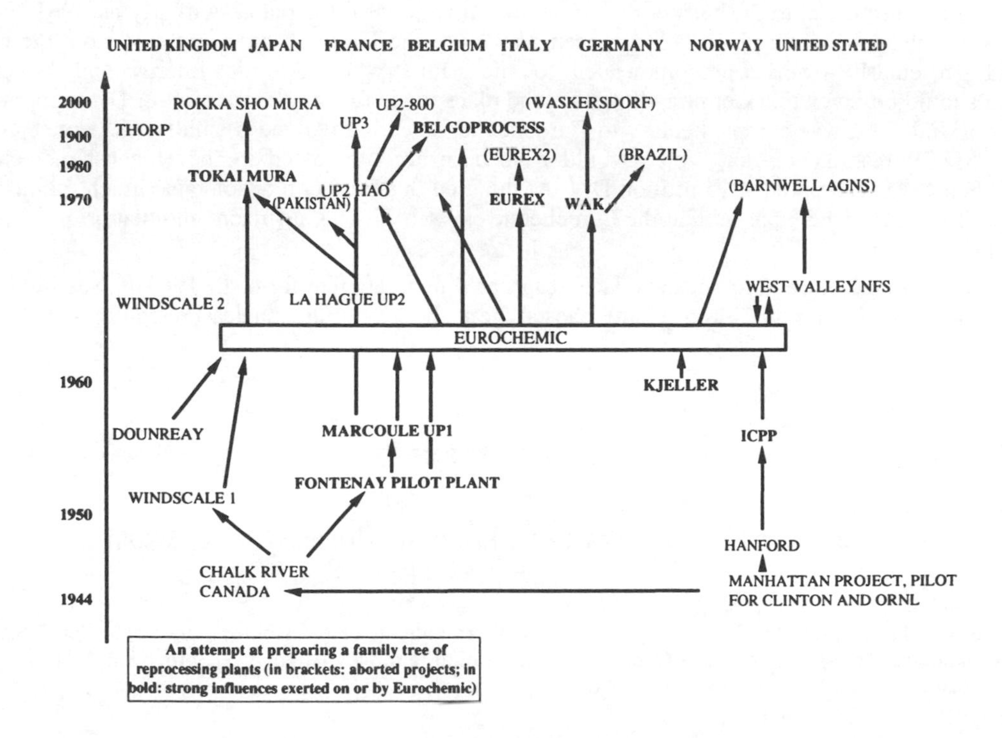

Figure 159. An attempt at a Eurochemic family tree.

Eurochemic apparently had a greater influence on Barnwell. In fact Allied Chemicals recruited Earl Shank as Director of Engineering following his trip to Mol. However he very soon felt ill at ease: for one thing he was homesick for Belgium but he also found it difficult to tolerate the authority of the Commercial Division over the Design Office[20]. Earl Shank therefore left the United States after two years to return to Belgium. He remained friendly with Teun Barendregt, who after leaving Eurochemic made a career as Managing Director of Comprimo. The former Technical Advisor to Eurochemic therefore joined the Belgian subsidiary of Comprimo which had just been set up in Antwerp.

In fact Barnwell was never commissioned following the Carter abandonment of reprocessing[21] and no reprocessing plant has been opened since, which is of course the most important factor in explaining Eurochemic's low influence in the United States.

However Eurochemic provided direct inspiration for projects in Italy and Germany.

The German case is a good example of how national industrial know-how can be acquired by participating in an international undertaking like Eurochemic.

The emergence of the Karlsruhe pilot plant known as Wiederaufarbeitungsanlage Karlsruhe (WAK), as the first step in Germany's strategy for acquiring reprocessing technology, was in fact closely linked to Eurochemic.

Eurochemic and Germany: from international apprenticeship to national acquisition of an advanced technology[22]

Eurochemic, WAK, the government and German industry

German participation in Eurochemic: the temporary victory of the international approach 1956-1962

The German chemical industry showed its interest as early as 1956 when Hoechst established a radiochemicals laboratory near Frankfurt[23]. German participation in Eurochemic had not caused the country to give up the idea of building a German pilot plant. It is true that Eurochemic provided the quickest and cheapest means of gaining access to American and French data[24], but the international pilot plant was insufficient for national industry to gain a mastery of all the necessary knowledge and experience. The difficulties the Ministry for Atomic Energy experienced in persuading private enterprise to buy shares demonstrates how the international project received more support from governments than from private interests.

The latter soon returned to the business of promoting a national pilot plant. As early as 1961, when the detailed plans for the Eurochemic plant had just been drawn up, the DAtK[25], under pressure from the chemical industry and particularly Hoechst, recommended to the Ministry for Atomic Energy and Water[26] that consideration should be given to a German pilot plant and plans made for the necessary R & D programme. The project in its initial form was complementary to Eurochemic, since it involved a small unit[27] for reprocessing fuel from the MZFR research reactor, which would have been accommodated in the Hoechst radiochemicals laboratory. Its cost was estimated at 15 million DM. At the time this project did not raise much enthusiasm on the part of the authorities who believed that the Eurochemic experiment was sufficient and that priority should be given to reactor research[28].

The project was therefore postponed and later reappeared in a different form. In 1963 it was broadened to encompass light water reactor fuels and a site chosen near the Karlsruhe nuclear research centre[29]. In the

---

20 "The obsession with cash flow resulted in cell design so poor that it became very difficult to envisage any subsequent intervention".

21 See Part IV Chapter 1.

22 Sources: SCHÜLLER W. (1985), pp. 178-181, WEINLÄNDER W., HUPPERT K. L., WEISHAUPT M. (1991), TEBBERT H., ZÜHLKE P. (1970).

23 See Part I Chapter 2.

24 SCHÜLLER W. (1985), p. 178: [Eurochemic] "hat es uns ermöglicht, den Anschluss an das in den USA und in Frankreich vorhandene Know-how auf dem Gebiet der Wiederaufarbeitung zu finden": [Eurochemic] "has given us access to American and French know-how in the field of reprocessing".

25 Arbeitskreis III/2, Fachkommission III.

26 Bundesministerium für Atomkernenergie und Wasserwirtschaft.

27 Technikumsanlage.

28 This was a field in which the chemical industry had less interest.

29 The plant is 10 kilometres from the city of Karlsruhe, on a 200 x 200 metre site near the nuclear centre, whose waste treatment plant and services it utilises. In this connection the resemblance to Eurochemic is striking.

meantime, Karl Winnacker and Ludwig Küchler had promoted the notion that a state-funded German pilot plant was an essential step in providing the broad base of experience[30] necessary for reaching the industrial scale[31].

Emergence of a German pilot plant: 1963-1971

This time the Hoechst initiative had a positive outcome, because on 20 May 1963 the Ministry for Scientific Research[32] decided to initiate a study on a pilot installation for reprocessing oxide fuels, and provided 3 million DM for the preliminary study and associated research programmes. The IGK, an association of three German industrial architects – Leybold, Lurgi and Uhde – participating in Eurochemic, was charged with the preliminary study, which was submitted in the summer of 1964.

On 18 December 1964 saw the creation in Frankfurt of the company which was to be the operator of WAK, the Gesellschaft zur Wiederaufarbeitung von Kernbrennstoffen (GWK). At the time it had three shareholders: Hoechst with 50% of the capital, Gelsenkirchener Bergwerke AG (GBAG, subsequently Gelsenberg AG) with 25%, and Nukem GmbH with 25%. Hoechst later sold half of its shares to Bayer AG.

The sum of 60 million DM was set aside in the budget for the Gesellschaft für Kernforschung mbH – Geschäftsbereich Versuchsanlagen (GfK/V), the public company which owned the plant.

The project made progress on paper until the beginning of 1966, when a crisis occurred which almost brought it to an end. There were two causes. First, there were differences of opinion between the main contractor and IGK on the question of costs. Then criticism was voiced about the PUREX process chosen as the basis of the project. In fact General Electric had proposed both to the BMwF and to Hoechst to build an installation using the revolutionary Aquafluor process[33] in Germany. Some officials in the Ministry for Research and in Industry were tempted by this technological poker game. However in May 1966 the differences over funding were overcome and the Aquafluor approach rejected in favour of the PUREX process.

Construction work began in January 1967 just as reprocessing was commencing at Eurochemic. The first reprocessing campaign began on 7 September 1971.

Eurochemic and WAK

Eurochemic, the learning centre for German reprocessing before the commissioning of WAK

Throughout the period, Eurochemic experience was transferred directly home by the Germans who, as shareholders in the company, had access to all information and who used the company as a training centre for some of their engineers and technicians (Table 119).

The number of German managers and technicians at Eurochemic went up from 4 to 13 between the end of 1961 and the end of 1966, to 12 in 1968 and then to 13 in 1969, before falling rapidly. There were three in 1971 and only one in 1972, when WAK had been started up. In percentage terms, the Germans accounted for 8% of Eurochemic staff up to 1968 and about 30% of managers and technicians up to 1969. Thereafter the percentages fell to reach a minimum in 1972 with 2% and 10% of the workforce respectively.

Mention may be made of two individual cases to illustrate the close links that had been established between the two plants at a high level.

Two WAK directors had important posts at Eurochemic for a number of years. One was Walter Schüller, who was a member of the design and research team in 1959, where he was in charge of the Civil Engineering Section before heading the Nuclear Services Section attached to the Technical Division. At Eurochemic he was regarded as the "Hoechst man"; he was present throughout the preparatory phase and most of the construction period until his return to Germany in 1965. He became technical director of WAK in 1970 and in 1985 chaired the firm's steering committee.

The second was Wilhelm Heinz, who in 1964 was Head of the plant's analytical laboratory and who became Head of the Operations Division during the period of reprocessing until the changes in 1972.

During the design and construction period, German firms were able almost immediately to reinvest the know-how acquired at Eurochemic.

---

30 "Die breite Erfahrungbasis".

31 AtW (1967), p. 194.

32 BMwF: Bundesministerium für wissenschaftliche Forschung.

33 It will be recalled that this process produced uranium hexafluoride which could be directly recycled in the enrichment plants. This was perhaps an attempt to try out, on a pilot scale, the laboratory process developed by General Electric and which was to be transferred vainly to the industrial scale in the United States at the Morris plant in the early 1970s.

Table 119. **Evolution in the number of German staff at Eurochemic**

	Managers (categories A and B)	All staff (ABCD)	% managers	% all staff
1961	4	15	20.6	11.5
1962	n.a.	n.a.	n.a.	n.a.
1963	9	26	34.6	9.6
1964	n.a.	n.a.	n.a.	n.a.
1965	n.a.	n.a.	n.a.	n.a.
1966	13	29	26.1	7.7
1967	12	26	25.8	7.1
1968	13	26	30.3	7.3
1969	10	17	28.3	5.2
1970	6	12	21.3	3.9
1971	3	9	15.7	3.0
1972	1	4	7.7	1.7
1973		4		1.7
1974		4		2.4
1975		4		2.4
1976		4		2.3
1977		5		2.6
1978		6		3.1
1979		6		2.9
1980		6		2.9
1981		6		3.0
1982		7		3.1

Reinvestment and development of industrial and engineering design know-how

WAK design and construction was entrusted almost entirely to German contractors[34]. The 118 firms listed by the *Atomwirtschaft* journal as WAK suppliers include one Dutch firm, the subsidiary of an American company making highly specialised equipment (Baird Atomic Europe N.V. of the Hague), one Austrian firm and one Swiss company. The list includes the German firms who contributed to the construction of Eurochemic, occupying leading positions on the basis of that experience. For example this was the situation of IGK formed from Uhde, Leybold and Lurgi which became the industrial architect for the whole plant. Civil engineering was entrusted to an association involving Eurochemic partners such as Wayss & Freytag KG and Hochtief AG. The suppliers of machines include in particular Essener Apparatenbau GmbH[35], which supplied the conditioning for the transport of plutonium, evaporators and condensers. However most of the German WAK contractors had not been involved in the construction of Eurochemic. Thus the technological seeding process took place in two stages, first by transfer, then by dissemination.

Two comparable and complementary pilot plants

With WAK, and its participation in Eurochemic, Germany effectively had two pilot reprocessing plants, which were in competition as well as being complementary, and fairly close geographically. Architecturally the relationship is direct. From the technical standpoint, the plants supplement one another.

Physically WAK was a smaller installation than Eurochemic (Figures 160 and 161) with the compact design of the main building clearly inspired by Eurochemic. The ponds (164a) adjoin the cells as they do at Mol. However unlike the situation at Eurochemic, none of the cell walls give on to the outside of the building. To the west of the cells there are service rooms and chemical supply areas (155-156). The laboratories are in the building, separated from the cells by the service corridor and the working area.

From the technical standpoint, some aspects of WAK copy Eurochemic while others make it a complementary installation[36]. The separation of the projects in time is the reason for the specialisation of WAK. It was a question of building a functional plant for reprocessing fuel from light or heavy water reactors (which in Germany in the mid-1960s were regarded as the two possible approaches for developing domestic nuclear power) and thus a way of trying out different approaches, the choice of which remained open.

---

34 AtW (1970), 2, p. 96.

35 Builder of the first Eurochemic dissolver. This particular case is examined in more detail below.

36 AtW (1967), p. 196-197.

Schematic diagram of WAK site

1	Building
2	Final product store
3	Solid and liquid waste store
4	Power station
5	Chemicals store
6	Workshops and offices
7	Stack

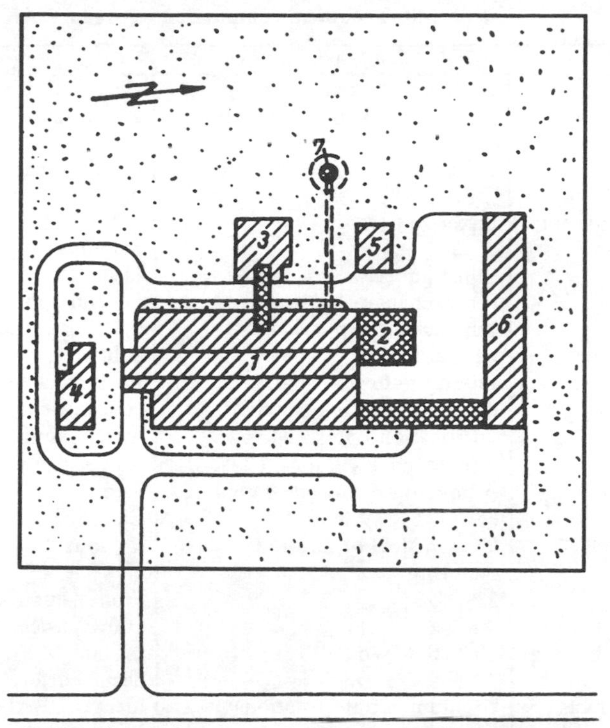

Figure 160. Outline sketch of the WAK site. 1. Treatment building. 2. Final product storage. 3. Solid and liquid waste storage. 4. Power plant. 5. Chemical store. 6. Workshops and offices. 7. Stack (Source: AtW (1967), p. 197).

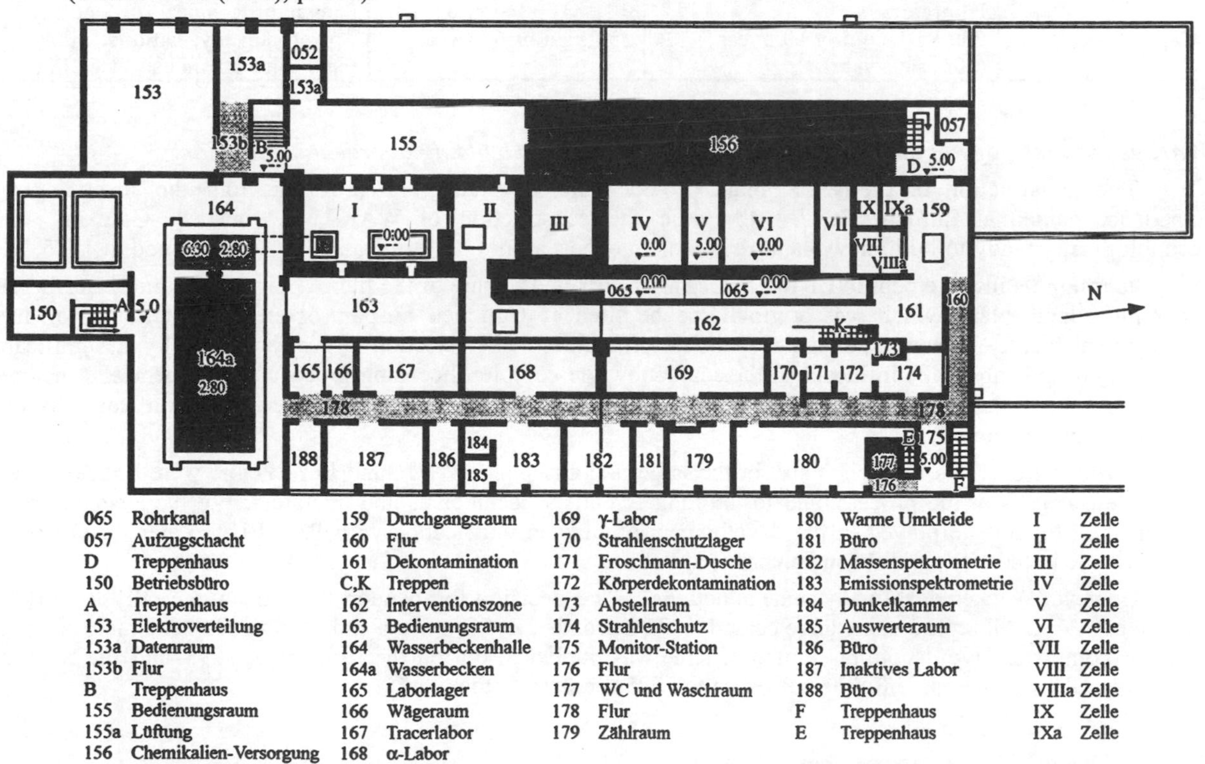

Figure 161. Ground floor plan of the WAK treatment building. The ponds used for storage and mechanical decladding adjoin the active cells, which are few in number, of unequal size and entirely surrounded by service areas. The plant is extremely compact and smaller than Eurochemic (Source: EITZ A. W., RAMDOHR H., SCHÜLLER W. (1970) p. 76).

Table 120. **Technical comparison between WAK and Eurochemic in 1970**[37]

	WAK (in 1970)	Eurochemic (in 1970)	Remarks
Head-end of process	Mechanical shearing (chop and leach) and chemical dissolving	Chemical decladding and dissolving	The choice for WAK related to the type of fuel reprocessed and feedback of Eurochemic experience
Extraction, U/Pu separation and decontamination	PUREX in 2 cycles (30% TBP) mixer-settlers	PUREX, variable number of cycles (2 or 3) (30% TBP), pulse columns	The choice of mixer-settlers at WAK as a supplement to the Eurochemic pulse columns
Tail-end of process	Silica gel (U) Ion exchange (Pu) Ion exchange (planned for neptunium)	Idem Several approaches tried out (ion exchange, pulse columns)	
Liquid waste management	Intermediate storage of concentrated liquid wastes, vitrification envisaged for disposal in a salt mine (Asse II)	Intermediate storage of concentrated liquid wastes, solidification research in progress	
Information transmission	Automatic pneumatic sampling (THOREX system)	Idem Electrical	Investigation of a cheaper method
Type of fuel	UO_2 in steel or zircaloy cladding	Various	WAK is specialised
Maximum enrichment	3% U235	From natural uranium to 100% enriched uranium	See above
Maximum burn-up	20 000 MWd/tU	25 000 MWd/t (initially designed for MWd/t)	Burn-up increases with reactor experience
Final products	Uranium nitrate solution (450g/l) Pu nitrate (250g Pu/l) (neptunium planned for the laboratory)	Idem et UF_4 (SFU) Pu oxide (neptunium planned for the laboratory)	Recovery of Np-237 as an intermediate product for Pu-238 (for isotopic batteries) No final purification of the Pu (source of technical difficulties)
Capacity	200 kg UO_2/d or 40 t/year (200 days/year)	350 kg/d or 60 t/year	Lower capacity, national pilot plant with no industrial calling

WAK as the first stage of the project – ultimately aborted – for building a large national plant

The construction of the WAK plant provided an opportunity for profiting from the technological experience gained at Eurochemic. By the time the construction of WAK was finished, Germany was technologically ready to build new plants and even to export a pilot unit to Brazil as it was planned in 1975.

Through its involvement in UNIREP, Germany obtained a share of the future European market, and began planning a large plant which was originally to be sited at Gorleben as part of an integrated nuclear fuel management facility. However, owing to political difficulties – the rejection of nuclear power by a significant proportion of the public, the influence of the coal lobby, the complex licensing procedures in a federal structure, and so on – the project was without a home for several years until the Wackersdorf site in Bavaria was selected, purchased and developed.

During these years of uncertainty for the large national plant, WAK and Eurochemic were both used for testing new aspects of the process, and for training scientists, technicians and operators. We have already seen the part the Belgian site played in the development of German vitrification with PAMELA. Later we shall see that of Lahde in the testing of the maintenance system.

However Wackersdorf finally came to nothing[38]. The decision to abandon the project naturally led to the closure of WAK. All reprocessing there ceased on 31 December 1990, only a month after Eurochemic ceased to exist. A comprehensive decommissioning scheme was published during the summer of 1993 and the sum of 77 million DM was set aside for the period up to 1995 for starting the work[39].

---

37 Source for WAK: AtW (1970) p. 83.

38 The reasons why the project was abandoned are complex. The German chemical industry had left the project, and were replaced by the electricity suppliers, who were more interested in cheaper fuel reprocessing than in funding a technically ambitious project. Against the general background of increasing ecological sensitivity, the continuing legal battles and the fears of proliferation were like millstones round the neck of reprocessing, none of which encouraged the federal government to support the project.

39 AtW (1993), p. 573.

The operation of WAK and Eurochemic. Common problems: partly resolved at WAK while it lasted

Drawing parallels between two plants is always a risky business. Eurochemic and WAK were two plants of comparable size commissioned five years apart by teams who had a number of points in common. Eurochemic ceased its activities after eight years of reprocessing; WAK continued for 19 years during which time it reprocessed 207 tonnes of fuel, coming almost equally from light water and heavy water reactors. The history of operations at WAK reveals comparable initial problems, very largely ascribable to the general state of technology at the design stage and thus generation-linked.

Differences between the head-ends of the processes make comparisons difficult, but WAK did experience a dissolver corrosion leak after nine years of operation. This incident interrupted plant operations from May 1980 to October 1982. As in the Mol plant, problems were experienced at an early stage with the evaporators and condensers, in the plutonium cycle and the intermediate uranium cycle, during the first few years of operation in 1974 and 1978.

Like Eurochemic, WAK found that concentrating plutonium was a complicated process, but by providing its customers with plutonium nitrate managed to avoid all Eurochemic's problems with the operations of precipitation and purification[40].

At WAK the MUF in terms of uranium was higher than at Eurochemic, but lower for plutonium owing to the absence of the nitrate conversion stage. The figures were 0.9% for uranium compared with 0.22% at Eurochemic, and 1.1% for plutonium against 1.8% at Eurochemic.

The reprocessing of 207 tonnes of fuel by WAK produced 73 m^3 of highly radioactive liquid waste concentrate, or 0.352 m^3 per tonne of uranium reprocessed, which it is planned will be vitrified in the PAMELA facility at Mol. These figures are to be compared with the 65 m^3 produced by the reprocessing of LEU at Mol, resulting from the reprocessing of 181.3 tonnes of LEU or 0.358 m^3/tU. These very similar figures demonstrate once again how parallel were the techniques used in the two plants.

However WAK was able, during its 19 years of operation – much longer than Eurochemic's life – to resolve certain problems, notably those of dose and gaseous emissions.

WAK was able substantially to reduce the average doses.

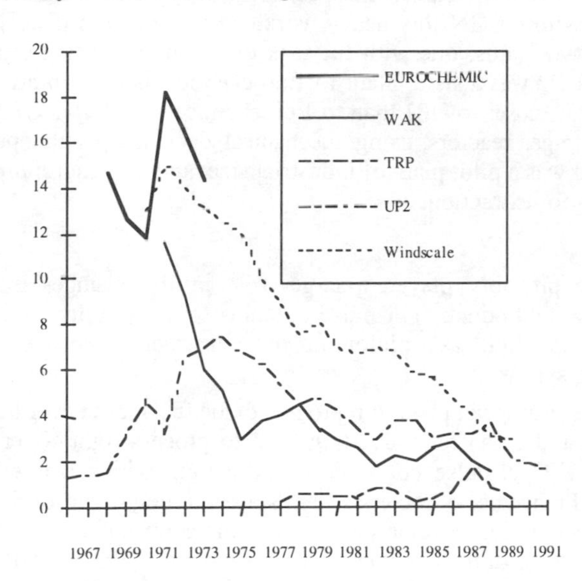

Figure 162. Variation in annual mean doses in certain reprocessing plants from 1967 to 1992 (mSv/person/year)

The above graphs, prepared from the statistics given in Part III Chapter 4, show how worker safety, starting from relatively high levels comparable to those at the Mol plant, around 12 mSv/year, moved towards the levels found in present-day plants. This was achieved in two ways. The first involved technical and standardisation measures, with continuous improvements to monitoring methods and the management of work

---

40 Unlike WAK, Eurochemic did not have a fuel fabrication plant nearby, and had to supply its plutonium in solid form. Calcination of the oxalate is a tricky operation if a recoverable form of plutonium is required.

in controlled areas. The second was simple but costly in terms of wages and salaries: it involved increasing the number of people working in controlled areas[41]. Starting with an initial workforce of 250, WAK ended up with 430, of whom 340 were working in controlled areas. Contract workers were also occasionally used. Eurochemic was never able to use this kind of approach to the same extent during its short period of reprocessing[42].

The other major step forward at WAK concerned gaseous emissions. As time went by, the number of monitoring systems was increased and filtration methods improved. Initially the situation was comparable with that of Eurochemic, with alpha and beta emissions and those from krypton-85, iodine-131 and tritium being monitored. Later on iodine-129, strontium-90, carbon-14 and plutonium-241 were also monitored. Special trapping methods were developed and applied in Germany, particularly for iodine-131, and used at WAK. These are today one of the poles of excellence in the German nuclear industry. The Japanese, in deciding what technologies to import for Rokka Shô (see below) decided on German technology for iodine removal.

Comparative study of Eurochemic and WAK suggests that the idea of restarting the Mol plant, even though it was politically and economically impossible, was technically well founded. It also reveals the commonality of problems at the two plants which were of the same generation and had virtually the same purpose.

Eurochemic's influence was also felt in Japan, which is paradoxical to say the least since Japan was never a member of the company and joined the ENEA (whereupon the Agency changed its name to NEA) only at the end of the reprocessing period. The explanation for this lies in the role of France and SGN. With only two exceptions, France made little use of the experience gained at Eurochemic for its own development. However it did use its Eurochemic experience for exporting its technologies to Japan.

Eurochemic, France and Japan

Eurochemic had very little influence on the French reprocessing plants

UP2 and Eurochemic "ring fenced" within SGN

Like West Valley, UP2 and Eurochemic were built contemporaneously. A degree of cross-fertilization between the two plants might have been expected, especially as they both had the same industrial architect, but this did not happen. Indeed within SGN this heavy workload was most probably responsible for the lack of communication between the two teams, one with the task of building the first establishment at La Hague, the other the international plant. UP2 was a sister plant to Eurochemic, sharing an ancestor – UP1 – and the same design bureau. However UP2 is closer to UP1 than to Eurochemic. Like UP1, UP2 was a plant specialising in reprocessing fuel from graphite-gas reactors, using mechanical decladding and operating with mixer-settlers[43]. Eurochemic on the other hand was a pilot plant of industrial size, and also multipurpose; finally it used chemical decladding and pulse columns for extraction.

Limited influence

Eurochemic experience probably played a larger role in the changes made to UP1 in the 1960s, particularly in the rather hesitant introduction of pulse columns to the plutonium concentration process using a solvent, in the use of uranium IV nitrate as a plutonium reducing agent instead of ferrous sulphamate, or in the simplification of the sampling systems.

Eurochemic was the first European plant to reprocess oxide fuels, but nevertheless provided information to its French shareholders (SGN and CEA) for converting UP2 to process light water reactor fuels. It also exerted influence in the adoption by UP3 of pulse columns for extraction, which marked a substantial technological departure from the traditional French use of mixer-settlers since the beginning of the 1950s. André Redon, who had acquired considerable experience in developing pulse column control practices at Eurochemic since the time of the pilot unit, was in fact director of the Contracting Department for UP3 and UP2-800 in the early days of the project from 1980 to 1983. However although the extraction and control principles applied were the same, the type of columns used was not the same as in Belgium. They were higher capacity systems with an annular design for criticality prevention at larger volumes.

In the beginning, France's role at Eurochemic was primarily one of a supplier of equipment and know-how. Until the plant was started up, French engineers and technicians accounted for between a third and a

---

[41] This was also used to a considerable extent at Tokai Mura. Financial reasons prevented Eurochemic from using this simple means of "spreading the dose".

[42] During the reprocessing period, there were never more than a few dozen contract workers at any one time. Indeed the company's policy tended rather to reduce the size of the workforce (see Part III Chapter 4).

[43] For a description of UP2, see DUBOZ M. (1965).

half of Eurochemic staff[44]. The French technical presence was again substantial during the industrial phase of reprocessing subsequent to 1971, with André Redon, not only with a view to fulfilling contracts as quickly as possible, but also to take a close look at the reprocessing of light water reactor fuels, with the conversion of UP2 in mind. Indeed relations between the two plants were good, unlike the situation with Marcoule.

This low overall influence is also related to the difference between the resources committed to national and international projects: vast in the former case, relatively modest in the latter. Cumulative French R & D investment in reprocessing has been evaluated at 8.23 billion francs (today's values) for the period 1970 to 1987[45], not counting the capital and running costs of the industrial installations[46]. This amount can be compared with the total cost of Eurochemic from 1956 to 1990, which was a little under 6 billion FF (1990 values), of which France contributed a quarter, about 1.5 billion FF.

However it is also linked to nationalism and the feeling of technical superiority which formed the cultural basis of a significant number of CEA engineers, traits which COGEMA inherited when it was established in 1976[47].

At COGEMA, Eurochemic seems to have had a rather negative image, particularly in the commercial sector. The international undertaking was seen as partly responsible for the downward spiral in reprocessing prices at the end of the 1960s, with the reputation of not being very safe, a "plutonium sieve". For example Jacques Couture believes that if there was any influence, it was rather "an example of what not to do", both as regards the commercial results and the form of international co-operation[48]. However this negative image does not seem to be shared by those members of the CEA and SGN who were directly involved in Eurochemic, nor by the engineers in the radioactive materials division at the CEN in Fontenay-aux-Roses (DGR). Perhaps this is an example of the conventionally opposite criteria applied by the "marketing men" of COGEMA and the CEA's "researchers".

In any event, Eurochemic's role in France was very different from the one it played in Italy and Germany. But it was thanks to Eurochemic that France became an exporter of nuclear reprocessing technology for the first time at the end of the 1950s.

French participation in Eurochemic gave SGN access to the Japanese market in the 1960s, just when European orders for reprocessing had dried up[49]. The Tokai Mura pilot plant provided SGN with the first commercial spin-off of its participation in Eurochemic. Even today, reprocessing is one of the rare examples of Franco-Japanese trade that shows a profit[50].

SGN and the construction of the TRP[51] Tokai-Mura pilot plant. A second experiment in international co-operation in reprocessing[52]

The beginning of Japanese reprocessing and the role of Franco-Japanese nuclear co-operation. 1956-1963

Reprocessing came into being in Japan in parallel with the establishment of its nuclear institutions which, as in the European countries, is itself inseparable from the Atoms for Peace policy[53].

---

44 See Annex.

45 Delange M. (1985), p. 565.

46 COGEMA apparently invested 45 billion FF (today's values) in La Hague between 1979 and 1989.

47 As regards the creation of COGEMA, the advent of a management team from the Military Applications Division (DAM) from the CEA and Marcoule apparently tended to reinforce this culture.

48 Jacques Couture is comparing Eurochemic with UP3 which was designed and built exclusively under French management for reprocessing foreign fuels. In his eyes this formula considerably simplifies decision-making and the sharing of responsibilities. See Part V Chapter 2.

49 UP2 and Eurochemic were completed in 1966. At that time SGN was experiencing a period of "short commons" in the nuclear field, so it resumed its engineering work in conventional chemistry. Michel Lung remembers having followed projects of this kind, completed in two or three years, a pleasant change from the much slower progress that prevailed in the nuclear sector.

50 This influence continues today with contracts to build parts of the Rokka Shô plant (at a cost estimated to be 3 billion FF).

51 Tokai Reprocessing Plant.

52 Japanese sources: MIYAHARA K. YAMAMURA O., TAKAHASHI K. (1991), YAMANOUCHI T., OMACHI S., MATSUMOTO K. (1987).

French sources: discussions with Michel Lung and André Redon, BLUM R. (1963), CURILLON R., VIENOT J. (1968), LEFORT G., MIQUEL P., RUBERCY M. de (1968), LUNG M., COIGNOT M. (1978).

53 A bilateral agreement with the United States was signed in 1955.

In January 1956 an Atomic Energy Commission (JAEC) was set up. In September the JAEC entrusted this task to a public company, the Atomic Fuel Corporation (AFC)[54]. However it was not until 1959 that an Advisory Committee was formed with the task of laying down guidelines for the development of reprocessing technology. A "survey team" was established, which visited overseas plants and drew up recommendations. It was decided to build a pilot plant using the advanced technologies employed by other countries, another instance of a classic Japanese strategy. However at that time the reprocessing project was not of primary urgency. As a massive importer of energy products, the country's immediate objective was to acquire know-how in nuclear power, and reprocessing was placed on the agenda only in 1963 when the programme of installing foreign reactors was already well advanced.

A contract had been signed with the United Kingdom in 1958 for the importation of a reactor and fuel. This led on 26 October 1963 to the start-up of an experimental 12 MWe power reactor, boiling water moderated and cooled, loaded with 2.6%-enriched uranium oxide and installed at Tokai Mura on the Pacific Coast, about 100 kilometres to the north of Tokyo. The first Japanese power reactor, JAPCO 1, came into service on the same site in July 1966; this was a magnox reactor of British manufacture with an installed capacity of 160 MWe. Subsequent reactors were American and included both General Electric BWRs and Westinghouse PWRs.

Contact had also been made with Canada for the supply of fuel. A widened programme for scientific and technical co-operation in the nuclear field had been set up with France in the early 1960s.

The first Franco-Japanese technical assistance contract was signed in 1960 between the CEA and the Japan Atomic Research Institute (JAERI), a public research body, for the construction of a radioisotope laboratory. The following year, the deputy head of the CEA plutonium chemistry section took part, as an expert, in a fuel reprocessing study for JAERI at the request of the IAEA. In February and March 1963, Franco-Japanese nuclear co-operation was formalized through the establishment of two national Steering Committees responsible for implementation. The French committee was made up of 17 members from the CEA, EdF and industry – through the ATEN – together with two senior civil servants, one from the Ministry of Foreign Affairs, the other from the Ministry for Scientific Research, Atomic Affairs and Space. These committees were supposed to promote the circulation of information, facilitate contacts and exchanges, suggest subjects of common interest, and plan the corresponding programmes. The committees were structured into four working groups, with group n° 3 being concerned with the fuel cycle. These structures enabled the Japanese to gain a better understanding of French achievements, particularly in the reprocessing field. Japanese engineers visited France and some of them also made the journey to Mol.

SGN and Tokai. The influence of experience gained at Eurochemic

In 1963 a project was initiated for the construction of a Japanese pilot reprocessing plant. The tendering procedure put three engineering companies in competition – one American (Weinrich and Dart), one British (Nuclear Chemical Plant Ltd., linked to the UKAEA) and one French (SGN) for producing the preliminary design for the pilot plant which was to be built beside the Tokai reactors[55].

There was fierce competition between Nuclear Chemical Plant and SGN over the contract for the complete design[56] which was to follow: it was awarded to SGN. The contract signed by AFC in December 1965 received government approval on 22 February 1966. The basic design work was carried out in 1966 and 1967 at a cost of 10 million FF.

The French firm's experience of international co-operation through Eurochemic carried considerable weight in the Japanese decision[57]. At Mol, SGN had learned to work with non-specialist engineering contractors. The Japanese wanted not only to build a pilot plant in their country but also to acquire the know-how.

During the contract negotiations covering the Basic Design Study, the Japanese had made it a requirement that the successful contractor should work with a local firm. The aim of this was of course to learn about the techniques of reprocessing engineering. However SGN was able to choose its own partner. It considered two[58]: Chioda, a subsidiary of Mitsubishi, and the Japan Gasoline Corporation (JGC), a medium-sized firm which had no links to any major groups. By opting for JGC, Jacques Viénot, a graduate of the prestigious École centrale des arts et manufactures, who was in charge of the project at the time, believed that SGN would have greater freedom of negotiation.

---

54 In October 1967 the AFC was absorbed by another public body, the Power Reactor and Nuclear Fuel Development Corporation (PNC).

55 AtW (1965), p. 646.

56 Also called the conceptual design or Basic Design Study.

57 CURILLON R., VIENOT J. (1968) p. 160, note 3.

58 Communication from Michel Lung.

Thus the objectives of the structural design studies[59] entrusted by Japan to SGN were identical with those it had received for Eurochemic: "the detailed design scheme must be such as to allow the widest possible involvement of Japanese industry and that the general construction work may in principle be entrusted to a Japanese company of substance and quality but without experience in the requirements of nuclear chemistry"[60].

The detailed design work took 16 months in 1967-1968[61]. JGC was essentially charged with the civil engineering, pipework and earthquake design. Engineers exchanged visits between Paris and Yokohama, where the two companies had their design offices[62]. This fruitful co-operation resulted in the two companies concluding a joint-venture agreement in 1970, with a view to the construction of the plant and the waste treatment unit, the more traditional sections being supervised directly by PNC. The construction contract was signed in December 1970.

Eurochemic and Tokai: Inspiration, feedback of experience and common problems

Construction of the plant took from April 1971 to September 1974. Between 10 and 12 SGN engineers were permanently seconded to Japan, under the direction of Michel Lung, who had represented SGN at Eurochemic during cold testing and commissioning of the Mol plant[63]. The contractors were exclusively Japanese and the equipment manufactured locally, except for a few special systems such as dissolvers, fission product evaporators, mixer-settlers and shears.

However it took three years for the plant to go active, owing to the contemporary worldwide uncertainties about reprocessing, and the need for the Japanese to obtain United States agreement to reprocess fuel that had been enriched on the other side of the Pacific[64].

From the technical standpoint, SGN's experience at Eurochemic was largely reinvested in Tokai, even though the plant had different objectives, being intended to reprocess exclusively light water reactor fuels. In fact the design for Tokai – a pilot unit of comparable size to Eurochemic – was produced by an engineering team consisting of a mixture of engineers and technicians who had worked on Eurochemic and on UP2.

The following Table 121[65] is an attempt to summarise the main contributions – a two-way process since there was feedback of experience – of Eurochemic, UP2 and HAO (the High Activity Oxide head-end grafted on UP2) to Tokai. It shows the influence of Eurochemic.

Table 121. **Influence of Eurochemic on the characteristics of the Tokai plant**

TRP : features	Inspired by Eurochemic	Inspired by UP2	Inspired by HAO
Dissolver	yes		
Essentially mixer-settlers		yes	
A few pulse columns	yes		
$U(NO_3)_4$ for Pu reduction	yes (purchase of licence)		
Bituminisation of intermediate activity liquid	yes		
Shearing			yes
Storage ponds	yes (feedback of experience)		
No purification of uranium on silica gel	yes (feedback of experience)		

59 Or Detailed Design.

60 CURILLON R., VIENOT J. (1968), p. 161.

61 LUNG, M., COIGNOT M. (1978), p. 318-319.

62 At the same time, AFC arranged for tests to be carried out by the CEA in France in conjunction with SGN, under bilateral international agreements, in order to verify the performance of the process. This work was carried out in the CYRANO cell of the CEN at Fontenay-aux-Roses. The intended Japanese fuels had characteristics similar to the snowflake elements. It is possible that the shipment of some of these to Eurochemic was related to this work. With regard to these aspects, see LEFORT G., MIQUEL P., RUBERCY M. de (1968).

63 Michel Lung arrived in Japan at the beginning of 1971 and followed the construction and testing of the plant until the eve of active commissioning in 1977. On his return to France he became commercial director of the nuclear branch of SGN, a post he held until 1989. In the meantime SGN had severed its links with Saint Gobain to become a joint subsidiary of COGEMA (two-thirds) and TECHNIP (one-third), an engineering company working mainly in the petroleum field. Thus the synergy between the nuclear and oil sectors had shown up long before the recent share swap between Total and COGEMA. Since 1989 Michel Lung has been working as a consultant engineer for FORATOM in Brussels. He is responsible for nuclear lobbying with the European Union.

64 See Part IV Chapter 1.

65 Sources: Interview with Michel Lung and ETR 318.

The plant's first campaign began on 22 September 1977; it is still in service in 1994. By 1990, it had reprocessed nearly 510 tonnes of fuel including 320 tonnes from BWRs and 183 tonnes from PWRs, the rest from the Fugen reactor[66]. Some 3400 kilogrammes of plutonium nitrate had been extracted; after conversion this was used in Fugen and in the Joyo and Monju experimental prototype fast reactors.

The operational history of the plant illustrated on Figure 163 shows that the plant experienced the same kind of evaporator problems as Eurochemic. It also had to deal with dissolver leaks and substantial plutonium MUF[67]

Not all Tokai experience has been reinvested in the Rokka Shô plant[68] now being built in northern Honshu. There are no links between the public operator of Tokai and the main contractor for Rokka Shô, Mitsubishi Heavy Industries. However SGN is still maintaining a good presence in Japan, without JGC. Thus the transfer of technology continues, particularly for the new dissolvers[69].

Table 122. **The mixture of world technologies at Rokka Shô[70]**

Sector	Developers	Notes
Main plant: principal process	SGN-MHI, Hitachi, Toshiba, SMM	Franco-Japanese consortium for engineering design
Main plant: reduced pressure evaporation	BNFL-Hitachi	Anglo-Japanese consortium
Main plant: iodine filtration	KEWA-Hitachi	German-Japanese consortium
U and Pu conversion units	PNC-MMC	Public-private Japanese consortium
Vitrification of high activity liquid wastes	PNC-IHI	Public-private Japanese consortium
Spent fuel storage unit	PNC-Hitachi, Toshiba, MHI	Public-private Japanese consortium

Abbreviations used above:

IHI: Ishikawajima-Harima Heavy Industries.

MHI: Mitsubishi Heavy Industries.

MMC: Mitsubishi Materials Corporation.

SMM: Sumitomo Metal Mining Company.

Thus SGN enhanced Eurochemic's influence overseas. The indistrial pilot plant built 400 kilometres from Paris helped SGN penetrate the Japanese nuclear market by the middle of the 1960s. Relying first on its domestic know-how, but also buoyed up by its Belgian and Japanese experience, and further reinforced by links forged through UNIREP with Germany and the United Kingdom from the beginning of the 1970s, the firm subsequently diversified its partners, naturally beyond the Rhine and the Channel, but also in the United States[71] and in various Third World countries[72]. This international experience certainly influenced the decision taken by COGEMA in 1989 to make its subsidiary the principal centre of its EURISYS engineering design network[73].

---

66 A heavy water moderated advanced thermal reactor (ATR).

67 See Part III Chapter 4.

68 Rokka Shô reprocessing plant.

69 See below.

70 TOYOTA M. et al (1991), p. 43.

71 Notably for research into the vitrification of fission products.

72 In the affair of the Pakistan plant, SGN supplied the detailed design and commenced civil engineering work on the plant until this was stopped by the political authorities (see Part IV Chapter 1). SGN was also responsible for systems at Tarapur and a small laboratory in Taiwan.

73 Notably by absorbing its reprocessing subsidiary USSI. See the COGEMA 1992 annual report, p. 26-27. For the role of USSI in EURODIF, see DAVIET J. P. (1993), pp. 233-251. As regards EURISYS, representing about 5000 people and with a turnover of 5 billion FF, and which is attempting to structure into a network the 50-odd companies that worked on the design of UP3 and who are now having to cope with a reduction in their nuclear workload, see the course report submitted by Sylvie Jehanno and Julien Jenoudet on 27 June 1994 to the Paris École des mines, *Comment fédérer des sociétés d'études de haute technologie: SGN et le réseau EURISYS [How to Federate High Technology Research Companies: SGN and the EURISYS Network]* in the optional third year project "Scientific Management".

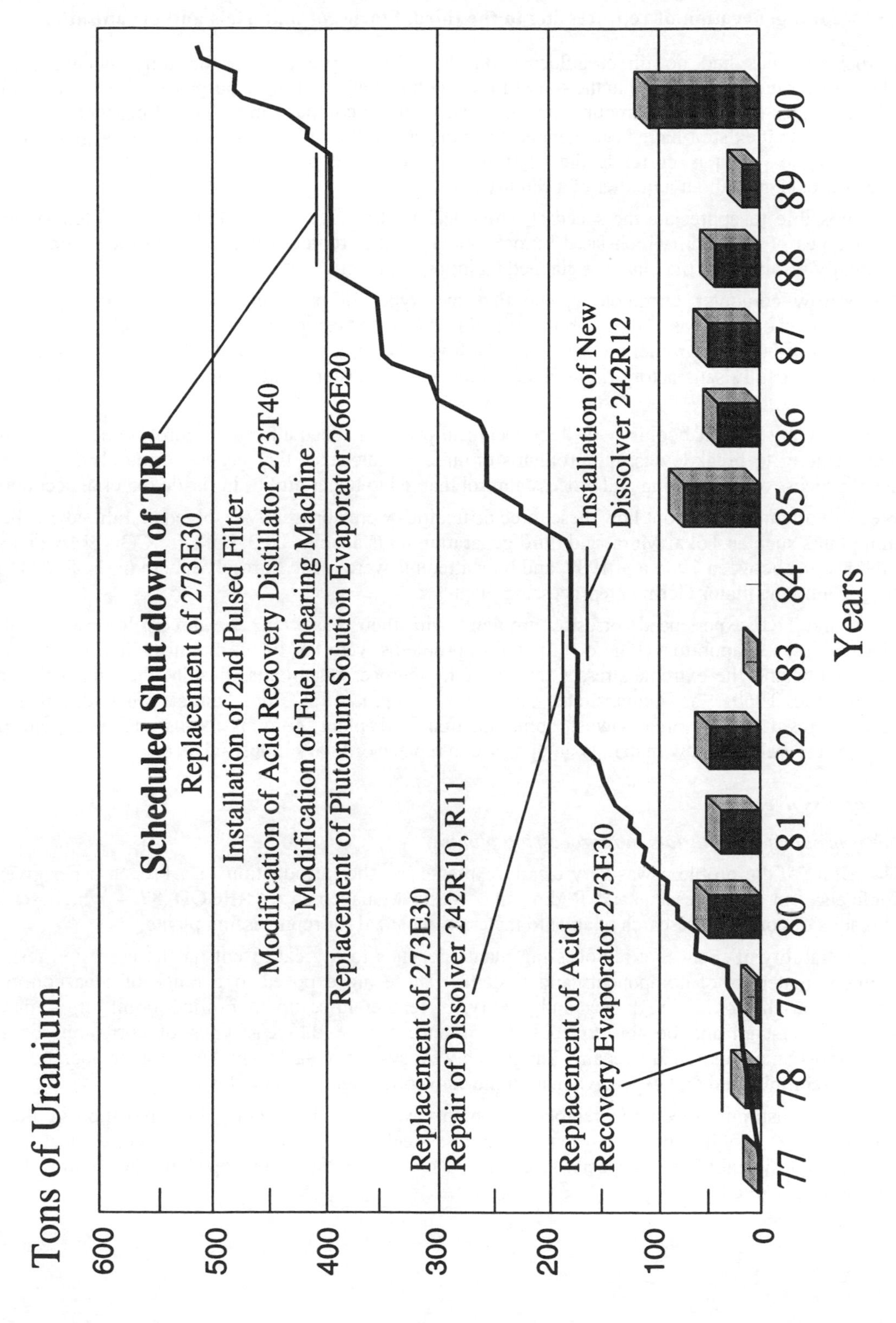

Figure 163. Principal phases in the operation of the Tokai Mura pilot plant between 1977 and 1990 (Source: MIYAHARA K. et al (1991), p. 52).

From the second generation of reprocessing to the third. Stories of materials and organisation

Eurochemic has had no direct influence on the design principles of the third generation, and its accumulated experience was diluted in the story of the second. However for the purposes of the history of nuclear chemical techniques it is worth attempting, on particular points, a comparison between Eurochemic and the third generation plants that exist or have been planned since the end of the 1970s. Pride of place is given to the UP3 plant at La Hague which is currently the most recent operational plant[74] and also had the same industrial architect as Eurochemic, about a quarter of a century later.

It is possible to appreciate the speed at which technical changes were taking place by considering how three major types of problem, which faced Eurochemic and other reprocessing plants of the second generation, were apparently resolved for the plants or planned facilities of the third generation.

To borrow computer terminology, the first two types of problem concerned plant and equipment ("hardware") and showed up as extensive contamination due to corrosion and the – relatively – extensive doses to workers; the third type of problem relates to the "software" area, the absence of which at Eurochemic made it difficult to set up a fully satisfactory fissile material accounting system and to analyse operational problems in real time.

The following section begins with three technical problems encountered at Eurochemic: the corrosion of evaporators leading to breakdowns, the problems of direct maintenance that generated unduly high doses, and problems of process data gathering and process control that led to uncertainties in fissile material accounting.

We shall attempt to find out how these three difficulties were apparently resolved or limited in the second generation plants such as Tokai Mura, and third generation units such as UP3, built by COGEMA at La Hague from SGN designs between 1984 and 1989, and how attempts were made to resolve them in the R & D projects gravitating around the major German reprocessing plant.

WAK and TRP experienced corrosion problems with their dissolvers; the three pilot plants had similar difficulties with their evaporators (Figure 164). These problems were the direct manifestation of the inability of materials to withstand the extreme stresses that exist in reprocessing plants, despite the undeniable progress made during the 1960s, as demonstrated by the development of Eurochemic's first dissolver. These shortcomings, resulting in breakdowns, contamination and plant commissioning delays, were powerful incentives for research into new materials, in this case to replace ordinary industrial steels.

The issue of dissolvers

The general problem of materials in reprocessing plants

The nature of the problem was very clearly expressed in the introduction to a French paper given at the Paris Conference on Reprocessing and Radioactive Waste Management – RECOD 87 – which set out the principal results of corrosion research relative to materials intended for reprocessing plants[75]:

"The reliability of spent fuel reprocessing plants depends to a great extent on the corrosion resistance of the constituent materials of components and systems. These are exposed to a range of subazeotropic nitric solutions[76], of variable oxidising power and acidity, at temperatures up to boiling point. It is particularly important to understand and be able to control resistance to the different types of corrosion, because the procedures for decontamination and repair – and possibly for system replacement – necessitate operations that are infinitely more complicated and expensive than in the conventional chemical industry".

Thus the constraints specific to reprocessing plants were a great incentive for innovation in the field of special steels such as very low carbon austenitic stainless steels. At Eurochemic, the systems most exposed to corrosion made considerable use of special steels and alloys successfully developed in the 1960s, despite the special problems involved.

---

74 THORP came into service on 27 March 1994 with fuel from Heysham, AtW (1994), May, p. 335.

75 LEDUC M., PELRAS M., SANNIER J., TURLUER G., DEMAY R. (1987), p. 1173.

76 Azeotropic: characteristic of a homogeneous liquid mixture of two components having a constant boiling point at a given pressure, the vapour of which having the same composition as the mixture. This applies for example to nitric acid in an aqueous solution containing 6% of nitric acid by weight. A change in pressure alters the composition of the azeotrope, which differentiates it from a pure substance, in ANGENAULT J. (1991), p. 49.

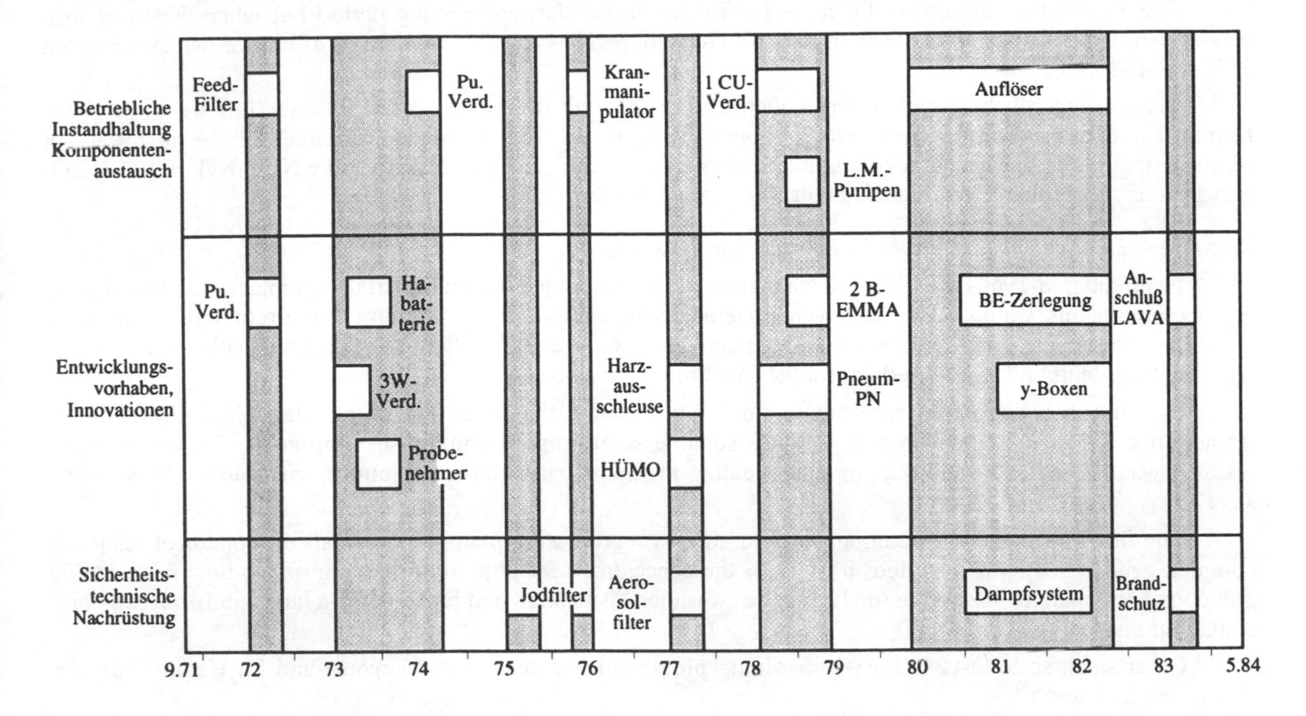

Figure 164. Periods of reprocessing (white) and shutdown for work on the system (cross-hatched) in WAK between September 1971 and May 1984. The shutdowns are divided into three main categories, from the bottom of the diagram upwards: improvements to safety systems; developments and innovations; shutdowns for maintenance and replacing equipment. The pilot plant was not intended to reprocess large quantities of fuel and experienced many operational problems (Source: SCHÜLLER W. (1984), p. 441).

The strengths and weaknesses of new materials used at Eurochemic: dissolvers and evaporators

For the first dissolver[77] Eurochemic had recommended the use of Ni-O-Nel, after conducting corrosion research on specimens supplied by the British firm, Higgins, the only one producing it at the time. Laboratory tests carried out by H. Dreissigacker at Eurochemic led to the definition and patenting by the company of a particular type of Ni-O-Nel known as CS.M2 or Incoloy 825 modified[78], for the construction of active parts. The other parts of the system had to be made using two other types of special steel, AISI 304 L and INCONEL 825, and all these materials had to be welded together in a fault-free manner.

The tendering procedure for the dissolver resulted in the contract being awarded to a subsidiary of Mannesmann AG, Essener Apparatenbau GmbH. However this firm was not capable itself of manufacturing the alloy and therefore turned to the British firm which had provided the specimens on which Dreissigacker had worked. However the raw material supplied by Higgins proved to be outside specifications; a further delivery was necessary which resulted in an 11-month delay in the final delivery to Mol.

The assembly procedure called for special precautions both in routine metalwork and in welding. The very low carbon content excluded routine hot forming. The operations of drawing and pressing[79] had to be carried out in a controlled atmosphere using inorganic materials.

Special welding procedures had to be developed since the traditional techniques caused microcracks[80] in the CS.M2. Finally, the acceptable tolerances were very narrow and called for special inspections. The same

---

[77] Sources: DREISSIGACKER H., FICHTNER H., REINARTZ W. (1966), and a written communication from Rudolf Rometsch.

[78] For the development of CS.M2, see DREISSIGACKER H. in *Chemical and Process Engineering,* May 1964. Ni-O-Nel is no longer regarded as a steel because its iron content is very low.

[79] "Zieh und Pressvorgänge".

[80] Mikrorisse.

material was used for the second dissolver but the contract was awarded to an Italian firm based in Milan, Siai Lerici, Cormano[81].

The Eurochemic dissolvers functioned normally during the reprocessing period but when the third unit was dismantled it showed traces of corrosion which might perhaps have led to the type of breakdown experienced at WAK and TRP.

In any event all three pilot plants suffered problems with their evaporators. The reasons were obviously technical and did not result from wrong choices being made. The inadequate strength of the materials in reprocessing plants thus encouraged research on austenitic steels[82] and special alloys like Ni-O-Nel, but also into non-ferrous substitute materials such as titanium and zirconium.

Zirconium metallurgy comes to the aid of the reprocessing plant

From the mid-1960s it was in France that research was carried out on materials to replace stainless steel in reprocessing plants. Comparative tests were done involving special steels, titanium and zirconium. In the early 1970s, zirconium was considered as a suitable material for supplementing the use of special steels in the French reprocessing plants and was used there for the first time[83].

Not only was zirconium resistant to boiling nitric acid thanks to the oxide film which forms on its surface, but industry learned how to produce suitable semi-finished and finished products. However further research was needed to develop appropriate welding methods, particularly for joining zirconium and stainless steel[84].

The first unit made of zirconium to be used in a reprocessing plant was a small evaporator of relatively complex geometry. It was installed in 1971 in the concentrator for plutonium-bearing oxalic liquors in a UP2 glovebox and remained in service for 12 years. It was then dismantled and replaced by a larger unit but was still contained inside gloveboxes.

Other small-scale tests followed, involving plutonium evaporators at Marcoule and La Hague, but also dissolvers, and so on[85].

When UP3 was built and UP2 extended, full-scale units were installed using greater thicknesses of zirconium. Some 100 tonnes of zirconium were used to produce the internals for evaporators principally but also for boilers, distillation columns and dust removal systems for the vitrification off-gases. Also 500 metres of zirconium tubing was installed in the two plants, some with connection welding to special stainless steel tubes, which are still the commonest materials used in piping[86].

The UP3 dissolver and dealing with dissolving problems in reprocessing plants. Formal elegance and sophisticated hardware

The UP3 dissolver broke new ground, turning away from the second generation of plants and managed partly or totally to resolve two substantial problems in the operation of reprocessing plants: the abundant solid wastes arising from dissolution[87] and the disparity between traditional, batchwise – and hence discontinuous – dissolving, and continuous extraction in pulse columns.

It was therefore essential, in one way or another, to turn both dissolving and the separation of hulls and insoluble fines into continuous processes. The traditional approach was to agitate the solutions. It was out of the

---

[81] Essener Apparatenbau had put in a bid but the Italian firm's price was lower. The Eurochemic management, knowing that the Italian firm had not previously welded this material, insisted on high *per diem* delay penalties in the contract. The Italian firm agreed, so long as a premium would be payable in the event of early delivery. They were one day early...

[82] For Rokka Shô for example, the results of co-operation between JAERI and the NKK Corporation's research centre are described in KIUCHI K., YAMANOUCHI N., KIKUCHI M. (1991), pp. 1054-1059. Research aimed at improving the performance of type 304L stainless steels was carried out jointly by Nippon Steel and Hitachi, see ONOYAMA M., NAKATA M., HIROSE Y., NAKAGAWA Y. (1991), pp. 1066-1071.

[83] SIMONNET J., DEMAY R., BACHELAY J. (1987); BERNARD C., MOUROUX J. P., DECOURS J., DEMAY R., SIMONNET J. (1991). Titanium proved brittle under certain conditions. The production of fuel cladding using zirconium and its alloys (zircaloy) had been routine for a long time, but sheet metal work in zirconium raised difficult problems.

[84] It is necessary to use special processes such as "explosion welding", involving the insertion of a sheet of tantalum between the other two metals, or "eutectic melting" which has to be done under vacuum at about 1000°C.

[85] List in BERNARD C., MOUROUX J. P., DECOURS J., DEMAY R., SIMONNET J. (1991), p. 572.

[86] Naturally enough, progress also continued at the same time in the steel industry.

[87] The same problem arose in Eurochemic's chemical decladding.

question to operate mechanical stirrers in a hostile environment. Accordingly gas injection was used, and higher temperatures, both of which caused additional corrosion. The problem was tackled in the UP3 dissolver using a simple and clever principle: the entire dissolver was moved in order to agitate the solution. The application of this principle involved a model that was rustic at the very least and therefore very suitable to the difficult environment: a water mill wheel with perforated buckets[88]. The dissolver developed for UP3, illustrated in Figure 165, functions as follows[89]:

"A Pelton wheel with perforated buckets is partly immersed in the solution. The chopped off fuel sections fall into a submerged bucket, the wheel rotates slowly and in steps, and by the time the bucket containing the cladding sections emerges from the solution, they must have been in it long enough to guarantee that the fuel contained is dissolved. After spending a certain time above the solution allowing the liquid entrained by the hulls to drain away, the hulls are unloaded during the subsequent rotational steps and disposed of.

One of the key features adopted is that the wheel is suspended on rollers at the top of the fixed structure and supporting the inside of the rim. This allows a hollow centre, the axis of rotation being virtual, allowing the installation of fixed central structures for filling the buckets and collecting the hulls as they fall out. This central structure, suspended from the lid, ensures continuity between the loading and unloading chutes integral with the tank."

The elegant fashion[90] in which these problems were resolved made UP3 – apart from being an object of pride for SGN and COGEMA – into a technological model since Rokka Shô also adopted the same principle.

The UP3 bucket dissolver apparently works effectively. The superior metallurgy of the zirconium brought about a considerable reduction in what was the primary cause of incidents in the second generation plants, leaks due to corrosion.

The combination of zirconium and austenitic stainless steels means that reliable and technically suitable containment materials are now available for the reprocessing of light water reactor fuels. Appreciable progress has also been made in reducing total dose, through developments in plant maintenance principles and systems.

Troubles with maintenance

The advantages and drawbacks of direct maintenance at Eurochemic

Eurochemic opted for direct maintenance because of experimental needs and cost considerations, but also because robotisation and remote handling techniques were still in a very embryonic state[91].

Of course the positive side to this was the relative ease of entry to repair breakdowns and implement the adaptations and modifications so essential in a pilot plant. The choice of a direct maintenance approach also made it possible for the installation to be decommissioned almost entirely by hand[92].

The major drawback was that before work could be done in the cells the installation had to be decontaminated, a process that depended largely on the dexterity of the operators. Also the exclusively manual maintenance operations together with the relatively frequent operational incidents were the main causes of the doses received on the site.

Although these characteristics were acceptable in a pilot plant designed in the early 1960s, they were already at the time not commensurate with the needs of a production establishment, which of course was not Eurochemic's primary purpose. As time went by they became unacceptable for their radiological impact on the personnel, even though it was possible by organisational means to ensure that the legal exposure limits were not exceeded[93].

---

88 The Japanese, for their part, had thought of an Archimedean screw. See OECD/NEA (1993a), p. 123, Fig. 5.11.

89 LORRAIN B., SAUDRAY D., TARNERO M. (1987), p. 1472.

90 However one must point out that elegant design does not preclude problems. Initially a stainless steel prototype was built, then the parts were made of zirconium. Inspection revealed welding problems and they all had to be replaced.

91 In France, it was only at the beginning of the 1970s that the first remote dismantling operations were tried out at Marcoule.

92 The remote maintenance approach used in the first generation of plants makes them extremely difficult to decommission today by other means, for which they were not designed. As a result the "entombing" process, used for example at the damaged Chernobyl reactor, might well be used for decommissioning abandoned plants of that period, like Hanford.

93 And even though towards the end of reprocessing at Eurochemic a net reduction in dose resulted. See Part IV Chapters 1 and 2.

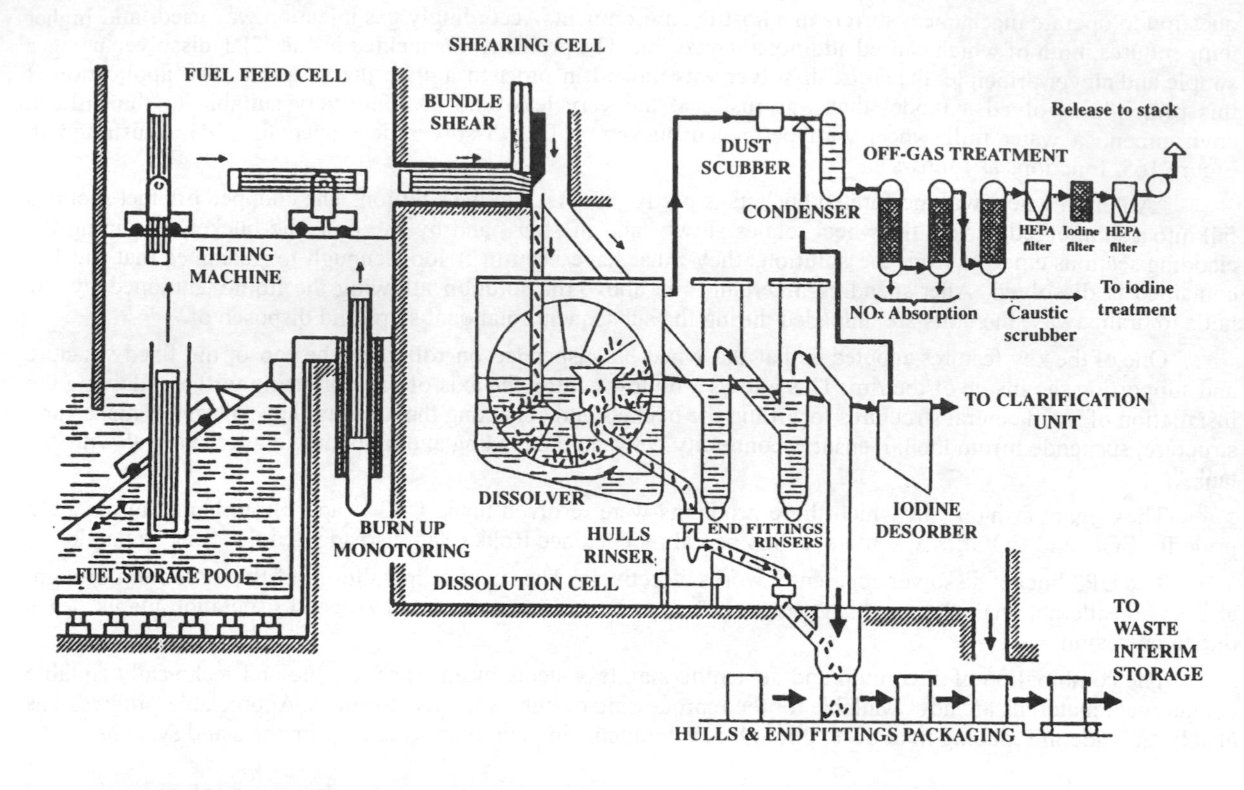

Figure 165. Diagram of the UP3 decladding and dissolving system (Source: PONCELET J. F., HUGELMANN D., SAUDRAY D., MUKOHARA S., CHO A. (1991), p. 95).

Thus a combination of production and radiological requirements were a powerful incentive for technical advances in remote maintenance[94]. It resulted in the development of new procedures and techniques, three examples of which are given below.

At Tokai Mura, a multipurpose maintenance robot was developed for use on the pilot plant dissolver. At UP3 the remote maintenance concept was integrated into the very design of the active parts of the plant. And then in Germany, a full-scale pilot experiment showed that the modular design of the installation planned for Wackersdorf was based upon the constraints of maintenance designed from the outset not to deliver dose to personnel.

Robotisation of the remote inspection of the Tokai Mura dissolvers (Figure 166)[95]

Operation of the Japanese pilot plant had been interrupted on several occasions as a result of contamination caused by corrosion[96]. In April 1982, a leak was found on one of the plant's two dissolvers and a few months later, in February 1983, the second dissolver was affected in the same way. The leaks did not occur in the basic material of the dissolvers, but in the weld metal which revealed "pinholes". A special remote repair technique using a robot had to be developed. This was achieved for the first time in the world in September-October 1983. However it was decided to install a third dissolver, designed and built in such a way as to avoid any welds that might come into contact with radioactive materials.

The robots developed for repairing the two defective dissolvers were subsequently adapted for the remote inspection of the repairs carried out. On the basis of this experience the Japanese developed a second generation of smaller robots with multiple interchangeable heads.

---

94 It should be noted that the second generation plants tested principles that were subsequently adopted in the third generation. At ITREC, the Italian plant resulting from bilateral co-operation with the United States (for reprocessing thorium fuels from Elk River) which was opened in 1971, already made use of the principles of modular design ("racks") and remote manipulation.

95 NAITO S., SUMIYA A., OHTAKA K., FURUKAWA H., TACHIHARA T., OKAMOTO H. (1991).

96 History of the plant in YAMANOUCHI T., OMACHI S., MATSUMOTO K. (1987).

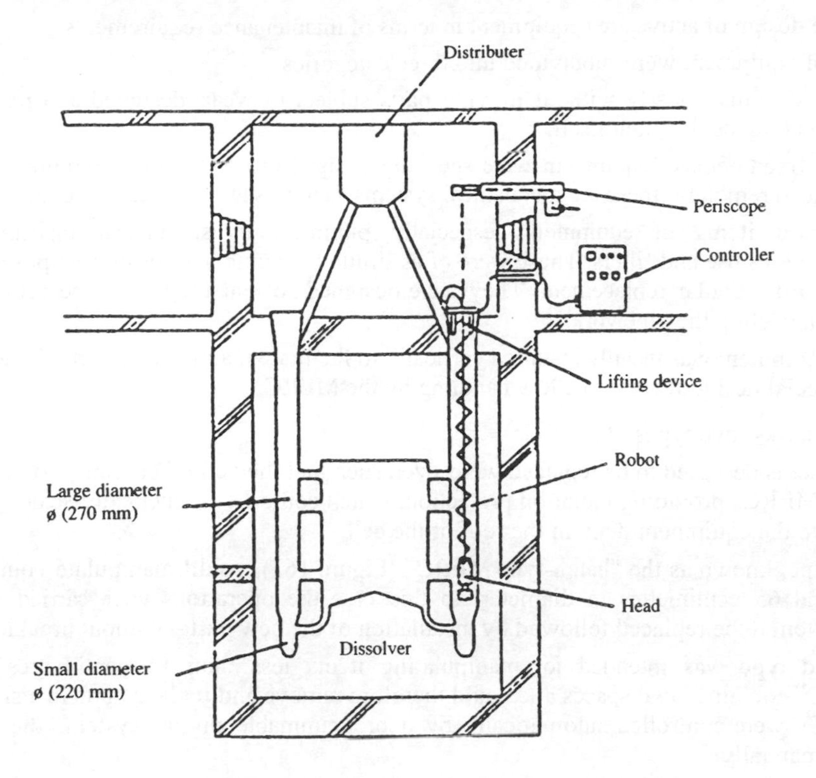

Figure 166. Basic principle of robot inspection of one of the dissolvers at the Tokai Mura plant (Source: NAITO S., SUMIYA A. et al. (1991), p. 161).

Today therefore, the dissolvers at Tokai Mura are maintained by remotely-controlled robots. Current research is concerned with remote maintenance of the robots themselves, which handle most but not all operations. Direct maintenance on the robots is sometimes still necessary.

Research is also being carried out into how to reduce the costs of robotisation, which are still very high, thus precluding in any event the gradual robotisation of an entire plant originally designed for direct maintenance[97].

In fact the "cure" applied at Tokai Mura is more expensive than the "preventive medicine" built into the maintenance design of UP3.

Maintenance at UP3: Standardisation of components and replacement machines[98]

Indirect maintenance: a prerequisite in the plant design

The approach adopted at UP3 was to integrate the idea of remote maintenance for the active parts of the plant at the design stage with the objective of reducing dose. "The very firm internal radiation protection targets applied in the design of these future plants necessitate advanced automation of all operations that carry any risk of irradiating workers together with the systematic adoption of remote dismantling systems for equipment installed on circuits carrying radioactive fluids"[99].

As far as maintenance was concerned, this approach led to two kinds of complementary action being taken at UP3. The first was to standardize the most vulnerable components and their location in the plant. The second was the development of mobile systems known as MERC[100], used for replacing defective active components remotely[101], both during preventive maintenance operations and when foreseeable incidents occurred.

---

97 Indeed there is no provision for the complete robotisation of a new plant.

98 IZQUIERDO J. J., CHAUVIRE P., PLESSIS L. (1991).

99 Extract from the COGEMA house magazine Cogémagazine, quoted without reference by the RGN in its first issue of 1982, p. 77.

100 Mobile Equipment Replacement Cask.

101 The activity threshold was set at 50 mCi/m^3.

Classification and design of active area equipment in terms of maintenance requirements

All items of equipment were subdivided into three categories:

- welded items, mostly without moving parts subject to wear, designed and manufactured to last as long as the plant itself;
- large fixed items of equipment were specially designed to allow their component parts to be replaced remotely (these are mechanical systems such as saws or overhead cranes);
- standard items of equipment, especially pumps, valves, nuclear instrumentation for continuous use and filters. These were of as limited a number of design as possible to allow standard modular replacement. They were designed so that they could be removed without disconnecting the pipework.

Standard equipment was usually installed vertically to facilitate maintenance from the top of the cells[102]. It incorporated a special design feature to allow handling by the MERC.

The maintenance casks: two types of MERC

The plant items designed to be replaced were sized such that they could be transported in one of two types of stainless steel MERCs providing radiation protection, which could be hermetically sealed on the access door immediately above the equipment item, in the roof of the cell.

The first type, known as the "hatch-type MERC" (Figure 167), could manipulate equipment items up to 2 metres high and 65 centimetres in diameter. In this case the operations were carried out in two stages – removal of the item to be replaced followed by installation of the new part – without breaking the cask seal.

The second type was intended for manipulating items less than 15 centimetres in diameter. The "revolving MERC" contained two spaces and could therefore remove and replace parts in a single operation. The hatch-type MERCs were controlled automatically by a programmable digital system, the revolving MERCs being controlled manually.

The advantages of the MERCs were that they obviated the need for workers to enter the cells, thus limiting external contamination, and that they reduced the amounts of secondary wastes such as plastic bags, gloves and rags which direct operations inevitably produce. The equipment removed was placed directly into waste containers which were then concreted (Figure 168).

During maintenance operations at UP3, workers entered the working areas only to operate the rotating MERCs.

Human intervention at Wackersdorf for the purposes of maintenance, had the plant been built, would have been confined to a control room. All maintenance would have been carried out remotely and the active parts of the plant fully modularised.

Wackersdorf (Figure 169)[103] *and the FEMO cells. Plans for a modular plant with no direct maintenance*[104]

The LAHDE-FEMO experiments: a contribution to the Wackersdorf feasibility study[105]

In 1979 a co-operation agreement was concluded between KfK and DWK for full-scale tests on the FEMO-Technik[106], or modular remote manipulation technique, which was the basic concept of the German reprocessing

[102] Except for the steam ejectors. The ventilation system, as in the directly maintained plants, puts these cells at a slight vacuum with respect to the working areas.

[103] With regard to Wackersdorf it must be pointed out that besides the planned arrangements for maintenance, extremely stringent measures were taken not only for providing protection against external attack – which resulted in a project for placing the entire plant in a containment vessel – but also in terms of waste production. The undertaking's own intentions were certainly reinforced by outside pressure. Michel Lung believes that the modular plans for Wackersdorf went beyond what was currently possible technically. Thus the project would have involved a kind of gamble, anticipating the rate of technical development, comparable with that which existed when the basic principles of computer control installed at La Hague were adopted (see below).

[104] SCHEUTEN G. H. (1985), Die deutsche Wiederaufarbeitungsanlage, Vortrag, am 21 may 1985 anlässlich der Jahrestagung Kerntechnik 1985, p. 9-10. Stencilled copy.

[105] PAMELA was another contribution.

[106] "Fernbedienungsgerechte Modultechnik".

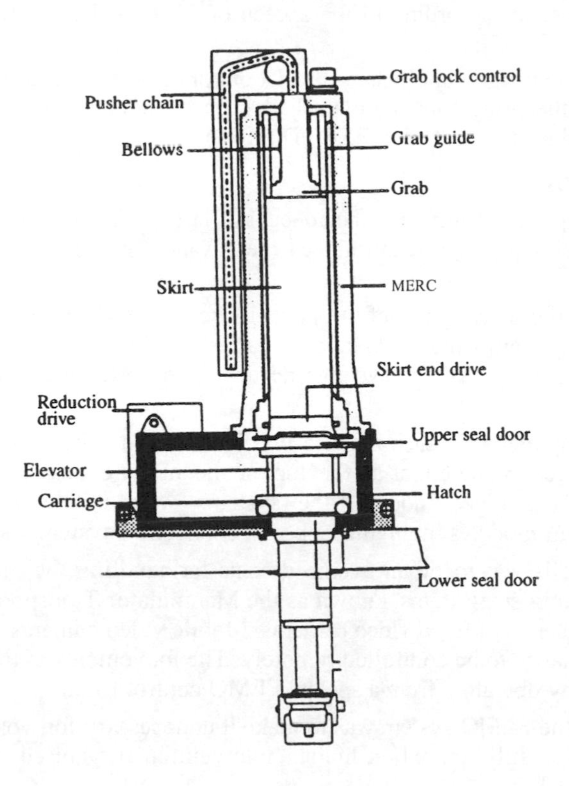

Figure 167. Schematic diagram of a hatch-type mobile equipment replacement cask (MERC) used at UP3 (Source: BETIS J. (1993), p. 273).

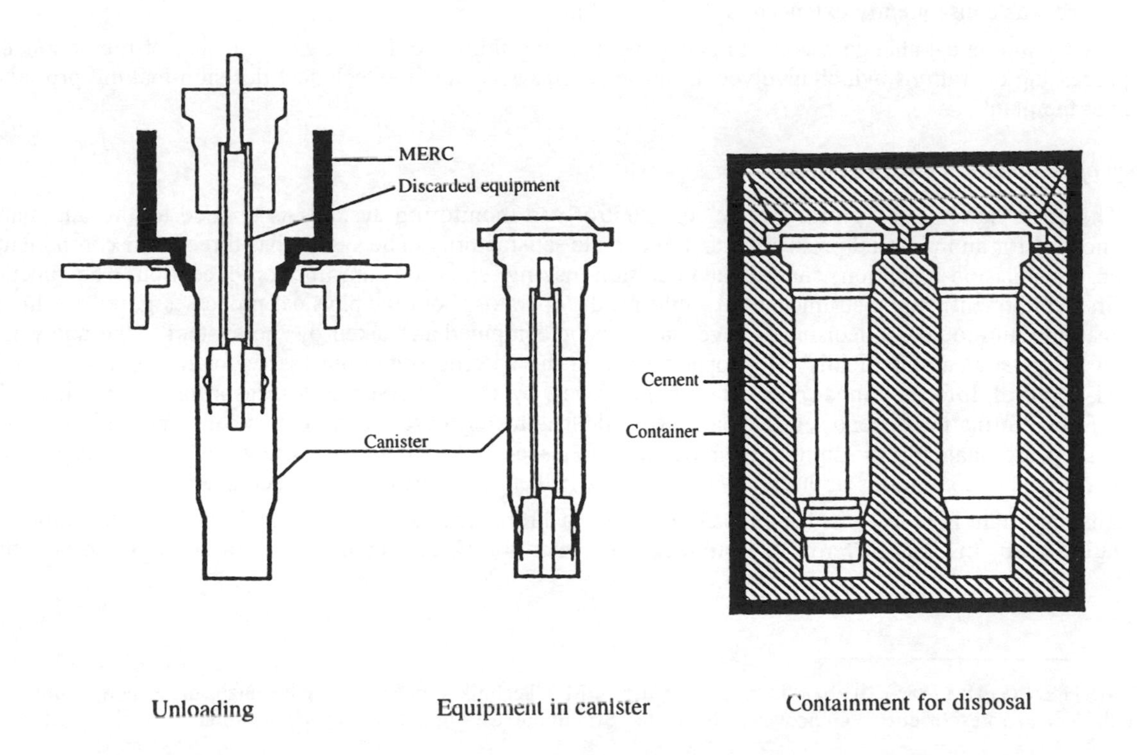

Figure 168. Diagram showing the direct concreting of used equipment collected by the UP3 MERCs (Source: IZQUIERDO J. J., CHAUVIRE P. (1991), p. 1116).

plant. The objective of this system, according to the speech of the President of DWK in 1985, was to reduce operating dose[107].

In 1981 the demonstration programme began in a laboratory at Lahde near Osnabrück. It was completed early in 1990 by which time the project for the plant had been abandoned. The project was reviewed at the Sendai Conference in 1991 and was published in RECOD (1991)[108].

The FEMO system (Figure 170)

The German reprocessing plant returned to the concept of large cells served by an overhead crane used in the first generation of reprocessing plants, but introduced the advances made in remote manipulation and video techniques.

In fact the plan was for the active parts of the plant to be enclosed in two large cells. Eighty modules measuring 3 x 3 metres and 12 metres high[109] were to be arranged along the walls of the two cells, leaving a corridor in the middle between them. All the systems needed for reprocessing were distributed between these 80 modules.

The arrangement of components in each module was based upon criteria of durability and accessibility. Plant items with a short life were to be either at the top of the module or at the corridor side, to facilitate automatic replacement. Within a given module, the systems were to be welded, the connections and disconnections between different modules involving a special connection system known as "jumpers".

Above these two large cells was to be an overhead crane for handling the modules, and a programmable manipulation system for replacing small items, known as the Manipulator Transport System (MTS). The entire unit was to be equipped with a series of fixed video cameras. Mobile video cameras could be located anywhere in the cells to allow plant maintenance to be controlled remotely. The movements of the cameras and manipulation systems were to be controlled by operators from a special FEMO control room[110].

Clearly the objective of the FEMO system was to make it unnecessary for workers to enter the radioactive cells. However the approach was still centred on human intervention, organised into three-man manipulation groups.

The plan was that the remote handling experience acquired with the FEMO cell at Lahde would be used for the decommissioning of WAK which, in any event, would be done differently from Eurochemic where direct maintenance was consequently extended to the dismantling.

The technological change was even more marked for the third problem mentioned, that of the conduct of the reprocessing operations, which involved the information system and the technical decision-making procedure applied at the plant.

History of automatic control systems

The term "automatic control system" or "control and monitoring system" is applied to the automated system needed for an industrial production unit to operate satisfactorily. The system has three main components: emission-transmission-reception; analysis and decision-making; emission-transmission-execution. The function of the first component – the instrumentation – informs the operator about the physical processes going on in the plant, enabling him to take a decision or have that decision computed and taken by the system – the analysis – and then to transmit automatically or by operator action the specific orders necessary for this decision to be executed – control. Information is collected from the process by sensors, essentially measuring instruments. The quality of the information depends closely on the specific features of these instruments, their type and precision, whether they are analogue or digital, their transmission speed, the time lapse between two measurements – whether these are continuous or discrete – and finally whether the information is processed or not.

It is important here to resist the temptation to be anachronistic about sources. At Eurochemic the approach to operating problems was in terms of instrumentation, monitoring and control, not in terms of information

---

107 SCHEUTEN G. H. (1985), p. 10: "Der Einsatz der FEMO-Technik eröffnet den Weg dazu, die schon heute weit unterhalb der vorgeschrieben Grenzwerte liegende Strahlenbelastung des Betriebspersonals noch weiter zu minimieren".

108 LEISTER P., OESER H. R., FRITZ P. (1991).

109 These dimensions were to allow new modules to be produced in the factory and delivered by truck.

110 Only three people were involved in each operation: a "supervisor" and two operators known as the "eye" and "hand".

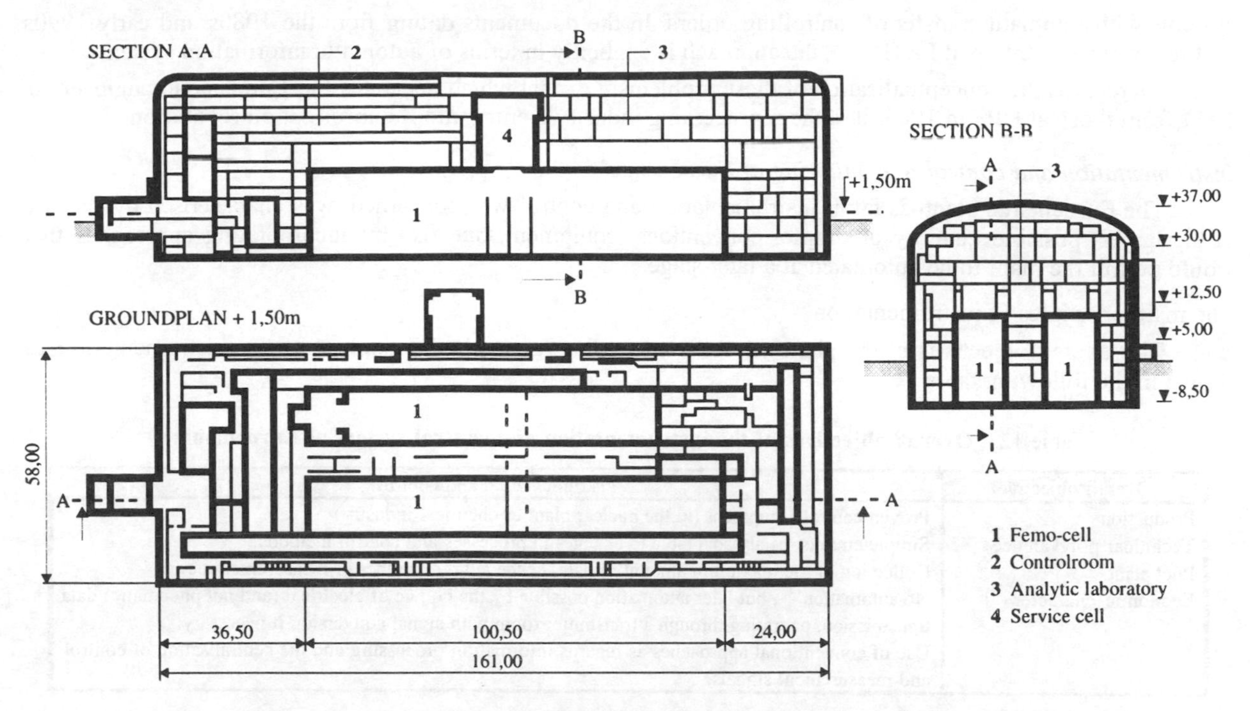

Figure 169. Floor plan and sectional view of the planned reprocessing plant at Wackersdorf. State of the project in 1987. The plant is extremely compact, enclosed in a concrete containment to protect it against crashing aircraft, and is built around two large FEMO cells (Source: HILPERT H. J. (1987), p. 1350).

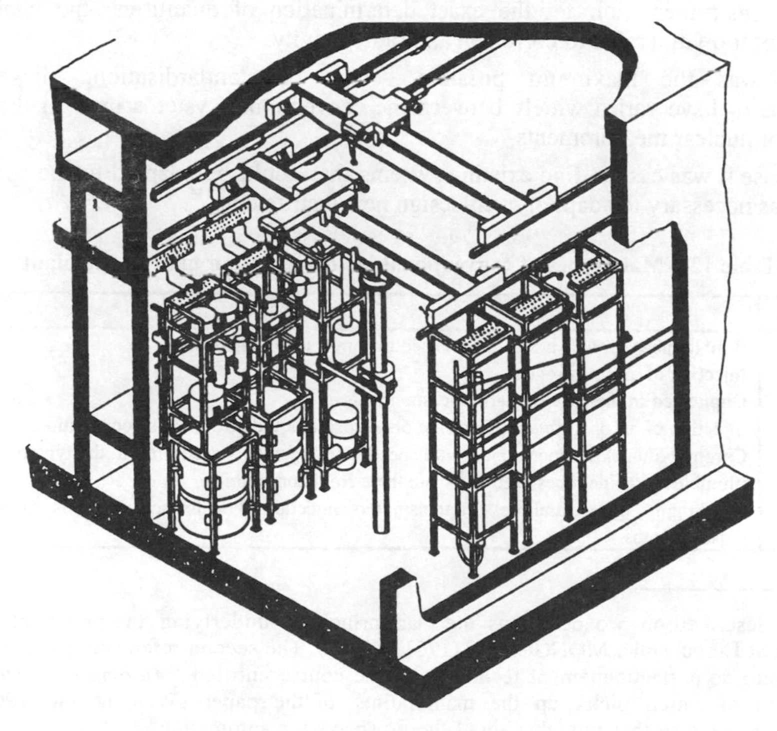

Figure 170. Sectional view of one of the Wackersdorf cells, showing the FEMO modules at either side of the central corridor, served by overhead cranes and MTS programmable manipulation systems (Source: MISCHKE J. (1984), p. 437).

systems with automatic transfer of controlling orders. In the documents dating from the 1980s and early 1990s that report what went on at La Hague, the approach is explicitly in terms of automatic information systems.

As regards the conceptualisation of these problems, i.e., that which separates the Eurochemic engineers of 1964 from those at UP3 in 1983, it is the reprocessing industry's entry into the information revolution.

Instrumentation and control & monitoring at Eurochemic[111]

The Eurochemic plant, as far as instrumentation and control was concerned, was characterised by the use of the largest possible quantity of simple, conventional equipment, but also by incorporating approaches that would permit the plant to be automated at a later stage.

The main principles of instrumentation

The overall objectives of the plant are reflected in the "general philosophy" of the instrumentation as shown in the following table[112].

Table 123. **Overall objectives of the instrumentation and control system at Eurochemic**

Overall objective	Instrumentation and control
Production	Proven control equipment (in the nuclear plant or chemical industry)
Technical polyvalence	Simple circuits, easily adaptable to changes in processes and control methods
Pilot plant	Collection of the maximum amount of data, hence a large number of measurement points
Economic character	No automation[113], but later automation possible by the choice of electrical (and not pneumatic) data transmission, operating through a transmitter room with signal conversion if necessary. Use of conventional approaches as regards information processing and the centralisation of control and measurement signals.

Classical chemical instrumentation for conventional measurements and equipment still at the experimental stage for the nuclear measurement systems

As regards the instrumentation for the exact determination of quantities, the choice fell on simple equipment that was not too vulnerable to corrosion and radioactivity.

An objective was the maximum possible degree of standardisation. However contemporary instrumentation seems to have varied widely between the conventional systems used in the chemical industry and the instruments for nuclear measurements.

In the former case it was easy to find existing systems that could be adapted to the specific requirements. In the latter case it was necessary to adapt or even design new instruments[114].

Table 124. **Main types of conventional instrumentation used in the plant**

Quantities measured.	Instruments
Volume, level and density	Bubble tubes (or subsurface tubes) used in conventional chemistry. Injection of air or nitrogen. Duplicated in the case of alarm measurements.
Pressure	Injection of air or nitrogen (for active or corrosive liquids), otherwise conventional pressure gauges.
Temperature	Chromel-alumel thermocouples, with specific arrangements according to the type of measurements (thimbles or guide tubes used to isolate these from corrosion).
Flow	Diaphragms, differential pressure transmitters, injection of air for active liquids, calibrated air lift for pulse columns.

---

[111] Alain Mongon has described on two occasions the main principles underlying the policy of instrumentation and control and monitoring at Eurochemic. MONGON A. (1963), (1964). The second reference is the publication of a paper given at the Paris Centre de perfectionnement technique in the course entitled "*Automatic Systems in the Nuclear Industry*" the first part of which picks up the main points of the paper given at the 1963 Colloquium on Instrumentation, but with some further thoughts about the prospects for automation.

[112] MONGON A. (1963), pp. 2-3.

[113] This fact is also related to the production requirement since the automation of reprocessing plants was in its very early stages (see below).

[114] SVANSSON L. (1963), Studies of in-line instrumentation at Eurochemic, OECD/ENEA-EUROCHEMIC (1963b), session II, paper g. See also Part III Chapter 3. These developments are not specific to reprocessing but stem from overall changes in the methods of measurement in the chemical industry in the early 1960s, with individual sample analysis being replaced by continuous measurement.

The nuclear measurements fell into two categories: occasional measurements involving the taking of samples, and continuous measurements related in particular to the prevention of criticality and for safety purposes.

Samples were taken automatically – a tapping point being provided on each tank – and the solutions for analysis were fed to airtight sampling boxes where they were bottled for analysis in the laboratory (Figures 171 to 173). The system did not operate very well during the early days of operation.

Continuous in-line measurements were made using special detectors, the main quantities concerned being alpha and gamma radiation, uranium and plutonium concentrations, and the conductivity and pH of the solutions. It took a long time before these instruments gave satisfactory results.

An operating system based upon the electrical transmission of data to synoptic diagrams monitored by human operators

Measurement values from the plant, converted into electrical signals in the transmitter gallery, were fed to five control panels arranged along a single gallery on the sixth floor (Figures 174 and 175)[115]. Each synoptic diagram, i.e. illustrating the main parts of the process, the reagent feeds, the heating and cooling circuits, corresponded to a group of units in the plant:

- decladding, dissolving, clarification, adjustment;
- concentration of fission products;
- first and second cycle extraction, regeneration;
- general services, ventilation, health and safety data;
- final purification of uranium and plutonium, treatment of solvent from the first and second cycles.

Each panel carried small indicators, controls and warning lights actually on the synoptic diagram. The recorders, regulators and larger indicators were arranged underneath the synoptic diagram.

Operators at each panel worked the controls necessary for the process and followed the progress of current operations on the recorders and indicators. Audible and visible alarms – a klaxon and flashing red lights – were triggered "whenever a plant variable reached a dangerous level and required immediate intervention by the operator. These were preceded by early warnings produced by contacts on an indicator or recorder. A continuous white light showed that levels were normal".

The other units at the plant – storage of fission product solutions, final storage facilities, storage of non-active substances, reception and storage of fuel elements – had their own local control panels. A few particularly important signals or alarms were repeated on the panels in the main plant control corridor.

To sum up, part of the control and monitoring system at Eurochemic – the main unit – was centralised geographically with only a few data being transmitted from local panels in other parts of the plant to the control corridor. The plant was also characterised by the separation of unit functions for operational purposes. Human operators therefore played an essential role in the analysis of information, which was routed and centralised for the purpose on the synoptic diagram panels. The operators received some assistance from a system of indicators, but it was they who initiated the procedures through the various controls and who took the decisions as to what action should be taken when abnormal situations arose. Thus Eurochemic was a labour-intensive undertaking in the field of control and monitoring, as well as in that of maintenance.

Provision for automation to come later

In 1963-1964 the designers of the control and monitoring system were fully aware that reprocessing was moving towards greater automation, although at the time this goal was still largely in the minds of the planners.

As Alain Mongon put it in the introduction to the second part of his 1964 paper[116], "the fact that the subject of this conference is automation in a reprocessing plant should give the specialists in the field something to smile about, because automation at present is extremely limited", especially when automation is taken to mean "not only the interlocks necessary – either to protect workers or to ensure satisfactory operation of the process – and the collection of data and their presentation in a form directly usable by the operators, but also the automatic operation of the plant without human intervention".

---

[115] See Figure 174, showing the control and operating corridor in the ICPP (1953, photograph taken around 1958) and Figure 175 showing Eurochemic's synoptic diagram corridor in 1966.

[116] MONGON A. (1964), p. 156 for the first quotation, p. 150 for the second. In this connection see also SVANSSON L. (1963) in OECD/ENEA-EUROCHEMIC (1963b).

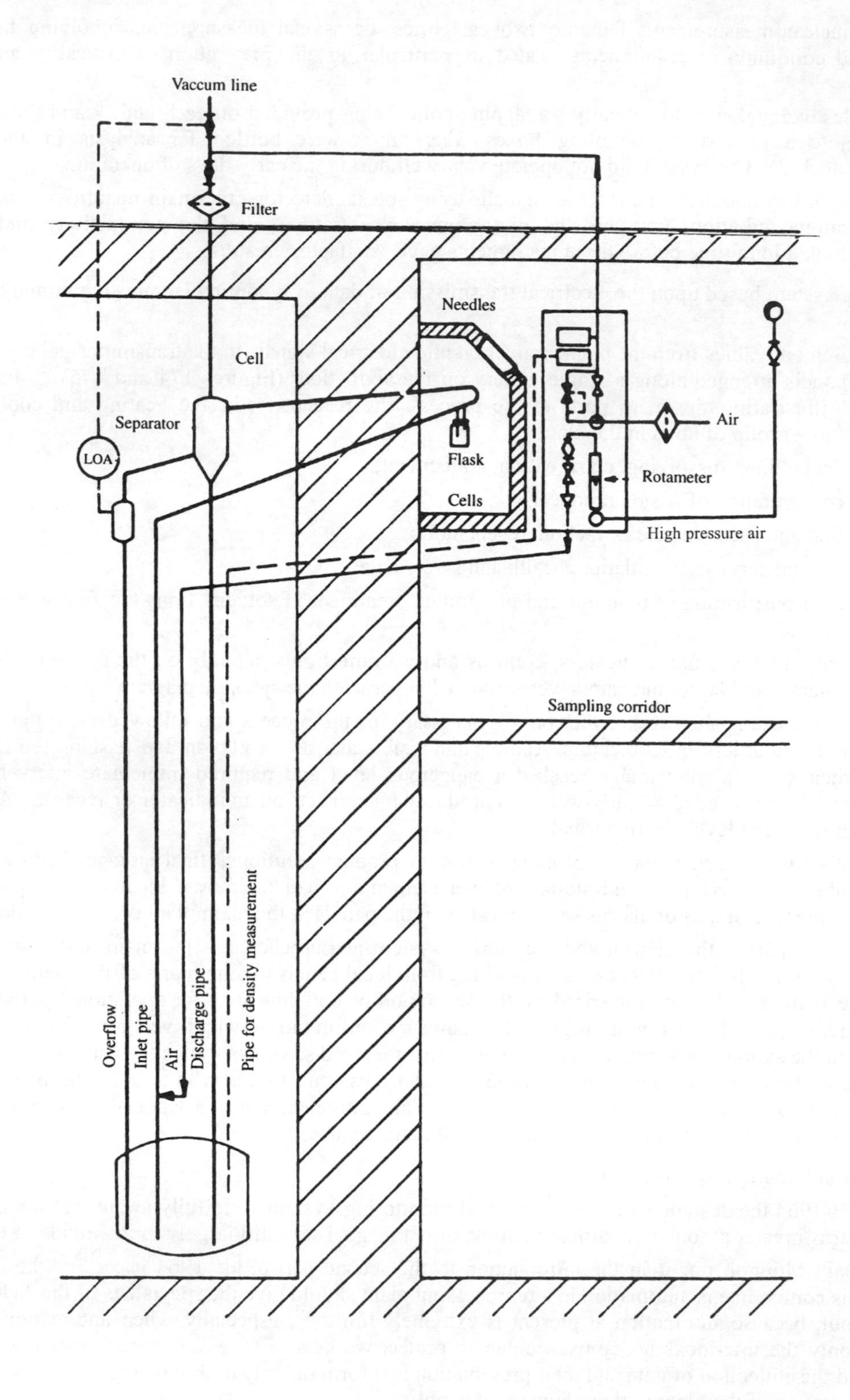

Figure 171. Schematic diagram of the sampling system in the plant (Source: RAE 1 (1963), p. 307).

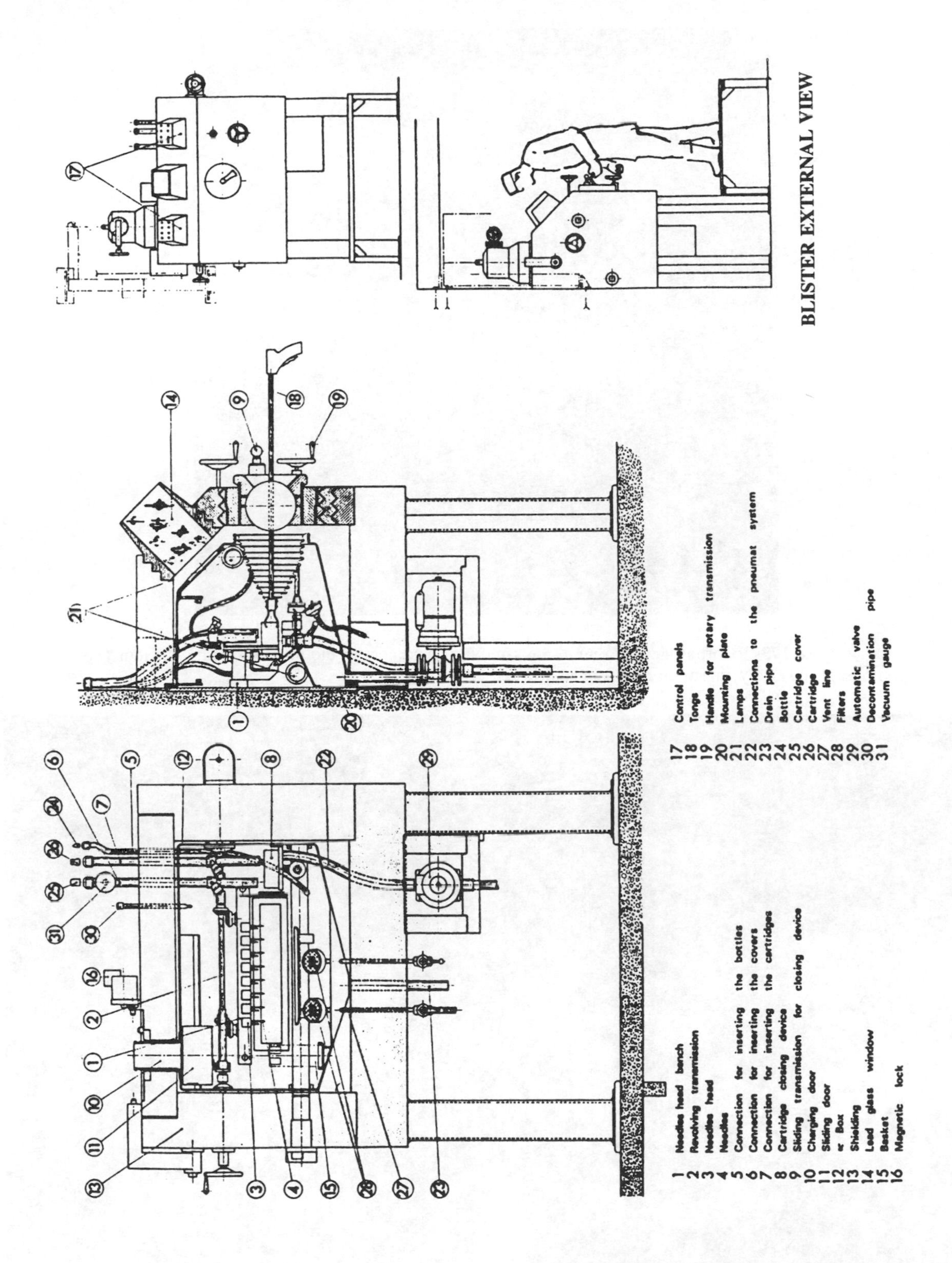

Figure 172. Diagrams of one of the sampling containers in the plant sampling corridor (Source: *Safety Analysis* (1965), Volume of Figures, VIII-9).

Figure 173. External view of one of the sampling containers in the plant sampling corridor (Source: photograph taken by the author in August 1992).

Figure 174. View of the principal corridor in the ICPP treatment building in 1958. Monitoring panels on the left, valve controls on the right (Source: HOGERTON J. F. ed. (1958), p. 94).

Figure 175. One of Eurochemic's control panel synoptic diagrams in operation (Source: Eurochemic slide, no date).

In the latter field in fact "the comprehensive automation of a reprocessing plant [...] is still no more than a fond hope in the minds of the instrumentation engineers, a hope that we want to come to fruition within a few years as progress is made with continuous measuring systems and computers".

Therefore Eurochemic's instrumentation and control and monitoring system left the door open for the subsequent automation of the plant. During the construction phase the intention was that the second extraction cycle would be automated at an early stage. The reasons for this choice were that[117] "to automate an entire unit requires a perfect understanding of the reactions involved in the process, the ability to measure or calculate all the necessary parameters, and operating conditions that are relatively constant and well-defined." The second cycle was in fact sufficiently unlike the stages making up the head-end of the process and the first cycle that were not very "constant and well-defined": this is why it was chosen.

However the changing financial position, as well as the polyvalent and experimental nature of reprocessing at Eurochemic, made it impossible to move on to this second stage. Automation at Eurochemic therefore remained as an "engineer's dream".

UP3: the information revolution reaches reprocessing

Twenty-five years later, in 1989, the UP3 reprocessing plant at La Hague came into service, with a largely automated control and monitoring system. For this stage to be reached however, significant progress had been needed in two fields: one specific to nuclear power and reflecting a continuous form of development, the other a technological breakthrough affecting all of industry.

First of all, the spent fuel being reprocessed was increasingly standardized as a result of the overwhelming success of light water reactors, and more particularly pressurised water reactors using oxide fuel clad in zircaloy[118]. Continuous progress had also been made with special instrumentation. The instrumentation at UP3[119] consists essentially of the same types as that of Eurochemic. The gas bubbling methods are still

---

[117] MONGON A. (1964), p. 157.

[118] Even though the geometrical configuration of the spent fuel sent for reprocessing is today still very diverse owing to the lifetime of reactors.

[119] GOMEZ J. (1983) and BERN J.-B., WAISS Y. (1983).

important for the measurement of volume, density and pressure, and thermocouples for temperatures. For flow measurement on the other hand, more significant progress has been made, leading to the development of new methods[120], such as the GEDEON – a submerged orifice flow generator – or the "measuring disk". Air lifts are no longer used for flow measurement but are used, as they were at Eurochemic, for transferring liquids. As regards on-line nuclear measurements, in what is now known as Nuclear Process Control, the systems have moved beyond the experimental stage of the 1960s, as regards gamma counters, neutron counters containing boron or helium 3 and the alpha analyser known as the ANPu[121].

The other major area of development was of course the revolution in information processing techniques, which should not be limited to the appearance and development of computers, which is to take a narrow view of the "data processing revolution". As we saw, Eurochemic had introduced a computer at the end of the 1960s for "number-crunching" in the analytical laboratory and the fissile material accounts section.

The emergence of microcomputers gave new impetus to the consideration of automatic control, tried out in France for example in the Marcoule pilot plant[122], which was completely refurbished in the mid-1980s and provided in particular with a centralised control room[123].

The emergence of computer networks, together with the materials revolution, the development of maintenance robots and significant advances in sensor technology, ushered in a new technical system, that of the third industrial revolution.

The principles of process control: a review of synoptic diagrams; feedback of experience. Centralisation and integration of information

The approach adopted for process control in the new plants at La Hague[124] was developed and introduced by SGN at the request of the plant owner, COGEMA, by a consortium of engineering companies consisting of the Société générale pour les techniques nouvelles, USSI, Technicatome and Technip. The specifically data processing aspects were dealt with by an Economic Interest Grouping (EIG) known as Engineering des systèmes informatisés d'automatisation (ESIA).

The approach[125] is based upon the centralisation of control and monitoring in a "single control room in a central building also containing the analytical laboratories, the facilities for preparing common reagents and the offices of the operations day shift [...]. This control room also accommodates the centralised radiation protection surveillance system".

The extent of the information problem can be illustrated by the planned list of measurement data to be provided by UP3: 3750 measurement signals, including 875 on a continuous basis, 400 000 digital signals[126], not counting the results of laboratory analyses. The expansion in the amount of information necessary in reprocessing plants had led not only to physical problems but also to operational difficulties.

In physical terms, the Eurochemic system described as "conventional" or "classic" in 1983 sources, no longer seems feasible. The physical connections – wiring, compressed air lines, and so on – would be too numerous and not flexible enough (any change to the process means that the whole system has to be rewired), and the indicator panels would become too long.

As far as operations are concerned, the "cascade" effect was one of the causes of plant shutdowns that lowered productivity[127].

---

[120] BERN J.-B., WAISS Y. (1983), p. 404-405.

[121] BERN J.-B., WAISS Y. (1983), p. 406-407.

[122] Commissioned in 1960.

[123] See MUS G., LINGER C. (1987). We may give some extracts from the abstract of the article, p. 1519: "The operating system has been completely recast: the control panels which had been situated locally near the different units have been assembled in a centralised control room; for this purpose, the measurement and actuation circuits have been replaced by new data acquisition and processing systems involving the use of numerical algorithms. Also certain units [...] are now managed and controlled by programmable automatic controllers. Finally, certain operations [...] are fully automatic".

[124] This is described in a series of articles published by RGN in September-October 1983, under the general title "*Contrôle-commande et instrumentation dans les usines de retraitement*", with an introduction by PEROT J. P. (1983), two articles on conventional instrumentation, GOMEZ J. (1983) and nuclear instrumentation, BERN J. B., WAISS Y. (1983) respectively, and two articles of which one covers the operating system, SILIE P. (1983), the other the control room, PIGNAULT J. (1983). All the articles were produced by SGN, principally the Operations-Instrumentation Section, part of the Process Department, itself included in the Process-Safety-Testing Division.

[125] PEROT J. P. (1983), p. 400-401.

[126] Of the ON-OFF type.

[127] SILIE P. (1983), p. 411.

"A 'cascade' effect arises from the simultaneous presence of a number of signals (sometimes a very large number), some of which serve no purpose, the others being induced by one or two important original signals. This leads to difficulties of interpretation, which can be risky for the operator, because faced with an abnormal and generally new situation in the status of the installation, he has to deduce what was the starting point of the installation [and] the fault which, through the dynamics of the process, produced the situation observed. Faced with this problem, as the process continues to develop and give off increasing alarms, the operator will shut the system down".

This kind of shutdown was recorded at Eurochemic. According to Emile Detilleux, the recorder charts were unwound on a table in the control room and quietly examined in an attempt to interpret the problem that had occurred. This kind of procedure, quite normal in a pilot plant, would be unacceptable in an industrial production unit.

Process control at UP3

"The development of digital technologies, with the advent of microprocessors, made it possible to envisage a digital control system for plants reprocessing irradiated fuels"[128]. This technological breakthrough resulted in the construction of an operating architecture on three levels, the overall approach being to process the information at each level in order to simplify and combine the data transmitted to the next higher level (Figure 176).

The immediate operational tasks were handled by automatic processing systems located in the plant and as close as possible to the sensors and actuators[129]. These systems transmit signals that have already been combined and receive instructions through the communication system located at the centralised control stations.

Each plant unit or group of units is operated through a "unit control computer" and the entire plant by a "supervision computer". Signals are combined by the machine and presented to the operator according to process requirements and no longer merely as a function of the physical reality of the installation as was the case with "conventional" synoptic diagrams. In practice what the operator sees on his console are functional diagrams or flow charts displaying synthesized information in binary form[130] which enable him to analyse the process with no danger of being submerged in information. He has access to a second level of display showing the detail of the installation "organised around conventional synoptic diagrams" if necessary.

All the centralised control stations are grouped in a single control room covering 1800 m^2, adjacent to the computer rooms and designed so that about 50 operators can work there on a shift basis[131].

Whereas at Eurochemic the control and monitoring panels were designed solely according to the technical requirements of the plant, the UP3 control room seeks to reflect, in the spatial layout of a computerised control room, many concepts of the organisation of work and the man-machine interface, all of which were embryonic at the time of Eurochemic. As summed up by J. Pignault[132], "the problem that the design study must resolve can [...] be put in the following way: having defined the processes to be used, find the organisation of the tasks and the technologies to be implemented whereby each operator will have a clear area of responsibility and an adequate workload in every phase of operation[133]; and thereby deduce the characteristics the control room should have".

La Hague and the Total Data Management System (TDMS)

However the objective of computerising reprocessing at La Hague went beyond process control, being to centralise all information in order to permit the real-time monitoring of changes in a large number of parameters and the integrated processing of all the data collected.

---

[128] SILIE P. (1983), introductory chapter p. 408.

[129] Two different kinds of processing system are used according to whether the information is in analogue or digital form, see SILIE P. (1983), p. 408. The analogue processing systems, exclusively in the chemical part of the plant, are able, through the presence of the microprocessor, automatically to correct the part of the process of which they are in control by following algorithms. The control process itself can be based upon multiple variables, whereas previously it used only one. The operator need then concern himself only with the detail of the settings.

[130] Of the ON-OFF type.

[131] PIGNAULT J. (1983) describes the general design methodology and the preliminary plans for the control room. Programmed visits to the site take in the UP3 control room. COGEMA allowed the author to visit the control room in operation on 25 May 1992.

[132] Ibid., p. 412.

[133] This idea is described earlier on the same page as the "criterion of flexibility whereby a team designed for routine operation can handle, with no significant increase in workload, the operation of the installations during startup, shutdown or in a degraded situation".

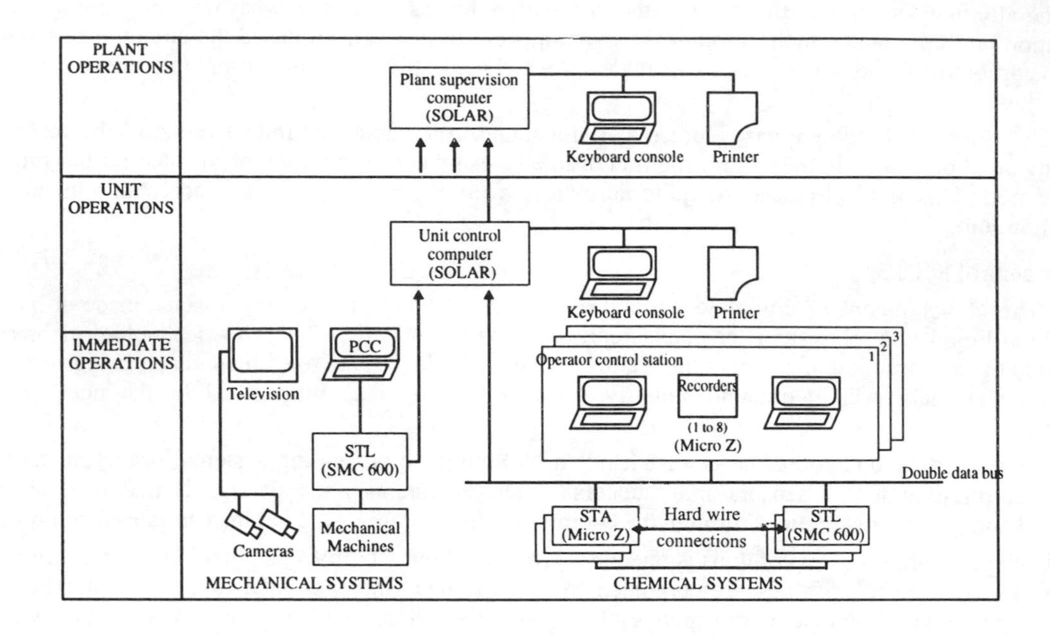

Figure 176. Schematic diagram showing the general organisation of the UP3 control systems (Source: SILIE P. (1983), p. 410).

Thus by the end of the 1980s the La Hague site received a Total Data Management System (TDMS), which manages the so-called HAGUENET network[134] over the entire site, which was commissioned during the summer of 1987[135].

The network carries data from the plant[136], covering the requirements of both operations and maintenance[137], technical documentation[138], personnel management, both traditional and in respect of radiation protection data[139]. The system also encompasses all the data processing documentation on the site[140] and the procedures for selective access to information.

The expected advantages, quite apart from making possible the rapid and frequent real-time processing of information itself more reliable in being emitted directly from the source systems, are improvements in both productivity and safety.

The conceptual advance[141] by comparison with Eurochemic as illustrated by the published sources is substantial, justifying the claim that this new technical generation evolved in less than one human generation.

---

[134] This is a network of the Ethernet type, the hardware consisting of 17 kilometres of optical fibres and 17 kilometres of coaxial cable linking 1600 nodes.

[135] Paper given at the RECOD Conference held at Sendai, Japan, in 1991, in CHABERT J., COIGNAUD G., PEROT J. P., FOURNIER W., SILVAIN B. (1991). See also CHABERT J., ROULLAND M. (1991).

[136] TMDS does not include the process control systems.

[137] As regards plant operations there are data relating to fissile material accounting, the analytical laboratory (150 000 sets of data a year), waste management, and inspection patrols. For maintenance, it is an expert system providing diagnostic help and a computerised spare parts management system known as SELAMO.

[138] Library of a million documents, used for computer-assisted design and for maintenance.

[139] 800 000 measurements a year.

[140] Software Maintenance Centre or SMC.

[141] It would be appropriate to supplement the conceptual study by a review of what actually happens on the plant.

Reprocessing and its human operators. Professional generations, the enterprise culture and the Eurochemic network

The final part of this review chapter begins with two simple questions.

In the 1950s reprocessing was a new activity in Europe. The engineers and technicians who designed, built and ran Eurochemic had not originally been specialists in reprocessing because there was no such thing. Hence it is not devoid of interest to see from where they came.

One function of Eurochemic was to train specialists. The company discharged this function both during its active life as a pilot reprocessing plant and during the waste management period. Reprocessing came to an end after eight years. It may be asked what became of the men[142] who worked there, after they left the company.

An exhaustive quantitative enquiry was neither possible nor desirable. However a qualitative assessment was done on the basis of scattered sources and information obtained during interviews. Use was also made of the list of participants at the colloquium on the "Eurochemic experience" held in June 1983 at BCMN in Geel.

Sources of expertise: Eurochemic staff before Eurochemic

The two principal sources of Eurochemic's researchers, engineers and technicians were the nuclear and chemical industries.

At the time, specialists with a foot in both camps, i.e., in the chemistry of radioactive materials, were in fact very scarce. Here Eurochemic was of course able to benefit from French know-how, since that country was the only one with specialists in reprocessing such as Yves Sousselier, André Redon and Michel Lung. The company also drew upon the resources of the Nuclear Chemistry Department of the Mol CEN, for example for Emile Detilleux. Dutch-Norwegian co-operation at Kjeller also provided a nucleus of specialists such as Teun Barendregt.

For most of these specialists, except for the French who were already experienced in building a large plant, the problem was to move from the level of the laboratory or very small unit to the industrial scale with its problems of commensurate size. We have already seen how important was the virtually continuous presence on the site of an American representative. Here it is worth stressing how important it was for a substantial number of these specialists to attend training courses in the United States, either for initial training – like Emile Detilleux at Brookhaven – or during the early stages of the project. A number of senior staff shared the experience of a learning trip to the United States.

However in the early 1960s there were very few specialists in nuclear chemistry, and the plant was built with researchers retrained from other areas, particularly chemistry, attracted by both the intellectual and career prospects in this new field, regarded then as having a vigorous and bright future. We have seen how much interest all the major chemical firms took in reprocessing at the time, with the seminal role of Du Pont de Nemours and that of ICI in the United Kingdom. The major contemporary names in European chemistry such as Saint-Gobain, Hoechst and Montecatini were present at the birth of Eurochemic. Chemists were heavily involved in the project, coming mainly from petrochemicals and pharmaceuticals.

Petrochemicals was the field of Teun Barendregt before he went to Kjeller; also Yves Sousselier graduated from the French École de chimie du pétrole. In fact there are quite a few links between petrochemicals and reprocessing. TBP extraction involves hydrocarbons, and petroleum engineers are accustomed to separation operations, even though a catalytic cracking tower and a pulse column have few points in common. Lubricating oils were refined by solvent extraction[143], and the oil industry used either mixer-settlers or extraction columns.

A typical example from the other branch of chemistry – pharmaceuticals – is the career of Rudolf Rometsch who, before going to Eurochemic, was a researcher in the laboratories of the Swiss chemicals and pharmaceuticals firm CIBA in Bâle where, as a physical chemist, he was responsible for introducing physical methods into chemistry research. He came into reprocessing as a result of two factors: first he had research experience with separation methods, particularly with counter-current extraction columns; secondly he had already had contacts both with the United States and with the nuclear field via the problem of licences for pharmaceuticals issued by the American Food and Drug Administration (FDA) which used methods of nuclear analysis in its laboratories.

---

[142] Primarily men, women having always been in a very small minority and restricted to office tasks, remote from the radioactive areas.

[143] Communication from Yves Sousselier, who worked in this sector at the Port Jérôme refinery by the Seine below Rouen when he was introduced by chance to Bertrand Goldschmidt, who was in fact looking for specialists in this type of operations.

After the initial period, the professional origins of people working at Eurochemic became more diverse. Training courses in nuclear chemistry became more frequent in the 1960s and students came to Eurochemic for two to three years to perfect their practical knowledge. The comings and goings amongst the senior staff caused management problems. Arrivals and departures were so frequent that the continuity of the research sometimes suffered. The progressive "Belgianisation" of the project certainly tended to stabilize the research teams, but at the expense of the training side. However the special position of Spaniards in the stable nucleus of staff deserves explanation. Their fairly strong presence was explained by the political situation in Spain during the 1960s and the early 1970s. It appears that most of the Spaniards at Eurochemic were opponents of the Franco regime. During their long stay in Belgium, some married local girls so when the Generalissimo died they did not return home.

Hiving off expertise: the Eurochemic staff after Eurochemic[144]

It was interesting to enquire what professional careers Eurochemic senior staff took up after leaving the company. Clearly the factors involved were many and varied and depended on individual, national, economic and other characteristics. The fact that most of the member countries did not subsequently set up their own national reprocessing plant did of course tend to amplify the extreme diversity of careers followed. However a certain logic can be identified from those professional careers that have been analysed.

Former Eurochemic staff can of course be found in their national reprocessing projects, both in Germany and in Italy, and most of the French reprocessors returned to that field at home.

More generally former Eurochemic staff took up posts in the agencies, national nuclear centres or nuclear firms who were not involved specifically in reprocessing, such as the Karlsruhe centre in Germany or that at Würenlingen in Switzerland. The waste management field is well represented: in Spain at ENRESA, in Switzerland at CEDRA/NAGRA or in Belgium at ONDRAF/NIRAS owing to the causal links between the two activities.

These career choices can be seen as following a sectoral approach, closely related to one of the company's objectives.

For a number of people, the stay at Eurochemic was just one step – sometimes the first – in a career in the international nuclear organisations. There are former Eurochemic staff can be found at OECD/NEA, at EURATOM and in its research establishments such as the Karlsruhe Transuranium Institute, the BCMN at Geel, and at the IAEA where Rudolf Rometsch recruited a number of his former Eurochemic colleagues into the Department of Guarantees he headed, which was responsible for developing and applying the inspection system under the NPT. They are also to be found in the international nuclear firms such as EURODIF, URENCO and Transnuklear.

Also some people left the nuclear field altogether or entered firms for whom the nuclear field was marginal, such as Comprimo. Some returned to their pre-Eurochemic origins, to chemistry at Rhône-Poulenc or Motor Columbus (now Colenco) or the Spanish fertilizer industry.

For some the link with the nuclear field is tenuous or nil. For example one former Eurochemic staff member has become the Director of a thermal power station and another the Sales Director for a major motor firm.

The Eurochemic "network"; the Eurochemic spirit[145]

Accordingly the end of European reprocessing led to a scattering of the people and skills built up in a number of sectors. However the very strong feeling of belonging to a community was not extinguished: on the contrary. Former staff call it the "Eurochemic family", "old Eurochemics", the "Eurochemic mafia". The "Eurochemic spirit" left an indelible mark on many of those involved in what they quite openly call the "Eurochemic adventure". There is no doubt that amongst the senior staff, Eurochemic generated an "enterprise culture". Most of those contacted stressed the bonds that persist. A great deal of information is exchanged about what has happened to former staff, and there are many letters and telephone calls.

The network built up by former Eurochemic staff has already been pointed out in the description of careers. It can be precisely located both geographically and institutionally by the photograph taken in 1983 at a major meeting of former staff at a time when the end of the company seemed close. It involved 150 people from the institutions shown in Table 125[146].

---

[144] For reasons of confidentiality, names are excluded.

[145] The relevant sources, apart from the 1983 Conference, are essentially incidental oral communications.

[146] The activities shown are those stated by the participants.

Table 125. **Affiliation of participants in the Seminar on the Eurochemic Experience held at Mol in June 1983**[147] (countries shown in decreasing number of participants; the number of people – if more than one – shown in brackets)

Country	Nuclear agency, nuclear centre or ministry	Company	Other and remarks
Belgique (40)	CEN/SCK (7), Department of Energy (4), Ministry for Public Health (2)	Belgonucléaire (11) Comprimo Belgium (8) EBES (2) UEEB (2) CORAPRO, SYNATOM, FBFC, Socertec, BBL	
Germany (34)	GfK (6) Ministry of the Interior, Ministry for Energy, BMFT	WAK (6) DWK (5) Uhde (5) Nukem (2) Gesellschaft für Reaktorsicherheit (2) KEWA, Kraftanlagen AG, Transnuklear, Quadrex	One private individual
International (30)	EURATOM (3) AEN/OCDE (2) Nordic Liaison Committee for Nuclear Energy	Eurochemic (19), URENCO (2) BCMN (2), European Transuranium Institute	
France (13)	CEN-FAR (3) CEA Marcoule (2), Council of State	SGN (4) USSI, Sofratome, Jacomex	
Italy (7)	ENEA (5)	EUREX Saluggia (2)	
Switzerland (5)	EIR (2) NAGRA	Motor Columbus Consulting Engineers (2)	
Denmark (4)	Risø National Lab (2) Danish Energy Agency,	Skaerbaekvaerket	
Netherlands (4)	Ministry of Housing, Energiecentrum Nederland (Petten)	Nucon (2)	
Japan (4)	PNC (3)	JGC	
Norway (3)	Institute for Energy Technology (Kjeller)	Norsk Hydro Research Center	One private individual
Sweden (2)	Studsvik Energiteknik AB	Swedish Nuclear Fuel Supply Co.	
Portugal (1)	Ministry for Industry and Technology		
Brazil (1)	NUCLEBRAS		
Austria (0)			
Spain (0)			
Turkey (0)			

[147] ETR 318, list of participants.

A further meeting took place in 1990 and another was held in June 1996 at La Hague.

These meetings provide opportunities not only for swapping memories but also for informal discussions and consultations about current problems with which some of the members are still involved.

This solidarity reflects the enterprise culture developed during 35 years of international co-operation. Former Eurochemic staff stress the difficulties as well as the pleasures of working with people of different nationalities. There were obviously language barriers, the answer to which was "Eurochemic English", but also difficulties arising from different ways of looking at problems. They all stress the delight of working together to find compromise solutions. In general the period they spent at Eurochemic has left them with very pleasant memories in both human and professional terms.

They regret that a European reprocessing undertaking never emerged. However they remain convinced that the international approach is still the best one for finding solutions to nuclear problems and so deplore the nationalistic approach that presently dominates.

These attitudes are the direct consequence of the feeling of belonging to an exceptional group in the history of reprocessing. One can also see in it the solidarity of a small group which was exposed – as "nucleocrats" and reprocessors – not only to attack from the antinuclear movements but, as participants in international co-operation, to a degree of contempt for this type of experiment from inside the nuclear world. It also demonstrates that the international experience was unique, with a varied outcome.

Thus Eurochemic is typical of a certain view of R & D and of international co-operation which, in many respects, belongs to a past that is irretrievably gone. However some features of international co-operation that were applied there may well yet have a future.

Chapter 2

Eurochemic and European nuclear co-operation
The economic and political perspective

Following our attempt to evaluate the technical inheritance of Eurochemic, it is important to consider its position in the institutional genealogy of European nuclear co-operation. In view of the many organisations and approaches used for international co-operation, this chapter will concentrate on what are considered to be the leading experiments in multilateral technical co-operation, i.e., those involving at least three partners and focused upon one technical object. The latter might be a large and essential research instrument, for example a particle accelerator and its auxiliary systems at CERN, an organisation which predates the Eurochemic project. The technical object might also be a production facility, for instance an industrial pilot such as Eurochemic, an enrichment plant like URENCO and EURODIF, or indeed fast breeder reactors for generating electricity such as NERSA or ESK. Notwithstanding this deliberate limitation of the field to be covered[1] we shall not exclude the background to the development of European international co-operation, particularly as regards the different sectors of the Community – ECSC, EURATOM and the EEC – or what is known about technical co-operation in other fields such as aviation and space[2].

The approach involving an international shareholding company governed by an international Convention has not been repeated since 1957 and Eurochemic is still the only "joint undertaking" of this type. However multilateral technical co-operation on a development project in the field of nuclear technology has taken other forms. It is important to determine why these developed, to consider the reasons for their differences and to try to discern some historical pattern in their emergence.

The morphological approach attempted by the lawyers does allow a first observation.

In May 1981 Pierre Strohl gave a paper to the Nancy Conference organised by the Société française pour le droit international on the "Unity and Diversity of Europe in International Relations"[3]. In an appendix he offered a schematic table setting out a selection of 15 international co-operative undertakings on technical subjects in the nuclear field, space, communications and air and rail transport, ranked in decreasing order of the extent of their integration in international public law. The part of the table covering the nuclear undertakings is given in Figure 177. Consideration of only the nuclear co-operative undertakings in the table, the construction of which was motivated by a legal type of approach, and their chronological listing between 1953 and 1973, suggests that there is a link between the date the seven undertakings mentioned were established and the extent to which they were integrated into international law, except as regards Halden and Dragon.

The earlier they were established, the stronger appears their international integration. Thus Eurochemic appears to be the institution most closely bound up with international law, second only to CERN.

This observation raises a number of questions that make it easier to understand the Eurochemic experience. For example, why did the 1950s appear to facilitate the development of highly integrated regional structures in the nuclear field? Why was there a decline in the 1960s and the early 1970s? Did the Eurochemic experience play any part in this decline? Did this trend continue after 1973? To what extent was the experience gained by Eurochemic taken up in the new forms of co-operation, and to what extent can lessons be drawn for the future of international co-operation in the nuclear field?

---

[1] Bilateral co-operative ventures are not included, even though the Franco-Belgian organisations SENA and SEMO, which manage the power plants at Chooz A and Tihange respectively, are joint undertakings of EURATOM. Multilateral co-operative ventures involving reactors, such as JEEP at Kjeller, Halden or Dragon – two other joint undertakings of the ENEA – have also been omitted owing to their low financial significance.

[2] See KOENIG C., THIETART R. A. (1988), KRIGE J. (1993a and b), CHADEAU E. dir. (1994).

[3] STROHL P. (1981), Appendix 2 and § 65 for an explanation of the classification used.

Date of creation	Name of undertaking	Purpose	Types of members	Legal status	Constitutive basis	Relevant legislation	Method of funding	Control by governments	Privileges and arrangements for settling disputes
1953	CERN	Fundamental research	Governments	Intergovernmental organisation Council	Treaty	Treaty and international law	Government contributions	Members of the Council	Usual privileges and immunities of an international organisation. International Court of Justice
1959	EUROCHEMIC	Construction and operation of a reprocessing plant, related R & D	Shareholders: governments, public and semi-public organisations, private companies	International shareholding company. General Assembly and Board of Directors	International Convention with Statutes annexed thereto	Convention, Statute and Belgian subsidiary law	Company's capital, government contributions, loans and receipts	Special Group of government representatives (OECD/NEA)	Immunities and privileges, Exemption from taxes and customs duties, Conciliation by the Special Group European Nuclear Energy Tribunal
1967	Institut Max von Laue Paul Langevin	Construction and operation of a high flux reactor, related R & D	Public research institutions, semi-public company	Non-commercial "Société civile" under French law. Steering Committee, Scientific Council	Treaty, private association agreement, registered in France	Statutes and French law	Government contributions, site made available free of charge by host organisation	Associates appointed by the government, appointment of members of the Steering Committee and changes to the Statutes, subject to approval	Facilities for recruiting staff, negotiations and arbitration between the governments concerned
1970	URENCO/ CENTEC	Industrial operation of the gas centrifuge enrichment process, related R & D	With government support, government agencies and commercial companies appointed by their governments as shareholders in the joint industrial undertaking	URENCO Ltd, company under English law (holding company providing marketing services), CENTEC GmbH, company under German law (execution of the R & D programme), Production companies: URENCO UK (under English law) URENCO Nederland (under Dutch law)	Based upon the Treaty of Almelo, formation of companies under the relevant national law	General provisions of the Treaty, national legislation	Company's capital, enrichment contract receipts, government contributions	Joint committee of government representatives exercises overall controll	No priviliges; arbitration procedure
1973	EURODIF SA	Construction and operation of a gaseous diffusion enrichment plant, associated R & D	Public and semi-public organisations, private companies (wholly or partially owned by governments)	Shareholding company under French law, Assembly of shareholders, Supervisory council, Board of Directors	Usual procedure for setting up a French company, articles of association	French law, articles of association (until 1980; since then, intergovernmental Convention dated 20 March 1980)	Company's capital, enrichment contract receipts	No special provisions	In the 1980 Treaty: specific tax exemptions. Arbitration under the rules of the International Chamber of Commerce
1958 and 1959	DRAGON and HALDEN (under the auspices of OEEC/ENEA and then OECD/NEA	R & D on experimental reactors	Governments, public organisations, private companies, Commission of the European Communities	No legal personality. Host organisation acting on behalf of the participants. Board of Directors representing the participants.	Private law agreement	Legislation of host country (United Kingdom and Norway)	Contributions by participants	No special provisions; regular reports to ENEA and NEA	No special provisions

* *Prepared from STROHL P. (1981), Appendix 2*

Figure 177. International organisations for scientific and technical co-operation in the nuclear field.

Consideration of whether Eurochemic was a success or a failure[4] can suggest an initial hypothesis. The success of the project shown by the signature of the Convention, the ratification of the Statutes enabling creation of the company, and the actual start-up of a reprocessing plant, must be considered against the general background of the development of international co-operation in the 1950s.

The collapse of international co-operation in reprocessing as seen in terms of the decision to halt reprocessing at Eurochemic is linked at least in part to changes in this background, which can be seen in the existing institutions and through the new forms of organisation created in the early 1970s.

However there is no reason why the approach used at Eurochemic should not serve as a basis for consideration of the co-operation that is necessary today in certain parts of the nuclear field and elsewhere.

The reasons for the success of the project

The project's success, as shown by the facts that the Convention was signed and ratified and the company set up, was based on a combination of favourable factors.

Eurochemic can be seen as the product of the golden age of European co-operation during the 1950s. Co-operation in reprocessing was made possible by the conjunction of special diplomatic and economic conditions, and the establishment of the project was facilitated by the institutional framework of its development and by the relative simplicity of contemporary decision-making.

Eurochemic benefited from an international context favourable to strong integrated co-operation

Alan S. Milward[5] and other historians of European economic integration such as Werner Bührer[6], who attempted to define the nature of European economic co-operation at the end of the 1940s and the beginning of the 1950s, emphasized that there was no contradiction at the time between a high degree of integration, expressed for example in the European Coal and Steel Community (ECSC) Treaty by the supranational powers of the High Authority, and the desire to affirm national power. For example the government of the Federal Republic of Germany saw the ECSC as a way of avoiding the humiliation that straightforward internationalisation of the Ruhr would have represented and, for the German steelmakers, it provided the way for rebuilding international markets. Integration in the OEEC helped speed up the economic reconstruction of Europe. Alan Milward examined how integration into the EEC had affected the "small countries" (Belgium, the Netherlands and Luxembourg) and stressed that in his view this policy had even saved these countries seriously weakened by the consequences of the Second World War.

In other words, although the integration movement received powerful impetus from the fact that the countries of Europe were relatively weak, it did not take place against their will but was rather to serve their interests. The driving force of European construction was apparently not the "European ideal" but the need to co-operate to overcome the post-1945 weakness of the individual countries.

Referring to the relative weakness of European nations as a factor to explain the multiplication of European co-operative organisations during the 1950s is particularly appropriate in the nuclear field[7].

The weakness was primarily economic and financial, when one considers the cost of major facilities, beginning with those of CERN. Financial problems lay behind the offers of French co-operation in the field of enrichment. Indeed weakness was not only a feature of the public organisations but was shared by some of the companies.

There were also scientific and technical shortcomings, since no research had been done during the war, at least in continental Europe, and because of the intense efforts needed to develop research during the period of reconstruction, owing especially to the holding back of information until the advent of the Atoms for Peace policy. As in the economic field, the creation of organisations for nuclear co-operation benefited from support from across the Atlantic, subject to a certain delay due to the rate of change in the general international situation.

---

4 As Hans-Joachim Braun points out with regard to the criteria for assessing technological failures, "...it is impossible to talk about 'success' or 'failure' in objective terms, and [...] the question 'success or failure for whom?' has always to be asked", BRAUN H. J. (1992), p. 216. One might also add the question of when the success or failure was evaluated.

5 See MILWARD A. S. (1984) and the papers given at EHESS in 1990-1991.

6 In particular BÜHRER W. (1986).

7 NAU H. R. (1975) compares the fertility of the period 1955-1958 with the sterility of the years 1968-1973 as regards the creation of nuclear R & D organisations. "In the first period, major actors [...] possessed unequal resources, and responded to the situation by the creation of new organisations", p. 617.

As far as reprocessing was concerned, the American aid that helped fill the "technical gap" followed the same institutional channel as Marshall Aid.

Finally, the weakness was institutional and legal. The international organisations helped fill the gap[8]. It is very striking to note that with the exception of France (but was not the strength of the CEA, sometimes regarded as a "state within a state" due to the inadequacy of existing government structures when the Order of October 1945 was passed?), national institutions in Europe were developed along with or even after the establishment of these organisations.

However two particular characteristics were equally indispensable to the success of the project.

A "purely civilian" form of co-operation, remote from the market

Co-operation between European countries, and between them and the United States, was made possible by the demilitarisation of reprocessing which became official at the first Geneva Conference in August 1955. This factor was not specific to reprocessing but was nonetheless highly significant for Eurochemic.

It was not fortuitous that the first major international nuclear institution, CERN, the foundations of which were laid in the middle of the Cold War, was focused on fundamental research into the physics of matter, a field far upstream of military applications.

As far as EURATOM was concerned, American support was made conditional on the abandonment of military applications. The "torpedoing" of the plans for an enrichment plant[9], a project on which France was particularly keen, was one expression of the American demilitarisation strategy that accompanied the Atoms for Peace policy of openness. Once EURATOM had abandoned the project, it received unceasing support from the United States[10].

In the space field too, demilitarisation made European co-operation possible[11]. In the early 1960s there was collaboration in ELDO and ESRO on civilian projects such as telecommunications satellite launchers.

As far as reprocessing was concerned, the ambiguity of the activity was also a factor in its success. Although it is not too difficult to demilitarise a technique, it is even easier to consider remilitarization if the situation should arise. This made it even more important to gain control of this technology through civilian European co-operation.

The remoteness of the market was also an important factor in the success of the project, and this explains, together with the previous factor, why of the two projects involving entirely comparable risks of proliferation – enrichment and reprocessing – it was the second that made progress. The fact that reprocessing was seen as the part of the fuel cycle for which viability seemed a remote prospect – or at least more remote than for enrichment – militated in favour of co-operation and even of association with European firms[12]. This probably goes some way towards explaining why private firms, apart from the industrial architects, were slow and mostly unenthusiastic about buying any shares.

However the declared objective was to promote a real European reprocessing industry, and Eurochemic's position as a "joint undertaking of the OEEC/ENEA" also played a significant role in the success of the project, notably by facilitating the decision-making process.

---

[8] Indeed it was this strategy that Jean Monnet had adopted when he decided that nuclear energy would be the lead sector in the "European recovery".

[9] See Part I Chapter 3.

[10] NAU H. R. (1975), p. 623 gives an excellent summary of the situation, quoting a secondary source, NIEBURG H. L. (1964), *Nuclear Secrecy and Foreign Policy*, Washington, p. 143: "At the same time [as the Atoms for Peace Plan], US officials hoped to pre-empt proliferation of strategic uses of atomic energy which might disrupt Western unity. Thus, in offering co-operation in nuclear research, the United States supported a proposal of the Monnet Action Committee to renounce military uses of atomic energy in Europe. When it became apparent that France would not accept this proposal, the United States, with British support, 'moved to find a basis for transforming EURATOM into a harmless regional effort to develop nuclear power reactors'".

[11] Demilitarisation of a technique, particularly in this case, is a political rather than a technical step. Some degree of ambiguity always remains, which may in fact explain the success of these instances of co-operation.

[12] At the risk of confusing the chronology, this was the "precompetitive stage" of development, when it was easier to establish co-operative arrangements between competing firms.

OEEC, the "nursery of co-operative ventures". Eurochemic, an undertaking "hatched" by an international organisation needing to move from a "dockyard approach" to a "market approach"

The "joint undertakings" of the OEEC were to form what would now be called "business nurseries" for the development of new technologies in Europe through "à la carte" co-operation.

The implicit strategy of the OEEC with regard to Eurochemic, its first joint undertaking, is not without some analogy with the "dynamic" characterising the "life cycle of major projects" analysed by Elie Cohen concerning the development of the telephone industry in France and more precisely the digital exchange. It is based upon the idea of a "dockyard industry" expressed by Jean-Jacques Salomon[13]. It also shows certain features of the national "major civilian programme" of which Dominique Finon sees the archetype in the development of fast breeder reactors in the United States and France[14].

According to Jean-Jacques Salomon, the operation of a dockyard industry is defined by the absence of any market sanctions or cost concerns, owing to its links with the public authorities (the original model being that of the defence industries, but which rapidly found civilian applications). Elie Cohen proposes to extract from this model an operational approach he calls a "dockyard approach" the characteristics of which are the mobilisation of "all financial, technical, political and regulatory resources", in order to "accomplish a technical-industrial project", "with no concern for commercial viability, for meeting consumers' expectations, or the delaying effects on industry of a unique specialisation"[15]. Once the purpose is defined, there are two possible scenarios. Either there is a change to an equipment supply approach, sanctioned by the emergence of a market and which, as appropriate, may lead to success, i.e., the change to a market approach (there are other possible scenarios: see the diagram below). Or, alternatively, the technical subject turns out to be a "white elephant", either because the project is a technical achievement with no apparent market or because the industrial costs are incompatible with those of the market".

The scenarios of Elie Cohen may be shown schematically as in Figure 178:

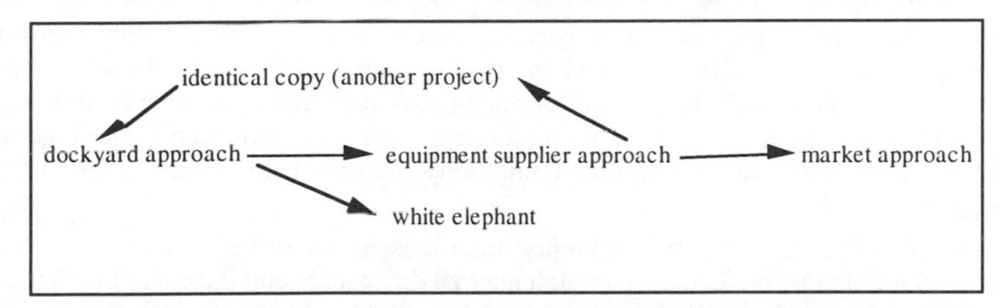

Figure 178. Scenarios in the life cycle of a major project according to Elie Cohen, pp. 176-180.

However the Eurochemic project differed from the "major project" through the international nature of governmental involvement, which put it in a fairly different position with respect not only to the industrial actors but also the national nuclear agencies, owing to the intervention of the international co-operation aspect, and to the fact of there being a different approach to costs.

The primary special feature of Eurochemic was that the project was led by an international organisation and not by a country, even if this does reflect the hybrid character of "major national programmes" combining industry and bureaucracy. The organisational culture of the OEEC had a profound influence on the way in which the company was created and its structures, with its Special Group, its restricted sub-groups working on specific problems, and the procedures for keeping members informed. The changes following the financial crisis of 1963-1964 enhanced these features still further.

At the outset however, the project embodied the hope that it would lead to a market approach. Its constitution allowed shares to be transferred to private undertakings with the minimum of constraints. Another expression of this concern was the limitation of the undertaking's privileges so as not to damage the future basis of competition. When the Statutes were being drawn up, few people doubted that the move to the market, enabling the public institutions to withdraw, would take place rapidly, and the project was kept open to private enterprise. But the actual rate of development of nuclear power and the very nature of the type of co-operation proposed quickly made this objective illusory.

---

[13] See COHEN E. (1992), p. 90 for the definition of the "dockyard approach", and p. 176-180 for the different scenarios for combining approaches, and SALOMON J.-J. (1986).

[14] FINON D. (1989), p. 9-10.

[15] COHEN E. (1992), p. 177.

The company was not indifferent to the problem of cost and the records are full of plans, counter-plans, assessments and scenarios, attempts to achieve savings, etc. which were not merely rhetorical. Here again the international dimension was of crucial significance, by allowing the pooling of effort and the mobilisation of significant funds. However the financial circuits were longer – since they involved governments – and they came more under the control of these governments, which made them relatively more transparent than those employed to bind together a government, its national nuclear agency and the relevant industry, especially when civilian and military interests are intermingled. The financial problems that were evident in the initial phases of the project stemmed essentially from early underestimates together with the political impossibility of halting a project once started, rather than from any indifference to cost.

This underestimate was itself related to two converging factors. First, the fuzziness of the criteria used for evaluating the costs of an industry that did not exist in Europe or was closely linked to military objectives, as in the United Kingdom and the United States. Secondly, the desire shared by the experts and senior civil servants to see the project carried out, which led during the crucial phase of political acceptance to the formulation of a project on the basis of the amount of funding that was acceptable rather than that which was technically necessary[16].

The final factor leading to success was undeniably the relative simplicity of the contemporary decision-making process on nuclear issues, due mainly to the national institutional and legal vacuum and the lack of any democratic participation in decision-making in the scientific and technical fields.

To put it simply, the OEEC drew up the overall framework of co-operation, the scientific and industrial experts formulated the proposals, the international bureaucracy worked out an "à la carte" project, the national bureaucracy conducted the necessary political bargaining as regards the site and the sharing out of the capital, and opened up the lines of credit. The undertaking was then launched and nothing could stop it until it was carried through. The financial problems that soon appeared had little influence in the face of the political need to continue co-operation, this time "whatever the cost", since it was imperative to settle the problems of waste.

Hence there was no initial indifference to the question of costs as there is in the "dockyard approach", but rather a kind of confinement within an international co-operative approach[17] which had to be pursued until the objectives were achieved. Here therefore the contractual role of the Convention and the Statutes which laid down the subject of co-operation and its minimum duration, was essential. The influence of the legal dimension, for governments seeking integration in an international community, appears here as one of the main factors in the dynamics of the project.

This substantial constraint, related to the highly ambitious nature of the project, may help explain the unique character of the Eurochemic project, even though internal difficulties and the changing face of international co-operation in the 1960s were in fact to make its repetition impossible.

The changing face of European co-operation since the 1960s, together with the limitations and institutional imperfections of the project that emerged, essentially explain the failure of European co-operation in reprocessing

However the conditions that presided over the success of the project – success demonstrated by the fact that the plant was actually started up – were largely responsible for the failure of the undertaking, which led to the decision to halt reprocessing. However there were also external causes for the failure, related – to use a biological metaphor – to the undertaking's "genetic programme" and the effect of the changes in the environment.

It is appropriate first to consider the changes in European co-operation in the 1960s through the general pattern of co-operation in EURATOM, and by analysing specific examples from the head-end, centre and tail-end of the fuel cycle.

Secondly it is important to understand why the undertaking was unable to withstand the changes in its environment.

---

16 This approach was fundamental in the recasting of the different schemes for the plant and generated friction amongst the experts concerned. Thus the fourth preliminary project stemmed from a reformulation of the final version of the third, notably lacking the part devoted to waste management, the financing of which was thus implicitly postponed. This made the project economically "presentable".

17 The termination of international co-operation would also have raised a problem of image. It is recalled that the abandonment considered at the end of 1963 was rejected on the grounds that it would have been "an international scandal".

European co-operation in the 1960s. Institutional passivity and changing conditions

Institutions concerned with co-operation multiplied during the 1940s and 1950s. The 1960s were less fertile. Of course the first reason for this is trivial: when a task is done it is done. However when the economic law of converting a supply market into a renewal market is applied to international organisations it takes no account of how existing organisations are changing. The crisis affecting EURATOM is highly illuminating with regard to the new climate characterising European international co-operation, where the concept of "a Europe of nation states" has prevailed over that of "a federal Europe" in the functioning of the European Community since the Luxembourg Compromise of January 1966, which put an end to the crisis begun by the "empty chair policy".

Co-operation in the nuclear field supplies an early example in the shape of the 1961 confrontation between Etienne Hirsch, former colleague and friend of Jean Monnet, who had succeeded Louis Armand as President of the EURATOM Commission, and General de Gaulle, President of the French Republic. This confrontation is reported by Hirsch in his memoirs[18].

Etienne Hirsch had asked the CEA to apply the measures provided for by the Treaty and confirmed by the EURATOM-United States Agreement on security control. His request was refused. He therefore sought an interview with the French President before embarking on the legal procedures provided for by the Treaty. The interview took place on 17 March 1961. The decision not to renew the appointment of the President of the Commission was taken by General de Gaulle at the next meeting of the cabinet.

During the conversation, the President of the Republic clearly indicates the change which had taken place: "Charles de Gaulle: 'The time when Mr. Monnet gave orders has gone'.

Etienne Hirsch: 'You know very well that Mr. Monnet never had any powers other than those of persuasion'.

Charles de Gaulle: 'I'm not criticising. That was perfectly normal because, at the time, governments were weak. The situation has changed. With the improvement in the economic situation – because today that's the source of power – we now have strong governments. I want co-operation between the three main governments'.

Etienne Hirsch: (indicates surprise).

Charles de Gaulle: 'You are not going to talk to me of Belgium and the others? I repeat that I want this co-operation in the cultural field and in military matters. But this is co-operation between governments who remain sovereign to decide freely on vital issues'".

The replacement of Etienne Hirsch by Pierre Châtenet who had been de Gaulle's Minister for the Interior after a diplomatic career at the United Nations and NATO, put an end to any pretence at independence on the part of the President of the Commission, and prepared the way for the disappearance of the whole structure. With the advent of Pierre Châtenet, as remarked in the biographical note on him given in the journal *Atomwirtschaft*[19], the Commission would cease "to be the driving force of any general European policy" in order to "focus on practical problems". In fact this appointment was the preliminary to the merger of the Community executives decided upon in 1965 and carried out in July 1967, which had the result of eliminating the EURATOM Commission and the ECSC High Authority, which were absorbed into the Common Market Commission. This put an end to any specific nuclear dimension in the European Community[20]. By a cruel twist, EURATOM was killed off in 1967 as an independent organisation through the "package deal" approach which had led to its creation 10 years earlier[21].

This same development had the effect in the ENEA of halting any further attempts to create joint undertakings until October 1970[22].

The growing powers of European governments then resulted in the emergence or development of national R & D structures, sometimes emerging – as in Germany – from those originally set up to grasp the opportunities offered by the Atoms for Peace policy, and by the development of "à la carte" co-operative ventures

[18] HIRSCH E. (1988), p. 171 for the quotation. This is a transcript of notes taken by Etienne Hirsch after his interview with Charles de Gaulle. Etienne Hirsch confirms "the exactness of the meaning, but not that of the actual text" (p. 169).

[19] AtW (1962) p. 405.

[20] In the beginning this merger had been sought by the "Federalists" in order to strengthen the powers of the Community executive.

[21] For the idea of the "package deal", see SCHEINMAN L. (1967).

[22] On 14 October 1970, ENEA created its fourth joint undertaking in conjunction with IAEA and the FAO: the "International Food Irradiation Project". This small project was established on the premises of the Karlsruhe nuclear centre and came to an end in December 1981. No new NEA joint undertakings have emerged since.

managed by national institutions within precise limits, as H. R. Nau shows very clearly in the second part of the article already mentioned, devoted to the problem of R & D co-operation in the period 1968-1973[23].

Governments' powers increased in parallel with economic growth, which speeded up private companies' move towards independence, at the same time as it made them more competitive. This development too was a factor unfavourable to international co-operation, particularly for an undertaking like Eurochemic under the auspices of an international organisation.

New forms of technical co-operation with industrial objectives in the 1970s

The new forms of co-operation that developed in the early 1970s were very different from that adopted for Eurochemic. We may quote three examples, relating to various stages of the fuel cycle: enrichment through URENCO at the beginning of the 1970s, the development of fast breeder reactor power plants with NERSA-ESK in 1974, and the reprocessing industry with the financial arrangements for the UP3 plant at La Hague defined in 1978.

A joint industrial undertaking of the 1970s, the Uranium Enrichment Corporation (URENCO)[24]: trilateral "à la carte" co-operation

One reason for the failure of the negotiations about a joint isotopic separation plant during the preparations for the Treaty of Rome, apart from American pressure, was the disagreement between France and the Federal Republic of Germany about the technical process to be used. The former favoured gaseous diffusion, a technique used at the time in the American plants and characterised by large-scale units, high capital and running costs, especially as regards energy. Germany preferred processes such as jet nozzle separation[25] which enabled cheaper and simpler plants to be used.

While France went ahead alone and built its gaseous diffusion separation plant at Pierrelatte[26], research continued during the 1960s into other processes, like the ultracentrifuge, that used less energy and could be built in small units[27]. Germany, the United Kingdom and the Netherlands conducted independent research of their own in this field. In the autumn of 1968 the three countries began negotiations with a view to co-ordinating their efforts, but it took 18 months before the Treaty of Almelo was signed on 4 March 1970. The reasons why the negotiations took so long were, first of all, the special legal problems and, secondly, the need to build a "customised" system that would represent a compromise between the differing aspirations of the three partners. The Treaty of Almelo was ratified by the three national parliaments and came into force on 19 July 1971. According to Article 2 §2 of the Treaty, the aim of the contracting parties was to "facilitate the establishment and operation of joint industrial undertakings with the task of building uranium enrichment facilities [...] and to operate them with a view to utilising this process on a commercial basis"[28].

At first sight the legal arrangements for the URENCO and Eurochemic projects have a number of parallel features, since they both involve international agreements enabling the creation and development of joint undertakings for industrial and commercial purposes.

However the form of co-operation for URENCO reflects the search for a compromise between the aspirations of the different partners and, although there is a pooling of resources, there is no common ownership. The various degrees of co-operation within URENCO vary according to the remoteness of the market.

The arrangements are complex (Figure 179) and involve a joint R & D structure, CENTEC, located in Germany, a marketing (and subsequently, co-ordinating) company, URENCO Limited established in the United Kingdom, together with two plants to be built in the United Kingdom and the Netherlands; a third plant in

---

[23] NAU H. R. (1975), pp. 630-645. The development of public R & D policies ultimately led to the downgrading of the status of nuclear research. In Germany, the scope of the Atomic Ministry set up in 1955 has been remarkably diversified. By the end of the 1950s it embraced water, then space in 1962 and computers in 1967. NAU H. R. (1975), p. 634.

[24] This section is based upon SCHMIDT-KÜSTER W. J. (1970), ASYEE J. (1985), PIROTTE O., GIRERD P., MARSAL P., MORSON S. (1988), pp. 193-195.

[25] See above: Part I Chapter 3.

[26] See DAVIET J.-P. (1993), pp. 181-207.

[27] This system used a large number of small centrifuges rotating at very high speed; the spinning drums contain uranium hexafluoride and the lighter U235 isotope concentrates at the centre while the heavier U238 isotope migrates outwards due to centrifugal force.

[28] Quoted by PIROTTE O. et al. (1988), p. 193.

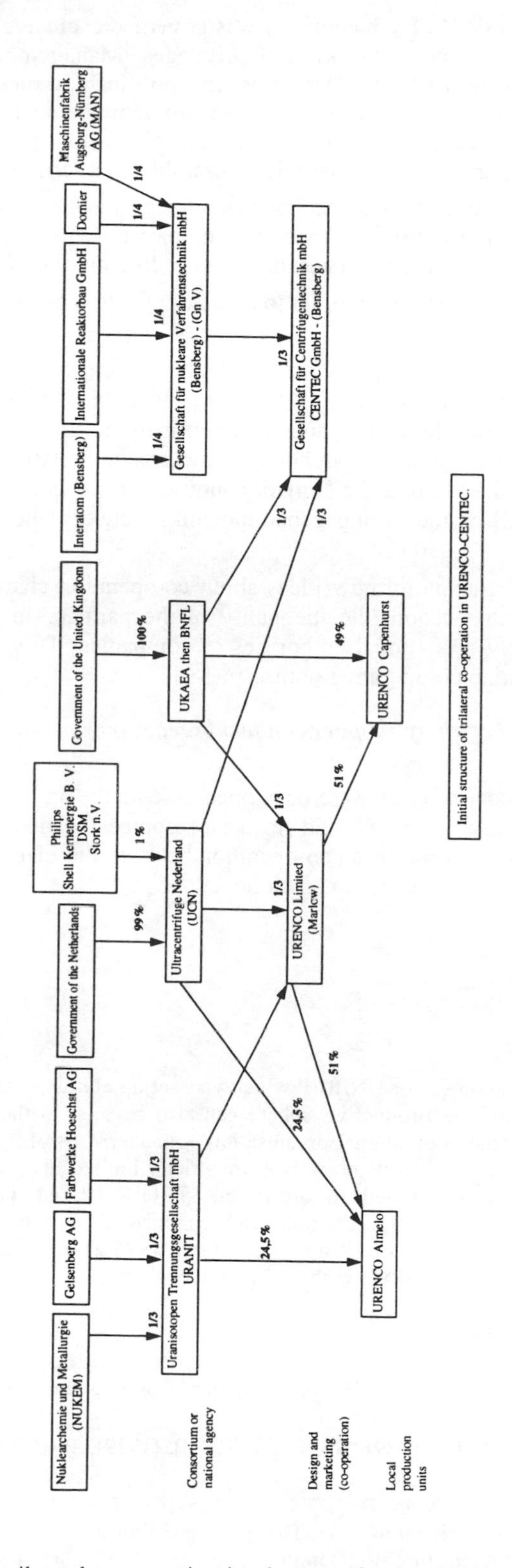

Figure 179. Structure of trilateral co-operation in ultracentrifuge enrichment between 1970 and 1974.

Germany was envisaged for a later date[29]. Each company was governed exclusively by the law of the host country and enjoyed no fiscal advantages nor any other kind of privileges. Management and recourse to the financial markets was also handled on a national basis. Decisions on "political" issues such as safety and guarantees required the unanimous agreement of the governments, other problems being discussed in a Joint Committee[30] with rotating chairmanship and secretariat, "thus avoiding the need to set up a separate bureaucratic administration"[31], in other words, any structure that might resemble an international organisation.

The arrangements for the ultracentrifuge contracts reinforced the principle of strict contractual equality between the partners, adopting a "fair return" approach by comparison with the industrial shareholdings whose structures precisely reflect the differing degrees of acceptance or utilisation of co-operation.

Research and development was pooled in order to avoid duplication of research, as well as was marketing in a cartel approach[32].

As regards the local production units, there are substantial differences between URENCO Capenhurst, in which BNFL has the biggest shareholding through its direct and indirect participation in URENCO Limited, and URENCO Almelo. In the latter, the Netherlands and Germany hold equal numbers of shares. The Dutch government's agreement to remain a minority shareholder compensated in part for the absence of any production site in Germany; it also reflected the fact that the "small countries" still wanted more intense co-operation than did the "large" countries, as well as the strong economic links between the two countries that bestride the Rhine[33].

URENCO is thus a good example of how ideas about co-operation changed during the 1960s, with the search for compromise taking into account the inequality of the partners in building a "made-to-measure" co-operative structure, run entirely by national authorities or companies. This kind of scheme is diametrically opposite to Eurochemic's egalitarian co-operative optimism.

Co-operation between Europe's electricity suppliers on fast breeder power reactors[34]: national dominance and "fair return" in NERSA

In Europe, experimental fast breeder reactors were developed essentially on a national and isolationist basis[35]. However the sheer extent of technical, scientific and financial resources needed to move from the pilot stage to the commercial reactor gave impetus to European co-operation between the electricity suppliers who had to bear the cost and who, led by EdF, did not wish to do so alone[36].

---

29 This was the Gronau plant.

30 "Gemeinsamer Ausschuss".

31 ASYEE J. (1985), p. 39.

32 Exactly the same approach was adopted for UNIREP which was set up almost at the same time. The hiatus between the levels of R & D co-operation and the productive sphere can also be seen in the space field. The European Space Agency (ESA) set up in 1975 is a true international organisation – in Pierre Strohl's classification it comes just after CERN in decreasing order of integration – but its objectives are strictly limited to R & D. Its contribution "ends once the product it has developed is demonstrated in orbit" in STROHL P. ed. (1985), p. 111, contribution by G. Lafferranderie, legal adviser to ESA. The production and commercial aspects are managed through structures governed by national law, like ARIANESPACE, set up on 26 March 1980 as a limited liability company under French law, with French interests dominating in 1985 with 59% of the shares, its chief executive at the time being the Director-General of the CNES.

33 The effects of ancient cultural differences in Europe are striking when one considers the development in the 1970s of the two co-operative – soon to be in competition – enrichment projects; after 1973 the "Germanic troika" co-operating within URENCO confronted the "Latin countries" forming the hard centre of co-operation within EURODIF.

34 This section is based upon FINON D. (1989), SAITCEVSKY B. (1979), PIROTTE O., GIRERD P., MARSAL P., MORSON S. (1988), pp. 162-178.

35 For France from 1956 to 1970, see FINON D. (1989), pp. 145-168. The RAPSODIE programme was launched in 1961 and the experimental reactor went critical in 1967. The programme for a small pilot PHENIX was started in 1965 and the reactor diverged in 1973. According to Dominique Finon, the French fast breeder programme is a perfect example of a major technological project bringing forth a white elephant. The section Dominique Finon devotes to Superphénix focuses on the national aspects and on the relationships between the CEA, EdF and the French government. This approach means that the different aspects of international co-operation are dealt with piecemeal between the CEA, pp. 188-190, and EdF, pp. 194.

36 On the strategy of cost dilution by EdF, see FINON D. (1989), p. 195. The establishment of co-operation enabled the Superphénix project to benefit from EURATOM loans, see PIROTTE O., GIRERD P., MARSAL P., MORSON S. (1988), p. 276 for the period 1981-1985.

In 1969, discussions began between EdF, RWE and ENEL in the framework of the Union of European Electricity Producers (UNIPEDE)[37]. On 16 July 1971 the three French, German and Italian electricity producers published a declaration expressing their intent jointly to build a 1200 MWe fast breeder reactor. The project was taken further by the signature of an agreement on 23 December 1972 providing for the construction of two prototype plants in turn, one in France as a sequel to Phénix, the other in Germany, in continuation of the SNR 300 at Kalkar. The principle of a "fair return" as regards contracts was made explicit and co-operation was distributed between two parallel organisations, the Société centrale nucléaire européenne à neutrons rapides (NERSA; European Fast Reactor Power Plant Company, and the Europaïsche Schnellbrüter Kernkraftwerkgesellschaft (ESK) (Figure 180).

The structure of the original holdings throws light on the co-operative approach adopted.

In France[38] NERSA was set up on 8 July 1974 as a company under French law with its headquarters in Paris, after a change in French law allowed a limited breach in the national monopoly for electricity production; a decree placed the company under the economic and financial control of the government[39].

The essentially national character of the undertaking was reinforced by its structure: a 16-man Steering Committee with six members, including the Chairman, from EdF; a Management Board of three members with a French Chairman; a specific and sparse operational structure with a Directorate overseeing three divisions and based upon the Equipment Division of EdF, which allowed foreign engineers to enter for the purpose.

As regards R & D and the commercial dissemination of knowledge French interests also predominated in other complex structures[40], so the organisation enshrined the country's technological lead in this sector.

As regards civilian reprocessing too, French domination in Europe made it possible to fund a new plant at La Hague using foreign capital. The basis for co-operation involved a strict separation between the technical and commercial aspects, once again an approach diametrically opposed to that used at Eurochemic.

French technology financed by foreign contracts: co-operation with separation of tasks at UP3 – La Hague[41]

In the reprocessing field, the ultimate in co-operation was achieved with the 1978 launch by Georges Besse, then in charge of COGEMA, of the UP3 project which was aimed to meet European reprocessing demand up to the year 2000. The customers undertook to pay for the construction of a new plant at La Hague, which would belong to and be managed exclusively by COGEMA, in return for having their fuel reprocessed.

At a press conference on 2 September 1981, Georges Besse announced the detail of the reprocessing contracts that had been concluded with foreign companies. Since the creation of UNIREP, contracts at COGEMA had gone through three generations, their changing pattern showing first the uncertainty about reprocessing costs and, secondly, the increasing importance of foreign fuel to French reprocessing. The first type of contracts in the initial period from 1971 to 1974 covered 514 tonnes of spent fuel from fifteen reactors, including that of Mühleberg; they kept the main principles adopted by Eurochemic and Windscale: firm prices, guaranteed delivery dates, freedom of the reprocessor to dispose of the wastes, obviously aiming to profit from their conversion. However the offer of firm prices with retention of wastes proved not to be feasible and resulted in renegotiation. From 1977 onwards, a second type of contract was instituted, which continued until 1979 and covered 713 tonnes of spent fuel from some twenty reactors. COGEMA's objective was "to cover nearly 20% of the cost of extending UP2-800 and similarly reduce the financial contribution of COGEMA and EdF".

The main features of these contracts were much higher prices, with revision clauses, the elimination of guaranteed delivery dates, and all wastes being returned to the fuel producers. In 1978 plans were announced to build the new plant, UP3, to be paid for by advances on reprocessing. COGEMA's strategy is clearly set out in the following passage[42]:

---

37 According to Dominique Finon, the CEA would have firmly opposed any British participation, FINON D. (1989), p. 194, by proposing a very costly "entry ticket".

38 In 1975 RWE set up an international consortium, SBK, of which it held 68.85% of the shares, the remainder being distributed between the Belgian company SYNATOM (14.75%), the Dutch company SEP (14.75%) and the British Central Electricity Generating Board (CEGB) (1.65% giving right of access to the technology). Hence RWE did not contribute more than a third to SNR2. This clearly shows a difference in national strategy. For RWE the cost dilution approach extends as far as losing majority control, but not for EdF.

39 Law of 23 December 1972 permitting the creation of undertakings exercising an activity of European importance as concerns electricity, in France, and the Prime Minister's decree of 13 May 1974.

40 See charts in PIROTTE O., GIRERD P., MARSAL P., MORSON S. (1988), pp. 166 and 169.

41 The technical side of this new plant was described in Part V Chapter 1. Here we focus on the commercial and financial aspects, using a COGEMA article published by RGN in 1981 in "RGN actualités", pp. 379-381.

42 RGN (1981), p. 379.

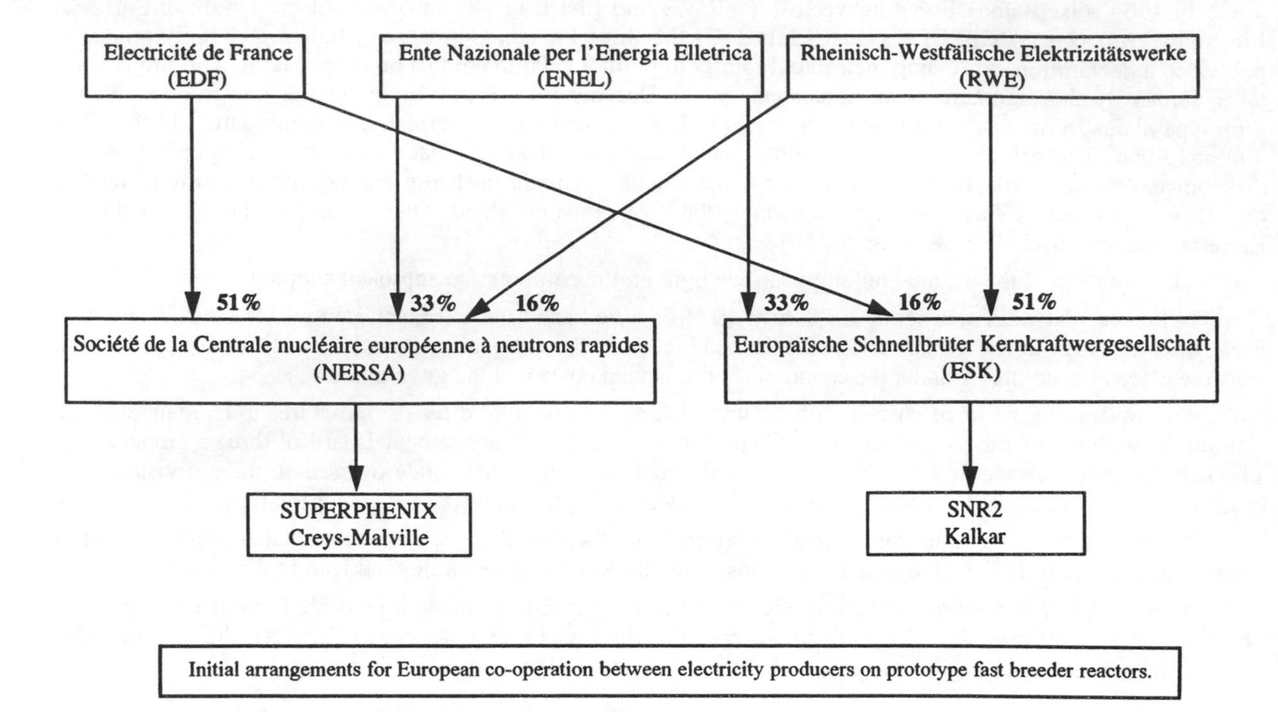

Figure 180. Trilateral co-operative structure between electricity producers in the programme to build prototype fast breeder reactors.

"In 1976, as a result of [*sic*] the expansion of the French nuclear power programme and the foreseeable development of French reprocessing requirements, it became clear that the planned extension to the existing plant (the UP2-800 extension to the UP2 plant) would be inadequate and that consideration must be given to commissioning a new, large plant (UP3-A) by the beginning of the 1990s and to raising total capacity to about 1600 t/year. The strategy adopted has therefore been to introduce this capacity of 1600 t/year somewhat earlier in order to provide a safety margin. However so as to avoid undue expenditure at too early a stage, allowance has been made for the fact that the overseas electricity producers were prepared to accept special financial conditions in return for having access to a certain reprocessing capacity"[43].

To this end, a third type of contract was negotiated globally with the foreign electricity producers concerned. An agreement was reached in 1978, but before contracts could be concluded it was necessary for international relations to match the progress made by trade and industry. The principles involved were as follows:

COGEMA undertook to build a new plant which for the first ten years of operation would be entirely devoted to reprocessing 6000 tonnes of spent fuel of foreign origin. The capital cost – 45 billion FF – was to be entirely covered by the clients[44] up to 1995.

Deliveries were to be made according to a schedule, with reprocessing invoiced on a "cost plus" basis.

All the wastes had to be returned to the producer along with the final products of reprocessing, i.e. the uranium and plutonium. This return would be guaranteed by an exchange of letters between the governments concerned.

In the event of dispute, the International Chamber of Commerce would be asked to arbitrate.

---

[43] The fact that these conditions were accepted must be seen in the light of the legal obligations on the electricity producers, at a time when only France appeared capable of reprocessing and when there was undeniable international uncertainty about the whole sector. Georges Besse and COGEMA were able to exploit these needs and these hazards, enabling French reprocessing to make a considerable quantitative and qualitative leap forward.

[44] According to the 1981 figures, Japan and Germany would contribute over 50%: Japan (2200 tonnes), Germany (2141 tonnes), Sweden (672 tonnes), Belgium (398 tonnes), Switzerland (469 tonnes), Netherlands (120 tonnes).

The contracts would also provide for very high compensation in the event of non-completion.

In this way the new plant would be funded and provided with work until the year 2000.

As the text makes clear "the foreign customers will have no access either to the technology or to the know-how"[45]. Thus technical co-operation reached a low point in the case of UP3, collaboration being limited to the financial and commercial aspects. There is no doubt that the outstanding technical sophistication of the plant[46] stems from the financial freedom obtained through the negotiation of these principles.

These three examples of co-operation on industrial projects show how the pattern of European co-operation was transformed in the 1960s and the early 1970s. The structures offered by the international organisations were not used. Neither NEA nor EURATOM were involved in the management of the projects, even though the European Community did make several unavailing attempts to bring together the "prodigal sons of European enrichment"[47]. There was a preference for specific agreements concluded between partners directly involved and along national lines.

This new pattern of international co-operation in nuclear engineering, the impact of which on the organisations set up during the period has already been examined, had a weakening effect on the pre-existing organisations, such as EURATOM. Naturally it also affected Eurochemic, where it added to the internal problems of reprocessing and of the company.

The effect of these changes on the international climate amplified the specific problems of European co-operation on reprocessing. The failure of the company stands out in contrast to the success of the project

The vulnerability of the existing co-operative structure

Financial problems and the lack of interest on the part of industry

The success of the project stemmed from deliberate underestimates of costs and the low capacity of the plant, which prevented the success of the undertaking, particularly as the delays to the schedule themselves brought about an increase in costs[48].

The project was under-capitalised at the outset, raising funding problems as soon as construction began and bringing about the first serious crisis in co-operation in 1963-1964. The reasons why co-operation did not break down completely at that time are complex and numerous. The main cause is very probably that stopping co-operation at the time, although not unthinkable[49], had a very significant symbolic aspect and that a project once decided upon must necessarily go on to its conclusion, in this case the first fifteen years laid down by the Convention[50].

The way in which this crisis was resolved was the first U-turn in the substance of the project. Co-operation was safeguarded through the automatic inclusion of Eurochemic funding in the OECD budget, but this departed from the original plan for a rapid move to industrial participation. Eurochemic now became an international public research and services organisation, an appendage of the OECD, even before the advent of the commercial problems of the early operations.

Moreover it is fairly clear that although the chemical industry in member countries continued to be interested in reprocessing, it was less attracted by the international framework offered by Eurochemic and preferred to develop in parallel its own experiments with national government aid. SGN saw Eurochemic as a tiny market compared with La Hague, built by the CEA, whose resources ultimately came from the government; for Hoechst, the transfer of technology had functioned since 1961 and the plans for a national pilot plant were

---

45 RGN (1981), p. 381.

46 See Part V Chapter 1.

47 For these attempts, see PIROTTE O., GIRERD P., MARSAL P., MORSON S. (1988), pp. 196-200.

48 It will be recalled that the Convention of December 1957 provided for the plant to be completed before 1961.

49 Although it might have been unthinkable in the 1950s. This could be one of the reasons for the loose provisions relative to the termination of the joint undertaking in Eurochemic's Statutes, which are legally very precise and very detailed in other respects.

50 Only a few international organisations disappeared like SDN which foundered during the Second World War but was replaced by the UN. It was not until the beginning of the 1970s that the first international organisations set up after the Second World War, ELDO and ESRO, disappeared, the causes being essentially technical, even if the actual timing of the decision to terminate their life was political, see KRIGE J. (1993b). The collapse of the Soviet bloc, accompanied by or leading to the disappearance of MEAC and the Warsaw Pact, has recently demonstrated how weak are regional international organisations dominated by a single country.

financed from public federal funds[51]. The situation appeared to be the same in Italy. However as far as the "small countries" were concerned, continued co-operation seemed more desirable, allowing the learning process to continue until it was possible to embark on purely national projects.

The diversification of reprocessing to highly enriched uranium fuels and its adaptation to new reactor characteristics raised costs still further, and the start-up of the plant merely revealed new financial needs. The company's new approach and the funding mechanism set up made it possible to complete construction and to commence the active operation of the plant.

Of course the technical success of this start-up again raised hope that commercial revenue would reduce the burden on member states. However these hopes were compromised by two factors.

The combination of viability problems and technical difficulties in the early 1970s

The primary difficulty arose from the low level of prices in a sector where there was no large commercial market. The problem of determining reprocessing prices in the 1960s would call for more detailed research, but they were certainly remote from economic reality, probably because it was impossible to calculate them through lack of experience. Eurochemic initially aligned itself on the prices declared by NFS, but was then forced to reduce them as it was undercut by Britain on the external market, both for fuel that Windscale could reprocess and for those that the British plant hoped to be able to reprocess in the near future. The deflationary spiral was made worse by the increased number of national projects, with which nuclear power generation did not keep up.

It seems that the move towards more realistic prices did not take place until the 1970s[52]. In any event, the unduly low prices were another obstacle to the project's earning reasonable income from reprocessing.

The second factor was more cyclical: it concerned the rate of reprocessing at Eurochemic, the slow increase in capacity and the attendant difficulties, and the fact that the plant was shut down for modifications between May 1970 and February 1971. These changes were made necessary by the various problems encountered during the early years. Thus one might consider that a further phase of co-operation had come to an end. Hence the closure of the pilot plant was better justified than in 1963-1964 when construction was in progress. By 1970 the experience of the first few years of operation has been added to that acquired during construction. The end of the 15-year period provided for in the Convention was approaching. However the decision to shut down the plant was made possible by the changed attitude to international co-operation.

Eurochemic as a "victim" of the 1960s change in the attitude to co-operation

In the history of European co-operation, the fundamental aspect of the early 1970s was not so much the decision to halt reprocessing at Mol for reasons of overcapacity – the UNIREP argument after the FORATOM report – as the decision not to adopt any multilateral co-operation formula in the industrial operation of the new plants it was planned to commission to meet the expansion – tardy but still hoped for – in nuclear power generation.

This approach resembles that which led to the establishment of URENCO, apart from the fact that there was still co-operation on R & D on the new ultracentrifuge process, whereas at UNIREP there was little more than commercial collaboration[53], since the interests of the three countries concerned were fairly divergent: awaiting better conditions for the existing French and British facilities, to obtain the assurance of being able to build their plant in Germany.

Thus the alliance between the three major European powers led to the sacrifice of multilateral co-operation. The small countries had no choice but to submit, and did so with a more or less good grace. The paradox is

---

[51] The French and German chemical industries did not break away from reprocessing until after 1975, and then did it in different ways. Saint-Gobain gradually sold its shares to the CEA (and then to COGEMA) and to the engineering (mainly oil) company Technip, which helped maintain an industrial approach in French reprocessing. The German chemical firms, led by Hoechst, handed control of the reprocessing interest to the electricity suppliers who adopted a mainly commercial approach to the problem, causing them to prefer using French reprocessing services, which was one of the reasons why domestic reprocessing was abandoned in Germany.

[52] Historical comparisons are difficult, owing to changes in fuel types and the reprocessing industry, quite apart from the effects of inflation. Simply as an illustration, let us recall that the first contracts Eurochemic signed with the CEA varied between \$20 and \$50/kg. In May 1994 COGEMA, in an estimate of German costs for reprocessing after the year 2000 at its La Hague plant, including transport but not final waste management, made an offer of 1960 DM/kg, or about \$1230/kg, a price increase of a factor of at least 25 over 30 years in constant money terms. Source for the 1994 French offer: DEVEZEAUX DE LAVERGNE J. G. (1994), p. 353, Table 2, Note 1. It is very probable that the discussion of changes to German nuclear legislation, which for COGEMA meant the threat of a reduction in or even abandonment of reprocessing in favour of direct storage was the reason behind the appearance of the article in the German journal.

[53] However exchanges of technical information were planned.

that this abandonment did not have the result of eliminating the co-operative structure or draining it of all substance.

It is necessary to explain why the joint undertaking continued in being for fifteen years after reprocessing had ended. The explanation of this long swan song, which was also a second life, was a combination of technical constraints and certain political features. The nature of European co-operation in Eurochemic from 1975 was in fact very different from what had gone before.

Changes in international co-operation at Eurochemic at the end of the 1970s

The post-reprocessing survival of the undertaking is linked to the fuzziness of the provisions in the Convention and the Statutes, together with the constraints of running a nuclear establishment.

Failure of the attempted "copycat project"[54] for co-operation on wastes

NEA tried to respond to the decision to halt reprocessing by initiating on the site an international radioactive waste management programme, a healthy sector of international co-operation on nuclear R & D. For example in the early 1980s[55], although EURATOM accounted for only 5% of R & D in the Community countries, the proportion was 25% in waste management. This field of activity had in fact so far been fairly marginal: a great deal remained to be done, and it was also remote both from the original industrial and commercial objectives of Eurochemic and from that of an international public service, which is what the undertaking became in the mid-1960s. What was being attempted was a further transformation, the creation from the existing joint undertaking of an international structure providing a service in the public interest, and going beyond the European framework, just as the ENEA had been transformed into the NEA.

However the project failed, and the partners began to search together for a way of bringing their co-operation to an end.

Forced collaboration aimed at terminating the joint undertaking

Reprocessing came to an end at the beginning of 1975. The overall approach to a disengagement from Eurochemic was found in 1978 through the signature of another Convention, but it took a further twelve years – until November 1990 – for the company to disappear completely. This long lapse of time was the result of the specific nature of the management of a nuclear undertaking, the fuzziness of the provisions in the Convention and Statutes concerning the dissolution of the company, the influence of the situation in Belgium and the legal and regulatory complexity of the decision-making processes concerned with nuclear facilities.

The problem of wastes was fundamental. The storage approach in the contemporary state of technology, together with the inevitable radioactive contamination of equipment – even in the absence of incidents – made it impossible simply to "lock the door". The question had to be settled before it was possible to consider disengagement. In the early 1970s it was not known how to stabilize the high activity liquid wastes that contained most of the radioactivity accumulated in a reprocessing plant. It was therefore necessary to carry out an R & D project up to the stage of vitrifying the liquid wastes generated during the short period of operation of the plant, which although modest in volume nevertheless contained most of the residual radioactivity.

The PAMELA project was entirely in line with the new forms of co-operation that have been a feature of European nuclear technology since the 1960s: national control of an international project, in this case by Germany. The difference by comparison with the Almelo plant lies in the fact that PAMELA was a German pilot unit on Belgian territory, at least until 1986 when it was agreed that it should be Belgianised.

The fuzziness of the provisions for termination of the company set out in Article 31 of the company Statutes relating to liquidation, and above all Article 32 on the principle of an agreement with the host country, necessitated negotiations with Belgium and made the future of international co-operation largely dependent on the ability of one of its member countries to reach a decision. There was no longer any overall collective policy, but there was "hands-on" control in order to bring about as quickly as possible the accomplishment of a waste management programme initiated in the early 1970s and the execution of a Convention signed at the end of the same period.

Unfortunately these mandatory negotiations took place at a time when it was difficult for Belgium to take clear decisions about nuclear policy and how this should embrace reprocessing. The Convention, which required the establishment of a Belgian takeover organisation, was further delayed by the wide differences that appeared within the country's decision-making authorities on energy questions. Finally, the financial calculations

---

[54] In the sense used by COHEN E. (1992), p. 178, see above. But here the change of subject also changes the meaning of the repeat process.

[55] STROHL P. (1981), p. 16 § 29.

essential to the termination of the international undertaking were hindered by the sheer novelty of the new activities, which had to be evaluated even though there was no previous practical experience. Nonetheless the prospect of the wastes being made safe was crucial for the continuation of operations and hence of the funding.

Thus following the success of the project as such, the progress of international co-operation at Eurochemic passed through three stages:

- a phase of sharing and acquiring an activity new to Europe, which was supposed to lead to an independent new organisation, but ended in fact in a financial fiasco resulting in the intergovernmental organisation taking over (1959-1964);
- a phase of acquiring know-how and providing an international reprocessing service in the public sector which came to an end as a result of changes in European co-operation (1964-1974);
- finally a phase of forced co-operation to bring about international disengagement, accompanied by the execution of a task in the public interest concerned with radioactive waste management, ending with Belgium gaining a nuclear site which today temporarily contains all the country's wastes, until a final storage solution can be found (1975-1990).

In concluding this discussion of the life cycle of an international organisation it is important to consider whether the highly original formula attempted by Eurochemic could not provide a basis for the preparation of new forms of co-operation needed to resolve certain contemporary burning questions.

The institutional inheritance of Eurochemic

Attempting to foretell the future is as risky for the historian as for anyone else, but the formula used for an international undertaking such as Eurochemic can be a subject for reflection, and may perhaps provide inspiration for resolving a number of current problems in the nuclear and other fields.

Other tasks for an international shareholding company working under the auspices of an international organisation. Possible applications in non-nuclear sectors

Such a formula would have definite advantages for pre-market R & D in fields requiring substantial input of funds on problems of common interest, such as the spread of AIDS, and where concrete results are needed rapidly. In fact the international combination of public and private partners can facilitate co-ordination between these two and between firms, introducing savings by avoiding duplication of research and making it possible to speed up the exchange of information; the pooling of public funds can create the conditions of strict equality as regards access to public support, and avoid wasting energy in settling conflicts concerned with commercial problems.

A specialised international technical company that closely combines the public sector with private enterprise could also play a role in organising "North-South" transfers of technology by relating them better to the mechanisms of international aid.

The aid organised by the international institutions is frequently criticised for using a "scatter-gun" approach and sometimes for being irrelevant, both held up as reasons for its being ineffective. Similar criticism is directed at technology transfers from private sources which serve only the interests of companies in the "North" and have led to the emergence of dual structures in the countries of the South.

For example to improve water supplies and urban sewage treatment in the large cities of the "South", international companies could be set up with precise objectives, funded from aid budgets and mobilising technical skills to carry out a clearly limited project that would result in local production units operating independently.

Using this kind of structure, problems with a transfrontier dimension could be handled in a transparent manner

As concerns conventional wastes for example, the developed countries are tending at present to stiffen regulations and to fall back on national approaches, which sometimes cannot be applied locally, at least in the short or medium term. Those involved are pressured into finding solutions that are on the fringes of legality or sometimes illicit, and transfer the problems beyond their own borders, particularly in circumstances that increase the hazard for local populations.

There is no doubt that the use of a common structure for industrial waste management, closely involving all concerned and covering the entire cycle of co-operation, concluded for a given period – so as not to hinder the slow adaptation of legislation – would make it possible to reduce the risks to the benefit of all.

However there is one sector in which the establishment of an international joint undertaking could make an immediate contribution to resolving a serious problem. This brings us back into the field of Eurochemic and is the sector of fissile materials and their civilian use.

A modest proposal to create an international company to manage the stocks of plutonium resulting from the planned reduction in strategic armaments[56]

The collapse of the USSR has moved the world on from the long "post-war" period and has radically changed the balance of military power. Application of the START I and II agreements and the execution of unilateral American and Russian commitments will result in the disarming of several thousand nuclear warheads, which by the year 2003 will generate military surplus stocks estimated at several hundred tonnes of highly enriched uranium (HEU) and some 50 tonnes of plutonium.

In 1992, the Committee on International Security and Arms Control (CISAC), an offshoot of the American National Academy of Sciences (NAS) that had existed since 1980, was asked by General Brent Scowcroft, National Security Adviser to President Bush, to look into the security questions raised by this new situation. The CISAC set up a plutonium group under the direction of Wolfgang K. H. Panovsky, Professor and Honorary Director at the Stanford Linear Accelerator Centre who was already monitoring the situation; the group prepared a report which was published in 1994.

The group focused on the problem of plutonium, believing that the HEU were relatively easier to manage since they could be diluted in LEU fuel intended for light water reactors. The United States has agreed to buy back 500 tonnes of Russian HEU diluted in LEU at a cost of $11.9 billion over the next twenty years under certain conditions[57]. However the problem of how to manage this fuel over the period still remains.

The Executive Summary of the report[58] points out that the very existence of this surplus constitutes a hazard, and that there are no institutional structures to deal with it. The report makes four recommendations, the third one being primarily technical in nature while the other three tend to be more political. A solution based upon Eurochemic's experience, naturally making allowance for its shortcomings, should enable some of the problems raised by the CISAC/NAS report to be resolved in an effective and internationally equitable manner.

From the technical standpoint, plutonium in a military usable form must be stored for as short a time as possible under conditions of safety and security comparable to those applying to weapons. Accordingly the plutonium must be processed in industrial facilities providing high levels of safety, security and protection both to the public and to the workers involved. Two immediate technical approaches are under consideration: the fabrication of mixed-oxide (MOX) fuel elements which, after use in the reactor, are to be stored without prior reprocessing or vitrified together with the fission products. For the future, the report proposes a third solution – burial in deep geological formations – but highlights the difficulties of administrative co-ordination and public acceptability, typified by the example of Yucca Mountain in the United States.

From the political standpoint, management of these stocks should form part of a global move to develop fuel cycles more proof against proliferation with the long term aim of entirely eliminating stocks of plutonium, currently estimated at 1100 tonnes and expected to reach 1600 to 1700 tonnes by the year 2000[59]. To be strictly practical, storage must form part of a new fissile material regime negotiated on a mutual basis, encompassing the disclosure of all weapon and fissile material stocks, with the cessation of production and stock reductions being handled in a negotiated and controlled manner. Moreover storage must be safe, under international surveillance and with the guarantee that withdrawals of fissile materials will be used for civilian purposes only[60].

These proposals could be implemented in practice by establishing a structure of the Eurochemic type.

An international company under the auspices of the IAEA, financed partly by the Agency and partly from the sale of its services, could be charged with managing the stocks both in the United States and in Russia, as well as with the processing of the plutonium and highly enriched uranium present in fuels or vitrified wastes. The United States and Russia would be majority shareholders with equal holdings, with other nuclear countries wishing to take part in world disarmament welcome to join. The company would be open to inspection by a large number of nuclear and non-nuclear countries through special mechanisms of the IAEA. Existing storage

---

56 The following section is based upon a reading of CISAC/NAS (1994), the publication of which was pointed out by Jean-Jacques Salomon, who stressed its value.

57 CISAC/NAS (1994), p. 5.

58 "Executive Summary", pp. 1-18.

59 CISAC/NAS (1994), p. 29. Over 75% of the stocks are made up of plutonium contained in nuclear reactor fuel, most of it in unreprocessed fuels that have left the reactor in the last fifteen years.

60 CISAC/NAS (1994), p. 1-2.

facilities would be internationalised and international industrial processing facilities[61] would be built on these sites.

A European contribution to such a structure would be both valuable and indispensable. The European technological lead in the fabrication of MOX fuels and that of France as regards vitrification, together with technical skills and surveillance other than by the principal partners could rapidly contribute to reducing, in a transparent manner, the substantial hazards generated by the consequences of the Second World War, to the benefit of world peace.

A structure of this kind, which is feasible today in the new order of international relations, would go some way to accomplishing the "reasonable utopia" proposed in 1946 in the Lilienthal-Acheson report, putting to an end to this "realistic folly", this perverse decline in reason constituted by the arms race and the "balance of terror".

---

61 Consideration could be given to taking over national establishments but this would raise problems including that of the acceptability of transporting fissile materials, particularly plutonium.

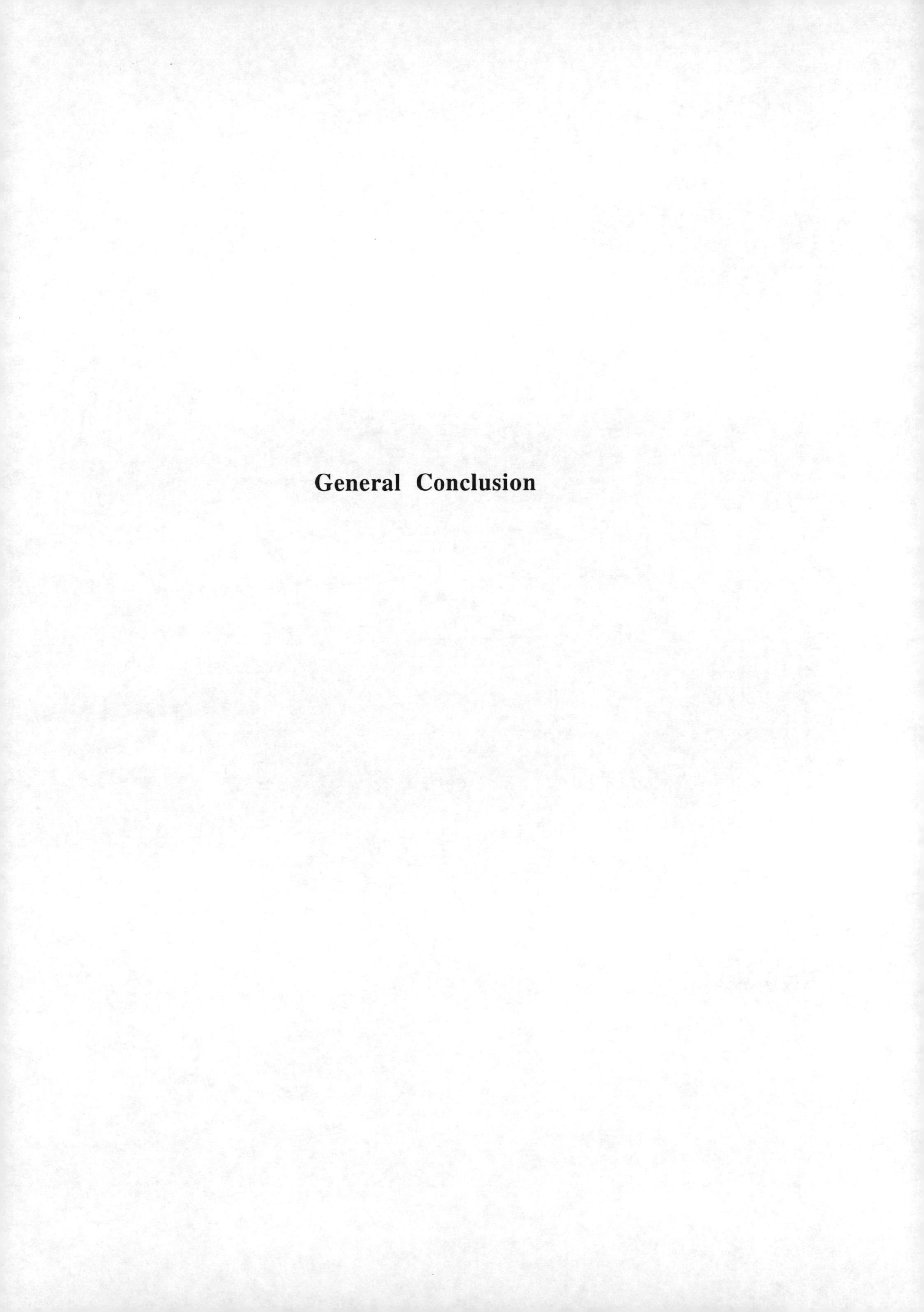

General Conclusion

General Conclusion

In this story of the Eurochemic company we have laid down the first milestones of a history of civilian nuclear technology and of European co-operation concerned with the tail-end of the fuel cycle.

The reprocessing industry, emerging from the demilitarisation of an industry that played a key part in the production of the plutonium bomb, was expanded in Europe under the Atoms for Peace programme and resulted in an intense transfer of technology. This transfer was geographical. The industry crossed the Atlantic and was one aspect of the adoption of American methods by the European economic actors in the aftermath of the Second World War. However the transfer was also sectoral. The radioactive materials chemical industry borrowed a great deal from petroleum chemistry and pharmaceuticals. However it also had to invent its own ways of dealing with its own specificities, notably as regards protecting workers against radiation.

The story of the life of Eurochemic, which was very brief, is a sort of microcosm of the problems that can face a nuclear undertaking and an international co-operative organisation.

Eurochemic was a pure product of the nuclear euphoria that gripped Europe in the second half of the 1950s.

The birth of Europe's nuclear industry was closely bound up with the contemporary perception of the regional "energy problem" and the search for a national replacement for coal. The need to ensure energy independence, regarded as one attribute of a government's power, was a powerful factor in the rise of the nuclear industry. From the outset therefore, reprocessing formed an integral part of the these development plans. By recycling the fissile uranium not "burned" in the power plants, it held out hope of utilising substantially more of the potential energy content of the nuclear fuel. By recovering the plutonium created by fission, possibly in fast breeder reactors, reprocessing appeared as one of the ways in which countries with a managed "fuel cycle" could escape from energy constraints.

This race for national power was strongly supported by countries looking to transform the situation of weakness in which they had been left by the Second World War. They relied heavily on the international organisations being created at the time. However because of lack of experience, the relevant costs, construction times and technical problems were largely underestimated. The circumstances governing decision-making and the completion of projects were more straightforward than they are today in that fewer actors were involved.

Eurochemic was one of the products of this period of abundance, optimism and close co-operation between public and private experts, national and international civil servants, engineering companies and the European chemical and metallurgical industries.

Apart from this desire on the part of countries for energy independence, there was another factor which gave considerable impetus to research into reprocessing: the "unspoken military aspect", which took two forms, opposite in nature but simultaneously present in co-operation.

For certain countries the acquisition of civilian reprocessing technology meant that one day they could consider making it military because, to use an expression widespread in the reprocessing field, "military isotopes don't wear uniforms". The attraction of the civilian technology enabling the fuel cycle – a long one – to be closed, at a time when nuclear power generation was only just taking off, can be attributed to the desire to implement a policy of energy independence in the very long term, but also to the desire to get into the club of atomic powers ready for the day when new members would be welcomed – which was not, in a Europe shaped by the Cold War, an absurdity.

At the same time other countries embarked upon European co-operation in reprocessing in the belief that mutual control was the best way of preventing proliferation, an approach initiated at Eurochemic ten years before the Non-Proliferation Treaty was signed.

However the true rate of development of nuclear power, the "languid years" of the 1960s, was to have a serious effect on the European reprocessing plant. The plans laid down in the 1950s, proposing the development of nuclear power as a way of ensuring energy independence, had been wrong about the role of oil and its low immediate cost for the economic actors. The "acceptable dependency" which was to feed the high rate of growth in Europe in the 1960s, was to considerably slow down the expansion of nuclear generation until the oil shocks of the 1970s, by which time the first industrial projects demonstrated that this new technique required a great deal of time and further progress if it was to be tamed. Thus Eurochemic found itself ahead of European needs, building up early and polyvalent experience in civilian reprocessing.

Thus Eurochemic was the first European plant to reprocess high burn-up oxide fuels from light water reactors. It perfected the reprocessing methods used in the most modern plants, contributed to resolving technical, material and organisational problems, and trained the scientists and technicians who were to be the seed corn for the whole of nuclear Europe.

However when the major nuclear power programmes were launched in Western Europe after the first oil shock, the political basis which had made European co-operation possible had disappeared. The reprocessing market had been shared out in the early 1970s by the three major regional industrial and nuclear powers. Thus the end of reprocessing activities at Eurochemic came only one year after the first oil shock.

The life of the company after 1975 was entirely focused on resolving the problem of managing its own wastes. Because of its short-lived reprocessing activities and its international character, Eurochemic was once again "ahead" of the other reprocessing plants and faced an unprecedented situation.

There was never any question of leaving the Mol plant just as it was after the shutdown. The participating countries assumed their responsibilities, some with a bad grace, and funded a programme to place the installation in a safe condition, the first of its kind to be applied to a reprocessing plant. The programme encompassed the partial decontamination of the buildings and stabilisation of the wastes produced by the nuclear activities on the site. Specific procedures were adopted for each type of waste: some of these were technical approaches developed by other organisations, others involved original aspects of applied research, as in the case of the alpha-contaminated solid wastes or the vitrification of the high activity liquid wastes. Although there is no doubt that the costs of this programme to the participating countries were reduced owing to the Belgian desire to resume activities on the site, thus sparing the member countries the cost of comprehensive decommissioning (the short-term completion of which is today very hypothetical) the history of Eurochemic demonstrates the extent of the financial burden generated by shutting down a nuclear undertaking of this kind, especially as the problem of the final storage of stabilized wastes still remains.

Of course one must be careful about relating this situation to the sites in operation today, with the savings achievable through automation, differences in design and methods of financing. The fact that Eurochemic was in operation for such a short time, the relevant economics, the experimental nature of the waste management techniques, and the substantial use of manpower in the dismantling operations, all contributed to inflating, for the company, the ratio between the running costs and the costs of making the site safe following the shutdown. It is clear however that the special requirements resulting from the nature of the work with radioactive substances and its legal and regulatory context – fortunately stringent – have substantial financial consequences that must be taken into account as soon as construction begins, and monitored during operations.

Of course this "lesson" taught by the history of Eurochemic must be applied not only to the nuclear field but to all undertakings that have a strong impact on the environment, if we are not to pass on to future generations costs that will not directly benefit them at all. In this context Eurochemic's "lead" supplements that of the European nuclear sector as a whole which, through its specificities, its peculiar hazards but also its close links with the public authorities in the democracies, has been subjected to constraints that prefigured those which are today being applied to other "heavy industries", at least in the countries that are more developed in economic and political terms.

ANNEXES

ANNEX 1

The Convention and Statute of 20 December 1957[1]

Convention on the Constitution of the European Company for the Chemical Processing of Irradiated Fuels (Eurochemic)

THE GOVERNMENTS of the Federal Republic of Germany, the Republic of Austria, the Kingdom of Belgium, the Kingdom of Denmark, the French Republic, the Italian Republic, the Kingdom of Norway, the Kingdom of The Netherlands, the Portuguese Republic, the Kingdom of Sweden, the Swiss Confederation and the Turkish Republic;

CONSIDERING that, in accordance with a Decision taken by the Council of the Organisation for European Economic Co-operation on 18th July, 1956, a Study Group has been formed among a number of Member countries of that Organisation, interested in the constitution of a joint undertaking for the chemical processing of irradiated fuels;

CONSIDERING that, on the basis of the studies made by the Study Group, the Government of the Federal Republic of Germany, the Government of the Republic of Austria, the Government of the Kingdom of Belgium, the Government of the Kingdom of Denmark, the Commissariat à l'Énergie Atomique in Paris, the Comitato Nazionale per le Ricerche Nucleari in Rome, the Government of the Kingdom of Norway, the Government of the Kingdom of The Netherlands, the Junta de Energia Nuclear in Lisbon, Aktiebolaget Atomenergi in Stockholm, the Government of the Swiss Confederation and the Government of the Turkish Republic have agreed to join in constituting a joint undertaking under the registered name of "The European Company for the Chemical Processing of Irradiated Fuels (EUROCHEMIC)";

CONSIDERING that this Company, both as regards its composition and its aims, has an international character and is in the general interests of the countries taking part;

CONSIDERING that the object of this Company is to carry out any research or industrial activity connected with the processing of irradiated fuels and the use of products arising therefrom, to contribute to the training of specialists in this field and thus to promote the production and peaceful uses of nuclear energy by the Member countries of the Organisation for European Economic Co-operation and, in furtherance of this object, to build before 1961 and operate a plant for the chemical processing of irradiated fuels and a research laboratory;

DESIROUS, in those circumstances, to give this Company all the support which it requires;

RECOGNISING that the constitution of the Company and its operations should be facilitated by special measures taken by the governments of the countries taking part, without, however, the facilities accorded to the Company constituting a precedent for other joint undertakings which might later be set up;

HAVE AGREED as follows:

Part I

Article 1

a) A joint undertaking under the registered name of "The European Company for the Chemical Processing of Irradiated Fuels (EUROCHEMIC)" (hereinafter referred to as the "Company") shall be constituted.

b) The constitution of the Company shall take place, in accordance with the provisions of the Statute annexed to the present Convention (hereinafter referred to as the "Statute"), after the signature of the Statute and upon the coming into force of the present Convention.

---

1 This edition is based, as regards the Convention and Statutes in their original version, on the first report of the Steering Committee for Nuclear Energy to the OEEC Council, March 1958, SCNE (1958) pp. 53-70. As concerns the modifications made to the Statute during the life of the company, it is based on the copy held by the secretary of the Board of Liquidators, updated to 15 April 1985.

Article 2

a) The Company shall be governed by the present Convention, by the Statute and, residuarily, by the law of the State in which its Headquarters are situated, insofar as the present Convention or the Statute do not derogate therefrom.

b) The Company shall possess juridical personality. It shall have power to do any act connected with its objects and in particular to conclude contracts, to acquire and dispose of movable and immovable property and to institute legal proceedings.

c) The character of public interest shall be recognised, in accordance with national laws, as regards the acquisition of the immovable property necessary for the establishment of the installations of the Company. The procedure of expropriation for reasons of public interest may be introduced by the Government in question in accordance with national law, with a view to acquiring such property in the absence of amicable agreement.

Article 3

The Governments party to the present Convention will take all action necessary, within their competence, to facilitate the Company doing any act connected with its objects, and in particular with processed fuels and recovered products.

Article 4

a) The provisions of the present Convention do not affect the rights and obligations resulting from the Treaty instituting the European Atomic Energy Community (EURATOM) signed at Rome on 25 March, 1957.

b) Contracts relating to source materials or special fissionable materials consigned from or destined for countries not members of the European Atomic Energy Community (EURATOM) shall benefit from the exceptions provided for in Article 75 of the said Treaty[2].

Article 5

a) The security control provided for by the Convention of 20th December, 1957, on the Establishment of a Security Control in the Field of Nuclear Energy shall be applicable to the operations of the Company and to its products and shall be exercised in accordance with the provisions of that Convention and of the Agreement provided for in Article 16 *a)* thereof[3].

Article 6

a) The installations and archives of the Company shall be inviolable. The property and assets of the Company, together with the materials despatched to it or by it, shall be immune from all administrative forms of requisition, expropriation or confiscation.

b) The property and assets of the Company may not be seized or be the subject of measures of enforced execution, except by an order of a court. Nevertheless, the installations and the materials necessary for the Company's activity may not be seized or be the subject of measures of enforced execution.

---

2 Quoted from ECSC-EEC-EAEC (1987), pp. 663 *et seq.* This article is included in the final section of Chapter VI concerning procurement – and in particular the Agency – and is entitled "Special Provisions":

"The provisions of this Chapter shall not apply to commitments relating to the processing, conversion or shaping of ores, source materials or special fissile materials and entered into,

a) ...

b) ...

c) by a person or undertaking and an international organisation or a national of a third State, where the material is processed, converted or shaped inside the Community and is then returned either to the original organisation or national or to any other consignee likewise outside the Community designated by such organisation or national.

The persons and undertakings concerned shall, however, notify the Agency of the existence of such commitments and, as soon as the contracts are signed, of the quantities of material involved in the movements...

The materials to which such commitments relate shall be subject in the territories of the Member States to the safeguards laid down in Chapter VII (Safeguards). The provisions of Chapter VIII (Property Ownership) shall not, however, be applicable to special fissile materials covered by the commitments referred to in subparagraph *c)*."

3 SCNE (1958), pp. 174-175: "An agreement shall be entered into between the Organisation and the European Atomic Energy Community (EURATOM) defining the arrangements under which the control established by the present Convention shall be carried out, within the territory to which the Treaty instituting the European Atomic Energy Community (EURATOM) signed at Rome on 25th March, 1957, applies, by the competent bodies of EURATOM by delegation from the Agency in order to attain the objectives of the present Convention. Proposals to this effect shall be submitted to the European Commission set up by the said Treaty as soon as it is constituted in order that such an agreement may be reached with the minimum delay."

c) The provisions of the present Article shall not prevent the competent authorities of the Headquarters State or of other countries where installations and archives of the Company are situated from having access to the installations and archives of the Company in their respective territories in order to ensure the execution of judicial decisions or regulations for the protection of public health and the prevention of accidents.

Article 7

a) The Company shall be exempt in the Headquarters State from all fees and taxes whether fiscal or quasi-fiscal at its constitution and when its capital is subscribed or increased, its shares are issued, as well as from various formalities which its activities may require in the Headquarters State. Similarly, it shall be exempt from all fees and taxes on its dissolution or winding up.

b) The Company shall be exempt in the Headquarters State as well as in other countries where its installations are situated from fees and taxes payable upon the acquisition of immovable property and from inscription and registration fees.

c) The Company shall be exempt in the Headquarters State from all direct taxes which might be imposed as regards its property, assets and income.

d) The Company shall be exempt from taxes of an exceptional or discriminatory nature levied by the Headquarters State, such as *ad hoc* capital levy or any tax not payable by other companies engaged in similar activities.

e) The exemptions laid down in the present Article shall not apply to any fee or tax charged in respect of any public utility service.

Article 8

a) Such raw materials, capital equipment and scientific and technical material as are necessary for the installations and for the operations of the Company shall, subject to the provisions of Article 9, be exempt from all customs duties or charges of like effect and from all import restrictions.

b) Products thus imported must not be resold on the territory of the country into which they were imported, except under conditions agreed with the Government of that country.

c) The import and export of fissionable materials destined for the Company as well as materials produced or recovered by the Company which are destined for the countries whose Governments are party to the present Convention and which are shareholders or have nationals who are shareholders in the Company (hereinafter referred to as "countries taking part") shall be exempt from all customs duties or charges of like effect and from all restrictions.

Article 9

a) The Company may, for the fulfilment of its objects, acquire, hold, and use currency of the Contracting Parties to the Agreement of 19th September, 1950, for the Establishment of a European Payments Union and of other countries whose Governments are party to the present Convention. The Governments party to the present Convention shall grant to the Company, where appropriate, any necessary authority in accordance with the procedure laid down in the regulations and agreements applicable.

b) The Governments party to the present Convention shall grant to the Company as freely as possible, any authority necessary to enable it to acquire, hold, and use currency not covered by paragraph *a)* of the present Article.

Article 10

a) The Company may recruit, without let or hindrance, technical, clerical and skilled manual staff from among the nationals of the countries taking part.

b) In particular, the Headquarters State will not apply provisions on immigration or alien registration in such a manner as to impede the recruitment or repatriation of skilled staff from the other countries taking part, except when this would be contrary to public policy, national security or public health.

c) Persons employed by the Company

- *i)* shall have the right, at the time of first taking up their position in the country in question, to import free of duty their furniture and effects from the country where they last resided or of which they are nationals, and to re-export free of duty their furniture and effects on the termination of their employment, subject in both cases to any conditions deemed requisite by the Government of the country in which the said rights are exercised;

ii) shall have the right to import free of duty any motorcar owned by them for their personal use obtained in the country where they last resided or of which they are nationals, under the normal internal market conditions prevailing in that country and to re-export the same free of duty, subject in both cases to any conditions deemed requisite by the Government of the country in which the said rights are exercised.

PART II

Article 11

a) The Company shall report each year to the Governments of the countries taking part on its development and financial position.

b) The reports of the Company shall be submitted to a Special Group of the Steering Committee of the European Nuclear Energy Agency (hereinafter referred to as the "Special Group") composed of representatives of the Governments of the countries taking part.

Article 12

a) The Special Group shall consider any problems of common interest to the Governments party to the present Convention which may be raised by the operations of the Company and shall propose the measures found necessary in that connection.

b) If it subsequently appears that the legislative provisions applied in the Headquarters State or in any other country taking part may give rise to difficulties in the operations carried out by the Company in pursuance of its objects, the Special Group shall propose measures for resolving such difficulties in the spirit of the present Convention.

c) Proposals formulated by the Special Group under the present Article shall be adopted by a simple majority.

Article 13

a) Any director or shareholder of the Company may submit to the Special Group difficulties arising in connection with

i) the processing of fuel consigned from the countries taking part or the allocation of the products recovered;

ii) the use of the resources of the Company for the development of research;

iii) the communication of the results of research.

b) When such application is made to the Special Group it will take a decision by a three-quarters majority of its members which will be binding upon the Company.

Article 14

a) The approval of the Special Group shall be required for amendments to the Statute concerning

– the headquarters of the Company (Article 2);

– its objects (Article 3);

– the conditions for the admission of new shareholders (Article 8);

– the adoption of decisions of the General Assembly (Article 15);

– the composition of the Board of Directors (Article 18);

– the adoption of decisions of the Board of Directors (Article 23);

– knowledge and patents (Article 26);

– the interim period (Article 27).

b) The approval of the Special Group shall be required for decisions of the Company concerning

i) the extension of the period fixed for the duration of the Company;

ii) the conclusion of contracts relating to the processing of fuel consigned from countries not taking part or to the delivery of special fissionable materials to such countries;

iii) the construction by the Company and the fixing of the site of any new plant and any extension of the existing plant leading to a new plant of large size.

c) In the case of transfer of shares or of subscription rights to any person who is not of the same nationality as the transferor, the choice of the transferee shall be subject to the approval of the Special Group. However, the Special Group shall not have the power to prevent the transfer of shares by a Government, which has declared its intention to give notice under Article 18 *a)* of the present Convention, or by shareholders nationals of such a Government, to Governments party to the present Convention or their nationals.

d) Decisions taken by the Special Group under the present Article shall be adopted unanimously by its members.

Article 15

a) The approval of the Special Group shall also be required for

i) amendments to the provisions of the Statute other than amendments contemplated in Article 14;

ii) any increase or reduction of capital which would result in changing the distribution of capital between the shareholders.

b) Decisions taken by the Special Group under the present Article shall be adopted by a three-quarters majority of its members.

Article 16

Any dispute arising between Governments party to the present Convention concerning the interpretation or application thereof shall be examined by the Special Group and in the absence of friendly settlement, may be submitted by agreement between the Governments concerned to the Tribunal established by the Convention of 20th December, 1957, on the Establishment of a Security Control in the Field of Nuclear Energy[4].

Article 17

a) The present Convention shall be concluded for a period of fifteen years. It will be automatically extended for periods of five years, if, at the end of the preceding period, the Company is still in existence.

b) However the continuation in force of all or part of the provisions of Article 7 and paragraphs *a)* and *b)* of Article 8 of the present Convention beyond the first period of five years shall be subject to a decision of the Special Group adopted unanimously by its members which will fix the duration of such further period.

c) The present Convention shall cease to be in force upon the completion of the period of the winding-up of the Company.

Article 18

a) A Government party to the present Convention which is not or is no longer a shareholder and has no national who is or continues to be a shareholder in the Company may, insofar as it is concerned, after a period of fifteen years, terminate the application of the present Convention by giving three months' notice to the Secretary-General of the Organisation for European Economic Co-operation.

b) Nevertheless, should this country be the Headquarters State, or a country in which any installation of the Company is situated, the present Convention shall not be terminated insofar as that country is concerned, unless the headquarters or installation of the Company shall have been transferred to another country.

Article 19

a) The Government of any Member or Associate country of the Organisation for European Economic Co-operation which is not a Signatory to the present Convention may accede thereto, provided that it becomes a party to the Convention of 20th December, 1957, on the Establishment of a Security Control in the Field of Nuclear Energy, by notification addressed to the Secretary-General of the Organisation.

b) The Government of any other country which is not a Signatory to the present Convention may accede thereto, provided that it becomes a party to the Convention of 20th December, 1957, on the Establishment of a Security Control in the Field of Nuclear Energy, by notification addressed to the Secretary-General of the Organisation and with the unanimous assent of the Special Group. Such accession shall take effect as from the date of such assent.

Article 20

a) The present Convention shall be ratified. Instruments of ratification shall be deposited with the Secretary-General of the Organisation for European Economic Co-operation.

4 Part III, Articles 2 to 15, in SCNE (1958) pp. 173-174. Article 16 has never been used.

b) The present Convention shall come into force when it has been ratified by the Government of the Headquarters State and when that part of the authorised capital allotted under Article 4 of the Statute to the Governments which have deposited their instruments of ratification, or to nationals of those Governments, amounts to 80 per cent of the capital of the Company.

c) For each Signatory ratifying thereafter, the present Convention shall come into force upon the deposit of its instrument of ratification.

Article 21

The Secretary-General of the Organisation for European Economic Co-operation shall give notice to all Governments party to the present Convention and the Company of the receipt of any instrument of ratification or accession, or of any notice of withdrawal. He shall also notify them of the date on which the present Convention comes into force.

IN WITNESS WHEREOF, the undersigned Plenipotentiaries, duly empowered, have signed the present Convention.

DONE in Paris, this twentieth day of December Nineteen Hundred and Fifty Seven, in the French, English, German, Italian and Dutch languages, in a single copy which shall remain deposited with the Secretary-General of the Organisation for European Economic Co-operation by whom certified copies will be communicated to all Signatories.

Annex

Statute of the European Company for the Chemical Processing of Irradiated Fuels (Eurochemic)[5]

Part I

NAME, OBJECTS, HEADQUARTERS, DURATION AND CAPITAL

Article 1

There is hereby constituted a joint undertaking under the name of "The European Company for the Chemical Processing of Irradiated Fuels (EUROCHEMIC)" which shall take the form of a joint stock company governed by the International Convention on the Constitution of the said Company (hereinafter referred to as the "Convention"), by the present Statute and, residuarily, by the law of the State in which its headquarters are situated.

Article 2

The headquarters of the Company shall be at Mol (Belgium).

The Company is established for a duration of fifteen years.

Article 3

The Company will build before 1961 and operate a plant and a laboratory for the processing of irradiated fuels; it will also ensure the development of techniques and the training of specialists in this field.

The Company will carry out any research or industrial activity with a view to enabling Member countries of the Organisation for European Economic Co-operation[6] to process the fuels used in their nuclear reactors under economic conditions.

When the quantity of irradiated fuels which Member countries of the Organisation for European Economic Co-operation wish to send for processing in a joint installation seems likely to exceed the capacity of this plant, the Company should examine means to meet the demand of these countries under economic conditions.

Article 4

The authorised capital of the Company shall be 20 million European Payments Union units of account. It shall be divided into 400 shares, each of the nominal value of 50,000 units of account, initially subscribed and allotted in full as follows:

The Government of the Federal Republic of Germany	68 shares	3,400,000
The Government of the Republic of Austria	20 shares	1,000,000
The Government of the Kingdom of Belgium	44 shares	2,200,000
The Government of the Kingdom of Denmark	22 shares	1,100,000
The Commissariat à l'Énergie Atomique in Paris	68 shares	3,400,000
The Comitato Nazionale per le Ricerche Nucleari in Rome[7]	44 shares	2,200,000
The Government of the Kingdom of Norway	20 shares	1,000,000
The Government of the Kingdom of The Netherlands	30 shares	1,500,000
La Junta de Energia Nuclear in Lisbon[8]	6 shares	300,000
Aktiebolaget Atomenergi in Stockholm[9]	32 shares	1,600,000

---

5 The text of December 1957 is shown in normal characters, with subsequent deletions in brackets. *Subsequent modifications and additions are shown in bold italic characters.* Substantial changes are dated at the beginning of the Articles, the others in notes.

6 The Organisation for European Economic Co-operation (OEEC) was transformed into the Organisation for Economic Co-operation and Development (OECD) by the Convention on the Organisation for Economic Co-operation and Development which came into force on 30 September 1961.

7 This organisation was dissolved under the law of 11.8.1960; its rights and liabilities were taken over by the Comitato Nazionale per l'Energia Nucleare (CNEN). In 1982 (law n° 84 of 6.3.1982), the CNEN was reorganised and became the Comitato Nazionale per la Ricerca e per lo Sviluppo dell'Energia Nucleare e delle Energie Alternative (ENEA).

8 The Junta ceased to exist in accordance with the decree-laws n° 548/77 of 31.12.1977 and 361/79 of 1.9.1979. In implementation of Order n° 126/78 of 31.5.1878, Portuguese participation in Eurochemic was then through the Energy Directorate of the Ministry for Industry and Technology.

9 After 5.6.1978 this organisation was renamed Studsvik Energiteknik AB with its head office in Nyköping.

The Government of the Swiss Confederation	30 shares	1,500,000
The Government of the Turkish Republic	16 shares	800,000

Article 4 bis (1959)[10]

The authorised capital is increased to 21.5 million European Payments Union units of account, divided into 430 shares, each of a nominal value of 50,000 units of account. The 30 new shares (1,500,000) are allotted to the Junta de Energia Nuclear of Madrid.

Article 4 ter (1963)[11]

The authorised capital is increased to 28.95 million European Monetary Agreement units of account, divided into 579 shares, each of a nominal value of 50,000 units of account. The 149 new shares, fully subscribed, are allotted as follows:

The Government of the Federal Republic of Germany	***32 shares***	***1,600,000***
The Government of the Republic of Austria	***5 shares***	***250,000***
The Government of the Kingdom of Belgium	***36 shares***	***1,800,000***
The Government of the Kingdom of Denmark	***5 shares***	***250,000***
The Commissariat à l'Énergie Atomique in Paris	***32 shares***	***1,600,000***
The Junta de Energia Nuclear in Madrid	***8 shares***	***400,000***
The Government of the Kingdom of Norway	***4 shares***	***200,000***
The Government of the Kingdom of The Netherlands	***9 shares***	***450,000***
Aktiebolaget Atomenergi in Stockholm[12]	***10 shares***	***500,000***
The Government of the Swiss Confederation	***8 shares***	***400,000***

The shares allotted to the Government of the Kingdom of Belgium under the terms of the present Article shall not be taken into account in any decisions which may be taken in regard to the coverage of any eventual operating losses which the Company may incur, and shall confer no right to any possible profits.

Article 4 quater (1964)[13]

The authorised capital is increased to 35.75 million European Monetary Agreement units of account, divided into 712 shares, each of a nominal value of 50,000 units of account and into 6 shares, each of a nominal value of 25,000 units of account.

The new shares, fully subscribed, that is 133 shares, each of a nominal value of 50,000 units of account, and the 6 shares, each of a nominal value of 25,000 units of account, are allotted as follows:

The Government of the Federal Republic of Germany	***28 shares of 50,000***	
	1 share of 25,000	***1,425,000***
The Government of the Republic of Austria	***5 shares of 50,000***	***250,000***
The Government of the Kingdom of Belgium	***22 shares of 50,000***	***1,100,000***
The Government of the Kingdom of Denmark	***5 shares of 50,000***	
	1 share of 25,000	***275,000***
The Commissariat à l'Énergie Atomique in Paris	***28 shares of 50,000***	
	1 share of 25,000	***1,425,000***
The Junta de Energia Nuclear in Madrid	***10 shares of 50,000***	
	1 share of 25,000	***525,000***
The Government of the Kingdom of Norway	***4 shares of 50,000***	
	1 share of 25,000	***225,000***
The Government of the Kingdom of the Netherlands	***10 shares of 50,000***	
	1 share of 25,000	***525,000***

---

10 This article was inserted by decision of the General Assembly on 28 July 1959, approved by the Special Group on 29 October 1959.

11 This article was inserted by decision of the General Assembly on 18 June 1963 and approved by the Special Group on 27 June 1963.

12 From 5.6.1978 this organisation was renamed Studsvik Energiteknik AB with its head office in Nyköping.

13 This article was inserted by decision of the General Assembly on 1 July 1964 and approved by the Special Group on 1 July 1964.

Aktiebolaget Atomenergi in Stockholm[14]	*12 shares of 50,000*	*600,000*
The Government of the Swiss Confederation	*9 shares of 50,000*	*450,000*

Article 4 quinquies (1975-1982)[15]

The authorised capital is reduced to 33.275 million European Monetary Agreement units of account, divided into 663 shares, each of a nominal value of 50,000 units of account and into 5 shares, each of a nominal value of 25,000 units of account. The 49 shares, each of a nominal value of 50,000 units of account and one of a nominal value of 25,000 units of account, allotted to the government of the Kingdom of the Netherlands in accordance with Articles 4, 4 ter, and 4 quater, are cancelled following their transfer to the Company.

Article 4 sexies (1980-1982)[16]

The authorised capital is reduced to 32.475 million European Monetary Agreement units of account, divided into 647 shares, each of a nominal value of 50,000 units of account and into 5 shares, each of a nominal value of 25,000 units of account. The 16 shares, each of a nominal value of 50,000 units of account, allotted to the government of the Turkish Republic in accordance with Article 4, are cancelled following their transfer to the Company.

Article 5

The shares of the Company shall be paid up to 20 per cent when the Company is constituted. The General Assembly shall have power to decide when other portions shall be paid up, according to the Company's needs and the progress of its work, bearing in mind the objects as laid down in Article 3.

If, within six months of the date on which the Convention comes into force, any Signatory is not in a position to ratify it, the General Assembly will be summoned in order to decide upon measures to be taken to ensure that the whole of the capital is subscribed.

Article 6

The shares shall be registered in the name of the holder.

They shall not be transferable except with the agreement of the General Assembly. Nevertheless, the General Assembly shall not have the power to prevent the transfer of shares by a shareholder to a person having the same nationality provided that the Government having jurisdiction over such a person has given its approval.

If, however, a Government declares its intention to give notice under Article 18 *a)* of the Convention, the transfer of shares by this Government or by its nationals to Governments party to the Convention or their nationals cannot be prevented by the General Assembly.

The Company shall maintain a Share Register in which the names and addresses of the shareholders shall be entered. The Company shall recognise as shareholders only those whose names are entered in this register.

Article 6 bis (1969)[17]

Beneficiaries' shares shall be attributed to shareholders who contribute to the operating expenses of the Company and to shareholders who are nationals of a Government that makes such contributions and who are designated by it, on the basis of one share for each payment of 5000 European Monetary Agreement units of account.

Beneficiaries' shares shall be attributed on 1st July 1969 for an amount equal to that of contributions paid up to that date and subsequently on 1st July of each year for an amount equal to that of contributions paid during the preceding year. The fractions of contributions below 5000 EMA u/a shall not give a right to beneficiaries' shares. Contributions which have not given the right to a beneficiary's share shall be taken into account in subsequent attributions.

---

[14] From 5.6.1978 this organisation was renamed Studsvik Energiteknik AB with its head office in Nyköping.

[15] This article was inserted by decision of the General Assembly on 11 September 1975 and approved by the Special Group on 30 June 1982.

[16] This article was inserted by decision of the General Assembly on 2 July 1980 and approved by the Special Group on 30 June 1982.

[17] By decision of the General Assembly on 26 June 1969, approved by the Special Group on 16 December 1969, Article 6 bis and the second paragraph of Article 30 were inserted, and Articles 14, 15 (paragraphs 1 to 3), 18 (paragraph 5), 23 (paragraph 2), 28 (paragraph 2) and 31 (paragraphs 2 and 3) were modified.

Beneficiaries' shares shall not be transferable.

The company shall maintain a Register of Beneficiaries' Shares in which the names and addresses of shareholders of such shares shall be entered.

Article 7

The capital of the Company may be increased, by the creation of new shares representing holdings in money or in kind, or reduced, by a vote of the General Assembly. When the capital is increased each shareholder shall be entitled to subscribe for a number of the new shares in proportion to the total number of shares registered in his name at the time of the increase, subject to the provisions of Article 8. Should any shareholder not exercise his right to subscribe, such right may, with the approval of the General Assembly, be transferred to another shareholder, without the General Assembly being able to prevent such transfer in the case provided in the second sentence of the second paragraph of Article 6.

The General Assembly shall lay down the conditions under which new shares may be issued and the rules governing the paying up of such shares in kind.

Article 8

Any Government or any person deriving his right to shares from or through a Government party to the Convention may be admitted as a shareholder in the Company by a decision of the General Assembly either by a transfer of shares or by subscription to an increase in the capital of the Company. In this case, all or a part of the new shares will be allotted for the new shareholder by decision of the General Assembly.

PART II

GENERAL ASSEMBLY

Article 9

The bodies of the Company shall be the General Assembly and the Board of Directors who will carry out their duties subject to the powers reserved to the Special Group established by the Convention

Article 10

The General Assembly shall be composed of all the shareholders of the Company. A representative of the European Nuclear Energy Agency[18] and a representative of the European Atomic Energy Community (EURATOM) shall take part in the General Assembly in an advisory capacity.

It shall be the superior body of the Company, and shall have the following powers:

1. To appoint the members of the Board of Directors and their Alternates and fix the remuneration of the members of the Board of Directors.
2. To appoint the Auditors.
3. To amend the present Statute.
4. To decide upon the paying up of new portions of the capital.
5. To decide upon any increase or reduction in the authorised capital.
6. To make any decision as to the transfer of shares or subscription rights.
7. To declare the extension of the period fixed for the duration of the Company.
8. To declare the dissolution of the Company.
9. To appoint the liquidators.
10. To approve the rules of management referred to in Article 21.
11. To consider the auditors' report, to examine and approve the management report, balance-sheet and profit and loss account, to decide on the allocation of the net profit and to receive the Directors' report on their conduct of the Company's business.
12. To approve the annual report to the Governments of the countries taking part.
13. To fix the maximum amount which may be borrowed during a given period.
14. To decide any other question reserved by law to it or submitted to it by the Board of Directors.

[18] This title was changed to "OECD Nuclear Energy Agency" by a decision of the OECD Council on 17 May 1972.

Article 11

The first meeting of the General Assembly shall be convened by the Secretary-General of the Organisation for European Economic Co-operation[19] within one month of the coming into force of the present Statute.

An ordinary meeting of the General Assembly shall be held each year and shall be convened by the Board of Directors within six months of the date to which the accounts are made up.

Article 12

Extraordinary meetings shall be convened:

1. On the decision of the General Assembly or of the Board of Directors;
2. At the request of the Special Group provided for in Article 11 of the Convention;
3. At the request of the Board of Auditors;
4. At the request of one or more shareholders whose shares together amount to at least one-tenth of the authorised capital. Such request shall be made in writing and shall specify the purpose for which the meeting is to be held.

The method of summoning an extraordinary meeting and the procedure thereat shall be in the same form as those of an ordinary meeting.

Article 13

The shareholders shall be summoned to a meeting of the General Assembly by registered letter at least two weeks before the date of the meeting.

The summons shall specify the business to be transacted at the meeting and, if such business includes any amendment to the present Statute (sub-clauses 3, 5, 7 and 8 of Article 10), the purport of such amendment shall be fully set out.

No decision shall be made on any matter not specified in the notice summoning the meeting, except in the case of a proposal made at the meeting to summon an extraordinary meeting of the General Assembly.

The General Assembly shall meet at the headquarters of the Company, unless the Board of Directors otherwise decides.

Article 14 (in part, 1969)[20]

The number of votes held by shareholders at a meeting of the General Assembly shall be proportional to the nominal value of all the shares (registered in their respective names. Each share shall carry the right to one vote.) ***and all the beneficiaries' shares registered in their respective names. Each share with a nominal value of 50,000 European Monetary Agreement units of account shall give the right to ten votes, each share with a nominal value of 25,000 units of account shall give the right to five votes and each beneficiary share shall give the right to one vote.***

Article 15 (in part, 1969)[21]

On a meeting of the General Assembly being summoned, it shall be entitled to proceed to business as soon as a majority of the shares ***and beneficiaries' shares*** are represented. Should this quorum not be present at the first session, a further session shall be convened upon at least two weeks' notice, and such a session shall be entitled to proceed to business whatever may be the number of shares represented.

The General Assembly shall take its decisions by the majority vote of the shares ***and beneficiaries' shares*** represented.

Decisions in the case of the powers set out in sub-clauses 3 to 8 and 13 of Article 10 shall require a majority of two-thirds (of the authorised capital) ***of the shares and beneficiaries' shares.***

Voting shall take place by show of hands, unless a shareholder asks for a secret ballot.

---

[19] Organisation for Economic Co-operation and Development (OECD) after 30 September 1961.

[20] Article 14 was modified by decision of the General Assembly on 26 June 1969, approved by the Special Group on 16 December 1969.

[21] Article 15 (paragraphs 1 to 3) was modified by decision of the General Assembly on 26 June 1969, approved by the Special Group on 16 December 1969.

Article 16

The Chairman of the Board of Directors, or, should he be unable to attend, one of the Vice-Chairmen, or, in default thereof, one of the Directors appointed by the Board shall be Chairman of the meetings of the General Assembly.

The General Assembly shall, by show of hands, appoint two tellers. It shall also appoint a Secretary, who need not necessarily be a shareholder.

Article 17

The discussions and decisions of the General Assembly shall be recorded in Minutes.

The Minutes shall be signed by the Chairman of the meeting, the tellers and the Secretary.

Copies or extracts shall be signed by the Chairman or one of the Vice-Chairmen of the Board.

PART III

BOARD OF DIRECTORS

Article 18

The Board of Directors shall be responsible for managing the business of the Company.

The Board of Directors shall consist of (15) ***16*** Directors. The Directors and their Alternates shall be appointed by the General Assembly regardless of nationality. A representative of the European Nuclear Energy Agency[22] and a representative of the European Atomic Energy Community (EURATOM) shall take part in the sessions of the Board of Directors in an advisory capacity[23].

Each shareholder or group of shareholders holding at least 5 per cent of the Company's (shares) ***initial capital*** shall be entitled to a seat on the Board of Directors and shall propose to the General Assembly the appointment of a Director and an Alternate[24].

The Directors and their Alternates shall be appointed for a period of three years. They may be re-elected. After the first period of three years, one-third of the Board will be replaced each year. To that end, at the meeting of the General Assembly following the expiry of the Company's third financial year, lots will be drawn to determine which Directors shall retire at the end of the Company's fourth and fifth financial years.

(All the Directors shall have an equal vote.) ***Each Director shall have a right of vote proportional to the votes of the shareholder or the group of shareholders who have proposed his appointment as determined by Article 14 of the present Statute[25].***

Article 19

The Directors and their Alternates shall be elected at an ordinary meeting of the General Assembly. The same procedure shall be followed upon a casual vacancy, unless a shareholder requests the vacancy be filled forthwith. In that event, the Board of Directors shall immediately summon an extraordinary meeting of the General Assembly to elect a new Director.

Article 20

The Chairman and Vice-Chairmen of the Board of Directors shall be appointed annually by the Board of Directors. They may be re-appointed. The Board shall appoint a Secretary who may not necessarily be one of its members.

Should the Chairman be unable to attend, one of the Vice-Chairmen or, in default, the eldest Director present at the meeting shall take the chair of the Board.

---

[22] "OECD Nuclear Energy Agency" by a decision of the OECD Council of 17 May 1972.

[23] The second paragraph of Article 18 was modified by decision of the General Assembly on 28 July 1959, approved by the Special Group on 29 October 1959.

[24] The third paragraph of Article 18 was modified by decision of the General Assembly of 28 July 1959, approved by the Special Group on 29 October 1959.

[25] Article 18 (paragraph 5) was modified by decision of the General Assembly on 26 June 1969, approved by the Special Group on 16 December 1969.

Article 21

The Board of Directors shall have power to determine any matter not coming within the competence of another body of the Company.

The Board of Directors shall have power to delegate all or any part of the management of the Company to one or more of its members or to third persons, who need not necessarily be Directors. It shall draw up rules of management which shall define the rights and duties of the Board of Directors, its delegates and the Managerial Staff.

In these rules, which must be approved by the General Assembly, the Board of Directors must, however, reserve to itself the right of determining the following matters:

1. Composition of the Managerial Staff, and the establishment of conditions of appointment and dismissal for members thereof including the acceptance of their resignation.
2. Appointment of the Directors and persons not on the Board of Directors (members of the Managerial Staff and signing clerks) empowered to sign on behalf of the Company.
3. Appointment of the Managing Director of the Company.
4. Negotiation of loans, in whatever form, subject to any limits imposed by the General Assembly.
5. Conclusion of contracts relating to the processing of irradiated fuels or to the allocation of the special fissionable materials recovered.
6. Conclusion of contracts relating to patents, rights of provisional protection, or utility models owned by the Company.
7. Necessary arrangements for the exercise of security control and any arrangement whatsoever with the European Nuclear Energy Agency[26].
8. Construction by the Company and the fixing of the site of any new plant and any extension of the existing plant leading to a new plant of large size.
9. Preparation of the management report, the annual report to the Governments of the countries taking part, the annual balance-sheet and the substance of any proposal to be submitted to the General Assembly. (It shall cause the accounts to be audited by accountants who have no part in the management of the Company)[27].

Article 22

The Board of Directors shall meet on the summons of the Chairman or one of the Vice-Chairmen as often as business requires and at least once every three months. Members of the Board shall be summoned by registered letter which shall specify the business to be transacted and which shall be despatched at least eight days before the date of meeting.

The Chairman must summon the Board on the written request of a Director, specifying the matter to be considered at the meeting. In this event, such meeting shall be held not later than two weeks after the receipt of the letter of request.

The summons shall indicate the place of meeting.

A Director who is unable to attend a meeting, and whose Alternate is also unable to attend, may vote in writing or may appoint another Director or Alternate expressly empowered to vote on his behalf, as his proxy. No Director or Alternate can act as proxy for more than one of his colleagues.

In urgent cases, decisions may be taken by letter or telegram, unless any of the Directors requests that a meeting should be summoned for the purpose.

Article 23[28]

The Board of Directors has power to hold discussions and take decisions only if it has been regularly summoned and if the majority of the Directors is present or represented by Alternates or by proxy.

Decisions of the Board shall be taken by a majority ***of the votes*** of the Directors present or represented by Alternates or by proxy. Should the votes be equally divided, the Chairman of the meeting has a second or

---

[26] "OECD Nuclear Energy Agency" by a decision of the OECD Council on 17 May 1972.

[27] Last sentence deleted by decision of the General Assembly on 25 February 1960, approved by the Special Group on 1 June 1960.

[28] The decision of the General Assembly of 26 June, 1969 to amend this Article was approved by the Special Group on 16 December, 1969.

casting vote. Exceptionally, for decisions on the matters enumerated in sub-clauses 3 to 8 of Article 21, a two-thirds majority ***of votes*** is required.

Article 24

The discussions and decisions of the Board of Directors shall be recorded in Minutes.

The Minutes shall be signed by the Chairman of the meeting and by the Secretary.

Copies or extracts shall be signed by the Chairman or one of the Vice-Chairmen.

Article 25

The remuneration of the Directors shall be fixed by the General Assembly.

PART IV

KNOWLEDGE AND PATENTS

Article 26

a) The shareholders shall be informed of the results of the scientific research and of the information obtained from the activities of the Company, except for knowledge obtained by the Company and which is not freely at its disposal. However, this obligation shall not prevent the Company from taking the necessary steps to ensure the protection of its inventions.

b) The results and information mentioned in the preceding paragraph shall be circulated through reports to the shareholders, who may, in addition, send trainees to the installations of the Company; the shareholders shall be responsible for the remuneration of trainees. The Board of Directors shall prepare rules for the admission of trainees. The maximum number of trainees for each shareholder shall be determined, taking into account his participation in the Company's capital.

c) The shareholders shall be entitled to acquire non-exclusive licences under patents, rights of provisional protection, or utility models belonging to the Company. They are also entitled to obtain sub-licences of licences of which the Company has the right to grant sub-licences. The conditions for these licences and sub-licences shall be fixed for all the interested shareholders without discrimination.

d) The communication of knowledge obtained from the Company by shareholders as well as the granting by shareholders, who are licensees of the Company, of sub-licences to third persons shall be subject to the consent of the Board of Directors. Nevertheless, the Board of Directors shall not have the power to prevent the communication of knowledge or the granting of a sub-licence by a Government or a public institution to an enterprise which undertakes to exploit, in its country, the knowledge or the invention in question.

e) The employees and trainees of the Company may not, without authorisation, communicate knowledge which they may have acquired concerning the work of the Company.

PART V

ACCOUNTS. LIQUIDATION

Article 27

The Company shall take over rights and obligations assumed by, and will reimburse for any special expenditure incurred by, the Organisation for European Economic Co-operation[29] in pursuance of the objects of the Company.

Article 28

The accounts of the Company shall be audited by a Board of three auditors elected by the General Assembly for a period of three years. ***The auditors may be re-elected. One of the three seats on the Board shall be renewed each year. The period of office of the first auditors elected shall be fixed, by ballot, at one, two and three years respectively***[30]**.**

---

[29] This title was changed to "OECD Nuclear Energy Agency" by a decision of the OECD Council on 17 May 1972.

[30] The first paragraph of Article 28 was modified by decision of the General Assembly on 25 February 1960, approved by the Special Group on 1 June 1960.

Any shareholder or group of shareholders representing 20 per cent of the authorised capital ***as well as any holder of beneficiaries' shares or group of holders of beneficiaries' shares representing 20 per cent of the total amount of beneficiaries' shares*** may require appointment of an additional auditor[31].

The auditors shall, in particular, be responsible for ascertaining whether the balance-sheet and the profit and loss account tally with the books, whether the latter are carefully kept and whether the Company's assets and the financial management are in conformity with the rules governing the Company under Article 1.

In the execution of their work, the auditors shall be entitled to consult the Company's books of account and all relevant papers. The balance-sheet and the profit and loss account must be submitted to them at least thirty days before the date of the ordinary meeting of the General Assembly.

They shall report in writing and submit their proposals to the meeting of the General Assembly at which a decision on the accounts is to be taken.

Article 29

The accounts and balance-sheet must be closed at the end of each calendar year.

The balance-sheet must be drawn up in conformity with the recognised principles of sound business management.

Article 30

Out of the balance remaining after the deduction of depreciation, a sum amounting to 5 per cent shall firstly be allocated to the ordinary reserve fund, until the latter attains one-fifth of the authorised capital already paid up. The reserve fund can only be drawn on to cover deficits.

The net profit shall be distributed between shareholders in proportion to the nominal value of shares and the amount of beneficiaries' shares of which they are registered holders. To this effect the amount of beneficiaries' shares shall be calculated on the basis of a lump sum of 5000 EMA u/a per share[32].

Article 31

In case of dissolution of the Company, the Company shall go into liquidation, and shall, from that time, be deemed to exist for the purpose of liquidation.

It shall be wound up by ***a Board of*** Liquidators appointed by the General Assembly. (Any shareholder or group of shareholders representing 20 per cent of the authorised capital may require the appointment of an additional liquidator. The liquidators shall have full power to realise the assets of the Company.) ***Each holder of beneficiaries' shares shall be entitled to one seat on the Board of Liquidators and shall propose to the General Assembly the appointment of a liquidator and an alternate. Each liquidator shall have a right of vote proportional to the number of votes attributed according to Article 14 to the shares and beneficiaries' shares held by the nationals of his country[33].***

Once the liabilities of the Company have been satisfied and the shares repaid, the balance remaining shall be distributed between the shareholders in proportion to the nominal value of the shares ***and the amount of the beneficiaries' shares*** of which they are registered holders. ***To this effect the amount of beneficiaries' shares shall be calculated on the basis of a lump sum of 5000 EMA u/a per share[34].***

Article 32

Upon the liquidation of the Company an agreement shall be concluded with the Government of the Headquarters State, and possibly with the Governments of countries in which installations of the Company are situated, as regards the possible taking over of all or part of the installations as well as the storage and control of radioactive wastes.

---

[31] Article 28 (paragraph 2) was modified by decision of the General Assembly on 26 June 1969, approved by the Special Group on 16 December 1969.

[32] Article 30 (paragraph 2) was inserted by decision of the General Assembly on 26 June 1969, approved by the Special Group on 16 December 1969.

[33] This paragraph was modified by decision of the General Assembly on 30 June 1982 and approved by the Special Group on the same day.

[34] Article 31 (paragraphs 2 and 3) was modified by decision of the General Assembly on 26 June 1969 and approved by the Special Group on 16 December 1969.

On 3 June 1964 a further modification was proposed to give priority to the reimbursement of the new shares in Articles 4 ter and 4 quater; however this proposal was rejected as it would have been against the interests of the private shareholders. AG/64(2) et AG/M/64(2).

PART V1

FINAL CLAUSES

Article 33

Correspondence addressed to shareholders shall be forwarded by registered letter.

Publication in the *Moniteur Belge* shall be deemed to be official notification by the Company of the matter published.

In the case of any other matter to be published, the Board of Directors shall decide the means of publication and, if necessary, shall designate newspapers or other publications for this purpose.

Article 34

Any amendment made to the present Statute shall be notified to the Government of the Headquarters State.

Article 35

The present Statute shall come into force at the same time as the Convention.

DONE in Paris, this twentieth day of December Nineteen Hundred and Fifty Seven, in the French, English, German, Italian and Dutch languages, in a single copy which shall remain deposited with the Secretary-General of the Organisation for European Economic Co-operation[35] by whom certified copies will be communicated to all shareholders having subscribed to the present Statute.

[35] Organisation for Economic Co-operation and Development (OECD) after 30 September 1961.

ANNEX 2

Summary of the company's balance sheets

A. Data for the interim period: May 1958-May 1959[36]

Data in millions of Belgian francs (current values)

Table 126. **Changes in staff numbers**

Date	17/05/58	1/06/58	1/07/58	28/06/58	1/10/58	14/2/59	May 1959
Number of employees	9.5	14	16	20	29	39	42

Table 127. **Cumulative expenditure in millions of BF (current values)**

Date	14/05/58	16/06/58	30/09/58	14/02/59	15/05/59
Total	0.309	0.932	4.100	10.016	15.866
Personnel costs	0.292	0.790	2.477	5.360	7.918
Equipment and services	0.017	0.090	0.981	2.092	3.190
Startup costs			0.167	1.203	2.071

B. Summary of annual balance sheets 1959-1990[37]

Money values in millions of BF (current values)

Table 128. **From the creation of the company to inauguration**

Year	1959	1960	1961	1962	1963	1964	1965
Net assets	1083.95	1089.10	1084.08	1132.49	1628.55	2024.54	2074.61
Turnover		40.6	54.05	79.1	98.22	142.09	224.8
Losses on year	31.22	33.67	44.44	65.75	86.07	54.48	91.2
Nuclear services							
Capital grants received							
OECD contributions						77.77	116.5
Staff costs	15.60	22.53	36.19	52.47	68.75	84.97	99.44
Workforce	37	75	131	195	271	292	360

36 Source: Progress Reports 1958-1959.

37 Sources: 1958-1959: Progress Reports. 1959: Management Report of 15 June 1960. 1960-1990: Management Reports. For Belgoprocess, Annual Reports of ONDRAF and Belgoprocess.

Table 129. **The reprocessing period**

Year	1966	1967	1968	1969	1970	1971	1972	1973	1974
Net assets	2016.84	2056	2081	2103	2117.71	2127.9	2187.2	2161.38	2247.99
Turnover	296.11	410	412	420	417.99	397.92	329.31	567.82	373.92
Losses on year	117.9	240	194	218.4	113.88	165.24	73.91	228.21	28.66
Nuclear services	16.38	69.1	39.4	58.28	28.45	104.74	80.61	71.26	32.84
Capital grants received						70	100	64	120
OECD contributions	150	75.7	155	120	223.75	105	145	155.16	282.23
Staff costs	112.03	122	123	124.1	122.38	128.1	132.72	151.6	184.26
Workforce	378	364	354	326	310	301	234	234	170

Table 130. **The waste management period**

Year	1975	1976	1977	1978	1979	1980	1981	1982 [38]	1983	1984
Net assets	2454	2612	592.7	736.21	827.93	967.3	942	966.33 [39]	1416.44	1636.4
Turnover	466	384	732.3	474.13	612.6	591.84	699.02	[40]	595.13	753.17
Losses on year	57.4	0.01								
Nuclear services	1.29	2.14	0							
Capital grants received (1977 capital subsidies)	253	392	117.6	168.5	289.5	361	332	718.07	634.69	1036.6
OECD contributions 1977: other operating income and contributions	446	345	351.4	346.37	489.62	415.28	493.24	574.70 [41]	279.98	53.89
Staff costs	228	190	180.7	223.88	254.67	296.05	346.38	367.07	393.61	518.8
Workforce	170	175	190	193	208	205	200	226	226	159

Table 131. **After the transfer to Belgoprocess up to the dissolution of the company**

Year	1985	1986	1987	1988	1989	1990
Net assets	1493	4321	1790	1151.122	655.40	40.32 [42]
Turnover	714	1681	632.1	314.063	62.03	150.48
Losses on year						
Nuclear services						
Capital subsidies	971	0				
Other operating income and contributions	504	504	501.9	482.96	445.61	420
Staff costs [43]						
Eurochemic staff	5	3	3	3	3	2
Belgoprocess staff	178	175	169	182	274	297

38 Incomplete data owing to the onset of liquidation in July 1982.

39 Total at the end of 1982. The total at the commencement of liquidation was 1351 million BF.

40 808.05 on 27 July, 276.05 on 31 December.

41 Period from 1 January to 27 July.

42 Balance after liquidation.

43 The staff costs are not given for reasons of confidentiality.

ANNEX 3

Workforce and breakdown by category and nationality Eurochemic (1959-1984)

Table 132. **Total staff numbers from 1959 to 1986**
(as from 1985 for Belgoprocess)

Year	1959	1960	1961	1962	1963	1964	1965	1966	1967	1968	1969
Staff numbers	37	75	131	195	271	292	360	378	364	354	326
Year	1970	1971	1972	1973	1974	1975	1976	1977	1978	1979	1980
Staff numbers	310	301	234	234	170	170	175	190	193	208	205
Year	1981	1982	1983	1984	*1985*	*1986*	*1987*	*1988*	*1989*	*1990*	
Staff numbers	200	226	226	159	*178*	*175*	*169*	*182*	*274*	*297*	

Table 133(a, b, c, d). **Eurochemic staff by category and nationality at 31 December of the year shown**

Note on the meaning of letters used:

A: former "Director" category + I; then > 480

B: former categories II and II bis, + draughtsmen in III; then 265-470

C: former category III except for draughtsmen; then 145-260

D: former category IV, then 100-142

Categories I, II, III, IV: 1961

Categories I, II, II bis, III, IV: 1963

Categories with four levels of indexes: 1966-1967

In 1969, categories III and IV merged in 100-260, shown as CD

After 1969, more than 10 salary levels in three groups (named by the author with a view to statistical continuity): A: 10 and beyond; B: 8 and 9; CD: 1 to 7

Definitions of new categories in 1970:

10: Directors, university graduates and equivalent;

8 and 9: Senior technicians, senior administrative assistants;

1 to 7: Workmen, process workers, craftsmen, operators, technicians, analysts, draughtsmen, guards, typists, clerks, secretaries, supervisors, administrative assistants

Table 133(a)	1961				1963				1966			
	A	B	C	D	A	B	C	D	A	B	C	D
Austria		1				3			1	4		
Belgium	1	5	35	23	1	9	55	78	6	24	96	89
Denmark		2			2	2	1		2	1		
France	5	2	1	4	5	3	3	9	5	7	18	
Germany	1	3	8	3	5	4	9	8	3	10	16	
Ireland											1	
Italy		3	6			6	4	1	2	1	10	
Netherlands	1	2	4			6	6	1	2	12	25	
Norway		1		1	1	4		1	1	1	2	
Portugal						1	2				2	1
Spain		3	5		1	2	17		3	1	15	1
Sweden		3			2	4				1	1	
Switzerland	2		1		2	1		2	2	2		
Turkey	1	1				1	1				2	
United Kingdom			2	1		2		6	1	1	5	1
Total	*11*	*26*	*62*	*32*	*19*	*48*	*98*	*106*	*28*	*65*	*193*	*92*

Table 133(b)	1967				1968				1969		
	A	B	C	D	A	B	C	D	A	B	CD
Austria	1	4			1	3				1	
Belgium	6	26	126	48	6	26	143	41	4	25	178
Denmark	1	1				1				1	
France	6	6	18	1	3	7	16	1	1	7	17
Germany	3	9	14		3	10	13		2	8	7
Ireland											
Italy	2	1	11		2		7		1		7
Netherlands	1	10	27		1	10	21		1	11	19
Norway	1	1	2		1	1	3		1		1
Portugal			2				2	1			3
Spain	2	2	17		2	2	16		2	1	17
Sweden		1	1							3	
Switzerland	2	2			2				1		
Turkey			1				1				
United Kingdom	1		7		1		7		1		6
Total	*26*	*63*	*226*	*49*	*22*	*60*	*229*	*43*	*14*	*57*	*255*

Table 133(c)	1970				1971				1972 and 1973			
	A	B	CD	ABCD	A	B	CD	ABCD	A	B	CD	ABCD
Austria		1		1		1	0	1	0	1	0	1
Belgium	7	20	179	206	4	19	183	206	5	23	135	163
Denmark	1	0	0	1	0	0	0	0	0	0	0	0
France	3	3	12	18	4	3	11	18	3	3	6	12
Germany	3	3	6	12	2	1	6	9	1	0	3	4
Ireland		0	0	0	0	0	0	0	0	0	0	0
Italy		0	8	8	0	0	8	8	0	0	5	5
Netherlands	6	6	19	31	3	5	20	28	2	5	14	21
Norway	1	0	2	3	1	0	1	2	1	0	0	1
Portugal	0	0	2	2	0	0	2	2	0	0	2	2
Spain	1	1	18	20	1	1	17	19	0	1	17	18
Sweden	2	0	0	2	1	1	0	2	1	1	0	2
Switzerland	0	0	0	0	0	0	0	0	0	0	0	0
Turkey	0	0	0	0	0	0	0	0	0	0	0	0
United Kingdom	0	0	6	6	0	0	6	6	0	0	5	5
Total	*24*	*34*	*252*	*310*	*16*	*31*	*254*	*301*	*13*	*34*	*187*	*234*

Table 133(d)	1974	1975	1976	1977	1978	1979	1980	1981	1982 to 1984
	ABCD	ABCD	ABCD	ABCD	ABCD	ABCD	ABCD	ABCD	ABCD
Austria	0	0	0	0	0				
Belgium	118	122	127	137	140	157	156	152	175
Denmark	0	0	0	0	0				
France	7	5	5	4	4	6	5	4	5
Germany	4	4	4	5	6	6	6	6	7
Ireland	0	0	0	0	0				
Italy	4	4	4	5	6	4	3	3	4
Netherlands	16	16	16	19	19	18	18	18	18
Norway	0	0	0	0	0				
Portugal	0	0	0	0	0				
Spain	15	14	14	15	14	13	13	13	13
Sweden	1	1	1	1	1	1	1	1	1
Switzerland	0	0	0	0	0				
Turkey	0	0	0	0	0				
United Kingdom	5	4	4	4	4	4	3	3	3
Total	*170*	*170*	*175*	*190*	*194*	*209*	*205*	*200*	*226* [44]

[44] Theoretical takeover of 170 employees; 159 accepted, 14 foreigners refused and were dismissed by Eurochemic with redundancy payments. Thirty-one took early retirement, 22 left the company of their own accord.

Table 134. **Doses received by staff working in controlled areas**

	Total			Eurochemic staff			Outside staff		
Year	*Number of persons*	*Collective dose (mSv)*	*Average (mSv/year/ person)*	*Number of persons*	*Collective dose (mSv)*	*Average (mSv/year/ person)*	*Number of persons*	*Collective dose (mSv)*	*Average (mSv/year/ person))*
1969	258	3765	14.6	252	3633	14.4	6	132	22.0
1970	268	3396	12.7	255	3153	12.4	13	243	18.7
1971	268	3152	11.8	238	2827	11.9	30	325	10.8
1972	235	4270	18.2	213	3963	18.6	22	307	14.0
1973	225	3715	16.5	206	3411	16.6	19	304	16.0
1974	187	2672	14.3	171	2468	14.4	16	204	12.8
1975	169	1566	9.3	152	1457	9.6	17	109	6.4
1976	175	1462	8.4	153	1256	8.2	22	206	9.4
1977	180	1746	9.7	160	1539	9.6	20	207	10.4
1978	212	2250	10.6	160	1320	8.3	52	930	17.9
1979		1963	9.3		1456			507	
1980		1264	6.0		1051			213	
1981		1141.1	5.4		928			213.1	
1982	217	798	3.7	174	707	4.1	43	91	2.1
1983	212	677	3.2	172	629	3.7	40	48	1.2
1984	203	1005.3	5.0	174	923.2	5.3	29	82.1	2.8

N.B.: The average doses for 1979 to 1981 were calculated on the assumption that the total number of people working in controlled areas had not changed since 1978.

ANNEX 5

Membership of Eurochemic committees

Members of the Study Group (preparatory period: October 1956-December 1957)[45]

Classification by country and in alphabetical order

Country or institution	*NAME*	*Initial*	*Affiliation*
AUSTRIA	GEHR	W.	Permanent delegate to the OEEC
	HOHN	H.	Technical University of Vienna
	LANG	Friedrich	Federal Chancellery
	PAHR	W.	Federal Chancellery
BELGIUM	BOVY	R. (Mlle)	CEN, Mol
	d'HONDT	Maurice	CEN, Mol
	de BAERDEMAKER	A.	Permanent delegate to the OEEC
	DEVADDER	Y.	Ministry of Foreign Affairs
	GAUDY	P.	Ministry of Finance
	LANNOY	J.	Société de la propriété industrielle
	OUTERS	L.	Ministry of Communications
	SYMON	E.	CEN, Brussels
	van der SPEK	Jean	Belgonucléaire
DENMARK	GOMARD	Bernhard	Atomic Energy Commission
	HOEYER	P.	Permanent delegate to the OEEC
	JACOBSEN	J.C.	Atomic Energy Commission
	PEDERSEN	O.	Atomic Energy Commission
	SKYTTE-JENSEN	B.	Atomic Energy Commission
	von BÜLOW	H.	Atomic Energy Commission
FRANCE	BRESSON	J.J. de	Ministry of Foreign Affairs
	DARD	J.M.D.	CEA
	FAUGERAS	P.	CEN Fontenay-aux Roses
	GAJAC	R.	Service de la propriété industrielle
	GOLDSCHMIDT	Bertrand	CEA
	GOURRIER	J.	CEA
	MERCEREAU	F.P.	CEA
	MONIN	J.	Contrôleur de l'armée
	REGNAUT	Pierre	CEA
	SARTORIUS	Robert	CEA
	SOUSSELIER	Yves	CEA
GERMANY	COSTA	Hermann	Bundesministerium für Atomkernenergie
	FRANTA	R.	Ministry of Justice

45 Sources: SCNE (1958) and archives.

	GÖTTE	Hans	Farbwerke Hoechst, Frankfurt
	HAEDRICH	H.	Ministry of Foreign Affairs
	JONAS	H.	Bayer/ Leverkusen
	KROPFF	B.	Ministry of Justice
	MARTIUS	H. von	Ministry of Atomic Affairs
	MEIBOM	H. von	Ministry of the Interior
	PAULY	W.	Ministry of Foreign Affairs
	PFANNER	K.	Ministry of Justice
	POHLAND	Erich	Ministry of Atomic Affairs
	SPRUNG	A.	Ministry of Finance
	WALLENBERG	H. von	Permanent delegate to the OEEC
	WOHLFAHRT	E.	Ministry of Justice
ITALY	ALBONETTI	Achille	Permanent delegate to the OEEC
	CAPROGLIO	P.	AGIP Nucleare, Milan
	CATALANO	N.	Avvoccatura dello stato
	CERRAL	E.	CNRN
	ORSONI	L.	SORIN, Milan
	PIZZI	L.	SORIN, Milan
	QUOIANI	M.	AGIP Nucleare, Rome
	ROLLIER	M.	CNRN
	SCHWEINICHEN	J. von	SORIN, Milan
	VALLAURI	G.	SORIN, Turin
NETHERLANDS	BARENDREGT	T.J.	JENER, Kjeller
	BENDIEN	G.W.	Ministry of Foreign Affairs
	GOUVERNE	M.W.	Ministry of Economic Affairs
	HOOGWATER	J.H.W.	Ministry of Economic Affairs
	SAUVEPLANNE	J.G.	Ministry of Foreign Affairs
	SCHIERBEEK	P.	Ministry of Economic Affairs
	van der WILLINGEN	A.	Permanent delegate to the OEEC
NORWAY	KOREN LUND	L.J.	JENER, Kjeller
	LIE	A.	Permanent delegate to the OEEC
	LOCHEN	E.	Ministry of Foreign Affairs
	MELIEN	R.	Ministry for Industry
	SOLLI	O.	Ministry of Foreign Affairs
PORTUGAL	do CARMO ANTA	M. (Mrs.)	Junta de Energia Nuclear
	SOTTOMAYOR	J.M.	Permanent delegate to the OEEC
SWEDEN	ALER	B.	A.B. Atomenergi
	HAAKANSON	H.	Atomic Energy Board
	HAEFFNER	E.	A.B. Atomenergi
	RAMEL	S.	Permanent delegate to the OEEC
	SVENKE	Erik	A.B. Atomenergi
SWITZERLAND	BIERI	R.	Department of Finance
	CAMPICHE	S.	Federal Policy Department
	DIEZ	E.	Department of Finance
	GRÄNACHER	Charles	CIBA, Basle
	ROMETSCH	Rudolf	CIBA, Basle
	WALTHARD	F.	Permanent delegate to the OEEC
	ZOELLY	H.	Policy Department
TURKEY	BERKI	K.	Permanent delegate to the OEEC
	KORKUD	S.	Permanent delegate to NATO

UNITED KINGDOM	BRAYNE	R.C.L.	Permanent delegate to the OEEC
	DARWIN	H.G.	Deputy Legal Advisor, Foreign Office
	GILLAMS	John	UKAEA, Risley
	KAVANAGH	M.	UKAEA, Risley
	STEWART	J.C.C.	UKAEA, Risley
	WILSON	W.W.	Permanent delegate to the OEEC
UNITED STATES	BISHOP	A.S.	USAEC European Liaison Office, Paris
	QUINN	J.R.	USAEC European Liaison Office, Paris
	SCOTT	E.W.	USRO, Paris
Secretary of Working Party in 1956-1957	SAELAND	Einar	OEEC consultant

Members of the interim Study Group, in alphabetical order of country or institution

Country or Institution	*NAME*	*Initial*	*Affiliation*
AUSTRIA	GEHR	W.	Permanent delegate to the OEEC
	LANG	Friedrich	Federal Chancellery
	ORLICEK	Adelbert	Ministry for Transport and Electricity
BELGIUM	BUYSE		Ministry for Economic Affairs (Director-General)
	d'HONDT	Maurice	CEN, Mol
	de BAERDEMAKER	A.	Permanent delegate to the OEEC
	de HEEM	Louis	Centre d'études nucléaires
	De MERRE	Marcel	Belchim (Director)
	FONTAINE		Permanent delegate to the OEEC
	HALTER		Ministry of Public Health (Director-General)
	LEROUX	A.	Belchim
	ROBILIART	Herman	Belgonucléaire
	van der MEULEN	J.	Ministry of Economic Affairs
	van der SPEEK	J.	Belgonucléaire
DENMARK	BOHR	Erik	Kryolitselkabet Oeresund A.S.
	GOMARD	Bernhard	Atomic Energy Commission
	JACOBSEN	C.F.	Atomic Energy Commission
	LOFT	Per	Atomic Energy Commission
ENEA	HUET	Pierre	Advisor, then Director-General
	PERRET	Roland	ENEA Secretariat
	SAELAND	Einar	ENEA Deputy Director
	SCHMID	F.	OEEC consultant
	STROHL	Pierre	ENEA Secretariat
	WEINSTEIN	Jerry L.	ENEA Secretariat
	WOLFF	Karlfritz	ENEA Secretariat
EURATOM	DIERKENS	F.	EURATOM
	FOCH	R.	EURATOM
	STIJKEL	E.	EURATOM
	VINCK	W.	EURATOM
EUROCHEMIC	HAEFFNER	E.	Director, Design and Research Bureau
	SOUSSELIER	Yves	Chairman, Management Committee
FRANCE	ASTY	Jacques	CEA
	FINKELSTEIN	A.	CEA

	GOLDSCHMIDT	Bertrand	CEA
	GRANDGEORGE	René	Saint-Gobain (Managing Director)
	REGNAUT	Pierre	CEA
	SARTORIUS	Robert	CEA
	SOUSSELIER	Yves	CEA
	TARANGER	Pierre	CEA
GERMANY	BROCKE	W.	Vereinigung Deutscher Elektrizitätswerke, Bonn
	COSTA	Hermann	Bundesministerium für Atomkernenergie
	GÖTTE	Hans	Farbwerke Hoechst, Frankfurt
	HAEDRICH	H.	Ministry of Foreign Affairs
	POHLAND	Erich	Ministry for Atomic Affairs
	WALLENBERG	H. von	Permanent delegate to the OEEC
ITALY	ALBONETTI	Achille	Permanent delegate to the OEEC
	CACCIARI	Alberto	CNRN
	IPPOLITO	Felice	CNRN
	ROLLIER	M.	CNRN
NETHERLANDS	BARENDREGT	T.J.	JENER, Kjeller
	CRAMER	J.W.C.	Ministry of Economic Affairs
	HARDEMANN		Ministry of Foreign Affairs
	KRAMER	E.L.	Ministry of Economic Affairs
	Le POOLE	J.	Ministry of Social Affairs
	RIEMENS	E.	Ministry of Foreign Affairs
	RUTTEN		Ministry of Foreign Affairs
	van der WILLINGEN	A.	Permanent delegate to the OEEC
NORWAY	SAELAND	Einar	JENER Lillestrøm
	TERJESEN	Sven G.	Norges Tekniske Hoegskole
PORTUGAL	SOTTOMAYOR	J.M.	Permanent delegate to the OEEC
SPAIN	GUTIERREZ-JODRA	José Luis	Junta de Energia Nuclear (Director of pilot projects and industrial chemical-metallurgical projects)
SWEDEN	HAEFFNER	E.	A.B. Atomenergi
	SVENKE	Erik	A.B. Atomenergi
SWITZERLAND	CAMPICHE	S.	Federal Policy Department
	GRÄNACHER	Charles	CIBA (Managing Director)
	ROMETSCH	Rudolf	CIBA
	SCHENK	M. von	Permanent delegate to the OEEC
	WALTHARD	F.	Permanent delegate to the OEEC
TURKEY	BERKI	K.	Permanent delegate to the OEEC
	INAN	Kamuran	Ministry of Foreign Affairs

Table 135. **Members and alternates of the interim Board of Management**

Source : Archives

Austria	LANG	Friedrich	Deputy Director, Bundeskanzleramt
Austria Alternate	ORLICEK	Adelbert	Deputy Director, Federal Ministry of Transport and Electricity
Belgium 1	van der MEULEN	J.	Director-General, Ministry for Economic Affairs
Belgium 1 Alternate	HATRY	P.	Ministry for Economic Affairs
Belgium 2	LEROUX	A.	Société belge de l'azote et des produits chimiques du Marly, Liège
Belgium 2 Alternate	De MERRE	Marcel	Société Belge de chimie nucléaire (Belchim), Brussels
Denmark	BOHR	Erik	Kryolitselskabel Oeresund A/S, Copenhagen
Denmark Alternate	GOMARD	Bernhard	Atomenergikommissionen, Copenhagen
EURATOM	STIJKEL	A.G.	Director-General, Economic and Industry Division, EURATOM
France 1	GOLDSCHMIDT	Bertrand	Director, Chemistry Department, CEA
France 1 Alternate	ASTY	Jacques	Director of Administration, CEA
France 2	GRANDGEORGE	René	Managing Director, Saint-Gobain
France 2 Alternate	SARTORIUS	Robert	Deputy Director, Chemistry Department, CEA
Germany 1	POHLAND	Erich	Bundesministerium für Atomkernenergie
Germany 1 Alternate	COSTA	Hermann	Bundesministerium für Atomkernenergie
Germany 2	GÖTTE	Hans	Radiochemisches Laboratorium, Hoechst
Germany 2 Alternate	BROCKE	W.	Vereinigung Deutscher Elektrizitätswerke, Bonn
Italy 1	IPPOLITO	Felice	Secretary-General, CNRN
Italy 1 Alternate	CAGLIOTI	V.	CNRN
Italy 2	ALBONETTI	Achille	Permanent delegate to the OEEC
Italy 2 Alternate	CACCIARI	Alberto	CNRN
Netherlands	van AKEN	J.S.A.J.M.	Director, Chemical Division, State Mines Limbourg, Geleen
Norway	TERJESEN	Sven G.	Institut for Kjemiteknikk, Norges Tekniske Hoegskole, Trondheim
Spain	OYARZUN	Roman	Permanent delegate to the OEEC
Spain Alternate	GUTIERREZ-JODRA	José Luis	Junta de Energia Nuclear
Sweden	SVENKE	Erik	Head of Industrial Division, A.B. Atomenergi, Stockholm
Switzerland	GRÄNACHER	Charles	Managing Director, CIBA, Basle
Turkey/ Portugal Alternate	SOTTOMAYOR	J.M.	Permanent delegate to the OEEC
Turkey/Portugal	INAN	Kamuran	Permanent delegate to NATO, Paris

Table 136. **Directors of Eurochemic from March 1958 to July 1982 and liquidators from July 1982 to November 1990**

Italics: Chairmen, Board of Directors, Board of Liquidators

IBD: Interim Board of Directors

Nota bene: Portugal and Turkey shared one seat until 1980, when Turkey withdrew

	Austria	Belgium 1	Belgium 2	Denmark	France 1	France 2
(IBD) 28/03/1958	Lang F.	van der Meulen J.	Robillard H.		Goldschmidt B.	
(IBD) 25/10/1958	"	"	Leroux A.	Bohr E.	"	Grandgeorge R.
1959	"	"	"	"	"	"
1960	Steinwender B.	Buyse M.	"	"	"	"
1961	"	"	"	"	"	"
1962	"	"	"	"	"	"
1963	"	"	"	"	"	"
1964	"	Coessens A.	"	"	"	de Vaissiere A.
1965	"	"	"	"	"	"
1966	"	"	"	"	"	"
1967	"	"	"	"	"	"
1968	"	"	"	"	"	Mabile J.
1969	"	"	de Bie E.	"	"	"
1970	"	"	"	"	"	"
1971	"	"	"	Bastrup-Birk E.	"	Sornein J.
1972	"	"	"	"	"	"
1973	"	"	"	"	"	"
1974	"	"	"	"	"	"
1975	"	"	"	"	"	Boussard R.
1976	"	Frerotte M.	Maricq E.R.	"	"	"
1977	"	"	"	"	Lefevre J.	Perrot F.
1978	"	"	"	"	"	"
1979	"	"	"	"	"	"
1980	"	"	"	"	"	"
1981	"	"	"	"	"	"
1982	Musyl E.	"	"	"	"	"
1983	"	"	End	"	"	End
1984	"	"		"	"	
1985	"	"		"	"	
1986	"	"		"	"	
1987	"	"		"	"	
1988	Wagner G.	"		Beck A.	"	
1989	"	"		"	"	
1990	"	"		"	"	

(see following page)

Directors of Eurochemic from March 1958 to July 1982 and liquidators from July 1982 to November 1990 (continuation of Table 136)

	Italy 1	Italy 2	Germany 1	Germany 2	Netherlands	Norway
(IBD) 28/03/1958	Albonetti A.	Ippolito F.	Pohland E.	Götte H.	Van Aken J.A.S.M.	
(IBD) 25/10/1958	"	"	"	"	"	Terjesen S.G.
1959	"	"	"	"	"	"
1960	"	"	Costa H.	"	"	"
1961	"	"	"	"	"	"
1962	"	"	"	"	"	"
1963	"	Zifferero M.	Schulte-Meerman W.	"	Hoekstra J.	"
1964	"	End	*Schulte-Meerman W.*	"	"	"
1965	"		"	"	"	"
1966	"		"	"	"	"
1967	"		"	"	Barendregt T.J.	"
1968	"		Haunschild H.H	"	"	"
1969	"		"	"	"	"
1970	"		Schmidt-Küster W.J.	Schlitt A.E.	"	"
1971	"		*Schmidt-Küster W.J.*	"	"	"
1972	"		"	"	"	"
1973	"		"	"	"	"
1974	"		"	"	"	"
1975	"		"	"	"	"
1976	"		"	"	Netherlands withdraw	"
1977	"		Randl R.P.	"		"
1978	"		"	Heinz W.		"
1979	"		"	"		"
1980	"		"	"		"
1981	"		"	"		"
1982	"		"	"		"
1983	"		"	End		"
1984	"		"			"
1985	"		"			"
1986	"		"			"
1987	"		"			"
1988	"		"			"
1989	"		"			"
1990	"		"			"

(see following page)

Directors of Eurochemic from March 1958 to July 1982 and liquidators from July 1982 to November 1990 (continuation of Table 136)

	Portugal	Spain	Sweden	Switzerland	Turkey
(IBD) 28/03/58	Sottomayor J.M.			Gränacher C.	Berki
(IBD) 25/10/58		Oyarzun R.	*Svenke E.*	“	Inan
1959		Gutierrez-Jodra L.	“	“	“
1960		“	“	“	Erikan B.
1961	“	“	“	“	“
1962	“	“	“	“	“
1963	“	“	“	“	Turgay
1964	“	“	“	“	“
1965	“	“	“	“	“
1966	“	“	“	“	Erben T.
1967	“	“	“	“	“
1968	“	*Gutierrez-Jodra L.*	“	“	“
1969	Marquez-Videira A.	“	“	“	“
1970	“	“	“	“	“
1971	“	“	“	Pictet J.M.	Okonsar K.
1972	“	“	“	“	Öymer
1973	“	“	Nøjd L.A.	“	“
1974	Carreira-Pich H.	“	“	“	Carkaci J.
1975	“	“	“	“	Afsar K.P.
1976	“	“	“	“	“
1977	“	“	“	“	“
1978	“	“	“	“	“
1979	“	“	“	“	“
1980	“	“	“	“	Turkey withdraws
1981	“	Lopez-Perez B.	“	“	
1982	“	“	*Nøjd L.A.*	“	
1983	“	“	“	“	
1984	“	“	“	“	
1985	“	“	“	“	
1986	“	“	“	“	
1987	“	“	“	“	
1988	“	“	“	“	
1989	“	Rodriguez-Parra M.	“	“	
1990	“	“	“	“	

Table 137. **Members of the Management Committee in 1958 and of Eurochemic's Technical Committee from 1959 to 1984**

	Austria	Belgium	Denmark	ENEA/NEA	EURATOM (advisory)	France	Germany
1958		D'Hondt M.				*Sousselier Y.*	
1959	Orlicek A.	"		Saeland E		"	
1960	"	"		"		"	
1961	"	"	Jacobsen C.F.	"	Ramadier C.	"	
1962	"	"	"	"	"	"	
1963	"	"	"	"	"	"	
1964	"	"	"	"	"	"	Giese K.
1965	"	"	"	"	"	"	"
1966	"	"	"	"	Baruffa A.	"	"
1967	"	"	"		"	"	Dreissigacker H.L.
1968	Bildstein H.	"	"		"	"	"
1969	"	"	Klitgaard J.		"	"	"
1970	"	"	"		"	"	Schüller W.
1971	"	"	Jacobsen C.F.	Olivier J.-P.	"	"	"
1972	"	"	"	"	"	"	Randl R.P.
1973	"	"	"	"	"	"	"
1974	"	"	"	"	Orlowski S.	"	"
1975	"	"	"	"	"	"	"
1976	"	"	Singer K.	"	"	"	"
1977	"	"	"	"	"	"	"
1978	"	"	"	"	Simon R.	"	"
1979	"	"	"	"	"	"	"
1980	"	"	"	"	"	"	"
1981	"	"	"	"	"	"	"
1982	"	Tonon P.	"	"	"	"	"
1983	"	"	"	"	"	"	"
1984	"	"	"	"	"	"	"

(see following page)

Members of the Management Committee in 1958 and of Eurochemic's Technical Committee from 1959 to 1984 (continuation of Table 137)

	Italy	Netherlands	Norway	Spain	Sweden	Switzerland
1958	Rollier M.	Barendregt T.J.				Rometsch R.
1959	Zifferero M.	"			Haeffner E.	
1960	"	"			"	
1961	"	X	Terjesen S.G.	Gutierrez-Jodra L.	"	
1962	"		"	"	"	
1963	"		"	"	"	
1964	"		"	"	"	
1965	"		"	"	"	
1966	"	Hoekstra J.	"	"	Larson A.	von Gunten H.
1967	"	Barendregt T.J.	"	"	"	"
1968	"	"	"	Lopez-Perez B.	"	"
1969	"	"	"	"	"	"
1970	"	"	"	"	"	"
1971	"	"	"	"	Hultgren A.	"
1972	"	"	"	"	"	"
1973	"	"	"	"	"	"
1974	"	"	"	"	"	"
1975	"		"	"	"	"
1976	"		"	"	"	"
1977	"		"	"	"	"
1978	"		"	"	"	"
1979	"		"	"	"	"
1980	"		"	"	"	"
1981	Cao S.		"	"	"	"
1982	Rolandi G.		"	"	"	"
1983	"		"	"	"	"
1984	"		"	"	"	"

Table 138. **Members of the Special Group in alphabetical order of country (except for the Netherlands, Portugal, Turkey and EURATOM, which are at the end of the list)**

(Italics: Chairmen)

	Austria	Austria	Austria	Belgium	Belgium	Belgium	Belgium
29/10/59	R. Polaczek	B. Steinwender		J. van der Meulen	J. Errera		
1/06/60	“	T. Klestil		E. Symon	“	P. Gaudy	H. Gillain de Bois
18/05/61	“	“	R. Polaczek	“	P. Bassette	“	
13/04/62	R. Hladik		“	S. Halter	J. Errera	“	R. Bataille
10/05/63	B. Steinwender	A. Orlicek				“	“
26/06/63	R. Hladik			G. Barthelemy	“	“	“
26/11/63	“		“	F. Leemans		“	“
24/02/64	“		“	R. Depovere		“	“
17/04/64	“		“	“		“	
1/07/64	“		“	“		“	“
23/11/65	E. Musyl		“	“	A. Coessens		“
24/03/66	B. Steinwender			“	“		
8/06/66		M. Graff	“		“	“	“
17/05/67		“	“		*A. Coessens*	“	
29/02/68	“				“	“	
23/04/69		M. Lutterotti	“		“	“	
16/12/69	“				“	“	
20/01/71		H. Bildstein	“		“	“	
8/11/72	“			P. Dejonghe	“		
2/10/73			“	“	“		
21/11/73	“			“	“		
4/06/74	“			“	“	S. Herpels	
10/07/74	“			“	“	J. Bousse	
8/07/75	“			“	“	M. Geuens	
14/04/78	E. Musyl			“	M. Frérotte	L. de Clerck	
28/06/78	“			“	“		
30/06/82	“			“	“		
9/04/86	“			“	“	E. Detilleux	

(see following page)

Members of the Special Group (continuation of Table 138)

	Denmark	Denmark	France	France	France
29/10/59	H. von Bülow	F. Juul	J. Renou	F. de Chocqueuse	
1/06/60	"	H.H. Koch		"	
18/05/61	"	J. Mandrup-Christensen		"	
13/04/62	"	A. Dybdal		"	
10/05/63	"		B. Goldschmidt	"	
26/06/63		O. Heyman		"	
26/11/63	"			"	
24/02/64	"		"	"	
17/04/64	"		"	"	Y. Sousselier
1/07/64	"			"	"
23/11/65		P.J. Snare		"	
24/03/66	E. Bastrup-Birk			"	
8/06/66	"		A. Finkelstein	"	A. Lagardère
17/05/67	"		B. Goldschmidt	"	Y. Sousselier
29/02/68	"			"	"
23/04/69	"			"	"
16/12/69	"			"	
20/01/71	"		"	"	J. Mabile
8/11/72	"		"	J. Sornein	Y. Sousselier
2/10/73	"			"	F. Perrot
21/11/73	"		"	"	"
4/06/74	"			"	
10/07/74	"			"	
8/07/75			R. Boussard	J. Lefevre	"
14/04/78	"			"	"
28/06/78	"			"	"
30/06/82	"		F. de Puybaudet	"	"
9/04/86	"			"	"

(see following page)

Members of the Special Group (continuation of Table 138)

	Germany	Germany	Germany	Italy	Italy	Italy
29/10/59	H.H. Haunschild	V. Rhamm				
1/06/60	H.Costa			A. Baroni		
18/05/61	"			M. Zifferero	S. Pittori	
13/04/62	"				A. Albonetti	C.C Bertoni
10/05/63	C.Zelle	K. Giese		M. Zifferero	"	Catalano
26/06/63	C. Zelle	W. Schulte-Meermann		"	"	F. Starace
26/11/63	"	"		"	"	S. d'Andrea
24/02/64	"	"		"	"	"
17/04/64	"	"			"	"
1/07/64	"	"			"	
23/11/65	W. Ungerer	"	H.A. von Rohr	A. Baglio	F. Starace	
24/03/66	M. Schreiterer	"		M. Sabelli	"	"
8/06/66	"	W. Ungerer	"	"	"	"
17/05/67	C. Zelle	W. Böcker		"	"	
29/02/68	M. Schreiterer	D. Taube	"	"	F. Conte	"
23/04/69	H.H. Haunschild			"	F. Starace	M. Gigliarelli-Fiumi
16/12/69			"		"	G. Licata
20/01/71	"				A. Albonetti	
8/11/72		R. Loosch		M. Zifferero		M. Gigliarelli-Fiumi
2/10/73					"	
21/11/73		"			"	
4/06/74		"			"	
10/07/74					"	
8/07/75	W. Sandtner				"	
14/04/78	"	R.P Randl			"	
28/06/78	"	"			"	
30/06/82		"		P. Longo		
9/04/86		"		"		

(see following page)

Members of the Special Group (continuation of Table 138)

	Norway	Norway	Norway	Spain	Spain	Spain
29/10/59	J.C. Hauge	H. Storhaug				
1/06/60	"	O. Kasa	E. Jensen	J.M. Otero de Navascues		
18/05/61		O. Solli		"	A. Duran	
13/04/62		"		"	"	F. del Castillo
10/05/63	S. Terjesen	"	O. Mosnesset	"		L. Gutierrez-Jodra
26/06/63	*J.C. Hauge*			"	"	
26/11/63	"			"	"	
24/02/64	"				R. Oyarzun	"
17/04/64	"		"	"		
1/07/64			"	"	A. Duran	
23/11/65	J.C. Hauge				"	
24/03/66						"
8/06/66	"				"	
17/05/67		J. Otterbech				
29/02/68		S. Terjesen				"
23/04/69	"			"		
16/12/69						"
20/01/71	"					"
8/11/72		"	V.O. Eriksen			"
2/10/73		"		B. Lopez-Perez		
21/11/73		"				"
4/06/74	N.G. Aamodt				J.L. Xifra	"
10/07/74		"			"	"
8/07/75		"				"
14/04/78		"				"
28/06/78		"		"		
30/06/82		"		"		
9/04/86	K. Mansika	"		"		

(see following page)

Members of the Special Group (continuation of Table 138)

	Sweden	Sweden	Switzerland	Switzerland	Switzerland
29/10/59			*J. Burckhardt*	S. Campiche	E. Stadelhofer
1/06/60	A. Edelstam		“	H. Bär	“
18/05/61	H. Hakanson		“		“
13/04/62	“		*U. Hochstrasser*		
10/05/63	E. Svenke	N.O. Hasslev	“	R. Lempen	G. Bodmer
26/06/63				S. Campiche	“
26/11/63	“		U. Hochstrasser		
24/02/64	“	J. Martenson	“	H. Voegeli	
17/04/64	H. Hakanson	“	“	“	
1/07/64	G. von Sydow	B. Aler	“		
23/11/65		“	“		
24/03/66	E. Svenke	P.G. Hasselmark			W. Pfister
8/06/66		B. Aler	“	“	
17/05/67	“			S. Lagger	“
29/02/68	“				“
23/04/69		H. Brynielson			“
16/12/69	“		J.M. Pictet		
20/01/71	*E. Svenke*		“		
8/11/72	“	A. Larsson	“		
2/10/73	“	L.A. Nöjd	“		
21/11/73		“	“		
4/06/74		“	“		
10/07/74	T. Eckered		“		
8/07/75		“	*J.M. Pictet*		
14/04/78	B.C. Johanson		“		
28/06/78		“	“		
30/06/82		“	“		
9/04/86	C. Danielsson	“	“		

(see following page)

Members of the Special Group (continuation of Table 138)

	Netherlands	Netherlands	Netherlands	Netherlands	Portugal
29/10/59	C. Rutten	J. Boekstal	F.J. Hardeman		Nogueira da Costa A.
1/06/60	S. Meijer	"	"	R. de Muralt	J.M. Sottomayor
18/05/61	"	A.A.v.d. Voort	"	"	"
13/04/62	"	"	"	"	"
10/05/63	"	M.M. Stillerroer	L.M.J. v.d. Winkel		"
26/06/63	"		"	H.K. Mani	"
26/11/63		A. Kruyt	"		"
24/02/64	"	C.H.A. Plug	"	"	"
17/04/64	"		"		F. de Paiva
1/07/64	*S. Meijer*		"	"	Nogueira da Costa A. and J.M. Sottomayor
23/11/65	*"*		"		
24/03/66	*"*	G.W. Bendien		R. Fruin	J.M. Sottomayor
8/06/66	*"*		"		
17/05/67	E. Beelaerts van Blokland				"
29/02/68	A. Kruyt				
23/04/69	W. Hellema				" and F. Marques-Videira
16/12/69	"				A.N. Da Costa
20/01/71	P.G. van der Lande		"		F. Marques-Videira
8/11/72	"	T.J. Barendregt			"
2/10/73	"	"			"
21/11/73	"	"			H. Carreira-Pich
4/06/74	"	"			"
10/07/74	"	A. Kruyt	"		"
8/07/75	"	R. Bosscher			"
14/04/78					"
28/06/78					"
30/06/82					"
9/04/86					"

(see following page)

Members of the Special Group (conclusion of Table 138)

	Turkey	EURATOM	EURATOM
29/10/59	K. Inan		
1/06/60	"		
18/05/61	B. Erikan	R. Foch	
13/04/62		"	M. Amory
10/05/63		"	C. Ramadier
26/06/63	B. Turgay		M. Amory
26/11/63	"		"
24/02/64	"	"	
17/04/64	"	E.R. von Geldern	"
1/07/64	"	C. Ramadier	"
23/11/65	"	H. van der Loos	"
24/03/66	"	"	"
8/06/66		R. Foch	"
17/05/67			
29/02/68		D. Blin	
23/04/69	F. Mutluay	R. Foch	
16/12/69			
20/01/71	B. Turgay	"	
8/11/72		"	
2/10/73	S. Carkaci		
21/11/73			
4/06/74	"	"	
10/07/74	" and S. Karaka	"	
8/07/75			
14/04/78	K.P. Afsar, S. Akilic, M. Kayalar		
28/06/78	M. Kayalar		
30/06/82		S. Orlowski	
9/04/86			

ANNEX 6

Table 139. **Reprocessing plants currently in service or under construction**[46]

Country (classified by region and in alphabetical order)	Name, location and date of first startup	Stated capacity in t/year	Output in 1992 (t/year)	Type of fuel and process	Remarks
France	UP1 Marcoule 1958	400	410	NUGG, PUREX	
	UP2 La Hague 1966 (UP2-800)	400 (800)	220	LWR PUREX	Extended to 800 t/year in 1994.
	UP3 La Hague 1989	800	448	LWR PUREX	
Italy	EUREX Saluggia 1971	10	0	LWR, HEU, PUREX	Included, but no longer in service.
United Kingdom	Dounreay 1958	8	2	HEU PUREX	
	B205 Windscale-Sellafield 1964	1500	?	MAGNOX PUREX	
	THORP Windscale-Sellafield 1994	850	n.a.	LWR PUREX	
Russia	RT1 Cheliabinsk 1971	400	120	VVER-440, naval reactors,FBR, MTR, aqueous extraction using TBP	Converted military plant.
	Tomsk 7 Tomsk ?	?	?	Plutonium-producing reactors	Military plant.

(see following page)

[46] Sources: NAS (1994), pp. 235-244; World Nuclear Industry Handbook 1994, Annual Supplement to Nuclear Engineering International, pp. 133-134. CHALIAND G., JAN M. (1993) for China. This source indicates which installations are on a laboratory scale: Dimona and Nahal Soreq in Israel, Rawalpindi and Islamabad in Pakistan, Yongbyon in North Korea and Pelindaba in South Africa are quoted here for information. Information on plants around the world is incomplete and sometimes contradictory, particularly as regards military plants. These uncertainties are reflected on the maps given in Annex 8.

Reprocessing plants currently in service or under construction (continuation of Table 139)

Country (classified by region and in alphabetical order)	Name, location and date of first startup	Stated capacity in t/year	Output in 1992 (t/year)	Type of fuel and process	Remarks
	RT2, Krasnoyarsk n.a.	(1500) (x2)	n.a.	VVER-1000	First section under construction. Completion of project uncertain.
India	Trombay 1965	50	?	Oxide PUREX	Second plant since 1985.
	Tarapur 1977	100	?	Oxide PUREX	
	Kalpakkam 1 1986	125	?	Oxide PUREX	
	Kalpakkam 2 n.a.	(1000)	n.a.	Oxide PUREX	Extension under construction.
China	Baotou ?	?	?	?	Military plant.
	Subei-Jiuquan ?	?	?	?	Military plant.
Japan	Tokaï Mura 1978	200	?	Oxide PUREX	Stated capacity reduced to 90 t.
	Rokka Shô Mura n.a.	(800)	n.a.	Oxide PUREX	Under construction since 1993. Completion schedule postponed until after 2000.
United States	ICRP Idaho Falls 1953(ex-ICPP)	?	?	?	The former ICPP modified. "has largely closed down"[47] (?).
	Rockwell Hanford ?	2400	?	Oxide PUREX	Military plant "has largely closed down" (?).
	Savannah River Site 1954	2700	?	Oxide PUREX	Military plant "has largely closed down" (?).
Argentina	Ezeiza n.a.	(5)	n.a.	Oxide PUREX	Under construction. Completion uncertain.
Brazil	Sao Paulo n.a.	?	n.a.	Oxide PUREX	Under construction.

47 NAS (1994), p. 243.

ANNEX 7

Table 140(a, b, c). **Nuclear power generation in Eurochemic member countries from 1960 to 1990**

Source: OECD/IEA (1990, 1991, 1992).

Note: five member countries – Austria, Denmark, Norway, Portugal and Turkey – have not produced any nuclear electricity.

For the purposes of comparison, the figures for generation in the United Kingdom, the United States and the European OECD countries and finally the OECD as a whole have been added.

Generation is expressed in GWh.

Table 140a	1960	1961	1962	1963	1964	1965	1966	1967	1968	1969
Region or country										
Belgium	0	0	4	47	51	0	6	94	61	22
France	146	271	473	466	655	1053	1606	2921	3539	4956
Germany	0	24	100	56	104	117	265	1225	1766	4937
Italy	0	0	0	323	2401	3510	3863	3152	2576	1679
Netherlands	0	0	0	0	0	0	0	0	28	315
Spain	0	0	0	0	0	0	0	0	83	829
Sweden	0	0	0	0	0	18	45	50	23	61
Switzerland	0	0	0	0	0	0	0	0	0	535
Eurochemic countries	146	295	577	892	3211	4698	5785	7442	8076	13334
% (nuclear/total electricity generation)	0.04	0.07	0.12	0.17	0.57	0.78	0.90	1.10	1.10	1.67
United Kingdom	*2224*	*2567*	*3720*	*6365*	*8163*	*15135*	*20217*	*23277*	*26190*	*29125*
OECD Europe	*2370*	*2862*	*4297*	*7257*	*11374*	*19833*	*26002*	*30719*	*34266*	*42459*
United States	*554*	*1808*	*2429*	*3437*	*3576*	*3913*	*5907*	*7646*	*13405*	*14903*
OECD	*2924*	*4670*	*6726*	*10787*	*15101*	*23874*	*32681*	*39118*	*49589*	*58991*

Table 140b	1970	1971	1972	1973	1974	1975	1976	1977	1978	1979
Region or country										
Belgium	57	0	11	76	148	6784	10036	11939	12513	11407
France	5711	9336	14592	14741	14695	18248	15778	17986	30483	39960
Germany	6030	5812	9137	11755	12276	21398	24262	36050	35942	42291
Italy	3176	3365	3626	3142	3410	3800	3807	3385	4428	2628
Netherlands	368	405	326	1108	3277	3335	3872	3710	4060	3489
Spain	923	2523	4751	6545	7223	7544	7555	6524	7649	6700
Sweden	56	90	1465	2111	2054	11969	15993	19913	23781	21039
Switzerland	1766	1391	3840	6310	7067	7761	7939	8114	8395	11805
Eurochemic countries	18087	22922	37748	45788	50150	80839	89242	107621	127251	139319
% (nuclear/total electricity generation)	2.11	2.52	3.85	4.32	4.59	7.40	7.50	8.86	9.95	10.32
United Kingdom	*26012*	*27548*	*29378*	*27997*	*33617*	*30338*	*36155*	*40021*	*37224*	*38308*
OECD Europe	*44099*	*50470*	*67127*	*73785*	*83767*	*111177*	*125397*	*150302*	*167739*	*184369*
United States	*23324*	*40552*	*57813*	*89167*	*121251*	*183512*	*202570*	*265936*	*292987*	*270464*
OECD	*73060*	*103289*	*141650*	*187913*	*239413*	*332384*	*379462*	*474239*	*551241*	*560497*

Table 140c	1980	1981	1982	1983	1984	1985	1986	1987	1988	1989	1990
Region or Country											
Belgium	12549	12859	15664	24106	27743	34601	39394	41967	43102	41217	42722
France	61251	105326	108919	144261	191234	224100	254155	265520	275521	303931	314081
Germany	43700	53631	63577	65833	92577	125902	119580	130515	145082	149390	147159
Italy	2208	2707	6804	5783	6887	7024	8758	174	0	0	0
Netherlands	4200	3658	3897	3589	3711	3899	4216	3556	3675	4019	3502
Spain	5186	9568	8771	10661	23086	28044	37458	41271	50466	56126	54273
Sweden	26488	37679	39045	41004	50926	58561	69951	67385	69424	65603	68185
Switzerland	14346	15330	15133	15710	18440	22558	22581	23003	22792	22836	23636
Eurochemic countries	169928	240758	261810	310947	414604	504689	556093	573391	610062	643122	653558
% (nuclear/total electricity generation)	12.33	17.15	18.55	21.26	27.00	31.35	33.76	33.37	34.55	35.45	35.02
United Kingdom	*37023*	*37969*	*43972*	*49928*	*53979*	*61095*	*59079*	*55238*	*63456*	*71734*	*65747*
OECD Europe	*213973*	*293392*	*322558*	*378595*	*487450*	*584843*	*634231*	*648275*	*693072*	*733946*	*738520*
United States	*266183*	*289034*	*299739*	*311298*	*347292*	*406712*	*438880*	*482586*	*558591*	*561116*	*611473*
OECD	*600779*	*710313*	*763064*	*852794*	*1021216*	*1211654*	*1312683*	*1395880*	*1513189*	*1557803*	*1625151*

ANNEX 8

Geographical data

Figures 181 a, b and c

Legend common to the three maps (see following pages)

	REPROCESSING LABORATORY	PILOT REPROCESSING	REPROCESSING PLANT
IN OPERATION	■	■	■
PLANNED	□	□	□
CLOSED	⊡	⊡	⊡
NOT BUILT		⧇	⧇
BUILT BUT NEVER STARTED UP			▲

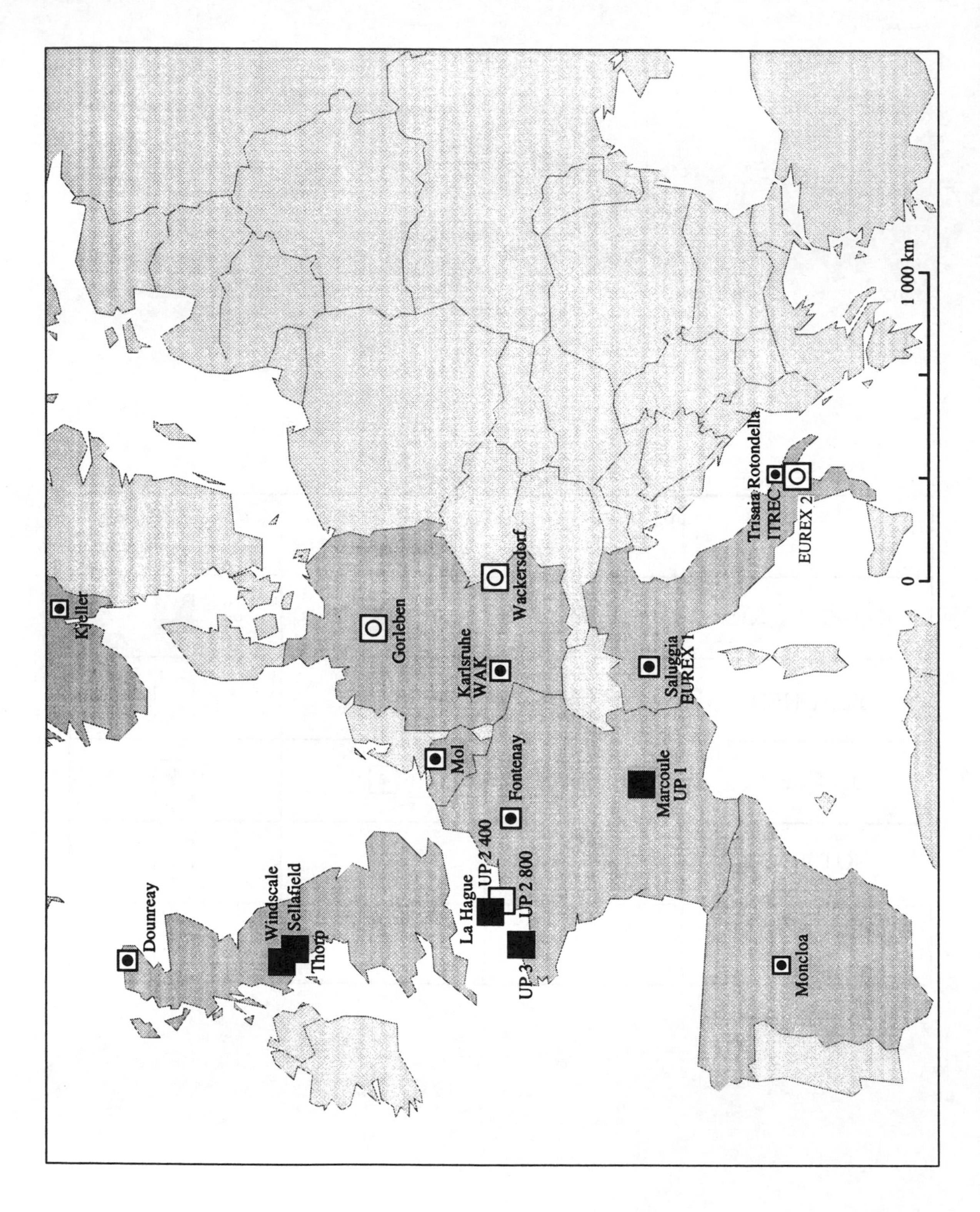

Figure 181a. Reprocessing installations in Europe.

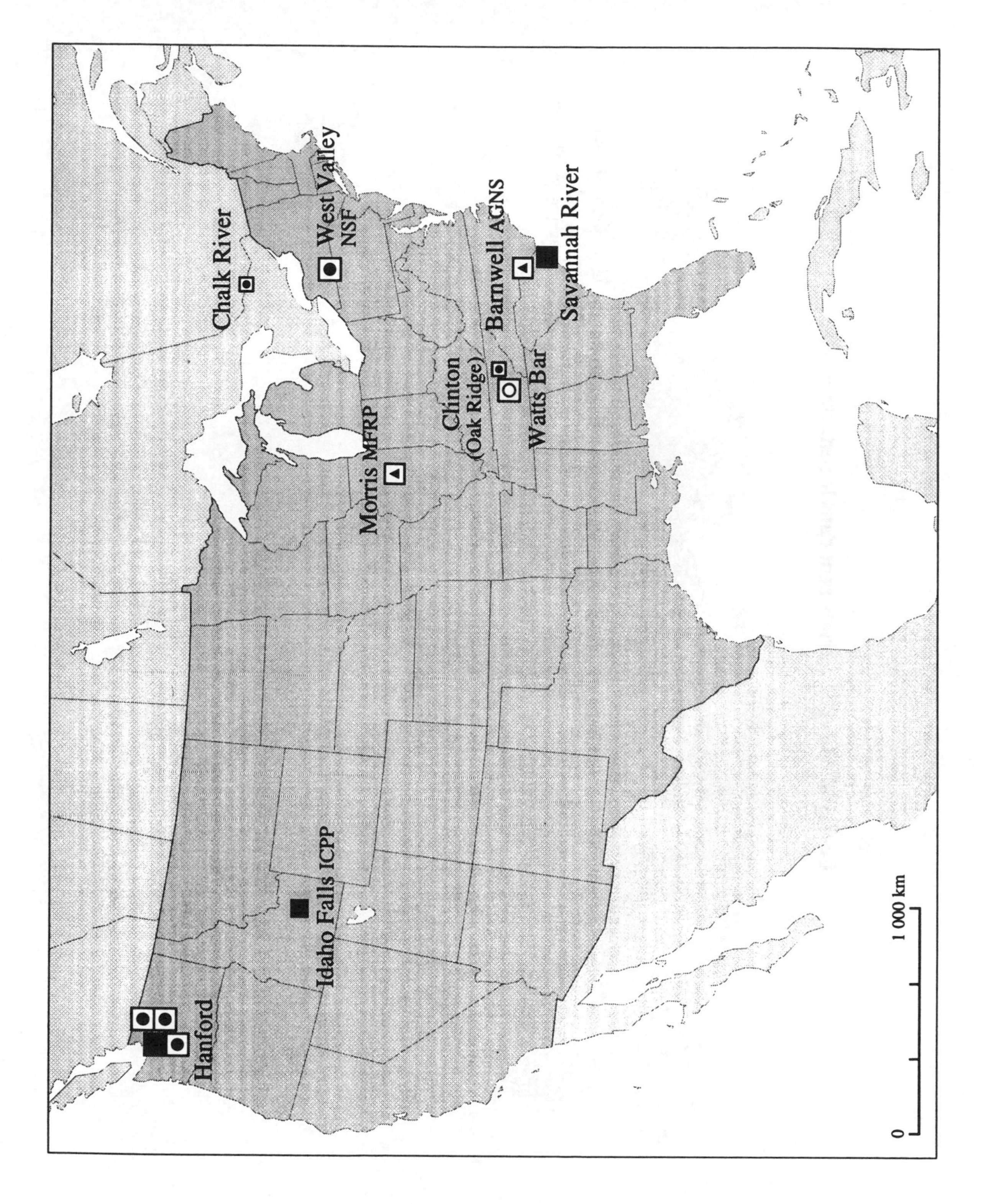

Figure 181b. Reprocessing installations in North America.

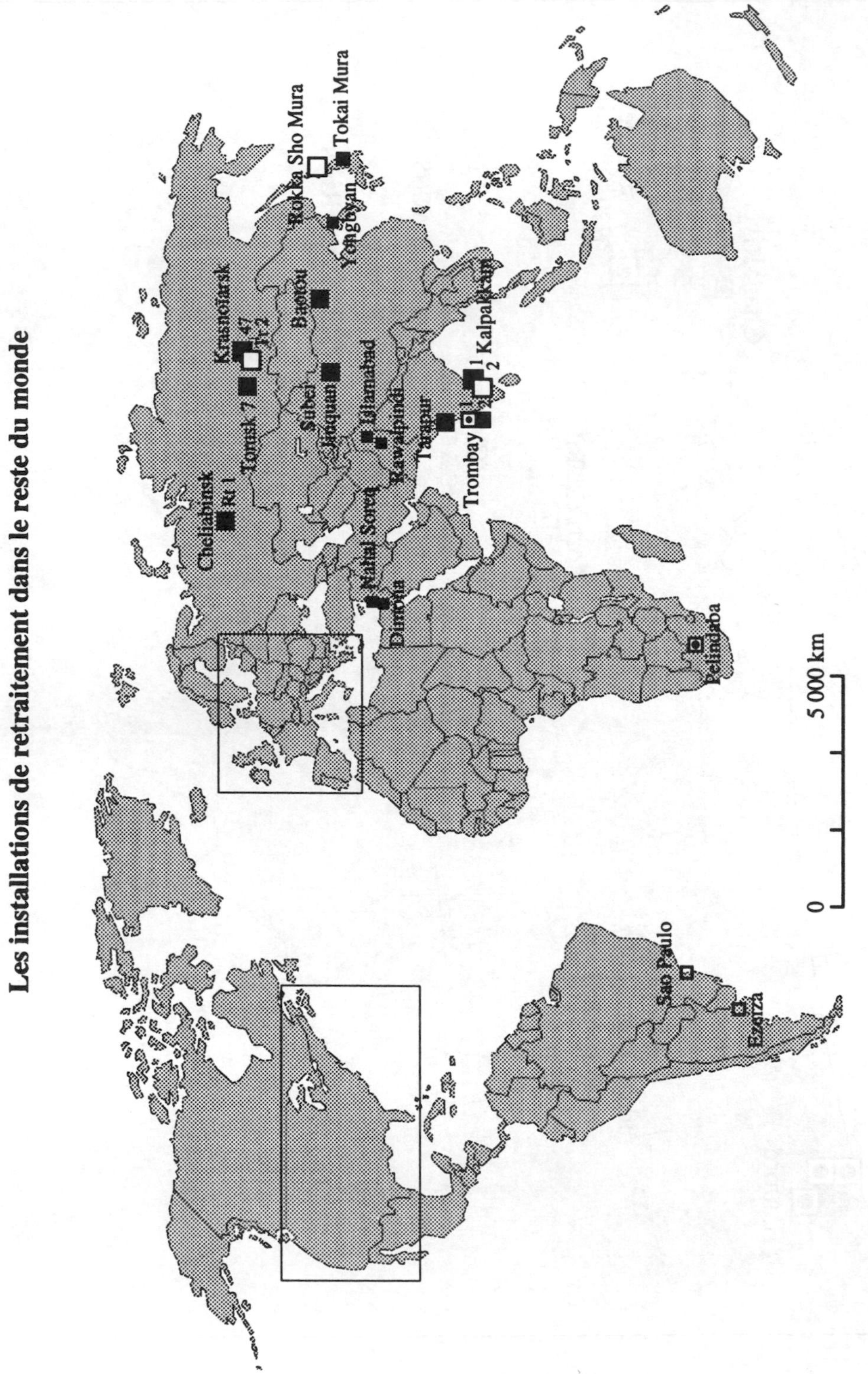

Les installations de retraitement dans le reste du monde

181c. Reprocessing installations in the rest of the world.

ANNEX 9

Chemical flow charts of the Eurochemic plant at the end of the 1960s

(See Figures 182 to 185 and Table 141)

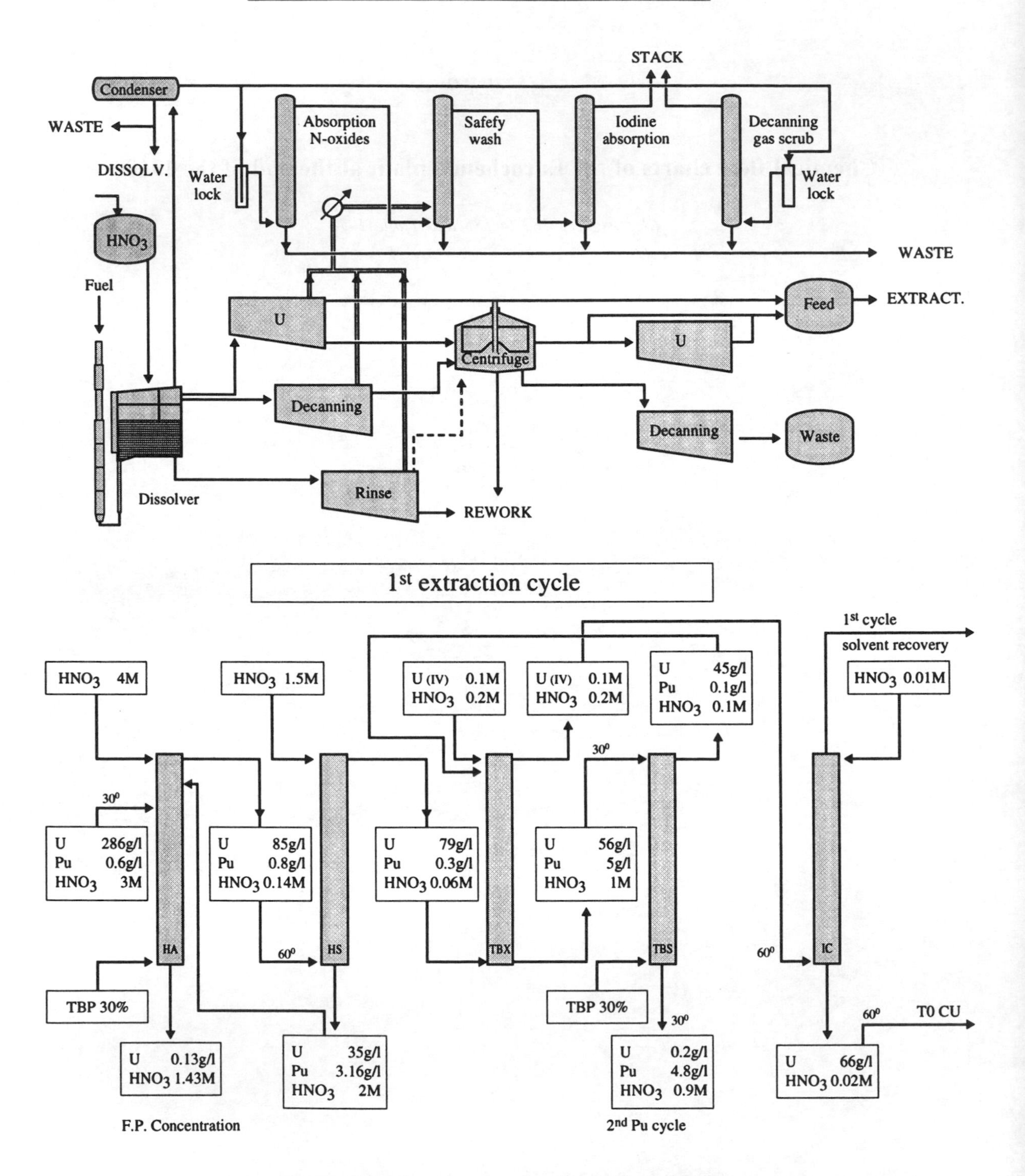

Figure 182. Chemical flow sheets produced in 1968 for the entire process (Source : BARENDREGT T. (1968), Figures 15 to 21 (I)).

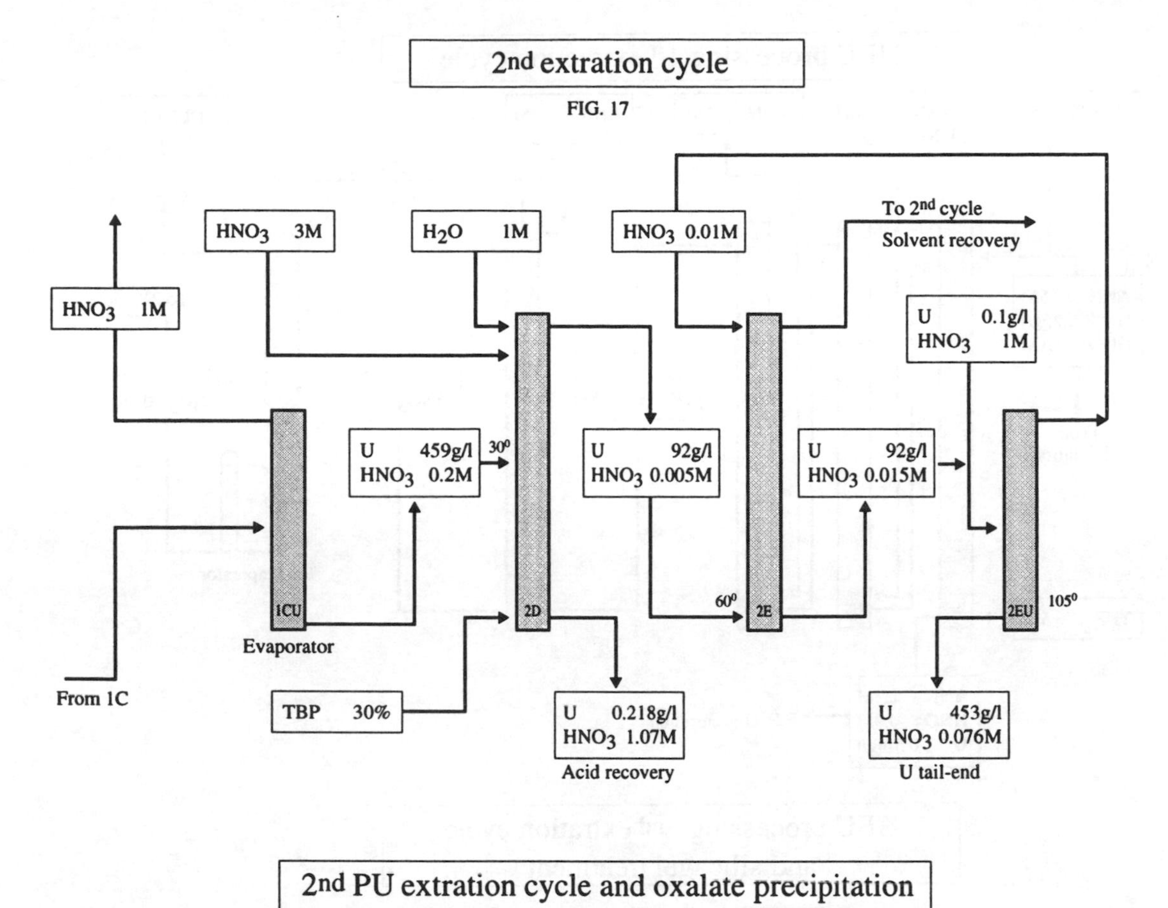

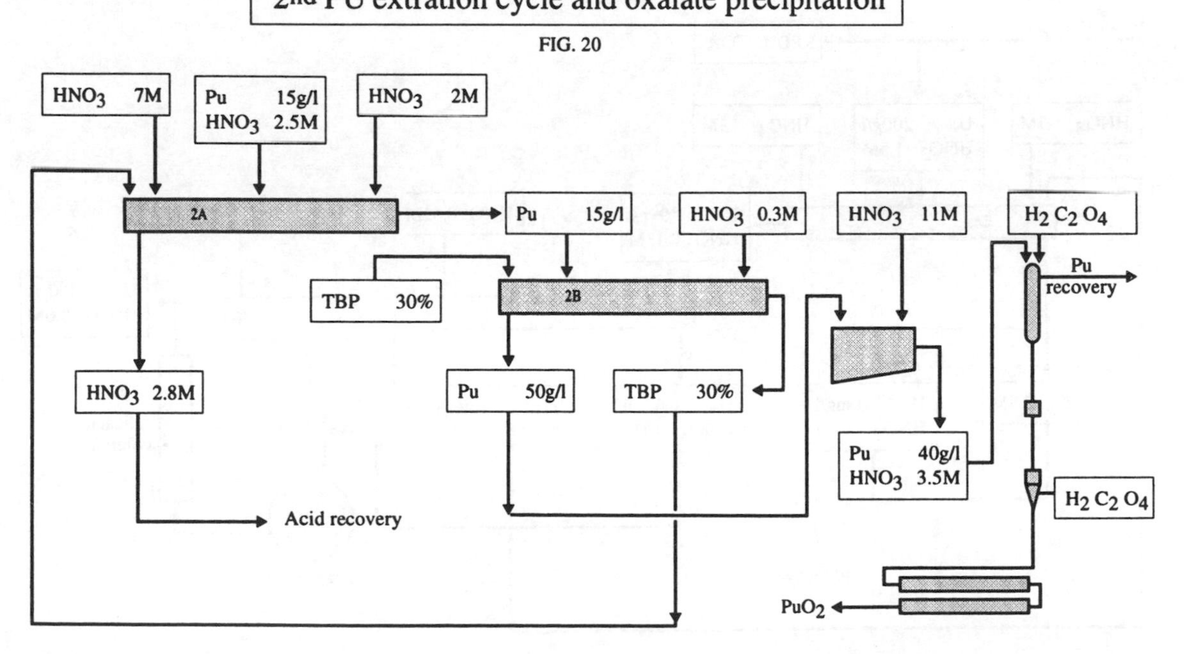

Figure 183. Chemical flow sheets produced in 1968 for the entire process (Source : BARENDREGT T. (1968), Figures 15 to 21 (II)).

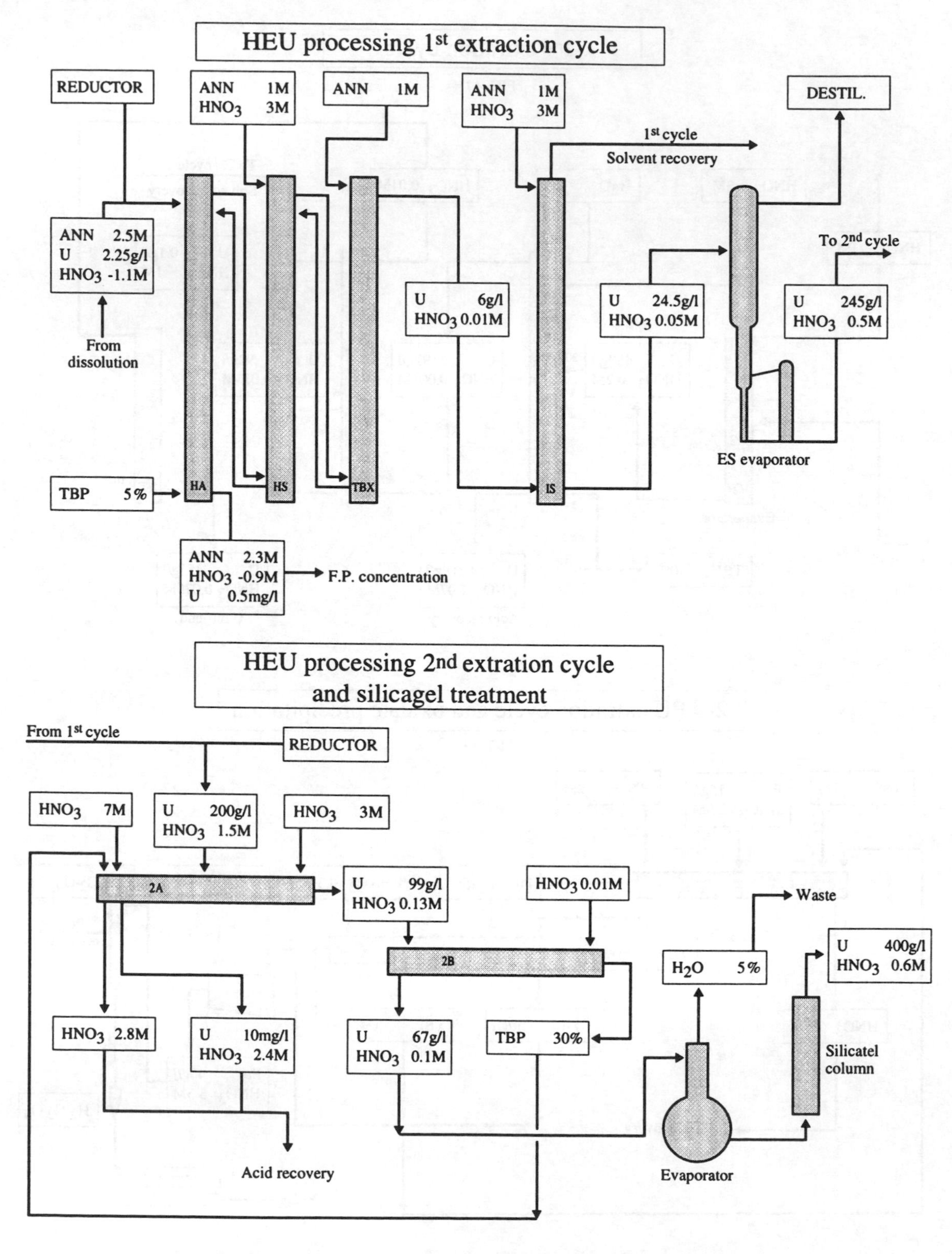

Figure 184. Chemical flow sheets prepared in 1968 for the entire process (Source : BARENDREGT T. (1968), Figures 15 to 21 (III)).

Fission product concentration

FIG. 21

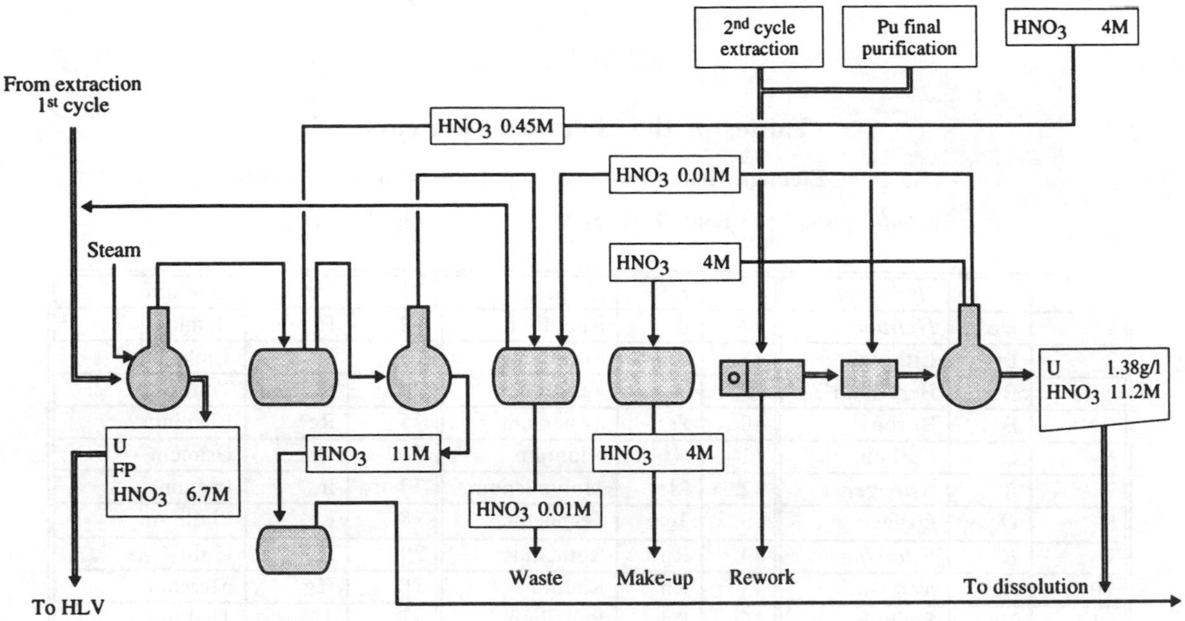

Figure 185. Chemical flow sheets prepared in 1968 for the entire process.
(Source : BARENDREGT T. (1968), Figures 15 to 21 (IV)).

Table 141. **Dissolving irradiated fuel and cladding in the head-end of the Eurochemic plant**
(Source : BARENDREGT T. (1968), Table 1, no page numbers)

Materials	Chemicals	Products	
		Solutions	***Gas***
Al	NaOH	$NaAlO_2 - NaOH$	H_2
Mg	H_2SO_4	$MgSO_4 - H_2SO_4$	H_2
Zr	$NH_4F - NH_4NO_3$	$(NH_4)_2ZrF_6$	NH_2
Steel, Cr	H_2SO_4	$Fe(CrNi)SO_4 - H_2SO_4$	H_2
Ni			
U	HNO_3	$UO_2(NO_3)_2 - HNO_3$	Oxides N_2 I 131
U, Mo	HNO_3	$UO_2(NO_3)_2 - HNO_3$ $H_2MoO_4H_2O$ prec.	Oxides N_2 I 131
U, Al	$HNO_3 - Hg^-$	$UO_2(NO_3)_2 - Al(NO_3)_3$	Oxides N_2 I 131
UO_2	HNO_3	$UO_2(NO_3)_2$	Oxides N_2 I 131

ANNEX 10

Tables of the chemical elements

Table 142a. **Elements listed in increasing order of atomic number**

Italic: gases **Bold: liquid at 20°** Underlined: artificial

1	*H*	*Hydrogen*	36	*Kr*	*Krypton*	71	Lu	Lutetium
2	*He*	*Helium*	37	Rb	Rubidium	72	Hf	Hafnium
3	Li	Lithium	38	Sr	Strontium	73	Ta	Tantalum
4	Be	Beryllium	39	Y	Yttrium	74	W	Tungsten
5	B	Boron	40	Zr	Zirconium	75	Re	Rhenium
6	C	Carbon	41	Nb	Niobium	76	Os	Osmium
7	*N*	*Nitrogen*	42	Mo	Molybdenum	77	Ir	Iridium
8	*O*	*Oxygen*	43	Tc	Technetium	78	Pt	Platinum
9	*F*	*Fluorine*	44	Ru	Ruthenium	79	Au	Gold
10	*Ne*	*Neon*	45	Rh	Rhodium	80	**Hg**	**Mercury**
11	Na	Sodium	46	Pd	Palladium	81	Tl	Thallium
12	Mg	Magnesium	47	Ag	Silver	82	Pb	Lead
13	Al	Aluminium	48	Cd	Cadmium	83	Bi	Bismuth
14	Si	Silicon	49	In	Indium	84	Po	Polonium
15	P	Phosphorous	50	Sn	Tin	85	At	Astatine
16	S	Sulphur	51	Sb	Antimony	86	*Rn*	*Radon*
17	*Cl*	*Chlorinee*	52	Te	Tellurium	87	**Fr**	**Francium**
18	*Ar*	*Argon*	53	I	Iodine	88	Ra	Radium
19	K	Potassium	54	*Xe*	*Xenon*	89	Ac	Actinium
20	Ca	Calcium	55	Cs	Caesium	90	Th	Thorium
21	Sc	Scandium	56	Ba	Barium	91	Pa	Proactinium
22	Ti	Titanium	57	La	Lanthanum	92	U	Uranium
23	V	Vanadium	58	Ce	Cerium	93	Np	Neptunium
24	Cr	Chromium	59	Pr	Praseodymium	94	Pu	Plutonium
25	Mn	Manganese	60	Nd	Neodymium	95	Am	Americium
26	Fe	Iron	61	Pm	Promethium	96	Cm	Curium
27	Co	Cobalt	62	Sm	Samarium	97	Bk	Berkelium
28	Ni	Nickel	63	Eu	Europium	98	Cf	Californium
29	Cu	Copper	64	Gd	Gadolinium	99	Es	Einsteinium
30	Zn	Zinc	65	Tb	Terbium	100	Fm	Fermium
31	Ga	Gallium	66	Dy	Dysprosium	101	Md	Mendelevium
32	Ge	Germanium	67	Ho	Holmium	102	No	Nobelium
33	As	Arsenic	68	Er	Erbium	103	Lr	Lawrencium
34	Se	Selenium	69	Tm	Thulium	104	Unq	Unnilquadium
35	**Br**	**Brominee**	70	Yb	Ytterbium			

Table 142b. **Elements listed in alphabetical order of symbols**

Ac	Actinium	*H*	*Hydrogen*	Pu	Plutonium
Ag	Silver	*He*	*Helium*	Ra	Radium
Al	Aluminium	Hf	Hafnium	Rb	Rubidium
Am	Americium	**Hg**	**Mercury**	Re	Rhenium
Ar	*Argon*	Ho	Holmium	Rh	Rhodium
As	Arsenic	I	Iodine	*Rn*	*Radon*
At	Astatine	In	Indium	Ru	Ruthenium
Au	Gold	Ir	Iridium	S	Sulphur
B	Boron	K	Potassium	Sb	Antimony
Ba	Barium	*Kr*	*Krypton*	Sc	Scandium
Be	Beryllium	La	Lanthanum	Se	Selenium
Bi	Bismuth	Li	Lithium	Si	Silicon
Bk	Berkelium	Lr	Lawrencium	Sm	Samarium
Br	**Bromine**	Lu	Lutetium	Sn	Tin
C	Carbon	Md	Mendelevium	Sr	Strontium
Ca	Calcium	Mg	Magnesium	Ta	Tantalum
Cd	Cadmium	Mn	Manganese	Tb	Terbium
Ce	Cerium	Mo	Molybdenum	Tc	Technetium
Cf	Californium	*N*	*Nitrogen*	Te	Tellurium
Cl	*Chlorine*	Na	Sodium	Th	Thorium
Cm	Curium	Nb	Niobium	Ti	Titanium
Co	Cobalt	Nd	Neodymium	Tl	Thallium
Cr	Chromium	*Ne*	*Neon*	Tm	Thulium
Cs	Caesium	Ni	Nickel	U	Uranium
Cu	Copper	No	Nobelium	Unq	Unnilquadium
Dy	Dysprosium	Np	Neptunium	V	Vanadium
Er	Erbium	*O*	*Oxygen*	W	Tungsten
Es	Einsteinium	Os	Osmium	*Xe*	*Xenon*
Eu	Europium	P	Phosphorous	Y	Yttrium
F	*Fluorine*	Pa	Proactinium	Yb	Ytterbium
Fe	Iron	Pb	Lead	Zn	Zinc
Fm	Fermium	Pd	Palladium	Zr	Zirconium
Fr	**Francium**	Pm	Promethium		
Ga	Gallium	Po	Polonium		
Gd	Gadolinium	Pr	Praseodymium		
Ge	Germanium	Pt	Platinum		

Sources, bibliography and references

Abbreviations and coding of references to the Archives, official document series, journals, etc.

Archives and official document series

Acts, vol., year, Act n°	Acts of OEEC/OECD Council of Ministers
C(year) n°	OEEC/OECD Council
CA(year) n°	Eurochemic Board of Directors
CL(year) n°	Eurochemic Board of Liquidators
ENB n°(year)	Eurochemic News Bulletin
ENEAR/OECD(year)	Report of the European Nuclear Energy Agency (1960 to 1972)
ENEAR/OEEC(year)	Report of the European Nuclear Energy Agency before 1960
ETR n°(year)	Eurochemic Technical Report
EUROCHEMIC(year) n°	Study Group: interim period
EUROCHEMIC/CA(year) n°	Interim Board of Directors
GA(year) n°	General Assembly of Eurochemic
NE(year) n°	Special Committee for Nuclear Energy (29/2/1956)
NE/EUR(year) n°	Special Group (Eurochemic) of the SCNE, then of the ENEA and the NEA
NE/EUR/WP(year) n°	Working Group of the SCNE Special Group
NEAR/OECD(year)	Report of the Nuclear Energy Agency after 1972
RAE (1970-1974)	Summary Report on Eurochemic Activities from 1970 to 1974
RAE(year)	Annual Eurochemic Activity Report (from 1967 to 1969, then from 1975 to 1984)
RAE 1	First Eurochemic Activity Report 1959-1961
RAE 2	Second Eurochemic Activity Report 1962-1964
RAE 3	Third Eurochemic Activity Report 1965-1966
SCNE(1958)	Report of the Steering Committee for Nuclear Energy (SCNE)
SEN (year) n°	Study Group of the ENEA
SEN/CHEM(year) n°	Study Group on Chemical Reprocessing (September 1956)
SEN/CHEM/WP(year) n°	Technical Sub-group of the Working Party

Journals, etc.

AtW	Die Atomwirtschaft, Düsseldorf since 1956
EN	Energie nucléaire, 1957-1973, Paris
RGN	Revue générale nucléaire, Paris, since 1974
NEW	Nuclear Europe (1982-1989) then Nuclear European Worldscan, Geneva

Sources

Eurochemic Archives [1]

Prior to the signature of the Convention and Statute of 20 December 1957

Documents of the OEEC Council.

Documents of the Steering Committee for Nuclear Energy.

Documents of the Eurochemic Study Group.

Interim period (December 1957-July 1959)

Documents of the Steering Committee for Nuclear Energy.

Documents of the Interim Study Group.

Documents of the Interim Board of Directors.

Documents of the Eurochemic Management Committee.

From 1959 to 1990

Documents of the Eurochemic Special Group of the Steering Committee for Nuclear Energy of the ENEA and then the NEA.

Documents of the Board of Directors of the Eurochemic Company until 27 July 1982, followed by those of the Board of Liquidators up to 28 November 1990.

Specific documents (in chronological order)

LECLERQ-AUBRETON Y., ASYEE J., NORAZ M. (1968), Preliminary Study of an International Financial Nuclear Energy Company, FINUCLEAIRE, unpaginated, dated 20 October 1968.

MEMORANDUM OF UNDERSTANDING (1970), Understanding Between Eurochemic and the UKAEA, 15 October 1970.

CONVENTION (1978) Convention Between the Government of the Kingdom of Belgium and the Eurochemic Company on Takeover of the Installations and Execution of the Obligations of the Company, 24 July 1978, in Dutch, French and English.

Collection of the Conventions, Protocol and "Lump Sum Agreement" of 1978, 1981, 1983, 1986, in Dutch, French and English.

VAN MERKSEN Baron (1979), Une entreprise de l'AEN (OCDE), vingt ans d'existence d'Eurochemic, course report.

Memorandum on the Takeover of the Company's Personnel by a Belgian Entity, Mol, 20 May 1983.

CA (07)/M/1, Copy of the Minutes of the 150th Meeting of the Board of Directors held in Mol on 1st April 2007 (probably dated 1982).

VIEFERS W. (1986), Incidents at Eurochemic, Description and Evaluation, unpublished report of April 1986.

VIEFERS W. (not dated), Waste Treatment Developments, Practices and Experience at Eurochemic, unpublished report probably dating from 1985.

Agreement between the CIEMAT and EUROCHEMIC, Madrid, 20 January 1988.

Verbal sources

The following kindly agreed to reply to the author's questions[2]:

Jacques Couture, Contracts Division, COGEMA, (12/11/1991),Vélizy.

Maurits Demonie, technician, then engineer at Eurochemic and later at Belgoprocess from 1962 (21/4/1992), Mol.

Emile Detilleux, Managing Director of Eurochemic from 1973 to 1984 (8/01/1992, 11/12/1992), Brussels.

Louis Geens, engineer at Eurochemic then Belgoprocess (22/04/1992), Mol.

Pierre Huet, Director-General of ENEA from 1959 to 1963, Legal Advisor to Eurochemic from 1963 to 1990 (19/02/1992), Paris.

---

[1] The Eurochemic archives, following their consultation for this book, will be transferred to the archives of the European Union in Florence under an OECD/EU Convention.

[2] This list includes only the most senior staff as concerns the history of Eurochemic.

Yves Leclercq-Aubreton, Managing Director of Eurochemic from 1969 to 1972 (3/6/1994), Paris.

Michel Lung, engineer at SGN (29/4 and 5/05/1994), Issy-Les-Moulineaux, Mareil-Marly.

Oscar Martinelle, Financial Director of Eurochemic (4/12/1991), Mol.

André Mongon, electrical and instrumentation engineer 1962-1964 (15/5/1992), Paris.

André Redon, engineer at the CEA, Technical Director of Eurochemic 1972-1974 (10/6/1992), Paris.

Rudolf Rometsch, Managing Director of Eurochemic from 1964 to 1968 (8/01/1992), Brussels.

Walter Schüller, Section Head at Eurochemic, Director of the Karlsruhe pilot plant, (30/05/1994), Weinheim.

Earl Shank, engineer, US advisor to Eurochemic from 1962 to 1969 (19/4/1994), Bruges.

Yves Sousselier, Chairman of the Eurochemic Technical Committee from 1957 to 1990 (08/01/1992), Brussels.

Pierre Strohl, Deputy Director-General of the OECD/NEA from 1982 to 1992. (17/12/1991, 23/01/1992, 04/02/1992), Paris.

Film and photographic sources

Eurochemic Archives

Eurochemic had 16 mm films made; these were viewed:

Three 16-minute films were made in 1966: *Construction of a Plant, The PUREX Process, Operating the Plant.*

Vitrification of Wastes by PAMELA, 14 minutes, 1983.

The archives include photographs and slides. A selection was made with the help of Willem Drent, former Head of the Eurochemic Documentation Section.

Belgoprocess

The firm has a library of slides and photographs. A selection was made with the help of Willem Drent.

Printed sources

Activity reports

These were published at intervals from 1961 to 1966 (3 issues, covering the periods 1959-1961, 1962-1964, and 1965-1966 respectively), annually from 1966 to 1969, a summary volume for the period 1970-1974 based on unpublished annual reports. CA (71)2, CA(72)1, CA(73)1, CA(74)2, and CA(75)2, then annually again from 1975 to 1984.

Each report was presented to the Board of Directors by the Managing Director. Reports were bilingual up to 1968 and in English only thereafter.

Technical reports (ETR)

Eurochemic Technical Reports.

A bibliography with abstracts is given in ETR 300 and 300 Supplement (1980, 1990). The following publication is of particular importance:

ETR 318: DRENT W., DELANDE E. (1984), Proceedings of the Seminar on Eurochemic Experience, June 9-11, 1983, Mol, as well as

HUMBLET L. (1987), Performance Assessment of the Eurochemic Reprocessing Plant.

Internal Bulletin (ENB)

Eurochemic News Bulletin: house magazine published from August 1960 to December 1966 (18 issues).

Other Eurochemic publications

In chronological order.

BARENDREGT T. et al. (1964), Conception de l'usine de retraitement d'Eurochemic, July 1964.

Safety Analysis (November 1965), 3 volumes of text and 1 volume of figures.

Staff manual (1966).

Inauguration of Eurochemic, Programme de la cérémonie, brochure published by the OECD for 7 July 1966.

DETILLEUX E. (1966), Le laboratoire de développement industriel d'Eurochemic, son rôle, ses moyens, July 1966, introductory brochure.

LOPEZ-MENCHERO E. (1968), Eurochemic Processes for the Treatment of Radioactive Effluents from a Multipurpose Nuclear Fuel Aqueous Processing Plant, May 1968, brochure.

BARENDREGT T. (1968), Conception de l'usine de retraitement d'Eurochemic, May 1968. Updated version of the 1964 booklet (see above).

Publicity brochure, no date (probably 1970), 4 pp.
Working regulations (October 1982).

Press articles and outside publications by staff and others involved with the undertaking

Eurochemic

BAZILE F. (1965), Eurochemic, editorial in Energie nucléaire, and Situation du projet, EN, pp. 317-320.

BUSEKIST O. von (1980), Der Werdegang der Eurochemic, eine Bilanz, AtW, May. French translation given in: BUSEKIST O. von (1982), Eurochemic, échec ou exemple, Consensus, n°2, p. 2-14.

BUSEKIST O. von, DETILLEUX E., OLIVIER J.P. (1983), Fuel Reprocessing and Radioactive Waste Management, Twenty Years of Experience in the Nuclear Energy Agency and the Eurochemic Company, IAEA (1983), Vol. 3, pp. 83-100.

DETILLEUX E. (1969), Le dégainage chimique: expérience acquise à Eurochemic, EN, pp. 292-297.

DETILLEUX E. (1978), Mise à l'arrêt des installations d'Eurochemic: programme, évolution et enseignements, RGN, 1978, 3, pp. 177-186.

DREISSIGACKER H., FICHTNER H., REINARTZ W. (1966), Bau des Dissolvers für die Eurochemic Wiederaufarbeitungsanlage, AtW, April 1966, pp. 170-173.

ENERGIE NUCLEAIRE (1959), La Société de chimie industrielle, La Société des amis de la maison de la chimie ont reçu le Conseil d'Administration d'Eurochemic, pp. 185-189 (Speech).

FREROTTE M. (1988), L'avenir des installations d'Eurochemic, Belgoprocess ou Belgowaste?, Intercommunale, 1st trimester 1988, Brussels.

HILD W., DETILLEUX E., GEENS L. (1980), Some Recommandations for Plant Design and Layout Resulting from the Decontamination of the Eurochemic Reprocessing Plant and the Partial Decommissioning of Some of its Facilities, NEA/OECD (1980), pp. 197-208.

HUET P. (1958), L'AEEN et la Société Eurochemic, contribution à l'étude des sociétés internationales, Annuaire français de droit international (AFDI), Paris.

MONGON A. (1964), Instrumentation et automatisme dans une usine de retraitement de combustibles irradiés, EN, pp. 150-158.

OSIPENCO A. (1980), Occupational Radiation Exposure at the Eurochemic Reprocessing Plant During Normal Operation and Intervention Periods, Occupational Radiation Exposure in Nuclear Fuel Facilities, IAEA, Vienna 1980, pp. 439-450.

OSIPENCO A., DETILLEUX E., FERRARI P. (1980), Décontamination et démantèlement partiel de l'usine de retraitement Eurochemic. Enseignements relatifs à la radioprotection et à la gestion des déchets, OECD/NEA (1980), pp. 255-262.

POHLAND E. (1958), Aufarbeitung bestrahlter Kernbrennstoffe, technische und wirtschaftliche Probleme, AtW, October 1958, pp. 385-388.

POHLAND E., BARENDREGT T. (1961), Die Aufarbeitung bestrahlter Kernspaltstoffe in der Eurochemic-Anlage, AtW, March 1961, pp. 149-153.

POHLAND E., STROHL P. (1962), Die Zusammenarbeit der Eurochemic mit der europaïschen Industrie beim Bau der Eurochemic-Anlage, AtW, Aug./Sept., pp. 403-409

ROMETSCH R. (1966), Wiederaufarbeitung abgebrannter Kernbrennstoffe in Europa – Aufgaben der Eurochemic, AtW, August-September 1966, pp. 423-427.

SOUSSELIER Y. (1958), Eurochemic, premier exemple de collaboration européenne atomique, EN, pp. 180-182.

SOUSSELIER Y. (1962), Eurochemic, buts et organisation de la Société, EN, pp. 510-516.

STROHL P. (1961), Problèmes juridiques soulevés par la constitution et le fonctionnement de la Société Eurochemic, Annuaire français de droit international, pp. 569-591, Paris.

STROHL P. (1991), Les aspects internationaux de la liquidation de la Société Eurochemic, Annuaire français de droit international, Paris, pp. 727-738.

PAMELA

DEMONIE H., WIESE H. (1990), Operation of the PAMELA High-level Waste Vitrification Facility, paper presented at the Jahrestagung Kerntechnik 1990, s.l.

DWK (1985), PAMELA, ein Projekt wird verwirklicht, Scherrer-Druck, Hanover.

DWK (sans date), Verglasungsanlage PAMELA in Mol/Belgien, introductory brochure of 27 pages.

HEIMERL W. (1975), HAW-Glaspartikel in Metallmatrix; ein endlagerfähiges Verfestigungsprodukt für hochradioaktive Lösungen, AtW, July-August 1975, pp. 347-349.

HÖHLEIN G. (1985), PAMELA: Advanced Technology for Waste Solidification, NEW, 1985, 2, pp. 16-17.

KAUFMANN F., WIESE H. (1987), Vitrification of High Level Waste, RECOD (1987), pp. 279-288.

KUHN K.-D., WIESE H., DEMONIE M. (1991), HLLW Conditioning in the PAMELA Vitrification Plant, RECOD (1991), pp. 273-277.

LUTZE W., CLOSS K.-D., TITTEL G., BRENNECKE P., KUNZ W. (1994), Vitrified HLW and Spent Fuel Management, AtW, 1994, 2, pp. 123-127.

TITTMANN E., SCHEFFLER K. (1987), PAMELA's LEWC Campaign, s.l.

ALONA

SWENNEN R., CUYVERS H., VAN GEEL J., WIECZOREK H. (1984), The Treatment of Combustible Alpha-wastes at Eurochemic Using Acid Digestion and a Plutonium Recovery Process. First Results of the Active Operation of a Demonstration Plant, American Nuclear Society, Idaho Section ed., Fuel Reprocessing and Waste Management, Jackson, Wyoming, August 26-29, 1984, Proceedings Vol. 1, pp. 549-563.

WIECZOREK H., OSER B. (1986), Entwicklung und aktive Demonstration des Verfahrens der Naßveraschung brennbarer plutoniumhaltiger Festabfälle, KfK Nachrichten, Vol. 18, n°2, pp. 77-82.

Belgoprocess

Annual reports of Belgoprocess and ONDRAF, 1987 to 1990.

DE CONINCK, DEMONIE M., CLAES J. (1986), Design and Preliminary Experience of Interim Storage Facility for Vitrified Waste, NEW, 1986, 9, pp. 27-29.

TEUNCKENS L. (1993), Decommissioning of Final Product Storage Buildings at the Ex-Eurochemic Reprocessing Plant, NEW, 1993, 11-12, p. 60.

Other printed sources

Documents of the OEEC, then OECD

Acts of the Council (since 1948)

Nuclear Energy Agency

Report of the Steering Committee for Nuclear Energy of the OEEC (March 1958), abbreviated as SCNE (1958).

Annual Reports of the European Nuclear Energy Agency of the OEEC, then OECD (1958 to 1971).

Annual Reports, OECD Nuclear Energy Agency (1972 to 1990).

OEEC (1955), Some Aspects of the European Energy Problem, also called the "Armand Report", June, Paris.

OEEC (1956a), Possibilities for Action in the Field of Nuclear Energy, also called the "Nicolaïdis Report", January, Paris.

OEEC (1956b), Europe and its Growing Energy Needs, also called the "Hartley Report", May, Paris.

OEEC (1956c), Joint Action by the OEEC Countries in the Field of Nuclear Energy, September, Paris.

Other organisations

CEA (1962), Eurochemic 1958-1962, Origins, Structures, Perspectives, Report by the Research Section of the Finance and Accounts Division.

ECAE (1958), Report on the Situation of the Nuclear Industries in the Community, Brussels.

ECMT(1955), Eurofima, European Company for the Financing of Railway Rolling Stock, constitution, Berne.

EURATOM Bulletin (1968), General Report on the Community's Nuclear Policy, VII/1968.

HOSTE A., JAUMOTTE A. (1976a), Commission d'évaluation en matière d'énergie nucléaire, Rapports de synthèse, March 1976, Kingdom of Belgium, Ministry for Economic Affairs, Brussels.

HOSTE A., JAUMOTTE A. (1976b), Commission d'évaluation en matière d'énergie nucléaire, Rapports techniques, Vol. VIII, Le cycle du combustible, Kingdom of Belgium, Ministry for Economic Affairs, Brussels.

HOSTE A., JAUMOTTE A. (1982), Commission d'évaluation en matière d'énergie nucléaire, Rapport final, éléments d'actualisation, Kingdom of Belgium, Ministry for Economic Affairs, Secretary of State for Energy, Brussels.

SYBELPRO (1983), Syndical Agreement, dated 13 December 1982.

Memoirs

HIRSCH E. (1988), Ainsi va la vie, Lausanne.

MARJOLIN R. (1986), Le travail d'une vie, Mémoires 1911-1986, Laffont, Paris.

MONNET J. (1976), Mémoires, Fayard, Paris (Published in the Livre de Poche series).

Bibliography

Section 1
Nuclear science and technology; history of nuclear science and technology

Reprocessing

General and international aspects of reprocessing

BERN J.B., WAISS Y. (1983), Contrôle en ligne dans les usines de retraitement, RGN, 1983, 5, pp. 406-408.

CHAUVE M. et al. (1986), Construction materials for spent fuel reprocessing plants, NEW, 1986, 2, pp. 19-21.

CHAYES A., LEWIS W.B. ed. (1977), International Arrangements for Nuclear Fuel Reprocessing, Cambridge (Mass.).

DELANGE M. (1985b), Operating Experience with Reprocessing Plants, AtW, 1985, 1, pp. 24-28.

GOMEZ J. (1983), Utilisation de l'instrumentation conventionnelle dans les usines de retraitement, RGN 1983, 5, pp. 401-405.

LANDRY J.W.(1955), Appareils d'échantillonnage de haute activité pour usines de traitement radiochimique, UN (1955), P/549, pp.635-638.

LOWRANCE W.W. (1977), Nuclear Futures for Sale: Issues Raised by the West-German Brazilian Nuclear Agreement, CHAYES A., LEWIS W.B. ed.(1977), pp. 201-221.

NUCLEAR ENGINEERING INTERNATIONAL (1993), Reprocessing Versus Direct Disposal: Weighing up the Costs, special edition "Spent Fuel Management and Transport", pp. 12-14.

OECD/ENEA-EUROCHEMIC (1963a), Aqueous Reprocessing Chemistry for Irradiated Fuels, La chimie du retraitement par voie aqueuse des combustibles irradiés, Brussels Symposium 1963, Paris.

OECD/ENEA-EUROCHEMIC (1963b) Colloquy [sic] on Instrumentation in Plants for the Chemical Reprocessing of Irradiated Fuels / Colloque sur l'instrumentation dans les usines de traitement chimique des combustibles irradiés, Paris, 22-23 November 1963, Paris.

OECD/NEA (1977), Nuclear Fuel Reprocessing in the OECD Countries, Paris.

OECD/NEA (1986), Management of Irradiated Nuclear Fuel, Experience and Options, Paris.

PEROT J.P. (1983), La conduite du procédé dans les usines de retraitement de combustibles irradiés, RGN, 1983 ,5, pp. 398-401.

PRATT H.R.C. (1956), Caractéristiques du matériel pour extraction liquide-liquide utilisé pour le traitement de substances radioactives, UN (1956), P/765, pp. 597-605.

RECOD (1987), International Conference on Nuclear Fuel Reprocessing and Waste Management, Proceedings of the conference held at the Palais des Congrès, Porte Maillot, Paris, from 23 to 27 August 1987. Sponsor: SFEN. Co-sponsor: European Nuclear Society, American Nuclear Society, Atomic Energy of Japan, 4 vols., 1628 p.

RECOD (1991), The Third International Conference on Nuclear Fuel Reprocessing and Waste Management, Proceedings of the conference held at Sendai from 14 to 18 April 1991. Sponsors: Atomic Energy Society of Japan, Japan Industrial Forum. Co-sponsors: ANS, ENS, JAERI, PNC. 2 vols., 1154 p.

REGNAUT P. (1957), L'extraction du plutonium, EN, pp. 196-206.

SCHÜLLER W. (1983), Betriebserfahrung mit der Wiederaufarbeitung, eine internationale Übersicht, AtW, 1983, 2, pp. 80-85.

UN (1956), Proceedings of the International Conference on the Peaceful Uses of Atomic Energy, Volume IX, Materials: Testing and Chemical Processing, Geneva.

UN (1958), Proceedings of the Second International Conference on the Peaceful Uses of Atomic Energy, Volume VIII, Irradiated Fuels and Radioactive Materials; Radiation Protection, Geneva.

UN (1965), Proceedings of the Third International Conference on the Peaceful uses of Atomic Energy, Volume X, Nuclear Fuels-I. Fabrication and Reprocessing, Volume XIV, Environmental Aspects of Atomic Energy and Waste Management, New York.

UN-IAEA (1972), Peaceful Uses of Atomic Energy, Proceedings of the Fourth International Conference on the Peaceful Uses of Atomic Energy, Volume VIII, New York, Vienna.

USAEC (1955), Selected Reference Material, United States Atomic Energy Program, Volume VI, Chemical Processing and Equipment, Geneva.

USAEC (1957), Symposium on the Reprocessing of Irradiated Fuels, held at Brussels, Belgium, May 20-25, 1957, Report TID-7534, Oak Ridge (Tenn.).

Reprocessing: regional and national aspects; installations

China

WANG D.Y., CHEN M. (1987), Plan for the Civil Reprocessing Pilot Plant of China, RECOD (1987), pp. 81-85.

YUN-QING J., ZHONG-YAO W. (1991), Some Aspects of a Civil Reprocessing Pilot Plant, RECOD (1991), pp. 69-73.

Europe

AtW (1979), Abgabe radioaktiver Stoffe aus Wiederaufarbeitungsanlagen in der EG 1972-1976, AtW, 1979, 12, pp. 612-613.

FORATOM (1970), The Future of Reprocessing in Europe, a FORATOM Study Prepared by a Groupe of Experts, February 1970.

GAUDERNACK B. et al. (1965), Preliminary Study of a Fuel Reprocessing Pilot Plant, UN (1965), P/704, pp. 247-251.

SCHLITT A. (1975), Zur Situation der Wiederaufarbeitung in Europa, AtW, July-August 1975, pp. 335-338.

France

BARBE A. (1990), LWR Fuel Reprocessing in France, AtW, 1990, 8-9, pp. 433-435.

CARLE R. (1987), Intervention lors de RECOD Paris, RECOD (1987), pp. 1559-1565.

DELANGE M. (1985a), Développement, expérience et innovation dans le retraitement, RGN, 1985, 6, pp. 559-568.

DEVEZEAUX DE LAVERGNE J.G. (1994), Wirtschaftlichkeit von Wiederaufarbeitung und Recycling, AtW, May 1994, pp. 350-353.

GLOAGEN A., LENAIL B. (1991), Policy in France Regarding the Back-end of the Fuel Cycle: Reprocessing/Recycling Route, RECOD (1991), pp. 7-11.

LEDUC M., PELRAS M., SANNIER J., TURLUER G., DEMAY R. (1987), Etudes de corrosion sur les matériaux destinés aux usines de retraitement, RECOD (1987), pp. 1173-1180.

LEWINER C.,GLOAGUEN A. (1988), The French Reprocessing Program, AtW, 1988, 5, pp. 227-229.

SCHAPIRA J.P., ZERBIB J.C. (1987), The "Castaing Exercise": an Original Contribution to the Technological Evaluation of the Back-end of the Fuel Cycle, RECOD (1987), pp. 1101-1108.

SIMONNET J., DEMAY R., BACHELAY J. (1987), Development and Use of Zirconium in Reprocessing Equipment, RECOD (1987), pp. 1215-1218.

STRAUCH G. (1987), Die genehmigten Radioaktivitätsabgabewerte für französische Kernanlagen, AtW, 1987, 1, pp. 53-55.

SYROTA J. (1993), Why Reprocess? The French View, Nuclear Engineering International, special edition "Spent Fuel Management and Fuel Transport", pp. 6-7.

Châtillon pilot plutonium extraction plant

FAUGERAS P., REGNAUT P. (1957), L'usine-pilote d'extraction du plutonium de Châtillon. Résultats d'exploitation, EN, pp. 123-137.

ATTILA (Fontenay-aux-Roses)

BOURGEOIS M., COCHET-MUCHY B. (1968), Le retraitement des combustibles irradiés par voie sèche, EN, pp. 192-200.

Marcoule

CEA (1963), La participation de l'industrie à la création et au fonctionnement du Centre de Marcoule, EN, pp. 329-334.

RODIER J., ESTOURNEL R., BOUZIGUES H., CHASSANY J. (1963), Le travail en milieu radioactif et ses problèmes, EN, pp. 291-301.

ROUVILLE M. de (1963), Le centre de production de plutonium de Marcoule, EN, pp. 213-220. Presented in a special edition devoted to Marcoule, n°4, June 1963.

TARANGER P. (1956), Marcoule, Age nucléaire, 1956, 2, pp. 35-39.

UP1

ALLES M. (1963), L'installation de dégainage, EN, pp. 257-262.

AUCHAPT P., GIRAUD J.-P., TALMONT X. (1968), Applications de l'extraction par solvant à la concentration et à la purification du plutonium; I. Marcoule: utilisation du phosphate tributylique, EN, pp. 181-186.

CURILLON R., COEURE M. (1958), Problèmes posés par la construction d'une usine d'extraction de plutonium: quelques solutions appliquées à Marcoule, UN (1958), P/1174.

CURILLON R., COEURE M. (1963), L'ensemble industriel d'exploitation de plutonium. L'usine – son évolution – ses annexes, EN, pp. 271-281.

DOUTRELUINGNE P. (1987), Avancées technologiques au service Laboratoires de Marcoule, RECOD (1987), pp. 885-891.

FAUGERAS P., CHESNE A. (1965), Le traitement des combustibles irradiés – amélioration et extension du procédé utilisant les solvants, UN (1965), P/65, pp. 295-306.

FERNANDEZ N. (1963), La section de traitement des effluents. Ses problèmes, son fonctionnement, ses résultats, EN, pp. 282-290.

FRUCHARD Y. (1968), Prévention de la criticité dans une usine de traitement des combustibles irradiés, EN, pp. 445-455.

GALLEY R. (1958), Les problèmes posés par l'usine de plutonium de Marcoule et leurs solutions, EN, 1958, pp. 2-20. English translation in: GALLEY R. (1959), Problems Connected with the Plutonium Factory at Marcoule and Their Solution, Journal of the Brirish Nuclear Energy Conference, 1959, 1, pp. 1-11.

JOUANNAUD C. (1963), L'usine d'extraction du plutonium et son exploitation, EN, pp. 263-270.

JOUANNAUD C. (1964), Expérience de six années de fonctionnement de l'usine de retraitement de Marcoule, UN (1964), P/67.

MICHEL P.M. (1987), Retraitement et opinion publique. Analyse de 30 ans de relations entre l'usine de Marcoule et son environnement, RECOD (1987), pp. 1121-1129.

RODIER J., BOUZIGUES H., BOUTOT P. (1962), La décontamination du matériel dans le centre de production du plutonium de Marcoule, EN, pp. 352-361.

APM (Marcoule pilot unit)

CALAME-LONGEAN A., REVOL G., ROUX J.P., RANGER G. (1987), Rénovation et extensions pour la R & D de l'atelier-pilote de Marcoule, RECOD (1987), pp. 115-121.

FARUGGIA J.M., MALTERRE G., TACHON M., POLYDOR G. (1987), Remise en état après 20 ans d'exploitation et redémarrage de l'atelier pilote de retraitement Marcoule, RECOD (1987), pp. 1405-1419.

GUILLET H. (1963), L'atelier-pilote de traitement des combustibles irradiés, EN, pp. 322-325.

MUS G., LINGER C. (1987), Conduite automatique des procédés de l'atelier pilote, RECOD (1987), pp. 1519-1526.

NIEZBORALA F., CALAME A. (1963), La télémanipulation à l'atelier-pilote de traitement des combustibles irradiés, EN, pp. 326-328.

UP1 and UP2

AUCHAPT P. et al. (1971), Adaptation des usines françaises de retraitement à de nouveaux types de combustibles irradiés, UN (1971), A/CONF.49/P/601.

La Hague

BASTIEN-THIRY H, JUSTIN F. (1988), Safety Engineering Achievements in Handling Casks at La Hague, NEW, 1988, 10, pp. 25-26.

BATHELLIER A., GRIENEISEN A., PLESSY L. (1968), Applications de l'extraction par solvant à la concentration et à la purification du plutonium II. La Hague: utilisation de la trilaurylamine, EN, pp. 186-191.

RGN (1981b), Mise en actif des "Nouvelle piscines de La Hague", RGN, 1981, 3, p. 260.

RGN-actualités (1981), Retraitement: lumières sur les contrats avec l'étranger, RGN, 1981, 4, pp. 379-381.

RGN (1982), Travaux d'amélioration à l'atelier Moyenne Activité Plutonium (MAPu), RGN, 1982, 5, p. 472.

RGN(1986), Le déchargement à sec des combustibles irradiés inauguré à La Hague, RGN, 1986, 5, p. 457.

UP2 (NUGG)

DELANGE M., CHAMBON M., PATIGNY M., TEXIER P. (1970), Mises au point industrielles réalisées à l'usine UP2 de La Hague de 1967 à 1970, EN, pp. 94-105.

DUBOZ M.(1965), Le Centre de La Hague, EN, pp. 228-245.

UP2 (LWR) and HAO

DELANGE M. (1987), LWR Spent Fuel Reprocessing at La Hague: Ten Years On, RECOD (1987), pp. 1423-1438.

DESVAUX J.L., FOURNIER W., SEUGNET J., MARTIN J.P. (1987), Maintenance in a Reprocessing Plant, RECOD (1987), pp. 171-181.

GUE J.P., ISAAC M., LESAUVAGE M., MALET G., COURTOIS G. (1987), Principales caractéristiques des déchets solides de haute activité récupérés en tête du procédé PUREX, RECOD (1987), pp. 607-610.

PATIGNY P., PAGERON D., SALOM J. (1987), Lessons from La Hague on Process Chemistry in LWR Spent Fuel Reprocessing, RECOD (1987), pp. 1593-1600.

RGN (1983), Bilan de 73 mois de retraitement des combustibles irradiés de réacteurs à eau légère, RGN , 1, pp. 58-59.

UP2-800 and UP3

BAILLIF L., BONNET C., DABAT R., DOMAGE M. (1987), Spent Fuel Unloading and Storage at La Hague, RECOD (1987), pp. 341-347.

BERN J.B., CHABERT J., LE GRAND F. (1987), Process Control and Monitoring in the New Reprocessing Plants at La Hague, RECOD (1987), pp. 985-992.

BERNARD C., MOUROUX J.P., DECOURS J. et al. (1991), Zirconium-made Equipement for the New La Hague Reprocessing Plants, RECOD (1991), pp. 570-575.

CHABERT J., COIGNAUD G., PEROT J.P., FOURNIER W., SILVAIN B. (1991), Total Data Management System for the
La Hague Spent Fuel Reprocessing Plants, RECOD (1991), pp. 77-82.

CHENEVIER F., BERNARD C., GIRAUD J.P. (1987), Design and Construction of the New Reprocessing Plants at La Hague, RECOD (1987), pp. 97-102.

UP3

BERNARD P., LAMARQUE G., et al. (1991), Process Nuclear Monitoring at UP3, RECOD (1991), pp. 125-130.

CHABERT J., ROULLAND M. (1991), Computer-aided Maintenance at La Hague UP3, NEW, 1991, 5-6, pp. 20-21.

DEVILLERS C., BASTIEN-THIRY H., DUBOIS G. (1987), Safety of the New Reprocessing Plants at La Hague, RECOD (1987), pp. 1311-1319.

DREYFUS G.P., RICHTER R. (1987), Maintenance Design in the New Reprocessing Plant at La Hague, RECOD (1987), pp. 1255-1262.

FOURNIER W., HUGELMANN D., DALVERNY G., LEUDET A. (1991), UP3 Plant First Reprocessing Campaigns, RECOD (1991), pp. 25-31.

GUEZENEC J., BASTIEN-THIRY H., CHAUMETTE A.M. (1987), Safety of Spent Fuel Handling in the D-pool at La Hague and Design of the Associated Control System, RECOD (1987), pp. 1329-1337.

IZQUIERDO J.J., CHAUVIRE P., PLESSIS L. (1991), The MERC Maintenance System, RECOD (1991), pp .1111-1116.

LAMARQUE G., FREJAVILLE G., LAVERGNE J. (1987), Mesures en lignes et contrôle de procédé dans UP3, RECOD (1987), pp. 1007-1014.

LEDERMANN P. (1994), Operating UP3. Three Years of Experience, Nuclear Engineering International, January 1994, pp. 46-49.

LORRAIN B., SAUDRAY D., TARNERO M. (1987), Dissolveur continu rotatif pour combustible à eau légère. Essais du prototype, RECOD (1987), pp. 1471-1481.

LUNG M. (1984), SGN Designs the Plants of Tomorrow, NEW, 1984, 4, 35-36.

PIGNAULT J.(1983), Conception d'une salle de conduite centralisée: application aux extensions de l'établissement de
La Hague, RGN, 1983, 5, pp. 411-414.

PRADEL P., DALVERNY G., MOULIN J.-P., GINISTY C. et al. (1991), The Organic Waste Treatment in UP3-La Hague, RECOD (1991), pp. 1101-1106.

RICAUD J.L., ROUX P., VIALA M. (1991), France's New UP3 Reprocessing Plant: Commissioning and First Year's Operation, NEW, 1991, 1-2, pp. 20-21.

RGN (1981), Retraitement: lumières sur les contrats avec l'étranger, RGN, 1981, p. 379-381.

RGN (1988), UP3 s'apprête au démarrage, RGN, 1988, 6, pp. 514-515.

RGN (1992), D'UP2-400 à UP3: les avancées technologiques., RGN, 1992, 2, pp. 156-159.

SILIE P. (1983), Les systèmes de conduite centralisés des usines de retraitement, RGN, 1983, 5, pp. 408-411.

SYROTA J. (1992), UP3, nouvelle installation de retraitement à La Hague, RGN, 1992, 4, pp. 318-323.

UP3 and Rokka-shô

BERNARD C., MIQUEL P., VIALA M. (1991), Advanced PUREX Process for the New Reprocessing Plants in France and Japan, RECOD (1991), pp. 83-88.

DRAIN F., BOULLIS B., HUGELMANN D., OHTOU Y. (1991), Extraction Process Technology for the New Reprocessing Plants in France and Japan, RECOD (1991), pp. 89-94.

PONCELET J.F., HUGELMANN D., SAUDRAY D., MUKOHARA S., CHO A. (1991), Head-end Process Technology for the New Reprocessing Plants in France and Japan, RECOD (1991), pp. 95-99.

AT1 La Hague and TOR Marcoule

AMAURY P., TALMONT X. (1967), AT1 – Maillon du programme d'étude de la filière surgénératrice française, EN, pp. 113-119.

PATARIN L. (1987), Le démarrage de TOR et l'expérience acquise en retraitement, RGN, 1987, 3, pp. 251-255.

Germany

HOSSNER R. (1990), Wiederaufarbeitung 1999 – Verträge jetzt, AtW, 1990, 1, p. 17.

SOLFRIAN W. (1974), Wer gehört zu wem in der Deutschen Atomindustrie; Beteiligungen und Verflechtungen ; Uranversorgung, Urananreicherung, Brennelementfertigung, Wieder-aufarbeitung, AtW, June 1974, pp. 309-315.

WAK

AtW (1970), Mit wichtigen Lieferungen an der WAK beteiligte Firmen, AtW, February 1970, pp. 95-96.

AtW (1986 to 1991), "Besondere Vorkommnisse" in der WAK, AtW, 1986, 12, pp. 626-627; AtW, 1987, 12, pp. 607-608; AtW, 1989, 2, pp. 98-100; AtW, 1990, 4, pp. 203-204; AtW, 1991, 5, pp. 250-252.

SCHÜLLER W. (1984), Betriebserfahrungen mit der Wiederaufarbeitungsanlage Karlsruhe, AtW, 1984, 8-9, pp. 438-444.

SCHÜLLER W. (1985), Zwanzig Jahre GWK/WAK. Bilanz des Erreichten und Ausblick auf zukünftige Aufgaben, AtW, 1985, 4, pp. 178-181.

SCHÜLLER W., HUPPERT K.L., HOFFMANN W. (1975), Betriebserfahrungen mit der WAK; Folgerungen für die weitere technologische Entwicklung, AtW, July-August 1975, pp. 342-346.

WEINLÄNDER W., HUPPERT K.L., WEISHAUPT M. (1991), Twenty Years of WAK Reprocessing Pilot Plant Operation, RECOD (1991), pp. 55-63.

WILLAX H.O. (1987), HLWC storage at WAK, RECOD (1987), pp. 315-318

WILLAX H.O., WEISHAUPT M. (1987), Reprocessing Plant Karlsruhe (WAK). Plant Experience and Future Tasks, RECOD (1987), pp. 203-209.

Project for a major German reprocessing plant

GASTEIGER R. (1987), Safety Design of the Wackersdorf Reprocessing Plant, RECOD (1987), pp. 1291-1307.

HILPERT H.J. (1987), Floor Response Spectra of the Main Process Building of a Reprocessing Plant Against Earthquake, Airplane Crash and Blast, RECOD (1987), pp. 1347-1350.

ISSEL W., KNOCH W. (1975), Zur Auslegung einer großen Wiederaufarbeitungsanlage, AtW, July-August 1975, pp. 339-342.

KOCH G., OCHSENFELD W., SCHMIEDER H. (1975), Überlegungen zum Fließschema einer Wiederaufarbeitungs-Großanlage, AtW, March 1975, pp. 123-127.

LEISTER P., OESER H.R. (1991), Remote Handling Technology for Reprocessing Plants: Lessons Learnt from LAHDE, RECOD (1991), pp. 163-168.

MISCHKE J. (1984), Konzept und Stand der Wiederaufarbeitungsanlage WA-350, AtW, 1984, 8-9, pp. 434-438.

NEGIDIUS N. (1988a), Zum Rechtsstreit um die Wiederaufarbeitungsanlage Wackersdorf, AtW, 1988, 2, p. 75 and 3 p. 125.

NEGIDIUS N. (1988b), Wackersdorf, Back to Square One?, AtW, 1988, 9, p. 410.

SCHEUTEN G.H. (1985), Die deutsche Wiederaufarbeitungsanlage, AtW, 1985, pp. 362-365.

SCHMIDT-KÜSTER W.-J. (1974), Das Entsorgungssystem im nuklearen Brennstoffkreislauf, AtW, July 1974, pp. 340-345.

WEINLÄNDER W. (1988), Germany's Wackersdorf Reprocessing Plant, NEW, 1988, 8-9, pp. 15 *et seq.*

WILLAX H.-O., KUHN K.-D. (1987), Betriebliche Erprobung neuer Verfahren und Komponenten in der WAK, AtW, 1987, 2, pp. 90-94.

ZÜHLKE P. (1974), Wiederaufarbeitung an der Schwelle zur wirtschaftlichen Nutzung, AtW, July 1974, pp. 346-351.

India

NAIR M.K.T., PRASAD A.N. (1991), Back-end of the Nuclear Fuel Cycle – Indian Experience, RECOD (1991), pp. 64-68.

Trombay

SETHNA H.N., SRINIVASAN N. (1964), Fuel Processing at Trombay, UN (1964), P/786.

Italy

EUREX and ITREC

CALLERI G., CAO S et al. (1971), Italian Progress Report on Reprocessing of Irradiated Fuel, UN (1971), A/CONF.49/P/182.

ENERGIA NUCLEARE (1961), La prima fase del PCUT, Energia Nucleare, 8, 4, p. 305.

ENERGIA NUCLEARE (1963), Firmato l'accordo USAEC-CNEN per il PCUT, Energia Nucleare, 10, 1, p. 57.

ENERGIA NUCLEARE (1968), La prima fase del PCUT, Energia Nucleare, 15, 6, p. 361-363.

ENERGIA NUCLEARE (1969), Prossima l'intrata in servizio del impianto EUREX, Energia Nucleare, 16, 8, p. 475.

ENERGIA NUCLEARE (1971), L'impianto EUREX-1 del CNEN per il trattamento del combustibile irragiato ha iniziato con successo la sua attività, Energia Nucleare, 18, 1, p. 10.

ENERGIA NUCLEARE (1975), In funzione il secondo impianto del CNEN per il trattamento del combustibile nuclcare, Energia Nucleare, 22, 8-9, p. 408 (ITREC).

POZZI F., RISOLUTI P., ROLANDI G. (1987), Technical and Commercial Aspects of European MTR Reprocessing, RECOD (1987), pp. 211-223.

SIMEN F. (1986), Italy's Policy on Back-end of the Fuel Cycle, NEW, 1986, 1, pp. 24-25.

Japan

OYAMA A. (1991), Nuclear Fuel Cycle Policy in Japan, RECOD (1991), pp. 3-6.

MUNAKATA K., NABESHIMA M., KOJIMA Y., TANAKA C. (1991), Prediction of Pulsed Column Behaviour During Off-standard Operation Using Numerical Calculation Code "DYNAC", RECOD (1991), pp. 617-622.

Tokai Reprocessing Plant (TRP)

CURILLON R., VIENNOT J. (1968), L'ensemble industriel de traitement des combustibles irradiés au Japon, EN, pp. 160-168.

LEFORT G., MIQUEL P., RUBERCY M. de (1968), Etude et expérimentation du schéma des extractions de l'usine japonaise, EN, pp. 169-180.

LUNG M., COIGNAUD M. (1978), L'usine de retraitement de Tokai-Mura, RGN, 1978, 4, pp. 314-319.

MIYAHARA K., YAMAMURA O., TAKAHASHI K. (1991), The Operational Experience at Tokai Reprocessing Plant, RECOD (1991), pp. 49-54.

NAITO S., SMIYA A., OHKATA K. et al. (1991), Remote Inspection of Repaired Dissolvers and Improvement of the Remote Maintenance Robots, RECOD (1991), pp. 157-162.

TAKASAKI K., EBANA M., NOMURA T. (1991), Radiation Control System at Tokai Reprocessing Plant, RECOD (1991), pp. 219-223.

YAMANOUCHI T., OMACHI S., MATSUMOTO K. (1987), Decadal Operational Experience of the Tokai Reprocessing Plant, RECOD (1987), pp. 195-202.

Rokka-shô Reprocessing Plant (RRP)

ATOMS IN JAPAN (1993), Construction Begins at Rokka-shô Reprocessing Plant, Aiming at Fuel Recycling, Atoms in Japan, April, p. 29-30.

BERNARD C., LENAIL B., REDON A., RUNGE S. (1989), Rokka-shômura Reprocessing Plant: France-Japan Cooperation, NEW, 1989, 11-12,pp. 33-34.

NAKAJIMA N., NAKAYA I., FUKAYA T., KIYOKAWA T. (1991), Manufacturing Experience of the Rotary Dissolver for HEDT [IIcad end Demonstration Test], RECOD (1991), pp. 1146-1149.

NOMURA A. (1987), Outline of the JNFS Reprocessing Plant Project, RECOD (1987), pp. 89-97.

ONOYAMA M., NAKATA M., HIROSE Y., NAKAGAWA Y. (1991), Development of a High Performance Type 304L Stainless Steel for Nuclear Fuel Reprocessing Facilities, RECOD (1991), pp. 1066-1071.

TOYOTA M. (1991), Outline of Rokka-shô Reprocessing Plant Project, RECOD (1991), pp. 38-43.

TSUYUKI T., KIYOKAWA T., OTODA T. (1991), Head-end Demonstration Test – Outline of the Test Program, RECOD (1991), pp. 141-145.

TSUYUKI T., KOMATSU K., OCHI K. et al. (1991), Extraction Demonstration Test – Outline of General Program and Seismic Test, RECOD (1991), pp. 146-150.

Norway

Kjeller

BARENDREGT T.J., KOREN LUND L. (1958), L'installation de retraitement de combustible du JENER, UN (1958), P/585.

GAUDERNACK B., LINDLAND K.P., JOSEPH C.J. (1965), Operational Experience from the Kjeller Reprocessing Pilot Plant, UN (1965), p. 253-259.

OOYEN J. van, ESCHRICH H. et al. (1965), Extraction Studies on Selected Problems in Reprocessing, UN (1965), P/758, pp. 409 *et seq.*

United Kingdom

ALLARDICE R.H. (1990), Nuclear Fuel Reprocessing in the United Kingdom, AtW, 1990, 8-9, pp. 436-439.

Dounreay

BOYLE J.G. et al. (1972), Operating Experience with the UK Fuel Reprocessing Plants at Windscale and Dounreay, UN (1972), A/CONF.49/P/492.

BUCK C. et al. (1958), Traitements chimiques aux installations de l'UKAEA à Dounreay, UN (1958), P/82, pp. 25-46.

Windscale-Sellafield

Windscale 1 and 2

COOTE J.A., ANDERSON R.W. (1991), Occupational Radiation Exposure Control at the Nuclear Fuel Reprocessing Plant at Sellafield, RECOD (1991), pp. 224-229.

CORNS H. et al. (1965), The New Separation Plant Windscale: Design of Plant and Plant Control Methods, UN (1965), P/161.

HUGHES T.G. et al. (1972), Development, Design and Operation of the Oxide Fuel Reprocessing Plant at the Windscale Works of BNFL, UN (1972), A/CONF.49/P/491.

WARNER B.F. et al. (1965), The Development of the New Separation Plant, Windscale, UN (1965), P/160, pp. 224-231.

THORP (Thermal Oxide Reprocessing Plant)

CHAMBERLAIN L.N. (1988), Sellafield in the 1990s, NEW, 1988, 5, pp. 22-23.

NUCLEAR ENGINEERING INTERNATIONAL (1993), THORP Awaits Starting Gun, February 1993, pp. 40-42.

NUCLEAR ENGINEERING INTERNATIONAL (1993), While THORP Lies Idle, BNFL News is Bad, May 1993, p. 14.

United States

SCHWENNESEN J.L. (1958), Examen sommaire de la conception et de l'exploitation des usines de traitement du combustible nucléaire, UN (1958), P/514.

THIRIET M. (1962), Coûts de retraitement des combustibles irradiés aux Etats-Unis, analysis and comments an a study by B. Manowitz, Nucleonics, February 1962, EN, pp. 217-221.

ICPP

LEMON R.B., REID D.G. (1956), Marche d'une usine de traitement radiochimique à entretien non télécommandé, UN (1956), P/543.

NATIONAL REACTOR TEST STATION (1955), Chemical Processing of Reactor Fuel Elements at the Idaho Chemical Processing Plant, USAEC (1955), Vol. 6, pp. 1-44.

ORNL

CULLER F.L., BLANCO R.E. (1965), Advances in Aqueous Processing of Power Reactor Fuels, UN (1965), P/249, pp. 316-328.

Metal Recovery and Thorex (Oak Ridge)

BRUCE F.R., SHANK E.M. et al. (1958), Expérience acquise pendant le fonctionnement de deux installations pilotes de traitement chimique, UN (1958), P/536.

NFS, MFRP, Barnwell

SINCLAIR E.E. et al. (1972), Existing and Projected Plants and Processes for Thermal Reactor Fuel Recovery, UN (1972), A/CONF.49/P/065.

West Valley (NFS)

ENERGIE NUCLEAIRE (1962), Première usine privée de retraitement des combustibles irradiés aux Etats-Unis, EN, pp. 615-617.

Barnwell

LARSON H.J. (1987), The Demise of the Barnwell Nuclear Fuel Plant. Impact of the Public, Press and Politics, RECOD (1987), pp. 1109-1114.

USSR

CHEVTCHENKO V.B. et al. (1958), Quelques particularités du traitement des éléments de combustibles irradiés de la première centrale nucléaire de l'URSS, UN (1958), p. 47-49.

DZEKUN E.G. et al. (1991), Commercial Reprocessing of Spent WWER-440 Fuel, RECOD (1991), pp. 44-48.

NIKIPELOV B.V. (1991), USSR Nuclear Fuel Cycle Industry, its Status and Outlook, RECOD (1991), pp. 18-21.

SEMENOV B. (1983), Les orientations de la politique soviétique en matière de retraitement et de gestion des déchets, Extract from the IAEA Bulletin reproduced in RGN, 1983, 5, pp. 420-421.

Activities "downstream" of reprocessing: management of solid and liquid wastes, decommissioning and dismantling of reprocessing facilities

Management of solid and liquid wastes: general and internatinal aspects

BOYASIS J.P., DONATO A., MERGAN L. (1987), The Conditioned Reprocessing Waste Returns. An Overview of the Question, RECOD (1987), pp. 633-644.

OECD/NEA-IAEA (1973), Management of Radioactive Wastes Produced in the Processing of Spent Fuel, Proceedings of the Symposium, 27 November-1 December 1972, Paris.

OLIVIER J.P. (1976), Le rejet en mer des déchets radioactifs, RGN, 1976, 6, pp. 513-519.

OLIVIER J.P. (1988), La contribution de l'AEN en matière de gestion des déchets radioactifs, RGN, 1988, 4, pp. 343-345.

ORLOWSKI S. (1976), Le programme de la commission des communautés européennes en matière de déchets radioactifs, RGN, 1976, 6, pp. 520-526.

SCHAPIRA J.P. (1991), Les déchets nucléaires, un problème mondial, Problèmes politiques et sociaux, n° 649, 1 February 1991.

SCHAPIRA J.P. (1990), Une nouvelle stratégie pour le plutonium, La Recherche, 226, November 1990, pp. 1434-1443.

SOUSSELIER Y., PRADEL J. (1972), La gestion des déchets radioactifs et leur stockage à long terme, UN – IAEA (1972), Vol. 10, p. 445-462.

SOUSSELIER Y. (1978), Les déchets radioactifs et leur gestion, RGN 1978, 3, pp. 160-165.

SOUSSELIER Y. (1981), Les déchets radioactifs: des programmes nationaux à la coopération communautaire, RGN, 1981, 1, pp. 24-27.

SOUSSELIER Y. (1984), Les déchets radioactifs: évolution, bilan et perspectives, RGN, 1984, 1, pp. 64-68.

TEILLAC J. (1988), Les déchets nucléaires, PUF, Paris

WARNECKE E. (1994), Disposal of Radioactive Waste – a Completing Overview, Kerntechnik, 59, 1-2, pp. 65-71.

High activity

KRAUSE H. (1982), Die Endkonditionierung hochaktiver Spaltproduktlösungen, AtW, 1982, 12, pp. 618-621.

SOMBRET C. (1987), Industrial Vitrification Processes for High Level Liquid Wastes: Potential and Limitations, RECOD (1987), pp. 1608-1614.

STRUXNESS E.G., BLOMEKE J.O. (1958), Procédés de traitement et d'élimination définitive de déchets radioactifs de natures différentes, UN (1958), P1073, pp. 498-512.

WATSON L.C. et al. (1958), Elimination des produits de fission par incorporation à un verre, UN (1958), pp. 485-493.

Intermediate and low activity

ALEXANDRE D., JORDA M., SAAS A., BAUDIN G. (1987), Immobilisation Matrices for Low and Intermediate Level Radioactive Wastes, RECOD (1987), pp. 811-815.

Special aspects (by country and/or installation)

Belgium

DETILLEUX E., DECAMPS F., BIESEMANS E. (1994), Radioactive Waste Disposal in Belgium, Kerntechnik, 59, 1-2, pp. 14-17.

CEN/SCK

DEJONGHE P., D'HONDT M. (1957), Quelques études en rapport avec le traitement chimique d'effluents radioactifs, EN, 1957, pp. 27-33.

DEJONGHE P. et al. (1965), Asphalt Conditioning and Underground Storage of Concentrates of Medium Activity, UN (1965), pp. 343-350.

DEJONGHE P. et al. (1958), Le traitement des résidus radioactifs dans les laboratoires de Mol, UN (1958), P1676, pp. 527-535.

France

BARBE A., PECH R. (1991), Cost Estimation of the Decommissioning of Nuclear Fuel Cycle Plants: Application to Reprocessing Plants, RECOD (1991), pp. 953-959.

BONNIAUD R., BRODSKY, BURKO, COHEN P. (1963), Traitement des solutions de produits de fission par vitrification, EN, pp. 427-430.

DELAUNAY H. (1992), La gestion des déchets du cycle du combustible, RGN, 1992, 5, pp. 389-393.

LEFILLATRE G. (1976), Conditionnement dans le bitume des déchets radioactifs de moyenne activité. L'expérience française, RGN, 1976, 6, pp. 496-500.

MONCOUYOUX J.P., BOEN R., PUYOU M., JOUAN A. (1991), New Vitrification Techniques, RECOD (1991), pp. 307-311.

RODIER J., LEFILLATRE G., SCHEIDHAUER J. (1964), Enrobage par le bitume des boues radioactives, EN, p. 81-88.

La Hague

ALEXANDRE D., CHOTIN M., LE BLAYE G. (1987), Vitrification of Fission Product Wastes: Industrial Experience and Construction of the New Vitrification Units at La Hague, RECOD (1987), pp. 289-296.

BARBE A., BAUDIN G., BOUDRY J.C., TCHEMITCHEFF E. (1991), Management of Low and Medium Activity Waste at La Hague, RECOD (1991), pp .399-403.

BOUDRY J.-C. (1988), La gestion des déchets de La Hague, traitement, conditionnements, contrôle, RGN, 1988, 4, pp. 300-305.

FOURNIER W., HUGONY P., SOMBRET C. et al. (1991), Start-up of Commercial HLW Vitrification Facilities at La Hague, RECOD (1991), pp. 278-282.

JOUAN A., TERKI M., PACAUD F., MONCOUYOUX J.P. (1987), Research and Development Status of the French Vitrification Process, RECOD (1987), pp. 669-676.

RGN (1989), Démarrage de R7 à La Hague, RGN, 1989, 6, p. 469-470.

Marcoule

FERNANDEZ N. (1969), Enrobage bitumineux des boues de traitement des effluents radioactifs. Réalisation industrielle, EN, pp. 357-365.

RODIER J., LEFILLATRE G., CHARBONNEAUX M. (1965), L'incinération des résidus radioactifs et ses aspects industriels, EN, pp. 25-37.

WORMSER G.,RODIER J., ROBIEN E. de, FERNANDEZ N. (1965), Améliorations apportées aux traitements des résidus radioactifs, UN (1965), Vol. 14, pp. 219-226.

AVM (Marcoule Vitrification Unit)

BERNAUD C. (1958), Quelques problèmes industriels posés par le traitement des effluents liquides radioactifs: solutions particulières à Marcoule, UN (1958), P/1178, pp. 536-539.

BONNIAUD R. (1976), La vitrification continue des produits de fission, RGN, 1976, 6, pp. 490-495.

BONNIAUD B., KERSALE M., ROZAND M. (1968), Atelier pilote de vitrification des produits de fission à Marcoule, EN, pp. 201-209.

GIRAUD J.P., LE BLAYE (1973), Conception d'un atelier industriel de vitrification des produits de fission et du stockage de verres de très haute activité, EN, pp. 41-50.

JOUAN A., PAPAULT C. et al. (1978), L'atelier de vitrification continue des produits de fission de Marcoule, RGN, 1978, 4, pp. 302-306.

ROZE B., COSTA J.C., CHEVILLARD H. (1987), Démantèlement du pont de la cellule de vitrification de l'AVM, RECOD (1987), pp. 231-244.

STEL (Liquid waste treatment unit)

DUCOS O., DELAFONTAINE G., MISTRAL J.P., PUIGREDO M. (1987), La station de traitement des effluents liquides de Marcoule, RECOD (1987), pp. 139-145.

DUCOS O., DELAFONTAINE G., SEYFRIED P., et al. (1991), New Evaporation Facility for Liquid Waste Treatment at Marcoule [EVA], RECOD (1991), pp. 112-116.

Germany (excluding PAMELA)

BRINKERT M. (1992), Die "Transnuklear-Affäre" und ihre Auswirkungen, AtW, 1992, 3, pp. 129-132.

GASTEIGER R., HÖHLEIN G. (1975), Behandlung radioaktiver Abfälle aus Wiederaufarbeitungsanlagen, AtW, July-August 1975, pp. 349-353.

JANBERG K., WEH R. (1991), Wiederaufarbeitung deutscher Brennelemente im Ausland: Rücknahme und Verbleib radioaktiver Abfälle, AtW, 1991, 1, pp. 40-43.

ODOJ R., FILSS P., WOLF J. (1988), Unregelmäßigkeiten bei der Deklaration von Abfallfässern durch Transnuklear/SCK Mol, AtW, 1988, 4, pp. 175-178.

ROSER T. (1983), Que faire des déchets radioactifs? L'état de réalisation du programme allemand, RGN, 1983, 3, pp. 232-234.

ROSER T. (1988), L'affaire Transnuklear, RGN, 1988, 2, pp. 176-177

ROSER T. (1991), Das Ende einer Affäre [Transnuklear], AtW, 1991, 1, p. 47.

WARNECKE E., HOLLMANN A. (1992), Aufkommen radioaktiver Abfälle in Deutschland, AtW, 1992, 2, p. 94-100.

India (Trombay)

CHINOY A.R. et al. (1965), Management of Radioactive Wastes at Trombay, UN (1965), Vol. 14, pp. 260-267.

United Kingdom

HOWDEN M., MOULDING T.L.J. (1987), Progress in the Reduction of Liquid Radioactive Discharges from the Sellafield Site, RECOD (1987), pp. 1045 *et seq.*

OPENSHAW S. et al. (1989), Britain's Nuclear Waste, Safety and Siting, London.

SHEIL A.E. (1991), BNFL's Decommissioning and Decommissioning Developement Programmes at Sellafield, RECOD (1991), pp. 960-964.

United States

BROWN R.E. (1958), Expérience acquise à Hanford et Savannah River dans l'enfouissement des déchets radioactifs [de faible activité], UN (1958), P/1767, pp. 545-551.

BURCH W.D., CROFF A.G., RAWLINS J.A. (1991), A New Look at Actinide Recycle, RECOD (1991), pp. 321-328.

PILKEY O.H. (1958), L'expérience des Etats-Unis d'Amérique dans la conception et l'exploitation des installations de stockage pour résidus de haute activité (Hanford, Savannah River, Idaho), UN (1958), P/389, pp. 472-484.

Hanford

ANDERSON C.R. (1955), Installations de stockage des effluents à haute activité. Conception et fonctionnement, UN (1955), P/552, pp. 734-742.

BROWN R.E. (1955), Décharge terrestre des déchets liquides, UN (1955), P/565, pp. 763-769.

GILLETTE R. (1973), Radiation Spill at Hanford: The Anatomy of an Accident, Science, 24 August 1973, Vol. 181, pp. 728-730.

ICPP

LOEDING J.W. et al. (1958), Conversion des résidus du traitement des combustibles en solides par la méthode du lit fluidifié en vue de leur élimination, UN (1958), pp. 513-526.

West Valley

KNABENSCHUH J.L. (1987), Startup of the West Valley High Level Waste Conditioning and Vitrification Systems B30(61), RECOD (1987), pp. 263-277.

USSR

COCHRAN T.B., NORRIS R.S., SUOKKO K.L. (1993), Radioactive Contamination at Chelyabinsk-65, Russia, Annual Review of Energy and the Environment, Vol. 18, pp. 507-528, Palo Alto (Cal.).

Decommissioning and dismantling

BRADBURY D. (1992), Decommissioning of Civil Nuclear Facilities: a World Review, THOMAS S., BERKHOUT F. ed.(1992), pp. 755-760.

COUTURE J. (1975), Y a-t-il une crise mondiale dans le retraitement des combustibles?, RGN, 1, pp. 31-34.

CREGUT A. (1978), Le déclassement des installations nucléaires, RGN, 1978, 3, pp. 166-172.

CREGUT A., LURIE R. (1985), L'expérience française en matière de déclassement, RGN, 1985, 2, pp. 133-138.

OECD/NEA (1980), Decommissioning of Nuclear Installations: Requirements at the Design Stage, report on an NEA Specialists' Meeting, 17-19 March 1980, Paris.

OECD/NEA (1981a), Decontamination Methods for the Decommissioning of Nuclear Installations, Paris.

OECD/NEA (1981b), Cutting Techniques Used in the Decommissioning of Nuclear Installations, Paris.

OECD/NEA (1991), Decommissioning of Nuclear Installations, an Analysis of Variations in Decommissioning Costs, Paris.

SCHALLER K.H. (1985), Le déclassement des installations nucléaires dans la Communauté européenne, RGN, 1985, 2, pp. 144-145.

SURREY J. (1992), Ethics of Nuclear Decommissioning, THOMAS S., BERKHOUT F. ed.(1992), pp. 632-640.

Safety and radiation protection

General

ANDERSON R. (1989), Environmental Safety and Health Issues at U.S. Nuclear Weapons Production Facilities, 1946-1988, Environmental Review, Fall-Winter 1989, pp. 69 *et seq.*

BIRRAUX C. (1992), Le contrôle de la sûreté et de la sécurité des installations nucléaires, Economica, Paris.

BIRRAUX C. (1994), Sur le contrôle de la sûreté et de la sécurité des installations nucléaires, Office parlementaire d'évaluation des choix scientifiques et technologiques, Rapport de l'Assemblée Nationale n°1008 et du Sénat n°280, Vol. 1, Conclusions du rapporteur.

BLANC D. (1991), Sûreté de l'énergie nucléaire, PUF, Paris.

BOIRON P., BOUCHEZ H. (1986), Radioprotection Facilities: a Decade of Progress, NEW, 1986, 12, pp. 11-15.

CHASSARD-BOUCHAUD C. (1993), Environnement et radioactivité, PUF, Paris.

LEFORT M. (1980), Les radiations nucléaires, PUF, Paris.

NENOT J.C. (1991), Les nouvelles recommandations de la CIPR. La publication 60 RGN, 1991, 4, pp. 303-305.

NENOT J.-C.(1991), Les accidents nucléaires et radiologiques. Le bilan en 1991, RGN, 1991, 6, pp. 469-474.

OECD/NEA (1993a), The Safety of the Nuclear Fuel Cycle.

OECD/NEA (1993b), Radiation Protection on the Threshold of the 21st Century; Proceedings of an NEA Workshop, Paris, 11-13 January 1993.

PHARABOD J.P., SCHAPIRA J.P (1988), Les jeux de l'atome et du hasard, les grands accidents nucléaires de Windscale à Tchernobyl, Calmann-Lévy, Paris.

TUBIANA M., BERTIN M. (1989), Radiobiologie et radioprotection, PUF, Paris.

In reprocessing plants

France

HENRY P., CLECH C., LAFFAILLE C. (1987), Radioprotection et dosimétrie dans les usines de retraitement, RECOD (1987), pp. 1079-1085.

Germany: WAK

KRAUT W., WICHMANN H.P., WILLAX H.O. (1987), Monitoring for Internal Contamination in the WAK Reprocessing Plant, RECOD (1987), pp. 1073-1077.

La Hague

BETIS J. (1993), Pratique de l'optimisation de la radioprotection sur le centre de retraitement des combustibles irradiés de La Hague, RGN, 1993, 4, pp. 270-275.

DELANGE M. (1984), 25 ans de sûreté et radioprotection dans les usines de retraitement, RGN, 1984, 1, pp. 58-63.

Fissile material accounting and systems of guarantees

BAHM W., BAUMGÄRTEL G., SEIFERT R. (1990), Fortgeschrittene Bilanzierungsmethoden zur internationalen Kernmaterialüberwachung, AtW, 1990, 10, pp. 461-464.

BERKHOUT F., FEIVESON H. (1993), Securing Nuclear Materials in a Changing World, Annual Review of Energy and the Environment, Vol. 18, pp. 631-665, Palo Alto (Cal.).

LAURENT J.P., REGNIER J., TALBOURDET Y., DE JONG P. (1991), Safeguards Implementation in UP3 Reprocessing Plant, RECOD (1991), pp. 524-527.

Fuel cycle

AtW (1972), Die Brennstoffkreislaufindustrie in der Europaïschen Gemeinschaft, AtW, March 1972, pp. 172-174.

BAUMIER J. (1989), Géopolitique et économie du cycle du combustible nucléaire, CEA, Paris.

FERNET P., PANTLEON M. (1969), Die europaïsche Atomindustrie und ihr Markt, 5. Brennelemente für Kernreaktoren, AtW, Februar 1969, pp. 80-84.

IAEA (1977), Regional Nuclear Fuel Cycle Centres, Vol. 1 and 2.

INFCE (1980), International Nuclear Fuel Cycle Evaluation, summary of proceedings, IAEA Vienna.

OECD/NEA (1978a), Nuclear Fuel Cycle Requirements and Long Term Supply Considerations, Paris.

WALKER W. (1992), The Back-end of the Nuclear Fuel Cycle, in KRIGE J. ed.(1992), Choosing Big Technologies, History and Technology, 1992, 1-4, pp. 189-201.

Also see the book by Jean-Pierre Daviet on EURODIF, DAVIET J.P. (1993), below.

Other aspects of nuclear science and technology

ANGELIER J.P. (1983), Le nucléaire, la Découverte Paris.

BLANC D. (1987), La chimie nucléaire, PUF, Paris.

CARLE R. (1993), L'électricité nucléaire, PUF, Paris.

CHELET Y. (1961), L'énergie nucléaire, Seuil, Paris.

COLOMEZ G., MAS P. (1981), Rôle et utilisation des réacteurs de recherche en France, RGN, 1981, 6, pp. 529-536.

CUNEY M, LEROY J., PAGEL M. (1992), L'uranium, PUF, Paris.

DUPUY G. (1982), Radioactivité Energie nucléaire, PUF, Paris

ERTAUD A. (1979), Réacteurs d'avant-hier et d'après-demain, RGN, 1979, 3, pp. 253-263.

FAUGERAS P. (1958), Le plutonium, production et caractère chimique, L'Age nucléaire, pp. 327-352.

GUERON J. (1982), L'énergie nucléaire, PUF, Paris.

IAEA (1983), Nuclear Power Experience, Vienna.

LEWINER C. (1988), Les centrales nucléaires, PUF, Paris.

MACKENZIE D., SPINARDI G. (1994), Tacit Knowledge, Weapons Design, and the Uninvention of Nuclear Weapons. Draft of March 1994 provided by Dominique Pestre.

NEEL L. préface (1982), Vivre avec le nucléaire, Hachette Pluriel, Paris.

OEEC/ENEA (1958), Industry and Nuclear Energy, papers given during the Second Information Meeting on Nuclear Energy for Managers, Amsterdam, 24 to 28 June 1957, Paris.

OEEC/ENEA (1959-1960), Industry and Nuclear Energy, Stresa Conference, 11 to 14 May 1959, Paris.

OEEC/EPA (1957), Industry and Nuclear Energy, papers given during the First Information Meeting on Nuclear Energy for Managers, Paris, 1 to 5 April 1957, Paris.

OECD/NEA (1978b), Symposium on International Co-operation in the Nuclear Field: Review and Prospects (20th Anniversary of the Agency), Paris.

OECD/NEA (1993c), Spin-off Technologies Developed Through Nuclear Activities, Paris.

OECD/NEA (1993d), Skilled Personnel for the Nuclear Industry; Supply and Demand, Paris.

REBOUD H. (1958), Les réacteurs de puissance. Principes et classifications, L'Age nucléaire, n°12, pp. 251-254.

SOUSSELIER Y. (1960), Demain l'atome, PUF, Paris.

SYNDICAT CFDT DE L'ENERGIE ATOMIQUE (1980), Le dossier électronucléaire, Paris.

History of chemistry, chemical engineering and the chemical industry

BENSAUDE-VINCENT B., STENGERS I. (1993), Histoire de la chimie, La Découverte, Paris.

DAVIET J.P. (1989), Une multinationale à la française, Saint-Gobain 1665-1989, Fayard, Paris.

HOUNSHELL D.A., SMITH J.K. Jr. (1988), Science and Corporate Strategy. Du Pont R & D, 1902-1980, Cambridge University Press, Cambridge (Mass.), Chapter 16.

PACHE P. (1970), Les activités du groupe Péchiney dans le cycle du combustible nucléaire, EN, pp. 318-322.

REGNAUT P. (1968), L'industrie chimique française et l'énergie nucléaire, EN, pp. 147-149.

RENAUD J. (1970), Activités nucléaires de la Société Ugine-Kuhlmann, Energie nucléaire, pp. 313-317.

General history and philosophy of science and technology

BRAUN H.J. ed.(1992), Symposium on Failed Innovations, Introduction, Social Studies of Science, SAGE, London and other locations, Vol. 22 (1992), pp. 213-230.

CHADEAU E. et al. (1992), Concorde, la puissance du rêve, le rêve de la puissance, Big Science: les Grands projets scientifiques du XXe siècle, Cahiers de Science et Vie n°9.

CHADEAU E. dir. (1994), Ariane: le défi technologique, Paris.

BRAUN H.J., KAISER W. (1992), Propyläen Technikgeschichte, KÖNIG W. ed.(1990-1992), Bd 5, 1914-1990, Ullstein Propyläen, Frankfurt, Berlin. See in particular KAISER W., Technisierung des Lebens seit 1945, Chapitre Die Problematik der Kernenergie als neue Primärenergiequelle, pp. 285-339.

DAUMAS M. ed.(1978), Histoire générale des techniques, PUF, Paris, T 4 et 5. In particular J. GUERON (1978), Energie nucléaire, pp. 244-305, et DAUMAS M. (1978), Grande industrie chimique, pp. 493-715.

DEBEIR J.C., DELEAGE P.P., HEMERY D. et al. (1986), Les servitudes de la puissance, une histoire de l'énergie, Flammarion, Paris, particularly chapters 8 and 9, pp. 299-360.

FERNE G. dir. (1993), Science, pouvoir et argent, la recherche entre marché et politique, Autrement, Science et société, n°7, January 1993, Paris.

GALISON P., HEVLY B. ed.(1992), Big Science: the Growth of Large Scale Research, Stanford, Stanford University Press.

GILLE B. (1978), Histoire des techniques, Gallimard, Paris

GOFFI J.Y. (1992), La philosophie de la technique, PUF, Paris.

KERVERN G.Y., RUBISE P. (1991), L'archipel du danger. Introduction aux cindyniques, Economica, Paris.

KRIEGER W. (1987), Zur Geschichte der Technologiepolitik und Forschungsförderung in der BRD, eine Problemskizze,Vierteljahreshefte für Zeitgeschichte, 1987/2, pp. 247-271.

KRIGE J. ed.(1992), Choosing Big Technologies, History and Technology, 1992, 1-4.

KRIGE J. (1993a), The Launch of ELDO, ESA HSR-7 (History Study Report n°7), March 1993.

KRIGE J. (1993b), Some New Thoughts on the Launch of ELDO and its Subsequent "Failure", paper presented at the Conference on Technological Change, Oxford, 8-11 September 1993.

LEROY A., SIGNORET J.P. (1992), Le risque technologique, PUF, Paris.

SALOMON J.-J. (1970), Science et politique, Seuil, Paris.

SALOMON J.-J. (1986), Le Gaulois, le cow-boy et le samouraï. La politique française de la technologie, Economica, Paris.

SALOMON J.-J. (1992), Le destin technologique, Balland, Paris.

TATON R. ed.(1964), Histoire générale des sciences, T.3 Vol.2, Paris.

WILLIAMS T.I. ed.(1978), A History of Technology. Vol. VI, The Twentieth Century c. 1900 - c. 1950, Part 1; Vol. VII, The Twentieth Century c. 1900 - c. 1950, Part 2, Oxford. Completed by SINGER C. et al. (1954 to 1958) in five volumes.

Section 2
History of nuclear energy and nuclear policy

DAMIAN M. (1992), Nuclear Power, the Ambiguous Lessons of History, THOMAS S., BERKHOUT F. ed.(1992), pp. 596-607.

DUMOULIN M., GUILLEN P., VAÏSSE M. dir. (1994), L'énergie nucléaire en Europe. Des origines à EURATOM. Actes des journées d'études de Louvain-La-Neuve (18 and 19 November 1991), Peter Lang, Collection Euroclio, Berne.

LA GORCE P.M. de dir. (1992), L'aventure de l'atome, Flammarion, Paris.

LECLERCQ J. (1986), L'ère nucléaire, le monde des centrales nucléaires, Hachette, Paris.

NELKIN D. (1981), The Atom Besieged, Cambridge (Mass.), in particular Chapter I, The Political Context of the Nuclear Controversy.

THOMAS S., BERKHOUT F. ed.(1992), The First 50 Years of Nuclear Power: Legacy and Lessons, special edition of Energy Policy, Vol. 20, n°7 and 8, July and August 1992.

VAÏSSE M. (1992), La coopération nucléaire en Europe (1955-1958), Etat de l'historiographie, Storia delle relazioni internazionali, VIII, 1992/1-2, pp. 201-213.

Section 3
International nuclear relations

BARBIER C. (1990), Les négociations franco-germano-italiennes en vue de l'établissement d'une coopération militaire nucléaire au cours des années 1956-1958, Revue d'Histoire diplomatique, 1-2 /1990, pp. 81-113

BLUM R. (1963), La coopération France-Japon dans le domaine nucléaire, EN, pp. 481-484.

CISAC/NAS (1994), Management and Disposition of Excess Weapons Plutonium, Committee on International Security and Arms Control (CISAC), National Academy of Sciences, National Academy Press, Washington.

CONZE E. (1990), La coopération franco-germano-italienne dans le domaine nucléaire dans les années 1957-1958, un point de vue allemand, Revue d'Histoire diplomatique, 1-2/1990, pp. 115-132.

GOLDSCHMIDT B. (1962), L'aventure atomique, Fayard, Paris.

GOLDSCHMIDT B. (1967), Les rivalités atomiques 1939-1966, Fayard, Paris.

GOLDSCHMIDT B. (1980), Le complexe atomique, Fayard, Paris

GOLDSCHMIDT B. (1983), La politique de non-prolifération: historique et résultats, RGN, 1983, 1, pp. 11-13, and debate pp. 18 to 21.

GOLDSCHMIDT B. (1987), Pionniers de l'atome, Stock, Paris.

GOLDSCHMIDT B. (1992), La France et la non-prolifération, Relations Internationales, n°69, Le nucléaire dans les relations internationales 2, spring 1992, pp. 41-49.

HATEM F. et SALAÜN F. (1990), Electronucléaire, le tournant de l'internationalisation, Economie prospective internationale, n°41, pp. 69-86.

HELMREICH J.E. (1986), Gathering Rare Ores, the Diplomacy of Uranium Acquisition, 1943-1954, Princeton University Press, Princeton.

LUXO A. (1978), Le transfert de technologie et l'exportation nucléaire, RGN, 1978, 4, pp. 320-326.

NAU H.R. (1974), National Politics and International Technology; Nuclear Reactor Development in Western Europe, John Hopkins University Press, Baltimore (Maryland), London.

NUTI L. (1990), Le rôle de l'Italie dans les négociations trilatérales 1957-1958, Revue d'Histoire diplomatique, 1-2/1990, pp. 132-158.

PESTRE D. (1991), Autour de la création du CERN: Physiciens, administrateurs et bureaucraties d'Etat en Europe vers 1950, PECHANSKI D. ed. (1991), Histoire politique et sciences sociales, Brussels.

SCHWARZ H.P. (1992), Adenauer, le nucléaire et la France, Revue d'Histoire diplomatique, 4/1992, pp. 299-311.

SOUTOU G.H. (1993), Les accords de 1957 et 1958: vers une communauté stratégique et nucléaire entre la France, l'Allemagne et l'Italie?, Matériaux pour l'Histoire de notre temps, n°31, April-June 1993, pp. 1-12, Nanterre.

STAMM-KUHLMANN T. (1992), EURATOM, ENEA und die nationale Kernenergiepolitik in Deutschland, Berichte zur Wissenschaftsgeschichte, 1992, Bd. 15, Heft 1, pp. 39-49.

VAÏSSE M. (1991), Un dialogue de sourds: les relations nucléaires franco-américaines de 1957 à 1960, Relations internationales, n°68, Le nucléaire dans les relations internationales. 1 winter 1991, pp. 407-423.

VAÏSSE M. (1990), Autour des "accords Chaban-Strauß", 1956-1958. Revue d'Histoire diplomatique, 1-2/1990, pp. 77-79.

VAÏSSE M. (1991), Avis de recherche, l'histoire de l'armement nucléaire, Vingtième siècle, October-December 1991, pp. 93-94.

WIESE M. (1978), Der Nuclear Non Proliferation Act of 1978, Enstehungsgechichte, Gesetzesinhalt und Schlußfolgerungen, Atom und Strom, 24, Heft 4.

Section 4
International nuclear organisations

General

STROHL P. (1970), European Joint Undertakings in Nuclear Energy – Objectives and Structures, IAEA Seminar on the Development of Nuclear Law, Bangkok, 6-11/4/1970, stencilled copy.

STROHL P. (1982), La coopération internationale dans le domaine de l'énergie nucléaire, Europe et pays de l'OCDE, SOCIETE FRANCAISE POUR LE DROIT INTERNATIONAL, L'Europe dans les relations internationales, unité et diversité, proceedings of the Nancy colloquium 21 to 23 May 1981, Pédone, Paris.

STROHL P. (1983), Les projets internationaux de recherche et développement dans le domaine de l'énergie nucléaire, expérience et perspectives, IAEA, Nuclear Power Experience, Vol. 5, pp. 511-531, Vienna.

Monographs (by organisations)

CERN

HERMANN A., KRIGE J., MERSITS U., PESTRE D.(1987 and 1990), History of CERN, Amsterdam.

KRIGE J. (1992), Le pouvoir du CERN, Big science: les grands projets scientifiques du XXe siècle, le CERN, Les cahiers de science et vie, n°12, pp. 76-80.

PESTRE D. (1986), La naissance du CERN, le comment et le pourquoi, Relations internationales, n°46, summer 1986, pp. 209-226.

PESTRE D. (1992), Qui fit le CERN, et Plus la machine est grande, mieux c'est, Big science: les grands projets scientifiques du XXe siècle, le CERN, Les cahiers de science et vie, n°12, pp. 16-28 et 68-75, Paris.

DRAGON

SHAW E.N. (1983), Europe's Nuclear Power Experiment: History of the OECD Dragon Project, Pergamon, Oxford.

EURATOM

DEUBNER C. (1977), Die Atompolitik der westdeutschen Industrie und die Gründung von EURATOM, Frankfurt/Main.

GUILLEN P. (1985), La France et la négociation du traité EURATOM, Relations Internationales, n°44, winter 1985, pp. 391-412, Paris.

PIROTTE O., GIRERD P., MARSAL P., MORSON S. (1988), Trente ans d'expérience EURATOM, Bruylant, Brussels.

POLACH J.G. (1964), EURATOM, its Background, Issues and Economic Implications, New York.

SCHEINMAN L. (1967), EURATOM: Nuclear Integration in Europe, International Conciliation, n°563, May 1967, New York.

WEILEMANN P. (1983), Die Anfänge der Europaïschen Atomgemeinschaft, Nomos, Baden-Baden.

WOLFF J.M. (1989), Une tentative d'intégration nucléaire européenne: EURATOM à l'époque du premier programme de recherche et d'enseignement (1958-1962), memoir submitted with a view to obtaining the DEA at the EHESS in September 1989.

EURODIF

DAVIET J.P. (1993), EURODIF, Histoire de l'enrichissement de l'uranium, 1973-1993, Anvers, Fonds Mercator/EURODIF S.A., Bruges.

NERSA-SBK

SAITCEVSKY B. (1979), Creys-malville: les accords de coopération européenne entre producteurs d'électricité, RGN, 1979, n°6, pp. 597-598.

OECD/NEA

ADKINS B.M. (1968), Dix ans de coopération européenne pour le développement de l'énergie nucléaire, EN, pp. 35-38.

URENCO

SCHMIDT-KÜSTER W.J. (1970), Zusammenarbeit bei der Urananreicherung mit Gaszentrifugen. Der deutsch-britisch-niederländische Vertrag von Almelo, AtW, June 1970, pp. 290-292.

ASYEE J. (1985), URENCO/CENTEC, une entreprise industrielle commune créée par un traité, STROHL P. dir. (1985), pp. 35-41.

Section 5
Histories of national developments (excluding reprocessing and waste management)

Belgium

DAMME R. van den (1979), L'énergie nucléaire en Belgique, RGN, 1979, 5, pp. 478-485.

ENERGIE NUCLEAIRE (1973), Le cycle du combustible en Belgique, EN, p. 209. Schéma émanant du CEN/SCK.

GAUBE M., SMISSAERT G. (1992), The Belgian Nuclear Industry, NEW, 1992, 5-6, p. 39 *et seq.*

NUCLEAR ENGINEERING INTERNATIONAL (1994), Datafile: Belgium, Nuclear Engineering International, February 1994, pp. 32-37.

VANDERLINDEN J. dir.(1994), Un demi-siècle de nucléaire en Belgique, book published by the Belgian Nuclear Society, the "French community in Belgium", and the FNRS, at the European Inter-University Press, Brussels. Drafts.

Canada

BOTHWELL R. (1988), Nucleus. L'histoire de la Société de l'énergie atomique du Canada Limitée, Quebec.

EGGLESTON W. (1965), Canada's Nuclear Story, Toronto, Vancouver.

France

ASSOCIATION DES AMIS DE LOUIS ARMAND (1986), Louis Armand, 40 ans au service des hommes, Paris.

BAUDOUÏ R. (1992), Raoul Dautry. Le technocrate de la République, Balland, Paris.

BAVILLE J.N. (1964), La participation industrielle dans le développement de l'énergie nucléaire en France, EN, pp. 349-354.

BELTRAN A. et al. (1984), Histoire de l'EdF, Paris, particularly Chapter 12.

BOITEUX M. (1969), Position of the EdF Regarding the Choice of Reactor System, typed summary of part of the press conference given by Mr. Boiteux on the occasion of the press visit to St. Laurent-des-Eaux on 16 October 1969, EN, pp. 519-521.

BOITEUX M. (1993), Haute tension, Odile Jacob, Paris, particularly Chapter 14. La bataille des filières nucléaires, pp. 137-156.

CEA (1951 to 1954), Rapport sur l'activité et la gestion du CEA au Président du Conseil et au Ministre des finances, published annually, from 1951 to 1954

CEA (1946-1950), Rapport d'activité du CEA du 1er janvier 1946 au 31 décembre 1950, Paris, Imprimerie Nationale, 1952.

CEA (year), annual reports since 1955.

COHEN S. (1988), Les pères de la bombe atomique française, L'Histoire n°117, December 1988, pp. 18-26.

DAVIS M. (1988), Guide de l'industrie nucléaire française, Paris.

DUVAL M., MONGIN D. (1993), Histoire des forces nucléaires françaises depuis 1945, PUF, Paris.

FINON D. (1984), La crise du Plutonium civil, La Recherche, 15, p. 884 *et seq.*, June 1984.

FROST R.L. (1991), Alternating Currents, Nationalized Power in France 1946-1970, Cornell University Press, Ithaca, London.

MONGIN D. (1993), Aux origines du programme atomique français, Matériaux pour l'histoire de notre temps, n°31, April-June 1993, pp. 13-212, Nanterre.

SOUTOU G.H. (1989), Die Nuklearpolitik der vierten Republik, Vierteljahreshefte für Zeitgeschichte, 37(1989) 4, pp. 605-610, Munich.

SOUTOU G.H. (1991), La logique d'un choix: le CEA et le problème des filières électro-nucléaires, 1953-1969, Relations internationales, n°68, Le nucléaire dans les relations internationales. 1 winter 1991, pp. 351-377.

SCHEINMAN L. (1965), Atomic Energy Policy in France Under the Fourth Republic, Princeton University Press, Princeton, New Jersey.

TEISSIER DU CROS H. (1987), Louis Armand, visionnaire de la modernité, Paris.

VAÏSSE M. (1992), Le choix atomique de la France, Vingtième Siècle, October-December 1992, pp. 21-30.

WEART S. (1979), Scientists in Power, Harvard University Press, French translation (1980), La grande aventure des atomistes français, Fayard, Paris. The French edition does not contain notes or bibliography.

Germany

ECKERT M. (1989), Die Anfänge der Atompolitik in der Bundesrepublik Deutschland, Vierteljahreshefte für Zeitgeschichte, 37(1989) 1, pp. 115-143, Munich.

ECKERT M. (1989) Kernenergie und Westintegration: die Zähmung des Westdeutschen Nuklearnationalismus, HERBST L. et al. ed. (1990), Vom Marshallplan zur EWG, Munich, pp. 313-334.

FRANK Sir C. intr. (1993), Operation Epsilon: the Farm Hall Transcripts, Institute of Physics Publishing, London.

MANDEL H. (1978), Le programme électronucléaire de la RFA, Annales des Mines, May-June 1978, pp. 199-206.

MÜLLER W.D. (1990), Geschichte der Kernenergie in der Bundesrepublik Deutschland, Anfänge und Weichenstellungen, Schäffer Verlag für Wirtschaft und Steuern Gmbh, Stuttgart (Kerntechnische Gesellschaft).

RADKAU J. (1983), Aufstieg und Krise der deutschen Atomwirtschaft, 1945-1975, Rowohlt, Hamburg.

RADKAU J. (1988), Das überschätzte System. Zur Geschichte der Strategie- und Kreislauf-Konstrukte in der Kerntechnik, Technikgeschichte, 56(1988) 3, pp. 207-215.

WALKER M. (1990), Legenden um die deutsche Atombombe, Vierteljahreshefte für Zeitgeschichte, 1990/1, pp. 45-74.

WALKER M. (1993), Selbstreflexionen deutscher Atomphysiker. Die Farm-Hall Protokolle und die Entstehung neuer Legenden um die deutsche Atombombe, Vierteljahreshefte für Zeitgeschichte, 1993/4, pp. 519-542.

WINNACKER K., WIRTZ K. (1975), Das unverstandene Wunder: Kernenergie in Deutschland, Econ Verlag, Düsseldorf. French translation (1977), Atome, illusion ou miracle, Paris, Presses Universitaires de France.

India

BHARGAVA G.S. (1992), Nuclear Power in India: the Cost of Independence, THOMAS S., BERKHOUT F., ed.(1992), pp. 735-743.

HART D. (1983), Nuclear Power in India. A Comparative Analysis, George Allen & Unwin, London.

Japan

LI YANG (1966), L'énergie nucléaire au Japon, EN, pp. 418-420.

Sweden

LINDSTRÖM S. (1992), The Brave Music of a Distant Drum: Sweden's Nuclear Phase Out, THOMAS S., BERKHOUT F. ed.(1992), pp. 623-631.

Swizterland

DÄNIKER G. (1991), Le projet de défense nucléaire de la Suisse des années 50 et 60, Relations internationales, n°68, Le nucléaire dans les relations internationales. 1, winter 1991, pp. 345-349.

FAVEZ J.C. (1992), Le nucléaire et la politique extérieure de la Suisse: le cas du Traité de non-prolifération, Relations internationales, n°69, Le nucléaire dans les relations internationales. 2, spring 1992, pp. 51-62.

HUG P.(1991), La genèse de la technologie nucléaire en Suisse, Relations internationales, n°68, Le nucléaire dans les relations internationales. 1, winter 1991, pp. 325-344.

RGN-actualités (1980), Ce que fut l'accident de Lucens, rapport définitif de la commission d'enquête, RGN 1980, 1, pp. 113-114.

SCHWEIZERISCHE GESELLSCHAFT DER KERNFACHLEUTE (1992), Die Geschichte der Kerntechnik in der Schweiz. Die ersten 30 Jahren, 1939-1969, Olynthus Verlag. Epreuves.

United Kingdom

CHESHIRE J. (1992), Why Nuclear Power Failed the Market Test in the UK, THOMAS S., BERKHOUT F., ed.(1992), pp. 744-754.

GOWING M. (1964), Britain and Atomic Energy 1939-1945, McMillan Co, London.

GOWING M. (1974), Independence and Deterrence. Britain and Atomic Energy 1945-1952, Vol. 1 Policy Making, Vol. 2, Policy Execution, The McMillan Press Ltd, London. See in particular Vol. 2, pp. 402-423, on chemical separation plants.

HALL T. (1983), Nuclear Politics, the History of Nuclear Power in Britain, London.

United States

CANTELON P.L., HEWLETT R.G., WILLIAMS R.C. ed. (1991), The American Atom. A Documentary History of Nuclear Policies from the Discovery of Fission to the Present, Philadelphia, University of Pennsylvania Press.

DERIAN M. (1969), Evolution récente de la structure de l'industrie nucéaire américaine, EN, pp. 439-442.

DUNCAN F., HOLL J.M. (1983), Shippingport, the Nation's first Atomic Power Station, Department of Energy, Washington.

FERMI L. (1957), Atoms for the World: US Participation in the Conference on the Peaceful Uses of Atomic Energy, University of Chicago Press.

GROVES L. (1962), Now It Can Be Told, the Story of the Manhattan Project, New York.

HACKER B.C. (1987), The Dragon's Tail: Radiation Safety in the Manhattan Project, 1942-1946, University of California Press, Berkeley.

HEILBRON J.L., SEIDEL R.W. (1989), Lawrence and His Laboratory, A History of the Lawrence Berkeley Laboratory, Vol. 1, University of Califormia Press, Berkeley, Los Angeles.

HEWLETT R.G., ANDERSON O.E. (1962), History of the U.S. Atomic Energy Commission, Vol. 1, The New World, 1939-1946, University Park, Pennsylvania State University Press.

HEWLETT R.G., DUNCAN F. (1969), History of the U.S. Atomic Energy Commission, Vol. 2, Atomic Shield, 1947-1952, University Park, Pennsylvania State University Press.

HEWLETT R.G., HOLL J.M. (1989), Atoms for Peace and War, 1953-1961. Eisenhower and the Atomic Energy Commission, University of California Press, Berkeley and Los Angeles.

HOGERTON J.F. ed.(1958), Atoms for Peace, a Pictorial Survey, s.l.

LEBLANC N.J. (1987), Du projet Manhattan à Hiroshima: histoire d'une décision, Relations Internationales, n°49, spring 1987, pp. 71-93.

MAZUZAN G.T., WALKER J.S. (1984), Controlling the Atom, the Beginnings of Nuclear Regulations, 1946-1962, University of California Press, Berkeley. For the sequel, see WALKER J.S. below.

McCAFFREY D.P. (1990), The Politics of Nuclear Power, a History of the Shoreham Nuclear Power Plant, Kluwer, Dordrecht.

ORNL REVIEW (1992), ORNL, The First 50 Years, ORNL Review, 25, 3/4. Chapter 1, Wartime Laboratory.

RADVANYI P. et BORDRY M. (1992), Les multiples chemins d'un projet démesuré, Le projet Manhattan, histoire de la première bombe atomique, Big Science, Les grands projets scientifiques du XXe siècle, Cahiers de Sciences et Vie n°7, pp. 32-52.

SEABORG G.T. (1958), The Transuranium Elements, Yale University Press, New Haven.

SMYTH H.DW. (1948), Atomic Energy for Military Purposes, the Official Report on the Development of the Atomic Bomb, Under the Auspices of the United States Government, 1940-1945, Princeton.

WALKER J.S. (1992), Containing the Atom: Nuclear Regulation in a Changing Environment, 1963-1971, University Of California Press, Berkeley, Los Angeles.

USSR

HOLLOWAY D. (1981), Entering the Nuclear Arms Race: the Soviet Decision to Build the Atomic Bomb, 1939-1945, Social Studies of Science 11.

HOLLOWAY D. (1994), Stalin and the Bomb: The Soviet Union and Atomic Energy, 1939-1956, New Haven Conn., Yale University Press.

Section 6
Comparative history

FINON D. (1989), L'échec des surgénérateurs, autopsie d'un grand programme, PUG, Grenoble.

JASPER J.M. (1992), Gods, Titans and Mortals: Patterns of State Involvement in Nuclear Development, THOMAS S., BERKHOUT F. ed.(1992), pp. 653-659.

STOFFAES C. (1992), Choix énergétiques, choix de société: trois modèles sociopolitiques face au nucléaire, Annales des Mines, Réalités industrielles, October 1992.

Section 7
General history

Historiography

CHAUVEAU A. et TETARD P. ed.(1992), Questions à l'histoire du temps présent, proceedings of the Round Table held at the Centre d'histoire de l'Europe du vingtième siècle, FNSP, Complexe, Brussels.

SCHWARZ H.P. (1983), Die Europaïsche Integration als Aufgabe der Zeitgeschichtsforschung, Forschungsstand und Perspektiven, Vierteljahreshefte für Zeitgeschichte, 31/4, pp. 555-572.

International relations: history and law

General

BOSSUAT G. (1992a), L'Europe occidentale à l'heure américaine, 1945-1952, Complexe, Brussels.

BOSSUAT G. (1992b), La France, l'aide américaine et la construction européenne 1944-1954, Comité d'Histoire économique et financière de la France, Paris.

BROMBERGER M. and S. (1968), Les coulisses de l'Europe, Paris.

BÜHRER W. (1986), Ruhrstahl und Europa. Die Wirtschaftsvereinigung Eisen- und Stahlindustrie und die Anfänge der europaïschen Integration, 1945-1952, Munich. Report on lecture given by WOLFF J.M. (1989), Le mouvement social, N°148, July-September 1989, pp. 112-114.

BUSSY E. de, et al. (1971), Approches théoriques de l'intégration européenne, Revue française de sciences politiques, Vol. XXI, n°3, June 1971, Paris.

ECSC-EEC-EAEC (1987), Treaties Setting up the European Communities, Luxembourg.

GERBET P. (1983), La construction de l'Europe, Imprimerie Nationale, Paris.

GIRAULT R., FRANK R., THOBIE J. (1993), La loi des géants, Masson, Paris.

GIRAULT R., LEVY-LEBOYER M. dir. (1993), Le Plan Marshall et le relèvement économique de l'Europe, proceedings of the colloquium held at Bercy the 21, 22 and 23 March 1991, Comité pour l'Histoire economique et financière de la France, Ministère des finances, Paris.

KRILL H.H. (1968), Die Gründung der Unesco, Vierteljahreshefte für Zeitgeschichte, 1968, pp. 247-279.

MELANDRI P. (1975), Les Etats-Unis et le défi européen 1955-1958, Paris.

MELANDRI P. (1980), Les Etats-Unis face à l'unification de l'Europe 1945-1954, Paris.

MILWARD A.S. (1984), The Reconstruction of Western Europe, 1947-1952, London.

MILWARD A.S. (1992), The European Rescue of the Nation State, Methuen, London.

NAU H.R. (1975), Collective Response to R & D Problems in Western Europe: 1955-1958 and 1968-1973, International Organization, Vol. 29, n°2, summer 1975, pp. 617-653.

POIDEVIN R. ed.(1980), Histoire des débuts de la construction européenne, March 1948-May 1950, Bruylant, Brussels.

SCHWABE K. ed.(1988), Die Anfänge des Schuman-Plans 1950/51, proceedings of the colloquium held at Aix-la-Chapelle, 28-30 May 1986, Baden-Baden, Brussels, Milan, Paris.

SERRA E. ed.(1989), Il rilancio dell'Europa e i trattati di Roma, proceedings of the colloquium held in Rome, 25-28 March 1987, Baden-Baden, Brussels, Milan, Paris.

STROHL P. (1965), Les organisations européennes, introduction, historique, bilan et perspectives, Juris-classeur de droit international, fascicule 150, Editions techniques, Paris.

STROHL P. dir. (1985), The International Undertakings for Technical Co-operation; Legal Aspects, Review and Prospects, Round Table held on 27 April 1985, organised by SFDI, OECD/NEA, OECD/IEA and ESA, Paris.

VAÏSSE M. (1990), Les relations internationales depuis 1945, Paris.

WEISS F. (1994), Die schwierige Balance. Österreich und die Anfänge der Westeuropaïschen Integration 1947-1957, Vierteljahreshefte für Zeitgeschichte, Heft 1, pp. 71-94.

History of the OEEC, the OECD and the OEEC/EPA

CAREW A. (1987), Labour Under the Marshall Plan. The Politics of Productivity and the Marketing of Management Science, Manchester University Press, Manchester.

LA DOCUMENTATION FRANCAISE (1948), La loi américaine de coopération économique de 1948, Notes documentaires et études, n°938, 29 June 1948, Paris.

LA DOCUMENTATION FRANCAISE (1959), L'AEP, Notes et études documentaires n°2604.

OEEC (1958), The EPA, Activities and Prospects, Review of the First Four Years.

STROHL P. (1959), L'OECE, Juris-classeur de droit international, fascicules 160-A et 160-B, Editions techniques, Paris.

STROHL P., REYNERS P. (1970), OCDE, Juris-classeur de droit international, fascicules 160-A et 160-B Editions techniques, Paris.

History and economics of organisations

CHANDLER Jr. A.D. (1962), Strategy and Structure. Chapters in the History of the Industrial Enterprise, Cambridge Mass. French translation (1989) Stratégies et structures de l'entreprise, Editions d'Organisation, Paris.

CHANDLER Jr. A.D. (1977), The Visible Hand, the Managerial Revolution in American Business, Cambridge Mass. French translation (1988) La main visible des managers, une analyse historique, Economica, Paris.

COHEN E. (1992), Le colbertisme High Tech, économie des Telecom et du grand projet, Hachette, Paris.

CROZIER M. (1963), Le phénomène bureaucratique, Seuil, Paris.

FRIDENSON P. (1989), Les organisations, un nouvel objet, Annales ESC n°6, pp. 1461-1477.

KOENIG C.,THIETART R.A. (1988), Managers, Engineers and Government, the Emergence of the Mutual Organization in the European Aerospace Industry, Technology in Society, X, 1988, pp. 45-69.

MENARD C. (1993), L'économie des organisations, La Découverte, Paris.

WATELET H. (1990), Vers un approfondissement factuel et théorique en histoire des entreprises, Revue belge d'histoire contemporaine, XXI, 1990, 1-2, pp. 143-161.

WEE H. van der (1984), Der gebremste Wohlstand. Wiederaufbau, Wachstum, Strukturwandel, 1945-1980, DTV, Munich.

Section 8
Reference works

Periodicals and bibliographical works

Bulletin signalétique d'histoire des sciences et des techniques.

Technology and Culture, Current Bibliography of the History of Technology.

Isis, Current Bibliography. Collection in WITHROW M. ed.(1976), Isis Cumulative Bibliography, 1913-1975, London and NEU J. (1989), Isis Cumulative Bibliography, 1976-1985, History of Science Society, Boston.

Statistical references

OECD/IEA (1990-1991-1992), Energy Statistics, 4 vols., 1960-1992, Paris.

OECD/NEA (annual since 1983), Nuclear Energy Data, Paris.

Chronologies

NEW (1983-1993), Nuclear Status Report for Western Europe, published annually. Became the World Nuclear Status Report in 1989.

UKAEA (1984), The Development of Atomic Energy, 1939-1984, Chronology of Events, Bournemouth (Dorset).

Atlas

CHALIAND G., JAN M. (1993), Atlas du nucléaire civil et militaire, Payot, Paris.

Dictionaries and guides

ANGENAULT J. (1991), La chimie, dictionnaire encyclopédique, Dunod, Paris.

CEA (1964), Dictionnaire des sciences et des techniques nucléaires, PUF, Paris.

CLASON W.E. (1970), Elsevier's Dictionary of Nuclear Science and Technology, Elsevier, Amsterdam, London, New York.

GMELIN (1974), Gmelin Handbook of Inorganic Chemistry, Transurane, Vol. A1, Berlin.

GMELIN (1982), Gmelin Handbook of Inorganic Chemistry, Uranium, Vol. A4, Irradiated Fuel Reprocessing, Berlin.

NUCLEAR ENGINEERING INTERNATIONAL (1993), Fuel Cycle Facilities, Nuclear Engineering Handbook 1994, pp. 124-135.

OECD/NEA (1983), Nuclear Energy Glossary, English-French, Paris.

PASCAL P. (1956, 1960, 1962, 1967, 1970), Nouveau traité de chimie minérale, Masson, Paris. Tomes 1 and 15.

Bibliographical references

ALEXANDRE D., CHOTIN M., LE BLAYE G. (1987), Vitrification of Fission Product Wastes: Industrial Experience and Construction of the New Vitrification Units at La Hague, RECOD (1987), pp. 289-296.

ALLES M. (1963), L'installation de dégainage, *EN*, pp. 257-262.

ANGENAULT J. (1991), *La chimie, dictionnaire encyclopédique,* Dunod, Paris.

ASSOCIATION DES AMIS DE LOUIS ARMAND (1986), *Louis Armand, 40 ans au service des hommes*, Paris.

ASYEE J. (1985), URENCO/CENTEC, une entreprise industrielle commune créée par un traité, STROHL P. dir. (1985), pp. 35-41.

AtW (1986 to 1991), "Besondere Vorkommnisse" in der WAK, *AtW*, 1986, 12, pp. 626-627; *AtW*, 1987, 12, pp. 607-608; *AtW*, 1989, 2, pp. 98-100; *AtW*, 1990, 4, pp. 203-204; *AtW*, 1991, 5, pp. 250-252.

AtW (1979), Abgabe radioaktiver Stoffe aus Wiederaufarbeitungsanlagen in der EG 1972-1976, *AtW*, 1979, 12, pp. 612-613.

BARBE A., PECH R. (1991), Cost Estimation of the Decommissioning of Nuclear Fuel Cycle Plants: Application to Reprocessing Plants, RECOD (1991), pp. 953-959.

BARBIER C. (1990), Les négociations franco-germano-italiennes en vue de l'établissement d'une coopération militaire nucléaire au cours des années 1956-1958, *Revue d'Histoire diplomatique*, 1-2 /1990, pp. 81-113.

BARENDREGT T. et al. (1964), *Conception de l'usine de retraitement d'Eurochemic*, July 1964.

BARENDREGT T.J., KOREN LUND L. (1958), L'installation de retraitement de combustible du JENER, UN (1958), P/585.

BAUDOUÏ R. (1992), *Raoul Dautry. Le technocrate de la République*, Balland, Paris.

BERN J.B., WAISS Y. (1983), Contrôle en ligne dans les usines de retraitement, *RGN*, 1983, 5, pp. 406-408.

BERNARD C., MOUROUX J.P., DECOURS J. et al. (1991), Zirconium-made Equipement for the New La Hague Reprocessing Plants, RECOD (1991), pp. 570-575.

BETIS J. (1993), Pratique de l'optimisation de la radioprotection sur le centre de retraitement des combustibles irradiés de La Hague, *RGN*, 1993, 4, pp. 270-275.

BLANC D. (1991), *Sûreté de l'énergie nucléaire*, PUF, Paris.

BLUM R. (1963), La coopération France-Japon dans le domaine nucléaire, *EN*, pp. 481-484.

BOITEUX M. (1969), Position of the EdF Regarding the Choice of Reactor System, typed summary of part of the press conference given by Mr. Boiteux on the occasion of the press visit to St. Laurent-des-Eaux on 16 October 1969, *EN*, pp. 519-521.

BOITEUX M. (1993), *Haute tension*, Odile Jacob, Paris, particularly Chapter 14. La bataille des filières nucléaires, pp. 137-156.

BOSSUAT G. (1992a), *L'Europe occidentale à l'heure américaine*, 1945-1952, Complexe, Brussels.

BOSSUAT G. (1992b), *La France, l'aide américaine et la construction européenne 1944-1954*, Comité d'Histoire économique et financière de la France, Paris.

BOTHWELL R. (1988), *Nucleus. L'histoire de la Société de l'énergie atomique du Canada Limitée*, Quebec.

BRAUN H.J. ed.(1992), Symposium on Failed Innovations, Introduction, *Social Studies of Science*, SAGE, London and other locations, Vol. 22 (1992), pp. 213-230.

BRAUN H.J., KAISER W. (1992), *Propyläen Technikgeschichte,* KÖNIG W. ed.(1990-1992), Bd. 5, 1914-1990, Ullstein Propyläen, Frankfurt, Berlin. See in particular KAISER W., Technisierung des Lebens seit 1945, Chapitre Die Problematik der Kernenergie als neue Primärenergiequelle, pp. 285-339.

BROMBERGER M. and S. (1968), *Les coulisses de l'Europe,* Paris.

BRUCE F.R., SHANK E.M. et al. (1958), Expérience acquise pendant le fonctionnement de deux installations pilotes de traitement chimique, UN (1958), P/536.

BÜHRER W. (1986), *Ruhrstahl und Europa. Die Wirtschaftsvereinigung Eisen- und Stahlindustrie und die Anfänge der europaïschen Integration, 1945-1952*, Munich. Report on lecture given by WOLFF J.M. (1989), *Le mouvement social*, N°148, July-September 1989, pp. 112-114.

BUSEKIST O. von (1980), Der Werdegang der Eurochemic, eine Bilanz, *AtW*, May. French translation given in: BUSEKIST O. von (1982), Eurochemic, échec ou exemple, *Consensus*, n°2, p. 2-14.

CALLERI G., CAO S et al. (1971), Italian Progress Report on Reprocessing of Irradiated Fuel, UN (1971), A/CONF.49/P/182.

CANTELON P.L., HEWLETT R.G., WILLIAMS R.C. ed.(1991), *The American Atom. A Documentary History of Nuclear Policies from the Discovery of Fission to the Present*, Philadelphia, University of Pennsylvania Press.

CAREW A. (1987), *Labour Under the Marshall Plan. The Politics of Productivity and the Marketing of Management Science*, Manchester University Press, Manchester.

CEA (1951 to 1954), *Rapport sur l'activité et la gestion du CEA au Président du Conseil et au Ministre des finances*, published annually, from 1951 to 1954.

CFDT (1980), *Le dossier électronucléaire*, Paris.

CHABERT J., COIGNAUD G., PEROT J.P., FOURNIER W., SILVAIN B. (1991), Total Data Management System for the La Hague Spent Fuel Reprocessing Plants, RECOD (1991), pp. 77-82.

CHABERT J., ROULLAND M. (1991), Computer-aided Maintenance at La Hague UP3, *NEW*, 1991, 5-6, pp. 20-21.

CHADEAU E. dir. (1994), *Ariane: le défi technologique*, Paris.

CHALIAND G., JAN M. (1993), *Atlas du nucléaire civil et militaire,* Payot, Paris.

CHANDLER Jr. A.D. (1962), *Strategy and Structure. Chapters in the History of the Industrial Enterprise*, Cambridge Mass. French translation (1989) *Stratégies et structures de l'entreprise*, Editions d'Organisation, Paris.

CHANDLER Jr. A.D. (1977), *The Visible Hand, the Managerial Revolution in American Business*, Cambridge Mass. French translation (1988), *La main visible des managers, une analyse historique*, Economica, Paris.

CHAYES A., LEWIS W.B. ed.(1977), *International Arrangements for Nuclear Fuel Reprocessing*, Cambridge (Mass.).

CISAC/NAS (1994), *Management and Disposition of Excess Weapons Plutonium*, Committee on International Security and Arms Control (CISAC), National Academy of Sciences, National Academy Press, Washington.

COCHRAN T.B., NORRIS R.S., SUOKKO K.L. (1993), Radioactive Contamination at Chelyabinsk-65, Russia, *Annual Review of Energy and the Environment*, Vol. 18, pp. 507-528, Palo Alto (Cal.).

COHEN E. (1992), *Le colbertisme High Tech, économie des Telecom et du grand projet*, Hachette, Paris.

COHEN S. (1988), Les pères de la bombe atomique française, *L'Histoire* n°117, December 1988, pp. 18-26.

CONZE E. (1990), La coopération franco-germano-italienne dans le domaine nucléaire dans les années 1957-1958, un point de vue allemand, *Revue d'Histoire diplomatique*, 1-2/1990, pp. 115-132.

COUTURE J. (1975), Y a-t-il une crise mondiale dans le retraitement des combustibles?, *RGN*, 1, pp. 31-34.

CREGUT A. (1978), Le déclassement des installations nucléaires, *RGN*, 1978, 3, pp. 166-172.

CREGUT A., LURIE R. (1985), L'expérience française en matière de déclassement, *RGN*, 1985, 2, pp. 133-138.

CURILLON R., COEURE M. (1963), L'ensemble industriel d'exploitation de plutonium. L'usine – son évolution – ses annexes, *EN*, pp. 271-281.

CURILLON R., VIENOT J. (1968), L'ensemble industriel de traitement des combustibles irradiés au Japon, *EN*, pp. 160-168.

DAVIET J.P. (1989), *Une multinationale à la française, Saint-Gobain 1665-1989,* Fayard, Paris.

DAVIET J.P. (1993), *EURODIF, Histoire de l'enrichissement de l'uranium, 1973-1993*, Anvers, Fonds Mercator/EURODIF S.A., Bruges.

DE CONINCK, DEMONIE M., CLAES J. (1986), Design and Preliminary Experience of Interim Storage Facility for Vitrified Waste, *NEW*, 1986, 9, pp. 27-29.

DEBEIR J.C., DELEAGE P.P., HEMERY D. et al. (1986), *Les servitudes de la puissance, une histoire de l'énergie*, Flammarion, Paris, particularly Chapters 8 and 9, pp. 299-360.

DELANGE M. (1985a), Développement, expérience et innovation dans le retraitement, *RGN*, 1985, 6, pp. 559-568.

DEMONIE H., WIESE H. (1990), Operation of the PAMELA High-level Waste Vitrification Facility, paper presented at the Jahrestagung Kerntechnik 1990, s.l.

DETILLEUX E. (1966), *Le laboratoire de développement industriel d'Eurochemic, son rôle, ses moyens*, July 1966, introductory brochure.

DETILLEUX E. (1978), Mise à l'arrêt des installations d'Eurochemic: programme, évolution et enseignements, *RGN*, 1978, 3, pp. 177-186.

DEVEZEAUX DE LAVERGNE J.G. (1994), Wirtschaftlichkeit von Wiederaufarbeitung und Recycling, *AtW*, May 1994, pp. 350-353.

DREISSIGACKER H., FICHTNER H., REINARTZ W. (1966), Bau des Dissolvers für die Eurochemic Wiederaufarbeitungsanlage, *AtW*, April 1966, pp. 170-173.

DUBOZ M.(1965), Le Centre de La Hague, *EN*, pp. 228-245.

DUMOULIN M., GUILLEN P., VAÏSSE M. dir. (1994), *L'énergie nucléaire en Europe. Des origines à EURATOM.* Actes des journées d'études de Louvain-La-Neuve (18 and 19 November 1991), Peter Lang, Collection Euroclio, Berne.

DUVAL M., MONGIN D. (1993), Histoire des forces nucléaires françaises depuis 1945, PUF, Paris.

DWK (1985), *PAMELA, ein Projekt wird verwirklicht,* Scherrer-Druck, Hanover.

ECKERT M. (1989), Die Anfänge der Atompolitik in der Bundesrepublik Deutschland, *Vierteljahreshefte für Zeitgeschichte*, 37(1989) 1, pp. 115-143, Munich.

ECKERT M. (1989), *Kernenergie und Westintegration: die Zähmung des Westdeutschen Nuklearnationalismus,* HERBST L. et al. ed. (1990), Vom Marshallplan zur EWG, Munich, pp. 313-334.

ECSC-EEC-EAEC (1987), *Treaties Setting up the European Communities*, Luxembourg.

EGGLESTON W. (1965), *Canada's Nuclear Story*, Toronto, Vancouver.

ETR 318: DRENT W., DELANDE E. (1984), *Proceedings of the Seminar on Eurochemic Experience*, June 9-11, 1983, Mol.

FAUGERAS P., CHESNE A. (1965), Le traitement des combustibles irradiés – amélioration et extension du procédé utilisant les solvants, UN (1965), P/65, pp. 295-306.

FERMI L. (1957), *Atoms for the World: US Participation in the Conference on the Peaceful Uses of Atomic Energy*, University of Chicago Press.

FERNANDEZ N. (1963), La section de traitement des effluents. Ses problèmes, son fonctionnement, ses résultats, *EN*, pp. 282-290.

FINON D. (1989), *L'échec des surgénérateurs, autopsie d'un grand programme*, PUG, Grenoble.

FORATOM (1970), *The Future of Reprocessing in Europe, a FORATOM Study Prepared by a Groupe of Experts,* February 1970.

FRANK Sir C. intr. (1993), *Operation Epsilon: the Farm Hall Transcripts*, Institute of Physics Publishing, London.

FREROTTE M. (1988), L'avenir des installations d'Eurochemic, Belgoprocess ou Belgowaste?, *Intercommunale*, 1st trimester 1988, Brussels.

FRIDENSON P. (1989), Les organisations, un nouvel objet, *Annales ESC* n°6, pp. 1461-1477.

FRUS, Foreign Relations of the United States.

GALLEY R. (1958), Les problèmes posés par l'usine de plutonium de Marcoule et leurs solutions, *EN*, 1958, pp. 2-20. English translation in: GALLEY R. (1959), Problems Connected with the Plutonium Factory at Marcoule and Their Solution, *Journal of the Brirish Nuclear Energy Conference*, 1959, 1, pp. 1-11.

GIRAULT R., FRANK R., THOBIE J. (1993), *La loi des géants*, Masson, Paris.

GIRAULT R., LEVY-LEBOYER M. dir. (1993), *Le Plan Marshall et le relèvement économique de l'Europe*, proceedings of the colloquium held at Bercy the 21, 22 and 23 March 1991, Comité pour l'Histoire economique et financière de la France, Ministère des finances, Paris.

GOLDSCHMIDT B. (1962), *L'aventure atomique*, Fayard, Paris.

GOLDSCHMIDT B. (1980), *Le complexe atomique*, Fayard, Paris.

GOLDSCHMIDT B. (1987), *Pionniers de l'atome*, Stock, Paris.

GOMEZ J. (1983), Utilisation de l'instrumentation conventionnelle dans les usines de retraitement, *RGN* 1983, 5, pp. 401-405.

GOWING M. (1964), *Britain and Atomic Energy 1939-1945*, McMillan Co, London.

GOWING M. (1974), *Independence and Deterrence. Britain and Atomic Energy 1945-1952,* Vol. 1 *Policy Making,* Vol. 2, *Policy Execution*, The McMillan Press Ltd, London. See in particular, Vol. 2, pp. 402-423, on chemical separation plants.

GROVES L.R. (1962), *Now It Can Be Told, the Story of the Manhattan Project*, New York.

GUILLEN P. (1985), La France et la négociation du traité EURATOM, *Relations internationales*, n°44, winter 1985, pp. 391-412, Paris.

HEILBRON J.L., SEIDEL R.W. (1989), *Lawrence and His Laboratory, A History of the Lawrence Berkeley Laboratory*, Vol. 1, University of California Press, Berkeley, Los Angeles.

HELMREICH J.E. (1986), *Gathering Rare Ores, the Diplomacy of Uranium Acquisition, 1943-1954*, Princeton University Press, Princeton.

HERMANN A., KRIGE J., MERSITS U., PESTRE D. (1987 and 1990), *History of CERN*, Amsterdam.

HERMET G. (1992), *L'Espagne au XXe siècle*, PUF, Paris.

HEWLETT R.G., ANDERSON O.E. (1962), *History of the U.S. Atomic Energy Commission*, Vol. 1, *The New World, 1939-1946,* University Park, Pennsylvania State University Press.

HEWLETT R.G., DUNCAN F. (1969), *History of the U.S. Atomic Energy Commission,* Vol. 2, *Atomic Shield, 1947-1952,* University Park, Pennsylvania State University Press.

HEWLETT R.G., HOLL J.M. (1989), *Atoms for Peace and War, 1953-1961. Eisenhower and the Atomic Energy Commission,* University of California Press, Berkeley and Los Angeles.

HILD W., DETILLEUX E., GEENS L. (1980), Some Recommandations for Plant Design and Layout Resulting from the Decontamination of the Eurochemic Reprocessing Plant and the Partial Decommissioning of Some of its Facilities, OECD/AEN (1980), pp. 197-208.

HILPERT H.J. (1987), Floor Response Spectra of the Main Process Building of a Reprocessing Plant Against Earthquake, Airplane Crash and Blast, RECOD (1987), pp. 1347-1350.

HIRSCH E. (1988), *Ainsi va la vie*, Lausanne.

HOGERTON J.F. ed.(1958), Atoms for Peace, a Pictorial Survey, s.l.

HOLLOWAY D. (1994), *Stalin and the Bomb: The Soviet Union and Atomic Energy, 1939-1956*, New Haven Conn., Yale University Press.

HOSTE A., JAUMOTTE A. (1976), *Commission d'évaluation en matière d'énergie nucléaire, Rapports de synthèse*, March 1976, Kingdom of Belgium, Ministry for Economic Affairs, Brussels.

HOSTE A., JAUMOTTE A. (1982), *Commission d'évaluation en matière d'énergie nucléaire, Rapport final, éléments d'actualisation*, Kingdom of Belgium, Ministry of Economic Affairs, Secretary of State for Energy, Brussels.

HOUNSHELL D.A., SMITH J.K. Jr. (1988), *Science and Corporate Strategy. Du Pont R & D, 1902-1980*, Cambridge University Press, Cambridge (Mass.). Chapter 16.

HUET P. (1958), L'AEEN et la Société Eurochemic, contribution à l'étude des sociétés internationales, *Annuaire français de droit international (AFDI)*, Paris.

HUG P.(1991), La genèse de la technologie nucléaire en Suisse, *Relations internationales*, n°68, *Le nucléaire dans les relations internationales.* 1, winter 1991, pp. 325-344.

IAEA (1980), *International Nuclear Fuel Cycle Evaluation, Summary of Results,* Vienna.

IZQUIERDO J.J., CHAUVIRE P., PLESSIS L. (1991), The MERC Maintenance System, RECOD (1991), pp. 1111-1116.

JOUANNAUD C. (1963), L'usine d'extraction du plutonium et son exploitation, *EN*, pp. 263-270.

KAUFMANN F., WIESE H. (1987), Vitrification of High Level Waste, RECOD (1987), pp. 279-288.

KERVERN G.Y., RUBISE P. (1991), *L'archipel du danger. Introduction aux cindyniques*, Economica, Paris.

KOENIG C.,THIETART R.A. (1988), Managers, Engineers and Government, the Emergence of the Mutual Organization in the European Aerospace Industry, *Technology in Society*, X, 1988, pp. 45-69.

KRAUSE H. (1982), Die Endkonditionierung hochaktiver Spaltproduktlösungen, *AtW*, 1982, 12, pp. 618-621.

KRIGE J. (1993a), *The Launch of ELDO,* ESA HSR-7 (History Study Report n°7), March 1993.

KRIGE J. (1993b), *Some New Thoughts on the Launch of ELDO and its Subsequent "Failure",* paper presented at the Conference on Technological Change, Oxford, 8-11 September 1993.

KRIGE J. ed.(1992), Choosing Big Technologies, *History and Technology,* 1992, 1-4.

KUHN K.-D., WIESE H., DEMONIE M. (1991), HLLW Conditioning in the PAMELA Vitrification Plant, RECOD (1991), pp. 273-277.

LECLERCQ J. (1986), *L'ère nucléaire, le monde des centrales nucléaires*, Hachette, Paris.

LECLERQ-AUBRETON Y., ASYEE J., NORAZ M. (1968), *Preliminary Study of an International Financial Nuclear Energy Company, FINUCLEAIRE,* unpaginated, dated 20 October 1968.

LEDUC M., PELRAS M., SANNIER J., TURLUER G., DEMAY R. (1987), Etudes de corrosion sur les matériaux destinés aux usines de retraitement, RECOD (1987), pp. 1173-1180.

LEFORT G., MIQUEL P., RUBERCY M. de (1968), Etude et expérimentation du schéma des extractions de l'usine japonaise, *EN*, pp. 169-180.

LEISTER P., OESER H.R. (1991), Remote Handling Technology for Reprocessing Plants: Lessons Learnt from LAHDE, RECOD (1991), pp. 163-168.

LEMON R.B., REID D.G. (1956), Marche d'une usine de traitement radiochimique à entretien non télécommandé, UN (1956), P/543.

LORRAIN B., SAUDRAY D., TARNERO M. (1987), Dissolveur continu rotatif pour combustible à eau légère. Essais du prototype, RECOD (1987), pp. 1471-1481.

LOWRANCE W.W. (1977), Nuclear Futures for Sale: Issues Raised by the West-German Brazilian Nuclear Agreement, CHAYES A., LEWIS W.B. ed.(1977), pp. 201-221.

LUNG M., COIGNAUD M. (1978), L'usine de retraitement de Tokai-Mura, *RGN*, 1978, 4, pp. 314-319.

MACKENZIE D., SPINARDI G. (1994), Tacit Knowledge, Weapons Design, and the Uninvention of Nuclear Weapons. Draft of March 1994 provided by Dominique Pestre.

MARJOLIN R. (1986), *Le travail d'une vie, Mémoires 1911-1986,* Laffont, Paris.

MELANDRI P. (1975), *Les Etats-Unis et le défi européen 1955-1958,* Paris

MELANDRI P. (1980), *Les Etats-Unis face à l'unification de l'Europe 1945-1954*, Paris.

MEMORANDUM OF UNDERSTANDING (1970), *Understanding Between Eurochemic and the UKAEA*, 15 October 1970.

MILWARD A.S. (1984), *The Reconstruction of Western Europe*, 1947-1952, London.

MISCHKE J. (1984), Konzept und Stand der Wiederaufarbeitungsanlage WA-350, *AtW*, 1984, 8-9, pp. 434-438.

MIYAHARA K., YAMAMURA O., TAKAHASHI K. (1991), The Operational Experience at Tokai Reprocessing Plant, RECOD (1991), pp. 49-54.

MONGIN D. (1993), Aux origines du programme atomique français, *Matériaux pour l'histoire de notre temps*, n°31, April-June 1993, pp. 13-212, Nanterre.

MONGON A. (1963), cf. AEEN/OCDE-EUROCHEMIC (1963b).

MONGON A. (1964), Instrumentation et automatisme dans une usine de retraitement de combustibles irradiés, *EN*, pp. 150-158.

MONNET J. (1976), *Mémoires,* Fayard, Paris (Published in the Livre de Poche series).

MÜLLER W.D. (1990), *Geschichte der Kernenergie in der Bundesrepublik Deutschland, Anfänge und Weichenstellungen,* Schäffer Verlag für Wirtschaft und Steuern Gmbh, Stuttgart (Kerntechnische Gesellschaft).

MUS G., LINGER C. (1987), Conduite automatique des procédés de l'atelier pilote, RECOD (1987), pp. 1519-1526.

NAITO S., SUMIYA A., OHTAKA K. et al. (1991), Remote Inspection of Repaired Dissolvers and Improvement of the Remote Maintenance Robots, RECOD (1991), pp. 157-162.

NAU H.R. (1974), *National Politics and International Technology; Nuclear Reactor Development in Western Europe,* John Hopkins University Press, Baltimore (Maryland), London.

NAU H.R. (1975), Collective Response to R & D Problems in Western Europe: 1955-1958 and 1968-1973, *International Organization*, Vol. 29, n°2, summer 1975, pp. 617-653.

NUTI L. (1990), Le rôle de l'Italie dans les négociations trilatérales 1957-1958, *Revue d'Histoire diplomatique*, 1-2/1990, pp. 132-158.

OECD/IEA (1990-1991-1992), *Energy Statistics*, 4 vols., 1960-1992, Paris.

OECD/NEA (1980), *Decommissioning of Nuclear Installations: Requirements at the Design Stage, report on an NEA Specialists' Meeting*, 17-19 March 1980, Paris.

OECD/NEA (1991b), *Decommissioning of Nuclear Installations, an Analysis of Variations in Decommissioning Costs*, Paris.

OECD/NEA (1993a), *The Safety of the Nuclear Fuel Cycle.*

OECD/NEA (1993b), *Radiation Protection on the Threshold of the 21st Century; proceedings of an NEA Workshop, Paris, 11-13 January 1993.*

OECD/NEA-IAEA (1973), *Management of Radioactive Wastes Produced in the Processing of Spent Fuel, proceedings of the symposium, 27 November-1 December 1972,* Paris.

OEEC (1955), *Some Aspects of the European Energy Problem*, also called the "Armand Report", June, Paris.

OEEC (1956a), *Possibilities of Action in the Field of Nuclear Energy*, also called the "Nicolaïdis Report", January, Paris.

OEEC (1956b), *Europe and its Growing Energy Needs*, also called the "Hartley Report", May, Paris.

OEEC (1956c), *Joint Action by OEEC Countries in the Field of Nuclear Energy*, September, Paris.

OEEC (1958), *The EPA, Activities and Prospects, Review of the First Four Years.*

OEEC/ENEA (1958), *Industry and Nuclear Energy,* papers given at the Second Information Meeting on Nuclear Energy for Managers, Amsterdam, 24 to 28 June 1957, Paris.

OEEC/EPA (1957), *Industry and Nuclear Energy*, papers given during the First Information Meeting on Nuclear Energy for Managers, Paris, 1 to 5 April 1957, Paris.

ONOYAMA M., NAKATA M., HIROSE Y., NAKAGAWA Y. (1991), Development of a High Performance Type 304L Stainless Steel for Nuclear Fuel Reprocessing Facilities, RECOD (1991), pp. 1066-1071.

ORNL REVIEW (1992), ORNL, The First 50 Years, *ORNL Review*, 25, 3/4. Chapter 1, Wartime Laboratory.

OSIPENCO A. (1980), Occupational Radiation Exposure at the Eurochemic Reprocessing Plant During Normal Operation and Intervention Periods, *Occupational Radiation Exposure in Nuclear Fuel Facilities,* IAEA, Vienna 1980, pp. 439-450.

OSIPENCO A., DETILLEUX E., FERRARI P. (1980), Décontamination et démantèlement partiel de l'usine de retraitement Eurochemic. Enseignements relatifs à la radioprotection et à la gestion des déchets, OECD/NEA (1980), pp. 255-262.

PEROT J.P. (1983), La conduite du procédé dans les usines de retraitement de combustibles irradiés, *RGN*, 1983, 5, pp. 398-401.

PESTRE D. (1986), La naissance du CERN, le comment et le pourquoi, *Relations internationales*, n°46, summer 1986, pp. 209-226.

PIGNAULT J. (1983), Conception d'une salle de conduite centralisée: application aux extensions de l'établissement de
La Hague, *RGN*, 1983, 5, pp. 411-414.

PIROTTE O., GIRERD P., MARSAL P., MORSON S. (1988), *Trente ans d'expérience EURATOM*, Bruylant, Brussels.

POHLAND E., STROHL P. (1962), Die Zusammenarbeit der Eurochemic mit der europaïschen Industrie beim Bau der Eurochemic-Anlage, *AtW*, Aug./Sept., pp. 403-409.

POLACH J.G. (1964), *EURATOM, its Background, Issues and Economic Implications*, New York.

PONCELET J.F., HUGELMANN D., SAUDRAY D., MUKOHARA S., CHO A. (1991), Head-end Process Technology for the New Reprocessing Plants in France and Japan, RECOD (1991), pp. 95-99.

RADKAU J. (1983), *Aufstieg und Krise der deutschen Atomwirtschaft, 1945-1975*, Rowohlt, Hamburg.

RADKAU J. (1988), Das überschätzte System. Zur Geschichte der Strategie- und Kreislauf-Konstrukte in der Kerntechnik, *Technikgeschichte*, 56(1988) 3, pp. 207-215.

RADVANYI P. et BORDRY M. (1992), Les multiples chemins d'un projet démesuré, *Le projet Manhattan, histoire de la première bombe atomique, Big Science, Les grands projets scientifiques du XXe siècle, Cahiers de Sciences et Vie* n°7, pp. 32-52.

RECOD (1987), *International Conference on Nuclear Fuel Reprocessing and Waste Management,* proceedings of the conference held at the Palais des Congrès, Porte Maillot, Paris, from 23 to 27 August 1987. Sponsor: SFEN. Co-sponsor: European Nuclear Society, American Nuclear Society, Atomic Energy of Japan, 4 vols., 1628 p.

RECOD (1991), *The Third International Conference on Nuclear Fuel Reprocessing and Waste Management*, proceedings of the conference held at Sendai from 14 to 18 April 1991. Sponsors: Atomic Energy Society of Japan, Japan Industrial Forum. Co-sponsors: ANS, ENS, JAERI, PNC. 2 vols., 1154 p.

RODIER J., ESTOURNEL R., BOUZIGUES H., CHASSANY J. (1963), Le travail en milieu radioactif et ses problèmes, *EN*, pp. 291-301.

ROUVILLE M. de (1963), Le centre de production de plutonium de Marcoule, *EN*, pp. 213-220. Presented in a special edition devoted to Marcoule, n°4, June 1963.

SAITCEVSKY B. (1979), Creys-malville: les accords de coopération européenne entre producteurs d'éléctricité, *RGN*, 1979, n°6, pp. 597-598.

SALOMON J.-J. (1986), Le Gaulois, le cow-boy et le samouraï. La politique française de la technologie, Economica, Paris.

SALOMON J.-J. (1992), *Le destin technologique*, Balland, Paris.

SCHEINMAN L. (1965), *Atomic Energy Policy in France Under the Fourth Republic,* Princeton University Press, Princeton, New Jersey.

SCHEINMAN L. (1967), EURATOM: Nuclear Integration in Europe, *International Conciliation*, n°563, May 1967, New York.

SCHEUTEN G.H. (1985), Die deutsche Wiederaufarbeitungsanlage, *AtW*, 1985, pp. 362-365.

SCHMIDT-KÜSTER W.J. (1970), Zusammenarbeit bei der Urananreicherung mit Gaszentrifugen. Der deutsch-britisch-niederländische Vertrag von Almelo, *AtW*, June 1970, pp. 290-292.

SCHÜLLER W. (1984), Betriebserfahrungen mit der Wiederaufarbeitungsanlage Karlsruhe, *AtW*, 1984, 8-9, pp. 438-444.

SCHÜLLER W. (1985), Zwanzig Jahre GWK/WAK. Bilanz des Erreichten und Ausblick auf zukünftige Aufgaben, *AtW*, 1985, 4, pp. 178-181.

SCHWEIZERISCHE GESELLSCHAFT DER KERNFACHLEUTE (1992), *Die Geschichte der Kerntechnik in der Schweiz. Die ersten 30 Jahren, 1939-1969*, Olynthus Verlag. Epreuves.

SCHWENNESEN J.L. (1958), Examen sommaire de la conception et de l'exploitation des usines de traitement du combustible nucléaire, UN (1958), P/514.

SEABORG G.T. (1958), *The Transuranium Elements*, Yale University Press, New Haven.

SERRA E. ed.(1989), *Il rilancio dell'Europa e i trattati di Roma*, proceedings of the colloquium held in Rome, 25-28 March 1987, Baden-Baden, Brussels, Milan, Paris.

SETHNA H.N., SRINIVASAN N. (1964), Fuel Processing at Trombay, UN (1964), P/786.

SHAW E.N. (1983), *Europe's Nuclear Power Experiment: History of the OECD Dragon Project*, Pergamon, Oxford.

SILIE P. (1983), Les systèmes de conduite centralisés des usines de retraitement, *RGN*, 1983, 5, pp. 408-411.

SIMONNET J., DEMAY R., BACHELAY J. (1987), Development and Use of Zirconium in Reprocessing Equipment, RECOD (1987), pp. 1215-1218.

SMYTH H.DW. (1948), *Atomic Energy for Military Purposes, the Official Report on the Development of the Atomic Bomb, Under the Auspices of the United States Government, 1940-1945,* Princeton.

SOUSSELIER Y. (1960), *Demain l'atome*, PUF, Paris.

SOUTOU G.H. (1991), La logique d'un choix: le CEA et le problème des filières électro-nucléaires, 1953-1969, *Relations internationales*, n°68, *Le nucléaire dans les relations internationales.* 1 winter 1991, pp. 351-377.

SOUTOU G.H. (1993), Les accords de 1957 et 1958: vers une communauté stratégique et nucléaire entre la France, l'Allemagne et l'Italie?, *Matériaux pour l'Histoire de notre temps*, n°31, April-June 1993, pp. 1-12, Nanterre.

STAMM-KUHLMANN T. (1992), EURATOM, ENEA und die nationale Kernenergiepolitik in Deutschland, *Berichte zur Wissenschaftsgeschichte*, 1992, Bd. 15, Heft 1, pp. 39-49.

STROHL P. (1959), L'OECE, *Juris-classeur de droit international*, Fascicules 160-A et 160-B, Editions techniques, Paris.

STROHL P. (1961), Problèmes juridiques soulevés par la constitution et le fonctionnement de la Société Eurochemic., *Annuaire français de droit international,* pp. 569-591, Paris.

STROHL P. (1982), La coopération internationale dans le domaine de l'énergie nucléaire, Europe et pays de l'OCDE, SOCIETE FRANCAISE POUR LE DROIT INTERNATIONAL, *L'Europe dans les relations internationales, unité et diversité*, proceedings of the Nancy colloquium 21 to 23 May 1981, Pédone, Paris.

STROHL P. (1983), Les projets internationaux de recherche et développement dans le domaine de l'énergie nucléaire, expérience et perspectives, IAEA, *Nuclear Power Experience*, Vol. 5, pp. 511-531, Vienna.

STROHL P. (1991), Les aspects internationaux de la liquidation de la société Eurochemic, *Annuaire français de droit international,* Paris, pp. 727-738.

STROHL P. dir. (1985), *The International Undertakings for Technical Co-operation; Legal Aspects, Review and Prospects, Round Table held on 27 April 1985 organised by SFDI, OECD/NEA and ESA*, Paris.

SVANSSON L. (1963), cf. AEEN/OCDE-EUROCHEMIC (1963b).

SWENNEN R., CUYVERS H., VAN GEEL J., WIECZOREK H. (1984), The Treatment of Combustible Alpha-wastes at Eurochemic using Acid Digestion and a Plutonium Recovery Process. First Results of the Active Operation of a Demonstration Plant, American Nuclear Society, Idaho Section ed., *Fuel Reprocessing and Waste Management,* Jackson, Wyoming, August 26-29, 1984, Proceedings Vol. 1, pp. 549-563.

SYBELPRO (1983), *Syndical Agreement,* dated 13 December 1982.

TAKASAKI K., EBANA M., NOMURA T. (1991), Radiation Control System at Tokai Reprocessing Plant, RECOD (1991), pp. 219-223.

TARANGER P. (1956), Marcoule, *Age nucléaire*, 1956, 2, pp. 35-39.

TEISSIER DU CROS H. (1987), *Louis Armand, visionnaire de la modernité*, Paris.

TEUNCKENS L. (1993), Decommissioning of Final Product Storage Buildings at the Ex-Eurochemic Reprocessing Plant, *NEW*, 1993, 11-12, p. 60.

TITTMANN E., SCHEFFLER K. (1987), PAMELA's LEWC campaign, s.l.

TOYOTA M. (1991), Outline of Rokka-shô Reprocessing Plant Project, RECOD (1991), pp. 38-43.

UKAEA (1984), *The Development of Atomic Energy, 1939-1984, Chronology of Events*, Bournemouth (Dorset).

UN (1956), *Proceedings of the International Conference on the Peaceful Uses of Atomic Energy,* Volume IX, *Materials: Testing and Chemical Treatment*, Geneva.

UN (1958), *Proceedings of the Second International Conference on the Peaceful Uses of Atomic Energy*, Volume VIII, *Irradiated Fuels and Radioactive Materials; Radiation Protection,* Geneva.

UN (1965), *Proceedings of the Third International Conference on the Peaceful Uses of Atomic Energy,* Volume X, *Nuclear Fuels-I. Fabrication and Reprocessing*, Volume XIV, *Environmental Aspects of Atomic Energy and Waste Management,* New York.

UN-IAEA (1972), *Peaceful Uses of Atomic Energy, Proceedings of the Fourth International Conference on the Peaceful Uses of Atomic Energy*, Volume VIII, New York, Vienna.

USAEC (1955), *Selected Reference Material, United States Atomic Energy Program,* Volume VI, *Chemical Processing and Equipment,* Geneva.

VAÏSSE M. (1990), Autour des "accords Chaban-Strauß", 1956-1958. *Revue d'Histoire diplomatique*, 1-2/1990, pp. 77-79.

VAÏSSE M. (1991), Avis de recherche, l'histoire de l'armement nucléaire, *Vingtième siècle*, October-December 1991, pp. 93-94.

VAÏSSE M. (1992), La coopération nucléaire en Europe (1955-1958), Etat de l'historiographie, *Storia delle relazioni internazionali*, VIII, 1992/1-2, pp. 201-213.

VANDERLINDEN J. dir. (1994), *Un demi-siècle de nucléaire en Belgique*, book published by the Belgian Nuclear Society, the "French community in Belgium", and the FNRS, at the European Inter-University Press, Brussels. Drafts.

VIEFERS W. (1986), *Incidents at Eurochemic, Description and Evaluation*, unpublished report from April 1986.

VIEFERS W. (not dated), *Waste Treatment Developments, Practices and Experience at Eurochemic*, unpublished report probably dating from 1985.

WALKER M. (1990), Legenden um die deutsche Atombombe, *Vierteljahreshefte für Zeitgeschichte*, 1990/1, pp. 45-74.

WALKER M. (1993), Selbstreflexionen deutscher Atomphysiker. Die Farm-Hall Protokolle und die Entstehung neuer Legenden um die deutsche Atombombe, *Vierteljahreshefte für Zeitgeschichte*, 1993/4, pp. 519-542.

WALKER W. (1992), The Back-end of the Nuclear Fuel Cycle, in KRIGE J. ed.(1992), Choosing Big Technologies, *History and Technology*, 1992, 1-4, pp. 189-201.

WEART S. (1979), *Scientists in Power*, Harvard University Press, French translation (1980), *La grande aventure des atomistes français*, Fayard, Paris. The French edition does not contain notes or bibliography.

WEILEMANN P. (1983), *Die Anfänge der Europaïschen Atomgemeinschaft*, Nomos, Baden-Baden.

WEINLÄNDER W., HUPPERT K.L., WEISHAUPT M. (1991), Twenty Years of WAK Reprocessing Pilot Plant Operation, RECOD (1991), pp. 55-63.

WEISS F. (1994), Die schwierige Balance. Österreich und die Anfänge der Westeuropaïschen Integration 1947-1957, *Vierteljahreshefte für Zeitgeschichte*, Heft 1, pp. 71-94.

WIECZOREK H., OSER B. (1986), Entwicklung und aktive Demonstration des Verfahrens der Naßveraschung brennbarer plutoniumhaltiger Festabfälle, *KfK Nachrichten*, Vol. 18, n°2, pp. 77-82.

WINNACKER K., WIRTZ K. (1975), *Das unverstandene Wunder: Kernenergie in Deutschland*, Econ Verlag, Düsseldorf. French translation (1977), *Atome, illusion ou miracle*, Paris, Presses Universitaires de France.

YAMANOUCHI T., OMACHI S., MATSUMOTO K. (1987), Decadal Operational Experience of the Tokai Reprocessing Plant, RECOD (1987), pp. 195-202.

Index

Names of individuals are italicised.

MAIN SALES OUTLETS OF OECD PUBLICATIONS
PRINCIPAUX POINTS DE VENTE DES PUBLICATIONS DE L'OCDE

ARGENTINA – ARGENTINE
Carlos Hirsch S.R.L.
Galería Güemes, Florida 165, 4° Piso
1333 Buenos Aires Tel. (1) 331.1787 y 331.2391
Telefax: (1) 331.1787

AUSTRALIA – AUSTRALIE
D.A. Information Services
648 Whitehorse Road, P.O.B 163
Mitcham, Victoria 3132 Tel. (03) 9210.7777
Telefax: (03) 9210.7788

AUSTRIA – AUTRICHE
Gerold & Co.
Graben 31
Wien I Tel. (0222) 533.50.14
Telefax: (0222) 512.47.31.29

BELGIUM – BELGIQUE
Jean De Lannoy
Avenue du Roi 202 Koningslaan
B-1060 Bruxelles Tel. (02) 538.51.69/538.08.41
Telefax: (02) 538.08.41

CANADA
Renouf Publishing Company Ltd.
1294 Algoma Road
Ottawa, ON K1B 3W8 Tel. (613) 741.4333
Telefax: (613) 741.5439

Stores:
61 Sparks Street
Ottawa, ON K1P 5R1 Tel. (613) 238.8985
12 Adelaide Street West
Toronto, ON M5H 1L6 Tel. (416) 363.3171
Telefax: (416)363.59.63

Les Éditions La Liberté Inc.
3020 Chemin Sainte-Foy
Sainte-Foy, PQ G1X 3V6 Tel. (418) 658.3763
Telefax: (418) 658.3763

Federal Publications Inc.
165 University Avenue, Suite 701
Toronto, ON M5H 3B8 Tel. (416) 860.1611
Telefax: (416) 860.1608

Les Publications Fédérales
1185 Université
Montréal, QC H3B 3A7 Tel. (514) 954.1633
Telefax: (514) 954.1635

CHINA – CHINE
China National Publications Import
Export Corporation (CNPIEC)
16 Gongti E. Road, Chaoyang District
P.O. Box 88 or 50
Beijing 100704 PR Tel. (01) 506.6688
Telefax: (01) 506.3101

CHINESE TAIPEI – TAIPEI CHINOIS
Good Faith Worldwide Int'l. Co. Ltd.
9th Floor, No. 118, Sec. 2
Chung Hsiao E. Road
Taipei Tel. (02) 391.7396/391.7397
Telefax: (02) 394.9176

CZECH REPUBLIC – RÉPUBLIQUE TCHÈQUE
Artia Pegas Press Ltd.
Narodni Trida 25
POB 825
111 21 Praha 1 Tel. (2) 242 246 04
Telefax: (2) 242 278 72

DENMARK – DANEMARK
Munksgaard Book and Subscription Service
35, Nørre Søgade, P.O. Box 2148
DK-1016 København K Tel. (33) 12.85.70
Telefax: (33) 12.93.87

EGYPT – ÉGYPTE
Middle East Observer
41 Sherif Street
Cairo Tel. 392.6919
Telefax: 360-6804

FINLAND – FINLANDE
Akateeminen Kirjakauppa
Keskuskatu 1, P.O. Box 128
00100 Helsinki

Subscription Services/Agence d'abonnements :
P.O. Box 23
00371 Helsinki Tel. (358 0) 121 4416
Telefax: (358 0) 121.4450

FRANCE
OECD/OCDE
Mail Orders/Commandes par correspondance :
2, rue André-Pascal
75775 Paris Cedex 16 Tel. (33-1) 45.24.82.00
Telefax: (33-1) 49.10.42.76
Telex: 640048 OCDE

Internet: Compte.PUBSINQ @ oecd.org

Orders via Minitel, France only/
Commandes par Minitel, France exclusivement :
36 15 OCDE

OECD Bookshop/Librairie de l'OCDE :
33, rue Octave-Feuillet
75016 Paris Tel. (33-1) 45.24.81.81
(33-1) 45.24.81.67

Dawson
B.P. 40
91121 Palaiseau Cedex Tel. 69.10.47.00
Telefax: 64.54.83.26

Documentation Française
29, quai Voltaire
75007 Paris Tel. 40.15.70.00

Economica
49, rue Héricart
75015 Paris Tel. 45.78.12.92
Telefax: 40.58.15.70

Gibert Jeune (Droit-Économie)
6, place Saint-Michel
75006 Paris Tel. 43.25.91.19

Librairie du Commerce International
10, avenue d'Iéna
75016 Paris Tel. 40.73.34.60

Librairie Dunod
Université Paris-Dauphine
Place du Maréchal-de-Lattre-de-Tassigny
75016 Paris Tel. 44.05.40.13

Librairie Lavoisier
11, rue Lavoisier
75008 Paris Tel. 42.65.39.95

Librairie des Sciences Politiques
30, rue Saint-Guillaume
75007 Paris Tel. 45.48.36.02

P.U.F.
49, boulevard Saint-Michel
75005 Paris Tel. 43.25.83.40

Librairie de l'Université
12*a*, rue Nazareth
13100 Aix-en-Provence Tel. (16) 42.26.18.08

Documentation Française
165, rue Garibaldi
69003 Lyon Tel. (16) 78.63.32.23

Librairie Decitre
29, place Bellecour
69002 Lyon Tel. (16) 72.40.54.54

Librairie Sauramps
Le Triangle
34967 Montpellier Cedex 2 Tel. (16) 67.58.85.15
Telefax: (16) 67.58.27.36

A la Sorbonne Actual
23, rue de l'Hôtel-des-Postes
06000 Nice Tel. (16) 93.13.77.75
Telefax: (16) 93.80.75.69

GERMANY – ALLEMAGNE
OECD Publications and Information Centre
August-Bebel-Allee 6
D-53175 Bonn Tel. (0228) 959.120
Telefax: (0228) 959.12.17

GREECE – GRÈCE
Librairie Kauffmann
Mavrokordatou 9
106 78 Athens Tel. (01) 32.55.321
Telefax: (01) 32.30.320

HONG-KONG
Swindon Book Co. Ltd.
Astoria Bldg. 3F
34 Ashley Road, Tsimshatsui
Kowloon, Hong Kong Tel. 2376.2062
Telefax: 2376.0685

HUNGARY – HONGRIE
Euro Info Service
Margitsziget, Európa Ház
1138 Budapest Tel. (1) 111.62.16
Telefax: (1) 111.60.61

ICELAND – ISLANDE
Mál Mog Menning
Laugavegi 18, Pósthólf 392
121 Reykjavik Tel. (1) 552.4240
Telefax: (1) 562.3523

INDIA – INDE
Oxford Book and Stationery Co.
Scindia House
New Delhi 110001 Tel. (11) 331.5896/5308
Telefax: (11) 332.5993

17 Park Street
Calcutta 700016 Tel. 240832

INDONESIA – INDONÉSIE
Pdii-Lipi
P.O. Box 4298
Jakarta 12042 Tel. (21) 573.34.67
Telefax: (21) 573.34.67

IRELAND – IRLANDE
Government Supplies Agency
Publications Section
4/5 Harcourt Road
Dublin 2 Tel. 661.31.11
Telefax: 475.27.60

ISRAEL – ISRAËL
Praedicta
5 Shatner Street
P.O. Box 34030
Jerusalem 91430 Tel. (2) 52.84.90/1/2
Telefax: (2) 52.84.93

R.O.Y. International
P.O. Box 13056
Tel Aviv 61130 Tel. (3) 546 1423
Telefax: (3) 546 1442

Palestinian Authority/Middle East:
INDEX Information Services
P.O.B. 19502
Jerusalem Tel. (2) 27.12.19
Telefax: (2) 27.16.34

ITALY – ITALIE
Libreria Commissionaria Sansoni
Via Duca di Calabria 1/1
50125 Firenze Tel. (055) 64.54.15
Telefax: (055) 64.12.57

Via Bartolini 29
20155 Milano Tel. (02) 36.50.83

Editrice e Libreria Herder
Piazza Montecitorio 120
00186 Roma Tel. 679.46.28
Telefax: 678.47.51

Libreria Hoepli
Via Hoepli 5
20121 Milano Tel. (02) 86.54.46
Telefax: (02) 805.28.86

Libreria Scientifica
Dott. Lucio de Biasio 'Aeiou'
Via Coronelli, 6
20146 Milano Tel. (02) 48.95.45.52
Telefax: (02) 48.95.45.48

JAPAN – JAPON
OECD Publications and Information Centre
Landic Akasaka Building
2-3-4 Akasaka, Minato-ku
Tokyo 107 Tel. (81.3) 3586.2016
Telefax: (81.3) 3584.7929

KOREA – CORÉE
Kyobo Book Centre Co. Ltd.
P.O. Box 1658, Kwang Hwa Moon
Seoul Tel. 730.78.91
Telefax: 735.00.30

MALAYSIA – MALAISIE
University of Malaya Bookshop
University of Malaya
P.O. Box 1127, Jalan Pantai Baru
59700 Kuala Lumpur
Malaysia Tel. 756.5000/756.5425
Telefax: 756.3246

MEXICO – MEXIQUE
OECD Publications and Information Centre
Edificio INFOTEC
Av. San Fernando no. 37
Col. Toriello Guerra
Tlalpan C.P. 14050
Mexico D.F.
Tel. (525) 606 00 11 Extension 100
Fax: (525) 606 13 07

Revistas y Periodicos Internacionales S.A. de C.V.
Florencia 57 - 1004
Mexico, D.F. 06600 Tel. 207.81.00
Telefax: 208.39.79

NETHERLANDS – PAYS-BAS
SDU Uitgeverij Plantijnstraat
Externe Fondsen
Postbus 20014
2500 EA's-Gravenhage Tel. (070) 37.89.880
Voor bestellingen: Telefax: (070) 34.75.778

NEW ZEALAND – NOUVELLE-ZÉLANDE
GPLegislation Services
P.O. Box 12418
Thorndon, Wellington Tel. (04) 496.5655
Telefax: (04) 496.5698

NORWAY – NORVÈGE
NIC INFO A/S
Bertrand Narvesens vei 2
P.O. Box 6512 Etterstad
0606 Oslo 6 Tel. (022) 57.33.00
Telefax: (022) 68.19.01

PAKISTAN
Mirza Book Agency
65 Shahrah Quaid-E-Azam
Lahore 54000 Tel. (42) 353.601
Telefax: (42) 231.730

PHILIPPINE – PHILIPPINES
International Booksource Center Inc.
Rm 179/920 Cityland 10 Condo Tower 2
HV dela Costa Ext cor Valero St.
Makati Metro Manila Tel. (632) 817 9676
Telefax: (632) 817 1741

POLAND – POLOGNE
Ars Polona
00-950 Warszawa
Krakowskie Przedmieácie 7 Tel. (22) 264760
Telefax: (22) 268673

PORTUGAL
Livraria Portugal
Rua do Carmo 70-74
Apart. 2681
1200 Lisboa Tel. (01) 347.49.82/5
Telefax: (01) 347.02.64

SINGAPORE – SINGAPOUR
Gower Asia Pacific Pte Ltd.
Golden Wheel Building
41, Kallang Pudding Road, No. 04-03
Singapore 1334 Tel. 741.5166
Telefax: 742.9356

SPAIN – ESPAGNE
Mundi-Prensa Libros S.A.
Castelló 37, Apartado 1223
Madrid 28001 Tel. (91) 431.33.99
Telefax: (91) 575.39.98

Mundi-Prensa Barcelona
Consell de Cent No. 391
08009 – Barcelona Tel. (93) 488.34.92
Telefax: (93) 487.76.59

Llibreria de la Generalitat
Palau Moja
Rambla dels Estudis, 118
08002 – Barcelona
(Subscripcions) Tel. (93) 318.80.12
(Publicacions) Tel. (93) 302.67.23
Telefax: (93) 412.18.54

SRI LANKA
Centre for Policy Research
c/o Colombo Agencies Ltd.
No. 300-304, Galle Road
Colombo 3 Tel. (1) 574240, 573551-2
Telefax: (1) 575394, 510711

SWEDEN – SUÈDE
CE Fritzes AB
S–106 47 Stockholm Tel. (08) 690.90.90
Telefax: (08) 20.50.21

Subscription Agency/Agence d'abonnements :
Wennergren-Williams Info AB
P.O. Box 1305
171 25 Solna Tel. (08) 705.97.50
Telefax: (08) 27.00.71

SWITZERLAND – SUISSE
Maditec S.A. (Books and Periodicals - Livres et périodiques)
Chemin des Palettes 4
Case postale 266
1020 Renens VD 1 Tel. (021) 635.08.65
Telefax: (021) 635.07.80

Librairie Payot S.A.
4, place Pépinet
CP 3212
1002 Lausanne Tel. (021) 320.25.11
Telefax: (021) 320.25.14

Librairie Unilivres
6, rue de Candolle
1205 Genève Tel. (022) 320.26.23
Telefax: (022) 329.73.18

Subscription Agency/Agence d'abonnements :
Dynapresse Marketing S.A.
38, avenue Vibert
1227 Carouge Tel. (022) 308.07.89
Telefax: (022) 308.07.99

See also – Voir aussi :
OECD Publications and Information Centre
August-Bebel-Allee 6
D-53175 Bonn (Germany) Tel. (0228) 959.120
Telefax: (0228) 959.12.17

THAILAND – THAÏLANDE
Suksit Siam Co. Ltd.
113, 115 Fuang Nakhon Rd.
Opp. Wat Rajbopith
Bangkok 10200 Tel. (662) 225.9531/2
Telefax: (662) 222.5188

TUNISIA – TUNISIE
Grande Librairie Spécialisée
Fendri Ali
Avenue Haffouz Imm El-Intilaka
Bloc B 1 Sfax 3000 Tel. (216-4) 296 855
Telefax: (216-4) 298.270

TURKEY – TURQUIE
Kültür Yayinlari Is-Türk Ltd. Sti.
Atatürk Bulvari No. 191/Kat 13
Kavaklidere/Ankara
Tel. (312) 428.11.40 Ext. 2458
Telefax: (312) 417 24 90

Dolmabahce Cad. No. 29
Besiktas/Istanbul Tel. (212) 260 7188

UNITED KINGDOM – ROYAUME-UNI
HMSO
Gen. enquiries Tel. (171) 873 8242
Postal orders only:
P.O. Box 276, London SW8 5DT
Personal Callers HMSO Bookshop
49 High Holborn, London WC1V 6HB
Telefax: (171) 873 8416
Branches at: Belfast, Birmingham, Bristol, Edinburgh, Manchester

UNITED STATES – ÉTATS-UNIS
OECD Publications and Information Center
2001 L Street N.W., Suite 650
Washington, D.C. 20036-4922 Tel. (202) 785.6323
Telefax: (202) 785.0350

Subscriptions to OECD periodicals may also be placed through main subscription agencies.

Les abonnements aux publications périodiques de l'OCDE peuvent être souscrits auprès des principales agences d'abonnement.

Orders and inquiries from countries where Distributors have not yet been appointed should be sent to: OECD Publications Service, 2, rue André-Pascal, 75775 Paris Cedex 16, France.

Les commandes provenant de pays où l'OCDE n'a pas encore désigné de distributeur peuvent être adressées à : OCDE, Service des Publications, 2, rue André-Pascal, 75775 Paris Cedex 16, France.

1-1996

OECD PUBLICATIONS, 2, rue André-Pascal, 75775 PARIS CEDEX 16
PRINTED IN FRANCE
(66 96 03 1) ISBN 92-64-14792-6 – No. 47107 1996